Volume I

Probability Theory and Mathematical Statistics

Proceedings of the
Fifth Vilnius Conference
June 25 - July 1, 1989

Edited by

**B. Grigelionis, Yu. V. Prohorov,
V. V. Sazonov and V. Statulevičius**

MOKSLAS
Vilnius, Lithuania

///VSP///
Utrecht, The Netherlands

VSP BV
Post Box 346
3700 AH Zeist
The Netherlands

MOKSLAS
Žvaigždžiu 23
Vilnius
Lithuania

First published in 1990

CIP-DATA KONINKLIJKE BIBLIOTHEEK, DEN HAAG
Probability theory and mathematical statistics:
Proceedings of the Fifth Vilnius Conference:
Vilnius, Lithuania, June 25 - July 1, 1989
ed. by B. Grigelionis... [*et al.*]
- Utrecht: VSP BV/ Vilnius: Mokslas.
ISBN 90-6764-128-6 (vol.I)
ISBN 90-6764-130-8 (set)
SISO 517 UDC 519.2(063)
Subject headings: probability theory/ mathematical statistics.

Typeset in Lithuania by Baltic Amadeus / Publishing Service Group of IMI, Vilnius
Printed in Lithuania by Spindulys, Kaunas

CONTENTS

PREFACE

The traditional Vilnius Conference on Probability Theory and Mathematical Statistics was held for the fifth time in 1989, June 26 – July 1. It was organized, as usual, by the Institute of Mathematics and Cybernetics of the Lithuanian Academy of Sciences, the Steklov Mathematical Institute of the USSR Academy of Sciences and the Vilnius University. The Organizing Committee was headed by the cochairmen Professor Yu.V. Prohorov (Moscow) and Professor V. Statulevičius (Vilnius). There were 744 participants at the Conference from 28 countries. The main directions of discussions covered asymptotic methods in probability and mathematical statistics, theory of Markov processes, stochastic analysis, martingales, stochastic physics and random fields. Quantum probability and computer statistics were included in the program for the first time. Two plenary lectures delivered by Professor A.A. Borovkov (USSR) and Professor K.R. Parthasarathy (India), as well as 446 talks, divided into 63 sectional sessions, were presented. Poster session included 172 communications.

The present Proceedings in two volumes contain the papers received by the Organizing Committee both from the invited speakers and authors of some selected short communications. We wish to thank most cordially all contributors to the success of the Vth Vilnius Conference and express hopes that these Proceedings will stimulate further progress in probability and mathematical statistics.

B. Grigelionis, Yu.V. Prohorov,
V.V. Sazonov, V. Statulevičius

Prob. Theory and Math. Stat., Vol. 1, pp. 1-27
B.Grigelionis et al. (Eds.)
1990 VSP/Mokslas

THE LOW DENSITY LIMIT AND THE QUANTUM POISSON PROCESS

L. ACCARDI and LU YUN GANG*
Centro Matematico V.Volterra, Dipartimento di Matematica,
Universita' di Roma II, Italy

ABSTRACT

We prove that, in the low density limit (z^2 = fugacity $\to 0$), the family of processes given by the collective Weyl operators and the collective coherent vectors (cf. a free Bose gas), converge to the Fock quantum Brownian motion over $L^2(\mathbf{R}, dt, K)$, where K is an appropriate Hilbert space (cf. Section 1). Moreover we prove that if we couple a nonrelativistic quantum system with a quadratic interaction (cf. formula (1.10) below) to a free Boson gas in the Fock state, then the matrix elements of the wave operator of the system at time t/z^2 in the collective coherent vectors converge to the matrix elements, in suitable coherent vectors of the quantum Brownian motion process, of a unitary Markovian cocycle satisfying a quantum stochastic differential equation ruled by some pure number process (i.e., no quantum diffusion part and only the quantum analogue of the purely discontinuous, or jump, processes). The explicit form of the equation is determined and the unitarity is proved.

1. INTRODUCTION

In the last two years our understanding of the weak coupling limit has drastically increased since it has been possible to acquire control of the limit not only of the reduced dynamics, but of the full quantum dynamics, both in the Schrödinger and in the Heisenberg form (Accardi et al., 1989, 1990a, 1990b; Accardi and Lu Yun Gang, 1989a, 1989b). The basic new idea in these papers was the introduction of some classes of "collective states"(collective coherent states or collective number states) and the study of the limit of the matrix elements of the Schrödinger as well as of the Heisenberg dynamics with respect to these collective states. Both in the classical and quantum case the results available in the low density case are much poorer than the corresponding ones for the weak coupling case: for example the existence of the Markovian semi-group has been proved only for a short time interval in a very special Fermion model (Dümcke, 1989a, 1989b).
Let H_0 and H_1 be complex separable Hilbert spaces interpreted respectively as the system Hilbert space and the one particle reservoir Hilbert space. Let

$$W(H_1) = \{W(f) : f \in H_1\} \tag{1.1}$$

* On leave of absence from Beijing Normal University

be the Weyl C^*–algebra on H_1; let H be a self–adjoint bounded below operator on H_1 and z, β positive real numbers interpreted respectively as density of the reservoir particles and inverse temperature. Define

$$Q_z := \left(1 + z^2 e^{-\beta H}\right)\left(1 - z^2 e^{-\beta H}\right)^{-1} = \coth(\beta H + \mu) \tag{1.2}$$

(with $z^2 = e^{\mu}$) and suppose that, for each z in an interval $[0, Z]$, Q_z is a self–adjoint operator on a domain D, independent on z. Denote φ_{Q_z} the mean zero gauge invariant quasi–free state on $W(H_1)$ with covariance operator Q_z, i.e.

$$\varphi_{Q_z}(W(f)) = \exp\left(-\frac{1}{2}\langle f, Q_z f\rangle\right), \qquad \forall f \in H_1 \tag{1.3}$$

and let $\{\mathcal{H}_{Q_z}, \pi_{Q_z}, \Phi_{Q_z}\}$ be the GNS – triple of $\{W(H_1), \varphi_{Q_z}\}$, so that

$$\langle \Phi_{Q_z}, \pi_{Q_z}(W(f))\Phi_{Q_z}\rangle = \varphi_{Q_z}(W(f)). \tag{1.4}$$

We shall write W_{Q_z} for $\pi_{Q_z} \circ W$. The *Fock representation* corresponds to the case $Q_z = 1$, i.e. $z = 0$. In this case the *GNS representation* will be simply denoted $\{\mathcal{H}, \pi, \Phi\}$. Let S_t be a unitary group on $B(H_1)$ (the one particle free evolution of the reservoir) and suppose that

$$S_t Q_z = Q_z S_t, \qquad \forall t \geqslant 0, \tag{1.5}$$

where the equality is meant on D. This implies that the second quantization of S_t, denoted $W(S_t)$, leaves φ_{Q_z} invariant hence it is implemented, in the GNS representation, by a unitary 1–parameter group $V_t^{(z)}$ whose generator $H^{(z)} =: H_R$ is called the *free Hamiltonian of the reservoir*. As in (Accardi et al., 1990a) we assume that there exists a non zero subspace K of H_1 (in all the examples it is a dense subspace) such that

$$\int_{\mathbf{R}} |\langle f, S_t g\rangle| dt < \infty, \qquad \forall f, g \in K. \tag{1.6}$$

Moreover, we suppose that $Q_z K \subseteq K$ and $e^{-tH_R} K \subseteq K$ for all $t \geqslant 0$.
Let be given a self–adjoint operator H_S on the system space H_0, called the system Hamiltonian. The total free Hamiltonian is defined to be

$$H_0 \otimes 1 + 1 \otimes H_1. \tag{1.7}$$

We define the interaction Hamiltonian V as in (Dümcke, 1989a), i.e., we fix two functions $g_1, g_0 \in K$ and define

$$\begin{aligned} V &:= i\left(D \otimes A^+(g_0)A(g_1) - D^+ \otimes A^+(g_1)A(g_0)\right) = \\ &= i \sum_{\varepsilon \in \{0,1\}} D_\varepsilon \otimes A^+(g_\varepsilon)A(g_{1-\varepsilon}) \end{aligned} \tag{1.8}$$

with the notations

$$D_0 = D, \qquad D_1 = -D^+ \tag{1.9}$$

and where D is a bounded operator on H_0 satisfying

$$\exp(-itH_S)\, D \exp(itH_S) = D. \tag{1.10}$$

Moreover we assume that g_0 and g_1 have disjoint energy spectra, i.e.,

$$\langle g_0, S_t g_1\rangle = 0, \qquad \forall t \in \mathbf{R}. \tag{1.11}$$

This condition has already been used in the literature on the weak coupling limit (cf. condition (2.5) of (Frigerio, 1988) and (Davies, 1974, §3)). As we shall prove in subsequent papers the assumptions (1.10), (1.12) are not essential, however they simplify several computations. With these notations, the total Hamiltonian is

$$H_{total} := H_S \otimes 1 + 1 \otimes H_R + V \tag{1.12}$$

and the wave operator at time t is defined by

$$U(t) := \exp\left(-itH^{(0)}\right) \exp(itH_{total}) \tag{1.13}$$

with

$$H^{(0)} := H_S \otimes 1 + 1 \otimes H_R. \tag{1.14}$$

Therefore

$$\frac{d}{dt}U(t) = \frac{1}{i}V(t)U(t); \qquad U(0) = 1, \tag{1.15}$$

where

$$\begin{aligned} V(t) &:= \exp(-itH_S \otimes 1)\, V \exp(itH_S \otimes 1) = \\ &= i \sum_{\varepsilon \in \{0,1\}} D_\varepsilon \otimes A^+(S_t g_\varepsilon) A(S_t g_{1-\varepsilon}). \end{aligned} \tag{1.16}$$

Moreover the solution of (1.15) is given by the iterated series

$$U(t) = \sum_{n=0}^{\infty} \int_0^t dt_1 \ldots \int_0^{t_{n-1}} dt_n V(t_1)\ldots V(t_n), \tag{1.17}$$

which is weakly convergent on the domain of the vectors of the form

$$u \otimes \Phi_{Q_z}\left(z \int_{S/z^2}^{T/z^2} S_u f\, du\right) := u \otimes W_{Q_z}\left(z \int_{S/z^2}^{T/z^2} S_u f\, du\right) \Phi_{Q_z}, \tag{1.17a}$$

where $u \in H_0$, $f \in K$, $S, T \in \mathbf{R}$. From Lemma (3.2) of (Accardi et al., 1990a), we know that the assumption (1.7) implies that the sesquilinear form $(\cdot|\cdot) : K \times K \longrightarrow$ $\longrightarrow C$ defined by

$$(f|g) := \int_{\mathbf{R}} \langle f, S_t g\rangle dt, \qquad f, g \in K \tag{1.18}$$

defines a pre-scalar product on K. We denote $\{K, (\cdot|\cdot)\}$, or simply K, the completion of the quotient of K by the zero $(\cdot|\cdot)$-norm elements.

Definition 1.1. Let $\mathcal{K}$ be a Hilbert space, T an interval in $\mathbf{R}$, $Q \geqslant 1$ be a self-adjoint operator on $\mathcal{K}$ and let

$$\{\mathcal{H}_Q, \pi_Q, \Phi_Q\} \tag{1.19}$$

denote the GNS representation of the CCR over $L^2(T, dt, \mathcal{K})$ with respect to the quasi-free state φ_Q on $W\big(L^2(T, dt, \mathcal{K})\big)$ characterized by

$$\varphi_Q(W(\xi)) = e^{-\frac{1}{2}\langle \xi, 1\otimes Q\xi\rangle}; \qquad \xi \in L^2(T, dt, \mathcal{K}). \tag{1.20}$$

Denote $\mathcal{D}$ the set of all vectors of the form $\pi(W(\xi))\Phi_Q = W_Q(\xi)\Phi_Q$ with $\xi \in L^2(T, dt, \mathcal{K})$. The stochastic process

$$\{\mathcal{H}_Q, \mathcal{D}, W_A(\chi_{(s,t]}\otimes f); \qquad (s,t] \subseteq T, f \in \mathcal{K}\} \tag{1.21}$$

is called the ***Q-quantum Brownian motion on*** $L^2(T, dt, \mathcal{K})$.
If $Q = 1$ we speak of the ***Fock Brownian Motion on*** $L^2(T, dt, \mathcal{K})$; if Q is the multiplication by a constant ($\beta \geqslant 1$), then we speak of the ***finite temperature quantum Brownian Motion***, in the terminology of (Hudson and Parthasarathy, 1984) or of the ***universal invariant quantum Brownian Motion*** in the terminology of (Hudson and Lindsay, 1989).
Sometimes, when no confusion can arise, we call quantum Brownian motion also the process

$$\{\mathcal{H}_Q, \mathcal{D}, A\left(\chi_{(s,t]}\otimes f\right), A^+\left(\chi_{(s,t]}\otimes f\right); \quad s,t \in T,\ f \in K\}, \tag{1.22}$$

where $A(\,\cdot\,)$, $A^+(\,\cdot\,)$ denote respectively the annihilation and creation fields. For the normalized coherent vectors we use the notation

$$W_Q\left(\chi_{[s,t]}\otimes f\right)\Phi_Q = \Phi_Q\left(\chi_{[s,t]}\otimes f\right).$$

Our main result in this paper is to prove that, the limit

$$\lim_{z\to 0}\left\langle u\otimes W\left(z\int_{S/z^2}^{T/z^2} S_u f du\right)\Phi, U(t/z^2)v\otimes W\left(z\int_{S'/z^2}^{T'/z^2} S_u f' du\right)\Phi\right\rangle \tag{1.23}$$

exists and is equal to

$$\langle u\otimes W(\chi_{[S,T]}\otimes f)\Psi, U(t)v\otimes W(\chi_{[S',T']}\otimes f')\Psi\rangle, \tag{1.24}$$

where $\{\mathcal{H}, W, \Psi\}$ is the Fock Brownian motion on $L^2(\mathbf{R}, dt, K)$ and $U(t)$ satisfies a quantum stochastic differential equation driven by purely discontinuous noises in the sense of (Frigerio and Maassen, 1990; Hudson and Parthasarathy, 1984), whose form is given by (6.1).

Acknowledgements
L. Accardi acknowledges support from Grant AFOSR 870249 and ONR N00014-86-K-0538 through the Center for Mathematical System Theory, University of Florida.

2. CONVERGENCE OF THE COLLECTIVE PROCESS

In this Section we show that, at a purely kinematical level, i.e. with $t = 0$ in (1.23), the low density limit coincides with the weak coupling limit and the limiting process is the Fock Brownian motion on $L^2(\mathbf{R}, dt, K)$ where K is equipped with the scalar product (1.18).
First recall from (Accardi et al., 1990a, Lemma 3.2) that for each $f, f' \in K, S, S'$, $T, T' \in \mathbf{R}$, one has

$$\lim_{z\to 0}\left\langle z\int_{S/z^2}^{T/z^2} S_u f\,du,\, z\int_{S'/z^2}^{T'/z^2} S_u f'\,du\right\rangle =$$

$$= \langle \chi_{[S,T]}, \chi_{[S',T']}\rangle_{L^2(\mathbf{R})}\int_{\mathbf{R}}\langle f, S_t f'\rangle dt =$$

$$= \langle \chi_{[S,T]}\otimes f, \chi_{[S',T']}\otimes f'\rangle_{L^2(\mathbf{R},dt,K)} \tag{2.1}$$

and it is easy to prove the following lemma:

LEMMA 2.1. *For each $n \in \mathbf{N}, \{f_k\}_{k=1}^n \subset K$, $\{S_k, T_k\}_{k=1}^n \subset \mathbf{R}$, $\{x_k\}_{k=1}^n \subset \mathbf{R}$,*

$$\lim_{z\to 0}\left\langle \Phi_{Q_z}, W\left(x_1 z\int_{S_1/z^2}^{T_1/z^2} S_u f_1\,du\right)\cdots W\left(x_n z\int_{S_n/z^2}^{T_n/z^2} S_u f_n\,du\right)\Phi_{Q_z}\right\rangle =$$

$$= \langle \Psi, W(x_1\chi_{[S_1,T_1]}\otimes f_1)\cdots W(x_n\chi_{[S_n,T_n]}\otimes f_n)\Psi\rangle \tag{2.2}$$

and the convergence is uniform for $\{x_k\}_{k=1}^n, \{S_k, T_k\}_{k=1}^n$ in a bounded set of $\mathbf{R}$.

3. THE BASIC ESTIMATES IN THE FOCK CASE

As in the weak coupling limit, our starting point will be the analysis of the quantities

$$\Delta_n^\varepsilon(z,t) := \int_0^{t/z^2} dt_1 \int_0^{t_1} dt_2 \cdots \int_0^{t_{n-1}} dt_n \left\langle W\left(z\int_{S/z^2}^{T/z^2} S_u f\,du\right)\Phi,\right. \tag{3.1}$$

$$\left. A^+(S_{t_1} g_{\varepsilon(1)})A(S_{t_1} g_{1-\varepsilon(1)})\cdots A^+(S_{t_1} g_{\varepsilon(1)})A(S_{t_n} g_{1-\varepsilon(n)})W\left(z\int_{S'/z^2}^{T'/z^2} S_u f'\,du\right)\Phi\right\rangle.$$

LEMMA 3.1. *For each $n \in \mathbb{N}$ and $\varepsilon \in \{0,1\}^n$*

$$\Delta_n^\varepsilon(z,t) =$$

$$= \sum_{d=0}^{n} \sum_{2 \leqslant q_1 < \cdots < q_d \leqslant n} \sum_{\substack{1 \leqslant p_1,\ldots,p_d \leqslant n \\ |\{p_h\}_{h=1}^d| = d, p_h < q_h, h=1,\ldots,d}} z^{2n-2d} \int_0^{t/z^2} dt_1 \int_0^{t_1} dt_2 \cdots \int_0^{t_{n-1}} dt_n =$$

$$= \left\langle W\left(z \int_{S/z^2}^{T/z^2} S_u f du \right)\Phi, W\left(z \int_{S'/z^2}^{T'/z^2} S_u f' du \right)\Phi \right\rangle \times$$

$$\times \prod_{h=1}^{d} \langle S_{t_{p_h}} g_{1-\varepsilon(p_h)}, S_{t_{q_h}} g_{\varepsilon(q_h)} \rangle \prod_{\alpha \in \{1,\ldots,n\} \setminus \{q_h\}_{h=1}^d} \int_{S/z^2}^{T/z^2} \langle S_u f, S_{t_\alpha} g_{\varepsilon(\alpha)} \rangle du \times$$

$$\times \prod_{\alpha \in \{1,\ldots,n\} \setminus \{p_h\}_{h=1}^d} \int_{S'/z^2}^{T'/z^2} \langle S_{t_\alpha} g_{1-\varepsilon(\alpha)}, S_u f' \rangle du \tag{3.2}$$

with the convention that if $d = 0$, then the sums over $q_1 < \cdots < q_d$ and $p_1, \ldots, p_d$ and the product over h, in (3.2) are equal to 1.

Proof. For each $n \in \mathbb{N}$, $\varepsilon \in \{0,1\}^n$, we want to bring the product

$$A^+(S_{t_1} g_{\varepsilon(1)}) A(S_{t_1} g_{1-\varepsilon(1)}) \ldots A^+(S_{t_n} g_{\varepsilon(n)}) A(S_{t_n} g_{1-\varepsilon(n)})$$

in Wick ordered form, i.e.

$$A^+(S_{t_1} g_{\varepsilon(1)}) A(S_{t_1} g_{1-\varepsilon(1)}) \ldots A^+(S_{t_n} g_{\varepsilon(n)}) A(S_{t_n} g_{1-\varepsilon(n)}) =$$

$$= \sum_{d=0}^{n} C(g,\varepsilon,d)(A^+ \ldots)^{n-d}(A \ldots)^{n-d}, \tag{3.3}$$

where $C(g,\varepsilon,d)$ is a product of some scalar products of the g_j and d is the number of scalar products arosen in the exchange of a creator with an annihilator at its left. Notice that the scalar product comes from the commutator $[A, A^+]$. Suppose that d pairs annihilators and creators are "used" to produce the scalar products and lebel these creators and annihilators by $2 \leqslant q_1 < \ldots < q_d \leqslant n$ and $1 \leqslant p_1, \ldots, p_d \leqslant n-1$ respectively, one obtains (3.2).

LEMMA 3.2. *For each $n \in \mathbb{N}$, $\varepsilon \in \{0,1\}^n$, denote*

$$\{i_1, \ldots, i_k\} = \{i : \varepsilon(i) = 1;\ i_1 < \ldots < i_k\} \subset \{1, \ldots, n\} \tag{3.4}$$

then the quantity $\Delta_n^\varepsilon(z,t)$, defined by (3.2) is equal to

$$\sum_{m=0}^{k\wedge(n-k)} \sum_{\substack{2\leqslant q_1<\cdots<q_m\leqslant n\\ \{q_h\}_{h=1}^m\subset\{i_h\}_{h=1}^k}} \sum_{m'=0}^{k\wedge(n-k)} \sum_{\substack{2\leqslant q'_1<\cdots<q'_{m'}\leqslant n\\ \{q'_h\}_{h=1}^{m'}\subset\{1,\ldots,n\}\setminus\{i_h\}_{h=1}^k}} \sideset{}{'}\sum_{(p_1,\ldots,p_m)} \sideset{}{'}\sum_{(p'_1,\ldots,p'_{m'})} z^{2n-2(m+m')}\times$$

$$\times\int_0^{t/z^2} dt_1 \int_0^{t_1} dt_2 \cdots \int_0^{t_{n-1}} dt_n \left\langle W\left(z\int_{S/z^2}^{T/z^2} S_u f du\right)\Phi, W\left(z\int_{S'/z^2}^{T'/z^2} S_u f' du\right)\Phi\right\rangle\times$$

$$\times\prod_{h=1}^{m}\langle S_{t_{p_h}}g_1, S_{t_{q_h}}g_1\rangle \prod_{h=1}^{m'}\langle S_{t_{p'_h}}g_0, S_{t_{q'_h}}g_0\rangle\times \tag{3.5}$$

$$\times\prod_{\alpha\in\{1,\ldots,n\}\setminus(\{q_h\}_{h=1}^m\cup\{q'_h\}_{h=1}^{m'})} \int_{S/z^2}^{T/z^2}\langle S_u f, S_{t_\alpha} g_{\varepsilon(\alpha)}\rangle du\times$$

$$\times\prod_{\alpha\in\{1,\ldots,n\}\setminus(\{p_h\}_{h=1}^m\cup\{p'_h\}_{h=1}^{m'})} \int_{S'/z^2}^{T'/z^2}\langle S_{t_\alpha} g_{1-\varepsilon(\alpha)}, S_u f'\rangle du,$$

where $\sum'_{(p_1,\ldots,p_m)}$ means the sum over all $\{p_h\}_{h=1}^m \subset \{1,\ldots,n\}\setminus\{i_h\}_{h=1}^k$ with $p_h < q_h$, $h = 1,\ldots,m$ and $\sum'_{(p'_1,\ldots,p'_{m'})}$ means the sum over all $\{p'_h\}_{h=1}^{m'} \subset \{i_h\}_{h=1}^k$ with $p'_h < q'_h$, $h = 1,\ldots,m'$.

Proof. The lemma follows from the condition

$$1-\varepsilon(p_h)\neq\varepsilon(q_h) \Longrightarrow \langle S_{t_{p_h}}g_{1-\varepsilon(p_h)}, S_{t_{q_h}}g_{\varepsilon(q_h)}\rangle = 0. \tag{3.6}$$

Now, proceeding as in (Accardi et al., 1990a), for each $n\in\mathbb{N}$ and $\varepsilon\in\{0,1\}^n$, in the expression $\Delta_n^\varepsilon(z,t)$, defined by (3.1), we separate the terms which will be relevant in the limit $z\to 0$ (type I terms) from those which will vanish in the limit (type II terms). More precisely we define

$$\Delta_n^\varepsilon(z,t) =: I_n^\varepsilon(z,t) + II_n^\varepsilon(z,t), \tag{3.7}$$

where

$$I_n^\varepsilon(z,t) := \sum_{m=0}^{k\wedge(n-k)} \sideset{}{'}\sum_{(q_1,\ldots,q_m)} \sum_{m'=0}^{k\wedge(n-k)} \sideset{}{'}\sum_{(q'_1,\ldots,q'_{m'})} z^{2n-2(m+m')}\times$$

$$\times\int_0^{t/z^2} dt_1 \int_0^{t_1} dt_2 \cdots \int_0^{t_{n-1}} dt_n \left\langle W\left(z\int_{S/z^2}^{T/z^2} S_u f du\right)\Phi, W\left(z\int_{S'/z^2}^{T'/z^2} S_u f' du\right)\Phi\right\rangle\times$$

$$\times\prod_{h=1}^{m}\langle g_1, S_{(t_{q_h}-t_{q_h-1})}g_1\rangle \prod_{h=1}^{m'}\langle g_0, S_{(t_{q'_h}-t_{q'_h-1})}g_0\rangle\times \tag{3.8}$$

$$\times \prod_{\alpha\in\{1,\dots,n\}\setminus(\{q_h\}_{h=1}^{m}\cup\{q'_h\}_{h=1}^{m'})} \int_{S/z^2}^{T/z^2} \langle S_u f, S_{t_\alpha} g_{\varepsilon(\alpha)}\rangle du \times$$

$$\times \prod_{\alpha\in\{1,\dots,n\}\setminus(\{q_h-1\}_{h=1}^{m}\cup\{q'_h-1\}_{h=1}^{m'})} \int_{S'/z^2}^{T'/z^2} \langle S_{t_\alpha} g_{1-\varepsilon(\alpha)}, S_u f'\rangle du$$

and

$$II_n^{\varepsilon}(z,t) = \Delta_n^{\varepsilon}(z,t) - I_n^{\varepsilon}(z,t) =:$$

$$=: \sum_{m=0}^{n-1} \sum_{2\leqslant\bar q_1<\cdots<\bar q_m\leqslant n} \sum_{(\bar q_1,\bar p_1,\dots,\bar q_m,\bar p_m)}' \Delta_n^{\varepsilon}(z,t,\{\bar q_h,\bar p_h\}_{h=1}^{m}), \tag{3.9}$$

where $\sum'_{(\bar q_1,\bar p_1,\dots,\bar q_m,\bar p_m)}$ means the sum for all $1\leqslant \bar p_1,\dots,\bar p_m\leqslant n$ with $|\{\bar p_h\}_{h=1}^{m}| = m$, $\bar p_h < \bar q_h$, $h=1,\cdots,m$ and $\bar p_h < \bar q_h - 1$, for some $h=1,\cdots,m$. From (3.7) and (3.8) one can deduce the form of $\Delta_n^{\varepsilon}(z,t,\{\bar q_h,\bar p_h\}_{h=1}^{m})$.

Lemma 3.3. *For each $n\in\mathbb{N}$, $\varepsilon\in\{0,1\}^n$,*

$$\lim_{z\to 0} II_n^{\varepsilon}(z,t) = 0. \tag{3.10}$$

Proof. For each $n\in\mathbb{N}$, $\varepsilon\in\{0,1\}^n$, since the coherent vectors are unit vectors, we have from (3.10)

$$|II_n^{\varepsilon}(z,t)| \leqslant \sum_{m=0}^{n-1} \sum_{2\leqslant\bar q_1<\cdots<\bar q_m\leqslant n} \sum_{(\bar q_1,\bar p_1,\dots,\bar q_m,\bar p_m)}' C_1^n \times$$

$$\times z^{2n-2m} \int_0^{t/z^2} dt_1 \int_0^{t_1} dt_2 \cdots \int_0^{t_{n-1}} dt_n \prod_{h=1}^{m} |\langle S_{t_{\bar p_h}} g, S_{t_{\bar q_h}} g\rangle|, \tag{3.11}$$

where $g = g_0$ or g_1 and

$$C_1 := \max_{F=f,f',G=g_0,g_1} \int_{-\infty}^{\infty} |\langle G, S_t F\rangle| dt. \tag{3.12}$$

Now let

$$h_0 := \min\{h:\ \bar q_h - \bar p_h \geqslant 2,\ h=1,\dots,m\}.$$

With the change of variables

$$t_{\bar q_h - 1}^{-} = \begin{cases} t_{\bar q_h-1}, & \text{if } \bar q_h > \bar q_{h-1}+1; \\ t_{\bar q_2-\alpha-1}, & \text{if } \bar q_h = \bar q_{h-1}+1 = \bar q_{h-2}+2 = \cdots = \\ & = \bar q_{h-\alpha}+\alpha > \bar q_{h-(\alpha-1)} + (\alpha+1), \end{cases} \tag{3.13}$$

$$t^-_{\bar{p}_h} = \begin{cases} t'_{\bar{q}_\alpha}, & \text{if } \bar{p}_h = \bar{q}_\alpha, \\ t_{\bar{p}_h}, & \text{if } \bar{p}_h \notin \{\bar{q}_h\}_{h=1}^m, \end{cases} \tag{3.14}$$

almost everywhere, therefore

$$|\Delta_n^\varepsilon(z,t,\{\bar{q}_h,\bar{p}_h\}_{h=1}^m)| \leqslant$$

$$\leqslant C_2^n \int_0^t dt_1 \int_0^{t_1} dt_2 \cdots \int_0^{t_{\bar{q}_1-2}} dt_{\bar{q}_1-1} \int_0^{t_{\bar{q}_1-1}} dt_{\bar{q}_1+1} \cdots$$

$$\cdots \int_0^{t_{\bar{q}_{h_0}-2}} dt_{\bar{q}_{h_0}-1} \int_{-t^-_{\bar{p}_{h_0}}/z^2}^{(t^-_{\bar{q}_{h_0}-1}-t^-_{\bar{p}_{h_0}})/z^2} dt'_{\bar{q}_{h_0}} |\langle g, S_{t'_{\bar{q}_{h_0}}} g\rangle| \cdots$$

$$\cdots \int_0^{t_{\bar{q}_m-2}} dt_{\bar{q}_m-1} \int_0^{t_{\bar{q}_m-1}} dt_{\bar{q}_m+1} \cdots \int_0^{t_{n-1}} dt_n. \tag{3.15}$$

Since $\bar{q}_{h_0} > \bar{p}_{h_0}+1$, $t^-_{\bar{q}_{h_0}-1} - t^-_{\bar{p}_{h_0}} < 0$ and therefore the right hand side of (3.15) tends to zero.

Having proved that the type II terms tend to zero, we shall now compute the limit of the type I terms.

LEMMA 3.4. *For each $n \in \mathbb{N}$, $\varepsilon \in \{0,1\}^n$,*

$$\lim_{z\to 0} I_n^\varepsilon(z,t) = \sum_{m=0}^{k\wedge(n-k)} \sum_{(q_1,\dots,q_m)}' \sum_{m'=0}^{k\wedge(n-k)} \sum_{(q'_1,\dots,q'_{m'})} (g_1|g_1)_-^m (g_0|g_0)_-^{m'} \times$$

$$\times \int_{0\leqslant t_n \leqslant\cdots\leqslant \widehat{t_{q'_{m'}}} \leqslant\cdots\leqslant \widehat{t_{q_m}} \leqslant\cdots\leqslant \widehat{t_{q'_1}} \leqslant\cdots\leqslant \widehat{t_{q_1}} \leqslant\cdots\leqslant t_1 \leqslant t} dt_1 \cdots \widehat{dt_{q_1}} \cdots \widehat{dt_{q'_1}} \cdots \widehat{dt_{q_m}} \cdots \widehat{dt_{q_{m'}}} \cdots dt_n \times$$

$$\times \prod_{\alpha\in\{1,\dots,n\}\setminus(\{q_h\}_{h=1}^m \cup \{q'_h\}_{h=1}^{m'})} (f|g_{\varepsilon(\alpha)})\chi_{[S,T]}(t_\alpha) \times$$

$$\times \prod_{\alpha\in\mathcal{C}\cap\{1,\dots,\bar{q}_1-2\}} (g_{1-\varepsilon(\alpha)}|f')\chi_{[S',T']}(t_\alpha) \times$$

$$\times (g_{1-\varepsilon(\bar{q}_{r_1})}|f')\chi_{[S',T']}(t_{\bar{q}_1-1}) \times$$

$$\times \prod_{\alpha\in\mathcal{C}\cap\{\bar{q}_{r_1}+1,\dots,\bar{q}_{r_1+1}-2\}} (g_{1-\varepsilon(\alpha)}|f')\chi_{[S',T']}(t_\alpha) \times$$

$$\times (g_{1-\varepsilon(\bar{q}_{r_2})}|f')\chi_{[S',T']}(t_{\bar{q}_1+1-1}) \times \cdots$$

$$\cdots \times \prod_{\alpha\in\mathcal{C}\cap\{\bar{q}_{r_{x-1}}+1,\dots,\bar{q}_{r_{x-1}+1}-2\}} (g_{1-\varepsilon(\alpha)}|f')\chi_{[S',T']}(t_\alpha) \times$$

$$\times (g_{1-\varepsilon(\bar{q}_{r_x})}|f')\chi_{[S',T']}(t_{\bar{q}_{x-1}-1})\times$$
$$\times \prod_{\alpha\in\mathcal{G}\cap\{\bar{q}_{r_x}+1,\dots,n\}} (g_{1-\varepsilon(\alpha)}|f')\chi_{[S',T']}(t_\alpha)\times$$
$$\times\langle W(\chi_{[S,T]}\otimes f)\Psi, W(\chi_{[S',T']}\otimes f')\Psi\rangle, \tag{3.16}$$

where $\widehat{t}$ means that the t is absent,

$$\mathcal{G} = \{1,\dots,n\}\setminus(\{q_h-1\}_{h=1}^{m}\cup\{q'_h-1\}_{h=1}^{m'})$$

and

$$\{\bar{q}_h\}_{h=1}^{m+m'} := \{q_h\}_{h=1}^{m}\cup\{q'_h\}_{h=1}^{m'}$$

with

$$\bar{q}_1 < \cdots < \bar{q}_{m+m'} \tag{3.17}$$

x and $\{r_h\}_{h=1}^{x}$ are uniquely determined by the prescription

$$\{\bar{q}_h\}_{h=1}^{m+m'} = \{\bar{q}_h\}_{h=1}^{r_1}\cup\{\bar{q}_h\}_{h=r_1+1}^{r_2}\cup\cdots\cup\{\bar{q}_h\}_{h=r_{x-1}+1}^{r_x} \tag{3.18}$$

and

$$\bar{q}_{r_y+1} = \bar{q}_{r_y+2}-1 = \bar{q}_{r_y+3}-2=\cdots=$$
$$= \bar{q}_{r_{y+1}} - (r_{y+1}-r_y-1) < \bar{q}_{r_{y+1}+1} - (r_{y+1}-r_y) \tag{3.19}$$

with

$$y = 0,1,\cdots,x-1 \qquad \textit{and} \qquad r_0 := 0.$$

Remark. The conditions (3.18), (3.19) mean that we split the ordered sequence (3.17) into ordered subsequences which are maximal with respect to the property that $\bar{q}_{h+1} = \bar{q}_h$ for all elements in the subsequence (a subsequence may contain a single element).

Proof. For each m, m', there exist unique x and $\{r_h\}_{h=1}^{x}$ such that $\{\bar{q}_h\}_{h=1}^{m+m'}$ satisfies (3.18) and (3.19). If in the expression of $I_n^\varepsilon(z,t)$ we put

$$t'_\alpha = \begin{cases} t_\alpha, & \text{if } \alpha\in\{1,\dots,n\}\setminus\{\bar{q}_h\}_{h=1}^{m+m'};\\ (t_{\bar{q}_h}-t_{\bar{q}_h-1})/z^2, & \text{if } \alpha\in\{\bar{q}_h\}_{h=1}^{m+m'}, \end{cases} \tag{3.20}$$

then $t_{\bar{q}_h} = z^2 t'_{\bar{q}_h} + t'_{\bar{q}_h-1}$, $h = 1,\dots,m+m'$. So $I_n^\varepsilon(z,t)$ includes three types of integrals

$$\int_{(S'-t'_\alpha)/z^2}^{(T'-t'_\alpha)/z^2} \langle F, S_u G\rangle du, \quad \int_0^{t'_{q\alpha}+z^2(\dots)}, \quad \int_{-t_{q\alpha}/z^2-(\dots)}^{0} . \tag{3.21}$$

In the limit $z\to 0$ the integrals of the form

$$\int_{(S'-t'_\alpha)/z^2}^{(T'-t'_\alpha)/z^2} \langle F, S_u G\rangle du$$

converge to $\chi_{[S',T']}(t'_\alpha)(F \mid G)$; the integrals of the form $\int_0^{t'_{q_\alpha}+z^2(\ldots)}$ converge to $\int_0^{t'_{q_\alpha}}$ and the integrals $\int_{-t_{q_\alpha}/z^2-(\ldots)}^{0}$ give rise to the factors $(g_j \mid g_j)_-$ and the corresponding integration variable disappears. The limiting procedure is justified by dominated convergence.

4. THE UNIFORM ESTIMATE

In Section 3, for each $n \in \mathbb{N}$, we have computed the limit, as $z \to 0$, of the n-th term of the iterated series. In this section, we shall prove the uniform estimates which allow to take the term by term limit of the series.
In the following, we shall use the notation

$$\|g_\varepsilon\|_-^2 := \int_{-\infty}^{0} |\langle g_\varepsilon, S_u g_\varepsilon\rangle| du, \qquad \varepsilon = 0, 1 . \tag{4.1}$$

The following Lemma is a further (with respect to (Accardi et al.,1990a)) generalization of the Pulé's inequality (i.e., Lemma (3.2) in (Pulé, 1974)).

LEMMA 4.1. *Let $f : \mathbb{R} \longrightarrow \mathbb{R}_+$ be a positive integrable symmetric function and let*

$$\mathcal{S}^0_{2m,n} :=$$

$$:= \{\varepsilon : \{1,\ldots,2m\} \longrightarrow \{1,\ldots,n\},\ \varepsilon(2) < \cdots < \varepsilon(2m) \quad and \quad \varepsilon(2h-1) < \varepsilon(2h)\}.$$

Then, for each $n \in \mathbb{N}$, $1 \leqslant m \leqslant n-1$, $2 \leqslant q_1 < \cdots < q_m \leqslant n$, $1 \leqslant p_1 < \cdots < p_m \leqslant n-1$, if $\{p_h, q_h\}_{h=1}^m = \{1,\ldots,n\}$ and $p_h < q_h$, $h = 1,\ldots,m$, one has

$$\lambda^{-2m} \int_0^t dt_1 \cdots \int_0^{t_{n-1}} dt_n \sum_{\sigma\in\mathcal{S}^0_m} \prod_{h=1}^m f\Big(\frac{t_{p_{\sigma(h)}} - t_{q_h}}{\lambda^2}\Big) \leqslant \frac{t^{n-m}}{(n-m)!}\Big[\int_{-\infty}^{0} f(t)dt\Big]^m, \tag{4.2}$$

where

$$\mathcal{S}^0_m := \{\sigma \in \mathcal{S}_m : \ p_{\sigma(h)} < q_h,\ h = 1,\ldots,m\}.$$

Proof. Let, for each $\sigma \in \mathcal{S}^0_m$,

$$\varepsilon_\sigma(j) = \begin{cases} q_h, & \text{if } j = 2h, \\ p_{\sigma(h)}, & \text{if } j = 2h-1, \end{cases} \tag{4.3}$$

then $\varepsilon_\sigma \in \mathcal{S}^0_{2m,n}$. Notice that for $\varepsilon \in \mathcal{S}^0_{2m,n}$, there exists a subset $\{r^\varepsilon_1,\ldots,r^\varepsilon_{2m-n}\} \subset \{1,\ldots,n\}$ such that

$$\{r^\varepsilon_1,\ldots,r^\varepsilon_{2m-n}\} \subset \{\varepsilon(2h-1)\}_{h=1}^m \cap \{\varepsilon(2h)\}_{h=1}^m.$$

For $\sigma \neq \sigma'$, one has $\varepsilon_\sigma \neq \varepsilon_{\sigma'}$, so

$$\lambda^{-2m} \int_0^t dt_1 \cdots \int_0^{t_{n-1}} dt_n \sum_{\sigma \in S_m^0} \prod_{h=1}^m f\left(\frac{t_{p_{\sigma(h)}} - t_{q_h}}{\lambda^2}\right) \leqslant$$

$$\leqslant \lambda^{-2m} \int_0^t dt_1 \cdots \int_0^{t_{n-1}} dt_n \sum_{\varepsilon \in S_{2m,n}^0} \prod_{h=1}^m f\left(\frac{t_{\varepsilon(2h-1)} - t_{\varepsilon(2h)}}{\lambda^2}\right). \tag{4.4}$$

Put

$$S_t^{(n)} := \{(t_1, \ldots, t_n),\ t \geqslant t_1 \geqslant \cdots \geqslant t_n \geqslant 0\}, \tag{4.5}$$

$$x_h^{(\varepsilon)} := \frac{t_{\varepsilon(2h-1)} - t_{\varepsilon(2h)}}{\lambda^2}, \qquad h = 1, \ldots, m, \tag{4.6}$$

$$y_h^{(\varepsilon)} := t_{\varepsilon(\delta_h^\varepsilon)}, \qquad h = 1, \ldots, n-m, \tag{4.7}$$

where $\delta_1^\varepsilon < \cdots < \delta_{n-m}^\varepsilon$ and $\{\varepsilon(\delta_h^\varepsilon)\}_{h=1}^{n-m} \cap \{\varepsilon(2h-1)\}_{h=1}^m = \varnothing$. So one has

$$0 \leqslant x_h^{(\varepsilon)} < \infty, \qquad \left(y_1^{(\varepsilon)}, \cdots, y_{n-m}^{(\varepsilon)}\right) \in S_t^{(n-m)}. \tag{4.8}$$

Therefore

$$\lambda^{-2m} \int_0^t dt_1 \cdots \int_0^{t_{n-1}} dt_n \sum_{\sigma \in S_m^0} \prod_{h=1}^m f\left(\frac{t_{p_{\sigma(h)}} - t_{q_h}}{\lambda^2}\right) \leqslant$$

$$\leqslant \int_{S_t^{(n-m)}} \int_{\mathbf{R}_+^m} \prod_{h=1}^m f(x_h) dy_1 \cdots dy_{n-m} dx_1 \cdots dx_m = \frac{t^{n-m}}{(n-m)!} \left[\int_{-\infty}^0 f(t)dt\right]^m. \tag{4.9}$$

COROLLARY 4.2. *Let $f : \mathbf{R} \longrightarrow \mathbf{R}_+$ be a positive integrable symmetric function, for each $n \in \mathbf{N}$, $1 \leqslant m \leqslant n-1$, $2 \leqslant q_1 < \cdots < q_m \leqslant n$, $1 \leqslant p_1 < \cdots < p_m \leqslant n-1$, $p_h < q_h$, $h = 1, \ldots, m$, one has*

$$\lambda^{-2m} \int_0^t dt_1 \cdots \int_0^{t_{n-1}} dt_n \sum_{\sigma \in S_m^0} \prod_{h=1}^m f\left(\frac{t_{p_{\sigma(h)}} - t_{q_h}}{\lambda^2}\right) \leqslant$$

$$\leqslant \frac{t^{n-m}}{(n-\bar{m})!(\bar{m}-m)!} \left[\int_{-\infty}^0 f(t)dt\right]^m, \tag{4.10}$$

where $\bar{m} := |\{p_h, q_h\}_{h=1}^m|$.

Proof. Denote for each $\sigma \in S_n^0$,

$$\{\alpha_h\}_{h=1}^{\bar{m}} := \{p_{\sigma(h)}, q_h\}_{h=1}^m, \quad \{\beta_h\}_{h=1}^{n-\bar{m}} := \{1, \ldots, n\} \setminus \{\alpha_h\}_{h=1}^{\bar{m}}$$

with

$$\alpha_1 < \alpha_2 < \cdots < \alpha_{\bar{m}}, \quad \beta_1 < \beta_2 < \cdots < \beta_{n-\bar{m}}.$$

Enlarging $t_{\beta_1 - 1}$ to t, $t_{\beta_2 - 1}$ to $t_{\beta_1}, \ldots, t_{\beta_{n-\bar{m}}-1}$ to $t_{\beta_{n-\bar{m}-1}}$ and $t_{\alpha_1 - 1}$ to t, $t_{\alpha_2 - 1}$ to $t_{\alpha_1}, \ldots, t_{\alpha_{\bar{m}}-1}$ to $t_{\alpha_{\bar{m}-1}}$, one has

$$\lambda^{-2m} \int_0^t dt_1 \cdots \int_0^{t_{n-1}} dt_n \sum_{\sigma \in S_m^0} \prod_{h=1}^m f\left(\frac{t_{p_{\sigma(h)}} - t_{q_h}}{\lambda^2}\right) \leqslant$$

$$\leqslant \lambda^{-2m} \sum_{\sigma \in S_m^0} \int_0^t dt_1 \cdots \int_0^{t_{\bar{m}-1}} dt_{\bar{m}} \prod_{h=1}^m f\left(\frac{t_{p_{\sigma(h)}} - t_{q_h}}{\lambda^2}\right) t^{n-\bar{m}} \frac{1}{(n-\bar{m})!}.$$

Notice that there $\{p_h, q_h\}_{h=1}^m$ is not the same as that in the left hand side of (4.10) but the order–relations is remained. So, using the Lemma 4.1, one gets (4.10).

LEMMA 4.3. *For each* $n \in \mathbb{N}$,

$$\sum_{\varepsilon \in \{0,1\}^n} |I_n^\varepsilon(z,t) + II_n^\varepsilon(z,t)| \leqslant$$

$$\leqslant 16^n \max_{0 \leqslant m \leqslant n-1} \frac{t^{n-m}}{(n-m)!} \left(\|g_0\|_- \bigvee \|g_1\|_-\right)^{2m} \times$$

$$\times \left[\max_{F=f,f';G=g_0,g_1} \int_{-\infty}^{\infty} |\langle F, S_t G\rangle| dt\right]^{2(n-m)} \tag{4.11}$$

Proof. Put

$$F(u) := \max_{\varepsilon \in \{0,1\}} |\langle g_\varepsilon, S_t g_{1-\varepsilon}\rangle|. \tag{4.12}$$

So, by Lemma (4.1) and Corollary (4.2),

$$\sum_{\varepsilon \in \{0,1\}^n} |I_n^\varepsilon(z,t) + II_n^\varepsilon(z,t)| \leqslant \sum_{k=0}^n \sum_{1 \leqslant i_1 < \cdots < i_k \leqslant n} \sum_{m=0}^{n-1} \binom{n}{m}^2 \times$$

$$\times \left[\max_{F=f,f';G=g_0,g_1} \int_{-\infty}^{\infty} |\langle F, S_t G\rangle| dt\right]^{2(n-m)} \times$$

$$\times \max_{m \leqslant n_0 \leqslant n} \frac{t^{n-m}}{(n-n_0)!(n_0-m)!} \left(\|g_0\|_- \bigvee \|g_1\|_-\right)^{2m} \leqslant$$

$$\leqslant \sum_{k=0}^n \sum_{1 \leqslant i_1 < \cdots < i_k \leqslant n} 8^n \max_{0 \leqslant m \leqslant n-1} \left(\frac{t^{n-m}}{(n-m)!} \left(\|g_0\|_- \bigvee \|g_1\|_-\right)^{2m} \times\right.$$

$$\times\left[\max_{F=f,f';G=g_0,g_1}\int_{-\infty}^{\infty}|\langle F,S_tG\rangle|dt\right]^{2(n-m)}\Bigg) \tag{4.13}$$

which easily implies (4.11).

COROLLARY 4.4. *For each $D\in B(H_0)$ if*

$$\left(\|g_0\|_-\bigvee\|g_1\|_-\right)^2<1/16\|D\| \tag{4.14}$$

then the series

$$\sum_{n=0}^{\infty}\sum_{\varepsilon\in\{0,1\}^n}(I_n^{\varepsilon}(z,t)+II_n^{\varepsilon}(z,t)) \tag{4.15}$$

converges absolutely and uniformly for $z>0$.

5. THE LOW DENSITY LIMIT AND ITS DERIVATIVE

In this Section, using the results of Sections 3 and 4 we derive the existence of the low density limit (1.23) and its explicit form. Using the latter, we deduce an integral equation for this limit.

THEOREM 5.1. *For each $u,v\in H_0$ and $D\in B(H_0)$ satisfying* (1.10) *and for each $g_0,g_1\in K$ satisfying* (1.11) *and*

$$\left(\|g_0\|_-\bigvee\|g_1\|_-\right)^2<1/16\|D\| \tag{5.0}$$

the low density limit

$$\lim_{z\to 0}\left\langle u\otimes W\left(z\int_{S/z^2}^{T/z^2}S_ufdu\right)\Phi,U(t/z^2)v\otimes W\left(z\int_{S'/z^2}^{T'/z^2}S_uf'du\right)\Phi\right\rangle \tag{5.1}$$

exists and is equal to

$$\sum_{n=0}^{\infty}\sum_{k=0}^{n}\sum_{1\leqslant i_1<\dots<i_k\leqslant n}\langle u,D_{\varepsilon(1)}\dots D_{\varepsilon(n)}v\rangle\times$$

$$\times\sum_{m=0}^{k\wedge(n-k)}\sum_{(q_1,\dots,q_m)}^{\prime}\sum_{m'=0}^{k\wedge(n-k)}\sum_{(q'_1,\dots,q'_{m'})}(g_1|g_1)_-^{m}(g_0|g_0)_-^{m'}\times$$

$$\times\int_{0\leqslant t_n\leqslant\dots\leqslant\widehat{t_{q'_{m'}}}\leqslant\dots\leqslant\widehat{t_{q_m}}\leqslant\dots\leqslant\widehat{t_{q'_1}}\leqslant\dots\leqslant\widehat{t_{q_1}}\leqslant\dots\leqslant t_1\leqslant t}dt_1\cdots\widehat{dt_{q_1}}\cdots\widehat{dt_{q'_1}}\cdots\widehat{dt_{q_m}}\cdots\widehat{dt_{q_{m'}}}\cdots dt_n\times$$

$$\times \prod_{\alpha\in\{1,\ldots,n\}\setminus(\{q_h\}_{h=1}^{m}\cup\{q'_h\}_{h=1}^{m'})} (f|g_{\varepsilon(\alpha)})\chi_{[S,T]}(t_\alpha)\times$$

$$\times \prod_{\alpha\in\mathcal{C}\cap\{1,\ldots,\bar q_1-2\}} (g_{1-\varepsilon(\alpha)}|f')\chi_{[S',T']}(t_\alpha)(g_{1-\varepsilon(\bar q_{r_1})}|f')\chi_{[S',T']}(t_{\bar q_1-1})\times$$

$$\times \prod_{\alpha\in\mathcal{C}\cap\{\bar q_{r_1}+1,\ldots,\bar q_{r_1+1}-2\}} (g_{1-\varepsilon(\alpha)}|f')\chi_{[S',T']}(t_\alpha)(g_{1-\varepsilon(\bar q_{r_2})}|f')\chi_{[S',T']}(t_{\bar q_1+1-1})\times$$

$$\cdots\cdots$$

$$\times \prod_{\alpha\in\mathcal{C}\cap\{\bar q_{r_{x-1}}+1,\ldots,\bar q_{r_{x-1}+1}-2\}} (g_{1-\varepsilon(\alpha)}|f')\chi_{[S',T']}(t_\alpha)(g_{1-\varepsilon(\bar q_{r_x})}|f')\chi_{[S',T']}(t_{\bar q_{x-1}-1})\times$$

$$\times \prod_{\alpha\in\mathcal{C}\cap\{\bar q_{r_x}+1,\ldots,n\}} (g_{1-\varepsilon(\alpha)}|f')\chi_{[S',T']}(t_\alpha)\langle W(\chi_{[S,T]}\otimes f)\Psi, W(\chi_{[S',T']}\otimes f')\Psi\rangle. \tag{5.2}$$

Proof. This follows easily from Lemmas 3.3, 3.4 and Lemma 4.5.
In order to find an integral equation satisfied by the limit (5.1), we proceed in analogy with the similar problem dealt with in (Accardi, 1990) and, for each $n \in \mathbb{N}$ we use the explicit form (1.16) of $V(t_1)$ in the expression

$$(-i)^n \int_0^{t/z^2} dt_1 \int_0^{t_1} dt_2 \cdots \int_0^{t_{n-1}} dt_n \Big\langle u\otimes W\left(z\int_{S/z^2}^{T/z^2} S_u f du\right)\Phi,$$

$$V(t_1)\cdots V(t_n)v\otimes W\left(z\int_{S'/z^2}^{T'/z^2} S_u f' du\right)\Phi\Big\rangle =$$

$$= \frac{1}{z^2}(-i)^{n-1}\int_0^{t} dt_1 \int_0^{t_1/z^2} dt_2 \int_0^{t_2} dt_3 \cdots \int_0^{t_{n-1}} dt_n \Big\langle u\otimes W\left(z\int_{S/z^2}^{T/z^2} S_u f du\right)\Phi,$$

$$(D\otimes A^+(S_{t_1/z^2}g_0)A(S_{t_1/z^2}g_1) - D^+\otimes A^+(S_{t_1/z^2}g_1)A(S_{t_1/z^2}g_0))\times$$

$$\times V(t_2)\cdots V(t_n)v\otimes W\left(z\int_{S'/z^2}^{T'/z^2} S_u f' du\right)\Phi\Big\rangle =$$

$$= \int_0^t dt_1 \int_{S/z^2}^{T/z^2} \langle S_u f, S_{t_1/z^2}g_0\rangle du \int_{S'/z^2}^{T'/z^2} \langle S_{t_1/z^2}g_1, S_u f'\rangle du \times$$

$$\times(-i)^{n-1}\int_0^{t_1/z^2} dt_2 \int_0^{t_2} dt_3 \cdots \int_0^{t_{n-1}} dt_n \times$$

$$\times\left\langle D^+ u \otimes W\left(z\int_{S/z^2}^{T/z^2} S_u f du\right)\Phi, V(t_2)\ldots V(t_n)v \otimes W\left(z\int_{S'/z^2}^{T'/z^2} S_u f' du\right)\Phi\right\rangle +$$

$$+\int_0^t dt_1 \int_{S/z^2}^{T/z^2} \langle S_u f, S_{t_1/z^2} g_1\rangle du \int_{S'/z^2}^{T'/z^2} \langle S_{t_1/z^2} g_0, S_u f'\rangle du \times$$

$$\times(-i)^{n-1}\int_0^t dt_1 \int_0^{t_1/z^2} dt_2 \int_0^{t_2} dt_3 \cdots \int_0^{t_{n-1}} dt_n \left\langle -Du \otimes W\left(z\int_{S/z^2}^{T/z^2} S_u f du\right)\Phi,\right.$$

$$\left. V(t_2)\cdots V(t_n)v \otimes W\left(z\int_{S'/z^2}^{T'/z^2} S_u f' du\right)\Phi\right\rangle +$$

$$+\int_0^t dt_1 \int_{S/z^2}^{T/z^2} \langle S_u f, S_{t_1/z^2} g_0\rangle du (-i)^{n-1}\frac{1}{z}\times$$

$$\times \int_0^{t_1/z^2} dt_2 \int_0^{t_2} dt_3 \cdots \int_0^{t_{n-1}} dt_n \left\langle D^+ u \otimes W\left(z\int_{S/z^2}^{T/z^2} S_u f du\right)\Phi,\right.$$

$$\left.[1\otimes A(S_{t_1/z^2} g_1), V(t_2)\cdots V(t_n)]v \otimes W\left(z\int_{S'/z^2}^{T'/z^2} S_u f' du\right)\Phi\right\rangle +$$

$$+\int_0^t dt_1 \int_{S/z^2}^{T/z^2} \langle S_u f, S_{t_1/z^2} g_1\rangle du (-i)^{n-1}\times$$

$$\times\frac{1}{z}\int_0^{t_1/z^2} dt_2 \int_0^{t_2} dt_3 \cdots \int_0^{t_{n-1}} dt_n \left\langle -Du \otimes W\left(z\int_{S/z^2}^{T/z^2} S_u f du\right)\Phi,\right.$$

$$\left.[1\otimes A(S_{t_1/z^2} g_0), V(t_2)\cdots V(t_n)]v \otimes W\left(z\int_{S'/z^2}^{T'/z^2} S_u f' du\right)\Phi\right\rangle =:$$

$$=: I_n(1,z,t) + I_n(2,z,t) + II_n(0,z,t) + II_n(1,z,t). \tag{5.3}$$

Now we analyze separately the contributions of the four type of terms (5.3) to the limit (5.1). From Section 4, we know that

$$\left| \int_0^{t_1/z^2} dt_2 \int_0^{t_2} dt_3 \cdots \int_0^{t_{n-1}} dt_n \left\langle D^+ u \otimes W\left(z \int_{S/z^2}^{T/z^2} S_u f du \right)\Phi, \right.\right.$$

$$\left.\left. V(t_2)\cdots V(t_n) v \otimes W\left(z \int_{S'/z^2}^{T'/z^2} S_u f' du\right)\Phi \right\rangle \right| \leqslant$$

$$\leqslant C\Big(16\|D\|\Big)^n \|u\|\|v\| \max_{0\leqslant m\leqslant n-1} \frac{t_1^{n-m}}{(n-m)!} \Big(\|g_0\|_- \bigvee \|g_1\|_-\Big)^{2m} \tag{5.4}$$

with a constant C. So, if $\Big(\|g_0\|_- \bigvee \|g_1\|_-\Big)^2 < 1/16\|D\|$, one has, by dominated convergence

$$\lim_{z\to 0} \sum_{n=1}^{\infty} I_n(1,z,t) = \sum_{n=1}^{\infty} \int_0^t dt_1 \lim_{z\to 0} \int_{S/z^2}^{T/z^2} \langle S_u f, S_{t_1/z^2} g_0\rangle du \times$$

$$\times \int_{S'/z^2}^{T'/z^2} \langle S_{t_1/z^2} g_1, S_u f'\rangle du\, (-i)^{n-1} \int_0^{t_1/z^2} dt_2 \int_0^{t_2} dt_3 \cdots \int_0^{t_{n-1}} dt_n \times$$

$$\times\left\langle D^+ u \otimes W\left(z \int_{S/z^2}^{T/z^2} S_u f du\right)\Phi, V(t_2)\ldots V(t_n) v \otimes W\left(z\int_{S'/z^2}^{T'/z^2} S_u f' du\right)\Phi\right\rangle. \tag{5.5}$$

Now, if $\Big(\|g_0\|_- \bigvee \|g_1\|_-\Big)^2 < 1/16\|D\|$, then the limit (5.1) is continuous for $u, v \in H_0$. So we can write this limit as $\langle u, G(t)\rangle$ with $G(t) \in H_0$ for each $t \geqslant 0$. With this notation (5.5) becomes

$$\lim_{z\to 0} \sum_{n=1}^{\infty} I_n(1,z,t) = \int_0^t dt_1 (f|g_0)\chi_{[S,T]}(t_1)(g_1|f')\chi_{[S',T']}(t_1)\langle D^+ u, G(t_1)\rangle. \tag{5.6}$$

Similarly, we obtain

$$\lim_{z\to 0} \sum_{n=1}^{\infty} I_n(2,z,t) = \int_0^t dt_1 (f|g_1)\chi_{[S,T]}(t_1)(g_0|f')\chi_{[S',T']}(t_1)\langle -Du, G(t_1)\rangle. \tag{5.7}$$

Now, we investigate $II_n(0,z,t)$. By definition for $n \geqslant 2$, put $s = t_1$, $s_1 = t_2, \ldots$, $s_n = t_{n+1}$,

$$II_n(0,z,t) = \int_0^t ds \int_{(S-s)/z^2}^{(T-s)/z^2} \langle S_u f, g_0\rangle du\, (-i)^n \frac{1}{z} \times$$

$$\times \int_0^{s/z^2} ds_1 \int_0^{s_1} ds_2 \cdots \int_0^{s_{n-1}} ds_n \Big\langle D^+ u \otimes W\left(z \int_{S/z^2}^{T/z^2} S_u f du\right) \Phi,$$

$$[1 \otimes A(S_{s/z^2} g_1), V(s_1) \cdots V(s_n)] v \otimes W\left(z \int_{S'/z^2}^{T'/z^2} S_u f' du\right) \Phi \Big\rangle. \tag{5.8}$$

We can write

$$[1 \otimes A(S_{s/z^2} g_1), V(s_1) \cdots V(s_n)] =$$

$$= \sum_{\varepsilon \in \{0,1\}^n} [1 \otimes A(S_{s/z^2} g_1), D_{\varepsilon(1)} \cdots D_{\varepsilon(n)} \otimes (I_n^\varepsilon + II_n^\varepsilon)] \tag{5.9}$$

and by Lemma 3.3

$$\lim_{z \to 0} \int_0^t ds \frac{1}{z} [A(S_{s/z^2} g_1), II_n^\varepsilon] = 0 \tag{5.10}$$

and moreover the following Lemma is valid.

LEMMA 5.2. *For each $n \in \mathbb{N}$, the limit*

$$\lim_{z \to 0} II_{n+1}(\varepsilon, z, t) \tag{5.11}$$

exists and is equal to

$$\lim_{z \to 0} z \int_0^{t/z^2} ds \int_{S/z^2 - s}^{T/z^2 - s} \langle S_u f, g_\varepsilon \rangle du \, (-i)^n \int_0^s ds_1 \int_0^{s_1} ds_2 \cdots \int_0^{s_{n-1}} ds_n \times$$

$$\times \Big\langle D_\varepsilon^+ u \otimes W\left(z \int_{S/z^2}^{T/z^2} S_u f du\right) \Phi, [1 \otimes A(S_s g_{1-\varepsilon}), V(s_1)] \times$$

$$\times V(s_2) \cdots V(s_n) v \otimes W\left(z \int_{S'/z^2}^{T'/z^2} S_u f' du\right) \Phi \Big\rangle. \tag{5.12}$$

Proof. It is enough to notice that other terms have the form of type II. Let us introduce, for simplicity the notation

$$\Phi_z(f, S, T) := W\left(z \int_{S/z^2}^{T/z^2} S_u f du\right) \Phi. \tag{5.13}$$

Then, using Lemma 5.2, formula (5.10) and the uniform estimate, we obtain that the limit

$$\lim_{z\to 0}\sum_{n=0}^{\infty} II_{n+1}(\varepsilon, z, t)$$

exists and is equal to

$$\lim_{z\to 0}\int_0^t dr\chi_{[S,T]}(r)(f, g_\varepsilon)\langle D_\varepsilon^+ u\otimes\Phi_z(f, S, T),$$

$$\frac{1}{z}[1\otimes A(S_{r/z^2}f_{1-\varepsilon}), U(r/z^2)]v\otimes\Phi_z(f', S', T')\rangle. \tag{5.14}$$

Clearly this limit is continuous in u (and in $D_\varepsilon^+ u$), hence it will have the form

$$\langle D_\varepsilon^+ u, K_\varepsilon(t)\rangle = \langle u, D_\varepsilon K_\varepsilon(t)\rangle \tag{5.15}$$

for some $K_\varepsilon(t)\in H_0$.
Our next step shall be to deduce an equation for $D_\varepsilon K_\varepsilon(t)$.

LEMMA 5.3. *Let $K_\varepsilon(t)$ be defined respectively by equations (5.14), (5.15). Then $K_\varepsilon(t)$ satisfies the equation*

$$\int_0^t \langle D_\varepsilon^+ u, K_\varepsilon(s)\rangle(f|g_\varepsilon)\chi_{[S,T]}(s)ds =$$

$$= \int_0^t ds(f|g_\varepsilon)\chi_{[S,T]}(s)(g_{1-\varepsilon}|g_{1-\varepsilon}) - \langle D_{1-\varepsilon}^+ D_\varepsilon^+ u, K_{1-\varepsilon}(s)\rangle +$$

$$+\int_0^t ds(f|g_\varepsilon)\chi_{[S,T]}(s)(g_{1-\varepsilon}|g_{1-\varepsilon}) - \chi_{[S',T']}(s)(g_\varepsilon|f')\langle D_{1-\varepsilon}^+ D_\varepsilon^+ u, G(s)\rangle. \tag{5.16}$$

Proof. It is clear that

$$[1\otimes A(S_{s/z^2}g_{1-\varepsilon}), V(s_1)] = i(D_{1-\varepsilon}\otimes 1)\langle S_{s/z^2}g_{1-\varepsilon}, S_{s_1}g_{1-\varepsilon}\rangle A(S_{s_1}g_\varepsilon),$$

therefore, bringing $A(S_{s_1}g_\varepsilon)$ to the right of the product $V(s_2)\cdot\ldots\cdot V(s_n)$, we obtain a commutator term plus a term coming from the action of $A(S_{s_1}g_\varepsilon)$ on the coherent vector $\Phi_z(f', S', T')$. Change of variable in the first term:

$$s_1 z^2 = r_1 \tag{5.17}$$

we see that the limit (5.14) is equal to

$$\lim_{z\to 0}\int_0^t ds(f|g_\varepsilon)\chi_{[S,T]}(s)\sum_{n=1}^{\infty}(-i)^{n-1}\int_0^s dr_1\int_0^{r_1/z^2}\int_0^{s_2} ds_3\cdots\int_0^{s_{n-1}} ds_n\times$$

$$\times\left\{\frac{1}{z^2}\langle g_{1-\varepsilon}, S_{(r_1-s)/z^2}g_{1-\varepsilon}\rangle\langle D^+_{1-\varepsilon}D^+_\varepsilon u\otimes\Phi_z(f,S,T)\rangle,\right.$$

$$\frac{1}{z}\left[1\otimes A(S_{r_1/z^2}g_\varepsilon), V(s_2)\right]V(s_3)\ldots V(s_n)v\otimes\Phi_z(f',S',T')\rangle+$$

$$+\langle D^+_{1-\varepsilon}D^+_\varepsilon u\otimes\Phi_z(f,S,T), V(s_2)\ldots V(s_n)v\otimes\Phi_z(f',S',T')\rangle\times$$

$$\left.\times\frac{1}{z^2}\langle g_{1-\varepsilon}, S_{(r_1-s)/z^2}g_{1-\varepsilon}\rangle\int\limits_{(S'-r_1)/z^2}^{(T'-r_1)/z^2}dv\langle g_\varepsilon, S_v f'\rangle\right\}. \tag{5.18}$$

When $z\to 0$, we obtain the right hand side of (5.16).
Now, by solving equation (5.16), we find the explicit form of $D_\varepsilon K_\varepsilon(t)$.

LEMMA 5.4. *In the above notations, denoting for each $\varepsilon\in\{0,1\}$*

$$D_g(\varepsilon) := (1-(g_\varepsilon|g_\varepsilon)_-(g_{1-\varepsilon}|g_{1-\varepsilon})_-D_\varepsilon D_{1-\varepsilon})^{-1} =$$

$$=\sum_{n=0}^{\infty}(g_\varepsilon|g_\varepsilon)^n_-(g_{1-\varepsilon}|g_{1-\varepsilon})^n_-(D_\varepsilon D_{1-\varepsilon})^n \tag{5.19}$$

one has

$$D_\varepsilon K_\varepsilon(t) = \chi_{[S',T']}(t)(g_{1-\varepsilon}|f')(g_\varepsilon|g_\varepsilon)_-(g_{1-\varepsilon}|g_{1-\varepsilon})_-D_g(\varepsilon)D_\varepsilon D_{1-\varepsilon}D_\varepsilon G(t)+$$

$$+\chi_{[S',T']}(t)(g_\varepsilon|f')(g_{1-\varepsilon}|g_{1-\varepsilon})_-D_g(\varepsilon)D_\varepsilon D_{1-\varepsilon}D_\varepsilon G(t). \tag{5.20}$$

Proof. First notice that, since f,S,T,u are arbitrary, then equation (5.16) is equivalent to

$$D_\varepsilon K_\varepsilon(t) = (g_{1-\varepsilon}|g_{1-\varepsilon})_-D_\varepsilon D_{1-\varepsilon}K_{1-\varepsilon}(t)+$$

$$+(g_{1-\varepsilon}|g_{1-\varepsilon})_-(g_\varepsilon|f')\chi_{[S',T']}(t)D_\varepsilon D_{1-\varepsilon}G(t). \tag{5.21}$$

Replacing $K_{1-\varepsilon}(t)$ by its expression (5.14) and with the notation

$$a_\varepsilon := (g_\varepsilon|g_\varepsilon)_- \tag{5.21a}$$

we obtain

$$D_\varepsilon K_\varepsilon(t) = (a_\varepsilon a_{1-\varepsilon}D_\varepsilon D_{1-\varepsilon})D_\varepsilon K_\varepsilon(t)+$$

$$+\Big\{(g_{1-\varepsilon}|f')\chi_{[S',T']}(s)(a_\varepsilon a_{1-\varepsilon}D_\varepsilon D_{1-\varepsilon})D_\varepsilon G(t)+$$

$$+(g_\varepsilon|f')\chi_{[S',T']}(s)(a_{1-\varepsilon}D_\varepsilon D_{1-\varepsilon})D_\varepsilon G(t)\Big\}. \tag{5.22}$$

Thus $D_\varepsilon K_\varepsilon(t)$ satisfies the operator equation

$$(1-T_\varepsilon)D_\varepsilon K_\varepsilon(t) = G_\varepsilon(t), \tag{5.23}$$

where $G_\varepsilon(t)$ is the term in braces in (5.22) and

$$T_\varepsilon := a_\varepsilon a_{1-\varepsilon}D_\varepsilon D_{1-\varepsilon}. \tag{5.24}$$

Notice that

$$\|T_\varepsilon\| \leqslant \|g_0\|^2\|g_1\|^2\|D\|^2$$

which is strictly less than 1 because of our assumption (5.0). This implies that

$$D_\varepsilon K_\varepsilon(t) = \sum_{n=0}^{\infty} T_\varepsilon^n G_\varepsilon(t) = (1 - T_\varepsilon)^{-1} G_\varepsilon(t). \tag{5.25}$$

From this, (5.21a) and (5.24), (5.20) immediately follows.
Up to now, one can get the following

THEOREM 5.5. *The low density limit*

$$\langle u, G(t)\rangle := \lim_{z\to 0} \left\langle u \otimes W\left(z \int_{S/z^2}^{T/z^2} S_u f du\right)\Phi, U_z(t/z^2)v \otimes W\left(z\int_{S'/z^2}^{T'/z^2} S_u f' du\right)\Phi\right\rangle \tag{5.26}$$

satisfies the integral equation

$$\langle u, G(t)\rangle = \langle u, G(0)\rangle + \int_0^t ds \sum_{\varepsilon\in\{0,1\}} \Big(\sum_{n=1}^{\infty} (f|g_\varepsilon)\chi_{[S,T]}(s)(g_\varepsilon|f')\chi_{[S',T']}(s)\times$$

$$\times (g_{1-\varepsilon}|g_{1-\varepsilon})_-^n (g_\varepsilon|g_\varepsilon)_-^{n-1} \langle (D_{1-\varepsilon}^+ D_\varepsilon^+)^n u, G(s)\rangle +$$

$$+ \sum_{n=1}^{\infty} (f|g_\varepsilon)\chi_{[S,T]}(s)(g_{1-\varepsilon}|f')\chi_{[S',T']}(s)\times$$

$$\times (g_{1-\varepsilon}|g_{1-\varepsilon})_-^{n-1}(g_\varepsilon|g_\varepsilon)_-^{n-1}\langle D_\varepsilon^+(D_{1-\varepsilon}^+ D_\varepsilon^+)^{n-1}u, G(s)\rangle\Big). \tag{5.27}$$

In the following, we shall use the notations

$$D_1(\varepsilon) := D_g(\varepsilon) D_\varepsilon \tag{5.28}$$

and

$$D_2(\varepsilon) := (g_{1-\varepsilon}|g_{1-\varepsilon})_- D_\varepsilon D_{1-\varepsilon} D_g(\varepsilon). \tag{5.29}$$

Then, (5.27) can be written as

$$\langle u, G(t)\rangle = \langle u, G(0)\rangle +$$

$$+ \int_0^t ds \sum_{\varepsilon\in\{0,1\}} \Big((f|g_\varepsilon)\chi_{[S,T]}(s)(g_\varepsilon|f')\chi_{[S',T']}(s)\langle D_2^+(\varepsilon)u, G(s)\rangle +$$

$$+(f|g_\varepsilon)\chi_{[S,T]}(s)(g_{1-\varepsilon}|f')\chi_{[S',T']}(s)\langle D_1^+(\varepsilon)u, G(s)\rangle\Big). \tag{5.30}$$

6. THE QUANTUM STOCHASTIC DIFFERENTIAL EQUATION

In this Section we identify the integral equation (5.30), satisfied by the low density limit (5.26), with the weak form of the quantum stochastic differential equation

$$U(t) = 1 + \int_0^t \sum_{\varepsilon \in \{0,1\}} \Big[D_1(\varepsilon) \otimes dN_s(g_\varepsilon, g_{1-\varepsilon}) + D_2(\varepsilon) \otimes dN_s(g_\varepsilon, g_\varepsilon)\Big] U(s) \quad (6.1)$$

on $H_0 \otimes \Gamma(L^2(\mathbf{R}) \otimes (K, (\cdot|\cdot)))$, where N is number process and for each $g, g' \in K$

$$N(g, g') := N_s(\chi_{[0,s]} \otimes |g\rangle\langle g'|). \quad (6.2)$$

Throughout the section we shall use freely the notations, definitions and results of (Hudson and Parthasarathy, 1984), with the only exception that we call "number process" the process called "gauge" in the paper and we denote it N rather than Λ.
Since D is bounded, it follows that the q.s.d.e. (6.1) has a unique solution $U(t)$ which is given by the iterated series (Hudson and Parthasarathy, 1984). Moreover, we have the following

THEOREM 6.1. *The solution of q.s.d.e.* (6.1) *is unitary.*

Proof. By Ito formula and (6.1), one gets

$$d(U^+(t)U(t)) = dU^+(t)U(t) + U^+(t)dU(t) + dU^+(t)dU(t) =$$

$$= U^+(t)\Big(D_1^+(1) \otimes dN_t(g_0, g_1) + D_1^+(0) \otimes dN_t(g_1, g_0) +$$

$$+ D_2^+(1) \otimes dN_t(g_1, g_1) + D_2^+(0) \otimes dN_t(g_0, g_0)\Big) U(t) +$$

$$+ U^+(t)\Big(D_1(1) \otimes dN_t(g_1, g_0) + D_1(0) \otimes dN_t(g_0, g_1) +$$

$$+ D_2(1) \otimes dN_t(g_1, g_1) + D_2(0) \otimes dN_t(g_0, g_0)\Big) U(t) +$$

$$+ U^+(t)\Big(D_1^+(1)D_1(1)(g_1|g_1) \otimes dN_t(g_0, g_0) + D_1^+(1)D_2(1)(g_1|g_1) \otimes dN_t(g_0, g_1) +$$

$$+ D_1^+(0)D_1(0)(g_0|g_0) \otimes dN_t(g_1, g_1) + D_1^+(0)D_2(0)(g_0|g_0) \otimes dN_t(g_1, g_0) +$$

$$+ D_2^+(1)D_1(1)(g_1|g_1) \otimes dN_t(g_1, g_0) + D_2^+(1)D_2(1)(g_1|g_1) \otimes dN_t(g_1, g_1) +$$

$$+ D_2^+(0)D_1(0)(g_0|g_0) \otimes dN_t(g_0, g_1) + D_2^+(0)D_2(0)(g_0|g_0) \otimes dN_t(g_0, g_0)\Big) U(t) \quad (6.3)$$

and

$$d(U(t)U^+(t)) = dU(t)U^+(t) + U(t)dU^+(t) + dU(t)dU^+(t) =$$

$$= \Big(D_1(1) \otimes dN_t(g_1, g_0) + D_1(0) \otimes dN_t(g_0, g_1) +$$

$$+D_2(1)\otimes dN_t(g_1,g_1)+D_2(0)\otimes dN_t(g_0,g_0)\Big)U(t)U^+(t)+$$

$$+U(t)U^+(t)\Big(D_1^+(1)\otimes dN_t(g_0,g_1)+D_1^+(0)\otimes dN_t(g_1,g_0)+$$

$$+D_2^+(1)\otimes dN_t(g_1,g_1)+D_2^+(0)\otimes dN_t(g_0,g_0)\Big)+$$

$$+\Big(D_1(1)\otimes 1\cdot U(t)U^+(t)D_1^+(1)(g_0|g_0)\otimes dN_t(g_1,g_1)+$$

$$+D_1(1)\otimes 1\cdot U(t)U^+(t)D_2^+(0)(g_0|g_0)\otimes dN_t(g_1,g_0)+$$

$$+D_1(0)\otimes 1\cdot U(t)U^+(t)D_1^+(0)(g_1|g_1)\otimes dN_t(g_0,g_0)+$$

$$+D_1(0)\otimes 1\cdot U(t)U^+(t)D_2^+(1)(g_1|g_1)\otimes dN_t(g_0,g_1)+$$

$$+D_2(1)\otimes 1\cdot U(t)U^+(t)D_1^+(0)(g_1|g_1)\otimes dN_t(g_1,g_0)+$$

$$+D_2(1)\otimes 1\cdot U(t)U^+(t)D_2^+(1)(g_1|g_1)\otimes dN_t(g_1,g_1)+$$

$$+D_2(0)\otimes 1\cdot U(t)U^+(t)D_1^+(1)(g_0|g_0)\otimes dN_t(g_0,g_1)+$$

$$+D_2(0)\otimes 1\cdot U(t)U^+(t)D_2^+(0)(g_0|g_0)\otimes dN_t(g_0,g_0)\Big). \tag{6.4}$$

The unitarity condition is therefore equivalent to

$$\begin{aligned}0 &= \big(D_2^+(0)+D_2(0)+(g_1|g_1)D_1^+(1)D_1(1)+(g_0|g_0)D_2^+(0)D_2(0)\big)\otimes dN_t(g_0,g_0)+\\ &+\big(D_2^+(1)+D_2(1)+(g_1|g_1)D_2^+(1)D_2(1)+(g_0|g_0)D_1^+(0)D_1(0)\big)\otimes dN_t(g_1,g_1)+\\ &+\big(D_1^+(1)+D_1(0)+(g_1|g_1)D_1^+(1)D_2(1)+(g_0|g_0)D_2^+(0)D_1(0)\big)\otimes dN_t(g_1,g_0)+\\ &+\big(D_1^+(0)+D_1(1)+(g_1|g_1)D_2^+(1)D_1(1)+(g_0|g_0)D_1^+(0)D_2(0)\big)\otimes dN_t(g_0,g_1)\end{aligned} \tag{6.5}$$

and

$$\begin{aligned}0 &= \big(D_2^+(0)+D_2(0)+(g_1|g_1)D_1(0)D_1^+(0)+(g_0|g_0)D_2(0)D_2^+(0)\big)\otimes dN_t(g_0,g_0)+\\ &+\big(D_2^+(1)+D_2(1)+(g_1|g_1)D_2(1)D_2^+(1)+(g_0|g_0)D_1^+(0)D_1(0)\big)\otimes dN_t(g_1,g_1)+\\ &+\big(D_1^+(0)+D_1(1)+(g_0|g_0)D_1(1)D_2^+(0)+(g_1|g_1)D_2(1)D_1^+(0)\big)\otimes dN_t(g_1,g_0)+\\ &+\big(D_1(0)+D_1^+(1)+(g_1|g_1)D_1(0)D_2^+(1)+(g_0|g_0)D_2(0)D_1^+(1)\big)\otimes dN_t(g_0,g_1).\end{aligned} \tag{6.6}$$

We shall prove that the coefficients of the noises $dN(g_i,g_j)$ $(i,j=0,1)$ are identically zero and this will imply unitarity.
First of all notice that from (6.5) and (6.6) the following algebraic identities easily are deduced

$$\begin{aligned}&D_1(0)D_1^+(0)=D_1^+(1)D_1(1); \quad D_2(0)D_2^+(0)=D_1^+(0)D_1(0);\\ &D_1(1)D_1^+(1)=D_1^+(0)D_1(0); \quad D_2(1)D_2^+(1)=D_1^+(1)D_1(1);\\ &D_1(0)D_2^+(1)=D_1^+(1)D_2(1); \quad D_2(0)D_1^+(1)=D_2^+(0)D_1(0);\\ &D_1(1)D_2^+(0)=D_1^+(0)D_2(0); \quad D_2(1)D_1^+(0)=D_2^+(1)D_1(1);\end{aligned} \tag{6.7}$$

for example,

$$D_1(0)D_1^+(0) = D_g(0)D_0D_0^+D_g^+(0) =$$

$$= \sum_{m,n=0}^{\infty} (g_1|g_1)_-^n (g_0|g_0)_-^n (g_1|g_1)_+^m (g_0|g_0)_+^m (D_0D_1)^n(-D_0D_1)[(D_0D_1)^+]^m =$$

$$= - \sum_{m,n=0}^{\infty} (g_1|g_1)_-^n (g_0|g_0)_-^n (g_1|g_1)_+^m (g_0|g_0)_+^m D_0(D_0D_1)^{n+m}D_1 =$$

$$= \sum_{m,n=0}^{\infty} (g_1|g_1)_-^n (g_0|g_0)_-^n (g_1|g_1)_+^m (g_0|g_0)_+^m (D_1)^+(D_0D_1)^{n+m}D_1 =$$

$$= D_1^+(1)D_1(1). \tag{6.8}$$

From these identities it follows that, if we prove that all the coefficients of the noises $dN_t(g_i, g_j)$ vanish in (6.5), then they will also vanish in (6.6). Let us show that the coefficient of $dN_t(g_0, g_0)$ in (6.5) vanishes, i.e.,

$$D_2^+(0) + D_2(0) + (g_1|g_1)D_1^+(1)D_1(1) + (g_0|g_0)D_2^+(0)D_2(0) = 0. \tag{6.9}$$

By (5.19), (5.21a), (5.28), (5.29) and (6.10) one has

$$a_\varepsilon D_2(\varepsilon) = D_g(\varepsilon) - 1. \tag{6.10}$$

If $(g_0|g_0)_- = 0$, the left hand side of (6.9) is equal to

$$(g_1|g_1)_+ D_1^+ D_0^+ + (g_1|g_1)_- D_0 D_1 + (g_1|g_1) D_1^+ D_1. \tag{6.11}$$

Since $D_0 = D$, $D_1 = -D^+$, $D_\varepsilon^+ = -D_{1-\varepsilon}$, (6.11) becomes

$$(g_1|g_1)_+ D_0 D_1 + (g_1|g_1)_- D_0 D_1 - (g_1|g_1) D_0 D_1 = 0. \tag{6.12}$$

If $(g_0|g_0)_- \neq 0$, the left hand side of (6.9) is equal to

$$\frac{1}{\bar{a}_0}(D_g^+(0) - 1) + \frac{1}{a_0}(D_g(0) - 1) + (a_1 + \bar{a}_1)D_1^+ D_g^+(1) D_g(1) D_1 +$$

$$+ \frac{a_0 + \bar{a}_0}{|a_0|^2}(D_g^+(0) - 1)(D_g(0) - 1), \tag{6.13}$$

so it is equal to zero if and only if

$$(a_0 + \bar{a}_0)D_g^+(0)D_g(0) - (a_0 D_g(0) + \bar{a}_0 D_g^+(0)) +$$

$$+ (a_1 + \bar{a}_1)|a_0|^2 D_1^+ D_g^+(1) D_g(1) D_1 = 0. \tag{6.14}$$

Again using (5.28) we have the identity

$$D_g(1)D_1 = \sum_{n=0}^{\infty} a_1^n a_0^n (D_1 D_0)^n D_1 = \sum_{n=0}^{\infty} a_1^n a_0^n D_1 (D_0 D_1)^n = D_1 D_g(0), \tag{6.15}$$

which gives

$$D_1^+ D_g^+(1) D_g(1) D_1 = D_g^+(0) D_1^+ D_1 D_g(0) = -D_g^+(0) D_0 D_1 D_g(0) =$$

$$= -D_g^+(0) D_g(0) D_0 D_1 = D_g^+(0) D_g(0) D D^+. \tag{6.16}$$

Moreover (5.19) implies that

$$a_0 D_g(0) + \bar{a}_0 D_g^+(0) = \frac{a_0}{1 + a_0 a_1 D D^+} + \frac{\bar{a}_0}{1 + \bar{a}_0 \bar{a}_1 D D^+} =$$

$$= \frac{(a_0 + \bar{a}_0) + |a_0|^2 (a_1 + \bar{a}_1) D D^+}{(1 + a_0 a_1 D D^+)(1 + \bar{a}_0 \bar{a}_1 D D^+)} =$$

$$= D_g^+(0) D_g(0) \Big((a_1 + \bar{a}_1) |a_0|^2 D D^+ + a_0 + \bar{a}_0 \Big). \tag{6.17}$$

From (6.17), (6.16) one immediately obtains (6.14). Similarly, one can get that the coefficients of $dN_t(g_0, g_1)$, $dN_t(g_1, g_1)$, $dN_t(g_1, g_0)$ vanish in (6.5), so U is unitary. We sum up our results in the following theorem

THEOREM 6.2. *For each $f, f', g_0, g_1 \in K$, $u, v \in H_0$, $D \in B(H_0)$ satisfying (1.11), $S, T, S', T' \in \mathbf{R}$, $t \geqslant 0$, if $\|g_0\|_- \vee \|g_1\|_- < 1/16\|D\|$ and $\langle g_0, S_t g_1 \rangle = 0$, $\forall t \geqslant 0$, then the low density limit*

$$\lim_{z \to 0} \left\langle u \otimes W\left(z \int_{S/z^2}^{T/z^2} S_u f du \right) \Phi, U(t/z^2) v \otimes W\left(z \int_{S'/z^2}^{T'/z^2} S_u f' du \right) \Phi \right\rangle \tag{6.18}$$

exists, where

$$\frac{d}{dt} U(t) = -i V(t) U(t)$$

and

$$V(t) = i \Big(D \otimes A^+(S_t g_0) A(S_t g_1) - D^+ \otimes A^+(S_t g_1) A(S_t g_0) \Big).$$

Moreover, the low density limit (6.18) is equal to

$$\langle u \otimes W(\chi_{[S,T]} \otimes f) \Psi, U(t) v \otimes W(\chi_{[S',T']} \otimes f') \Psi \rangle, \tag{6.19}$$

where $U(t)$ is the solution of q.s.d.e. (6.1) on $H_0 \otimes \Gamma(L^2(\mathbf{R}) \otimes (K, (\cdot|\cdot)))$, where

(i) Ψ is the vacuum of $\Gamma(L^2(\mathbf{R}) \otimes (K, (\cdot|\cdot)))$;

(ii) for each $g, g' \in K$, $N_s(g, g')$ is the number process

$$N_s(g, g') := N(\chi_{[0,s]} \otimes |g\rangle\langle g'|);$$

(iii) $D_1(\varepsilon)$, $D_2(\varepsilon)$ are given by (5.28), (5.29) respectively.

Proof. The theorem (5.1) has shown that the low density limit (6.18) exists. Now, we shall prove that it is equal to (6.19). Clearly, (6.19) is continuous for

$u, v \in H_0$, so, one can write (6.19) into $\langle u, F(t)\rangle$, where $F(t) \in H_0$. Hence, we have

$$\langle u, F(0)\rangle = \langle u \otimes W(\chi_{[S,T]} \otimes f)\Psi, v \otimes W(\chi_{[S',T']} \otimes f')\Psi\rangle = \\ = \langle u, G(0)\rangle. \tag{6.20}$$

Moreover,

$$\langle u, F(t)\rangle = \langle u, F(0)\rangle + \int_0^t \sum_{\varepsilon \in \{0,1\}} \langle u \otimes W(\chi_{[S,T]} \otimes f)\Psi,$$

$$\Big[D_1(\varepsilon) \otimes dN_s(g_\varepsilon, g_{1-\varepsilon}) + D_2(\varepsilon) \otimes dN_s(g_\varepsilon, g_\varepsilon)\Big] v \otimes W(\chi_{[S',T']} \otimes f')\Psi\rangle U(s). \tag{6.21}$$

Apply the theorem 4.3 of (Hudson and Lindsay, 1989) to (6.21), one obtains

$$\langle u, F(t)\rangle = \langle u, F(0)\rangle +$$

$$+ \int_0^t ds \sum_{\varepsilon \in \{0,1\}} \Big((f|g_\varepsilon)\chi_{[S,T]}(s)(g_\varepsilon|f')\chi_{[S',T']}(s)\langle D_2^+(\varepsilon)u, F(s)\rangle +$$

$$+ (f|g_\varepsilon)\chi_{[S,T]}(s)(g_{1-\varepsilon}|f')\chi_{[S',T']}(s)\langle D_1^+(\varepsilon)u, F(s)\rangle \Big). \tag{6.22}$$

Since (6.1) has unique solution, one knows that (6.22) has a unique solution. Therefore

$$\langle u, F(t)\rangle = \langle u, G(t)\rangle, \quad t \geqslant 0.$$

REFERENCES

Accardi, L. (1978). On the quantum Feynmann–Kac formula. *Rend. Sem. Mat. e Fisico, Milano* **48**, 135–180.

Accardi, L. (1990). Noise and dissipation in quantum theory. *Reviews of Modern Phys.* (to appear).

Accardi, L., Frigerio, A. and Lu Yun Gang (1989). The weak coupling limit for Fermion case. *Preprint.*

Accardi, L., Frigerio, A. and Lu Yun Gang (1990a). The weak coupling limit as the quantum functional central limit. *Commun. Math. Phys.* (to appear).

Accardi, L., Frigerio, A. and Lu Yun Gang (1990b). On the weak coupling limit: Langevin equation and finite temperature case. (to appear).

Accardi, L. and Lu Yun Gang (1989a). On the weak coupling limit without rotating approximation. *Preprint.*

Accardi, L. and Lu Yun Gang (1989b). On the weak coupling limit for non-linear interactions. *Preprint.*

Accardi, L. and Lu Yun Gang (1990). The low density for Boson Fock case. *Commun. Math. Phys.* (to appear).

Alicki, R. and Frigerio, A. (1989). Quantum Poisson noise and linear Boltzmann equation. *Preprint.*

Davies, E.B. (1974). Markovian master equation. *Commun. Math. Phys.* **39**, 91–110.

Dümcke, R. (1989a). The low density limit for N-level systems interacting with a free Bose or Fermi gas. *Commun. Math. Phys.* **97**, 311.

Dümcke, R. (1989b). The low density limit for N-level systems. *Lect. Notes in Math.* **1136**, 151–161.

Fagnola, F. (1989). A martingale characterization of quantum Poisson process. *Preprint.*

Frigerio, A. (1988). Quantum Poisson processes: physical motivations and applications. *Lect. Notes in Math.* **1303**, 107–127.

Frigerio, A. and Maassen, H. (1990). Quantum Poisson processes and dilations of dynamical semigroups. *Preprint.*

Grad, H. (1958). Principles of the kinetic theory of gases. In: *Handbuch der Physik.* V.12. Springer, Berlin–Heidelberg etc.

Hudson, R.L. and Lindsay, M. (1989). Uses of non–Fock quantum Brownian motion and a quantum martingale representation theorem. *Lect. Notes in Math.* **1136**, 276–305.

Hudson, R.L. and Parthasarathy, K.R. (1984). Quantum Ito's formula and quantum stochastic evolutions. *Commun. Math. Phys.* **93**, 301–323.

Pulé, J.V. (1974). The Bloch equation. *Commun. Math. Phys.* **38**, 241–256.

Prob. Theory and Math. Stat., Vol. 1, pp. 28–34
B.Grigelionis *et al.* (Eds.)
1990 VSP/Mokslas

SYSTEMS ASSISTING IN GATHERING AND ANALYSIS OF STATISTICAL DATA AND ADJUSTABLE TO SUBJECT DOMAINS

S.A. AIVAZYAN, Yu.N. BLAGOVESCHENSKY and L.D. MESHALKIN

Central Economics and Mathematics Institute of the USSR Academy of Science, Moscow, USSR
Moscow State University, Moscow, USSR
Research Computer Center of the USSR Health Ministry, Moscow, USSR

ABSTRACT

A new technology of the statistical research support is suggested. In comparison with traditional statistical packages it gives additional possibilities to primary user especially in a study design part. Some further development of applied statistics is needed to help statisticians to participate in the suggested technological process.

1. INTRODUCTION

Mass applications of the statistical methods in science, industry, medicine, geology, agriculture and so on leads to the fact that many statistical in their nature studies are fulfilled today without participation of professional statisticians. Software intellectualization could be an answer to this problem. There are some method–oriented statistical systems of different degree of complexity up to statistical expert system (MOSES) (Aivazyan, 1987a). The next step should be subject–oriented statistical expert system (SOSES). But their elaboration meets serious difficulties due to great diversity of concrete problems and looks rather impossible (Streitberg, 1988).

In the report a new technology of support of the statistical research is suggested. It is based on a new scheme of relations between system manager, statistician and primary user and on a new generation of the statistical software which could be adjusted by statistician to subject domain of his clients. We will call it as ABM–technology (ABM is the abbreviation from "adjusted by master") to emphasize the central role which a subject domain statistician plays in it.

In comparison with traditional use of statistical packages ABM–technology requires more effort from software producer and statistician. But it gives stronger support to user as the approach of the system adjustment to user's problems allows:

(i) to include more specific statistical models, methods and recommendations especially in a study design part;

(ii) to use subject-oriented and adapted to users statistical knowledge level dialogue;
(iii) to have archives of fulfilled investigations;
(iv) to use special measures for protection against non-competence including active search for internal difficulties in users understanding of research problems.

Fig. 1
Information flow under ABM-technology.

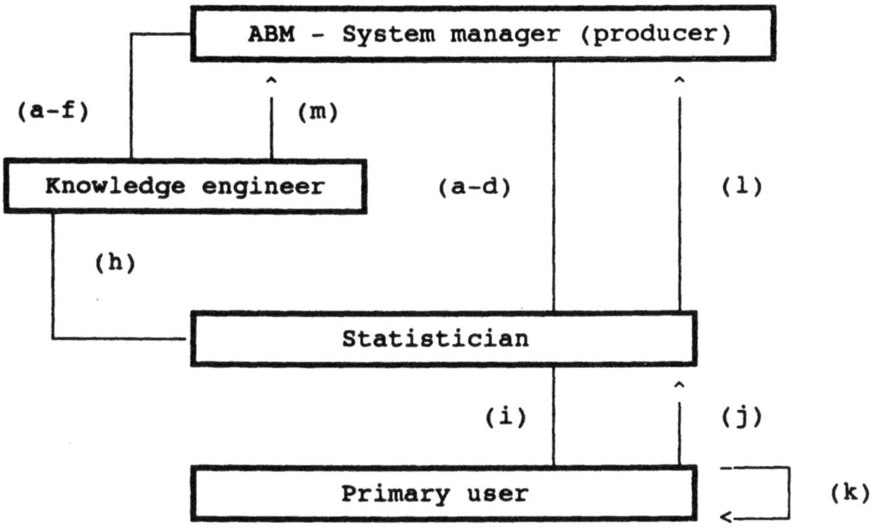

Notations:

- a – control system, which ensures adjustment to specific field of application;
- b – MOSES with free linquistic parameters;
- c – statistical phenomenological models of experiment conditions;
- d – statistical manual how to use (a,b,c);
- e – examles of informational support (IS) of SOSES;
- f – instruction for a knowledge engineer how to teach statistician to system adjustment and how to extract information from him for further system development;
- h – teaching course;
- i – SOSES designed by statistician;
- j – direct and indirect questions;
- k – information about finished research;
- l – demands for additional models and methods development;
- m – information about subject domain and statisticians in it.

The essence of ABM–technology is explained in sections 2–4. Some further development of applied statistics is needed to help statisticians to participate in the ABM–technology. These problems are described in section 5.

2. RELATIONS BETWEEN ABM–TECHNOLOGY PARTICIPANTS

There are four types of participants: ABM–system manager (producer), system knowledge engineer, subject domain statistician who serves several statistical work stations and primary user who works at one of these work stations with SOSES. It is desirable that homogeneous from the subject domain point of view users work at the same work station. This makes easy for statistical maintenance and archives keeping.
Information flow under ABM–technology is shown at Fig. 1. A knowledge engineer is a new person and represents a new type of activity in statistical methods application. He will work as intermediary between ABM–system manager and subject domain statisticians. Information he collected during teaching courses is vital for system survival.

3. USERS WORK UNDER ABM–TECHNOLOGY

ABM–technology is more friendly to user then traditional statistical packages. By means of dialogue a concrete investigation is connected with one of Basic Research Schemes (BRS) for each of which related subproblems, corresponding mathematical models and methods are picked out beforehand by statistician. However not every study is described completely by one of given set of BRS. If this is a case user calls directly or indirectly with system help for statistical consultation.
The principal advantage of ABM–technology is statistical assistance to all stages of the study beginning with introduction of summary variables, designing forms for data collection, organizing the collection procedure, warning against possible errors in research design. Other possibilities such as adaptation of language level of dialogue to the users level and protection against uncompetence are although important but auxiliary.
From the point of view of assistance given to user ABM–technology is equal to permanent consultation of the experienced statistician. In the same time ABM–technology unloads statistician from routine and helps him to concentrate on questions where his knowledge is used more efficiently.
A scheme of users work under ABM–technology is shown at Fig. 2. It starts with users identification and his statistical knowledge level estimation. This is an inevitable price of the individual approach. Then dialogue follows directed towards identification of the nearest to planned research BRS. Our experience of statistical consulting work indicates that the number of BRS in clinical medicine is significantly less than numbers of different statistical methods suggested in revolver type packages. A fragment of possible dialogue is given at Fig. 3. It is usefull to do BRS identification with some abundance in order to verify an internal consistence of users answers and the completeness of used set of BRS. During dialogue linguistical parameters are defined for futher use.
As one can see from Fig. 2 during solution of BRS subproblems data are entered,

mathematical models and statistical subroutines are used and one could reference to archives by using key words or the BRS type.
The work is ended by system recommendations to user and user's estimation of system performance. Recommendations should follow from comparison of users level of statistical knowledge with requirements of corresponding BRS to knowledge level, users questions to statistician during study fulfillment and study results. Users system estimates should be used for system development.
Let us summarize suggested methods of intellectual study support. First of all, it is bringing study together BRS and indication of corresponding subproblems. Secondly it is reference to archives. Thirdly it is support of adequate level of dialogue with use of subject domain language and shielding of suggested subproblems in dependence from users knowledge level. Lastly active search for moments of research where statistician interference would be useful.

Fig. 2
Scheme of users work under ABM-technology.

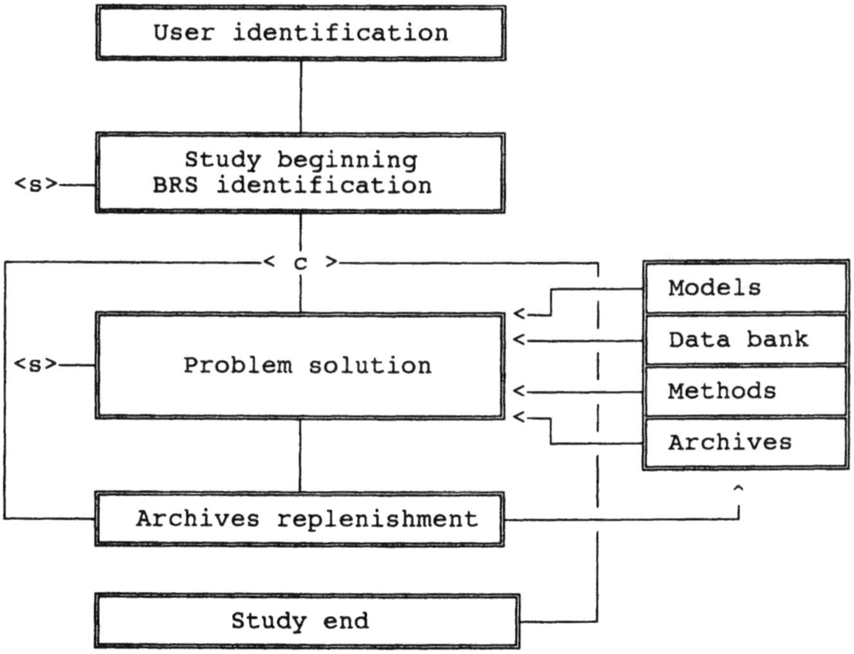

Notations: <s> – going out statistician;
<c> – counter of subproblems of BRS.

Fig. 3
An example of beginning of dialogue in medical applications.

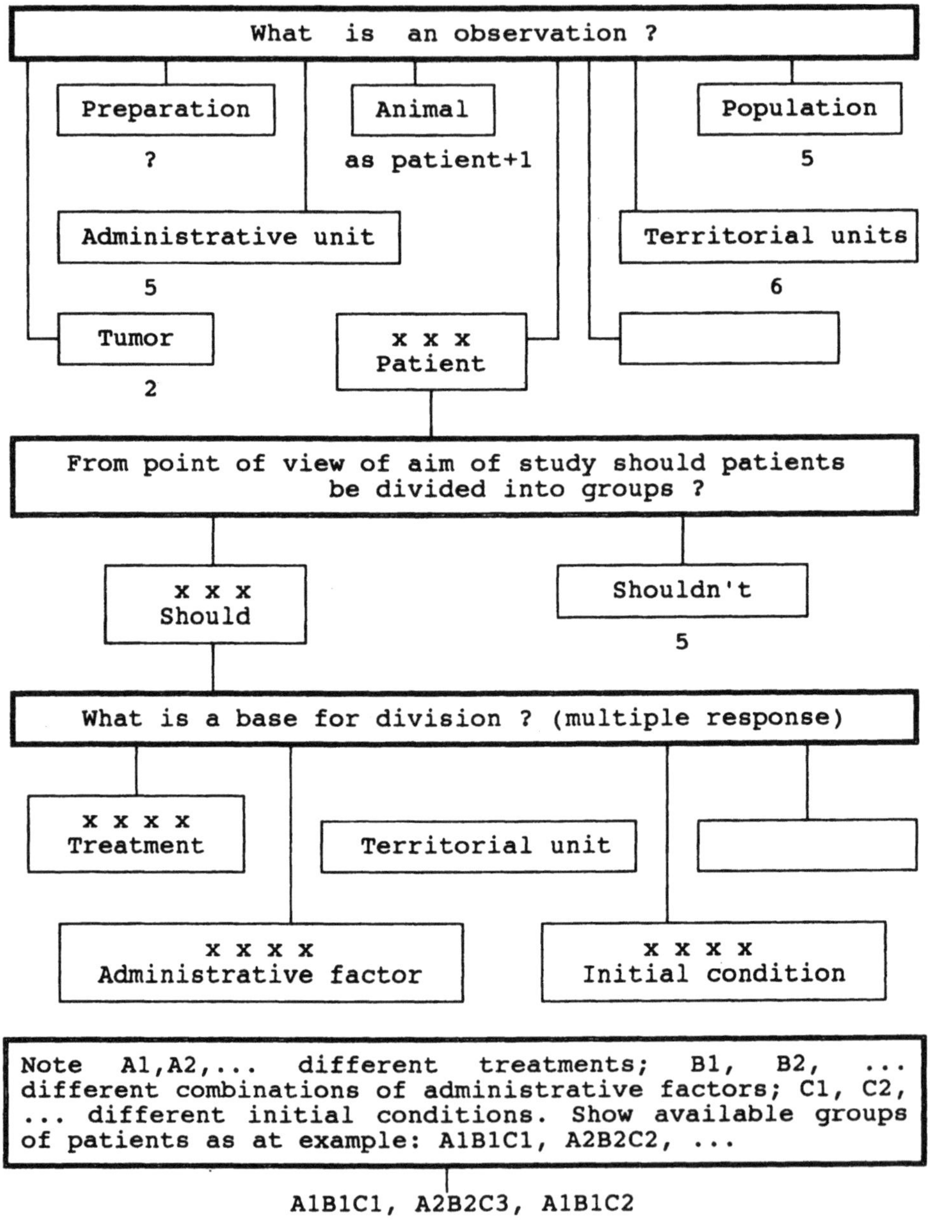

Notations: [ ] – question or instruction;

[ ] – possible answer;

[x x x] – actual answer.

The number of corresponding answer BRS are shown below answer.

4. STATISTICIAN WORK

The basic problems which should be solved by statistician under ABM– technology is to pick out study designes of his clients, generalized them into BRS, formulate corresponding subproblems and find methods appropriate for their solution. This is a serious task which is equivalent to creation of textbook on application of statistical methods in a given subject domain. To make easy the solution of these problems statistician has to be equipped with all auxiliary tools: examples of BRS and corresponding subproblems for relatively close subject domains, reach set of statistical subroutines and means of modeling of pseudorandom samples for comparative study of different statistical methods efficiency at condition of concrete BRS.

Other statisticians works such as adjustment of method of user testing, design of dialogue for BRS identification, design corresponding trapes for identification of unforeseen situation or nonadequate users responds, design of archives and, recommendation to user are less creative and good knowledge engineer could teach user to their fulfillment.

5. APPLIED STATISTICS PROBLEMS

As it was shown in section 3 the basic aid which ABM–technology gives to user is making easier the following chain of passages: concrete study→BRS→set of substantial problems→set of formal problems→set of concrete methods. In general the realization of this chain has to be served by statistics of given applied domain. But applied statistics could introduce concept of distance between concrete studies, elaborate language and means for their systematization and therefore help significantly to subject domain statistician. It is necessary also to introduce a dynamic under application domain scale of usefulness of statistical methods and put into operation the best methods, especially for multidimensional sample creation (Blagoveschensky and Meshalkin, 1986), automatic classification (Aivazyan, 1987b, 1989), general models of censored data (Blagoveschensky, 1987).

In conclusion one can remark that ABM–technology promote the social role of statistician. It is provided by:

– technical support of intellectual dialogue between user and statistician;

– availability of instrument for accumulation and representation of professional knowledge;

– solution of creative task during introduction of BRS and their elaboration;

– possibility of new statistical methods investigation and serious extend of user served number.

Therefore one can hope that statisticians will support ABM–technology if it will be well implemented, early mentioned methodological question will be solved and sufficient number of schools teaching ABM–technology will be available.

REFERENCES

Aivazyan, S.A. (1987a). Model– and method–oriented intelligent software for statistical data analysis. In: *Model-Oriented Data Analysis Systems.* Springer, Berlin—Heidelberg etc, pp. 153–158.

Aivazyan, S.A. (1987b). On constraction a general theory of automatic classification. In: *Proc. First World Congress Bernoulli Soc.,* VNU Science Press, Utrecht, pp. 111–122.

Aivazyan, S.A., Buchstuber, V.M., Enyukov, I.S. and Meshalkin, L.D. (1989). *Applied Statistics: Classification and Reduction of Dimensionality.* Finances and Statistics, Moscow.

Blagoveshensky, Yu.N. and Meshalkin, L.D. (1986). Multidimensional T–normal distributions In: *Abstr. First World Congress Bernoulli Soc.,* **1**, Nauka, Moscow, pp. 79.

Blagoveschensky, Yu.N. (1987). *Estimation and hypothesis testing on noncomplete samples.* Dr. Thesis. Vilnius State Univ., Vilnius.

Streitberg, B. (1988). On the non–existence of expert systems. Critical remarks on AI in statistics and discussion. *Statist. Software Newsletter* **2**, 55–74.

Prob. Theory and Math. Stat., Vol. 1, pp. 35–51
B.Grigelionis *et al.* (Eds.)
1990 VSP/Mokslas

ON THE MAXIMUM OF SUMS OF RANDOM VARIABLES AND THE SUPREMUM FOR STABLE PROCESSES

A.K. ALEŠKEVIČIENĖ
Institute of Mathematics and Informatics,
Akademijos 4, Vilnius 232600, Lithuania

ABSTRACT

Given a separable stable process with stationary independent increments, we consider the problem of finding the characteristic function and distribution of its supremum process.

Let

$$X_1, X_2, \dots \tag{1}$$

be a sequence of independent identically distributed random variables whose distribution function F belongs to the domain of attraction of a stable law V of index $0 < \alpha < 2$ and whose characteristic function is given by

$$v(t) = \exp\{i\gamma t - c|t|^\alpha(1 + i\beta \operatorname{sign} t\, \omega(t, \alpha))\}, \tag{2}$$

where $0 < \alpha < 2$, $c = c_1 + c_2 > 0$, $c_1 \geqslant 0$, $c_2 \geqslant 0$, $\beta = (c_1 - c_2)/(c_1 + c_2)$, $-1 \leqslant \beta \leqslant 1$, $-\infty < \gamma < \infty$,

$$\omega(t, \alpha) = \begin{cases} \operatorname{tg}(\pi\alpha/2), & \alpha \neq 1, \\ (2/\pi)\log|t|, & \alpha = 1; \end{cases}$$

and α, c_1, c_2 are the same constants present in the Levy formula for the logarithm of the characteristic function of the stable law V (see (Gnedenko and Kolmogorov, 1954)).

The Levy formula is given by

$$\log v(t) = i\gamma t + \int_{-\infty}^{0} \left(e^{itu} - 1 - \frac{itu}{1+u^2}\right) dM(u) + \int_{0}^{\infty} \left(e^{itu} - 1 - \frac{itu}{1+u^2}\right) dN(u), \tag{3}$$

where

$$M(u) = c_1|u|^{-\alpha}, \qquad N(u) = -c_2 u^{-\alpha}.$$

Write

$$S_0 = 0, \quad S_n = X_1 + \dots + X_n,\ n \geqslant 1, \quad M_n = \max(0, S_1, \dots, S_n).$$

The random walk $\{S_n\}$ generated by the S_k, $k = 1, 2, \dots$, drifts either to $-\infty$, or to $+\infty$, or oscillates. Drift to $-\infty$ (i.e. $S_n \to -\infty$ as $n \to \infty$, almost surely) occurs if and only if

$$\sum_{n=1}^{\infty} \frac{1}{n} \mathbf{P}\{S_n > 0\} < \infty,$$

or equivalently, if and only if $M = \lim_{n\to\infty} M_n$ is finite almost surely. In terms of the parameters of V we can say that drift to $-\infty$ takes place in two cases:

(1) F has mean $a = \mathbf{E}X_1 < 0$,

(2) $0 < \alpha < 1$, $\beta = 1$.

The distribution of M is given by the characteristic function

$$\mathbf{E}\exp\{itM\} = \prod_{k=1}^{\infty} \exp\left\{\frac{1}{k}\left[\mathbf{E}\exp\{it\max(0, S_k)\} - 1\right]\right\}$$

(Spitzer, 1956). Great attention has been paid to this distribution. For details see, for example (Borovkov, 1970; Bingham, 1973).

Drift to $+\infty$ (i.e., $S_n \to +\infty$, a.s.) occurs if and only if

$$\sum_{n=1}^{\infty} \frac{1}{n} \mathbf{P}\{S_n < 0\} < \infty.$$

In terms of the parameters of V drift to $+\infty$ takes place in two cases:

1) F has mean $a > 0$,

2) $0 < \alpha < 1$, $\beta = -1$.

In the case of drift to $+\infty$ the weak convergence theorem for M_n has been obtained in (Heyde, 1967; Bingham, 1973). According to this theorem, the limit behaviour of M_n is similar to that of S_n. For details see (Heyde, 1967; Bingham, 1973).

Oscillation (i.e., $\limsup S_n = +\infty$, $\liminf S_n = -\infty$, a.s.) occurs if and only if

$$\sum_{n=1}^{\infty} \frac{1}{n} \mathbf{P}\{S_n > 0\} = \infty, \qquad \sum_{n=1}^{\infty} \frac{1}{n} \mathbf{P}\{S_n < 0\} = \infty.$$

In terms of the parameters of V oscillation occurs if and only if

1) F has mean $a = 0$,

2) $0 < \alpha < 1$, $|\beta| < 1$.

In the case of oscillation the limit behaviour of $\{M_n\}$ is quite different from that of $\{S_n\}$. In this case a weak convergence theorem was obtained in (Heyde, 1969; Bingham, 1973).

In the sequel we shall deal only with the case of oscillation.

Let $\{Y(z), z \geqslant 0, Y(0) = 0\}$ be a separable stable process with stationary independent increments characterised by

$$\mathbf{E}\exp\{itY(z)\} = \exp\left\{ - zc|t|^{\alpha}(1 + i\beta \operatorname{sign} t \operatorname{tg}\frac{\pi\alpha}{2})\right\}, \tag{4}$$

where parameters c, α and β are the same as those in (2).
Suppose that there exists a monotone sequence of constants $\{B_n, n = 1, 2, \ldots\}$ such that $B_n^{-1} S_n$ converges in distribution to the stable law V with characteristic function (2) and parameters $c > 0$, $0 < \alpha < 1$, $\beta = 0$ if $\alpha = 1$, $|\beta| < 1$ if $0 < \alpha < 1$, $|\beta| \leqslant 1$ if $1 < \alpha < 2$. Then (see (Heyde, 1969))

$$\lim_{n\to\infty} \mathbf{P}\{B_n^{-1} M_n < x\} = W(x), \tag{5}$$

where

$$W(x) = \mathbf{P}\{\sup_{0\leqslant z\leqslant 1} Y(z) < x\}$$

and

$$\mathbf{P}\{\sup_{0\leqslant z\leqslant Z} Y(z) < x\} = W(xZ^{-1/\alpha}).$$

In connection with the parameter restrictions, we remind that the stable law with characteristic function (2) is one–sided if $0 < \alpha < 1$, $|\beta| = 1$ so that the random walk generated by the S_n does not oscillate. The case $\alpha = 1$, $\beta \neq 0$ (asymmetric Cauchy law) was excluded because of a normalization complication.
A number of works were devoted to study the function W itself. Let us look at the history of this problem.
As the distributions V are absolutely continuous (Heyde, 1969, p. 422), so are W: thus, W has a density p. Let h be the Laplace transform of p:

$$h(s) = \int_0^\infty e^{-sx} dW(x) = \int_0^\infty e^{-sx} p(x) dx.$$

Darling (Darling, 1956) and Heyde (Heyde, 1969) obtained the integral equation

$$\int_0^\infty e^{-x} h(sx^{1/\alpha}) dx = g(s), \quad s \geqslant 0, \tag{6}$$

where

$$g(s) = \exp\left\{ -\frac{\sin \pi\rho}{\pi} \int_0^\infty \frac{\log(1 + c_3 (sx)^\alpha) dx}{(x^2 + 2x \cos \pi\rho + 1)} \right\}, \tag{7}$$

$$\rho = \mathbf{P}\{X_1 > 0\} = \frac{1}{2} + (\pi\alpha)^{-1} \mathrm{arctg}(-\beta \mathrm{tg} \frac{\pi\alpha}{2}),$$

$$c_3 = c\left(1 + \beta^2 \mathrm{tg}^2 \frac{\pi\alpha}{2}\right)^{1/2}$$

and c is given in (2).
The Darling—Heyde integral equation (6) may be solved by Mellin transforms. Let us write

$$\tilde{h}(s) = \int_0^\infty x^{s-1} p(x) dx,$$

$$\tilde{g}(s) = \int_0^\infty x^{s-1} g(x)dx$$

for the Mellin transforms of p and g. Then,

$$\tilde{h}(s) = \tilde{g}(1-s)/\left(\Gamma(1-s)\Gamma(1-\alpha^{-1}+s\alpha^{-1})\right) \tag{8}$$

and, consequently,

$$p(x) = \frac{1}{2\pi i} \int_{c_0-i\infty}^{c_0+i\infty} \frac{\tilde{g}(1-s)x^{-s}}{\Gamma(1-s)\Gamma(1-\alpha^{-1}+s\alpha^{-1})} ds. \tag{9}$$

In principle (8) and (9) determine $\tilde{h}$ and p as well. But in a general case one cannot evaluate the integral in (7) to obtain g explicitly. Consequently, in a general case one cannot obtain p explicitly, too. Only in two special cases (symmetric Cauchy and spectrally negative) it is known how to solve the Darling—Heyde integral equation explicitly.

Darling (Darling, 1956) succeeded to simplify (9) for the case, when X_i have a symmetric Cauchy distribution with common density $\pi^{-1}(1+x^2)^{-1}$, $x > 0$ (in (2) $\alpha = 1$, $\beta = 0$). He obtained the following result:

$$p(x) = \frac{1}{\pi x^{1/2}(1+x^2)^{3/4}} \exp\left\{-\frac{1}{\pi}\int_0^x \frac{\log w}{1+w^2}dw\right\}, \quad x > 0.$$

Darling noted that for $x \to \infty$ $p(x) \sim \pi^{-1}x^{-2}$ and for $x \to 0$ $p(x) \sim \pi^{-1}x^{-1/2}$. Bingham (Bingham, 1973) solved the Darling—Heyde integral equation in a spectrally negative case. Let X_i themselves have a stable distribution V whose Levy—Khinchine measure gives no mass to $(0, \infty)$ (see (Gnedenko and Kolmogorov, 1954, §18)). X_i have this property if and only if $\beta = 1$ in (2). In this case X_i is said to be spectrally negative. Bingham (1973) considered the spectrally negative case when $1 < \alpha < 2$, $\beta \doteq 1$, and obtained the following results: the distribution of $\sup_{0 \leqslant z \leqslant 1} Y(z)$ is the Mittag—Leffler distribution with parameter $1/\alpha$, whose Laplace transform is given by

$$\mathbf{E}\exp\left\{-t \sup_{0\leqslant z\leqslant 1} Y(z)\right\} = \sum_{n=0}^{\infty}\left(-tc_3^{1/\alpha}\right)^n/\Gamma(1+n\alpha^{-1}), \quad t > 0;$$

the density of $\sup_{0\leqslant z\leqslant 1} Y(z)$ is given by

$$p_{1/\alpha}(x) = \frac{\alpha}{\pi c^{1/\alpha}} \sum_{n=1}^{\infty} \frac{(-1)^{n-1}}{n!} \sin\frac{n\pi}{\alpha} \Gamma(1+n\alpha^{-1}) \left(\frac{x}{c_3^{1/\alpha}}\right)^{n-1}, \quad x > 0.$$

With the help of a different method Heyde (Heyde, 1969) obtained another expression for W in a spectrally negative case. Heyde's result is given by

$$W(x) = 1 - G_{1/\alpha}(-\lambda \sec \frac{\pi\alpha}{2} x^{-\alpha}), \quad x \geqslant 0,$$

where $G_{1/\alpha}$ is the distribution function of the stable law whose characteristic function is given by

$$g_{1/\alpha}(t) = \exp\Big\{ - |t|^{1/\alpha}\Big(\cos \frac{\pi}{2\alpha} - i \sin \frac{\pi}{2\alpha} \mathrm{sign} t\Big)\Big\}.$$

In no other case (apart from symmetric Cauchy and spectrally negative cases) one can solve the Darling—Heyde integral equation (6) explicitly. However, Bingham (Bingham, 1973) obtained some results on the asymptotic behaviour of $h(s)$ and $W(x)$ directly from the equation (4). We quote here only the results on the asymptotic behaviour of $W(x)$:

1) if $x \to 0$, then

$$W(x) \sim x^{\alpha\rho} / (c_3^{\rho}\Gamma(1-\rho)\Gamma(1+\alpha\rho));$$

2) if $\alpha\rho < 1$ and $x \to \infty$, then

$$1 - W(x) \sim c_1 \Gamma(\alpha) \sin \pi\alpha\rho / (\pi x^{\alpha});$$

3) if $\alpha\rho = 1$ (spectrally negative case) and $x \to \infty$, then

$$1 - W(x) \sim c_4 (x c_3^{-\rho})^{-1/(2(1-\rho))} \exp\{-c_5 (x c_3^{-\rho})^{1/(1-\rho)}\},$$

where

$$c_4 = c_4(\rho) = (2\pi(1-\rho)\rho^{1/(1-\rho)})^{-1/2},$$
$$c_5 = c_5(\rho) = (1-\rho)\rho^{\rho/(1-\rho)}.$$

Note that always $\alpha\rho \leqslant 1$ (see (Bingham, 1973)).
From the above statements it follows that in the spectrally negative case the distribution function $W(x)$ has the moments of all orders, and the tail $1 - W(x)$ decreases exponentially fast. In no other case does $W(x)$ have a variance, and the tail $1 - W(x)$ behaves like a power of x.
In this paper we consider the problem of finding an expression for the characteristic function $w(t)$ as well as for density of the distribution W in terms of the parameters of the law V. At first we obtain an expression for $w(t)$.

THEOREM 1. *Let $\{Y(z), z \geqslant 0, Y(0) = 0\}$ be a separable stable process with stationary independent increments, defined by the characteristic function* (4) *with the parameters $0 < \alpha < 1$, $|\beta| < 1$, $c > 0$. Then,*

$$w(t) = v(t) + \psi_t \int_0^1 \exp\Big\{-cu|t|^{\alpha}\Big(1 + i\,\beta \mathrm{sign}\, t \,\mathrm{tg} \frac{\pi\alpha}{2}\Big)\Big[1 +$$
$$+ \sum_{l=1}^{\infty} \frac{1}{l!(l+1)!} \psi_t^l (1-u)^l\Big] du,$$

where

$$\psi_t = c_0|t|^\alpha(1 + i \operatorname{sign} t \operatorname{tg}\frac{\pi\alpha}{2}),$$

$$c_0 = \Gamma(1-\alpha)c_1 \cos\frac{\pi\alpha}{2}$$

and c_1 is the constant appearing in the Levy formula (3) for the logarithm of the characteristic function $v(t)$, and $v(t)$ is given by (2).

Proof. In order to simplify the calculations assume, without loss of generality, that the random variables X_i in sequence (1) are stable with the distribution V characterised by the characteristic function (2) with the parameters $0 < \alpha < 1$, $|\beta| < 1$ and $\gamma = 0$. Then

$$\lim_{n\to\infty} \mathbf{P}\{n^{-1/\alpha}S_n < x\} = V(x)$$

and, consequently (Heyde, 1969),

$$\lim_{n\to\infty} \mathbf{P}\{n^{-1/\alpha}M_n < x\} = W(x). \tag{10}$$

On the other hand, since $\mathbf{P}\{S_k \leqslant 0, k = 1,\dots,n\} \to 0$ as $n\to\infty$

$$\lim_{n\to\infty} \mathbf{P}\{n^{-1/\alpha}M_n < x\} = \lim_{n\to\infty} \mathbf{P}\{n^{-1/\alpha}\bar{S}_n < x\}, \bar{S}_n = \max_{1\leqslant k\leqslant n} S_k. \tag{11}$$

Basing on the relations (10), (11) we can conclude that $w(t) = \int e^{itx}dW(x)$ must coincide with the limit of the sequence of characteristic functions $w_n(tn^{-1/\alpha})$ determined by the distributions

$$W_n(xn^{1/\alpha}) = \mathbf{P}\{n^{-1/\alpha}\bar{S}_n < x\}, \quad n = 1,2,\dots,$$

i.e. the relation

$$w(t) = \lim_{n\to\infty} w_n(tn^{-1/\alpha}), \quad w_n(t) = \mathbf{E}\exp\{it\bar{S}_n\}, \tag{12}$$

must be valid. Suppose

$$\bar{\varphi}_n(t) = \mathbf{P}\{\bar{S}_n < 0\} - \int_{-\infty}^{0-} e^{itx}dW_n(x), \quad n \geqslant 1, \quad \bar{\varphi}_0(t) \equiv 0.$$

Then (Nagaev, 1970)

$$w_n(t) = \int_{-\infty}^{\infty} e^{itx}dW_n(t) = v^n(t) + \sum_{k=1}^{n-1} v^k(t)\bar{\varphi}_{n-k}(t). \tag{13}$$

When calculating the functions $\bar{\varphi}_k$ we shall make use of the following analogue of Spitzer identity (Spitzer, 1956, Nagaev, 1970)

$$1 + \sum_{n=1}^{\infty} \bar{\varphi}_n(t)z^n = \exp\left\{\sum_{k=1}^{\infty} \frac{z^k}{k} \int_{-\infty}^{0-} (1 - e^{itx})dV_k(x)\right\}, \tag{14}$$

$$V_k(x) = V^{*k}(x),$$

and the asymptotical property of stable distributions

$$V(x) = \frac{c_1}{|x|^\alpha}(1+\alpha_1(x)), \quad x<0,$$

$$1-V(x) = \frac{c_2}{x^\alpha}(1+\alpha_2(x)), \quad x>0, \tag{15}$$

where $\alpha_i(x) \to 0$ as $|x| \to \infty, i = 1,2$, and c_1, c_2 are the same constants from the Levy formula (3).
Assume $t_n = tn^{-1/\alpha}$. Let us calculate the integral

$$I_{k,n} = \int_{-\infty}^{0} (1-e^{it_n x})dV_k(x) = -it_n \int_{-\infty}^{0-} e^{ixt_n} V_k(x)dx. \tag{16}$$

Since $V(x)$ is a stable distribution with the parameters $0<\alpha<1$, $|\beta|<1$, $\gamma=0$ and $c = (c_1+c_2)\Gamma(1-\alpha)\cos(\pi\alpha/2)$ (here c_1 and c_2 are the same constants as in (15)) by virtue of (2) $V_k(x)$ is also a stable distribution with the parameters α, β, $\gamma=0$ and $kc = (kc_1+kc_2)\Gamma(1-\alpha)\cos(\pi\alpha/2)$. Then, according to (15)

$$V_k(x) = \frac{kc_1(1+\alpha_1(x))}{|x|^\alpha}, \quad x<0, \quad \alpha_1(x)\to 0 \text{ as } x\to-\infty. \tag{17}$$

Inserting (17) into (16) we get as $n \to \infty$

$$I_{k,n} = -it_n c_1 k \int_{-\infty}^{0-} e^{ixt_n}(1+\alpha_1(x))\frac{dx}{|x|^\alpha} =$$

$$= it_n c_1 k \int_{0+}^{\infty} (\cos x|t_n| - i\,\mathrm{sign}t \sin x|t_n|)(1+\alpha_1(x))\frac{dx}{x^\alpha} =$$

$$= \frac{ic_1 k}{n} t|t|^{\alpha-1} \int_{0+}^{\infty} (\cos x - i\,\mathrm{sign}t\sin x)\left(1+\alpha_1\left(\frac{xn^{1/\alpha}}{|t|}\right)\right)\frac{dx}{x^\alpha}. \tag{18}$$

Note that for $n \to \infty$ and any finite t

$$\int_0^\infty \frac{\sin x}{x^\alpha}\left(1+\alpha_1\left(\frac{xn^{1/\alpha}}{|t|}\right)\right)dx \sim \left(1+\alpha_1\left(\frac{n^{1/\alpha}}{|t|}\right)\right)\int_0^\infty \frac{\sin x}{x^\alpha}dx,$$

$$\int_0^\infty \frac{\cos x}{x^\alpha}\left(1+\alpha_1\left(\frac{xn^{1/\alpha}}{|t|}\right)\right)dx \sim \left(1+\alpha_1\left(\frac{n^{1/\alpha}}{|t|}\right)\right)\int_0^\infty \frac{\cos x}{x^\alpha}dx. \tag{19}$$

The proof of the relations (19) is analogous to that of the lemma 2.6.1 from (Ibragimov and Linnik, 1965). Therefore, it is not presented here.

Further, by the Euler formula (see, e.g., (Ibragimov and Linnik, 1965, Chapter 2, §6))

$$\int_0^\infty \frac{\sin x}{x^\alpha} dx = \Gamma(1-\alpha)\cos\frac{\pi\alpha}{2}, \quad \int_0^\infty \frac{\cos x}{x^\alpha} dx = \Gamma(1-\alpha)\sin\frac{\pi\alpha}{2}.$$

Hence and from the relations (18), (19) we obtain

$$I_{k,n} = \frac{ic_1 k}{n} t|t|^{\alpha-1}\left(\Gamma(1-\alpha)\sin\frac{\pi\alpha}{2} - \right.$$
$$\left. -i\Gamma(1-\alpha)\cos\frac{\pi\alpha}{2}\operatorname{sign}t\right)\left(1+\alpha_1\left(\frac{n^{1/\alpha}}{|t|}\right)\right)(1+\delta_n) =$$
$$= \frac{c_0 k|t|^\alpha}{n}\left(1+i\operatorname{sign}t\ \operatorname{tg}\frac{\pi\alpha}{2}\right)\left(1+\alpha_1\left(\frac{n^{1/\alpha}}{|t|}\right)\right)(1+\delta_n),$$

where $c_0 = c_1\Gamma(1-\alpha)\cos(\pi\alpha/2)$, $\delta_n \to 0$ for $n \to \infty$.
By inserting the obtained representation for $I_{k,n}$ into (14) we get

$$1+\sum_{k=1}^\infty z^k\bar\varphi_k(t_n) = \exp\left\{\frac{c_0|t|^\alpha}{n}\left(1+i\operatorname{sign}t\ \operatorname{tg}\frac{\pi\alpha}{2}\right)\left(1+\alpha_1\left(\frac{n^{1/\alpha}}{|t|}\right)\right)\frac{z}{1-z}\right\}. \tag{20}$$

Hence, denoting

$$\psi_t = c_0|t|^\alpha\left(1+i\operatorname{sign}t\ \operatorname{tg}\frac{\pi\alpha}{2}\right),$$
$$\varepsilon_n = \alpha_1(n^{1/\alpha}/|t|)+\delta_n+\delta_n\alpha_1(n^{1/\alpha}/|t|), \tag{21}$$

we get

$$\bar\varphi_k(t_n) = \frac{1}{n}\psi_t(1+\varepsilon_n)+\frac{1}{2!}\left(\frac{\psi_t}{n}(1+\varepsilon_n)\right)^2 C_k\left(\frac{z^2}{(1-z)^2}\right)+\cdots+$$
$$+\frac{1}{l!}\left(\frac{\psi_t}{n}(1+\varepsilon_n)\right)^l C_k\left(\frac{z^l}{(1-z)^l}\right)+\cdots+$$
$$+\frac{1}{(k-1)!}\left(\frac{\psi_t}{n}(1+\varepsilon_n)\right)^{k-1} C_k\left(\frac{z^{k-1}}{(1-z)^{k-1}}\right)+\frac{1}{k!}\left(\frac{\psi_t}{n}(1+\varepsilon_n)\right)^k,$$

where symbol $C_k(h(z))$ denotes the coefficient at z^k in the expansion into power series of the function $h(z)$.
Since

$$C_k\left(\frac{z^l}{(1-z)^l}\right) = \frac{1}{(l-1)!}(k-(l-1))(k-(l-2))\ldots(k-1) = C_{k-1}^{l-1},$$

then

$$\bar{\varphi}_k(t_n)=\frac{\psi_t}{n}(1+\varepsilon_n)+\frac{1}{2!}\left(\frac{\psi_t}{n}(1+\varepsilon_n)\right)^2C^1_{k-1}+\cdots+\frac{1}{l!}\left(\frac{\psi_t}{n}(1+\varepsilon_n)\right)^lC^{l-1}_{k-1}+\cdots+$$

$$+\frac{1}{(k-1)!}\left(\frac{\psi_t}{n}(1+\varepsilon_n)\right)^{k-1}C^{k-2}_{k-1}+\frac{1}{k!}\left(\frac{\psi_t}{n}(1+\varepsilon_n)\right)^k, \tag{22}$$

where C^l_k are binomial coefficients.
Finally, inserting (22) into (13) we conclude

$$w_n(t_n)=v^n(t_n)+\psi_t(1+\varepsilon_n)\frac{1}{n}\sum_{k=1}^{n-1}v^k(t_n)\Big\{1+$$

$$+\sum_{l=1}^{n-k-2}\frac{1}{l!(l+1)!}\left(\psi_t\Big(1-\frac{k}{n}\Big)(1+\varepsilon_n)\right)^l\Big(1-\frac{1}{n-k}\Big)\Big(1-\frac{2}{n-k}\Big)\cdots\Big(1-\frac{l}{n-k}\Big)+$$

$$+\frac{1}{(n-k)!}\left(\frac{\psi_t}{n}(1+\varepsilon_n)\right)^{n-k-1}\Big\}. \tag{23}$$

Let A be an arbitrarily large integer satisfying the inequality $A>|\psi_t|e$ (here t is finite). Then

$$|\psi_t|\sum_{l=A+1}^{n-k-1}\frac{1}{l!(l+1)!}\left(|\psi_t|\Big(1-\frac{k}{n}\Big)(1+\varepsilon_n)\right)^l\Big(1-\frac{1}{n-k}\Big)\cdots\Big(1-\frac{l}{n-k}\Big)=$$

$$=o_n(1). \tag{24}$$

Further, taking into account (2), (21) and inequality $x^\alpha e^{-x}<e^{-\alpha}\alpha^\alpha$, $x>0$, $\alpha>0$, it is easily seen that

$$|\psi_t|\sum_{k=n-\sqrt{n}}^{n-1}\frac{1}{n}v^k(t_n)\Big[1+\sum_{l=1}^{A}\frac{1}{l!(l+1)!}\left(|\psi_t|\Big(1-\frac{k}{n}\Big)(1+\varepsilon_n)\right)^l\Big]=o_n(1). \tag{25}$$

In accordance with the relations (2), (23)-(25)

$$w_n(t_n)=v(t)+\psi_t\sum_{k=1}^{n-\sqrt{n}}\frac{1}{n}v^k(t_n)\Big[1+$$

$$+\sum_{l=1}^{A}\frac{1}{l!(l+1)!}\left(\psi_t\Big(1-\frac{k}{n}\Big)(1+\varepsilon_n)\right)^l\Big(1-\frac{1}{n-k}\Big)\cdots\Big(1-\frac{l}{n-k}\Big)\Big]+o_n(1). \tag{26}$$

Making use of the inequalities

$$e^{-x/(1-x)}<1-x<e^{-x},\quad 0<x<1, \tag{27}$$

we obtain

$$\exp\left\{-\frac{1}{n-k}\right\} < \left(1-\frac{1}{n-k}\right)\left(1-\frac{2}{n-k}\right)\cdots\left(1-\frac{l}{n-k}\right) <$$

$$< \exp\left\{-\frac{1}{n-k}\frac{l(l+1)}{2}\right\}.$$

Thus, for $1 \leqslant k \leqslant n-\sqrt{n}$ and $1 \leqslant l \leqslant A$ we have

$$\left(1-\frac{1}{n-k}\right)\left(1-\frac{2}{n-k}\right)\cdots\left(1-\frac{l}{n-k}\right) = 1 + O\left(\frac{l^2}{n-k}\right). \qquad (28)$$

Since for $1 \leqslant k \leqslant n-\sqrt{n}$

$$\frac{1}{\sqrt{n-k}}|\psi_t|\sum_{l=1}^{A}\frac{l^2}{l!(l+1)!}\left(\psi_t\left(1-\frac{k}{n}\right)(1+\varepsilon_n)\right)^l = o_n(1),$$

and

$$\varepsilon_n A\sum_{l=1}^{A}\frac{1}{l!(l+1)!}\psi_t^l\left(1-\frac{k}{n}\right)^l = o_n(1)$$

from the relations (26), (28) it follows

$$w_n(t_n) = v(t) + \psi_t(1+\varepsilon_n)\sum_{k=1}^{n-\sqrt{n}}\frac{1}{n}v^k(t_n)\times$$

$$\times\left[1+\sum_{l=1}^{A}\frac{1}{l!(l+1)!}\left(\psi_t\left(1-\frac{k}{n}\right)(1+\varepsilon_n)\right)^l\right] + o_n(1) =$$

$$= v(t) + \psi_t\sum_{k=1}^{n-1}\frac{1}{n}v\left(t\left(\frac{k}{n}\right)^{1/\alpha}\right)\times$$

$$\times\left[1+\sum_{l=1}^{A}\frac{1}{l!(l+1)!}\psi_t^l\left(1-\frac{k}{n}\right)^l + o_n(1)\right]. \qquad (29)$$

Applying the Euler—McLauren summation formula (see, e.g. (Cramer, 1946, Chapter 2)) to the first sum of the righthand side of relation (29) and taking into account (2) we obtain

$$w_n(t_n) = v(t) + \frac{\psi_t}{n}\int_1^{n-1}\exp\left\{-\frac{k}{n}c|t|^\alpha\left(1+i\beta\operatorname{sign}t\ \operatorname{tg}\frac{\pi\alpha}{2}\right)\right\}\times$$

$$\times\left[1+\sum_{l=1}^{A}\frac{1}{l!(l+1)!}\psi_t^l\left(1-\frac{k}{n}\right)^l\right]dk + o_n(1) =$$

$$= v(t) + \psi_t \int_{1/n}^{1-1/n} \exp\Big\{ - cu|t|^\alpha \Big(1 + i\,\beta \operatorname{sign} t \operatorname{tg}\frac{\pi\alpha}{2}\Big) \Big\} \times$$

$$\times \Big[1 + \sum_{l=1}^{\infty} \frac{1}{l!(l+1)!} \psi_t^l (1-u)^l \Big] du + o_n(1).$$

Hence

$$\lim_{n\to\infty} w_n(t_n) = v(t) + \psi_t \int_0^1 \exp\Big\{ - cu|t|^\alpha \Big(1 + i\,\beta \operatorname{sign} t \operatorname{tg}\frac{\pi\alpha}{2}\Big) \Big\} \times$$

$$\times \Big[1 + \sum_{l=1}^{\infty} \frac{1}{l!(l+1)!} \psi_t^l (1-u)^l \Big] du,$$

$$\psi_t = c_0 |t|^\alpha \Big(1 + i \operatorname{sign} t \operatorname{tg}\frac{\pi\alpha}{2}\Big),$$

$$c_0 = \Gamma(1-\alpha) c_1 \cos\frac{\pi\alpha}{2}.$$

Consequently,

$$\lim_{n\to\infty} w_n(t_n) = w(t).$$

Theorem 1 is proved.

Theorem 2. *Let* $\{Y(z), z \geqslant 0, Y(0) = 0\}$ *be a separable stable process with stationary independent increments, defined by* (4), *and with the parameters* $1 < \alpha < 2$, $|\beta| < 1$, $c > 0$. *Then*

$$w(t) = v(t) + it \int_0^1 \exp\Big\{ - cu|t|^\alpha \Big(1 + i\,\beta \operatorname{sign} t \operatorname{tg}\frac{\pi\alpha}{2}\Big) \Big\} (1-u)^{1/\alpha - 1} du. \tag{30}$$

Proof. As in the proof of Theorem 1 we shall start from relation (12) and search for the limit $\lim w_n(tn^{-1/\alpha})$, $n \to \infty$.
First we shall show that if $V(x)$ is the stable law of index $1 < \alpha < 2$ and $a = \int x dV(x) = 0$, then

$$\bar{a}_k = \int_{-\infty}^{0-} x dW_k(x) = k^{1/\alpha - 1} v^-, \quad v^- = \int_{-\infty}^{0-} x dV(x). \tag{31}$$

In fact, since

$$\bar{a}_k = \frac{1}{k} \int_{-\infty}^{0} x dV_k(x)$$

(Nagaev, 1970), then

$$\bar{a}_k = k^{1/\alpha-1}\int_{-\infty}^{0} x dV_k(xk^{1/\alpha}) = k^{1/\alpha-1}\int_{-\infty}^{0-} x dV(x) = k^{1/\alpha-1}v^{-}.$$

Now we shall proceed to a more detailed study of the function $\bar{\varphi}_k(t)$. Let $t_n = tn^{-1/\alpha}$. Then

$$\bar{\varphi}_k(t) = \int_{-\infty}^{0-}(1-e^{it_n x})dW_k(x) = it_n\int_{-\infty}^{0} e^{it_n x}W_k(x)dx =$$

$$= it_n\bar{a}_k + it_n\int_{-\infty}^{0-}(\cos xt_n - 1)W_k(x)dx - t_n\int_{-\infty}^{0}\sin xt_n W_k(x)dx. \tag{32}$$

First we calculate the integral

$$it_n\int_{-\infty}^{0}(\cos xt_n - 1)W_k(x)dx =$$

$$= it_n|t_n|^{\alpha-1}\int_{-\infty}^{0}\frac{\cos u - 1}{|u|^\alpha}\left(\frac{u}{|t_n|}\right)^\alpha W_k\left(\frac{u}{|t_n|}\right)du. \tag{33}$$

Note that if $x \in [-Ak^{1/\alpha}, 0)$, where A will be defined later on, then

$$|x|^\alpha W_k(x) = o_k(1). \tag{34}$$

Really, taking into account (33) we find that for $-k^{1/(2\alpha)} \leqslant x < 0$

$$|x|^\alpha W_k(x) \leqslant |x|^{\alpha-1}\int_{-\infty}^{x}|u|dW_k(u) < |x|^{\alpha-1}|\bar{a}_k| = O\left(\frac{|x|^{\alpha-1}}{k^{1-1/\alpha}}\right) = O\left(k^{1/(2\alpha)-1/2}\right),$$

and for $x \in [-k^{1/\alpha}, -k^{1/(2\alpha)})$

$$|x|^\alpha W_k(x) = |x|^{\alpha-1}\int_{-\infty}^{x}|u|dW_k(u) = |x|^{\alpha-1}o(|\bar{a}_k|) = |x|^{\alpha-1}o(k^{1/\alpha-1}) = o_k(1).$$

Finally, by means of (31) we have

$$\int_{-\infty}^{x}|u|dW_k(u) = |\bar{a}_k|\varepsilon(|x|),$$

where $\varepsilon(u) \to 0$ as $u \to \infty$. Then for $x \in [-Ak^{1/\alpha}, -k^{1/\alpha})$

$$|x|^\alpha W_k(x) \leqslant |x|^{\alpha-1} \int_{-\infty}^{x} |u| dW_k(u) =$$

$$= |x|^{\alpha-1} |\bar{a}_k| \varepsilon(|x|) \leqslant A^{\alpha-1} \varepsilon(k^{1/\alpha}) = o_k(1); \quad A = (\varepsilon(k^{1/\alpha}))^{-1/2}.$$

Relation (34) is proved.
Further, note that if $x < 0$, then for any $k > 1$

$$W_k(x) \leqslant V(x).$$

Then, basing on the properties of the stable distribution $V(x)$ it can be concluded that when $x < -Ak^{1/\alpha}$ the function $|x|^\alpha W_k(x)$ is a bounded function. Hence and from the relations (33), (34) it follows

$$\left| t_n \int_{-\infty}^{0-} (\cos xt_n - 1) W_k(x) dx \right| \leqslant |t_n|^\alpha \left| \int_{-\infty}^{0-} \frac{\cos u - 1}{|u^\alpha|} \left(\frac{|u|^\alpha}{|t_n|} \right)^\alpha W_k \left(\frac{u}{|t_n|} \right) du \right| =$$

$$= o_k(|t_n|^\alpha) \int_{-Ak^{1/\alpha}|t_n|}^{0-} \frac{|\cos u - 1|}{|u|^\alpha} du + O(|t_n|^\alpha) \int_{-\infty}^{-Ak^{1/\alpha}|t_n|} \frac{|\cos u - 1|}{|u|^\alpha} du =$$

$$= o_k(|t_n|^\alpha) + O\left(|t_n|^\alpha \left(Ak^{1/\alpha} |t_n| \right)^{1-\alpha} \right) = o_k\left(\frac{|t|^\alpha}{n} \right) + O\left(\frac{|t_n|}{A^{1-\alpha} k^{1-1/\alpha}} \right) =$$

$$= o_k\left(\frac{|t|^\alpha}{n} \right) + O\left(\frac{|t|}{A^{\alpha-1} n^{1/\alpha} k^{1-1/\alpha}} \right). \tag{35}$$

Since $A = (\varepsilon(k^{1/\alpha}))^{-1/2}$ (35) means that

$$\left| t_n \int_{-\infty}^{0-} (\cos xt_n - 1) W_k(x) dx \right| = o_k\left(\frac{|t|^\alpha}{n} + \frac{|t|}{n^{1/\alpha} k^{1-1/\alpha}} \right). \tag{36}$$

Analogously, we find

$$\left| t_n \int_{-\infty}^{0-} \sin xt_n W_k(x) dx \right| = o_k\left(\frac{|t|^\alpha}{n} + \frac{|t|}{n^{1/\alpha} k^{1-1/\alpha}} \right). \tag{37}$$

Finally, from the relations (31), (32), (36) and (37) we obtain

$$\bar{\varphi}_k(t_n) = it_n v^- k^{1/\alpha-1} + o_k\left(|t_n|^\alpha + |t_n| k^{1/\alpha-1} \right). \tag{38}$$

Inserting (38) into (13) we find

$$w_n(t_n) = v(t) + it_n v^- \sum_{k=1}^{n-1} v^k(t_n)(n-k)^{1/\alpha-1}(1+o_{n-k}(1)) + t_n^\alpha \sum_{k=1}^{n-1} v^k(t_n) o_{n-k}(1).$$

However, since

$$\left| t_n \sum_{k=n-\sqrt{n}+1}^{n-1} v^k(t_n)(n-k)^{1/\alpha-1} o_{n-k}(1) \right| = O\left(n^{1/(2\alpha)}|t_n|\right) = O\left(\frac{|t|}{n^{1/(2\alpha)}}\right)$$

and

$$\left| t_n^\alpha \sum_{k=n-\sqrt{n}+1}^{n-1} v^k(t_n) o_{n-k}(1) \right| = O\left(\frac{|t|^\alpha}{\sqrt{n}}\right),$$

then

$$w_n(t_n) = v(t) + it_n v^- \sum_{k=1}^{n-1} v^k(t_n)(n-k)^{1/\alpha-1}(1+o_n(1)) +$$

$$+ o_n\left(|t_n|^\alpha \sum_{k=1}^{n-1} |v(t_n)|^k\right) + o\left(\frac{|t|}{n^{1/(2\alpha)}} + \frac{|t|^\alpha}{\sqrt{n}}\right) =$$

$$= v(t) + it_n v^- \sum_{k=1}^{n-1} |v(t_n)|^k (n-k)^{1/\alpha-1}(1+o(1)) + o_n(|t|^\alpha + |t| n^{-1/(2\alpha)}). \quad (39)$$

Hence, by virtue of (2) (as $\gamma = 0$)

$$w_n(t_n) = v(t) + \frac{itv^-}{n} \sum_{k=1}^{n-1} \exp\left\{ -c|t|^\alpha \frac{k}{n}\left(1 + i\beta \operatorname{sign} t \operatorname{tg}\frac{\pi\alpha}{2}\right)\right\} \times$$

$$\times \left(1 - \frac{k}{n}\right)^{1/\alpha-1} (1+o(1)) + o_n\left(|t|^\alpha + |t| n^{-1/(2\alpha)}\right).$$

Making use of the Euler—McLauren formula (see, e.g. (Cramer, 1946, Chapter 2)) we obtain

$$w_n(t_n) = v(t) +$$

$$+ \frac{itv^-}{n} \int_1^{n-1} \exp\left\{ -c|t|^\alpha \frac{x}{n}\left(1 + i\beta \operatorname{sign} t \operatorname{tg}\frac{\pi\alpha}{2}\right)\right\} \left(1 - \frac{x}{n}\right)^{1/\alpha-1} dx(1+o_n(1)) +$$

$$+ \frac{itv^-}{n} v(t_n)\left(1 - \frac{1}{n}\right)^{1/\alpha-1}(1+o_n(1)) + \frac{itv^-}{n} v^{n-1}(t_n) n^{1-1/\alpha}(1+o_n(1)) +$$

$$+O\left(\frac{|t|}{n^2}\int\limits_1^{n-1}\left[\left|v^x(t_n)\ln v(t)\right|+\left|v^x(t_n)\right|\left(1-\frac{x}{n}\right)^{1/\alpha-2}\right]dx\right)+$$

$$+o_n(|t|^\alpha+|t|n^{-1/(2\alpha)}).$$

Hence

$$w_n(t_n)=itv^-\int\limits_{1/n}^{(n-1)/n}\exp\left\{-cu|t|^\alpha\left(1+i\beta\operatorname{sign}t\ \operatorname{tg}\frac{\pi\alpha}{2}\right)\right\}(1-u)^{1/\alpha-1}du+$$

$$+o_n(|t|^\alpha+|t|)=itv^-\int\limits_0^1\exp\left\{-cu|t|^\alpha\left(1+\right.\right.$$

$$\left.\left.+i\beta\operatorname{sign}t\operatorname{tg}\frac{\pi\alpha}{2}\right)\right\}(1-u)^{1/\alpha-1}du\,(1+o_n(1))+o_n(|t|^\alpha+|t|).$$

Then for any finite t

$$\lim_{n\to\infty}w_n(t)=v(t)+itv^-\int\limits_0^1\exp\left\{-cu|t|^\alpha(1+\right.$$

$$\left.\left.+i\beta\operatorname{sign}t\operatorname{tg}\frac{\pi\alpha}{2}\right)\right\}(1-u)^{1/\alpha-1}du.$$

Theorem 2 is proved.

THEOREM 3. *Let $\{Y(z), z\geqslant 0, Y(0)=0\}$ be a separable stable process with stationary independent increments, determined by the characteristic function (4) with the parameters $1<\alpha<2$, $|\beta|<1$, $c>0$. Let*

$$W(x)=\mathbf{P}\{\sup_{0\leqslant z\leqslant 1}Y(z)<x\}.$$

Then

$$1-W(x)=\int\limits_x^\infty\left(p(y)+v^-\int\limits_0^1 u^{-2/\alpha}(1-u)^{1/\alpha-1}p'(yu^{-1/\alpha})du\right)dy,$$

where

$$v^-=\int\limits_{-\infty}^{0-}x\,dV(x),$$

$p(x)$ is a density of the stable distribution $V(x)$ defined by the characteristic function

$$v(t)=\exp\left\{-c|t|^\alpha\left(1+i\beta\operatorname{sign}t\ \operatorname{tg}\frac{\pi\alpha}{2}\right)\right\},$$

and

$$p'(y) = \frac{d}{dy}p(y).$$

Proof. According to Theorem 2 the characteristic function $w(t)$ of distribution $W(x)$ can be expressed by formula (30). Then

$$\frac{d}{dx}W(x) = \frac{1}{2\pi}\int_{-\infty}^{\infty} e^{-itx}w(t)dt =$$

$$= \frac{1}{2\pi}\int_{-\infty}^{\infty} e^{-itx}\Big[v(t) + itv^{-}\int_{0}^{1}\exp\Big\{-cu|t|^{\alpha}\Big(1+i\,\beta\mathrm{sign}\,t\ \mathrm{tg}\frac{\pi\alpha}{2}\Big)\Big\}(1-u)^{1/\alpha-1}du\Big]dt =$$

$$= p(x) + v^{-}\int_{0}^{1}(1-u)^{1/\alpha-1}p_u'(x)du, \tag{40}$$

where

$$p_u'(x) = \frac{d}{dx}p_u(x)$$

and $p_u(x)$ is the density of distribution $V_u(x)$, connected with $V(x)$ by this manner: if the characteristic function of the law $V(x)$ is expressed by the formula

$$v(t) = v(t,c) = \exp\Big\{-c|t|^{\alpha}\Big(1+i\,\beta\mathrm{sign}\,t\ \mathrm{tg}\frac{\pi\alpha}{2}\Big)\Big\},$$

then a characteristic function of the law $V_u(x)$ is given by the formula

$$v_u(t) = v(t,cu) = \exp\Big\{-cu|t|^{\alpha}\Big(1+i\,\beta\mathrm{sign}\,t\ \mathrm{tg}\frac{\pi\alpha}{2}\Big)\Big\}.$$

In other words, if $V(x)$ is a stable distribution with the parameters α, β, $\gamma = 0$, c then $V_u(x)$ is a stable distribution with the parameters α, β, $\gamma = 0$, cu, respectively. From (40) we find

$$1 - W(x) = \int_{x}^{\infty}\Big(p(y) + v^{-}\int_{0}^{1}(1-u)^{1/\alpha-1}p_u'(y)du\Big)dy. \tag{41}$$

Let us remind that the density

$$p(y) = p(y,\alpha,\beta,0,c)$$

of the stable distribution $V(x)$ with the parameters α, β, $\gamma = 0$, c satisfies the relation

$$p(y,\alpha,\beta,0,c) = c^{1/\alpha}p(yc^{-1/\alpha},\alpha,\beta,0,1)$$

(see (Ibragimov and Linnik, 1965, formula (2.3.1)). Then

$$p_u(y) = p(y,\alpha,\beta,0,cu) = u^{-1/\alpha}p(yu^{-1/\alpha},\alpha,\beta,0,c)$$

and, consequently,

$$p'_u(y) = \frac{d}{dy}p_u(y) = u^{-2/\alpha}p'(yu^{-1/\alpha},\alpha,\beta,0,c) = u^{-2/\alpha}p'(yu^{-1/\alpha}), \tag{42}$$

where

$$p'(y) = \frac{d}{dy}p(y) = \frac{d}{dy}p(y,\alpha,\beta,0,c). \tag{43}$$

Inserting (42) and (43) into (41) we obtain

$$1 - W(x) = \int_x^\infty \Big(p(y) + v^- \int_0^1 u^{-2/\alpha}(1-u)^{1/\alpha-1}p'(yu^{-1/\alpha})du\Big)dy.$$

Theorem 3 is proved.

REFERENCES

Baxter, G. and Donsker, M.D. (1957). On the distribution of the supremum functional for processes with stationary independent increments. *Trans. Amer. Math. Soc.* **85**, 73–87.

Bingham, N.H. (1973). Maxima of sums of random variables and suprema of stable processes. *Z. Wahrsch. verw. Geb.* **26**, 273–296.

Borovkov, A.A. (1970). Factorization identities and properties of the distribution of the supremum of sequential sums. *Theory Probab. Appl.* **15**, 359–403.

Cramer, H. (1946). *Mathematical Methods of Statistics.* Princeton Univ., Princeton.

Darling, D.A. (1956). The maximum of sums of stable random variables. *Trans. Amer. Math. Soc.* **83**, 164–169.

Gnedenko, B.V. and Kolmogorov, A.N. (1954). *Limit Theorems for Sums of Independent Random Variables.* Addison—Wesley, London.

Heyde, C.C. (1967). A limit theorem for random walk with drift. *J. Appl. Probab.* **4**, 144–150.

Heyde, C.C. (1969). On the maximum of sums of random variables and the supremum functional for stable processes. *J. Appl. Probab.* **6**, 419–429.

Ibragimov, I.A. and Linnik, Yu.V. (1965). *Independent and Stationary Related Random Variables.* Nauka, Moscow.

Nagaev, S.V. (1970). On the rate of convergence in a boundary problem. *Theory Probab. Appl.* **15**, 179–199.

Spitzer, F. (1964). *Principles of Random Walk.* Van Nostrand, New York.

Prob. Theory and Math. Stat., Vol. 1, pp. 52–56
B.Grigelionis *et al.* (Eds.)
1990 VSP/Mokslas

ON THE NECESSITY OF CRAMÉR'S CONDITION IN LOCAL LIMIT THEOREMS

N.N. AMOSOVA
Polytechnic Institute, 195251, Leningrad, USSR

ABSTRACT

Let $X_1, X_2, \dots$ be a sequence of independent identically distributed random variables for which $\mathbf{E}X_1 = 0$, $\mathbf{D}X_1 = \sigma^2 > 0$. We prove that Cramér's condition is necessary in local limit theorems for probabilities of large deviations.

In the present paper we consider the question of the necessity of Cramér's condition in local limit theorems for probabilities of large deviations. For a random variable X, Cramér's condition holds if there exists a positive constant A such that for $|h| < A$

$$\mathbf{E}\exp\{hX\} < \infty.$$

It is known (Petrov, 1975) that Cramér's condition is equivalent to each of the following two requirements:

1) There exists a positive constant a such that

$$\mathbf{E}\exp\{a \mid X_1 \mid\} < \infty.$$

2) There exist positive constants b and c such that for all $x > 0$

$$\mathbf{P}\{|X_1| \geqslant x\} \leqslant b\exp\{-cx\}.$$

Moreover, if $\mathbf{E}X_1 = 0$ then 1) and 2) are also equivalent to the following condition:

3) There exist positive constants g and T such that

$$\mathbf{E}\exp\{tX_1\} \leqslant \exp\{gt^2\}, \ \mid t \mid \leqslant T.$$

Let at first $X_1, X_2, \dots$ be a sequence of independent identically distributed integer-valued random variables. Denote $\mathbf{E}X_1 = m, \mathbf{D}X_1 = \sigma^2$ and assume that the maximal step of the distribution equal to one. Put $S_n = \sum_{i=1}^{n} X_i$. If Cramér's condition holds then (Richter, 1957) for $|x| > 1, x = o(\sqrt{n})$, $n \to \infty$,

$$\sigma\sqrt{n}\mathbf{P}(S_n = k) = \frac{1}{\sqrt{2\pi}}\exp\Big\{-\frac{x^2}{2} + \frac{x^3}{\sqrt{n}}\lambda\Big(\frac{x}{\sqrt{n}}\Big)\Big\}\Big(1 + O\Big(\frac{|x|}{\sqrt{n}}\Big)\Big), \ n \to \infty, \quad (1)$$

where $x = (k - nm)/(\sigma\sqrt{n})$ and $\lambda(t) = \sum_{k=0}^{\infty} c_k t^k$ is the well-known Cramér's series that converges for all sufficiently small t. The coefficients of this series depend only on the cumulants of X_1.

Let now $X_1, X_2, \dots$ be a sequence of independent identically distributed random variables. Suppose that $\mathbf{E}X_1 = 0, \mathbf{D}X_1 = 1$ and that the random variable X_1 has a bounded density $p(x)$.

Let $p_n(x)$ be the density of $S_n/\sqrt{n}$. If Cramér's condition holds, then (Richter, 1957) for $|x| \geqslant 1, x = o(\sqrt{n}),\ n \to \infty$,

$$p_n(x) = \frac{1}{\sqrt{2\pi}} \exp\Big\{-\frac{x^2}{2} + \frac{x^3}{\sqrt{n}}\lambda\Big(\frac{x}{\sqrt{n}}\Big)\Big\}\Big(1 + O\Big(\frac{|x|}{\sqrt{n}}\Big)\Big),\ n \to \infty, \tag{2}$$

where $\lambda(t)$ is the Cramér series.

Now the following question arises: Is Cramér's condition necessary for the asymptotics (1) and (2) in the domain $1 < |x| \leqslant \sqrt{n}/\psi(n)$ where $\psi(n)$ is an arbitrary function such that $\psi(n) \to \infty$ as $n \to \infty$?

We assume that the series $\lambda(t) = \sum_{k=0}^{\infty} c_k t^k$ converges for all sufficiently small t.

Let us formulate our basic results.

1. Let $X_1, X_2, \dots$ be sequence of independent identical distributed integer–valued random variables, $\mathbf{E}X_1 = m, \mathbf{D}X_1 = \sigma^2$ and assume that the maximal step of the distribution of X_1 is equal to one.

The following results are true (Amosova, 1987).

THEOREM 1. *For any function $\psi(n), \psi(n) \to \infty$ as $n \to \infty$ there exist positive constants δ and n_0 such that*

$$\mathbf{P}\{S_n = k\} < \frac{1}{\sigma\sqrt{n}} e^{-\delta x^2}$$

for all $n \geqslant n_0$ and all integer k in the domain $|x| \leqslant \sqrt{n}/\psi(n)$ if and only if Cramér's condition is satisfied.

THEOREM 2. *The relation (1) is fulfilled in the domain $1 < |x| \leqslant \sqrt{n}/\psi(n)$ for any function $\psi(n)$ such that $\psi(n) \to \infty$ as $n \to \infty$, where $x = (k - nm)/(\sigma\sqrt{n})$ and $\lambda(t) = \sum_{k=0}^{\infty} c_k t^k$ is some series that converges for all sufficiently small t, if and only if the Cramér's condition is satisfied.*

Remark 1. If (1) is fulfilled in the domain $1 < |x| \leqslant \sqrt{n}/\psi(n)$ for any function $\psi(n), \psi(n) \to \infty$ as $n \to \infty$, where $\lambda(t) = \sum_{k=0}^{\infty} c_k t^k$ is some series that converges for all sufficiently small t, then $\lambda(t)$ coincides with Cramér's series.

2. Let $X_1, X_2, \ldots$ be a sequence of independent identically distributed random variables. $\mathbf{E}X_1 = 0, \mathbf{D}X_1 = 1$ and assume that the distribution of X_1 have a bounded density $p(x)$. Denote by $p_n(x)$ the density of $S_n/\sqrt{n}$.

THEOREM 3. *The relation (2) is fulfilled in the domain* $1 \leqslant |x| \leqslant \sqrt{n}/\psi(n)$ *for any function* $\psi(n), \psi(n) \to \infty$ *as* $n \to \infty$, *where* $\lambda(t) = \sum_{k=0}^{\infty} c_k t^k$ *is some series converging for all sufficiently small* t, *if and only if Cramér's condition is true.*

Remark 2. If (2) is fulfilled in the domain $1 \leqslant |x| \leqslant \sqrt{n}/\psi(n)$ for any $\psi(n), \psi(n) \to \infty$ as $n \to \infty$, where $\lambda(t) = \sum_{k=0}^{\infty} c_k t^k$ is some series converging for all sufficiently small t, then $\lambda(t)$ coincides with Cramér's series.

THEOREM 4. *The relation*

$$p_n(x) = \frac{1}{\sqrt{2\pi}} \exp\Big(- \frac{x^2}{2}(1 + o(1))\Big), n \to \infty,$$

is fulfilled uniformly with respect to x *in the domain* $1 \leqslant |x| \leqslant \sqrt{n}/\psi(n)$ *for any function* $\psi(n), \psi(n) \to \infty$ *as* $n \to \infty$, *if and only if Cramér's condition is satisfied.*

THEOREM 5. *For any function* $\psi(n), \psi(n) \to \infty$ *as* $n \to \infty$, *there exist positive constants* δ *and* n_0 *such that*

$$p_n(x) \leqslant \frac{1}{\sqrt{2\pi}} e^{-\delta x^2} \tag{3}$$

for $n \geqslant n_0$ *in the domain* $1 \leqslant | x | \leqslant \sqrt{n}/\psi(n)$ *if and only if Cramér's condition is satisfied.*

The following one-sided analogue of Theorem 5 is true.

THEOREM 6. *For any function* $\psi(n), \psi(n) \to \infty$ *as* $n \to \infty$, *there exist positive constants* δ *and* n_0 *such that the relation* (3) *is fulfilled for* $n \geqslant n_0$ *in the domain* $1 \leqslant x \leqslant \sqrt{n}/\psi(n)$, *if and only if the following condition holds: there exists a positive constant* A *such that*

$$\int_0^\infty e^{hx} p(x) dx < \infty, \quad 0 \leqslant h < A.$$

Remark 3. In Theorems 3—6 the condition that there exists a bounded density $p(x)$ can be replaced by the condition that there exists n_0 such that the distribution function of $X_1 + \cdots + X_{n_0}$ is absolutely continuous with a bounded density.
We prove necessity in Theorems 1 and 2. In other theorems necessity may be proved analogously.
The following result is true.

LEMMA. *X_1 doesn't satisfy Cramér's condition if an only if*

$$\mathbf{P}(|X_1| = k) = e^{-k/g(k)} \tag{4}$$

for all $k > 0$ such that $\mathbf{P}(|X_1| = k) > 0$. Here $g(x)$ is a function such that $\lim\sup_{k\to\infty} g(k) = +\infty$.

Proof. Assume that random variable X_1 does not satisfy Cramér's condition and let the assertion of lemma be false. Then (4) holds with some function $g(x)$ such that $\lim\sup_{k\to\infty} g(k) = M < +\infty$. Then we have $g(k) < M + \varepsilon$ for some $\varepsilon > 0$ and for all sufficiently large $k > 0$. Put $0 < c < 1/(M + \varepsilon)$, then for sufficiently large N we have

$$\int_N^\infty e^{cx} dF(x) + \int_{-\infty}^{-N} e^{-cx} dF(x) \leqslant \sum_{k=N}^\infty e^{k(c-1/g(k))} \leqslant$$

$$\leqslant \sum_{k=N}^\infty e^{k(c-1/(M+\varepsilon))} = e^{(c-1/(M+\varepsilon))N}/(1 - e^{c-1/(M+\varepsilon)}) < +\infty.$$

Hence, Cramér's condition is satisfied that contradicts with the condition of lemma. Suppose now that the condition (4) holds with some function $g(x)$ such that $\lim\sup_{k\to\infty} g(k) = +\infty$. Then there exists a subsequence $\{k_n\}$ such that $\lim_{n\to\infty} g(k_n) = +\infty$. Then for all $c > 0$ and all sufficiently large $n > N$ we have $1/g(k_n) < \varepsilon < c$ and

$$\int_{-\infty}^{-k_N} e^{-cx} dF(x) + \int_{k_N}^\infty e^{cx} dF(x) \geqslant$$

$$\geqslant \sum_{n=N+1}^\infty e^{ck_n - k_n/g(k_n)} \geqslant \sum_{n=N+1}^\infty e^{k_n(c-\varepsilon)} = +\infty.$$

Hence, Cramér's condition fails. The Lemma is proved.
If (1) were satisfied, then for all sufficiently large n we should have

$$\mathbf{P}(S_n = k) < \frac{1}{\sigma\sqrt{n}} e^{-x^2/4},$$

where $x = (k - nm)/(\sigma\sqrt{n})$, $|x| > 1$, $x = o(\sqrt{n})$, $n \to \infty$. Assume now that X_1 does not satisfy the Cramér's condition.
Then from Lemma we have

$$\mathbf{P}(|X_1| = k) = e^{-k/g(k)}, \quad k > 0,$$

where $\lim\sup_{k\to+\infty} g(k) = +\infty$. Then there exists a sequence $\{k_n\}$ such that $\mathbf{P}(|X_1| = k_n) > 0$, $k_n \to \infty$, and $g(k_n) \to \infty$ as $n \to \infty$. Let $\gamma(n)$ be a function such that $\gamma(n) \to \infty$, $\gamma(n) = o(g(k_n))$, $\gamma(n) = o(k_n)$ as $n \to \infty$.

Put $N(n) = [k_n\gamma(n)]$, $\varphi(N,m) = [N(n)m] - N(n)m$. Here $[x]$ denotes the integral part of x. Then we have as $n \to \infty$.

$$\mathbf{P}(S_{N(n)} = k_n + [N(n)m]) + \mathbf{P}(S_{N(n)} = -k_n + [N(n)m]) \geqslant$$

$$\geqslant \mathbf{P}(S_{N(n)-1} = [N(n)m])\mathbf{P}(|X_1| = k_n) \geqslant \frac{1}{3\sigma\sqrt{N(n)}}e^{-k_n/g(k_n)} >$$

$$> \frac{1}{\sigma\sqrt{N(n)}}\Big(\exp\left\{-\alpha(k_n + \varphi(N,m))^2/N(n)\right\} +$$

$$+\exp\left\{-\alpha(-k_n + \varphi(N,m))^2/N(n)\right\}\Big) =$$

$$= \frac{2+o(1)}{\sigma\sqrt{N(n)}}\exp\left\{-\frac{\alpha k_n^2(1+o(1))}{N(n)}\right\}$$

for all $\alpha > 0$, because

$$\frac{\alpha k_n^2(1+o(1))}{N(n)} - \frac{k_n}{g(k_n)} =$$

$$= \frac{k_n^2}{N(n)}\left(\alpha(1+o(1)) - \frac{N(n)}{k_n g(k_n)}\right) \geqslant \alpha_1\frac{k_n^2}{N(n)}, \quad 0 < \alpha_1 < \alpha.$$

This implies that (1) cannot be true. Necessity in Theorems 1 and 2 is proved.

REFERENCES

Amosova, N.N. (1987). One complement for Richter's theorem. *Math. Notes* **42**, 120—123.

Petrov, V.V. (1975). *Sums of Independent Random Variables.* Springer, Berlin—Heidelberg etc.

Richter, W. (1957). The local limit theorem for large deviations. *Theory Probab. Appl.* **2**, 214—229.

Prob. Theory and Math. Stat., Vol. 1, pp. 57–67
B.Grigelionis *et al.* (Eds.)
1990 VSP/Mokslas

MARKOV RANDOM GRAPHS AND POLYGONAL FIELDS WITH Y–SHAPED NODES

T. ARAK and D. SURGAILIS
Tartu University, 202400 Tartu, Estonia
Institute of Mathematics and Cybernetics, 232600 Vilnius, Lithuania

ABSTRACT

A new class of Markov random graphs and polygonal fields on the plane is introduced, complementing the authors' earlier works.

INTRODUCTION

We define a new class of Markov random graphs and polygonal fields on the plane with Y–, Λ– and X–shaped nodes. The probability distribution of the random graph is proportional to

$$\prod_{\langle i,j\rangle} | z_i - z_j |^{-1} \, dz_1 \ldots dz_n,$$

where $z_1, \ldots, z_n \in T$ are vertices, $T \subset \mathbf{R}^2$ is a bounded convex set and the product is taken over edges. Our previous papers (Arak, 1986; Arak and Surgailis, 1988, 1989a, b) dealt with a different class of polygonal fields (without Y–shaped nodes), where the product Lebesgue measure was attributed to the lines forming the edges of the random graph.

We present here an exposition of the basic ideas of the construction together with comments and open problems. A complete proof of the Markov property of the random graph can be found in (Arak and Surgailis, 1989c). The second part of the paper is given to a dynamic construction of a family of stationary isotropic random graphs in $\mathbf{R}^2$ in terms of the equilibrium evolution of particle systems. A similar evolution of particle systems for polygonal fields without Y–shaped nodes was discussed in detail in (Arak and Surgailis, 1989a, b).

1. DEFINITIONS

Let $\mathcal{T}$ denote the class of all open connected bounded domains $T \subset \mathbf{R}^2$ with piecewise smooth boundary ∂T. Set $\bar{T} := T \cup \partial T$, $\mathcal{T}_{\text{conv}} := \{T \in \mathcal{T} : T \text{ is convex}\}$. A *segment* is any closed connected subset I of a line $\ell \subset \mathbf{R}^2$ of strictly positive length. We write $I[z_1, z_2]$ for the segment whose end points are $z_1, z_2 \in \mathbf{R}^2$,

$z_1 \neq z_2$, and $\ell[z_1, z_2]$ for the line which contains $I[z_1, z_2]$. Sometimes we shall denote an interval by $[\ell]$, indicating thus the line which it belongs to.
All lines $\ell \subset \mathbf{R}^2 = \{(t, y) : t, y \in \mathbf{R}\}$ will be parametrized as

$$\ell = \ell(p, \alpha), \quad (p, \alpha) \in \mathbf{R} \times [0, \pi), \tag{1}$$

where p is the signed length of the perpendicular to the line from the origin $0 \in \mathbf{R}^2$ and α is the angle this perpendicular makes with the t-axis.
Let G_T ($T \in \mathcal{T}$) be the class of all graphs γ whose vertices are points $z_i \in \bar{T}$, $i = 1, \ldots, n < \infty$, and whose edges are segments $I[z_i, z_j]$. Write $V(\gamma)$ and $I(\gamma)$ for the sets of vertices and edges of γ, respectively.
Given $T \in \mathcal{T}_{\text{conv}}$, denote by $\bar{T}_0^n (n \geqslant 1)$ the set whose elements are collections $(z)_n = (z_1, \ldots, z_n)$ of points $z_i \in \bar{T}$ such that $z_i \neq z_j$ $(i \neq j)$, none of three points z_i, z_j, z_k such that $\{z_i, z_j, z_k\} \not\subset \partial T$ lie on one line, and moreover, none of three lines $\ell[z_i, z_j]$, $i, j = 1, \ldots, n$, $i \neq j$, intersect at one point $z \in T \backslash (z)_n$. A graph $\gamma \in G_T$ with $V(\gamma) = (z)_n \in \bar{T}_0^n$ is called *admissible* if

$$\deg z_i = \begin{cases} 1, & z_i \in \partial T, \\ 2 \text{ or } 3, & z_i \in T, \end{cases} \tag{2}$$

$i = 1, \ldots, n$, where $\deg z$ is the degree of $z \in V(\gamma)$ (i.e., the number of incident edges). The class of all admissible γ's with $V(\gamma) = (z)_n$ will be denoted by $\Gamma_T(z)_n$. Clearly $\operatorname{card}\Gamma_T(z)_n < \infty$ for each $(z)_n \in \bar{T}_0^n$. Let $\Gamma_T(z)_n$, $n = 0, 1$, consist of the empty graph, and

$$\Gamma_T = \bigcup_{n=0}^{\infty} \bigcup_{(z)_n \in \bar{T}_0^n} \Gamma_T(z)_n.$$

A metrizable topology can be introduced in G_T and Γ_T such that $\gamma_n \to \gamma$ $(n \to \infty)$, $\gamma_n, \gamma \in G_T$ iff γ_n and γ are isomorphic for all n sufficiently large and all vertices of γ_n converge to the corresponding vertices of γ. Let $\mathcal{F}_T$ be the σ-field of Borel subsets of Γ_T.
Let $\gamma \in \Gamma_T$ be given. A *node* is either a vertex, or an inner intersection point of two segments of $\gamma \in \Gamma_T$, together with its infinitesimal neighborhood. To distinguish between a node and a point of $\bar{T}$, we shall denote the former by $z\cdot$. Write $N(\gamma)$ for the set of nodes of $\gamma \in \Gamma_T$. Then

$$N(\gamma) = N_I(\gamma) \cup N_\Lambda(\gamma) \cup N_Y(\gamma) \cup N_X(\gamma),$$

where $N_I(\gamma)$ (respectively, $N_\Lambda(\gamma)$, $N_Y(\gamma)$, $N_X(\gamma)$) is the set of I-shaped (respectively, Λ- , Y- , X-shaped) nodes. By definition, $z\cdot \in N_I(\gamma)$ (respectively, $z\cdot \in N_\Lambda(\gamma), z\cdot \in N_Y(\gamma)$) iff $z \in V(\gamma)$, $\deg z = 1$ (respectively, $\deg z = 2$, $\deg z = 3$), while X-shaped nodes correspond to the inner intersection points of edges. Write $z\cdot \in \tilde{N}_Y(\gamma)$ if $z\cdot \in N_Y(\gamma)$ and for any line $\ell \ni z$, the three incident segments do not lie in one of the two halfplanes bordered by ℓ. Set $\tilde{\Gamma}_T = \{\gamma \in \Gamma_T : N_Y(\gamma) = \tilde{N}_Y(\gamma)\}$.
Let us introduce the measure dz on $\bar{T}$ which is equal to the 1-dimensional Lebesgue measure $d_1 z$ on ∂T and to the 2-dimensional Lebesgue measure $d_2 z$ on T. Let $d^n(z)_n = dz_1 \ldots dz_n$ be the corresponding product measure on $\bar{T}_0^n$. Note that $\bar{T}^n \backslash \bar{T}_0^n$

is $d^n(z)_n$–negligible. For $T \in \mathcal{T}_{\text{conv}}$ consider the class Φ_T of all measurable functions $F : \Gamma_T \to \mathbf{R} \cup \{+\infty\}$ such that

$$Z_T \equiv Z_T(\Gamma_T) < +\infty,$$

where

$$Z_T(A) \equiv Z_{T,F}(A) = \sum_{n=0}^{\infty} \frac{1}{n!} \int_{\tilde{T}_0^n} d^n(z)_n \sum_{\gamma \in \Gamma_T(z)_n \cap A} e^{-F(\gamma)} \prod_{\langle i,j \rangle} | z_i - z_j |^{-1} \tag{3}$$

is the measure on Γ_T, not necessarily normalized, and the product is taken over the edges $I[z_i, z_j]$ of γ.
For any $F \in \Phi_T$ we define the probability measure $\mathbf{P}_T = \mathbf{P}_{T,F}$ on Γ_T by

$$\mathbf{P}_T(A) = Z_T(A)/Z_T, \quad A \in \mathcal{F}_T. \tag{4}$$

Comments

1. One would like to modify the above construction to include random graphs with T–shaped nodes, c.f. Comment 2 below. Another shortcoming of it is the fact that the restriction $\mathbf{P}_T \mid_{\Gamma_S}$ of $\mathbf{P}_T = \mathbf{P}_{T,F}$ to Γ_S with $S \in \mathcal{F}_{\text{conv}}, S \subset T$ is not necessarily a probability measure of the type $\mathbf{P}_{S,F'}$ with some $F' \in \Phi_S$. The reason is that $\mathbf{P}_T \mid_{\Gamma_S}$ assigns positive mass to graphs $\gamma \in \Gamma_S$ which are "untypical" for $\mathbf{P}_{S,F'}$ (see Fig. 1). The situation is different, however, if $\mathbf{P}_T$ is concentrated on the smaller set $\tilde{\Gamma}_T \subset \Gamma_T$.

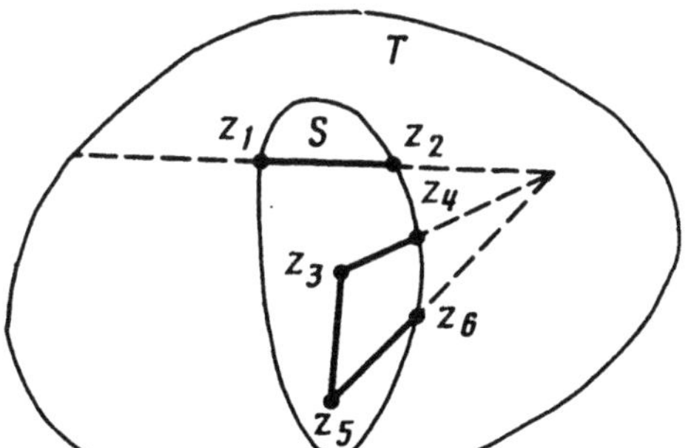

Fig. 1. *$\gamma \in \Gamma_S$ is typical for $\mathbf{P}_T \mid_{\Gamma_S}$ and untypical for $\mathbf{P}_S$ as the three lines $\ell[z_1, z_2], \ell[z_3, z_4]$ and $\ell[z_5, z_6]$ intersect at one point.*

2. In (Arak,1986; Arak and Surgailis, 1988,1989a) it was constructed another class of Markov random graphs and polygonal fields on the plane. There, the product Lebesgue measure was attributed to the lines forming the edges of the random graph. This construction includes T–shaped nodes but excludes Y–shaped ones

as the probability of three random lines intersecting at one point is zero. On the other hand, the two constructions yield the same probability measures on the set of admissible graphs with I–, Λ– and X–shaped nodes only (Arak and Surgailis, 1989c).

3. A direct verification of the condition $Z_{T,F} < +\infty$ (i.e., the existence of $\mathbf{P}_{T,F}$) seems difficult. For example, we do not know even if $\mathbf{P}_{T,0}$ exist for T's of sufficiently small diameter. On the other hand, the dynamic construction of Section 4 gives a sufficient condition for the existence of $\mathbf{P}_{T,F}$ (concentrated on $\tilde{\Gamma}_T$) for arbitrary $T \in \mathcal{T}_{\mathrm{conv}}$ (Corollary at the end of Section 4).

4. The constraint (2) on the maximal degree of an admissible graph is also connected with the divergence of $Z_{T,F}$. The increase of the maximal degree from 3 to 4 leads to the divergence of the summands of $Z_{T,F}$ unless F satisfies some hard–core condition which should violate the Markov property of $\mathbf{P}_{T,F}$ (Arak and Surgailis, 1989c).

5. The definitions (3), (4) can be carried over to (polygonal) random fields with finite number of values, as follows. Fix $T \in \mathcal{T}_{\mathrm{conv}}$ and a finite set $J \subset \mathbf{Z}$. Denote by $\tilde{\Omega}_T$ the set of all measurable functions $\omega : T \to J$ such that

$$\omega(z) = \lim_{\varepsilon \to 0} \operatorname{ess\,sup}\{\omega(z') : \mid z' - z \mid < \varepsilon\} \quad (\forall z \in T).$$

For any $(z)_n \in \bar{T}_0^n (n \geqslant 0)$ consider the class $\Omega_T(z)_n$ consisting of all functions $\omega \in \tilde{\Omega}_T$ for which there exists a graph $\gamma \in \Gamma_T(z)_n$ such that

$$\partial\omega = \bigcup_{I \in I(\gamma)} I \cap T,$$

where $\partial\omega \subset T$ is the set of discontinuity points of ω. Introduce the space of (polygonal) trajectories

$$\Omega_T = \bigcup_{n=0}^{\infty} \bigcup_{(z)_n \in \bar{T}_0^n} \Omega_T(z)_n$$

and the measure $W_{T,F}(A)$ on Ω_T by

$$W_{T,F}(A) = \sum_{n=0}^{\infty} \frac{1}{n!} \int_{\bar{T}_0^n} d^n(z)_n \sum_{\omega \in \Omega_T(z)_n \cap A} e^{-F(\omega)} \prod_{\langle i,j \rangle} \mid z_i - z_j \mid^{-1},$$

where $F : \Omega_T \to \mathbf{R} \cup \{+\infty\}$ is a (measurable) function. By a ***polygonal field*** on T we mean the probability measure

$$\mathbf{R}_{T,F}(A) = W_{T,F}(A)/W_{T,F} \ ,$$

where $W_{T,F} = W_{T,F}(\Omega_T) < +\infty$. Most of the results and comments of this paper on random graphs apply also to polygonal fields, in an appropriate formulation (Arak and Surgailis, 1989c).

2. THE MARKOV PROPERTY

Given $T \in \mathcal{T}_{\rm conv}, S \in \mathcal{T}, S \subset T$ and $\gamma \in \Gamma_T$, we define $\gamma = \pi_T^S \gamma \in G_S$ as follows. Let $I \in I(\gamma)$ and let $I' \subset I$ be the smallest segment containing $I \cap S$. Then $I' \in I(\gamma')$. Let $\mathcal{F}_T^S$ be the σ–field on Γ_T generated by the mapping $\pi_T^S : \Gamma_T \to G_S$. Let a probability measure **P** on $\Gamma_T(T \in \mathcal{T}_{\rm conv})$ be given. **P** will be called *Markov* if for any $S, U \in \mathcal{T}, S \cup U = T$, and any $A \in \mathcal{F}_T^S$ the following relation holds:

$$\mathbf{P}(A \mid \mathcal{F}_T^U) = \mathbf{P}(A \mid \mathcal{F}_T^{S \cap U}).$$

A random variable $F : \Gamma_T \to \mathbf{R} \cup \{+\infty\}$ will be called *additive* if for any $S, U \in \mathcal{T}, S \cup U = T$ there exist random variables $F_S, F_U : \Gamma_T \to \mathbf{R} \cup \{+\infty\}$ measurable with respect to $\mathcal{F}_T^S$ and $\mathcal{F}_T^U$ respectively, and such that

$$F(\gamma) = F_S(\gamma) + F_U(\gamma) \quad (\mathbf{P} - -a.s.).$$

THEOREM 1. *Let $F \in \Phi_T$ be additive. Then $\mathbf{P}_{T,F}$ is Markov.*

Comments

6. A rather general form of homogeneous additive functional is

$$F(\gamma) = \lambda L(\gamma) - \sum_{z \cdot \in N(\gamma)} h(z \cdot), \tag{5}$$

where $\lambda \in \mathbf{R}$ is a parameter, $L(\gamma)$ is the total length of all segments of γ and $h(z \cdot)$ is a function of the angular coordinates of $z \cdot \in N(\gamma)$ (i.e., the angles between the segments containing z, or the segment and the tangent to ∂T in the case when $z \in \partial T$) . A special case of (5) is considered in Sect. 3.
7. A crucial role in the Markov property of $\mathbf{P}_{T,F}$ is played by the product

$$\prod_{\langle i,j \rangle} \mid z_i - z_j \mid^{-1} .$$

Indeed, consider the conditional probability $\mathbf{P}_T(d\gamma_S \mid \gamma_{T \setminus S})$, where $\gamma_U = \pi_T^U \gamma$, $U \subset T$. Assume that γ contains an edge $I[z_1, z_2]$ with $z_1 \in S$ and $z_2 \in T \setminus S$ (see Fig. 2).

Fig. 2

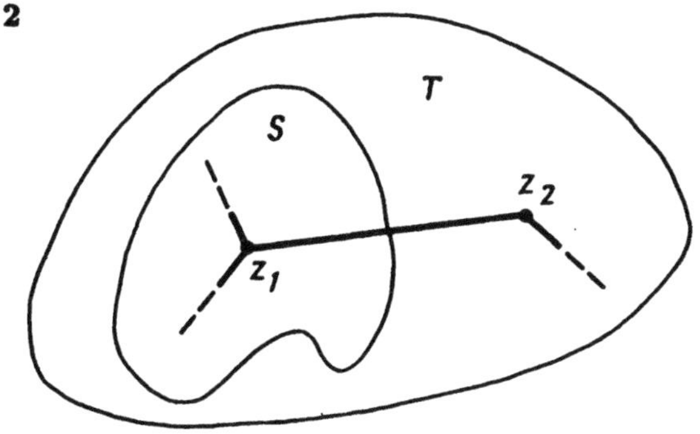

It is easy to understand that because of the Markov property of $\mathbf{P}_{T,F}$ the two vertices z_1 and z_2 must be conditionally independent provided we know the line $\ell = \ell[z_1, z_2]$ on which they lie.
Call a finite non-atomic measure $\mu = \mu(dz_1, dz_2)$ *primitive Markov* if it satisfies the above mentioned conditional independence property. Assume, moreover, that μ is homogeneous and isotropic, i.e.

$$\mu(dz_1, dz_2) = k(| z_1 - z_2 |)d_2z_1d_2z_2.$$

Then μ is primitive Markov iff

$$\mu(dz_1, dz_2) = \text{const} \mid z_1 - z_2 \mid^{-1} d_2z_1d_2z_2. \tag{6}$$

The above statement is equivalent to

$$d_2z_1d_2z_2 = \mid z_1 - z_2 \mid dp\, d\alpha d_1z_1d_1z_2, \tag{7}$$

where (p, α) are the coordinates (1) of $\ell = \ell[z_1, z_2]$ and d_1z is the element of length on ℓ.
A short proof of (7) can be given as follows. Fix α first, then $dp\,d_1z_1 = d_2z_1$ as dp and d_1z are the elements of length on perpendicular lines. Next, fix z_1, then

$$d\alpha d_1z_2 \mid z_1 - z_2 \mid = d_2z_2.$$

This proves (7).
One can prove, finally, that the Poisson field of segments $I = I[z_1, z_2]$ with mean $\mu = \mu(dz_1, dz_2)$ is Markov iff μ is primitive Markov. This fact seems to be closely related to the Markov property of $\mathbf{P}_{T,F}$ in Theorem 1.

3. MARKOV RANDOM GRAPHS IN $\mathbf{R}^2$

Write Γ for the inductive limit of the spaces Γ_T, $T \in \mathcal{T}_{\text{conv}}$ as $T \nearrow \mathbf{R}^2$. In other words, Γ is the set of all countable graphs $\gamma, V(\gamma) \subset \mathbf{R}^2$, such that $\pi^T_{\mathbf{R}^2}\gamma \in \Gamma_T$ for each $T \in \mathcal{T}_{\text{conv}}$.
Let $(\mathbf{P}_T)_T$ be a family of probability measures $\mathbf{P}_T$ on $\Gamma_T, T \in \mathcal{T}_{\text{conv}}$. The family $(\mathbf{P}_T)_T$ is said *consistent* if the equality

$$\mathbf{P}_T \mid_{\Gamma_S} = \mathbf{P}_S$$

holds for any pair $T, S \in \mathcal{T}_{\text{conv}}$ such that $T \supset S$. According to Kolmogorov's theorem, there exists a probability measure $\mathbf{P}$ on Γ such that $\mathbf{P} \mid_{\Gamma_T} = \mathbf{P}_T$ for each $T \in \mathcal{T}_{\text{conv}}$.
Let $b_i \geqslant 0, i = 0, 1$, and $q > 0$ be parameters such that

$$b_0 + \frac{\pi}{2}b_1 \leqslant 1. \tag{8}$$

Consider the additive functional $F(\gamma) = F_T(\gamma)$ (5), where

$$\lambda = q(2b_0 + \frac{3\pi}{8}b_1) \tag{9}$$

and $h(z\cdot)$ equals to

$$\begin{array}{ll} \log(q^{1/2}\sin\varphi) & \text{for } z\cdot\in N_I(\gamma), \\ \log(q(b_0 \;+\; b_1 H(\varphi))\sin\varphi) & \text{for } z\cdot\in N_\Lambda(\gamma), \\ \log(q^{1/2} b_1 \prod_{i=1}^{3}\sin\varphi_i) & \text{for } z\cdot\in \tilde{N}_Y(\gamma), \\ \log(1-b_0-b_1\sum_{i=1}^{2} H(\varphi_i)) & \text{for } z\cdot\in N_X(\gamma), \\ -\infty & \text{for } z\cdot\in N_Y(\gamma)\setminus\tilde{N}_Y(\gamma). \end{array} \tag{10}$$

Here

$$H(\varphi)=\frac{1}{2}(\sin\varphi+(\pi-\varphi)\cos\varphi)>0,\ \varphi\in(0,\pi), \tag{11}$$

and $\varphi, \varphi_i \in (0,\pi)$ are the angles shown in Fig. 3.

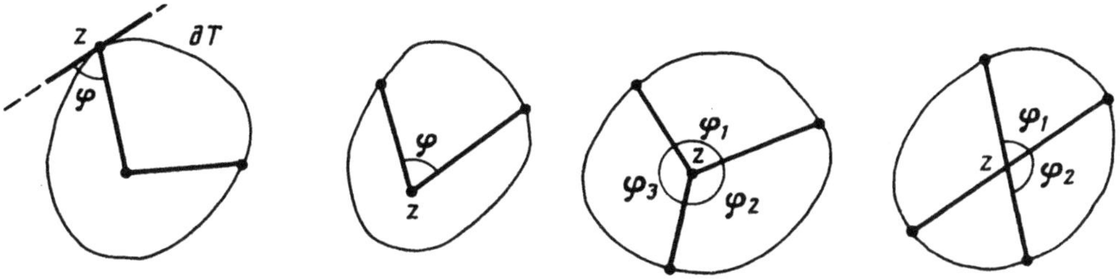

1) $z\cdot \in N_I(\gamma)$ 2) $z\cdot \in N_\Lambda(\gamma)$ 3) $z\cdot \in \tilde{N}_Y(\gamma)$ 4) $z\cdot \in N_X(\gamma)$

Fig. 3

Put also

$$d=q^2(\pi b_0+\frac{\pi^2}{4}b_1). \tag{12}$$

THEOREM 2. *Let $F(\gamma)$ (5) be defined by (8)–(11). Then:*

(i) *For each $T\in\mathcal{T}_{\text{conv}}$, $\mathbf{P}_{T,F}$ is well–defined and*

$$Z_{T,F}=\exp\{qL(\partial T)+dD(T)\},$$

where $D(T)$ is the area of T.

(ii) *The family $(\mathbf{P}_{T,F})_T$ is consistent. The corresponding probability measure $\mathbf{P}$ on Γ is stationary and isotropic.*

(iii) *For each line ℓ the point process $(r_j,\varphi_j)\in\ell\times(0,\pi)$ of the intersection points and angles of the random graph with the line ℓ is Poisson with intensity $q\,dr\,d\varphi\sin\varphi$.*

Comments

8. For $b_1 = 0$ Theorem 2 follows from Theorem 4.1 in (Arak and Surgailis, 1989a) and the relation between the classes of random graphs mentioned in Comment 2.
9. Theorem 2 does not exhaust all cases of random graphs of the form (4) with the properties (ii) and (iii).
10. According to the results of (Ambarzumyan,1973), random graphs with the property (iii) cannot have vertices of degree greater than 3.

4. A SKETCH OF THE PROOF OF THEOREM 2

The main idea of the proof of Theorem 2 consists in a dynamic description of $\mathbf{P}_{T,F}$ in terms of the equilibrium evolution of systems of 1-dimensional particles with death, birth and branching. For simplicity we shall discuss this evolution in a rectangular domain

$$T = \{(t_0, t_1) \times (a_-, a_+)\} \subset \mathbf{R}^2 = \{(t, y) : t, y \in \mathbf{R}\}.$$

Write

$$\partial T = \partial^0 T \cup \partial^+ T \cup \partial^- T \cup \partial^1 T,$$

where

$$\partial^0 T = \{(t_0, y) : a_- \leqslant y \leqslant a_+\}, \ \partial^{\pm} T = \{(t, a_{\pm}) : t_0 \leqslant t \leqslant t_1\},$$
$$\partial^1 T = \{(t_1, y) : a_- \leqslant y \leqslant a_+\}$$

The dynamics is defined in a$_1$)–a$_4$) below.

a$_1$) A *particle* is any pair (y, α), where $y \in [a_-, a_+]$ is position and $\alpha = -\operatorname{arcctg} v$ characterizes velocity $v \in \mathbf{R}$. Trajectory $(y_t)_{t \geqslant t_0}$ of a particle is completely determined by its initial position y_{t_0} and the evolution $(\alpha_t)_{t \geqslant t_0}$ which is the homogeneous jump-type Markov process on $(0, \pi)$ with the jump intensity:

$$q(\sin\alpha)^{-1}(b_0 + b_1 H(\pi - \mid \alpha - \beta \mid)) \sin \mid \alpha - \beta \mid d\beta, \quad \beta \in (0, \pi).$$

Moreover, the particle (y, α) can branch at any time $t > t_0$ into two particles $(y, \beta'), (y, \beta'')$ such that $0 < \beta' < \alpha < \beta'' < \pi$ with the branching intensity:

$$qb_1(\sin\alpha)^{-1} \sin(\beta'' - \alpha) \sin(\alpha - \beta') \sin(\beta'' - \beta') d\beta' d\beta''.$$

Branching is interpreted as death of the old particle and birth of the two new ones. Trajectory of a particle in the (t, y)-plane is a sequence of line segments whose lengths are independent and distributed exponentially with the parameter λ (9), which is equal to the full intensity of jumps and branching multiplied by $\sin\alpha$.

a$_2$) Consider four independent Poisson point processes:

$$(z_j^0, \alpha_j^0)_j, \ (z_j^+, \alpha_j^+)_j, \ (z_j^-, \alpha_j^-)_j, \ (z_j', \alpha_j', \alpha_j'')_j$$

on $\partial^0 T \times (0, \pi), \partial^+ T \times (0, \pi/2), \partial^- T \times (\pi/2, \pi)$ and $T \times (0, \pi)^2$ respectively, with intensities $q \sin\alpha d_1 z d\alpha$, $q \cos\alpha d_1 z d\alpha$, $-q \cos\alpha d_1 z d\alpha$ and

$$q^2(b_0 + b_1 H(\alpha'' - \alpha')) \sin(\alpha'' - \alpha') \mathbf{1}(\alpha'' > \alpha') d_2 z d\alpha' d\alpha'',$$

respectively. Write

$$z_j^0 = (t_j^0, y_j^0),\ z_j^+ = (t_j^+, y_j^+),\ z_j^- = (t_j^-, y_j^-),\ z_j' = (t_j', y_j') = (t_j'', y_j'').$$

The points $(t_j^0, y_j^0, \alpha_j^0)_j, (t_j^+, y_j^+, \alpha_j^+)_j, (t_j^-, y_j^-, \alpha_j^-)_j, (t_j', y_j', \alpha_j')_j, (t_j'', y_j'', \alpha_j'')_j$ will be considered as the birth coordinates of particles, with the t–coordinate standing for the time of birth.

a_3) The evolution of the system of particles is determined by the following conditions. Each of the born particles in a_2) undegoes independent evolution described in a_1) during its lifetime. This applies also to the evolution of the particles born at the moments of branching and collision (see a_4)). Death occurs at the moment of branching (see a_1)), the exit from T or the collision with some other moving particles (in the last case, the colliding particles can also survive with some probability; see below).

a_4) Let two particles (y, α) and (y, β) collide at $(t, y) \in T$ with $\alpha < \beta$. Then:

– with probability $b_0 + b_1 H(\beta - \alpha)$ the particles die without producing an offspring (Fig. 4, 1);

– with probability $b_1 \sin(\beta - \alpha') \sin(\alpha' - \alpha) d\alpha'$ a new particle $(y, \alpha'), \alpha' \in (\alpha, \beta)$ is born in their place (Fig. 4, 2);

– with probability $1 - b_0 - b_1(H(\beta - \alpha) + H(\pi - \alpha + \beta))$ both particles survive (Fig. 4, 3).

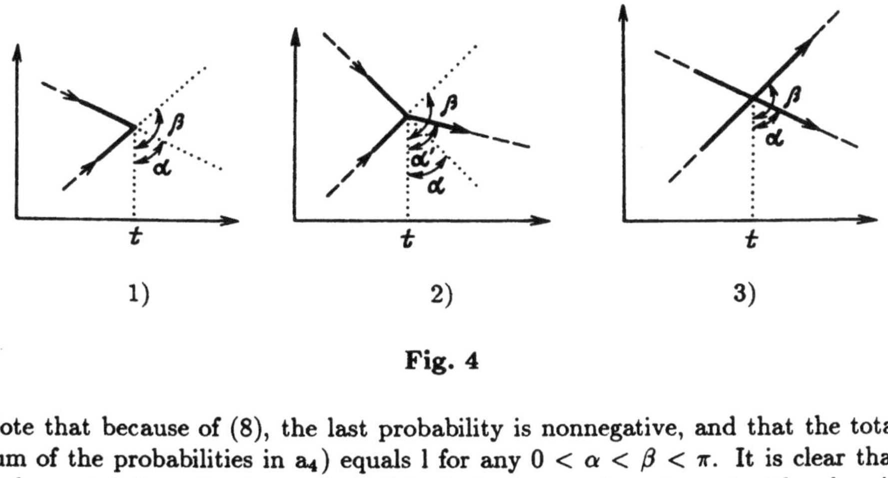

1) 2) 3)

Fig. 4

Note that because of (8), the last probability is nonnegative, and that the total sum of the probabilities in a_4) equals 1 for any $0 < \alpha < \beta < \pi$. It is clear that such an evolution of systems of particles induces a random element with values in $\tilde{\Gamma}_T$. Let $\mathbf{Q}_T$ be the probability measure on Γ_T induced by this random element.

LEMMA. $\mathbf{Q}_T = \mathbf{P}_{T,F}$.

The proof of the Lemma uses explicit formulae for the intensities and the probabilities governing our evolution, together with some elementary transformations. As an example, let us prove the relation

$$\mathbf{Q}_T(d\gamma) = \mathbf{P}_{T,F}(d\gamma) \tag{13}$$

for γ represented in Fig. 5.

Fig. 5

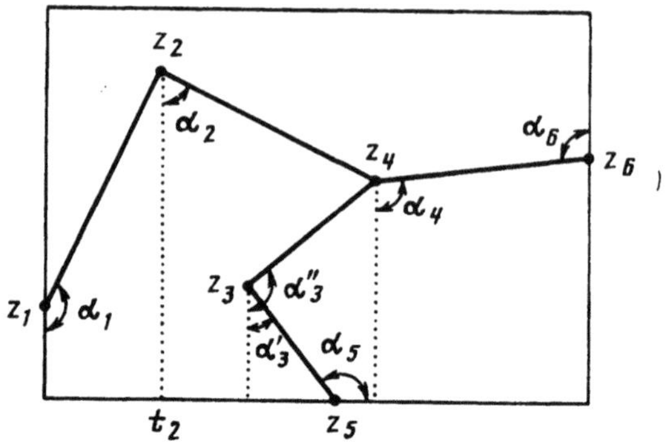

According to a₁)–a₄) and (9)–(12),

$$\mathbf{Q}_T(d\gamma) = \exp\{-qL(\partial T) - dD(T)\}qd_1z_1 \sin\alpha_1\, d\alpha_1 \times$$

$$\times \exp\{-\lambda L(\gamma)\} \sin^{-1}\alpha_1 dt_2 \exp\{h(z_2\cdot)\}d\alpha_2 \times$$

$$\times d_2z_3 q \exp\{h(z_3\cdot)\}d\alpha_3' d\alpha_3'' \exp\{h(z_4\cdot)\}q^{-1/2}\sin^{-1}(\alpha_3'' - \alpha_2)d\alpha_4.$$

Hence (13) follows from the elementary equalities:

$$d\alpha_1 dt_2 = d_2z_2 \mid z_1 - z_2 \mid^{-1} \sin\alpha_1,$$

$$d\alpha_2 d\alpha_3'' = d_2z_4 \mid z_2 - z_4 \mid^{-1} \mid z_3 - z_4 \mid^{-1} \sin(\alpha_3'' - \alpha_2),$$

$$d\alpha_3' = d_1z_5 \mid z_3 - z_5 \mid^{-1} \cos\alpha_3',$$

$$d\alpha_4 = d_1z_6 \mid z_4 - z_6 \mid^{-1} \sin\alpha_4$$

and $\cos\alpha_3' = \sin\alpha_5, \sin\alpha_4 = \sin\alpha_6$.

Theorem 2 follows from the Lemma in a similar way as Theorem 4.1 is derived from Lemma 4.1 in (Arak and Surgailis, 1989a). In particular, let $T' = (t_0, t_1') \times \times (a_-, a_+) \subset T$. Then

$$\mathbf{Q}_T \mid_{\Gamma_{T'}} = \mathbf{Q}_{T'} \tag{14}$$

as both $\mathbf{Q}_T$ and $\mathbf{Q}_{T'}$ can be constructed via the evolution of the same particle system: for $\mathbf{Q}_{T'}$ this evolution must be observed up to time t_1' only. From (14), the Lemma and the invariance of $\mathbf{P}_{T,F}$ with respect to the "time–reversal": $(t, y) \to (-t, y)$ we have at once that the distribution of the particle system at time t_1' coincides with its distribution at time t_0, which is Poisson.

COROLLARY. *Let $T \in \mathcal{T}_{\text{conv}}$ and let $F' : \Gamma_T \to \mathbf{R} \cup \{+\infty\}$ satisfy the condition:*

$$F'(\gamma) \geqslant F(\gamma)\ ,\ \gamma \in \Gamma_T\,,$$

where $F(\gamma) = F(\gamma; b_0, b_1, q)$ *is the functional from Theorem 2. Then* $F' \in \Phi_T$.

REFERENCES

Ambarzumyan, R. (1973). On random fields of segments and random mosaics on the plane. *Theory Probab. Appl.* **18**, 515–523.

Arak, T. (1986). A class of Markov fields with finite range. In: *Proc. Int. Congress of Mathematicians. Berkeley.* AMS, Reading, pp. 994–999.

Arak, T. and Surgailis, D. (1988). Polygonal Markov random fields. *Soviet Math. Dokl.* **302**, 521–523.

Arak, T. and Surgailis, D. (1989a). Markov fields with polygonal realizations. *Prob. Th. Rel. Fields.* **80**, 543–579.

Arak, T. and Surgailis, D. (1989b). Consistent polygonal fields. *Preprint.* Centro matematico V. Volterra, Universita Roma II.

Arak, T. and Surgailis, D. (1989c). Polygonal fields with Y-shaped nodes. *Preprint.* Inst. Math. and Cybern., Vilnius.

Prob. Theory and Math. Stat., Vol. 1, pp. 68–77
B.Grigelionis *et al.* (Eds.)
1990 VSP/Mokslas

PROBABILITY MEASURES ON QUANTUM LOGICS AND A NON–COMMUTATIVE CHOQUET THEORY

Sh.A. AYUPOV

Institute of Mathematics of Uzbek Academy of Sciences,
ul.F.Hodjaeva 29, Tashkent 700143, USSR

ABSTRACT

In the present paper we suggest a geometric approach to the notion of probability measure on projections of operator algebras in terms of the theory of affine function spaces on convex sets. Following (Alfsen and Shultz, 1976) we introduce the notion of projective units, P–projections and projective faces which generalize respectively self–adjoint projections p, the maps $a \longrightarrow pap$ and closed faces of state spaces of operator algebras.

INTRODUCTION

The well–known Gleason theorem and its generalizations for measures on projections of von Neumann algebras and Jordan Banach algebras are considered as the problem of extending a measure from the lattice of projective units to a linear functional on the space of all bounded affine functions on the given convex set. In its turn the last problem is equivalent to the problem of the existence of the so–called "barycenter" of a measure defined on the quantum logic of projective faces of the convex set. Therefore if we consider the last measure as a non commutative version of measures defined on extreme points of a simplex or of more general boundary measures on compact convex sets, then these problems can be interpreted as a "non commutative" or "quantum" Choquet theory (Alfsen, 1971; Asimov and Ellis, 1980).

In Section 1 we give preliminaries from the non commutative spectral theory for affine function spaces. In Section 2 we recall some notions and results from the Choquet theory. Section 3 is devoted to probability measures on the lattice (quantum logic) of projective faces of a spectral convex set. We introduce the notion of barycenter for such measures and study the so called Gleason sets and Jauch—Piron sets which generalize the notion of Choquet simplices.

1. SPECTRAL CONVEX SETS

Let A be an ordered vector space over the $\mathbf{R}$. A positive element $e \in A$ is called *order unit* if given any $a \in A$ there is $\lambda \in \mathbf{R}^+$ such that $-\lambda e \leqslant a \leqslant \lambda e$. It is clear that if A has an order unit then the cone A^+ of positive elements is a generating cone, i.e., $A = A^+ - A^+$. The space A with an order unit e is said to be ***Archimedean*** if $na \leqslant e$ for all $n = 1, 2, \ldots$ implies $a \leqslant 0$. In this case A admits the so-called ***order-unit norm***

$$\|a\| = \inf\{\lambda > 0\colon\ -\lambda e \leqslant a \leqslant \lambda e\}.$$

We shall use the term ***order unit space*** and the notation (A, e) to denote an Archimedean ordered vector space with a distinguished order unit e considered as a normed vector space in the order-unit norm.
A convex set B in a vector space E is ***radially compact*** if $B \cap L$ is a closed bounded segment for any line L passing through the origin of E. Recall also that a convex subset K of a hyperplane H not passing through the origin of E is said to be a ***base*** for the cone $C = \cup\{\lambda K,\ \lambda \geqslant 0\}$. We shall use the term ***base norm space*** and the notation (E, K) to denote an ordered vector space E for which E^+ is a generating cone with a base K such that $B = conv(K \cup -K)$ is radially compact, and E is considered as a normed space in the ***base norm***

$$\|x\| = \inf\{\lambda \geqslant 0\colon\ x \in \lambda B\},$$

where $convX$ denotes the convex hull of X in E.
Let now K be an arbitrary convex set in a locally convex space V. Denote by A or $A^b(K)$ the order-unit space of all bounded affine functions on K with the pointwise order and order unit $e = 1$. From now on we shall assume that K is embedded in V in such a way that (V, K) becomes a base-norm space and $A = A^b(K) \approx V^*$. If K is compact then such imbedding is always possible (the "regular embedding") (see (Alfsen, 1971, Chapter II, Section 2)), namely one can take $V = A(K)^*$, where $A(K)$ is the order-unit space of all continuous affine functions on K.

Definition. A positive projection $R\colon A \to A$ with norm 1 is called ***P-projection*** if there exists a unique positive projection $R'\colon A \to A$ with norm 1 such that

$$\begin{aligned} \mathrm{Im}^+R &= \mathrm{Ker}^+R', & \mathrm{Im}^+R^* &= \mathrm{Ker}^+R'^*, \\ \mathrm{Ker}^+R &= \mathrm{Im}^+R', & \mathrm{Ker}^+R^* &= \mathrm{Im}^+R'^*, \end{aligned}$$

where $R^*\colon V \to V$ is the dual map for R.
Denote by $\mathfrak{P}$ the set of all P-projections on A, and for $R, Q \in \mathfrak{P}$ put $R \leqslant Q$ if $\mathrm{Im}R \subseteq \mathrm{Im}Q$. Then $\mathfrak{P}$ is ordered and $0 \leqslant R \leqslant I$ for all $R \in \mathfrak{P}$, where 0 is the zero projection and I is the identity map. The projection R' in the definition of P-projection R is also a P-projection which is called ***the quasicomplement*** of R.
In the order-unit space (A, e) the elements of the form $u = Re$, $R \in \mathfrak{P}$, are called ***projective units***. The set $\mathfrak{U}$ of all projective units is equipped with the ordering induced from A and the complementation $u^\perp = e - u$.
Recall now that a subset G of the convex set K is called ***a face*** if $x, y \in K$, $\lambda \in (0, 1)$, $\lambda x + (1 - \lambda)y \in G$ imply $x, y \in G$. A face G is said to be ***exposed*** if $G = \{\rho \in K\colon\ a(\rho) = 0\}$ for some $a \in A^+$. If moreover a is a projective unit, say $a = Re$ for some $R \in \mathfrak{P}$, then G is called a ***projective face***. In other words projective faces are those which have the form

$$G = \mathrm{Im}^+R^* \cap K = F_R, \quad R \in \mathfrak{P}.$$

Denote by $\mathfrak{F}$ the set of all projective faces of K endowed with the inclusion ordering and the complementation

$$F^{\#} = F_{R'} = \mathrm{Im}^{+} R'^{*} \cap K.$$

THEOREM 1.1 (see (Alfsen and Shultz, 1976, Corollary 2.18)). *The sets* $\mathfrak{P}$, $\mathfrak{U}$, $\mathfrak{F}$ *are canonically order isomorphic. Specifically, we have a commutative diagram of order preserving maps*

$$\mathfrak{F} \xrightarrow{T_1} \mathfrak{P}, \qquad \mathfrak{P} \xrightarrow{T_2} \mathfrak{U},$$

$$\mathfrak{F} \xleftarrow{T_3} \mathfrak{U},$$

where $T_1\colon F_R \to R$, *i.e.*

$$T_1(F)(a) = \sup\{b \in A^{+}\colon\ b \leqslant a \text{ on } F, \quad b = 0 \text{ on } F^{\#}\} =$$

$$= \inf\{b \in A^{+}\colon\ b \geqslant a \text{ on } F, \quad b = 0 \text{ on } F^{\#}\},$$

$$T_2\colon R \to Re, \quad T_3\colon Re \to F_R = \{\rho \in K\colon\ (Re, \rho) = 1\},$$

here (a, ρ) *means* $a(\rho)$.

Now following (Bunce and Wright, 1985 b) we shall say that a convex set K is *projective* if its every exposed face is projective.

THEOREM 1.2 (see (Alfsen and Shultz, 1976, Corollary 12.5)). *If* K *is projective then* $\mathfrak{P}$, $\mathfrak{U}$ *and* $\mathfrak{F}$ *are mutually isomorphic complete orthomodular lattices (quantum logics).*

Recall that two P-projections $P, Q \in \mathfrak{P}$ (respectively projective faces $F, G \in \mathfrak{F}$ or projective units $u, v \in \mathfrak{U}$ are said to be *orthogonal* if $PQ = 0$ (respectively $G \subseteq F^{\#}$ or $u + v \leqslant e$). Two elements $a, b \in A^{+}$ are said to be orthogonal, denoted by $a \perp b$, if there is a projective face F such that $(a, \rho) = 0$, $(b, \sigma) = 0$ for all $\rho \in F$, $\sigma \in F^{\#}$.

Definition. A convex set K is called *spectral* if it is projective and any $a \in A = A^{b}(K)$ can be uniquely decomposed as $a = a_{+} - a_{-}$, where $a_{+}, a_{-} \in A^{+}$, $a_{+} \perp a_{-}$. It should be noted (Alfsen and Shultz, 1976) that any finite dimensional projective convex set is spectral. Moreover if K is a projective set and a convex body in a reflexive Banach space then it is spectral (Alfsen and Shultz, 1979, Proposition 2.5). Further examples of spectral convex sets are given by state spaces of JB- and C^{*}-algebras and by more general normal state spaces of JBW- and W^{*}-algebras (Alfsen and Shultz, 1978; Alfsen et al., 1980). Among spectral convex sets normal state spaces of JBW-algebras can be characterized as follows (Iochum and Shultz, 1983, Theorem 1.5).

THEOREM 1.3. *A spectral convex set* K *is affinely isomorphic to the normal state space of a* JBW*-algebra iff* K *is elliptic, i.e.*

$$P(Q - Q')P' = 0$$

for all $P, Q \in \mathfrak{P}$.

For different characterizations of state spaces of $JB-$ and C^*-algebras or normal state spaces of $JBW-$ and W^*-algebras see (Alfsen and Shultz, 1978; Alfsen et al., 1980).
If K is spectral then the spectral theorem is valid for $A = A^b(K)$. Recall (Alfsen and Shultz, 1976) that a family $\{e_\lambda,\ \lambda \in \mathbf{R}\}$ of projective units from A is called *spectral family* if

(i) $e_\lambda \leqslant e_\mu$ for $\lambda \leqslant \mu$;

(ii) $\inf e_\lambda = 0, \quad \sup e_\lambda = e$;

(iii) $e_\lambda = \inf\{e_\mu,\ \mu \gg \lambda\}$ for any $\lambda \in \mathbf{R}$.

THEOREM 1.4 (see (Alfsen and Shultz, 1976), Theorem 7.6). *If K is a spectral convex set then given any $a \in A$ there is a unique spectral family $\{e_\lambda\} \in \mathfrak{U}$, such that $a = \int \lambda\, de_\lambda$, where the right hand side is norm convergent Riemann—Stieltjes integral.*

Finally, recall one more example of a spectral convex set. A compact convex set K is called *Choquet simplex* if the base-norm space $(V, K) = (A(K)^*, K)$ is a vector lattice. In this case $A = A^b(K)$ is also a vector lattice. Every Choquet simplex is a spectral convex set. A closed face F of a convex set K is called *split face* if there is a face F' in K, such that K is the direct sum of F and F' ($K = F \oplus cF'$), i.e., any $\rho \in K$ can be uniquely represented as $\rho = \lambda\mu + (1-\lambda)\nu$ with $\lambda \in [0,1]$, $\mu \in F, \quad \nu \in F'$.

THEOREM 1.5 (see (Alfsen and Shultz, 1976), Section 5; (Asimov and Ellis, 1980), Theorem 7.2). *Let K be a spectral convex set. The following conditions are equivalent:*

(1) *K is Choquet simplex;*

(2) *any closed exposed face of K is a split face;*

(3) *the isomorphic lattices $\mathfrak{P}$, $\mathfrak{U}$, $\mathfrak{F}$ are Boolean algebras.*

Remarks. If K is the normal state space of a JBW-algebra J then $A^b(K) \approx$ J, moreover projective units of $A^b(K)$ correspond to projections of J, P-projections are maps $a \to pap$, etc. For details of the theory of JBW-algebras we refer the reader to (Hanche-Olsen and Stormer, 1984).

2. ELEMENTS OF CHOQUET THEORY

Let K be a compact convex set in a locally convex space E over $\mathbf{R}$. Denote by $A(K, E)$ the vector space of all restrictions to K of continuous affine real-valued functions on E, by $C_{\mathbf{R}}(K)$ the space of all continuous real-valued functions on K. Then it is clear that $A(K) \supseteq A(K, E)$ and $A(K)$ is uniformly closed in $C_{\mathbf{R}}(K)$. We shall denote *the extreme boundary* (the set of all extreme points) of K by the symbol $d_e K$. Given $f \in C_{\mathbf{R}}(K)$ define *the upper envelope of f* as

$$f^\frown = \inf\{a \in A(K, E):\ a \geqslant f\}.$$

In the present section we shall use the word *measure* to denote a real *Radon measure* μ on some compact Hausdorf space T (i.e., $\mu \in C_{\mathbf{R}}(T)^*$). Recall that the Radon measures on T correspond biuniquely to the Baire measures on T, i.e., to the real valued countably additive set functions. Specifically, the correspondence is given by

$$\mu(f) = \int f\, d\mu, \quad f \in C_{\mathbf{R}}(T).$$

If μ is a measure on a compact Hausdorf space T, then a function f from T into a locally convex space E is said to be *weakly μ-integrable* with μ-integral $y \in E$ if $p \cdot f$ is μ-integrable with

$$p(y) = \int (p \cdot f)\, d\mu \quad \text{for all} \quad p \in E^*.$$

Clearly weak integrals are unique whenever they exist, and we shall denote them by $y = \int f\, d\mu$.

Now let μ be a measure on a compact convex set K in a locally convex space E. By ((Alfsen, 1971), Proposition 1.2.1) the identity function $x\colon K \to E$ is weakly μ-integrable, and μ-integral $\int x\, d\mu(x)$ is called *the moment* of μ. If μ is a positive non-zero measure on K, then the point

$$x_\mu = \mu(K)^{-1} \int x\, d\mu(x)$$

is called *barycenter* of μ; it is in K since $\mu(K)^{-1}\mu$ is a probability measure. In the sequel we shall most often work with probability measures (i.e., positive and normalized: $\mu(K) = 1$). For such measures the following "barycentric formula" is valid

$$a(x_\mu) = \int a(x)\, d\mu(x) \quad \text{for all} \quad a \in A(K).$$

Thus any probability measure on K has a (unique) barycenter in K. The converse problem is more complicated: can any point of K be represented as the barycenter of some measure on K? The Choquet theory gives the affirmative answer to this question, moreover the corresponding measure can be found among the so-called boundary measures on K.

Recall that a probability measure μ on K is called *boundary measure* if $\mu(K - B_f) = 0$ for all $f \in C_{\mathbf{R}}(K)$, where $B_f = \{t \in K\colon \hat{f}(t) = f(t)\}$ is the boundary set of f, $\hat{f}$ is the upper envelope for f. In particular if μ is a boundary measure then its support $\operatorname{supp}(\mu) \subseteq \overline{d_e K}$. If K is metrizable then μ is boundary measure iff $\mu(K - d_e K) = 0$.

THEOREM 2.1 (Choquet—Bishop—de-Leeuw). *Every point in a compact convex set can be represented by a boundary probability measure.*

In general the representing boundary measure is not unique. For this case we have.

THEOREM 2.2 (Choquet). *A compact convex set K is Choquet simplex iff any point of K is the barycenter of a unique boundary probability measure.*

3. PROBABILITY MEASURES ON THE LATTICE OF FACES OF A SPECTRAL CONVEX SET

Let K be a spectral convex set with the complete orthomodular lattice (quantum logic) $\mathfrak{F}$ of its projective faces.

*Definition. A **probability measure** μ* on $\mathfrak{F}$ is a non negative function μ: $\mathfrak{F} \to R_+$ such that $\mu(\emptyset) = 0$, $\mu(K) = 1$ and, moreover, if $\{F_\alpha\}$ is a family of mutually orthogonal faces from $\mathfrak{F}$ then $\mu(\bigvee F_\alpha) = \sum \mu(F_\alpha)$.
Let us show that any point from K canonically defines a probability measure on $\mathfrak{F}$.

THEOREM 3.1. *Given any $t \in K$ the function*

$$m_t(F) = (t, \ \inf\{a \in A: \ \chi_F \leqslant a \leqslant e\})$$

is a probability measure on $\mathfrak{F}$, where $F \in \mathfrak{F}$, χ_F is the characteristic function of the face F.

Proof. By ((Alfsen and Shultz, 1976), Theorem 2.17) the map

$$T: \ F \to \inf\{a \in A: \ \chi_F \leqslant a \leqslant e\} \in \mathfrak{U} \subset A$$

is an order isomorphism between the lattices $\mathfrak{F}$ and $\mathfrak{U}$. Namely, T is the inverse map for the map T_3 in Theorem 1.1. Therefore $m_t(F) = (t, u)$, where $u = T(F) \in \mathfrak{U}$, i.e., if we identify the isomorphic logics $\mathfrak{F}$ and $\mathfrak{U}$, then m_t coincides with the restriction onto $\mathfrak{U}$ of the linear functional t on A. Since elements of K can be considered as *-weakly continuous states (positive normalized functionals) on A, m_t is an additive and *-weakly continuous, therefore completely additive, map from $\mathfrak{U}$ to $[0,1]$. Thus m_t is a probability measure on $\mathfrak{F}$.

Definition. Let μ be an arbitrary probability measure on the logic $\mathfrak{F}$. If there exists a point $t \in K$ such that $\mu = m_t$, then t is called ***barycenter*** of μ.
From the proof of the previous theorem and from spectral theorem in A it follows that if a barycenter of μ exists then it is unique. Considering μ as a probability measure on the logic $\mathfrak{U}$ by identifying $\mathfrak{F}$ and $\mathfrak{U}$ and denoting the barycenter of μ by t_μ we can see that in this case μ is the restriction from A to $\mathfrak{U}$ of the linear functional

$$\rho(a) = a(t_\mu), \quad a \in A.$$

Thus the existence of the barycenter for μ means that μ can be extended from the logic $\mathfrak{U}$ to a linear functional from $K \subset V \subset A^*$. Moreover, by the spectral theorem $\rho(A)$ can be represented as

$$\rho(a) = \int \lambda \, d\mu(e_\lambda),$$

where $\{e_\lambda\}$ is the spectral family of the element $a \in A$, i.e., $a = \int \lambda \, de_\lambda$.
Therefore

$$a(t_\mu) = \int \lambda \, d\mu(e_\lambda)$$

for any bounded affine function $a \in A = A^b(K)$. The last formula can be considered as an analogue of the barycentric formula from Section 2.

It is clear that the notion of a boundary measure on a compact convex set K differs from the notion of a probability measure on projective faces of K. Nevertheless both of them are generalizations of measures on the extreme boundary of a finite dimensional simplex, specifically, boundary measures generalize this notion in the framework of the classical measure theory, and measures on the logic of faces represent a non commutative or quantum generalization.

Let us show that in the both approaches to a measure on the extreme boundary of a simplex (i.e., as to a boundary measure and as to a measure on the lattice of faces) the notions of the barycenter coincide.

Let ν be a probability measure on the set $d_e K$ of all extreme points of a finite dimensional simplex K. Considering the extreme points of K as minimal faces, one can define a probability measure on the lattice $\mathfrak{F}$ of all (projective) faces of K. Namely, put

$$\mu(F) = \sum_{p \in F \cap d_e K} \nu(p), \quad F \in \mathfrak{F}.$$

It is clear that μ is a probability measure on the logic (Boolean algebra) $\mathfrak{F}$. We shall prove that the barycenter of the boundary measure ν (in the sense of Section 2) coincides with the barycenter of the measure μ on $\mathfrak{F}$ (in the sense of the definition above). Since K is a finite dimensional simplex the affine function

$$T(F) = \inf\{a \in A\colon\ \chi_F \leqslant a \leqslant e\}$$

is given by

$$T(F)(p) = \begin{cases} 1, & \text{if} \quad p \in F \cap d_e K; \\ 0, & \text{if} \quad p \in d_e K - F. \end{cases}$$

If x_ν is the barycenter of the boundary measure ν then by the barycentric formula from Section 2 we have

$$a(x_\nu) = \int a(x)\, d\nu(x), \quad a \in A(K) = A.$$

In particular

$$(x_\nu,\ \inf\{a \in A\colon\ \chi_F \leqslant a \leqslant e\}) = T(F)(x_\nu) =$$

$$= \int T(F)(x)\, d\nu(x) = \sum_{p \in d_e K} T(F)(p)\nu(p) = \sum_{p \in F \cap d_e K} \nu(p) = \mu(F),$$

i.e., x_ν is the barycenter of the measure μ on the lattice of faces of K. Therefore $x_\nu = t_\mu$.

Thus the notion of the barycenter for probability measures on the quantum logic $\mathfrak{F}$ of all projective faces of a spectral convex set K is a non commutative or quantum generalization of the barycenter for a boundary measure on a simplex. But unlike to the case of simplex or the case of boundary measures on general compact convex sets not all probability measures on $\mathfrak{F}$ have barycenters in K. This means that given an arbitrary probability measure on the orthomodular lattice $\mathfrak{U}$ of all projective units from A, it may not have an extension to a *-weakly continuous state on $A = A^b(K)$. In this connection we shall introduce the following important notion.

Definition. A spectral convex set K is said to be a *Gleason set* if any probability measure on the logic $\mathfrak{F}$ of its projective faces has a barycenter in K.
Since the barycenter is unique if it exists theorem 2.2 shows that Gleason sets are non commutative analogues of Choquet simplices.
Theorems from (Ayupov, 1987; Ayupov and Adizov, 1984; Bunce and Wright, 1985a) about measures on projection of W^*-algebras and JBW-algebras generalizing the well-known Gleason theorem imply

THEOREM 3.2. *The following convex sets are Gleason sets:*

a. *Choquet simplices;*

b. *normal state spaces of W^*-algebras without type I_2 direct summands;*

c. *normal state spaces of JBW-algebras without type I_2 direct summands.*

Proof. Consider the most general case c. As it was mentioned above any measure on the logic $\mathfrak{F}$ canonically defines a measure on the logic $\mathfrak{U}$ of projective units from $A = A^b(K)$. In our case given a JBW-algebra J we have $J \approx A$ and projective units of A correspond to idempotents of the Jordan algebra J.Therefore we have a probability measure on the lattice P of idempotents from J. By the generalization of Gleason theorem for JBW-algebras (Ayupov, 1985; Ayupov and Adizov, 1984; Bunce and Wright, 1985a) any probability measure μ on P can be extended to a normal state t on J, i.e., there is $t \in K$ (K is identified with the normal state space of J) such that $\mu = t$ restricted onto P. Identifying isomorphic logics $\mathfrak{U}$ and P, we have

$$\mu(F) = t(T(F)) = (t, \inf\{a \in A\colon \chi_F \leqslant a \leqslant e\}), \quad F \in \mathfrak{F}.$$

Thus t is the barycenter of μ.
In this connection the following questions seem to be interesting:
1.Is there any Gleason set not affinely isomorphic to the normal state space of a JBW-algebra?
2. Is it possible to describe all finite dimensional Gleason sets? (For $\dim K \leqslant 4$, see the remark below.)
Let us introduce one more class of spectral convex sets closely related to Gleason sets.
Recall that a probability measure μ on a complete quantum logic L is called Jauch—Piron state (Bunce et al., 1985; Ruttiman, 1977) if given any sequence $\{a_n\} \subset L$ with $\mu(a_n) = 0$ for all $n = 1, 2, \ldots$, one has $\mu(\bigvee a_n) = 0$.
A complete quantum logic L is said to be *Jauch—Piron logic* if any probability measure on L is Jauch—Piron state.

Definition. A spectral convex set K is called *Jauch—Piron set* if the quantum logic $\mathfrak{F}$ of all its projective faces, and therefore the logics $\mathfrak{P}$ and $\mathfrak{U}$ are Jauch—Piron logics.

THEOREM 3.3. *Let K be a Gleason set. Then K is Jauch—Piron set.*

Proof. Let μ be a probability measure on the logic $\mathfrak{U}$ of projective units from $A = A^b(K)$, $\{e_n\}$ be a sequence from $\mathfrak{U}$ with $\mu(e_n) = 0$ for all $n = 1, 2, \ldots$ Put $x = \sum 2^{-n} e_n$. Since K is a Gleason set, μ can be considered as a *-weakly continuous state on $A = A^b(K)$. Therefore, $\mu(x) = 0$. By spectral theorem there exists a sequence $\{p_n\}$ of projective units in A increasing to the support $r(x)$ of the element x, such that $x \geqslant \varepsilon_n p_n$ for some $\varepsilon_n > 0$, $n = 1, 2, \ldots$ Thus,

$$\mu(\vee e_n) = \mu(r(x)) = \sup \mu(p_n) = 0$$

because

$$\mu(p_n) \leqslant \varepsilon_n^{-1} \mu(x) = 0.$$

This means that the logic $\mathfrak{U}$ and therefore $\mathfrak{F}$ are Jauch—Piron logics.
Let us consider an example of spectral set which is not a Jauch—Piron (therefore not Gleason) set.

Example. Let K be the unit ball in a Hilbert space H ($\dim H \geqslant 2$). Then K is spectral; namely it is the state space of a spin factor— JBW-factor of type I_2. The proper projective faces of K are exactly the extreme points. Given any unit vector $h \in H$ put

$$\mu(F) = \begin{cases} 0, & \text{if} \quad (f, h) \geqslant 0; \\ 1, & \text{if} \quad (f, h) < 0; \end{cases}$$

where F is the proper face of K defined by the unit vector $f \in H$. Now, putting $\mu(\varnothing) = 0$, $\mu(K) = 1$, we obtain a probability measure μ on $\mathfrak{F}$. But μ is not a Jauch—Piron state because for $F_1, F_2 \in \mathfrak{F}$, $F_1 \neq F_2$ with $\mu(F_1) = \mu(F_2) = 0$, we have $\mu(F_1 \bigvee F_2) = \mu(K) = 1$. Therefore K is not a Jauch—Piron set.

Remark. For $\dim K \leqslant 4$ all possible spectral convex sets were described in (Ayupov and Yadgorov, 1989). In particular it was proved that for spectral convex sets with such low dimensions the following conditions are equivalent:

a. K is a Gleason set;

b. K is Jauch—Piron set;

c. K is a simplex.

REFERENCES

Alfsen, E.M. (1971). Compact convex sets and boundary integrals. *Ergebnisse. der Math.* **57**, Springer, Berlin—Heidelberg etc.

Alfsen, E.M. and Shultz, F.W. (1976). Non-commutative spectral theory for affine function spaces on convex sets. *Mem. Amer. Math. Soc.* **172**.

Alfsen, E.M. and Shultz, F.W. (1978). State spaces of Jordan algebras. *Acta Math.* **140**, 155—190.

Alfsen, E.M. and Shultz, F.W. (1979). On non-commutative spectral theory and Jordan algebras. *Proc. London Math. Soc.* **38**, 497—516.

Alfsen, E.M. Hanche-Olsen, H. and Shultz, F.W. (1980). State spaces of C^*-algebras. *Acta Math.* **144**, 267—305.

Asimov, L. and Ellis, A.J. (1980). *Convexity Theory and Its Applications in Functional Analysis*. Academic Press, London.

Ayupov, Sh.A. (1987). Jordan operator algebras. *J. Sov. Math.* **37**, 1422—1448.

Ayupov, Sh.A. and Adizov, A.A. (1984). Probability measures on projections of JBW–algebras. Dep. N 7822–84 VINITI, Moscow.

Ayupov, Sh.A. and Yadgorov, N.D. (1989). Spectral convex sets in finite–dimentional spaces. *Izvestiya AN UzSSR, ser. Fiz.-Mat. Nauk* **3**, 3—7.

Bunce, L.J., Navara, M., Ptak, P. and Wright, J.D.M. (1985). Quantum logics with Jauch—Piron states. *Quart. J. Math. Oxford* **36**, 261—271.

Bunce, L.J. and Wright, J.D.M. (1985a). Quantum measures and states on Jordan algebras. *Commun. Math. Phys.* **98**, 187—202.

Bunce, L.J. and Wright, J.D.M. (1985b). Quantum logic and convex geometry. *Commun. Math. Phys.* **101**, 87—96.

Hanche–Olsen, H. and Stormer, E. (1984). *Jordan Operator Algebras*. Pitman, London.

Iochum, B. and Shultz, F.W. (1983). Normal state spaces of Jordan and von Neumann algebras. *J. Funct. Analysis* **50**, 317—328.

Ruttiman, G.T. (1977). Jauch—Piron states *J. Math. Phys.* **18**, 189—192.

Prob. Theory and Math. Stat., Vol. 1, pp. 78-90
B.Grigelionis *et al.* (Eds.)
1990 VSP/Mokslas

CONVOLUTION SEMIGROUPS OF INSTRUMENTS IN QUANTUM PROBABILITY

A. BARCHIELLI and G. LUPIERI
Dipartimento di Fisica, Università di Milano, Via Celoria 16, 20133 Milano, Italy;
Instituto Nazionale di Fisica Nucleare, Sezione di Milano, Italy

ABSTRACT

Convolution semigroups of instruments arising in the study of continuous measurement processes in quantum probability, from a mathematical point of view are natural generalization of convolution semigroups of probability measures, the difference lying in the fact that an instrument is a measure with values in a generally noncommutative Banach algebra. An important open problem is to obtain a characterization of convolution semigroup of instruments, analogous to the Hunt theorem for probability measures. A solution of this problem is given in the case of finite-dimensional algebra, by using the probability operator approach. In the general case various continuity properties of the semigroups are introduced and relations between them are established.

1. INSTRUMENTS AND PROBABILITY OPERATORS

The notion of instrument (Davies, 1976) has been introduced in quantum theories in order to extract probabilities and to describe state changes due to measurements. An instrument is a certain kind of normalized positive-operator-valued measure and it is an analogue of a probability measure. By using this notion, a theory of continuous (in time) measurements in quantum mechanics has been developed (Barchielli, 1987, 1990; Barchielli et al., 1983; Barchielli and Lupieri, 1985; Davies, 1976). Convolution semigroups of instruments are the main ingredient in this theory. They have been studied by using Fourier transform techniques (Barchielli et al., 1983; Barchielli and Lupieri, 1985, 1990; Holevo, 1988; Parthasarathy, 1986) (see also the talk by L.V. Denisov at this conference) and a quantum analogue of the notion of probability operator (Barchielli, 1987, 1989, 1990; Barchielli and Lupieri, 1989). The aim of this paper is to present this second approach. In this section we give in detail the construction of the probability operator, because this presents some delicate points and has not been done in the previous papers. The second section is entirely devoted to some new results about the relations among the various continuity properties of semigroups of probability operators. In the last section we present some cases in which the general structure of the generators of such semigroups has been obtained.

Let $\mathcal{H}$ be a separable complex Hilbert space, $\mathcal{L}(\mathcal{H})$ the algebra of the bounded linear maps from $\mathcal{H}$ into itself, $\mathcal{M}$ a von Neumann subalgebra of $\mathcal{L}(\mathcal{H})$ and $\mathcal{M}_*$

the predual of $\mathcal{M}$. We denote by $\langle \varrho, a\rangle$, $\varrho \in \mathcal{M}_*$, $a \in \mathcal{M}$ the bilinear form giving the duality between $\mathcal{M}_*$ and $\mathcal{M}$. Let G be a locally compact topological Hausdorff group, $\mathcal{B}$ its Borel σ-algebra and e the neutral element of G.

Definition 1.1. We call *quasi-instrument* on $\mathcal{M}$ with value space $(G, \mathcal{B})$ a map J from $\mathcal{B}$ into $\mathcal{L}(\mathcal{M})$ such that

1.1.A) $J(B)$ is completely positive (Gorini et al., 1976; Lindblad, 1976; Takesaki, 1979, p. 192) and normal (Dixmier, 1969, p. 53), $\forall B \in \mathcal{B}$;

1.1.B) $J(\cdot)$ is weakly* σ-additive, i.e. the measure $\langle \varrho, J(\cdot)[a]\rangle$ is σ-additive, $\forall \varrho \in \mathcal{M}_*$, $\forall a \in \mathcal{M}$;

1.1.C) $\forall \varrho \in \mathcal{M}_*^+$, $\forall a \in \mathcal{M}^+$ (the positive cones of $\mathcal{M}_*$ and $\mathcal{M}$), the positive finite measure $\langle \varrho, J(\cdot)[a]\rangle$ is regular.

If we have also the normalization condition

1.1.D) $J(G)[\mathbf{1}] = \mathbf{1}$,

J is called an *instrument*.

Let us introduce now some Banach spaces (in all of them the norm is the usual supremum norm). $\mathcal{C}_0(G)$: the space of the continuous complex-valued functions on G, vanishing at infinity; $\mathcal{C}_c(G) \equiv \mathcal{C}(G_\infty)$: the space of the continuous complex-valued functions on the one-point compactification G_∞ of G; $\mathcal{C}_b(G)$: the space of the continuous bounded functions on G. In an analogous way we introduce the spaces $\mathcal{C}_0(G; \mathcal{M})$, $\mathcal{C}_c(G; \mathcal{M})$ and $\mathcal{C}_b(G; \mathcal{M})$, whose elements are now norm-continuous $\mathcal{M}$-valued functions.

Because a quasi-instrument is a w^*-σ-additive measure, one can define w^*-integrals. As in the classical case, this gives rise to an equivalent characterization of quasi-instruments. The proof of the following proposition (which we omit) is more or less standard and it is based on the classical Riesz representation theorem (Heyer, 1977, p. 21). For something similar see (Davies, 1976, p. 50, Theorem 1.2) and the construction of the convolution operator in the following.

PROPOSITION 1.2. *The formula*

$$\langle \varrho, \overline{J}(f)[a]\rangle = \int_G f(x)\langle \varrho, J(dx)[a]\rangle, \qquad \forall \varrho \in \mathcal{M}_*, \quad \forall a \in \mathcal{M}, \tag{1.1}$$

sets up a one-to-one correspondence between quasi-instruments J and the linear maps $\overline{J}$ from $\mathcal{C}_0(G)$ into $\mathcal{L}(\mathcal{M})$ such that

1.2.A) *$\overline{J}(f)$ is completely positive and normal if $f \geqslant 0$;*

1.2.B) *$\forall f \in \mathcal{C}_0(G)$, $\|\overline{J}(f)\| \leqslant M\|f\|$, for some constant M.*

Moreover, J is an instrument if and only if, $\forall \varrho \in \mathcal{M}_*^+$,

$$\sup\{\langle \varrho, \overline{J}(f)[\mathbf{1}]\rangle : f \in \mathcal{C}_0(G),\ 0 \leqslant f \leqslant 1\} = \langle \varrho, \mathbf{1}\rangle. \tag{1.2}$$

Let ω be a left invariant Haar measure on G and consider the Banach space $L^\infty(G, \omega; \mathcal{M}) = \mathcal{N}$ of all $\mathcal{M}$-valued essentially bounded weakly* locally ω-measurable functions on G. Then, $\mathcal{N}$ is a W^*-algebra under pointwise multiplication, and its predual is $\mathcal{N}_* = L^1(G, \omega; \mathcal{M}_*)$, the Banach space of all $\mathcal{M}_*$-valued Bochner ω-integrable functions on G (Sakai, 1971, p. 68, Theorem 1.22.13). The W^*-algebra

$\mathcal{N}$ is isomorphic to $\mathcal{M} \otimes L^\infty(G,\omega)$ (W^*–tensor product) and the algebra $\mathcal{N}_*$ to $\mathcal{M}_* \otimes L^1(G,\omega)$ (tensor product with respect to the greatest cross norm) (Sakai, 1971, p. 58, 59, 67, 68, Definition 1.22.10, Proposition 1.22.12). The W^*–algebra $\mathcal{N}$ can be seen as a von Neumann algebra of operators on $\mathcal{H} \otimes L^2(G,\omega)$. We denote by $\|\cdot\|_1$ the norms in $\mathcal{M}_*$, $L^1(G,\omega)$, $\mathcal{N}_*$ and by $\|\cdot\|$ all the other norms.

Let us consider a quasi–instrument J. $J(B)$ positive and normal is equivalent to positive and w^*–continuous and $J(B)$ w^*–continuous is equivalent to the existence of the preadjoint $J_*(B) \in \mathcal{L}(\mathcal{M}_*)$ (Sakai, 1971, p. 50–54). Then, $J_*(B)$, $B \in \mathcal{B}$, turns out to be a positive–operator–valued measure, strongly σ–additive (Davies, 1976; Ozawa, 1984).

Let $\varphi \in L^1(G,\omega)$, $\varphi \geqslant 0$; $\forall a \in \mathcal{M}^+$, $\forall \varrho \in \mathcal{M}_*^+$, let us set

$$I(\varrho, a; \varphi)(x) = \int\limits_{y\in G} \varphi(y^{-1}x)\langle \varrho, J(dy)[a]\rangle. \tag{1.3}$$

By Fubini theorem the previous integral exists ω–a.e. in G and defines a positive function belonging to $L^1(G,\omega)$. Moreover, by the positivity of J we have $I(\varrho, a; \varphi)(x) \leqslant \|a\| I(\varrho, \mathbf{1}; \varphi)(x)$, so that (1.3) exists for every a if it exists for $a \equiv \mathbf{1}$. Let now x_0 be a point such that $I(\varrho, \mathbf{1}; \varphi)(x_0)$ exists, a_α be an increasing net in $\mathcal{M}^+$ with least upper bound a and F be the set of all positive simple functions

$$f(y) = \sum_{i=1}^{n} f_i \chi_{B_i}(y)$$

such that $f(y) \leqslant \varphi(y^{-1}x_0)$. Then, by the fact that J is normal, we have that $I(\varrho, \cdot; \varphi)(x_0)$ is normal; indeed, by Theorem 3.13, p. 67, in (Rudin, 1970), we can express the integral in (1.3) by a supremum on simple functions and we have

$$\sup_\alpha I(\varrho, a_\alpha; \varphi)(x_0) = \sup_\alpha \sup_{f\in F} \sum_{i=1}^{n} f_i \langle \varrho, J(B_i)[a_\alpha]\rangle =$$

$$= \sup_{f\in F} \sup_\alpha \sum_{i=1}^{n} f_i \langle \varrho, J(B_i)[a_\alpha]\rangle = \sup_{f\in F} \sum_{i=1}^{n} f_i \langle \varrho, J(B_i)[a]\rangle =$$

$$= \int\limits_{y\in G} \varphi(y^{-1}x_0)\langle \varrho, J(dy)[a]\rangle = I(\varrho, a; \varphi)(x_0).$$

Because $I(\varrho, a; \varphi)(x_0)$ is also affine on $\mathcal{M}^+$, there exists (Dixmier, 1969) an element $I(\varrho; \varphi)(x_0)$ of $\mathcal{M}_*^+$ such that $\langle I(\varrho; \varphi)(x_0), a\rangle = I(\varrho, a; \varphi)(x_0)$, $\forall a \in \mathcal{M}^+$. Now, the $\mathcal{M}_*$–valued function $I(\varrho; \varphi)$ is weakly measurable by definition and, by the separability of $\mathcal{M}_*$, it is strongly measurable (Hille and Phillips, 1968, Theorem 3.5.3, Corollary 2, p.72, 73). Moreover, we have

$$\int\limits_G \omega(dx)\, \|I(\varrho; \varphi)(x)\|_1 = \int\limits_G \omega(dx)\, I(\varrho, \mathbf{1}; \varphi)(x) =$$

$$= \int_{y\in G} \langle \varrho, J(dy)[\mathbf{1}]\rangle \int_{x\in G} \omega\,(dx)\,\varphi(y^{-1}x) =$$

$$= \|\varphi\|_1\,\|J_*(G)[\varrho]\|_1 \leqslant \|\varphi\|_1\,\|\varrho\|_1\,\|J_*(G)\|;$$

therefore $I(\varrho;\varphi) \in L^1(G,\omega;\mathcal{M}_*)$ (Hille and Phillips, 1968, Theorem 3.7.4, p.80) and $I(\cdot;\cdot)$ is a bounded bilinear map.
Now we can extend $I(\cdot;\cdot)$ by linearity and continuity to the $\|\cdot\|_1$–closure of $\mathcal{M}_* \odot L^1(G,\omega)$ ($\odot \equiv$ algebraic tensor product), which is $\mathcal{N}_*$ (Takesaki, 1979, p. 254–256), and we obtain a bounded linear operator on $\mathcal{N}_*$ which we call $(T_J)_*$. This construction entitles us to write formally

$$(T_J)_*[\Phi](x) = \int_{y\in G} J_*(dy)[\Phi(y^{-1}x)], \qquad \forall\Phi \in \mathcal{N}_*. \tag{1.4}$$

We consider now the adjoint T_J of $(T_J)_*$, which belongs to $\mathcal{L}(\mathcal{N})$. We call T_J the ***convolution operator*** associated with the quasi–instrument J; if J is an instrument we call T_J ***probability operator***.

PROPOSITION 1.3. *The convolution operator T_J enjoys the following properties:*

1.3.A) *T_J is positive and $\|T_J\| = \|J(G)\|$;*

1.3.B) *T_J is the unique w^*–continuous operator on $\mathcal{N}$ whose restriction to the product elements $a\otimes f$, $a\in\mathcal{M}$, $f\in L^\infty(G,\omega)$, satisfies ω–a.e.*

$$\langle \varrho, T_J[a\otimes f](x)\rangle = \int_{y\in G} f(yx)\langle\varrho, J(dy)[a]\rangle, \qquad \forall\varrho\in\mathcal{M}_*; \tag{1.5}$$

1.3.C) *T_J leaves the spaces $\mathcal{C}_\varepsilon(G;\mathcal{M})$, $\varepsilon = 0, c, b$, invariant.*

Proof. Positivity follows trivially from the construction of T_J. Then, by positivity, $\|T_J\| = \|T_J[\mathbf{1}]\|$ and this last quantity is just $\|J(G)\|$.
To show equation (1.5) is a matter of straightforward computations. Moreover, as T_J has a preadjoint, it is w^*–continuous. As $\mathcal{M}\odot L^\infty(G,\omega)$ is w^*–dense in $\mathcal{N}$ (by definition of W^*–tensor product), T_J is the unique operator with the wanted properties.
Let us note that the weak* regularity 1.1.C implies strong regularity. Indeed, let $\mathcal{M}_*^1$ be the unit ball in $\mathcal{M}_*$ and $\mathcal{M}_*^{1+} = \mathcal{M}_*^1 \bigcap \mathcal{M}_*^+$; $\forall a\in\mathcal{M}^+$, we have

$$\|J(B)[a]\| = \sup_{\varrho\in\mathcal{M}_*^{1+}} \sup\{\langle\varrho, J(K)[a]\rangle : K\subset B,\ K \text{ compact}\} =$$

$$= \sup\{\sup_{\varrho\in\mathcal{M}_*^{1+}} \langle\varrho, J(K)[a]\rangle : K\subset B,\ K \text{ compact}\} =$$

$$= \sup\{\|J(K)[a]\| : K\subset B,\ K \text{ compact}\}.$$

Let now K be a compact neighborhood of e and $A \in \mathcal{C}_b(G; \mathcal{M})$. Let us consider the decomposition $A(zx) - A(zy) = B_1(z) - B_2(z) + iB_3(z) - iB_4(z)$ with $B_i(z) \geqslant 0$. Then, by the positivity of T_J, we have

$$\|T_J[A](x) - T_J[A](y)\| \leqslant \sum_{i=1}^{4} \|T[B_i](e)\| \leqslant$$

$$\leqslant \sum_{i=1}^{4} \|T_J[\mathbf{1} \otimes \|B_i(\cdot)\|](e)\| =$$

$$= \sum_{i=1}^{4} \sup_{\varrho \in \mathcal{M}_*^1} \left| \int_{z \in G} \|B_i(z)\| \langle \varrho, J(dz)[\mathbf{1}] \rangle \right| \leqslant$$

$$\leqslant 4 \sup_{\varrho \in \mathcal{M}_*^1} \left| \int_{z \in G} \|A(zx) - A(zy)\| \langle \varrho, J(dz)[\mathbf{1}] \rangle \right| \leqslant$$

$$\leqslant 4 \sup_{z \in K} \|A(zx) - A(zy)\| \|J(G)[\mathbf{1}]\| + 8\|A\| \|J(G \backslash K)[\mathbf{1}]\|.$$

This result and strong regularity imply that T_J maps $\mathcal{C}_b(G; \mathcal{M})$ into itself. A similar estimation on $\|T_J[A](x)\|$ shows that $T_J[A]$ vanishes at infinity if $A \in \mathcal{C}_0(G; \mathcal{M})$ and, therefore, T_J leaves invariant $\mathcal{C}_0(G; \mathcal{M})$. By the fact that T_J maps constants into constants, one has that T_J leaves invariant also $\mathcal{C}_c(G; \mathcal{M})$. By property 1.3.B we are entitled to write, $\forall A \in \mathcal{N}$,

$$T_J[A](x) = \int_{y \in G} J(dy)[A(yx)], \tag{1.6}$$

where this integral has to be understood in the sense of the construction above. Let us introduce now left (L_z) and right (R_z) translations by defining

$$L_z[f](x) = f(zx), \quad R_z[f](x) = f(xz), \qquad \omega\text{–a.e.}, \quad \forall f \in L^\infty(G, \omega). \tag{1.7}$$

Translations in $\mathcal{N}$ are denoted by the same symbols L_z and R_z.

THEOREM 1.4. *Given a map $T \in \mathcal{L}(\mathcal{N})$ there exists a unique quasi–instrument J such that T is the convolution operator associated with J, if and only if*

1.4.A) *T is completely positive and normal;*

1.4.B) *$T \circ R_z = R_z \circ T$, $\forall z \in G$.*

J is an instrument if and only if we have also

1.4.C) $T[\mathbf{1}] = \mathbf{1}$.

Proof. Let J be a quasi–instrument and T the associated convolution operator. Property 1.4.B follows from property 1.3.B. By definition T has a preadjoint and it is positive; therefore, it is also normal (Dixmier, 1969). For showing complete positivity one can use a result of (Ozawa, 1984) on instruments (see the proof in

(Barchielli, 1989)). If J is an instrument, property 1.4.C follows trivially from the definition of T and the normalization of J.
Let now $T \in \mathcal{L}(\mathcal{N})$ be a map satisfying properties 1.4.A and 1.4.B. For $x, x_0 \in G$, $A \in \mathcal{N}$, we have

$$\sup_{y\in G} \|T[A](yx) - T[A](yx_0)\| = \|(R_x - R_{x_0}) \circ T[A]\| =$$

$$= \|T_0(R_x - R_{x_0})[A]\| \leqslant \|T\| \|R_x[A] - R_{x_0}[A]\| =$$

$$= \|T\| \sup_{y\in G} \|A(yx) - A(yx_0)\|,$$

which implies that T transforms left uniformly continuous functions into left uniformly continuous ones.
Let us set now $\overline{J}(f)[a] := T[a \otimes f](e)$, $\forall a \in \mathcal{M}$, $\forall f \in \mathcal{C}_0(G)$. This quantity is well defined because of the continuity property shown above. It is easy to see that $\overline{J}(f)$ satisfies properties 1.2.A and 1.2.B and, by Proposition 1.2, it defines a quasi-instrument J. Property 1.4.B implies, $\forall f \in \mathcal{C}_0(G)$, $\forall a \in \mathcal{M}$, $\forall \varrho \in \mathcal{M}_*$,

$$\langle \varrho, T[a \otimes f](x)\rangle = \langle \varrho, T[a \otimes R_x[f]](e)\rangle = \int_{y\in G} R_x[f](y) \langle \varrho, J(dy)[a]\rangle =$$

$$= \int_{y\in G} f(yx) \langle \varrho, J(dy)[a]\rangle = \langle \varrho, T_J[a \otimes f](x)\rangle,$$

where T_J is the convolution operator associated with J. By property 1.4.A and Proposition 1.3, we have $T_J = T$. If 1.4.C holds, it is trivial to see that J is normalized.

2. SEMIGROUPS OF PROBABILITY OPERATORS: CONTINUITY PROPERTIES

Let T_{J_1} and T_{J_2} be two convolution operators. Also the operator $T_{J_1} \circ T_{J_2}$ satisfies properties 1.4.A and 1.4.B and, therefore, there is a unique quasi-instrument $J_1 * J_2$ associated with it, which we call the convolution of J_1 with J_2. Formally we can write, $\forall B \in \mathcal{B}$,

$$J_1 * J_2(B) = \int_G J_1(dy) \circ J_2(By^{-1}) = \int_G J_1(x^{-1}B) \circ J_2(dx). \tag{2.1}$$

The objects which we are interested in are one-parameter convolution semigroups of instruments. From now on $\{T_t,\ t \geqslant 0\}$ will be a one-parameter semigroup (i.e. $T_t \circ T_s = T_{t+s}$, $\forall t, s \geqslant 0$) of probability operators, such that $T_0 = \mathbf{1}$. By $\{J_t,\ t \geqslant 0\}$ we shall denote the associated convolution semigroup of instruments. Let us stress that $J_0(\cdot) = \delta_e(\cdot)\mathbf{1}$, where δ_x is the Dirac measure concentrated at x; moreover, it turns out that $J_t(G)$ is a so called (completely positive) quantum dynamical semigroup on $\mathcal{M}$ (Gorini et al., 1976; Lindblad, 1976).

THEOREM 2.1. *The following continuity properties are equivalent:*
2.1.A) *for any neighborhood V of e*

$$w^*\text{-}\lim_{t\downarrow 0} J_t(V) = \mathbf{1};$$

2.1.B) $\forall \varrho \in \mathcal{M}_*$, $\forall a \in \mathcal{M}$, $\forall f \in C_b(G)$, *the function* $\int_G f(x)\langle \varrho, J_t(dx)[a]\rangle$ *is continuous for* $t \in [0, +\infty)$;
2.1.C) T_{t*} *(the preadjoint of* T_t*) is strongly continuous for* $t \in [0, +\infty)$ *(on* $\mathcal{N}_*$*);*
2.1.D) $s\text{-}\lim_{t\downarrow 0} T_{t*} = \mathbf{1}$ *(on* $\mathcal{N}_*$*);*
2.1.E) $w^*\text{-}\lim_{t\downarrow 0} T_t = \mathbf{1}$ *(on* $\mathcal{N}$*);*
2.1.F) T_t *is weakly* continuous for* $t \in [0, +\infty)$ *(on* $\mathcal{N}$*).*

Proof. 2.1.C $\iff$ 2.1.D: see (Dynkin, 1965, p. 23, 1.3.A).
2.1.C $\iff$ 2.1.E. By proposition 1.23, p. 15, of (Davies, 1980), property 2.1.C is equivalent to the weak continuity of T_{t*} at $t = 0$; but this last property is the same as 2.1.E by the definition of adjoint, of the weak topology on $\mathcal{N}_*$ and of the weak* one on $\mathcal{N}$.
2.1.C $\Rightarrow$ 2.1.F. We have, $\forall \Phi \in \mathcal{N}_*$, $\forall A \in \mathcal{N}$,

$$|\langle \Phi, T_t[A]\rangle - \langle \Phi, T_{t_0}[A]\rangle| =$$

$$= |\langle (T_{t*} - T_{t_0*})[\Phi], A\rangle| \leqslant \|(T_{t*} - T_{t_0*})[\Phi]\|\|A\| \longrightarrow 0 \quad \text{as} \quad t \to t_0.$$

2.1.F $\Rightarrow$ 2.1.B. Let B be a basis of the filter of the neighborhoods of e in G, such that any element of B is compact. For any $V \in B$ let φ_V be a positive continuous function such that supp $\varphi_V \subset V$ and $\int_G \varphi_V(x)\omega(dx) = 1$. Let us set $\nu_V(dy) = \varphi_V(y)\omega(dy)$, $g_V(x) = \int_{y\in G} f(xy)\nu_V(dy)$, where f is any continuous bounded function on G. The net of measures ν_V converges to the Dirac measure δ_e in the weak topology of measures (Heyer, 1977, p. 24). By Theorem 1.5.5, p. 64, in (Heyer, 1977), $\|g_V - f\| \to 0$.
Let us set now $\mu_t(dx) = \langle \varrho, J_t(dx)[a]\rangle$, $\forall \varrho \in \mathcal{M}_*^+$, $\forall a \in \mathcal{M}^*$; then we have

$$\left|\int_G f(x)\mu_t(dx) - \int_G f(x)\mu_s(dx)\right| \leqslant \left|\int_G (f(x) - g_V(x))\mu_t(dx)\right| +$$

$$+\left|\int_G g_V(x)\mu_t(dx) - \int_G g_V(x)\mu_s(dx)\right| + \left|\int_G (f(x) - g_V(x))\mu_s(dx)\right| \leqslant$$

$$\leqslant 2\|f - g_V\|\|\varrho\|_1\|a\| + |\langle \varrho \otimes \varphi_V, (T_t - T_s)[a \otimes f]\rangle|.$$

The first term can be taken arbitrary small by the previous discussion and the second one tends zero as $t \to s$ by 2.1.F.
2.1.B $\Rightarrow$ 2.1.A. By taking $f(x) \equiv 1$, $\forall x \in G$, we have immediately

$$w^*\text{-}\lim_{t\downarrow 0} J_t(G) = \mathbf{1}. \tag{2.2}$$

Let now $f \in C_b(G)$ be such that $f(e) = 0$, $f(x) \geqslant \chi_{G\setminus V}(x)$ $\forall x \in G$. Then, $\forall \varrho \in \mathcal{M}_*^+$, $\forall a \in \mathcal{M}^+$, we have

$$0 \leqslant \langle \varrho, J_t(G\setminus V)[a]\rangle \leqslant \int_G f(x)\langle \varrho, J_t(dx)[a]\rangle =$$

$$= \int_G f(x)\langle \varrho, J_t(dx)[a]\rangle - f(e)\langle \varrho, a\rangle \longrightarrow 0 \qquad \text{as} \quad t \downarrow 0.$$

Together with (2.2), this is equivalent to 2.1.A.

2.1.A $\Rightarrow$ 2.1.D. Let $\mathcal{K}(G)$ be the space of the compact–support continuous complex–valued functions on G. By the fact that $\mathcal{M}_* \odot \mathcal{K}(G)$ is $\|\cdot\|_1$–dense in $\mathcal{N}_*$ (Takesaki, 1979, p. 254–256), it is enough to show that

$$\lim_{t\downarrow 0} \|T_{t*}[\varrho \otimes \varphi] - \varrho \otimes \varphi\|_1 = 0, \quad \forall \varrho \in \mathcal{M}_*^+, \quad \forall \varphi \in \mathcal{K}(G).$$

Let us first show point convergence. Let V be a compact neighborhood of e; $\forall x \in$ $\in G$, we have

$$\|T_{t*}[\varrho \otimes \varphi](x) - \varphi(x)\varrho\|_1 \leqslant \sup_{a \in \mathcal{M}_1} \left\{ \int_G |\varphi(y^{-1}x) - \varphi(x)|\,|\langle \varrho, J_t(dy)[a]\rangle| + \right.$$

$$\left. + |\varphi(x)|\,|\langle \varrho, (J_t(G) - 1)[a]\rangle| \right\} \leqslant \sup_{x\in G}\sup_{y \in V} |\varphi(y^{-1}x) - \varphi(x)|\,\|\varrho\|_1 +$$

$$+2\|\varphi\|\langle \varrho, J_t(G\setminus V)[1]\rangle + \|\varphi\|\,\|J_{t*}(G) - 1\|\,\|\varrho\|_1.$$

By 2.1.A and the strong continuity of $J_{t*}(G)$ (Davies, 1980, p.15, Proposition 1.23), the last two terms tend to zero. By the fact that a function in $\mathcal{K}(G)$ is (left and right) uniformly continuous, the first term can be made arbitrary small by choosing V small enough.

Let us set now $\operatorname{supp}\varphi = K$; we have

$$\|T_{t*}[\varrho \otimes \varphi] - \varrho \otimes \varphi\|_1 \leqslant \int_K \omega(dx)\|T_{t*}[\varrho \otimes \varphi](x) - \varphi(x)\varrho\|_1 +$$

$$+ \int_{G\setminus K} \omega(dx) \int_G |\varphi(y^{-1}x)|\langle \varrho, J_t(dy)[1]\rangle.$$

The first term vanishes with t by the previous result and the fact that $\omega(K) < +\infty$. The second term can be rewritten as

$$\int_G g(y)\langle \varrho, J_t(dy)[1]\rangle,$$

$$g(x) = \int\limits_{G\backslash K} \omega(dx)|\varphi(y^{-1}x)| = \int\limits_{K} \omega(dz)|\varphi(z)|\chi_{G\backslash K}(yz).$$

Then, g is a continuous bounded function and $g(e) = 0$. Finally, we have

$$0 \leqslant \int\limits_{G} g(y)\langle \varrho, J_t(dy)[\mathbf{1}]\rangle \leqslant \|g\|\langle \varrho, J_t(G\backslash V)[\mathbf{1}]\rangle + \sup_{y\in V}|g(y)|\ \|\varrho\|_1,$$

where V is a small neighborhood of e. By the continuity of g and 2.1.A, the statement is proved.

THEOREM 2.2. *The following continuity properties are equivalent:*
2.2.A) *for any neighborhood V of e*

$$s\text{–}\lim_{t\downarrow 0} J_t(V) = \mathbf{1};$$

2.2.B) *the restriction of T_t to $\mathcal{C}_c(G;\mathcal{M})$ is strongly continuous.*

Proof. 2.2.B $\Rightarrow$ 2.2.A. Let $f \in \mathcal{C}_c(G)$ be such that $f(e) = 0$, $f(x) \geqslant \chi_{G\backslash V}(x)$, V is a neighborhood of e. Then, we have

$$\|J_t(G\backslash V)\| \leqslant \sup_{\varrho\in\mathcal{M}_*^{1+}} \int\limits_{G} f(y)\langle \varrho, J_t(dy)[\mathbf{1}]\rangle \leqslant$$

$$\leqslant \sup_{\varrho\in\mathcal{M}_*^{1+}} \sup_{x\in G} \left| \int\limits_{G} f(yx)\langle \varrho, J_t(dy)[\mathbf{1}]\rangle - f(x)\langle \varrho, \mathbf{1}\rangle \right| \leqslant$$

$$\leqslant \|T_t[\mathbf{1}\otimes f] - \mathbf{1}\otimes f\| \longrightarrow 0 \qquad \text{as} \quad t\downarrow 0.$$

Moreover, $\forall a \in \mathcal{M}$,

$$\|J_t(G)[a] - a\| = \|T_t[a\otimes \mathbf{1}] - a\otimes \mathbf{1}\| \longrightarrow 0 \qquad \text{as} \quad t\downarrow 0.$$

The two results above are equivalent to 2.2.A.

2.2.A $\Rightarrow$ 2.2.B. Let V be a neighborhood of e; for any $a \in \mathcal{M}^+$ and any $f \in \mathcal{C}_c(G)$, we have

$$\|T_t[a\otimes f] - a\otimes f\| \leqslant \sup_{x\in G} \sup_{\varrho\in\mathcal{M}_*^1} \left| \int\limits_{G\backslash V} (f(yx) - f(x))\langle \varrho, J_t(dy)[a]\rangle \right| +$$

$$+ \sup_{x\in G} \sup_{\varrho\in\mathcal{M}_*^1} \left| \int\limits_{V} (f(yx) - f(x))\langle \varrho, J_t(dy)[a]\rangle \right| + \|f\|\ \|J_t(G)[a] - a\| \leqslant$$

$$\leqslant 2\|f\|\ \|J_t(G\backslash V)[a]\| + \sup_{x\in G}\sup_{y\in V}|f(yx) - f(x)|\ \|a\| + \|f\|\ \|J_t(G)[a] - a\|.$$

This shows the strong continuity of T_t on the product elements and by linearity on $\mathcal{M} \odot \mathcal{C}_c(G)$. By the fact that $\mathcal{M} \odot \mathcal{C}_c(G)$ is dense in $\mathcal{C}_c(G; \mathcal{M})$ (Takesaki, 1979, p. 254, Proposition 7.3) and that $\|T_t\| = 1$, we have strong continuity on the whole $\mathcal{C}_c(G; \mathcal{M})$.

Finally, we consider two other continuity properties:

2.3) for any neighborhood V of e

$$\lim_{t \downarrow 0} \|J_t(V) - \mathbf{1}\| = 0;$$

2.4) the semigroup T_t on $\mathcal{N}$ is norm continuous.

It is easy to show that 2.4 implies 2.3, so that we have the chain of implications among the continuity properties: 2.4 $\Rightarrow$ 2.3 $\Rightarrow$ 2.2 $\Rightarrow$ 2.1.

3. INFINITESIMAL GENERATOR

The infinitesimal generator of semigroups of probability operators has been studied under particular continuity conditions on T_t or under particular choices of G and $\mathcal{M}$. The simplest case is that of norm-continuous semigroups. The proof of the following theorem is given in (Barchielli, 1989).

THEOREM 3.1. *The generator K of a norm-continuous semigroup of probability operators $T_t \equiv \exp(Kt)$ can be written in a unique way as*

$$K[A](x) = T_\Gamma[A](x) - \frac{1}{2}\{\Gamma(G)[\mathbf{1}], A(x)\} + L[A(x)], \qquad \forall A \in \mathcal{N}, \tag{3.1}$$

where $\{a, b\} \equiv ab + ba$, Γ is a quasi-instrument on $\mathcal{M}$ with value space $(G, \mathcal{B})$ without mass at e, T_Γ is the associated convolution operator and L is the generator of a norm-continuous quantum dynamical semigroup on $\mathcal{M}$. Conversely, any bounded operator with the structure (3.1) is the generator of a norm-continuous semigroup of probability operators.

From now on we shall consider semigroups satisfying the continuity property 2.3.

PROPOSITION 3.2. *Let G be a discrete group. Then, property 2.3 implies the norm continuity of T_t (property 2.4). Therefore, the generator of T_t has the structure (3.1).*

Proof. Without any restriction we can take $A \in \mathcal{N}^+$; then, we have

$$\|T_t[A](x) - A(x)\| \leqslant \|J_t(\{e\})[A(x)] - A(x)\| +$$

$$+ \left\| \int_{G \setminus e} J_t(dy)[A(yx)] \right\| \leqslant \|A\| \, \{\|J_t(\{e\}) - \mathbf{1}\| + \|J_t(G \setminus e)\|\}.$$

By 2.3 and the fact that, for a discrete group, $\{e\}$ is a neighborhood of e, the quantities in curly brackets vanish as $t \downarrow 0$ and this implies

$$\lim_{t\downarrow 0} \|T_t - \mathbf{1}\| = 0.$$

Let us consider now a n–level quantum system, i.e. $\mathcal{H} = \mathbf{C}^n$, $\mathcal{M} = \mathcal{L}(\mathbf{C}^n) \equiv \equiv M(n; \mathbf{C})$ ($n \times n$ complex matrices) and assume that G is a Lie group of finite dimension $d \geqslant 1$. In (Barchielli and Lupieri, 1989) we have obtained the structure of the generator of T_t under the continuity hypotheses 2.2 (which in this case are equivalent to 2.1 and 2.3). We end the paper by presenting (without proofs) the main results of (Barchielli and Lupieri, 1989). The case of a Lie group G and a generic W^*–algebra $\mathcal{M}$ is presently under study in collaboration with A.S. Holevo and L.V. Denisov in Moscow.

Let $\Lambda(G)$ be the Lie algebra of G and let $\{Y_j,\ j = 1, \ldots, d\}$ be a basis in it. We denote by the same symbols Y_j the left (uniform) derivatives in the direction Y_j. We denote by $C_2(G)$ the Banach space of all functions in $\mathcal{C}_c(G)$ for which all first and second left (uniform) derivatives exist and by $C_2(G; \mathcal{M})$ the Banach space of the $n \times n$ matrices of C_2–functions with norm

$$\|A\|_2 = \|A\| + \sum_i \|Y_i[A]\| + \sum_{ij} \|Y_i \circ Y_j[A]\|. \tag{3.2}$$

Let $\{y_j,\ j = 1, \ldots, d\}$ be a set of canonical coordinates, associated with the basis $\{Y_j,\ j = 1, \ldots, d\}$ and extended to the whole G in such a way that $y_j \in \in C_2(G)$, $y_j(x) \in \mathbf{R}$, and let φ be a "Hunt function": $\varphi \in C_2(G)$, $\varphi(x) \in \mathbf{R}$, $\varphi(x) > 0$ for $x \neq e$, $\varphi(x) = \sum_i y_i^2(x)$ in a neighborhood of e and $\varphi(\infty) > 0$.

We denote by K_c the infinitesimal generator of the restriction of T_t to $\mathcal{C}_c(G; \mathcal{M})$ and by D_c the domain of K_c. Moreover, we introduce the *generating functional* N of T_t by

$$N[A] = \lim_{t\downarrow 0} \frac{1}{t}(T_t[A](e) - A(e)); \tag{3.3}$$

the domain D_N of N is the set of all $A \in \mathcal{C}_b(G; \mathcal{M})$ such that the limit (3.3) exists. By property 1.4.B we have

$$K_c[A](x) = N \circ R_x[A], \qquad \forall A \in D_c. \tag{3.4}$$

The first step is to show that the domain of N is sufficiently rich (Barchielli and Lupieri, 1989, Theorem 2.3 and Corollary 2.4).

THEOREM 3.3. $D_N \supset C_2(G; \mathcal{M})$.

Then, by exploiting the complete positivity of T_t, one obtains the structure of N (Barchielli and Lupieri, 1989, Theorem 2.5 and Theorem 3.3).

THEOREM 3.4. *The generating functional N of a semigroup T_t with the properties described above has necessarily the structure* ($\forall A \in C_2(G; \mathcal{M})$)

$$N[A] = L[A(e)] + \sum_{i,j=1}^{d} b_{ij}(C_j^* A(e) C_i - \frac{1}{2}\{C_j^* C_i, A(e)\} + D_j[A] C_i +$$

$$+C_j^* D_i[A] + \frac{1}{2} D_{ji}[A]) + \sum_{i=1}^{d} b_i D_i[A] +$$

$$+ \sum_{r,s=1}^{n^2-1} \int_{G_\infty^*} \gamma_{rs}(dx) \frac{1}{\varphi(x)} (F_s^* A(e) F_r - \frac{1}{2}\{F_s^* F_r, A(e)\}) +$$

$$+ \sum_{r,s=0}^{n^2-1} \int_{G_\infty^*} \gamma_{rs}(dx) \frac{1}{\varphi(x)} F_s^* (A(x) - A(e) - \sum_{i=1}^{d} D_i[A] y_i(x) \delta_{r0} \delta_{s0}) F_r, \qquad (3.5)$$

where L is the generator of a norm–continuous quantum dynamical semigroup on $\mathcal{M}$ (Gorini et al., 1976), $C_i \in \mathcal{M}$, $b_j \in \mathbf{R}$, $\{b_{ij}\}$ *is a real positive $d \times d$ matrix,*

$$D_i[A] = Y_i[A](e), \quad D_{ij}[A] = Y_i \circ Y_j[A](e), \quad G_\infty^* = G_\infty \setminus e,$$

$\{\gamma_{rs}\}$ *is a positive $n^2 \times n^2$ matrix of complex regular finite measures on G_∞ such that the integrals appearing in equation* (3.5) *exist*, $\{F_r,\ r = 0, 1, \ldots, n^2 - 1\}$ *is a basis in $\mathcal{M}$ such that* $F_0 = 1/\sqrt{n}$, $\mathrm{Tr}\ F_r = 0$, $r \geqslant 1$, $\mathrm{Tr}\ F_s^* F_r = \delta_{rs}$, $s, r = = 0, 1, \ldots, n^2 - 1$.

By equation (3.4) we have now a candidate to be the generator of a semigroup T_t of probability operators. We say that an operator K on $C_2(G; \mathcal{M})$ admits a Lévy—Khintchine (LK) representation if

$$K[A](x) = L[A(x)] + K_1[A](x) + K_2[A](x), \quad \forall A \in C_2(G; \mathcal{M}), \qquad (3.6)$$

$$K_1[A](x) = \sum_{r,s=1}^{n^2-1} \int_{G \setminus e} \gamma_{rs}(dz) \frac{1}{\varphi(z)} (F_s^* A(x) F_r - \frac{1}{2}\{F_s^* F_r, A(x)\}) +$$

$$+ \sum_{r,s=0}^{n^2-1} \int_{G \setminus e} \gamma_{rs}(dz) \frac{1}{\varphi(z)} F_s^* (A(zx) - A(x) - \sum_i Y_i[A](x) y_i(z) \delta_{r0} \delta_{s0}) F_r, \qquad (3.7)$$

$$K_2[A](x) = \sum_{ij} b_{ij} (C_j^* A(x) C_i - \frac{1}{2}\{C_j^* C_i, A(x)\} + Y_j[A](x) C_i +$$

$$+ C_j^* Y_i[A](x) + \frac{1}{2} Y_j \circ Y_i[A](x)) + \sum_i b_i Y_i[A](x). \qquad (3.8)$$

The various quantities appearing in these formulas have been introduced before, but now the measures γ_{rs} are defined on G and not on G_∞. Indeed, at the end, one proves that the fact that T_t is normal implies that the measures γ_{rs} appearing in equation (3.5) cannot have mass at infinity.
The last step is to show (by approximating the generator of T_t by bounded generators of the form (3.1')) that an operator of the form (3.6)—(3.8) is indeed the most general generator of a semigroup of probability operators (Barchielli and Lupieri, 1989, Lemma 4.2 and Theorem 4.3).

THEOREM 3.5. *Let K be a linear operator on $C_2(G;\mathcal{M})$ admitting a LK-representation. Then, there exist a unique w^*-continuous semigroup $\{T_t,\ t \geqslant 0\}$ of probability operators such that K is the restriction to $C_2(G;\mathcal{M})$ of its generator K_c. Conversely, any w^*-continuous semigroup of probability operators has a generator whose restriction to $C_2(G;\mathcal{M})$ admits a LK-representation.*

REFERENCES

Barchielli, A. (1987). Convolution semigroups in quantum probability. *Semesterbericht Funktionalanalysis Tubingen* **12**, 29–42.

Barchielli, A. (1989). Probability operators and convolution semigroups of instruments in quantum probability. *Probab. Theory Rel. Fields* **82**, 1–8.

Barchielli, A. (1990). Some Markov semigroups in quantum probability. *Lect. Notes in Math.* (to appear).

Barchielli, A., Lanz, L. and Prosperi, G.M. (1983). Statistics of continuous trajectories in quantum mechanics: Operator-valued stochastic processes. *Found. Phys.* **13**, 779–812.

Barchielli, A. and Lupieri, G. (1985). Quantum stochastic calculus, operator-valued stochastic processes, and continual measurements in quantum mechanics. *J. Math. Phys.* **26**, 2222–2230.

Barchielli, A. and Lupieri, G. (1989). A quantum analogue of Hunt's representation theorem for the generator of convolution semigroups on Lie groups. *Preprint* 1FUM 360/FT, Milan.

Barchielli, A. and Lupieri, G. (1990). Convolution semigroups in quantum probability and quantum stochastic calculus. *Lect. Notes in Math.* (to appear).

Davies, E.B. (1976). *Quantum Theory of Open Systems.* Academic Press, London.

Davies, E.B. (1980). *One-Parameter Semigroups.* Academic Press, London.

Dixmier, J. (1969). *Les algebres d'operateurs dans l'espace Hilbertien.* Gautier-Villars, Paris.

Dynkin, E.B. (1965). *Markov Processes.* Vol.I. Springer, Berlin–Heidelberg etc.

Gorini, V., Kossakowski, A. and Sudarshan, E.C.G. (1976). Completely positive dynamical semigroups of N-level systems. *J. Math. Phys.* **17**, 821–825.

Heyer, H. (1977). *Probability Measures on Locally Compact Groups.* Springer, Berlin–Heidelberg etc.

Hille, E. and Phillips, R.S. (1968). *Functional Analysis and Semigroups.* Amer. Math. Soc., Providence.

Holevo, A.S. (1988). A noncommutative generalization of conditionally positive definite functions. *Lect. Notes in Math.* **1303**, 128–148.

Lindblad, G. (1976). On the generators of quantum dynamical semigroups. *Commun. Math. Phys.* **48**, 119–130.

Ozawa, M. (1984). Quantum measuring processes of continuous observables. *J. Math. Phys.* **25**, 79–87.

Parthasarathy, K.R. (1986). On parameter semigroups of completely positive maps of groups arising from quantum stochastic differential equations. *Boll. Univ. Mat. Ital.* **5A**, 391–397.

Rudin, W. (1970). *Real and Complex Analysis.* McGraw-Hill, London.

Sakai, S. (1971). *C^*-algebras and W^*-algebras.* Springer, Berlin–Heidelberg etc.

Takesaki, M. (1979). *Theory of Operator Algebras.* I. Springer, Berlin–Heidelberg etc.

Prob. Theory and Math. Stat., Vol. 1, pp. 91–109
B.Grigelionis *et al.* (Eds.)
1990 VSP/Mokslas

STOCHASTIC POSTERIOR EQUATIONS FOR QUANTUM NONLINEAR FILTERING

V.P. BELAVKIN
MIEM, B.Vuzovski 3/12, 109028 Moscow, USSR

ABSTRACT

Quantum stochastic (QS) calculus for open Markovian system and nondemolition processes of continuous observation is developed. A natural integrability condition for the existence and uniqueness of the solution of linear operator-valued QS evolution equations is given. A stochastic linear filtering equation as a quantum generalization of the Zakai equation is derived for a general observable process, absolutely continuous with respect to the standard Wiener and Poisson processes. A unique solution of the quantum filtering problem is found in the form of a multiple stochastic integral by resolving the corresponding wave stochastic evolution equation describing the posterior dynamics of a quantum object under the complete nondemolition observation.

INTRODUCTION

The time evolution of a quantum object under a continuous observation can be effectively described in the framework of quantum stochastic (QS) calculus worked out in (Hudson and Parthasarathy, 1984). The basic idea for the description of the continuous quantum measurements is the nondemolition principle, formulated in (Belavkin, 1988a) as the commutativity of the observed process with respect to the present and future representation of the quantum object under observation. This principle giving no restrictions on the future observations, is consistent with the casuality principle and is necessary and sufficient for the existence of posterior conditional expectations for any initial quantum state. Moreover, as it follows from the reconstruction theorem (Belavkin, 1985), it can be satisfied for any type of quantum measurements in appropriate representation of the object under observation.

The QS calculus of the nondemolition processes (Belavkin, 1988a), generalizing the canonical output signals and fields introduced for a model in (Belavkin, 1980, 1987) and for a more general case in (Gardiner and Collet, 1985; Barchielli, 1988), helps to derive a stochastic filtering equation (Belavkin, 1989) for the posterior conditional expectations of any QS process of the object. In the Markovian case it gives a recursive equation for the posterior states, introduced in the context of quantum filtering for the general Markovian quantum dynamics in an operational approach in (Belavkin, 1983). The corresponding nonlinear stochastic equations for the posterior normalized density operator and wave function, obtained for the

canonical output processes in (Belavkin, 1988a,b), give quantum generalizations of the Stratonovich filtering equation in the Ito form (Stratonovich, 1966; Liptser and Shiryaev, 1974).
Here we give a direct derivation of the linear stochastic equations for the posterior density operator and wave function, corresponding to the Zakai filtering equation for the posterior state, normalized on the probability density with respect to a standard probability measure. We use a general QS calculus in the matrix form with an indefinite metrics, introduced in (Belavkin, 1988a) and based on a new definition of the QS integrals, given in (Belavkin, 1989). We define the corresponding multiple QS ξ integrals by an explicit formula generalizing (Lindsay and Maassen, 1987) and describe the natural sufficient conditions for the existence and uniqueness of the solution T_t of a stochastic operator–valued evolution equation as the multiple QS integral over the time ordered product of generators of this equation. We construct the solution for the general QS filtering equation as a conditional expectation of the stochastic processes $Z_t = T_t^* X T_t$, defined by the operators X of the quantum object. Another method of construction of the solutions in a particular case of equations of such kind is given in (Holevo, 1989).

1. MULTIPLE QS INTEGRALS AND QS EQUATIONS

Let $\mathcal{H}$ be a Hilbert space with the probability vector $\mathbf{h} \in \mathcal{H}, \| \mathbf{h} \| = 1$, of a quantum dynamical object described at any $t \in \mathbf{R}_+$ by a von Neumann algebra $\mathcal{B}$, i.e., a weakly closed subspace $\mathcal{B} \subseteq \mathcal{B}(\mathcal{H})$ of bounded operators in $\mathcal{H}$ with the identity operator $I \in \mathcal{B}$ and $X^*X \in \mathcal{B}$ for any $X \in \mathcal{B}$. We denote by $\mathcal{K} = \int^{\oplus} \mathcal{E}(v)\, dv$ the Hilbert integral of complex Euclidean spaces $\mathcal{E}(v), v \in V$, that is the space of all square integrable vector–functions $\mathbf{c} : v \mapsto \mathbf{c}(v) \in \mathcal{E}(v)$ with respect to the measure $dv = dt\, dx$ on the product $V = \mathbf{R}_+ \times \Delta$ of $\mathbf{R}_+ = \{t \geqslant 0\}$ and a Borel set Δ with measure dx. The probability vector space of quantum noise is the space $\mathcal{F} = \mathcal{F}(1)$, where

$$\mathcal{F}(\xi) = \int^{\oplus} \xi^{|\upsilon|} \mathcal{E}^{\oplus}(\upsilon) d\upsilon$$

is the integral of tensor products $\mathcal{E}^{\oplus}(\upsilon) = \bigoplus_{v \in \upsilon} \mathcal{E}(v)$ on the space Υ of all finite subsets $\upsilon \subset V$ with the measure $d\upsilon = \prod_{v \in \upsilon} dv$ defined as a sum over $n = | \upsilon |$:

$$\int \xi^{|\upsilon|} \|c(\upsilon)\|^2 \, d\upsilon = \sum_{n=0}^{\infty} \xi^n \int \cdots \int_{0 \leqslant t_1 < \cdots < t_n < \infty} \|c(v_1, \ldots, v_n)\|^2 \, dv_1 \ldots dv_n \equiv$$

$$\equiv \| c \|^2(\xi), \ \xi > 0.$$

Here $\mathbf{c}(v_1, \ldots, v_n) = c(\upsilon)$ for $\upsilon = \{v_1, \ldots, v_n\} \subset V$ are symmetrical functions on $v_i = (t_i, x_i)$ defining the Hilbert space $\mathcal{F}$ of square integrable tensor–functions $c : \upsilon \mapsto c(\upsilon) \in \mathcal{E}(\upsilon)$ as the Fock space $\Gamma(\mathcal{K})$ over the integral $\mathcal{K}$. The quantum system "object+noise" is described by the Hilbert space $\mathcal{G} = \mathcal{H} \otimes \mathcal{F}$ of square integrable vector–functions $\mathbf{a} = \int^{\oplus} \mathbf{a}(\upsilon)\, d\upsilon$ on $\upsilon \in \Upsilon$ with tensor–values $\mathbf{a}(\upsilon) \in \mathcal{H} \otimes \mathcal{E}^{\otimes}(\upsilon)$ called the probability amplitudes, if $\| \mathbf{a} \|^2 = \int \| \mathbf{a}(\upsilon) \|^2 \, d\upsilon = 1$.

Following (Belavkin, 1988a,b) we define the QS integral

$$\Lambda_t(\mathbf{D}) = \int_0^t d\Lambda_s(\mathbf{D})$$

for the table $\mathbf{D} = (D^\mu_\nu)^{\mu=-,0}_{\nu=0,+}$ of functions $D^\mu_\nu(u)$ on $u \in V$ with operator values

$$D^0_0(u) : \mathcal{G} \otimes \mathcal{E}(u) \to \mathcal{G} \otimes \mathcal{E}(u), \quad D^-_+(u) : \mathcal{G} \to \mathcal{G},$$

$$D^0_+(u) : \mathcal{G} \to \mathcal{G} \otimes \mathcal{E}(u), \quad D^-_0(u) : \mathcal{G} \otimes \mathcal{E}(u) \to \mathcal{G}, \tag{1.1}$$

as the sum $\Lambda_t(\mathbf{D}) = \sum \Lambda^\nu_\mu(t, D^\mu_\nu)$ of the operators

$$\Lambda^\nu_\mu(t, D) : \mathbf{a} \in \mathcal{G} \mapsto \Lambda^\nu_\mu(t, D)\mathbf{a},$$

$$[\Lambda^0_0(t, D^0_0)\mathbf{a}](v) = \sum_{u \in v^t} [D^0_0(u)\dot{\mathbf{a}}(u)](v \backslash u), \tag{1.2a}$$

$$[\Lambda^+_0(t, D^0_+)\mathbf{a}](v) = \sum_{u \in v^t} [D^0_+(u)\mathbf{a}](v \backslash u), \tag{1.2b}$$

$$[\Lambda^0_-(t, D^-_0)\mathbf{a}](v) = \int_{V^t} [D^-_0(u)\dot{\mathbf{a}}(u)](v)\, du, \tag{1.2c}$$

$$[\Lambda^+_-(t, D^-_+)\mathbf{a}]\,(v) = \int_{V^t} [D^-_+(u)\mathbf{a}](v)\, du, \tag{1.2d}$$

where $v^t = v \cap V^t$, $V^t = [0,t) \times \Delta$, $v \backslash u = \{v \in v \mid v \neq u\}$, $\dot{\mathbf{a}}(u) = \int^\oplus \mathbf{a}(v \sqcup u)\, dv$, $u \sqcup v = \{u, v \mid u \notin v\}$. The operator–functions called in the QS calculus (Hudson and Parthasarathy, 1984) the *gauge, creation, annihilation* and *time processes* respectively are densely defined in $\mathcal{G} = \int^\oplus \mathcal{H} \otimes \mathcal{E}^\otimes(v)\, dv$, if the operator–functions $D^\mu_\nu(u)$, $\mu = -, 0$; $\nu = 0, +$ are locally QS–integrable in the sense of (Belavkin, 1988b)

$$\| D^0_0 \|^{(\infty)}_t < \infty, \ \| D^+_0 \|^{(2)}_t < \infty, \ \| D^0_- \|^{(2)}_t < \infty, \ \| D^+_- \|^{(1)}_t < \infty, \tag{1.3}$$

where

$$\| D \|^{(p)}_t = \left(\int_{V^t} \| D(u) \|^p \, du \right)^{1/p}, \ p = 1, 2, \ldots, \infty$$

($\| D \|^{(\infty)} =$ ess sup $\| D(u) \|$).
Let us define the multiple QS–integral

$$\Lambda^\odot_t(B) = \sum_{n=0}^\infty \int \cdots \int_{0 \leqslant t_1 < \ldots < t_n < t} d\Lambda^{\odot n}_{t_1, \ldots, t_n}(B) \equiv \int_{\tau \subset [0,t)} d\Lambda^\odot_\tau(B)$$

for the operator-valued function $B(v)$ on the table $v = (v^\mu_\nu)^{\mu=-,0}_{\nu=0,+}$ of subsets $v^\mu_\nu \in \Upsilon$ with values

$$B\begin{pmatrix} v_0^- & v_+^- \\ v_0^0 & v_+^0 \end{pmatrix}: \mathcal{G} \otimes \mathcal{E}^{\otimes}(v_0^-) \otimes \mathcal{E}^{\otimes}(v_0^0) \to \mathcal{G} \otimes \mathcal{E}^{\otimes}(v_0^0) \otimes \mathcal{E}^{\otimes}(v_+^0)$$

as the operator in $\mathcal{G}$ with the action

$$[\Lambda_t^{\odot}(B)\mathbf{a}](v) = \sum_{v_0^0 \sqcup v_+^0 \subseteq v^t} \int_{\Upsilon^t}\int_{\Upsilon^t} \left[B(v)\dot{\mathbf{a}}(v_0^- \sqcup v_0^0)\right](v_-^0) dv_0^- dv_+^-, \tag{1.4}$$

where $v_+^0 = v \backslash v_0^0 \backslash v_+^0 = v \cap \overline{(v_0^0 \sqcup v_+^0)}$, $\Upsilon^t = \{v \in \Upsilon \mid v \subset V^t\}$ and $\dot{\mathbf{a}}(v_0) =$ $= \int^{\oplus} \mathbf{a}(v \sqcup v_0) dv$, i.e., $[\dot{\mathbf{a}}(v_0)](v) = \mathbf{a}(v \cup v_0)$ for almost all v $(v \cap v_0 = \varnothing)$. We shall say that the function B is locally QS-integrable (as the uniform inductive limit) if for any $t \in \mathbf{R}_+$ there are pairs $\eta^{\cdot} = (\eta^-, \eta^0)$, $\eta_{\cdot} = (\eta_0, \eta_+)$ of numbers $\eta^\mu, \eta_\nu > 0$, for which $\| B \|^{\eta_\cdot}_{\eta^\cdot}(t) < \infty$,

$$\| B \|^{\eta_\cdot}_{\eta^\cdot}(t) = \int_{\Upsilon^t} \left[\int_{\Upsilon^t}\int_{\Upsilon^t} \frac{(\eta_+)^{|v_+^0|}}{(\eta^-)^{|v_0^-|}} \operatorname*{ess\,sup}_{v_0^0 \in \Upsilon^t} \frac{(\eta_0)^{|v_0^0|}}{(\eta^0)^{|v_0^0|}} \times \| B(v) \|^2 \, dv_+^0 dv_0^- \right]^{1/2} dv_+^-. \tag{1.5}$$

THEOREM 1. *For any locally QS-integrable operator-function B the multiple QS-integral (1.5) is densely defined in $\mathcal{G}$ as (ξ^+, ξ_-)-continuous operator $\mathcal{H} \otimes \mathcal{F}(\xi^+) \to \mathcal{H} \otimes \mathcal{F}(\xi_-)$ with $\xi^+ > 1 + \eta^- + \eta^0$, $\xi_-^{-1} \geqslant 1 + \eta_0^{-1} + \eta_+^{-1}$:*

$$\| \Lambda_t^{\odot}(B) \|^{\xi_-}_{\xi^+} = \sup\{\| \Lambda_t^{\odot}(B)\mathbf{a} \| (\xi_-)/\| \mathbf{a} \| (\xi^+)\} \leqslant \| B \|^{\eta_\cdot}_{\eta^\cdot}(t), \tag{1.6}$$

as well as the formally conjugate operator

$$\Lambda_t^{\odot}(B)^* = \Lambda_t^{\odot}(B^*),\ B^*(v) = B(v^*)^*,\ v^* = \begin{pmatrix} v_+^0 & v_+^- \\ v_0^0 & v_0^- \end{pmatrix}$$

as a continuous operator $\mathcal{H} \otimes \mathcal{F}(1/\xi_-) \to \mathcal{H} \otimes \mathcal{F}(1/\xi^+)$ with the norm

$$\| \Lambda_t^{\odot}(B^*) \|^{1/\xi^+}_{1/\xi_-} = \| \Lambda_t^{\odot}(B) \|^{\xi_-}_{\xi^+}.$$

This operator has a QS differential $d\Lambda_t^{\odot}(B) = d\Lambda_t(\mathbf{D})$ in the sense

$$\Lambda_t^{\odot}(B) = B(\varnothing) + \Lambda_t(\mathbf{D}),\ D^\mu_\nu(s,x) = \Lambda_s^{\odot}(\dot{B}^\mu_\nu(s,x)),\ s < t, \tag{1.7}$$

with (ξ^+, ξ_-)-continuous derivatives $D^\mu_\nu(u)$, densely defined as in (1.4) for almost all $u = (s,x)$ by $\dot{B}^\mu_\nu(u)(v) = B(\mathbf{u}^\mu_\nu \sqcup v)$, where $\mathbf{u}^\mu_\nu$ denotes the table $(u^\varkappa_\lambda)^{\varkappa=-,0}_{\lambda=0,+}$ with $u^\varkappa_\lambda = \emptyset$ except $\varkappa = \mu, \lambda = \nu$ with single-point $u^\mu_\nu = \{u\}$, and $v = (v^\mu_\nu)$ has the components $v^\mu_\nu \in \Upsilon^s$ disjunctive almost everywhere with single-point tables $\mathbf{v} \in v$ defined as $\mathbf{v}^\mu_\nu$ if $v \in v^\mu_\nu$.
If $B(v)$ is defined for disjunctive (v^μ_ν), $v = \bigsqcup_{\mathbf{v} \in v} \mathbf{v}$, as a chronologically ordered

product $B(v) = \overleftarrow{\bigodot_{\mathbf{v}\in v}} L(\mathbf{v}) \equiv \mathbf{L}^{\overleftarrow{\odot}}(v)$ *by the recurrency* $B(\mathbf{u}\sqcup v) = L(\mathbf{u}) \odot B(v)$, $v^\mu_\nu \in \Upsilon^s$, *i.e.*

$$\dot{B}^\mu_\nu(s,x)(v) = (L^\mu_\nu(s,x)\otimes \widehat{1})B(v),\ \widehat{1}(v) = \bigotimes_{v\in v} I(v) \equiv I^\otimes(v) \tag{1.8}$$

with $B(\emptyset) = \widehat{I}$, *where* $B^\mu_0(u)B(v) = (B^\mu_0(u)\otimes I^\otimes(v^0_0 \sqcup v^0_+))\ (B(v)\otimes I(u))$, $B^\mu_+(u)B(v) = \left(B^\mu_+(u)\otimes I^\otimes(v^0_0\sqcup v^0_+)\right)B(v)$, $\mu = -,0$, *then the adapted process* $U_t = \Lambda^\odot_t(B)$ *satisfies the QS differential equation* $dU_t = d\Lambda_t(\mathbf{L})U_t$ *in the sense*

$$U_t = \widehat{I} + \Lambda_t(\mathbf{L}\odot U),\quad where\quad (\mathbf{L}\odot U)^\mu_\nu(s,x) = (L^\mu_\nu(s,x)\otimes\widehat{1})U_s, \tag{1.9}$$

$B^\mu_0(v)U_t = B^\mu_0(v)(U_t\otimes I(v))$, $B^\mu_+(v)U_t = B^\mu_+(v)U_t$, $\mu = -,0$, $\widehat{I} = I\otimes\widehat{1}$, *is the identity in* $\mathcal{G} = \mathcal{H}\otimes\mathcal{F}$.

Moreover, the solution $U_t = \Lambda^\odot_t(\mathbf{L}^{\overleftarrow{\odot}})$ *of the equation* (1.9) *with an integrable in the sense* (1.3) *function* $\mathbf{L}(u) = (L^\mu_\nu)(u)$ *is unique,* (ξ^+, ξ_-)*–continuous, if* $\xi^+\xi_-^{-1} > (1+ \| L^0_0 \|^{(\infty)}_t)^2$, *has the estimate*

$$\| U_t \|^{\xi_-}_{\xi^+} \leqslant \exp\left\{\int_{V^t}\left[\| L^-_+(u)\| + \left(\| L^-_0(u)\|^2 + \| L^+_0(u)\|^2\right)/2\varepsilon\right]du\right\}, \tag{1.10}$$

if $\xi^+ \geqslant 1+\varepsilon+\| L^0_0 \|^{(\infty)}_t$, $\xi_-^{-1} \geqslant 1+\varepsilon+\| L^0_0\|^{(\infty)}_t$, $\varepsilon > 0$, and *is isometric* $U^*_t U_t = I\otimes\widehat{1}$ *(unitary* $U^*_t = U^{-1}_t$*), if the triangular matrix* $\mathbf{S}(u) = [S^\mu_\nu(u)]^{\mu=-,0,+}_{\nu=-,0,+}$ with $S^\mu_\nu = L^\mu_\nu + I^\mu_\nu$, $L^\mu_\nu = 0$, *if* $\mu = +$ or $\nu = -$, $I^\mu_\nu(u) = I\otimes 1^\mu_\nu(u)$, $1^-_-(u) = 1 = 1^+_+(u)$, $1^0_0(u) = I(u)$, and $1^\mu_\nu(u) = 0$ *for* $\mu\neq\nu$, *is pseudoisometric* $\mathbf{S}^\star(u)\mathbf{S}(u) = [I^\mu_\nu(u)] \equiv I(u)$ *(pseudounitary:* $\mathbf{S}^\star(u) = \mathbf{S}^{-1}(u)$*) for almost all* $u\in V^t$ *under the pseudometrics*

$$\mathbf{g}(u) = \widehat{I}\otimes\begin{pmatrix}0 & 0 & 1\\ 0 & I(u) & 0\\ 1 & 0 & 0\end{pmatrix},$$

$$\mathbf{S} = \begin{pmatrix}1 & s^-_0 & s^-_+\\ 0 & s^0_0 & s^0_+\\ 0 & 0 & 1\end{pmatrix} \mapsto \mathbf{S}^\star = \begin{pmatrix}1 & s^{0*}_+ & s^{-*}_+\\ 0 & s^{0*}_0 & s^{-*}_0\\ 0 & 0 & 1\end{pmatrix} = \mathbf{g}\,\mathbf{S}^*\mathbf{g}.$$

Proof. Using the property

$$\int\sum_{\sqcup v_\nu = v} f(v_-, v_0, v_+)dv = \int\int\int f(v_-,v_0,v_+)\prod dv_\nu,$$

we obtain from the definition (1.4) for $\mathbf{a},\mathbf{c}\in\mathcal{G}$

$$\int \Big\langle \mathbf{c}(v) \mid \Big[\Lambda_t^{\odot}(B)\mathbf{a}\Big](v)\Big\rangle dv =$$

$$= \int_{\Upsilon^t} dv_+^- \int_{\Upsilon^t} dv_0^- \int_{\Upsilon^t} dv_+^0 \int_{\Upsilon^t} dv_0^0 \Big\langle \dot{\mathbf{c}}(v_0^0 \sqcup v_+^0) \mid B(v)\dot{\mathbf{a}}(v_0^- \sqcup v_0^0)\Big\rangle =$$

$$= \int_{\Upsilon^t} dv_+^- \int_{\Upsilon^t} dv_0^- \int_{\Upsilon^t} dv_+^0 \int_{\Upsilon^t} dv_0^0 \Big\langle B(v)^* \dot{\mathbf{c}}(v_0^0 \sqcup v_+^0) \mid \dot{\mathbf{a}}(v_0^- \sqcup v_0^0)\Big\rangle =$$

$$= \int \Big\langle [\Lambda_t^{\odot}(B^\star)\mathbf{c}](v) \mid \mathbf{a}(v)\Big\rangle dv.$$

This defines the formally conjugate operator $\Lambda_t^{\odot}(B)^*$ as QS–integral $\Lambda_t^{\odot}(B^\star)$ on $B^\star(v) = B(v^\star)^*$, and

$$\| \Lambda_t^{\odot}(B) \|_\xi^{1/\eta} = \sup_{\|\mathbf{a}\|(\xi)\leqslant 1} | \langle \mathbf{c} \mid \Lambda_t^{\odot}(B)\mathbf{a}\rangle | =$$

$$= \sup_{\|\mathbf{c}\|(\eta)\leqslant 1} | \langle \Lambda_t^{\odot}(B^\star)\mathbf{c} \mid \mathbf{a}\rangle | = \| \Lambda_t^{\odot}(B^\star) \|_\eta^{1/\xi} .$$

Let us estimate the integral $\langle \mathbf{c} \mid \Lambda_t^{\odot}(B)\mathbf{a}\rangle$ using the Schwarz inequality

$$\int \| \dot{\mathbf{c}}(v) \| \| \dot{\mathbf{a}}(v) \| (\eta_0/\eta^0)^{|v|/2} dv \leqslant \| \dot{\mathbf{c}} \| (1, \eta_0^{-1}) \| \dot{\mathbf{a}} \| (1, \eta^0)$$

and

$$\| \dot{\mathbf{a}} \| (\xi, \eta) = \left(\int\int \xi^{|\sigma|} \eta^{|\tau|} \| \mathbf{a}(\sigma \sqcup \tau) \|^2 \, d\sigma d\tau \right)^{1/2} = \| \mathbf{a} \| (\xi + \eta).$$

We have

$$| \langle \mathbf{c} \mid \Lambda_t^{\odot}(B)\mathbf{a}\rangle | \leqslant$$

$$\leqslant \int_{\Upsilon^t} dv_+^- \int_{\Upsilon^t} dv_0^- \int_{\Upsilon^t} dv_+^0 \int_{\Upsilon^t} dv_0^0 \| \dot{\mathbf{c}}(v_0^0 \sqcup v_+^0) \| \| B(v) \| \| \dot{\mathbf{a}}(v_0^- \sqcup v_0^0) \| \leqslant$$

$$\leqslant \int_{\Upsilon^t} dv_+^- \int_{\Upsilon^t} dv_0^- \int_{\Upsilon^t} dv_+^0 \| \dot{\mathbf{c}}(v_+^0) \| (1 + \eta_0^{-1}) \| \dot{\mathbf{a}}(v_0^-) \| (1 + \eta^0) \times$$

$$\times \operatorname*{ess\,sup}_{v_0^0 \in \Upsilon^t} \left(\frac{\eta_0}{\eta^0}\right)^{|v_0^0|/2} \| B(v) \| \leqslant$$

$$\leqslant \int_{\Upsilon^t} dv_+^- \| \mathbf{c} \| (1 + \eta_0^{-1} + \eta_+^{-1}) \| \mathbf{a} \| (1 + \eta^- + \eta^0) \times$$

$$\times \left[\int_{\Upsilon^t}\int_{\Upsilon^t} \frac{(\eta_+)^{|v_+^0|}}{(\eta^-)^{|v_0^-|}} \operatorname*{ess\,sup}_{v_0^0 \in \Upsilon^t} \left(\frac{\eta_0}{\eta^0}\right)^{|v_0^0|} \| B(v) \| \right]^{1/2} .$$

This gives

$$| \langle \mathbf{c} \mid \Lambda_t^{\odot}\mathbf{a}\rangle | \leqslant \| \mathbf{c} \| (\xi_-^{-1}) \| \mathbf{a} \| (\xi^+) \| B \|_{\eta}^{\eta \cdot} (t)$$

for $\xi^+ \geqslant 1 + \eta^- + \eta^0$, $\xi_-^{-1} \geqslant 1 + \eta_0^{-1} + \eta_+^{-1}$.
Using the definition (1.4) and the property

$$\int_{\Upsilon^t} \varphi(v)\, dv = \varphi(\emptyset) + \int_{V^t} du \int_{\Upsilon^t} \varphi(u \sqcup v)\, dv,$$

where $u = (s, x)$ one can obtain

$$\Big[\Lambda_t^{\odot}(B)\mathbf{a}\Big](v) - \Big[\Lambda_0^{\odot}(B)\mathbf{a}\Big](v) = \Big[(\Lambda_t^{\odot}(B) - B(\emptyset))\mathbf{a}\Big](v) =$$

$$= \int_{V^t} du \Bigg(\sum_{v_0^0 \sqcup v_+^0 = v^s} \int_{\Upsilon^s}\int_{\Upsilon^s} \Big[B(\mathbf{u}_+^- \sqcup v)\dot{\mathbf{a}}(v_0^- \sqcup v_0^0) +$$

$$+ B(\mathbf{u}_0^- \sqcup v)\dot{\mathbf{a}}(u \sqcup v_0^- \sqcup v_0^0)\Big](v_-^0) dv_+^- dv_0^- \Bigg) +$$

$$+ \sum_{u \in v^t} \Bigg(\sum_{v_0^0 \sqcup v_+^0 \subseteq v^s} \int_{\Upsilon^s}\int_{\Upsilon^s} \Big[B(\mathbf{u}_+^0 \sqcup v)\dot{\mathbf{a}}(v_0^- \sqcup v_0^0) +$$

$$+ B(\mathbf{u}_0^0 \sqcup v)\dot{\mathbf{a}}(u \sqcup v_0^- \sqcup v_0^0)\Big](v_-^0) dv_+^- dv_0^- \Bigg) =$$

$$= \int_{V^t} du \Big[\Lambda_s(\dot{B}_+^-(u))\mathbf{a} + \Lambda_s(\dot{B}_0^-(u))\dot{\mathbf{a}}(\dot{u})\Big](v) +$$

$$+ \sum_{u \in v^t} \Big[\Lambda_s(\dot{B}_+^0(u))\mathbf{a} + \Lambda_s(\dot{B}_0^0(u)\dot{\mathbf{a}}(u)\Big](v \backslash u).$$

Hence

$$\Lambda_t^{\odot}(B) - \Lambda_t^{\odot}(B) = \Lambda_-^+(t, D_+^-) + \Lambda_-^0(t, D_0^-) + \Lambda_0^+(t, D_+^0) + \Lambda_0^0(t, D_0^0),$$

where $\Lambda_\nu^\mu(t)$ are the QS–integrals (1.2) of operators $D_\nu^\mu(u)$, defined in (1.7). Due to inequality (1.6) we have

$$\int_{V^t} \| D_+^-(u) \|_{\xi^+}^{\xi_-} du \leqslant \int_{V^t}\int_{\Upsilon^t} \| \dot{B}_+^-(u \sqcup v) \|_{\eta}^{\eta \cdot} (s) dv du =$$

$$= \| B \|_{\eta}^{\eta \cdot} (t) - \| B_+^-(\emptyset) \|_{\eta \cdot}^{\eta \cdot} (t) \leqslant c,$$

where $c = \| B \|_{\eta}^{\eta \cdot} (t)$, $\dot{B}_+^-(v_+^-, v \backslash v_+^-) = B(v)$, $u = (s, x)$. In the same way

$$\int_{V^t} (\| D_0^-(u) \|_{\xi^+}^{\xi_-})^2 du \leqslant \int_{V^t} (\| \dot{B}_0^-(u) \|_{\eta \cdot}^{\eta \cdot} (s))^2 du \leqslant \eta^- (\| B \|_{\eta \cdot}^{\eta \cdot} (t))^2,$$

$$\int_{V^t}(\| D^0_+(u) \|^{\eta_-}_{\eta^+})^2 du \leqslant \int_{V^t}(\| \dot{B}^0_+(u) \|^{\eta_\cdot}_{\eta^\cdot} (s))^2 du \leqslant \eta_+^{-1}(\| B \|^{\eta_\cdot}_{\eta^\cdot} (t))^2,$$

and

$$\operatorname*{ess\,sup}_{u\in V^t} \| D^0_0(u) \|^{\xi_-}_{\xi^+} \leqslant \operatorname*{ess\,sup}_{u\in V^t} \| \dot{B}^0_0(u) \|^{\eta_\cdot}_{\eta^\cdot} (s) \leqslant (\eta^0\eta_0^{-1})^{1/2} \| B \|^{\eta_\cdot}_{\eta^\cdot} (t).$$

It proves the QS–integrability (1.3) of the derivatives $D^\mu_\nu(u)$ with respect to the (ξ^+,ξ_-)–norms.
If $B(v)$ satisfies the recurrency (1.8), then $\dot{B}^\mu_\nu(u) = L^\mu_\nu(u) \odot B$ and

$$D^\mu_\nu(u) = \Lambda^\odot_s(L^\mu_\nu(u)\odot U) = L^\mu_\nu(u) \odot \Lambda^\odot_s(B) = \Lambda^\mu_\nu(u) \odot U_s,$$

where $u = (s, x)$. Hence, $U_t = \Lambda^\odot_t(B)$ satisfies the QS equation $U_t = \hat{I} + \Lambda_t(\mathbf{L}\odot U)$. Conversely, let $U_t = \Lambda^\odot_t(B)$ satisfy QS equation (1.9). Then this equation written in terms of $\Lambda^\odot_t(B) - B(\varnothing)\otimes\hat{1} = \Lambda_t(\Lambda^\odot(\dot{\mathbf{B}}))$ in the form

$$\Lambda_t(\Lambda^\odot(\dot{B})) = \Lambda_t(L \odot \Lambda^\odot(B)) = \Lambda_t(\Lambda^\odot(L \odot B)),$$

is equivalent to the system of equations $\Lambda^\odot(\dot{B}^\mu_\nu) = \Lambda^\odot(L^\mu_\nu \odot B)$ or $\dot{\mathbf{B}} = \mathbf{L} \odot B$. But this is the recurrency (1.8) having the unique solution $B(v) = \left[\overleftarrow{\bigodot_{\mathrm{v}\in v}} L(\mathrm{v})\right]$ with $B(\varnothing) = \hat{I}$ in terms of the chronological product $\overleftarrow{\odot}$, with

$$\| B(v) \| = \| B^-_+(v^-_+) \| \cdot \| B^-_0(v^-_0) \| \cdot \| B^+_0(v^+_0) \| \cdot \| B^0_0(v^0_0) \|,$$

where $\| B(v) \| = \prod_{\mathrm{v}\in v} \| L(V) \|$. If the function $L(u)$ is locally QS–intergrable in the sence (1.3), then the chronological products $B(v)$ give a densely defined operator $U_t = \Lambda_t(B)$ due to

$$\| B \|^{\eta_\cdot}_{\eta^\cdot} (t) = \int_{\Upsilon^t} \| B^-_+(v) \| dv \Big(\int_{\Upsilon^t} (1/\eta^-)^{|v|} \| B^-_0(v) \|^2 dv \times$$

$$\times \int_{\Upsilon^t} (1/\eta_+^{-1})^{|v|} \| B^0_+(v) \| dv \sup_{v\in\Upsilon^t} \Big(\frac{\eta_0}{\eta^0}\Big)^{|v|} \| B^0_0(v) \|^2 \Big)^{1/2} =$$

$$= \sum_{n=0}^\infty \frac{1}{n!}\Big(\| L^-_+ \|^{(1)}_t \Big)^n \Big(\sum_{n=0}^\infty \frac{1}{n!} \Big(\frac{1}{\zeta^-} \int_{V^t} \| L^-_0(u) \|^2 du \Big)^n \times$$

$$\times \sum_{n=0}^\infty \frac{1}{n!} \Big(\frac{1}{\zeta_+^{-1}} \int_{V^t} \| L^0_+(u) \|^2 du \Big)^n \Big)^{1/2} \sup_n \Big((\frac{\eta_0}{\eta^0})^{1/2} \| L^0_0 \|^0_t \Big)^n =$$

$$= \exp\left\{\int_{V^t}\left(\| L_+^-(u) \| + \frac{1}{2}\left(\frac{1}{\zeta^-} \| L_0^-(u) \|^2 + \frac{1}{\zeta_+^{-1}} \| L_+^0(u) \|^2\right)\right)du\right\} < \infty$$

for any $t > 0, \zeta^-, \zeta_+ > 0$, and $(\eta^0\eta_0^{-1})^{1/2} \leqslant \| L_0^0 \|_t^{(\infty)} = \operatorname{ess\,sup}_{u\in V^t} \| L_0^0(u) \|$. In particular, for $\zeta^- = \varepsilon = \zeta_+^{-1}$, $\eta^0 = \| L_0^0 \|_+^{(\infty)} = \eta_0^{-1}$ we obtain (1.10). The condition $U_t^*U_t$ from $\mathbf{S}^*(u)\mathbf{S}(u) = \mathbf{I}(u)$ and $U_t^{-1} = U_t^*$ from $\mathbf{S}^{-1}(u) = \mathbf{S}^*(u)$ are proved in (Belavkin, 1988a).

2. QS REPRESENTATION OF A NONDEMOLITION PROCESS

Let $\mathbf{y}_t = [y_i(t)]$, $i = 1, \ldots, d$, be a classical stochastic vector–process

$$y_i(t,\omega) = q_i(t,\omega) + \int \varkappa_i \, dn_\varkappa(t,\omega), \quad \omega \in \Omega, \ t \in \mathbf{R}_+$$

defined on the product $(\Omega, \mathcal{A}, \nu) = (\Omega_0 \times \Omega_1, \mathcal{A}_0 \otimes \mathcal{A}_1, \nu_0 \cdot \nu_1)$ of the standard probability space $(\Omega_0, \mathcal{A}_0, \nu_0)$ of the d–dimensional Wiener process $\mathbf{w}_t = [w_i(t)]$, $t \in \mathbf{R}_+$, by $\mathbf{q}_t = \mathbf{w}_t + \mathbf{c}_t$ and the space $(\Omega_1, \mathcal{A}_1, \nu_1)$ of the Poisson stochastic vector process $\mathbf{n}_t = [n_\varkappa(t)]$, $\varkappa \in \Delta_1 = \{\varkappa \in \mathbf{R}^d \mid \varkappa \neq 0\}$ with intensity $\lambda_t = [t\lambda_\varkappa]$. We shall denote by $d\mathbf{n}_t$ the increments $[n_\varkappa(t+dt) - n_\varkappa(t)]$ of the process $\mathbf{n}_t$, and by $dn_{t\varkappa}$ the corresponding stochastic interger measure on $\mathbf{R}_+ \times \Delta_1$ with intensity measure $dt\, d\lambda_\varkappa$. Assuming that $\mathbf{c}(t) = \mathbf{q}(t) - \mathbf{w}(t)$ is a non-stochastic process, one can suppose that the vector–process $\mathbf{y}_t$ has the orthogonal increments defined as

$$d\mathbf{y}_t = d\mathbf{w}_t + \varkappa dm_{t\varkappa}, \quad dm_{t\varkappa} = dn_{t\varkappa} - dt\, d\lambda_\varkappa \tag{2.1}$$

even if either $\lambda(\Delta_1) = \int_{\varkappa\neq 0} d\lambda_\varkappa = \infty$ or $\operatorname{ess\,sup} \| \varkappa \| = \infty$.

We denote by $\Delta = \Delta_0 \sqcup \Delta_1$ the direct union $\Delta_0 \cup \Delta_1$, $\Delta_0 \cap \Delta_1 = \emptyset$ of the sets $\Delta_0 = \{1, \ldots, d\}$ and $\Delta_1 \subset \mathbf{R}^d$ with $dx = 1$ for $x = i \in \Delta_0$ and $dx = d\lambda_\varkappa$ for $x = \varkappa \in \Delta_1$ and by $dv = dt\, dx$ the product measure on $V = \mathbf{R}_+ \times \Delta$ defining the scalar product

$$\int \| \mathbf{c}(v) \|^2 \, dv = \int \left(\sum_{i=1}^d \| \mathbf{c}(t,i) \|^2 + \int_{\varkappa\neq 0} \| \mathbf{c}(t,\varkappa) \|^2 \, d\lambda_\varkappa \right) dt$$

on the Hilbert sum $\mathcal{K} = \mathcal{K}_0 \oplus \mathcal{K}_1$, where

$$\mathcal{K}_0 = \int^\oplus \bigoplus_{i=1}^d \mathcal{E}(t,i)dt, \quad \mathcal{K}_1 = \int^\oplus \int \mathcal{E}(t,\varkappa)d\lambda_\varkappa dt.$$

As the set Υ of all finite subsets $v \subset V_0 \sqcup V_1$ is identical to the set $\Upsilon_0 \times \Upsilon_1$ of all pairs (v_0, v_1) where $v_i = v \cap V_i$, $V_i = \mathbf{R}_+ \times \Delta_i$, and $dv = dv_0\, dv_1$ for $v = v_0 \sqcup v_1$, the Fock space $\mathcal{F} = \Gamma(\mathcal{K}_0 \oplus \mathcal{K}_1)$ is identical to the tensor product $\mathcal{F}_0 \otimes \mathcal{F}_1$ of $\mathcal{F}_i = \Gamma(\mathcal{K}_i)$.

Let us assume that $\dim \mathcal{E}(v) \geqslant 1$ almost everywhere on V. Then the Hilbert space of complex-valued stochastic integrals

$$l_t(\varphi) = \int_0^t \left(\sum_{i=1}^d \varphi(s,i)\, dw_{si} + \int_{\varkappa \neq 0} \varphi(s,\varkappa)\, dm_{s\varkappa} \right) = w_t(\varphi) + m_t(\varphi) \tag{2.2}$$

identified with $\mathcal{L}^2(V^t) = \{\varphi\colon V^t \to \mathbf{C} \mid \int_{V^t} |\varphi(u)|^2\, du < \infty\}$ due to

$$\int | l_t(\varphi) |^2 (\omega)\nu(d\omega) =$$

$$= \int_0^t \left(\sum_{i=1}^d |\varphi(s,i)|^2 + \int_{\varkappa \neq 0} |\varphi(s,\varkappa)|^2\, d\lambda_\varkappa \right) ds = \int_{V^t} |\varphi(u)|^2\, du$$

can be embedded into $\mathcal{K} = \int^\oplus \mathcal{E}(v)\, dv$ by the isometry

$$\mathbf{e}\colon \varphi \mapsto \int^\oplus \left(\bigoplus_{i=1}^d \varphi(s,i)\mathbf{e}_{si} \oplus \int^\oplus_{\varkappa \neq 0} \varphi(s,\varkappa)\mathbf{e}_{s\varkappa}\, d\lambda_\varkappa \right) ds = \int^\oplus \varphi(v)\mathbf{e}_v\, dv$$

defined as $\mathbf{e}(\varphi)(v) = \varphi(v)\mathbf{e}_v$ by a measurable family $\{\mathbf{e}_v\}$ of normalized almost everywhere vectors $\mathbf{e}_v \in \mathcal{E}(v)$, $\| \mathbf{e}_v \| = 1$. It yields the embedding $\mathbf{e}^\otimes(\varphi)(v) = = \varphi(v)\mathbf{e}_v^\otimes$ into $\mathcal{F} = \int^\oplus \mathcal{E}^\otimes(v)\, dv$,

$$\mathbf{e}^\otimes\colon \varphi \mapsto \int^\oplus \int^\oplus \varphi(v_0,v_1)\mathbf{e}^\otimes(v_0) \otimes \mathbf{e}^\otimes(v_1)\, dv_0\, dv_1 = \int^\oplus \varphi(v)\mathbf{e}^\otimes(v)\, dv,$$

of the Hilbert space $\mathcal{L}^2(\Omega, \mathcal{A}^t, \nu)$ of repeated stochastic integrals

$$Z_t = \int_{\Upsilon_0^t} \int_{\Upsilon_1^t} \varphi(v_0,v_1)\, dw_{v_0}\, dm_{v_1}, \quad dw_v = \prod_{u \in v} dw_u, \quad dm_v = \prod_{u \in v} dm_u, \tag{2.3}$$

identified with $\mathcal{L}^2(\Upsilon^t) = \{\varphi\colon \Upsilon^t \to \mathbf{C} \mid \int_{\Upsilon^t} |\varphi(v)|^2\, dv < \infty\}$ due to

$$\int | z_t(\omega) |^2\, \nu(d\omega) = \int_{\Upsilon_0^t} \int_{\Upsilon_1^t} |\varphi(v_0,v_1)|^2\, dv_0\, dv_1 = \int_{\Upsilon^t} |\varphi(v)|^2\, dv.$$

THEOREM 2. *The family $Y_t = [Y_i(t)]$ of the operators*

$$Y_i(t) = W_i(t) + \int \varkappa_i\, dM_\varkappa(t), \quad i = 1, \dots, d, \quad t \in \mathbf{R}_+, \tag{2.4}$$

defined by the quantum stochastic processes

$$W_i(t) = \Lambda_0^+(t, E_i) + \Lambda_-^0(t, E_i^*) \equiv 2\mathrm{Re}\Lambda_0^+(t, E_i),$$

$$M_\varkappa(t) = \Lambda_0^0(t, I_\varkappa) + 2\mathrm{Re}\Lambda_0^+(t, J_\varkappa), \quad \varkappa \in \Delta_1 \tag{2.5}$$

with $E_i = I \otimes \mathbf{e}(\delta_i)$, $J_\varkappa = I \otimes \mathbf{e}(\chi_\varkappa)$, $I_\varkappa = I \otimes i(\chi_\varkappa)$ *defined by* $e(\varphi) = \int^\oplus \varphi(v) e_v dv$, $e^*(\varphi) = \int^\oplus \varphi(v) e_v^* dv$, $i(\varphi) = \int^\oplus \varphi(v) I(v) dv$ *for*

$$\delta_i(t, x) = \begin{cases} 1, & x = i \in \Delta_0 \\ 0, & x \neq i, \end{cases}$$

$$\mathcal{X}_\varkappa(t, x) = \begin{cases} 1, & x \in \Delta_1^\varkappa = \{x \in \Delta_1 \mid x < \varkappa\} \\ 0, & x \notin \Delta_1^\varkappa \end{cases}$$

gives a normal faithful QS representation

$$\pi(z_t) = \int_{\Upsilon^t}\int_{\Upsilon^t} \varphi(v_0, v_1)\, dW_{v_0}\, dM_{v_1}$$

of the algebra $\mathcal{C}^t = \mathcal{L}^\infty(\Omega, \mathcal{A}^t, \nu)$ *of bounded integrals (2.3) defined as in (1.4) by the operators* $\widehat{z}_t = \Lambda_t^\odot(\widehat{c})$ *in* $\mathcal{F}$ *where* $\widehat{c} = \mathbf{j}^\otimes(\varphi)$ *is*

$$\widehat{c}(v) = \varphi(v_0, v_1)\mathbf{j}_v^\otimes, \quad v_0 = V_0 \cap (v_0^- \cup v_+^0), \quad v_1 = V_1 \cap (v_0^- \cup v_+^0 \cup v_0^0),$$

$$\mathbf{j}_v^\otimes = \bigotimes_{\mathbf{v} \in v} j(\mathbf{v}), \quad j_0^0(v) = \begin{cases} 0, & v \in V_0, \\ I(v), & v \in V_1 \end{cases},$$

$$j_+^0(v) = \mathbf{e}_v, \quad j_0^-(v) = \mathbf{e}_v^*, \quad j_+^-(v) = 0. \tag{2.6}$$

The probability measure μ_t *on* $(\Omega, \mathcal{A}^t)$ *induced by*

$$\int |\, z_t(\omega)\,|^2\; \mu(d\omega) = \|\; \pi(z_t)\mathbf{a}\; \|^2$$

for any probability amplitude $\mathbf{a} \in \mathcal{G}$, $\|\,\mathbf{a}\,\| = 1$, *is absolutely continuous with respect to the measure* ν, *corresponding to the separating vector–amplitude* $\mathbf{a}_\varnothing = \mathbf{h} \otimes 1_\varnothing$, $\mathbf{h} \in \mathcal{H}$, *where* $\mathbf{h} \in \mathcal{H}$ *is a probability vector* $\|\,\mathbf{h}\,\| = 1$, *and* $1_\varnothing$ *is the vacuum amplitude* $1_\varnothing(v) = 0$ *if* $v \neq \varnothing$, $1_\varnothing(\varnothing) = 1$.

Proof. Firstly, we show the commutativity of the process (1.6) using the canonical commutation relations for the basic processes (1.2) in the form (Belavkin, 1988a)

$$[\Lambda_r(\mathbf{D}_i),\; \Lambda_t(\mathbf{D}_k)] = \Lambda_{r\wedge t}([\mathbf{D}_i,\; \mathbf{D}_k]), \tag{2.7}$$

where

$$[L, J] = \begin{pmatrix} L_0^- J_0^0 - J_0^- L_0^0, & L_0^- J_+^0 - J_0^- L_+^0 \\ L_0^0 J_0^0 - J_0^0 L_0^0, & L_0^0 J_+^0 - J_0^0 L_+^0 \end{pmatrix}$$

Representing (2.6) as $Y_i(t) = \Lambda_t(\mathbf{D}_i)$, where $\mathbf{D}_i = \mathbf{J}_i + \mathbf{K}_i$,

$$L_i = \begin{pmatrix} E_i^* & 0 \\ 0 & E_i \end{pmatrix} = I \otimes \mathbf{l}_i, \quad K_i = \int \varkappa_i \, dJ_\varkappa, \quad J_\varkappa = \begin{pmatrix} J_\varkappa^* & 0 \\ I_\varkappa & J_\varkappa \end{pmatrix} = I \otimes \mathbf{j}_\varkappa$$

one obtains $[Y_i(r), Y_k(t)] = \Lambda_{r \wedge t}(0) = 0$ because $[\mathbf{D}_i, \mathbf{D}_k] = 0$ due to the relations $[L_i, L_k] = 0$, $[L_i, J_\varkappa] = 0$, $[J_i, J_\varkappa] = 0$ of the commutativity of the QS processes

$$W_i(t) = \Lambda_t(L_i) = I \otimes \widehat{w}_i(t), \quad M_\varkappa(t) = \Lambda_t(J_\varkappa) = I \otimes \widehat{m}_\varkappa(t).$$

Hence the von Neumann algebra $\mathcal{C} \subset \mathcal{B}(\mathcal{G})$ generated by the family $Y = \{Y_i(t)\}$ is abelian and has the form $\mathcal{C} = I \otimes \mathfrak{e}$, where $\mathfrak{e} \in \mathcal{B}(\mathcal{F})$ is the Abelian algebra generated on $\mathcal{F}$ by the processes

$$\widehat{w}_i(t) = \Lambda_t(\widehat{\mathbf{l}}_i), \quad \widehat{m}_\varkappa(t) = \Lambda_t(\widehat{\mathbf{j}}_\varkappa), \mathbf{l}_i = \begin{pmatrix} \mathbf{e}_i^* & 0 \\ 0 & \mathbf{e}_i \end{pmatrix}, \quad \mathbf{j}_\varkappa = \begin{pmatrix} \mathbf{i}_\varkappa & 0 \\ \widehat{\mathbf{1}}_\varkappa & \mathbf{i}_\varkappa \end{pmatrix}.$$

This algebra coincides with the algebra of bounded operators $Z_t = \Lambda_t^{\odot}(I \otimes \mathbf{j}^{\otimes}(\varphi))$ denoted as

$$Z_t = \int_{\Upsilon^t} \int_{\Upsilon^t} \varphi(v_0, v_1) \, dW(v_0) \, dM(v_1) = \Lambda_t^{\odot}(B), \quad t \in \mathbf{R}_+, \tag{2.8}$$

representing the repeated integrals (2.4) in the form (1.4) with

$$B(v) = I \otimes \varphi(\sigma_0^- \sqcup \sigma_+^0, \rho_0^- \sqcup \rho_+^0 \sqcup v_0^0) \widehat{1}(v_0^0) \otimes \mathbf{e}_{v_+^0}^{\otimes} \otimes \mathbf{e}_{v_0^-}^{\otimes *} = I \otimes \widehat{c}(v),$$

where $\sigma_0^- = v_0^- \cap V_0$, $\sigma_+^0 = v_+^0 \cap V_0$, $\rho_0^- = v_0^- \cap V_1$, $\rho_+^0 = v_+^0 \cap V_1$ and $v_+^- = \emptyset$, $v_0^0 \cap V_0 = \emptyset$, otherwise $v_+^- \neq \emptyset$ or $v_0^0 \cap V_0 \neq \emptyset \Rightarrow B(v) = 0$. Indeed, the bounded integral (2.3) with

$$\varphi(v_0, v_1) = \varphi(v_0)\varphi(v_1), \quad \varphi(v) = \prod_{v \in v} \varphi(v), \quad \mathrm{Re}\varphi = 0, \tag{2.9}$$

defining the Doleans–Dade (factorial) exponent in (2.2)

$$g_t(\varphi) =: \exp l_t(\varphi) :=: \exp\{w_t(\varphi) + m_t(\varphi)\} :=$$

$$= \exp\{w_t(\varphi) + m_t(\ln(1+\varphi)) - \frac{1}{2} \int_{V^t} \varphi(u)^2 \, du + \int_{V^t} (\ln(1 + \varphi(u) - \varphi(u)) \, du\}$$

as the solution $z_t = g_t(\varphi)$, $z_0 = 1$, of the stochastic equation $dz_t = dl_t(\varphi) z_t$ satisfies the product formula

$$z_t^* z_t =: \exp\{w(t, 2\mathrm{Re}\varphi) + m(t, |\varphi|^2 + 2\mathrm{Re}\varphi) \int |\varphi(u)|^2 \, du\} : .$$

On the other hand, the Wick (normal) chronological exponent

$$: \overleftarrow{\exp} \Lambda_t(C) := \Lambda_t^{\odot}(\mathbf{C}^{\overleftarrow{\odot}}), \quad \mathbf{C}^{\overleftarrow{\odot}}(v) = \overleftarrow{\bigodot_{\mathbf{v} \in v}} C(\mathbf{v}) = \bigodot_{\mathbf{v} \in v} C(\mathbf{v})$$

defined as the solution $Z_t = \Lambda_t^{\odot}(\mathbf{C}^{\odot})$ of the QS equation $dZ_t = d\Lambda_t(\mathbf{C})Z_t,\ Z_0 = \widehat{I}$, has for $\mathbf{C}(v) = I \otimes \mathbf{c}(v)$ the form : $\exp \Lambda_t(I \otimes \mathbf{c}) : = I \otimes \Gamma_t(\mathbf{c}),\ \ \Gamma_t(\mathbf{c}) = \Lambda_t^{\odot}(\mathbf{c}^{\otimes})$ due to $\Lambda_t^{\odot}(I \otimes \mathbf{c}) = I \otimes \Lambda_t^{\odot}(\mathbf{c})$ and

$$\mathbf{c}(v) = \mathbf{c}^{\otimes}(v),\quad \mathbf{c}^{\otimes}(v) = c_0^0(v_0^0) \otimes c_+^0(v_+^0) \otimes c_0^-(v_0^-),\quad c_\nu^\mu(v) = \bigotimes_{\mathrm{v}\in v} c_\nu^\mu(\mathrm{v}).$$

Let us find the product $\widehat{z}_t^* \widehat{z}_t,\ \ \widehat{z}_t = :\exp \Lambda_t(\mathbf{c}) := \widehat{g}_t(\varphi)$ for the operator $\Lambda_t(\mathbf{c}) = \widehat{w}_t(\varphi) + \widehat{m}_t(\varphi)$ corresponding to $\mathbf{c} = \mathbf{j}(\varphi)$, i.e., $\mathbf{c}(v) = \varphi(v)\mathbf{j}_v$ with $\mathbf{j}_v = (j_\nu^\mu(v))$, defined in (1.6). We use the formula

$$\Gamma_t(\mathbf{c})^*\Gamma_t(\mathbf{c}) = \Gamma_t(\mathbf{c}^* + \mathbf{c}^*\mathbf{c} + \mathbf{c}), \tag{2.10}$$

$$\mathbf{c} = \begin{pmatrix} c_0^- & c_+^- \\ c_0^0 & c_+^0 \end{pmatrix},\quad \mathbf{c}^* = \begin{pmatrix} c_+^{0*} & c_+^{-*} \\ c_0^{0*} & c_0^{-*} \end{pmatrix},\quad \mathbf{c}^*\mathbf{c} = \begin{pmatrix} c_+^{0*}c_+^0 & c_+^{0*}c_+^0 \\ c_0^{0*}c_0^0 & c_0^{0*}c_+^0 \end{pmatrix}$$

which follows directly from the definition (1.4) of $\widehat{z}_t = \Lambda_t^{\odot}(\mathbf{c})$ on $a(v) = \bigotimes_{\mathrm{v}\in v} \mathbf{a}(\mathrm{v}) = \mathbf{a}^{\otimes}(v)$ for $c = \mathbf{c}^{\otimes}$

$$Z_t a(v) = \mathbf{a}(v\backslash v^t) \otimes (c_+^0 + c_0^0\mathbf{a} + \mathbf{a})^{\otimes}(v^t) \exp\left\{ \int_{V^t} (c_0^-(u)\mathbf{a}(u) + c_+^-(u))\, du \right\}.$$

Taking into account that $\mathbf{c}^*\mathbf{c} = \mathbf{c}_0^*\mathbf{c}_0 + \mathbf{c}_1^*\mathbf{c}_1$, where $\mathbf{c}_i = \mathbf{j}_i(\varphi),\ \ \mathbf{j}_i(\varphi)(v) = \varphi_i(v)\mathbf{j}_v,\ \ \varphi_i \mid V_i = \varphi,\ \ \varphi_i(v) = 0$, if $v \notin V_i$, so that

$$\mathbf{c}_0 = \varphi_0\begin{pmatrix} \mathbf{e}^* & 0 \\ 0 & \mathbf{e} \end{pmatrix},\quad \mathbf{c}_0^*\mathbf{c}_0 = \mid \varphi_0 \mid^2 \begin{pmatrix} 0 & 1 \\ 0 & 0 \end{pmatrix},\quad \mathbf{c}_1 = \varphi_1\begin{pmatrix} \mathbf{e}^* & 0 \\ I & \mathbf{e} \end{pmatrix},\quad \mathbf{c}_1^*\mathbf{c}_1 = \mid \varphi_1 \mid^2 \begin{pmatrix} e^* & 1 \\ I & e \end{pmatrix},$$

we obtain

$$\mathbf{c}^* + \mathbf{c}^*\mathbf{c} + \mathbf{c} = 2\mathrm{Re}\varphi_0\begin{pmatrix} \mathbf{e}^* & 0 \\ 0 & \mathbf{e} \end{pmatrix} + (\mid \varphi_1 \mid^2 + 2\mathrm{Re}\varphi_1)\begin{pmatrix} \mathbf{e}^* & 0 \\ I & \mathbf{e} \end{pmatrix} + \mid \varphi \mid^2 \begin{pmatrix} 0 & 1 \\ 0 & 0 \end{pmatrix},$$

$$\widehat{z}_t^*\widehat{z}_t = :\exp\left\{ \widehat{w}_t(2\mathrm{Re}\varphi) + \widehat{m}_t(\mid \varphi_1 \mid^2 + 2\mathrm{Re}\varphi) + \int_{V_t} \mid \varphi(u) \mid^2\ du\widehat{1} \right\}:$$

for $\widehat{z}_t = :\exp\{\widehat{w}_t(\varphi) + \widehat{m}_t(\varphi)\}:$, where $\widehat{w}_t(\varphi) = \Lambda_t(j_0(\varphi)),\ \ \widehat{m}_t(\varphi) = \Lambda_t(j_1(\varphi))$. It proves that $\pi(z) = I \otimes \widehat{z}$ is a representation $\pi(z^*z) = \pi(z)^*\pi(z),\ \pi(1) = I \otimes \widehat{1}$, on the initial abelian subalgebra of linear combinations $z = \sum \lambda^i g_{t_i}(\varphi_i)$ the w^*–closure of which with bounded $g_t(\varphi)$ is the W^*–algebra $\mathfrak{e} = \mathcal{L}^{(\infty)}(\Omega, \mathcal{A}, \nu)$ of bounded stochastic integrals (2.3) $z \in \mathcal{L}^2(\Omega, \mathcal{A}, \nu), \| z \|^{(\infty)} < \infty$. Moreover due to w^*–continuity of π, it is a normal representation of $\mathfrak{e}$ onto the von Neumann algebra $\mathcal{C} \subset \mathcal{B}(\mathcal{G})$ generated by the Wick exponents $\pi(g(\varphi)) = I \otimes \Gamma(j(\varphi))$ with separating probability vector $\mathbf{a} = \mathbf{h} \otimes 1_\emptyset,\ \ \pi(z)\mathbf{a} = 0 \Rightarrow z = 0$ inducing the standard measure $\int \mid z(\omega) \mid^2 \nu(d\omega) = \| \pi(z)a \|^2 = \| \widehat{z}1_\emptyset \|^2$. It follows from the property

$$\pi(z)(\mathbf{h} \otimes 1_\emptyset)(v) = \int C\begin{pmatrix} \emptyset & v_+^- \\ \emptyset & v \end{pmatrix}\mathbf{h}\, dv_+^- = \mathbf{h} \otimes c_+^0(v) = \mathbf{h} \otimes \mathbf{e}^{\otimes}(\varphi)(v)$$

due to the definition (1.4) of $\Lambda_t^{\odot}(I \otimes c)$ for $c = \mathbf{j}^{\otimes}(\varphi)$ on $\mathbf{a}(\upsilon) = 0$ for $\upsilon \neq \emptyset$ and $\mathbf{a}(\upsilon) = \mathbf{h}$ for $\upsilon = \emptyset$, and isometricity of $\mathbf{e}^{\otimes}(\varphi) = \mathbf{j}^{\otimes}(\varphi)^0_+ \colon \| h \otimes \mathbf{e}^{\otimes}(\varphi) \| = = \int | \varphi(\upsilon) |^2 \, d\upsilon = \| \widehat{z} 1_{\emptyset} \|^2$.

Hence $\pi(z) = \int^{\oplus} \pi_\omega(z)\nu(d\omega)$, where $\pi_\omega(z) = I \otimes \widehat{z}_\omega$, $\widehat{z}_\omega = z(\omega)\widehat{1}_\omega$ for almost all $\omega \in \Omega$, where $\{\widehat{1}_\omega\}$ defines the orthogonal solution $\widehat{\mathfrak{c}} = \int^{\oplus} \mathbf{C}\widehat{1}_\omega \nu(d\omega)$ of the algebra of operators $\widehat{z}$ in $\mathcal{F}$. Due to the isomorphism of $\widehat{\mathfrak{c}}$ and C any normal state ρ on C as a positive normalized functional of the predual space $\mathfrak{c}_*$ corresponds to a positive normalized function $p \in \mathcal{L}^1(\Omega, \mathcal{A}, \nu) = \widehat{\mathfrak{c}}_*$, $p(\omega) = \rho \circ \pi_\omega$:

$$\rho(Z) = \int \omega(Z)p(\omega)\nu(d\omega), \quad \omega(Z) \equiv \pi_\omega^{-1}(Z).$$

It means the absolute continuity of the measure $\mu : z \mapsto \int z(\omega)\mu(d\omega) = \rho \circ \pi(z)$ with the probability density $p(\omega) = \mu(d\omega)/\nu(d\omega)$, which is $p(\omega) = \| \mathbf{a}_\omega \|^2$ in the case $\rho(Z^*Z) = \| Z\mathbf{a} \|^2$ for an arbitrary probability amplitude $\mathbf{a} \in \mathcal{G}$, $\| \mathbf{a} \| = 1$, in the representation $\mathbf{a} = \int^{\oplus} \mathbf{a}_\omega \nu(d\omega)$.

3. MARKOVIAN QS FILTERING AND POSTERIOR EQUATIONS

Let us consider an isometric QS evolution $\mathbf{a}(t) = U_t\mathbf{a}$ defined by equation (1.9) with the initial vector amplitude $\mathbf{a} = \mathbf{h} \otimes 1_{\emptyset}$, corresponding to a given probability vector $\mathbf{h} \in \mathcal{H}$ of a QS object, and the described representation $\pi = \int^{\oplus} \pi_\omega \nu(d\omega)$ of the classical process $\mathbf{y}_t$, defined by (2.1) with the standard measure ν. We are interested in the posterior evolution of the QS object under the continuous quantum filtering with respect to the measuring the process $\mathbf{y}_t$ as a nondemolition QS process $\widehat{\mathbf{Y}}_t = U_t^*\pi(\mathbf{y}_t)U_t$ commuting with the future of any QS process $\widehat{X}_t = = U_t^*XU_t$, defined by an operator $X \in \mathcal{B}$ in Heisenberg picture. It is described by the resolution $\mathbf{a}(t) = \int^{\oplus} \mathbf{a}_\omega(t)\nu(d\omega)$ giving the posterior state

$$\rho_t[X](\omega) = \langle \mathbf{a}_\omega(t) \mid X\mathbf{a}_\omega(t) \rangle, \quad \rho_t[I](\omega) = p_t(\omega) \tag{3.1}$$

normalized on the probability density $p_t(\omega) = \| \mathbf{a}_\omega(t) \|^2$ of the observed process up to a time instant $t \in \mathbf{R}_+$ with respect to ν.

Let us describe the stochastic linear equation for the posterior quantum state ρ_t as the quantum filtering equation (Belavkin, 1988a) generalizing the Zakai equation of the classical filtering theory (Stratonovich, 1966; Liptser and Shiryaev, 1974) for the process $\mathbf{y}$. We find it from the evolution equation for the generating functional

$$\mathcal{X}_t(\varphi)[X] = \langle U_t(h \otimes 1_{\emptyset})] \mid (X \otimes \widehat{g}_t(\varphi))U_t(h \otimes 1_{\emptyset}) \rangle$$

of the state on $\mathcal{B} \otimes \mathcal{C}^t$ at $t \in \mathbf{R}_+$ as the standard mean

$$\mathcal{X}_t(\varphi)[X] = \int \rho_t[X](\omega)g_t(\varphi)(\omega)\nu(d\omega)$$

of the product of the conditional expectation of X and the factorial exponent $z_t = g_t(\varphi)$ satisfying the Ito equation

$$z_t = 1 + \int_0^t \left(\sum_{i=1}^d \varphi(s,i)dw_{si} + \int_{\varkappa \neq 0} \varphi(s,\varkappa)dm_{s\varkappa} \right) z_s. \tag{3.2}$$

Let us denote by $X \in \mathcal{B}(\mathcal{H}) \to \Phi_t[X]$ the posterior map defining ρ_t for any $\mathbf{h} \in \mathcal{H}$ as $\rho_t[X] = \langle \mathbf{h} \mid \Phi_t[X]\mathbf{h}\rangle$ and $\Theta_t : X \mapsto \Theta_t[X] \in \mathcal{B}(\mathcal{H})$ the generating factorial map $\Theta_t(\varphi)[X]$ for a $\varphi \in \mathcal{L}^2(V^t)$ defining the corresponding $\mathcal{X}_t(\varphi)$ as $\mathcal{X}_t[X] = \langle \mathbf{h} \mid \Theta[X]\mathbf{h}\rangle$.

THEOREM 3. *Let $L(t)$, $H(t)$, R_{ti}, $F_{t\varkappa}$ be the functions*

$$H(t) = \frac{i}{2}\int (L_+^-(t,x) - L_+^-(t,x)^*)\,dx, \quad L(t) = \int^\oplus L_+^0(t,x)\,dx,$$

$$R_{ti} = e_{ti}^* L_+^0(t,i), \quad F_{t\varkappa} = L_+^0(t,\varkappa) + I \otimes e_{t\varkappa}, \tag{3.3}$$

with $\| H \|_t^{(1)} < \infty$, $\| L \|_t^{(2)} < \infty$, $\| R \|_t^{(2)} < \infty$, $\| F \|_t^{(\infty)} < \infty$. Then the generating map Θ is defined by the equation

$$\frac{d}{dt}\Theta_t[X] + \Theta_t\left[i[X,H(t)] + \frac{1}{2}\Big(L^*(t)[X,L(t)] + [L^*(t),X]L(t)\Big)\right] =$$

$$= \sum_{i=1}^{d} \Theta_t[R_{ti}^* X + XR_{ti}]\varphi(t,i) + \int_{\varkappa\neq 0} \Theta_t[F_{t\varkappa}^*(X\otimes I(t,\varkappa))F_{t\varkappa} - X]\varphi(t,\varkappa)\,d\lambda_\varkappa, \tag{3.4}$$

where $[X,H] = XH - HX$, $[L^*,H] = L^*(X\otimes I) - XL^* = [X,L]^*$ having for the initial $\Theta_0[X] = X \in \mathcal{B}$ the unique solution

$$\Theta_t(\varphi)[X] = \int \Phi_t[X](\omega)g_t(\varphi)(\omega)\nu(d\omega). \tag{3.5}$$

The posterior map Φ_t satisfies the Ito equation

$$\Phi_t[X] + \int_0^t \Phi_s\left[i[X,H(s)] + \frac{1}{2}\Big([L^*(s),X]L(s) + L^*(s)[X,L(s)]\Big)\right]ds =$$

$$= X + \int_0^t \left(\sum_{i=1}^{d} \Phi_s[R_{st}X + XR_{si}]\,dw_{si} + \int_{\varkappa\neq 0} \Phi_s[F_{s\varkappa}^*(X\otimes I(s,\varkappa))F_{s\varkappa} - X]\,dm_{s\varkappa}\right) \tag{3.6}$$

having the unique solution defined as the classical conditional expectation

$$\Phi_t[X](\omega) = \varepsilon[Z_t](\omega), \quad Z_t(\omega) = T_t^*(\omega)XT_t(\omega), \tag{3.7}$$

where $T_t(\omega) = \omega(\Lambda_t^{\odot}(\overleftarrow{\odot}\mathbf{D}))$ is a classical operator–valued stochastic integral representing on a standard space $(\Omega,\mathcal{A},\nu)$ a normal chronological exponent $\vdots\overleftarrow{\exp}\Lambda_t(\mathbf{D}) := \Lambda_t^{\odot}(\overleftarrow{\odot}\mathbf{D})$.

The stochastic propagator $\mathbf{T}$, satisfying the Ito equation

$$T_t + \int_0^t K(s)T_s\,ds = I + \int_0^t \left(\sum_{x\in\Delta_0} L_{sx}\,dw_{sx} + \int_{x\in\Delta_1} L_{sx}\,dm_{sx}\right)T_s \tag{3.8}$$

is defined in the case $\dim \mathcal{E}(t,x) = 1$ for almost all $x \in \Delta$ by $\Phi_t[X] = T_t^* X T_t$

$$K(t) = iH(t) + \int\limits_{x\in\Delta} \left(\frac{1}{2}L_{tx}^* L_{tx} + i\mathrm{Im}\ l_{tx}^* L_{tx}\right) dx + k(t),$$

$$L_{tx} = R_{tx} + l_{tx}, \quad x \in \Delta_0, \quad L_{tx} = s_{tx} F_{tx} - I, \quad x \in \Delta_1, \tag{3.9}$$

uniquely up to a $\mathcal{B}'$-valued functions l_{tx}, $\| l \|_{\Delta_0 t}^{(2)} < \infty$, $s_{tx} = I + l_{tx}$, $x \in \Delta_1$, and $k(t)$, $\| k \|_t^{(1)} < \infty$, satisfying the conditions

$$l_{tx}^* + l_{tx} = 0, \quad s_{tx}^* s_{tx} = I, \quad k^*(t) + k(t) = 0. \tag{3.10}$$

Proof. Let us find the QS equation for the process $\widehat{G}_t = U_t^*(X \otimes \widehat{g}_t)U_t$, where $X \in \mathcal{B}$ and $\widehat{g}_t(\varphi) = \Gamma_t(j(\varphi))$ is the normal exponent, representing in $\mathcal{F}$ the factorial exponent $z_t = g_t(\varphi)$. Due to (3.2) the operator $G_t = X \otimes \widehat{g}_t$ satisfies the QS equation

$$G_t - X \otimes \widehat{1} = \int_0^t d\Lambda_s(\mathbf{J}(\varphi))G_s = \int_0^t G_s\, d\Lambda_s(\mathbf{J}(\varphi)),$$

where $\mathbf{J}(\varphi) = I \otimes \mathbf{j}(\varphi)$ and $\mathbf{j}(\varphi)(v) = \varphi(v)[j_\nu^\mu(v)]$ is defined by (2.6). Applying the QS Ito formula (Hudson and Parthasarathy, 1984) in the form (Belavkin, 1988a,b)

$$d(\Lambda_t(\mathbf{D} - Z \otimes \mathbf{1})^* \Lambda_t(D - Z \otimes \mathbf{1})) = d\Lambda_t(D^* D - Z^* Z \otimes \mathbf{1})$$

to the product $\widehat{G}_t = U_t^* G_t U_t$, we have the QS equation

$$\widehat{G}_t - X \otimes \widehat{1} = \int_0^t U_s^*\, d\Lambda_s(\mathbf{S}^*(G \otimes \mathbf{f})\mathbf{S} - G \otimes \mathbf{1})U_s,$$

where $\mathbf{f} = \mathbf{1} + \mathbf{j}(\varphi)$. Using the property $\langle \mathbf{h} \otimes 1_\emptyset \mid \Lambda_\nu^\mu(t, D)\mathbf{h} \otimes 1_\emptyset \rangle = 0$, $(\mu,\nu) \neq (-,+)$, of the QS integrals (1.2), and the explicit form of the Lebesgue integral $\Lambda_-^+(t)$ we obtain the equation

$$\langle \mathbf{a}(t) \mid G_t \mathbf{a}(t)\rangle - \langle \mathbf{h} \mid X\mathbf{h}\rangle = \int_0^t \int \langle \mathbf{a}_s \mid (S^*(G \otimes f)S)_+^-(s,x)\mathbf{a}_s\rangle\, ds\, dx,$$

where $\mathbf{a}_s = U_s(\mathbf{h} \otimes 1_\emptyset)$, and

$$(\mathbf{S}^*(G \otimes \mathbf{f})\mathbf{S})_+^- = \sum_{\mu \leqslant \nu} L_\mu^{*-}(G \otimes f_\nu^\mu)L_\mu^\nu =$$

$$= \Big(L_+^{0*}(X \otimes j_0^0(\varphi))L_+^0 + L_+^{0*}(X \otimes e(\varphi)) +$$

$$+ (X \otimes e^*(\varphi))L_+^0 + L_+^{-*}X + XL_+^-\Big) \otimes \widehat{g}(\varphi).$$

It gives equation (3.4) for the generating map $\Theta_t : \mathcal{B} \mapsto \mathcal{B}(\mathcal{H})$ defined for a fixed $\varphi \in \mathcal{L}^2(V^t)$, $V^t = [0,t) \times \Delta$ with $\mathrm{Re}\,\varphi(x) = 0$ for $x \in \Delta_0$ and $\mathrm{Re}\ln(1+\varphi(x)) = 0$ for $x \in \Delta_1$ as a bounded operator by

$$\langle \mathbf{h} \mid \Theta_t(\varphi)[X]\mathbf{h}\rangle = \langle \mathbf{h}\otimes 1_\emptyset \mid U_t^*(X \otimes \hat{g}_t(\varphi))U_t(\mathbf{h}\otimes 1_\emptyset)\rangle, \quad \forall \mathbf{h} \in \mathcal{H}.$$

Now we prove that the solution of (3.4) can be represented as the expectation (3.5) of a non–anticipating stochastic operator–valued function on the probability space of the observed process y. Let $\mathcal{E}_0(t) = \bigoplus_{x\in X_0} \mathbb{C}e_{tx}$ be a complete orthogonal decomposition of the Hilbert space $\mathcal{E}_0(t) = \bigoplus_{i=1}^{d} \mathcal{E}(t,i)$ containing the family $\{e_{ti}\}$ of unit vectors $e_{ti} \in \mathcal{E}(t,i)$ indexed by a measurable subset $\{1,\dots,d\} \subseteq \Delta_0$ and $\mathcal{E}_0(t) = \int_{\Delta_1}^{\oplus} \mathbb{C}e_{tx}\,d\lambda_x$ be a complete disintegration of $\mathcal{E}_1(t) = \int^{\oplus} \mathcal{E}(t,\varkappa)\,d\lambda_\varkappa$ consistent in the sense $\int^{\oplus} \varphi(\varkappa)e_\varkappa\,d\lambda_\varkappa = \int^{\oplus} \varphi(\varkappa(x))\,d\lambda_x$ with the family $\{e_{t\varkappa}\}$ of the unit vectors $e_{t\varkappa} \in \mathcal{E}(t,\varkappa)$ indexed by a measurable factor–set $\Delta_1 \preccurlyeq X_1$ as a partition $X_1 = \bigsqcup_{\varkappa\in\Delta_1} X(\varkappa)$ of a space X_1 with the measure λ_x inducing $\lambda_\varkappa$ by $\varkappa(x) = \varkappa \Leftrightarrow x \in X(\varkappa)$ as $\int \varphi(\varkappa)\,d\lambda_\varkappa = \int \varphi(\varkappa(x))\,d\lambda_x$. Constructing $(\Omega, \mathcal{A}, \nu)$ as the product of the standard probability space $(\Omega_0, \mathcal{A}_0, \nu_0)$ for the Wiener vector process $\mathbf{w}_t = \{w_x(t) \mid x \in X_0\}$ and the space $(\Omega_1, \mathcal{A}_1, \nu_1)$ for the Poisson measure dn_{tx} with intensity $dt\,d\lambda_x$, one can represent Θ_t as

$$\Theta_t(\varphi)[X] = \int\!\!\int g_t(\varphi)(\omega_0,\omega_1)T_t(\omega_0,\omega_1)^* X T_t(\omega_0,\omega_1)\nu_0(d\omega_0)\nu_1(d\omega_1),$$

where

$$g_t(\varphi)(\omega) =: \exp\int_0^t \left(\sum_{x=1}^{d} \varphi(s,x)\,dw_{sx} + \int \varphi(s,\varkappa(x))\,dm_{sx}\right) ds :$$

$T_t(\omega)$, $\omega \in \Omega_0 \times \Omega_1$, is the solution of (3.8). Indeed, using the classical multiplication table for the stochastic differentials dw, dm having the only non–zero products

$$(dw_{xt})^2 = dt, \quad (dm_{xt})^2 = dm_{xt} + dt\,d\lambda_x,$$

we obtain from (3.2) and (3.8) by the Ito formula for $Z = zT^*XT$

$$Z_t - X = \int_0^t \left(\sum A_{sx}\,dw_{sx} + \int B_{sx}\,dm_{sx} - C(s)\,ds\right) +$$

$$+ \int_0^t \left(\sum_{x=1}^{d} \varphi(s,x)(Z_s\,dw_{sx} + A_{sx}\,ds) + \int_{x\in X_1} \varphi(s,\varkappa(x))(Z_s\,dm_{sx} + B_{sx}\,ds\,d\lambda_x)\right)$$

where

$$A_{tx} = z_t T_t^*(L_{tx}^* X + X L_{tx})T_t,$$

$$B_{tx} = z_t T_t^*(L_{tx}^* X + X L_{tx} + L_{tx}^* X L_{tx}) T_t,$$

$$C(t) = z_t T_t^* \Big(K(t)^* X + X K(t) + \sum L_{tx}^* X L_{tx} + \int L_{tx}^* X L_{tx}\, d\lambda_x \Big) T_t.$$

The expectation of this equation gives

$$\Theta_t[X] - X = \int_0^t \Big(\sum_{x=1}^{d} \varphi(s,x) A_{sx} + \int \varphi(s, \varkappa(x)) B_{sx}\, d\lambda_x) - C(s) \Big)\, ds$$

which is equivalent to the differential equation (3.4) iff

$$\sum_{x=1}^{d} (L_{tx}^* X + X L_{tx}) \varphi(t,x) = \sum_{i=1}^{d} (R_{ti}^* X + X R_{ti}) \varphi(ti), \quad \forall \varphi \in \mathcal{L}^2(V^t),$$

$$\int \Big(L_{tx}^* X + X L_{tx} + L_{tx}^* X L_{tx} \Big) \varphi(t, k(x))\, d\lambda_x =$$

$$= \int \Big(F_{t\varkappa}^* (X \otimes I(t,\varkappa)) F_{t\varkappa} - X \Big) \varphi(t,\varkappa)\, d\lambda_\varkappa.$$

$$K^* X + X K + \int L_x^* X L_x\, dx = i[X,H] + \frac{1}{2} \Big(L^*[X,L] + [L^*,X]L \Big).$$

This system of operator equations has solution (3.9) at (3.10) where $R_{tx} = e_{tx}^* L_+^0$, $F_{tx} = R_{tx} + I$ which is unique up to $l_{tx} \in \mathcal{B}'$, $s_{tx} \in \mathcal{B}'$, $k(t) \in \mathcal{B}'$ in the case $\Delta_0 = \{1, \ldots, d\}$, $\Delta_1 = \mathbf{R}^d \backslash \{0\}$, corresponding to $\dim \mathcal{E}(t,i) = 1$, $\dim \mathcal{E}(t,\varkappa) = 1$. According to Theorem 1, the solution of the equation (3.8) exists, is unique and can be represented in the space $\mathcal{G} = \mathcal{H} \otimes \mathcal{F}$ by the normal chronological exponent

$$: \overleftarrow{\exp}\{\Lambda_t(\mathbf{D})\} : = \Lambda_t^{\odot}(\mathbf{D}^{\overleftarrow{\odot}}), \quad [\mathbf{D}^{\overleftarrow{\odot}}](v) = \overleftarrow{\bigodot_{\mathbf{v} \in v}} D(\mathbf{v})$$

if the table $\mathbf{D} = (D_\nu^\mu)$ of $\mathcal{B}(\mathcal{H})$-valued functions

$$D_0^0(t,x) = 1_{X_1}(x) L_{tx}, D_+^0(t,x) + L_{tx}, D_0^-(t,x) = L_{tx}^*, D_+^-(t,x) = c(x) K(t),$$

where $1_{X_1}(x) = 1$, if $x \in X_1$, $1_{X_1}(x) = 0$, if $x \in X_0$, $\int_X c(x)\, d\lambda_x = -1$, satisfies the conditions (1.3). These conditions are satisfied due to $\| L_+^0 \|_t^{(2)} < \infty$ and $\| L_+^- \|^{(1)} < \infty$ iff

$$\| F \|_{t\Delta_1}^{(\infty)} = \operatorname{ess\,sup}\{ \| F_{sx} \| \mid s < t, x \in \Delta_1 \} < \infty,$$

$$\| l \|_{t\Delta_0}^{(2)} = \Big(\int_0^t \sum_{x=1}^{d} \| l_{sx} \|^2\, ds \Big)^{1/2} < \infty, \quad \| k \|_t^{(1)} = \int_0^t \| k(s) \|\, ds < \infty.$$

In order to obtain the posterior map Φ_t uniquely defined by the repesentation (3.5) of the map $\Theta_t(\varphi)$ as an expectation with respect to the observed process due to arbitrary φ one should take the conditional expectation (3.7) with respect to the σ-algebra $\mathcal{A}$ generated by the process y up to $t = \infty$, coinciding with $\varepsilon[Z_t \mid \mathcal{A}^t]$. The algebra $\mathcal{A}^t$ coincides with the algebra generated by factorial exponents $z_t(\omega) = g_t(\varphi)(\omega)$ with arbitrary $\varphi \in \mathcal{L}^2(V^t)$ due to the uniqueness of the decomposition (2.1) of the observed process into the continuous part $w(t)$ and the purely discontinuous part uniquely defining the Poisson process on the measurable space $\mathbf{R}^d \backslash \{0\}$ with non-zero jumps $\varkappa \neq 0$ of intensity $d\lambda_\varkappa \neq 0$.

REFERENCES

Barchielli, A. (1988). Input and output channels in quantum systems and quantum stochastic differential equations. In: *Quantum Probability and Applications III, Proc. Oberwolf. 1987,* ed. L.Accardi, W. von Waldenfeld. Springer, Berlin–Heidelberg etc, pp. 3751.

Belavkin, V.P. (1980). Quantum filtering of Markovian signals with quantum white noises. *Radiotechnika i Electronika* **25**, 1445–1453.

Belavkin, V.P. (1983). Theory of control of observable quantum systems. *Automatica and Remote Control* **44**, 178–188.

Belavkin, V.P. (1985). Reconstruction theorem for a quantum stochastic process. *Theor. Math. Phys.* **62**, 275–289.

Belavkin, V.P. (1987). Nondemolition measurement and control in quantum dynamical systems. In: *Information Complexity and Control in Quantum Physics. Proc. of CISM, Udine 1985,* ed. A.Blaquiere, S.Diner, G.Lochak. Springer, Wien–NewYork, pp. 311–336.

Belavkin, V.P. (1988a). Nondemolition measurements, nonlinear filtering and dynamic programming of quantum stochastic processes. *Lect. Notes Control Inform. Sci.* **121**, 245–265.

Belavkin, V.P. (1988b). Nondemolition stochastic calculus in Fock space and nonlinear filtering and control in quantum systems. In: *Stochastic Methods in Mathematics and Physics. Proc. XXIV Karpacz Winter School, 1988,* ed. R.Guelerak and W.Karwowski. World Scientific, Singapore, pp. 310–324.

Belavkin, V.P. (1989). Quantum stochastic calculus and quantum nonlinear filtering. *Preprint* N6, Centro Matematico V. Volterra, Universita degli studi di Roma.

Gardiner, C.W. and Collet, M.J. (1985). Input and output in damped quantum systems: Quantum statistical differential equations and the Master equation. *Phys. Rev.* **A31**, 3761–3774.

Holevo, A.S. (1989). Time–ordered exponentials in quantum stochastic calculus. *Preprint* N517, Universität Heidelberg, Heidelberg.

Hudson, R.L. and Parthasarathy, K.R. (1984). Quantum Ito's formula and stochastic evolution. *Comm. Math. Phys.* **93**, 301–323.

Lindsay, M. and Maassen, H. (1988). An integral kernel approach to noise. In: *Quantum Probability and Applications III, Proc. Oberwolf, 1987,* ed. L.Accardi, W. von Waldenfeld. Springer, Berlin–Heidelberg, pp. 192–208.

Liptser, P.Sh. and Shiryaev, A.N. (1974). *Statistics of Random Processes.* Nauka, Moscow.

Stratonovich, R.L. (1966). *Conditional Markovian Processes and Their Applications to the Theory of Optimal Control.* Moscow State Univ., Moscow.

Prob. Theory and Math. Stat., Vol. 1, pp. 110–117
B.Grigelionis et al. (Eds.)
1990 VSP/Mokslas

DIFFUSIONS AND PARABOLIC EQUATIONS IN PRINCIPAL BUNDLES

Ya.I.BELOPOLSKAYA
Leningrad Civil Engineering Institute, 2–Krasnoarmeiskaya 4,
Leningrad 198005, USSR

ABSTRACT

The investigation of smooth manifold valued stochastic processes comes to be interesting both due to intrinsic development of stochastic process theory and to problems arising in various fields of mathematics and physics such as partial differential equations, harmonic analysis, gauge theory and so on. A number of articles and some books devoted to this subject had been published up to now (Ikeda and Watanabe, 1981; Elworthy, 1982; Schwartz, 1984; Belopolskaya and Dalecky, 1989).
In the present paper we develop the approach of (Belopolskaya and Dalecky, 1989) to deal with stochastic processes valued in a principal bundle total space as well as their functionals. The investigation of these functionals shows the way to construct solutions of the Cauchy problem in sections of principal bundles.

1. Introduce briefly some differential geometry notions which will be crucial for our investigations. Their detailed description may be found in (Belopolskaya and Dalecky, 1989; Kobayashi and Nomizu, 1963; Eliassion, 1967).
Let X be a smooth manifold modelled on a real separable Hilbert spase B, TX, T^2X be correspondingly its first and second tangent bundles. The connection map K^x: $T^2X \to TX$ is defined locally by the relation

$$K^X(x,y,z,u) = \big(x, u + \Gamma^X_x(z,y)\big) \tag{1}$$

with Γ^X_x: $B \times B \to B$ being bilinear mapping which is called a *local connection coefficient*. The transformation rule for Γ^X_x under changing the chart is the following

$$\Gamma^{\psi}_{x^\psi}(y^\psi, z^\psi) = (\psi\circ\varphi^{-1})'\Gamma^{\varphi}_{x^\varphi}(y^\varphi, z^\varphi) - (\psi\circ\varphi^{-1})''(y^\varphi, z^\varphi), \tag{2}$$

$$x^\psi = (\psi\circ\varphi^{-1})x^\varphi,\ y^\psi = (\psi\circ\varphi^{-1})'y^\varphi.$$

Here (U,φ), (V,ψ) are charts of the atlas of X and $x \in U\cap V$.
Consider a principal bundle p: $\mathcal{P} \to X$ on X with structural group G acting on $\mathcal{P}$ from the right. We assume that G is a Lie group with Hilbert spase $\mathcal{G}$ standing for its Lie algebra. Denote $Ad(p)$: $\mathcal{G}^X \to X$ a vector bundle on X associated with p by adjoint representation Ad of G in $\mathcal{G}$. It means that $Ad(p)$ is a vector bundle

with a typical fibre $\mathcal{G}$ and a structural group $Ad(G)$. Let φ be a local section of p, φ: $U \to \mathcal{P}$. Consider the inverse image of the vertical tangent bundle $VT\mathcal{P}$ on $\mathcal{P}$ over φ

$$\begin{array}{ccc} \varphi^*(VT\mathcal{P}) & \longrightarrow & VT\mathcal{P} \\ \big\downarrow \varphi^*(\tau_p) & & \big\downarrow \tau_p \\ X & \xrightarrow{\varphi} & \mathcal{P} \end{array}$$

and notice that there is a vector bundle isomorphism between $\varphi^*(VT\mathcal{P})$ and $\mathcal{G}^X$. To construct a connection on $\mathcal{P}$ let us start with the definition of the connection map K^p: $T\mathcal{P} \to \mathcal{G}^X$ of the principal fibre bundle p. Locally K^p is given by the relation

$$K^p(x, g, y, A^0) = (x, Ad(g)\{A + \Gamma^p_x y\})$$

where

$$A^0(z) = \frac{d}{dt}(z \exp tA)|_{t=0} \quad (z \in \mathcal{P})$$

is a fundamental vector field on $\mathcal{P}$, generated by $A \in \mathcal{G}$, while Γ^p_x: $B \to \mathcal{G}$ is a $\mathcal{G}$–valued 1–form on X. In the sequel we shall need as well a connection map $K^{Ad(p)}$: $T\mathcal{G}^X \to \mathcal{G}^X$ which may be given locally in the form

$$K^{Ad(p)}(x, A, y, B,) = (x, B + [\Gamma^p_x y, A]).$$

Here $[\cdot,\cdot]$ stands for commutator in $\mathcal{G}$.
Given connection maps in τ_X: $TX \to X$, p and $Ad(p)$ one may define covariant derivatives ∇, ∇^p, $\nabla^{Ad(p)}$ for correspondent bundle sections. Namely,

$$\nabla y = K^X \circ Ty\colon\ TX \to TX, \quad \nabla^p \varphi = K^p \circ T\varphi\colon\ TX \to \mathcal{G}^X,$$

$$\nabla^{Ad(p)} \psi = K^{Ad(p)} \circ T\psi\colon\ TX \to \mathcal{G}^X$$

and

$$\nabla_z y(x) = y'(x)z + \Gamma^X_x(z, y(x)), \quad \nabla^p_z \varphi(x) = \varphi'(x)z + \varphi(x)\Gamma^p_x z,$$

$$\nabla^{Ad(p)}_z \Psi(x) = \Psi'(x)z + [\Gamma^p_x z, \Psi(x)]. \tag{3}$$

The above relations make it possible to define second order differential operators of the form

$$\nabla^{Ad(p)}_z \nabla^p_y \varphi\ , \ \nabla^{Ad(p)}_z \nabla^p_y \varphi - \nabla^p_{\nabla_z y}\varphi$$

for the section φ of p.
Given the connection maps K^X and K^p one may define the connection map $K^{\mathcal{P}}$: $T^2\mathcal{P} \to T\mathcal{P}$ on P which should respect its structure of principal fibre bundle total space.
In local coordinates it looks like

$$K^{\mathcal{P}}(x, g, y, A^0, z, B^0, u, C^0) = (x, g, u + \Gamma^X_x(z, y), \quad Ad(g)\{(\Gamma^p_x)'(z, y)+$$

$$+[\Gamma^p_x z, \Gamma^p_x y] - \Gamma^p_x \Gamma^X_x(z, y) + [\Gamma^p_x y, B] + [\Gamma^p_x z, A] + C\}). \tag{4}$$

Denote $\exp^{\mathcal{P}}$ the exponential map $\exp^{\mathcal{P}}: T\mathcal{P} \to \mathcal{P}$ corresponding to the above $K^{\mathcal{P}}$ and exp the group Lie exponential. We assume below that the connection map K^G on the Lie group G is left invariant and its local coefficient $\Gamma^G_g(\cdot,\cdot)$ at the unit element e is given by

$$\Gamma^G_e(A, A) = -A^2, \qquad A \in \mathcal{G}.$$

2. Let $(\Omega, \mathcal{F}, P)$ be a probability spase, $H_+ \subset H \subset H_-$ be a rigged Hilbert space, $w(t)$ be a canonical Wiener process in H_-. Denote $\varkappa: \Theta = X \times H \to X$ a trivial vector bundle on X and $L_{12}(\varkappa, \tau_X)$: $L_{12}(\Theta, TX) \to X$ a bundle of Hilbert–Schmidt operators on X. Given vector field $a_x(t)$ and operator field $A_x(t) \in L_{12}(H, T_xX)$ depending on the time parameter $t \in [0, T]$ one may consider a stochastic equation of Ito type

$$d\xi = \exp^X_{\xi(t)}(a_{\xi(t)}(t)dt + A_{\xi(t)}(t)dw) \tag{5}$$

having the following local representation

$$d\xi = a_{\xi(t)}(t)dt - \frac{1}{2}\mathrm{Tr}\Gamma^X_{\xi(t)}(A_{\xi(t)}(t)\cdot, \quad A_{\xi(t)}(t)\cdot)dt + A_{\xi(t)}(t)dw. \tag{6}$$

Remind that equation (5) may be treated as a section of Ito bundle which may be described in the following way. Given a local manifold $U \subset B$ and direct sum $B \oplus L_{12}(H, B)$ of the Hilbert space B and Hilbert–Schmidt operator space $L_{12}(H, B)$ define the local Ito bundle as the direct product $I(U) = U \times B \oplus \oplus L_{12}(H, B)$. If (V, φ), (U, ψ) stand for two charts of the given atlas on X, $U \cap V \neq \varnothing$, define a transformation rule

$$\Psi_{\varphi\psi}: I(U) \to I(V), \quad \Psi_{\varphi\psi} = (\psi \circ \varphi^{-1}, \Psi^0_{\varphi\psi}),$$

where

$$\Psi^0_{\varphi\psi}(b^\varphi, A^\varphi) = (b^\psi, A^\psi)$$

and

$$b^\psi = (\psi \circ \varphi^{-1})' b^\varphi + \frac{1}{2}\mathrm{Tr}(\psi \circ \varphi^{-1})''(A^\varphi, A^\varphi),, \quad A^\psi = (\psi \circ \varphi^{-1})' A^\varphi.$$

Given a tripple $(x, b, A) \in I(U)$ define a germ $\Re_x(b, A)$ of diffusion processes starting at zero time from the point x and not leaving the neighborhood U with probability 1 which coinside with solutions of Cauchy problems for stochastic differential equations

$$d\xi = b_{\xi(s)}(s)ds + A_{\xi(s)}(s)dw.$$

Here $b_x(s) \in B$, $A_x(s) \in L_{12}(H, B)$ are assumed to be smooth fields with support U_1, $U_1 \subset U$, satisfying the relations

$$b_x(0) = b, \qquad A_x(0) = A, \qquad x \in U_1.$$

Let $f = (\psi \circ \varphi^{-1})$: $U \to V$ be a local manifold *morphism* and define the transformation rule for $\mathfrak{R}_x(b, A)$ as

$$\mathfrak{R}(f): \mathfrak{R}_x(b, A) \to \mathfrak{R}_{f(x)}(b^f, A^f),$$

$$b^f = f'b + \frac{1}{2}\mathrm{Tr} f''(A, A), \qquad A^f = f'A. \tag{7}$$

It means that

$$f(\xi(t)) = \eta(t) \in \mathfrak{R}_{f(x)}(b^f, A^f)$$

and thus consides up to stochastic equivalence with the solution of the Cauchy problem

$$d\eta = b^f_{\eta(t)}(t)dt + A^f_{\eta(t)}(t)dw, \qquad \eta(s) = f(x)$$

where

$$b^f_{\eta(t)}(t) = f'(\xi(t))b_{\xi(t)}(t) + \frac{1}{2}\mathrm{Tr} f''(\xi(t))(A_{\xi(t)}(t)\cdot, A_{\xi(t)}(t)\cdot),$$

$$A^f_{\eta(t)}(t) = f'(\xi(t))A_{\xi(t)}(t).$$

Let $\mathfrak{R}_U(b, A) = \bigcup_{x \in U} \mathfrak{R}_x(b, A)$.

The bundle $\mathfrak{R}$: $\mathfrak{R}(I(X)) \to X$ is called the *bundle of stochastic differentials*, while its section is called *a stochastic differential equation* (Belopolskaya and Dalecky, 1989).

We also note that $\mathfrak{R}_x(b, A)$ defined above may be identified with an element of T^2X of the form

$$\mathfrak{R}_x(b, A) = (x, \frac{1}{\sqrt{2}}A_x dw, \frac{1}{\sqrt{2}}A_x dw, b_x dt + A_x dw)$$

since for f: $U \to V$ we have

$$T^2\ f\mathfrak{R}_x(b, A) = (x^f, \frac{1}{\sqrt{2}}f'A_x dw),$$

$$\frac{1}{\sqrt{2}}f'A_x dw, \qquad f'b_x dt + f'A_x dw + \frac{1}{2}\mathrm{Tr} f''(A_x\cdot, A_x\cdot)dt. \tag{8}$$

Comparing (7), (8) and (2) one sees that $-\ \Gamma^X_x(\cdot,\cdot)$ may stand for b_x as well as the element $a_x - \Gamma^X_x(\cdot,\cdot)$ with a_x being a vector field. Thus the stochastic equation localy has the form (6) wich justifies (5).

Due to the described above identification of T^2X with $\mathfrak{R}_x(b, A)$ one may consider

$$K^X \circ \mathfrak{R}_x(b, A) = (x, b_x dt + \frac{1}{2}\mathrm{Tr}\Gamma^X_x(A_x\cdot, A_x\cdot)dt + A_x dw).$$

This may be called *Stratonovich differential* since it concides with the classical variant of $d_s\xi$ for a special connection on X with local coefficients $\Gamma^X_x(A_x, A_x) = -A'_x A_x$. Thus locally Stratonovich differential equation has the form

$$d_s\xi = b_{\xi(t)}dt + \frac{1}{2}\mathrm{Tr}\Gamma^X_{\xi(t)}(A_{\xi(t)}\cdot, A_{\xi(t)}\cdot)dt + A_{\xi(t)}dw.$$

Consider next a stochastic differential equation on the total space of the principal fibre bundle p: $\mathcal{P} \to X$

$$d\gamma = \exp^{\mathcal{P}}_{\gamma(t)}(q_{\gamma(t)}(t)dt + Q_{\gamma(t)}(t)dw). \tag{9}$$

Assume that $\exp^{\mathcal{P}}$ corresponds to the connection $K^{\mathcal{P}}$ with local coefficients (4) while principal parts of the fields $q_\gamma \in T_\gamma\mathcal{P}$, $Q_\gamma \in L_{12}(H, T_\gamma\mathcal{P})$, $\gamma \in \mathcal{P}$, may be written in the form

$$q_\gamma = (a_x, g(b_x - \Gamma^p_x a_x)),$$
$$Q_\gamma = (A_x, g(B_x - \Gamma^p_x A_x)) \tag{10}$$

where $\gamma = (x, g)$, $b_x \in \mathcal{G}$, $B_x \in L_{12}(H, \mathcal{G})$, $a_x \in T_xX$, $A_x \in L_{12}(H, T_xX)$. In local coordinates equation (9) under above assumptions is the system of stochastic differential egations

$$d\xi = a_{\xi(t)}(t)dt - \frac{1}{2}\mathrm{Tr}\Gamma^X_{\xi(t)}(A_{\xi(t)}(t)\cdot, A_{\xi(t)}(t)\cdot)dt + A_{\xi(t)}(t)dw, \tag{11}$$

$$dg = g(t)\{b_{\xi(t)}(t)dt - \Gamma^p_{\xi(t)}a_{\xi(t)}(t)dt + B_{\xi(t)}(t)dw - \Gamma^p_{\xi(t)}A_{\xi(t)}(t)dw +$$
$$+\frac{1}{2}\mathrm{Tr}\{(\Gamma^p_{\xi(t)})'(A_{\xi(t)}(t)\cdot, A_{\xi(t)}(t)\cdot) - \Gamma^p_{\xi(t)}\Gamma^X_{\xi(t)}(A_{\xi(t)}(t)\cdot, A_{\xi(t)}(t)\cdot)\}dt +$$
$$+\mathrm{Tr}[\Gamma^p_{\xi(t)}A_{\xi(t)}(t)\cdot, B_{\xi(t)}(t)\cdot]dt\}. \tag{12}$$

It results from the general theory (Belopolskaya and Dalecky, 1989) that the following assertion is valid.

THEOREM 1. *Assume that all the fields $a_x(t)$, $A_x(t)$, $b_x(t)$, $B_x(t)$, $(\Gamma^p_x)'$, Γ^X_x are C_2–smooth and bounded. Then there exists a unique solution $\gamma(t)$ of equation (9) such that*

$$\gamma(s) = \gamma = (x, e). \tag{13}$$

Its component $\xi(t) = p(\gamma(t))$ is an X–valued diffusion process while the component $g(t)$ stands for a G–valued multiplicative functional of the process $\xi(t)$.

Given a solution $\xi(t)$ of (5) and a smooth section φ of the bundle p consider the G–valued process $\eta(t) = \varphi(\xi(t))g(t)$.
To derive the expression for its stochastic differential $d\eta$ we need some additional results concerning G–valued diffusion processes which we shall state here without proof. Remind that G is assumed to be equipped with the left invariant connection K^G with $\Gamma^G(A, A) = -A^2$ and $\exp^G$ denotes the corresponding exponential map. Consider the stochastic equation

$$dg = \exp^G_{g(t)}\left(k_{g(t)}(t)dt + K_{g(t)}(t)dw\right) \tag{14}$$

with $k_g(t) \in T_gG$, $K_g(t) \in L_{12}(H, T_gG)$ being left invariant fields. Then equation (14) is equivalent to the equation

$$dg = g(t)\exp(k(t)dt + K(t)dw) \tag{15}$$

Along with (15) consider yet another stochastic equation

$$df = f(t)\exp(l(t)dt + \mathcal{L}(t)dw) \tag{16}$$

with $l(t) \in \mathcal{G}$, $\mathcal{L}(t) \in L_{12}(H, \mathcal{G})$.

LEMMA 1. *Let $g(t)$ solves the equation (15), $f(t)$ solves the equation (16) and $f(s) = f,\ g(s) = e$. The process $f(t)g(t) = \eta(t)$ obeys the equation*

$$d\eta = \eta(t)\exp\{(k(t) + l(t))dt + (K(t) + \mathcal{L}(t))dw + \frac{1}{2}\mathrm{Tr}[\mathcal{L}(t)\cdot, K(t)\cdot]dt\}$$

which has locally the following form

$$d\eta = \eta(t)\{(k(t) + l(t))dt + (K(t) + \mathcal{L}(t))dw + \frac{1}{2}\mathrm{Tr}K^2(t)dt +$$

$$+ \frac{1}{2}\mathrm{Tr}\mathcal{L}^2(t)dt + \mathrm{Tr}\mathcal{L}(t)K(t)dt\}.$$

LEMMA 2. *If $f: X \to G$ is a diffeomorphism and $\xi(t) \in X$ is a process satisfying (5) then the stochastic differential of the G-valued process $\eta^0(t) = f(\xi(t))$ may be written as*

$$d\eta^0 = f'(\xi(t))a_{\xi(t)}(t)dt - \frac{1}{2}f'(\xi(t))\mathrm{Tr}\Gamma^X_{\xi(t)}(A_{\xi(t)}(t)\cdot, A_{\xi(t)}(t)\cdot)dt +$$

$$+ \frac{1}{2}\mathrm{Tr}f''(\xi(t))(A_{\xi(t)}(t)\cdot, A_{\xi(t)}(t)\cdot)dt + f'(\xi(t))A_{\xi(t)}(t)dw. \tag{17}$$

Remark. It would be nice to obtain an expression for $d\eta$ in (17) in terms of invariant differential operators instead of partial derivatives. Still it needs some correction to become invariant. Note that there is an analogous situation in dealing with sections of vector bundles (Belopolskaya and Dalecky, 1989).
Now we are ready to write down the stochastic differential of $\eta(t) = \varphi(\xi(t))g(t)$

$$d\eta = (\varphi'(\xi(t))a_{\xi(t)}(t)dt + \frac{1}{2}\mathrm{Tr}\varphi''(\xi(t))(A_{\xi(t)}(t)\cdot, (A_{\xi(t)}(t)\cdot)dt + \varphi'(\xi(t))A_{\xi(t)}(t)dw -$$

$$-\frac{1}{2}\varphi'(\xi(t))\mathrm{Tr}\Gamma^X_{\xi(t)}(A_{\xi(t)}(t)\cdot, A_{\xi(t)}(t)\cdot dt)g(t) + \varphi(\xi(t))g(t)(b_{\xi(t)}(t)dt -$$

$$-\Gamma^p_{\xi(t)}a_{\xi(t)}(t)dt + B_{\xi(t)}(t)dw - \Gamma^p_{\xi(t)}A_{\xi(t)}(t)dw + \frac{1}{2}\mathrm{Tr}\{(\Gamma^p_{\xi(t)})'(A_{\xi(t)}(t)\cdot, A_{\xi(t)}(t)\cdot) -$$

$$-\Gamma^p_{\xi(t)}\Gamma^X_{\xi(t)}(A_{\xi(t)}(t)\cdot, A_{\xi(t)}(t)\cdot)\}dt + \mathrm{Tr}[\Gamma^p_{\xi(t)}A_{\xi(t)}(t)\cdot, B_{\xi(t)}(t)\cdot]dt -$$

$$-\varphi(\xi(t))(\varphi^{-1}(\xi(t)))\{\mathrm{Tr}(\varphi'(\xi(t))A_{\xi(t)}(t)\cdot)g(t)(\Gamma^p_{\xi(t)}A_{\xi(t)}(t)\cdot)dt -$$

$$-\mathrm{Tr}(\varphi'(\xi(t))A_{\xi(t)}(t)\cdot)g(t)(B_{\xi(t)}(t)\cdot)dt\}.$$

Here φ^{-1} denotes the inverse element in G, $\varphi\varphi^{-1} = e$. Consider next the expression

$$d\alpha = \varphi(\xi(t))g^{-1}(t)dg\varphi^{-1}(\xi(t))d\varphi(\xi(t))\varphi^{-1}(\xi(t))$$

and compute the average of the difference $\eta^{-1}(t)d\eta - d\alpha$ at the point $t = s$. We obtain

$$\mathbf{E}_{s,x}\{d\eta\ \eta^{-1}(t) - d\alpha\}\mid_{t=s} = Ad(\varphi(x))\Big\{\frac{1}{2}\varphi^{-1}(x)\mathrm{Tr}\ \varphi''(x)(A_x(s)\cdot, A_x(s)\cdot) -$$

$$-\frac{1}{2}\varphi^{-1}(x)\varphi'(x)\mathrm{Tr}\Gamma^X_x(A_x(s)\cdot, A_x(s)\cdot) + \varphi^{-1}(x)\varphi'(x)a_x(s) + b_x(s)-$$

$$-\Gamma^p_x a_x(s) - \frac{1}{2}\mathrm{Tr}\{(\Gamma^p_x)'(A_x(s)\cdot, A_x(s)\cdot) - \Gamma^p_x\Gamma^X_x(A_x(s)\cdot, A_x(s)\cdot)\}+$$

$$+\mathrm{Tr}\{[\Gamma^p_x A_x(s)\cdot, B_x(s)\cdot] + \varphi^{-1}(x)(\varphi'(x)A_x(s)\cdot)(B_x(s)\cdot)-$$

$$-\varphi^{-1}(x)(\varphi'(x)A_x(s)\cdot)(\Gamma^p_x A_x(s)\cdot)+$$

$$+(\Gamma^p_x A_x(s)\cdot)\varphi^{-1}(x)(\varphi'(x)A_x(s)\cdot) - (B_x(s)\cdot)\varphi^{-1}(x)\varphi'(x)A_x(s)\cdot)\}\Big\}ds.$$

Note that the last expression may be presented in terms of covariant derivatives. Indeed

$$A\varphi(x) = \mathbf{E}_{s,x}\{d\eta\ \eta^{-1}(t) - d\alpha\}\ |_{t=s} = \frac{1}{2}\mathrm{Tr}\Big\{\nabla^{Ad(p)}_{A_x(s)}\nabla^p_{A_x(s)} -$$

$$-\nabla^p_{\nabla_{A_x(s)}A_x(s)}\Big\}\varphi(x) + \nabla^p_{a_x(s)}\varphi(x)+$$

$$+\varphi(x)\mathrm{Tr}\left[\varphi^{-1}(x)\nabla^p_{A_x(s)}\varphi(x), B_x(s)\right] + \varphi(x)b_x(s). \tag{18}$$

3. Assume for the moment that p is a trivial fibre bundle, that is $\mathcal{P} = X \times G$, $b_x(s) \equiv 0$, $B_x(s) \equiv 0$, and $a_x(s) \equiv a_x$, $A_x(s) \equiv A_x$ do not depend on s. Let φ be a diffeomorphism from X to G. Consider the Cauchy problem

$$\frac{\partial u}{\partial s} + \frac{1}{2}\mathrm{Tr}\left\{\nabla^{Ad(p)}_{A_x}\nabla^p_{A_x} - \nabla^p_{\nabla_{A_x}A_x}\right\}u + \nabla^p_{a_x}u = 0, \tag{19}$$

$$u_x(T) = \varphi(x) \tag{20}$$

with respect to diffeomorphism $u_x(s)$ from X to G depending on parameter $s \in [0, T]$.
Let X be a finite dimensional compact manifold and G be a finite dimensional Lie group. Denote G^0 the infinite dimensional Hilbert—Lie group formed by square integrable infinitely smooth morphisms from X to G. The relation

$$\mathfrak{B}\varphi = \frac{1}{2}\mathrm{Tr}\left\{\nabla^{Ad(p)}_{A_x}\nabla^p_{A_x} - \nabla^p_{\nabla_{A_x}A_x}\right\}\varphi + \nabla^p_{a_x}\varphi \tag{21}$$

defines a vector field on G^0 while

$$u_x(s) = \varphi(x)\ \overrightarrow{\exp}\int_s^t \mathbf{E}_{s,x}Ad(g^{-1}(\tau))\varphi^{-1}(\xi(\tau))d\varphi(\xi(\tau))$$

defines an evolution family in G^0 which satisfy the equation

$$\frac{du}{ds} + \mathfrak{B}u = 0.$$

The above considerations lead to the following assertion.

THEOREM 2. *Let* X *be a compact smooth manifold,* G *be a Lie group,* $a_x \in T_xX$, $A_x \in L_{12}(H, T_xX)$ *be smooth bounded fields and* $\varphi \in G^0$ *be a smooth map from* X *to* G. *Then there exists a unique solution* $u_x(s)$ *of parabolic equation (19) satisfying (20) which may be written as*

$$u_x(s) = \varphi(x) \overrightarrow{\exp} \int_s^t L_s(\tau)d\tau,$$

where

$$L_s(\tau)d\tau = \mathbf{E}_{s,x} Ad\big(g^{-1}(\tau)\big)\varphi^{-1}\big(\xi(\tau)\big)d\varphi\big(\xi(\tau)\big) \tag{22}$$

with $\gamma(t) = (\xi(t),\ g(t))$ *being a solution of the stochastic equation*

$$d\gamma = \exp^{\mathcal{P}}_{\gamma(t)}(q_{\gamma(t)}dt + Q_{\gamma(t)}dw)$$

whose coefficients locally have the form

$$q_\gamma = (a_x, -\Gamma^p_x a_x),\ Q_\gamma = (A_x, -\Gamma^p_x A_x).$$

The above result may be extended to nontrivial principal bundle and to nonzero fields b_x and B_x. In this way one may prove the existence and uniqueness of the solution of the Cauchy problem

$$\frac{\partial u}{\partial s} + \frac{1}{2}\mathrm{Tr}\left\{\nabla^{Ad(p)}_{A_x\cdot}\nabla^p_{A_x\cdot} - \nabla^p_{\nabla_{A_x\cdot}A_x\cdot}\right\}u+$$

$$+\nabla^p_{a_x}u + u\mathrm{Tr}\left[u^{-1}\nabla^p_{A_x\cdot}u, B_x\cdot\right] + ub_x = 0, \quad u_x(T) = \varphi(x)$$

and present the solution in the form

$$u_x(s) = \varphi(x) \overrightarrow{\exp} \int_s^t L_s(\tau)d\tau,$$

where $L_s(\tau)$ is given by (22) and $\gamma(t) = (\xi(t), g(t))$ is the solution of (11), (12).

REFERENCES

Belopolskaya, Ya.I. and Dalecky, Yu.L. (1989). Stochastic equations and differential geometry. Kluwer, Dordrecht

Eliasson, H. (1967). Geometry of manifolds of maps. *J. Diff. Geometry* 1, 32—62.

Elworthy, K.D. (1982). Stochastic Differential Equations on Manifolds. London Math. Soc., London.

Ikeda, N. and Watanabe, S. (1981). Stochastic Differential Equations and Diffusion Processes. North—Holland, Amsterdam.

Kobayashi, S. and Nomizu, K. (1963). Foundations of Differential Geometry. 1

Schwartz, L. (1984). Semimartingales and their Stochastic Calculus on Manifolds.

Prob. Theory and Math. Stat., Vol. 1, pp. 118-127
B.Grigelionis *et al.* (Eds.)
1990 VSP/Mokslas

ON MARKOV PROCESSES ASSOCIATED WITH A PROJECTIVE SEQUENCE OF HARMONIC SPACES

A.D. BENDIKOV
Rostov Civil Engineering Institute,
344022 Rostov-on-Don, Sotsialisticheskaya street, 162, USSR

ABSTRACT

Let E be a locally compact Hausdorff space with a countable base (semi-compact) and S be a hyperharmonic sheaf on E such that (E, S) is a P-harmonic space (Constantinesqu and Cornea, 1972). It is known (Bauer, 1969, 1984; Constantinesqu and Cornea, 1972; Meyer, 1963) that if $1 \in S^+(E)$ then there exists a standard Markov process on E such that the cone $S^+(E)$ coincides with the cone of excessive functions with respect to this process. Does this property survive under the procedure of projective limit? We give a positive answer to this question and consider the example of projective sequence generated by an elliptic differential operator on the infinite dimensional group $\mathbf{R}^n \times \mathbf{T}^\infty$. We also show that the class of elliptic harmonic groups is exhausted by this example.

1. PROJECTIVE SEQUENCES OF HARMONIC SPACES

Let $(E_n, r_{nm})_1^\infty$ be a projective sequence of semicompacts. We suppose that the maps $r_{nm}: \quad E_m \to E_n$, $n \leqslant m$, satisfy the conditions:

(A) $r_{nm}(E_m) = E_n$, $r_{mm} = id$, r_{nm} is a continuous and open map;

(B) for any compact $K \subset E_n$ the set $r_{nm}^{-1}(K)$ is a compact subset of E_m.
It follows from (A) and (B) that the set $E_\infty = \lim \text{proj } E_n$ is semi-compact.
Suppose that on each E_n a hyperharmonic sheaf S_n is defined such that $1 \in S_n^+(E_n)$ and (E_n, S_n) is a P-harmonic space. We shall assume that the following property is satisfied:

(C) for any $n \leqslant m$, open subset $V \subset E_n$ and a function $f \in S_n(V)$, the function $f \cdot r_{nm}$ belongs to $S_m(r_{nm}^{-1}(V))$; in this case we shall write:

$$S_n \cdot r_{nm} \subset S_m.$$

Definition 1.1. A sequence $(E_n, S_n, r_{nm})_1^\infty$ satisfying conditions (A), (B) and (C) will be called projective sequence of harmonic spaces.
In the sequel we use the following notations:

(i) B is the set of bounded Borel functions on E_∞;

(ii) B_n is the subset of B of cylindric functions $f \cdot r_n$, where r_n is the projection of E_∞ on E_n and f is a function on E_n;

(iii) $\widehat{B} = \bigcup_n B_n$;

(iv) C is the set of bounded continuous functions on E_∞;

(v) C_c is the subset of C of functions f such that supp f is a compact set;

(vi) $P = (P_t)$ is a sub-Markov semi-group on B;

(vii) $R = (R_s)$ is the resolvent of semi-group P;

(viii) $\mathcal{E}^P$ is the cone of P-excessive functions;

(ix) $\mathcal{E}^P_{\text{l.s.c.}}$ is the sub-cone of $\mathcal{E}^P$ of lower semicontinuous functions;

(x) $\mathcal{E}^P_n = \mathcal{E}^P \cap B_n$;

(xi) $S^+_n(E_\infty) = \{f = \widehat{f} \cdot r_n : \ \widehat{f} \in S^+_n(E_n)\}$;

(xii) $S^+_\infty(E_\infty) = \{f = \uparrow \lim f_n : \ f_n \in S^+_n(E_\infty), n = 1, 2, \ldots\}$.

THEOREM 1.1. *Let $(E_n,\ S_n, r_{nm})_1^\infty$ be a projective sequence of harmonic spaces. There exists a sub-Markov semi-group P on E_∞ such that:* 1) *for any $f \in C_c$:*

a) $P_t(f) \in C$;

b) $\lim_{t \downarrow 0} \sup_{E_\infty} |P_t(f) - f| = 0$;

c) *for any closed set F disjoined from the support of f,*

$$\lim_{t \downarrow 0} \sup_F \frac{1}{t} P_t(f) = 0;$$

2) *for any $f \in \widehat{B}$ and $s \geqslant 0$*

$$R_s(f) \in C;$$

3) a) $\mathcal{E}^P_n = S^+_n(E_\infty)$ *for any* $n = 1, 2, \ldots$;

b) $\mathcal{E}^P_{\text{l.s.c.}} = S^+(E_\infty)$.

We only outline the proof of this theorem. Let $P_n \subset S^+_n(E_n)$ be the cone of potentials on E_n. For any $n < m$ the following including is true:
1) $P_n \cdot r_{nm} \subset P_m$.
Fix $n \geqslant 1$ and a potential $p \in P_n \cap C(E_n)$. According to (Constantinesqu and Cornea, 1972, Proposition 10.2.2) there exists a unique resolvent $R^{(n)}$ on $B(E_n)$ such that:

(a) $R^{(n)}_0(1) = p$;

(b) $R^{(n)}_0 : \quad B^+(E_n) \to P_n \cap C(E_n)$.

If $n < m$, then $p \cdot r_{nm} \in P_m$ and consequently there exists a unique resolvent $R^{(m)}$ on $B(E_m)$ satisfying (a), (b). The following equality is true:
2) $R^{(n)}_s(f) \cdot r_{nm} = R^{(m)}_s(f \cdot r_{nm})$.

Let R be the unique resolvent on B such that for any $m \geqslant n$

$$R_s(f)\cdot r_m = R_s^{(m)}(f\cdot r_m).$$

It follows from (a) and (b) that

(c) $R_0(1) = p\cdot r_n$,

(d) $R_0 : \quad \widehat{B}^+ \to \bigcup_n P_n\cdot r_n$.

We pick for each $n = 1, 2, \dots$ a potential $p_n \in P_n \cap C(E_n)$ such that:

(e) p_n is a hyperstrict potential (Constantinesqu and Cornea,1972, p. 167);

(f) $p_{n+1} \leqslant p_n\cdot r_{n\ n+1}/2$,

and put

$$p = \sum p_n\cdot r_n, \quad R_0 = \sum R_0^n,$$

where R^n is the resolvent associated with potential p_n in the sence of (c).
The kernel R_0 maps the space B into itself and satisfies the complete maximum principle. According to (Meyer, 1966) there exists a resolvent R on B such that $R_s|_{s=0} = R_0$.
The resolvent R is generated by some sub–Markov semi–group P. The proof of this property and properties 1 and 2 of the theorem is a modification of the proof of the analogous theorem in (Constantinesqu and Cornea, 1972).
The proof of the property 3 of the theorem relies on the following result: let f be a continuous P–excessive function and denote

$$\widehat{f}_n(x) = \inf_{r_n^{-1}(x)} f, \qquad x \in E_n,$$

$$f_n = \widehat{f}_n\cdot r_n.$$

Then $\widehat{f}_n \in S_n^+(E_n)$ and $f = \uparrow \lim_{n\to\infty} f_n$.
Let X be a Markov process on E_∞ generated by the semi–group P from Theorem 1.1. It follows from 1) that X is a standard process with almost surely continuous trajectories.
A set of hyperharmonic sheafs S on E_∞ such that

1) S is closed under $\uparrow$ lim;

2) $S_n\cdot r_n \subset S$ for any $n = 1, 2, \dots$ and S has a minimal element S_∞.

It follows from 3) that S_∞ coincides with the sheaf of lower semi–continuous superharmonic functions of the process X.
In the general case (E_∞, S_∞) doesn't satisfy the axioms of P–harmonic space.

THEOREM 1.2. *The following properties hold true.*

Axiom of Positivity. Harmonic sheaf $H_\infty = S_\infty \cap (-S_\infty)$ *is non–degenerated at all points of* E_∞.

Weak Axiom of Convergence. H_∞ *is closed under the operation of uniform–limit.*

Axiom of Resolutivity. The resolutive sets (with respect to S_∞) (Constantinesqu and Cornea, 1972, § 1.2) *form a base in E_∞.*

Axiom of Completeness. A lower semi-continuous, lower finite function f on an open set U of E_∞ belongs to $S_\infty(U)$ if for any relatively compact resolutive set V (with respect to S_∞) with $\bar{V} \subset U$ we have $m^V(f) \leqslant f$ on V where m^V is the harmonic measure constructed with respect to S_∞ (it coincides with the hitting probability of the process X).

Remark 1.1. According to Theorem 1.2 (E_∞, S_∞) is a P-harmonic space iff the harmonic sheaf H_∞ possesses the Bauer convergence property: the limit function of any increasing sequence of H_∞-functions on any open set of E_∞ is an H_∞-function whenever it is locally bounded.
The Bauer convergence property is equivalent to any one of the following properties (see property 2 of Theorem 1.1):

2′) for any $f \in B$ and $s \geqslant 0$, $R_s(f) \in C$,

2″) $\mathcal{E}^P_{\text{l.s.c.}} = \mathcal{E}^P$.

2. EXAMPLES

Consider an infinite matrix $A = (a_{ij})_1^\infty$ and a vector $b = (b_i)_1^\infty$. We shall suppose that the following condition is fulfilled: for any $n = 1, 2, \ldots$

$$(a_{ij})_1^n > 0.$$

For a fixed number $p \geqslant 3$ and any $n = 1, 2, \ldots$ we consider the following objects:

$$E_n = \mathbf{R}^p \times \mathbf{T}^n,$$

where $\mathbf{R}^p$ is the p-dimensional Euclidean space and $\mathbf{T}^n \overset{*}{=} \mathbf{R}^n/2\pi\mathbf{Z}^n$ is the n-dimensional torus;

$$L_n = \sum_{i,j=1}^{p+n} a_{ij}\partial_i\partial_j + \sum_{i=1}^{p+n} b_i\partial_i;$$

S_n is the sheaf of lower semi-continuous functions u on E_n such that:

1) u is locally summable with respect to the Haar measure,

2) $L_n u \leqslant 0$ in the distribution sense,

$$H_n = S_n \cap (-S_n),$$

r_{nm} is the projection of E_m on E_n.
We have:

1) (E_n, S_n) for any $n = 1, 2, \ldots$ is a P-harmonic space (and even Brelot space (Constantinesqu and Cornea, 1972, Chapter 3),

2) $H_n = \{u \in C^2 : L_n(u) = 0\}$;

3) $(E_n, S_n; r_{nm})_1^\infty$ is a projective sequence of harmonic spaces (def. 1.1);

4) $E_\infty = \mathbf{R}^p \times \mathbf{T}^\infty$;

$$L_\infty = \sum_{i,j=1}^{\infty} a_{ij}\partial_i\partial_j + \sum_{i=1}^{\infty} b_i\partial_i;$$

$$H_\infty = \{u \in C : \quad L_\infty u = \infty \text{ in the distribution sense}\};$$

5) the associated Markov process X;

$$\widehat{E} = \mathbf{R}^p \times \mathbf{Z}^{(\infty)},$$

(m_t) is a semi–group of Gaussian measures on E_∞ such that the Fourier transform $\widehat{m}_t(n), n \in \widehat{E}$, of the measure m_t has the form

$$\widehat{m}_t(n) = \exp(-tl(n)),$$

$$l(n) = \sum_{k,j=1}^{\infty} a_{kj}n_k n_j - i\sum_{k=1}^{\infty} b_k n_k, n = (n_k)_1^\infty, \ i = \sqrt{-1};$$

X is a space–homogeneous Markov process on E_∞ such that:

$$\mathbf{P}(x_t \in V, x_0 = 0) = m_t(V), \qquad V \subset E_\infty.$$

THEOREM 2.1. *The following properties are equivalent:*

1) (E_∞, H_∞) *is an elliptic harmonic space* (Constantinesqu and Cornea, 1972, p. 66);

2) *the measure* $r = \int_0^\infty m_t dt$ *is absolutely continuous with respect to the Haar measure.*

Proof. See (Bendikov, 1987, Theorem 2.1.1).

Example 2.1. Let $A = \operatorname{diag}(a_k)_1^\infty$ with $a_k = k^s, s \geqslant 0$ and $b = 0$; then we have:

1) If $s = 0$, then the measure r is singular with respect to the Haar measure and consequently (E_∞, H_∞) is not an elliptic harmonic space. But it can be proved that (E_∞, H_∞) is a quasi–harmonic space (Stoica, 1980).

2) If $s > 0$, then the measure r is absolutely continuous with respect to the Haar measure and consequently (E_∞, H_∞) is an elliptic harmonic space. But if $s \leqslant 1$ then (E_∞, H_∞) is not a Brelot space.

3) If $s > 1$, then (E_∞, H_∞) is a Brelot space.

Proof. See (Bendikov, 1987, § 3.3).

In the general case we denote for any $i = 1, 2, \ldots$

$$d_i = \sum_{i=j} |a_{ij}|.$$

THEOREM 2.2.

1) *If the following conditions are fulfilled*

a) $\sup_{i \geqslant 1} d_i/a_{ii} < 1$,

b) *for any* $t > 0$ $\sum_{i=1}^{\infty} \exp(-ta_{ii}) < \infty$, *then* (E_∞, H_∞) *is an elliptic harmonic space.*

2) *Let in addition to condition* a) *the following conditions be fulfilled*

c) $\sum_{i=1}^{\infty} d_i/a_{ii} < \infty$;

d) $\sum_{i=1}^{\infty} 1/a_{ii} < \infty$. *Then* (E_∞, H_∞) *is a Brelot space.*

Proof.

1) Let m_t^s be a symmetric Gaussian measure on E_∞ such that

$$\widehat{m}_t^s(n) = \exp(-t \sum_{i,j=1}^{\infty} a_{ij} n_i n_j),$$

then

$$m_t(V) = m_t^s(V - b_t), \tag{2.1}$$

where $b_t = e(tb)$ and $e: \quad \mathbf{R}^\infty \to \mathbf{R}^p \times \mathbf{T}^\infty$ is a natural homomorphism. It follows from the definition of the measure r and (2.1) that it is sufficient to proof that the measure m_t^s is absolutely continuous with respect to the Haar measure for any $t > 0$. Let m_t^* be a Gaussian measure on E_∞ such that

$$\widehat{m}_t^* = \exp(-t \sum_{i=1}^{\infty} a_{ii} n_i^2).$$

It follows from condition a) (Bendikov, 1987, Lemma 2.3.1) that there exists a constant $c > 0$ such that for any Borel set $B \subset E_\infty$

$$m_t^s(V) \leqslant c m_t^*(V). \tag{2.2}$$

It follows from condition b) (Bendikov, 1987, Theorem 2.2.3) that for any $t > 0$ the measure m_t^* is absolutely continuous with respect to the Haar measure. Therefore by (2.2) the measure m_t^s is exactly the described one.

2) We have d) $\Rightarrow$ b), so (E_∞, H_∞) is an elliptic harmonic space. Let $r(x)$ be the excessive density of the measure r with respect to the Haar measure and (m_x^V) be the set of harmonic measures (H_∞–sweeping). Then for any open subset V of E_∞ such that $0 \notin V$ and for any $x \in V$

$$m_x^V(r) = r(x). \tag{2.3}$$

We shall show that $r(x)$ is locally bounded on the set $E_\infty \backslash \{0\}$. From this property and (2.3) it will follow that $r(x)$ is continuous on the set $E_\infty \backslash \{0\}$. According to (Bendikov, 1987, Theorem 2.1.2) the last property is equivalent to the assertion 2) of the theorem.

We have

$$r(x) = \int_0^\infty m_t(x)dt = \int_0^\infty m_t^*(x - b_t)dt \leqslant c_1 \int_0^\infty m_t^*(x - b_t)dt = \int_0^s \cdot + \int_s^\infty \cdot; \tag{2.4}$$

$$\int_s^\infty m_t^*(x - b_t)dt \leqslant \int_s^\infty m_t^*(0)dt \leqslant c_2 \int_s^\infty t^{-p/2}dt < \infty. \tag{2.5}$$

Fix an open set V such that $0 \in \bar{V}$ and pick $s > 0$ such that some neighbourhood of the neutral element has emply intersection with the set $V - \{b_t :\ 0 \leqslant t \leqslant s\}$.Let for example

$$\inf\{|x_1 - b_{1t}| : \quad x = (x_i) \in V,\ 0 \leqslant t \leqslant s\} > 0. \tag{2.6}$$

We can write

$$E_\infty = \mathbf{R}^1 \times (\mathbf{R}^{p-1} \times \mathbf{T}^\infty) = E_1 \times E_2,$$

$$m_t^* = m_t^{**} \otimes m_t^{***},$$

where m_t^{**} and m_t^{***} are the projections of the measure m_t^* on E_1 and E_2 respectively.
It follows from (2.6) that for some $c_3 > 0$

$$m_t^{**}(x_1 - b_{1t}) \leqslant \exp(-c_3/t),$$

for any $x \in V$ and $0 \leqslant t \leqslant s$ ($x = (x_1, x_2)$). For any $x_2 \in E_2$ and $t > 0$

$$m_t^{***}(x_2 - b_{2t}) \leqslant m_t^{***}(0).$$

It follows from condition d) (Bendikov, 1987, Lemma 2.2.2) that for any $c > 0$

$$\lim_{t \downarrow 0} \exp(-c/t) m_t^{***}(0) = 0,$$

consequently for some $c_4 > 0$ and for any $x \in V, 0 \leqslant t \leqslant s$,

$$m_t^*(x - b_t) \leqslant c_4. \tag{2.7}$$

By (2.4), (2.5) and (2.7) the function $r(x)$ is locally bounded on the set $E_\infty \backslash \{0\}$. The proof is finished.

3. ELLIPTIC HARMONIC GROUPS

Let E be a locally compact abelian group and S be a hyperharmonic sheaf on $E, H = S \cap (-S)$.

Definition 3.1. We say that (E, S) is a harmonic group if

1) the sheaf S is translation invariant;

2) (E, S) is a harmonic space.
We say also that (E, S) is an elliptic harmonic group if (E, S) is an elliptic harmonic space.

THEOREM 3.1. *If (E, S) is an elliptic harmonic group such that $1 \in H(E)$, then*

1) *E is a locally connected group;*

2) *the component E_0 of the neutral element is of the form*

$$E_0 \approx \mathbf{R}^n \times \mathbf{T}^m, \qquad n < \infty, \quad m \leqslant \infty;$$

3) *there exists a translation invariant elliptic differential operator $L = \sum a_{ij}\partial_i\partial_j + \sum b_i\partial_i$ such that*

$$H = \{u \in C, Lu = 0 \text{ in the distribution sense}\}.$$

Proof.

1) See (Constantinesqu and Cornea, 1972, p. 31).

2) See (Heyer, 1977, Theorem B, p. 19).

3) If $E_0 = \mathbf{R}^n \times \mathbf{T}^m (m < \infty)$, then 3) is true by Bony's theorem (Bony, 1970, Theorem 3.4). Let $m = \infty$. For any $k = 1, 2, \ldots$ we set:

$$E_k = \mathbf{R}^n \times \mathbf{T}^k,$$

r_k is the projection of E_0 onto E_k, $S_k(V) = \{u : \quad u \cdot r_k \in S(r_k^{-1}(V))\}$, for any open $V \subset E_k$.

It's proved that (E_k, S_k) for any $k = 1, 2, \ldots$ is an elliptic harmonic group. Consequently there exists translation invariant elliptic differential operator L_k on E_k (unique up to a multiplicative constant) such that

$$H_k = S_k \cap (-S_k) = \{u \in C^2 : \quad L_k u = 0\}.$$

We have for any $m \geqslant k$

$$H_k \cdot r_{km} \subset H_m,$$

so the sequence $\{L_k\}$ can be chosen such that

$$L_k(u) \cdot r_{km} = L_m(u \cdot r_{km}).$$

From this property it follows that the operator

$$L(u \cdot r_k) = L_k(u) \cdot r_k$$

is defined correctly on the set of cylindric smooth functions.

For any $k = 1, 2, \ldots$ denote:

m_1 the Dirac measure supported at the neutral element of E_k,

m_2 the Haar measure of the group T^∞,

$M_k = m_1 \otimes m_2$ (this is a measure on the group $E_k \times \mathsf{T}^\infty \approx E_0$).

The sequence $\{M_k\}$ converges weakly to the Dirac measure supported at the neutral element of the group E_0 as $k \to \infty$. Let $V = r_k^{-1}(\widehat{V})$ be an open cylindric subset of E_0 and $u \in H(V)$. For any $n \geqslant k$ denote

$$u_n = u * M_n,$$

$$\widehat{u}_n(x) = u_n(y), \qquad x \in V_n = r_n(V), \qquad y \in r_n^{-1}(x).$$

Then

$$\widehat{u}_n \in H_n(V_n), \qquad u_n = \widehat{u}_n \cdot r_n,$$

$$Lu_n = (L_n \widehat{u}_n) \cdot r_n = 0.$$

The sequence $\{u_n\}$ converges to u as $n \to \infty$ locally uniformly on the set V, so for any cylindric smooth function f

$$(Lu, f) = (u, L^* f) = \lim_{n \to \infty} (u_n, L^* f) = \lim_{n \to \infty} ((L_n \widehat{u}_n) \cdot r_n, f) = 0.$$

The proof is finished.

4. WEAK SOLUTIONS

Let E, L and H be the locally compact abelian group $\mathbf{R}^p \times \mathsf{T}^\infty$, the elliptic differential operator $\sum_{i,j=1}^{\infty} a_{ij} \partial_i \partial_j + \sum_{i=1}^{\infty} b_i \partial_i$ and the sheaf of functions $u \in C$ satisfying the equation $Lu = 0$ in the distribution sense respectively. When does H coincide with the sheaf of classical solutions of the equation $Lu = 0$? Here we consider the case $L = \sum_{k=1}^{\infty} a_k \partial_k^2$.

Put

$$G = \left\{ \mathbf{x} = (x_k) \in R^\infty \colon ||\mathbf{x}||^2 = \sum_{k=1}^{\infty} x_k^2 / a_k < \infty \right\},$$

and let e be the natural homomorphism of the group $\mathbf{R}^\infty$ onto the group E. For $x \in G, t \in \mathbf{R}^1$ we set $x_t = e(t \cdot x)$. Obviously $(x_t)_{t \in \mathbf{R}^1}$ is an one parameter sub-group of E.

Definition 4.1. We say that a function u defined in a neighbourhood of a point $x_0 \in E$ is differentiable at x_0 in the direction of a vector $x \in G$, if the function $t \to u(x_0 + x_t)$ is differentiable at $t = 0$. We denote by C_G^∞ the sheaf of functions that possess continuous derivatives of any order in the direction of any vector $x \in G$.

THEOREM 4.1. *Let one of the following conditions be fulfilled:*

$$\lim_{n\to\infty} a_n/n = \infty, \tag{4.1}$$

$$\sum_{n=1}^{\infty} 1/a_n < \infty. \tag{4.2}$$

Then $H \subset C_G^\infty$ and for any set V and function $u \in H(V)$ the series $\sum_{n=1}^{\infty} a_n \partial_n^2 u$ converges absolutely and locally uniformly on V.

Proof. (See (Bendikov, 1987, Theorem 3.3.2).

Remark 4.1. Theorem 3.3.2 in (Bendikov, 1987) contains a stronger result: any measure m such that $Lm = 0$ in the distribution sense on some open set V is absolutely continuous with respect to the Haar measure on V and its harmonic density belongs to C_G^∞.

Example 4.1. Let $a_k = k^s, s \geqslant 0$, and let r be the corresponding potential kernel (see example 2.1). We have $Lr = -\mathcal{E}_0$, where $\mathcal{E}_0$ is the Dirac measure at $x = 0$, in particular $Lr = 0$ on the set $E\backslash\{0\}$.

1) If $s = 0$, then the measure r is singular with respect to the Haar measure.

2) If $0 < s \leqslant 1$, then the measure r is absolutely continuous with respect to the Haar measure, but its excessive density is not contained in the set $H(E\backslash\{0\})$.

3) If $s > 1$, then the conditions of theorem 4.1 are satisfied and the corresponding density belongs to $C_G^\infty(E\backslash\{0\})$.

REFERENCES

Bauer, H. (1970). Harmonic spaces and associated Markov processes. In: *Potential Theory,* C.I.M.E., Roma, pp. 3–68.

Bauer, H. (1984). Harmonic spaces – A survey. *Comp. del Semin. di Mat. dell'Univ. de Bari,* 1–46.

Bendikov, A.D. (1987). Markov processes and partial differiantial equations on a group: the space–homogeneous case. *Uspekhi Matem. Nauk* **42**, 41–78.

Bony, J.M. (1970). Operateurs elliptiques degeneres associes aux axiomatiques de la theorie du potential. In: *Potential Theory,* C.I.M.E., Roma, pp. 69–119.

Constantinesqu, C. and Cornea, A. (1972). *Potential theory on harmonic spaces.* Die Grundl. der Math. Wiss. 158, Springer, Berlin–Heidelberg etc.

Heyer, H. (1977). *Probability measures on locally compact groups.* Springer, Berlin–Heidelberg etc.

Meyer, P.A. (1963). Brelot's axiomatic theory of the Dirichlet problem and Hunt's theory. *Ann. Inst. Fourier* **13**, 357–372.

Meyer, P.A. (1966). *Probability and Potentials.* Blaisdell, Waltham, Toronto–London.

Stoicǎ, L. (1980). *Local operators and Markov processes.* Lect. Notes in Math. 816, Springer, Berlin–Heidelberg etc.

Prob. Theory and Math. Stat., Vol. 1, pp. 128–140
B.Grigelionis *et al.* (Eds.)
1990 VSP/Mokslas

ESTIMATION OF THE TAIL OF THE SPECTRAL DISTRIBUTION BY MEANS OF HIGH LEVEL SOJOURN TIMES

S.M. BERMAN*
Courant Institute of Mathematical Sciences, New York University, USA

ABSTRACT

Let $F(\lambda)$ be a spectral distribution function of a real stationary Gaussian process. We assume that there exists a number α, $0 < \alpha < 2$, such that $F(\infty) - F(\lambda)$ is regularly varying of index $-\alpha$, for $\lambda \to \infty$. We present a method of estimating the parameter α based on the observed sojourn times of the sample function above specified high levels.

1. INTRODUCTION AND SUMMARY

Let $X_t, t \geqslant 0$, be a real stationary Gaussian process with mean 0 and covariance function $r(t) = \mathrm{E}X_t X_0$. For simplicity we take $r(0) = 1$. Let $F(\lambda)$ be the spectral distribution function in the representation of r:

$$r(t) = \int_{-\infty}^{\infty} e^{i\lambda t} dF(\lambda). \tag{1.1}$$

F is simmetric about 0 because r is real. Our main hypothesis is that there exists a number α, $0 < \alpha < 2$, such that

$$1 - F(\lambda) \textit{ is regularly varying of index } -\alpha, \textit{ for } \lambda \to \infty. \tag{1.2}$$

According to the classical theory of the domain of attraction of the symmetric stable law of index α, the condition (1.2) is equivalent to the following condition on r (Feller, 1971, Chapter IX):

$$1 - r(t) \textit{ is regularly varying of index } \alpha, \textit{for } t \to 0. \tag{1.3}$$

This work is about the estimation of α as an unknown parameter on the basis of observed sojourn times of the sample function above extreme levels. For $t > 0$ and $u > 0$, put

* This paper represents results obtained at the Courant Insitute of Mathematical Sciences, New York University, under the sponsorship of the National Science Foundation, Grant DMS 8801188, and the U. S. Army Research Office, Contract DAAL 03 86K 0127.

$$L_t(u) = \int_0^t \mathbf{1}_{[X_s > u]} ds. \tag{1.4}$$

Now take $u = u(t)$ as a function of t, and, for fixed $c, 0 < c < 1$, and an integer $m \geqslant 2$, define the sojourn times above the levels $c^j u(t),\ j = 0, 1, \ldots, m$:

$$L_{t,j} = L_t(c^j u(t)) = \int_0^t \mathbf{1}_{[X_s > c^j u(t)]} ds,\ j = 0, 1, \ldots, m. \tag{1.5}$$

We propose an estimator of α based on the observed values of $(L_{t,j})$ for $t \to \infty$. The main result is that under additional conditions on $r(t)$, stipulated for large $t > 0$, and which are independent of (1.3), the estimator of α is, for large m, asymptotically normally distributed with mean 0 and a specified variance. Approximate confidence intervals may be obtained in this way.

The use of the extremes of the process in the estimation of α is justified by an apparent connection between the high frequencies of the spectral distribution and the behavior of the sample functions at large values. Indeed, the stationary Gaussian process sample function, upon attaining a large value, tends to get pushed down almost immediately to a much smaller value (see (Berman, 1980, p. 513)). Thus the sample function tends to have an exaggerated oscillatory behavior near its large values. A simple, heuristic way of explaining this is as follows: We have $\mathbf{E}[X_{t+h} - X_t | X_t = x] = -x(1 - r(h))$ and $\mathrm{Var}[X_{t+h} - X_t | X_t = x] = 1 - r^2(h)$. If $x \to \infty$ and $h \to 0$ in such a way that $x(1 - r(h))$ approaches a positive limit μ, then the conditional displacement $X_{t+h} - X_t$ has limiting mean $-\mu$ and limiting variance 0, so , $X_{t+h} - X_t$ converges in conditional probability to $-\mu$ as the interval length h tends to 0.

The estimation procedure is as follows. We actually first estimate $1/\alpha$, and, from this, α itself. Take m in (1.5) to be an odd integer. For given c and j, we have, by Fubini's theorem and stationarity,

$$\mathbf{E}L_{t,j} = \int_0^t \mathbf{P}[X_s > c^j u(t)] ds = t\mathbf{P}[X_0 > c^j u(t)];$$

this does not depend on α. Furthermore, it is shown in (Berman, 1988), that, under (1.3),

$$\mathrm{Var} L_{t,j} \sim H_j(t) q_\alpha(t) c^{2j/\alpha}, \tag{1.6}$$

for $t \to \infty$, where $H_j(t)$ does not depend on α and $q_\alpha(t)$ does not depend on j. Define

$$L^*_{t,j} = \frac{L_{t,j} - \mathbf{E}L_{t,j}}{\sqrt{\mathrm{Var}\ L_{t,j}}},\quad j = 0, 1, \ldots, m. \tag{1.7}$$

Then, by the main result of (Berman, 1988), the random variables $L^*_{t,j}$ have standard normal limiting distributions. Furthermore, we shall prove in Theorem 3.1

below that the $m+1$ random variables (1.7) are also asymptotically independent. Noting the asymptotic form of Var $L_{t,j}$ in (1.6) we see that the random variables

$$c^{j/\alpha} L^*_{t,j} = \frac{L_{t,j} - \mathbf{E} L_{t,j}}{\sqrt{H_j(t) q_\alpha(t)}}, \tag{1.8}$$

are asymptotically independent with limiting $N(0, c^{2j/\alpha})$ distributions, $j = 0, 1, \ldots, m$.

Define the ratios

$$R_j = c^{1/\alpha} \frac{L^*_{t,2j+1}}{L^*_{t,2j}}, \tag{1.9}$$

$j = 0, 1, \ldots, (m-1)/2$. It is clear from (1.6) and (1.7) that R_j does not depend asymptotically on the parameter α. Furthermore, since the ratio of two standard normal independent random variables has a standard Cauchy density, it follows from the results cited after (1.7) above that $L^*_{t,2j+1}/L^*_{t,2j}$ are asymptotically mutually independent with standard Cauchy limiting distributions. Therefore, the random variables R_j in (1.9) are asymptotically independent with a common limiting Cauchy density with median 0 and scale parameter $c^{1/\alpha}$. From the theory of the estimation of the scale parameter of this density, we obtain our estimator of the parameter $1/\alpha$:

$$e(1/\alpha) = \frac{2}{(m+1)\log c} \sum_{j=0}^{(m-1)/2} \log |R_j|. \tag{1.10}$$

For fixed m, $e(1/\alpha)$ is our estimator of $1/\alpha$, justified by our deduction that the random variables R_j have limiting Cauchy distributions. Thus $e(1/\alpha)$ may be considered to be a good estimator for large t, that is, large enough so that the Cauchy assumption is reasonably justified. In the following theorem we analyze the limiting distribution of the estimator under the additional limit operation $m \to \infty$.

THEOREM 1.1. *Define the constant*

$$\sigma^2 = \frac{8}{\pi} \sum_{n=0}^{\infty} (-1)^n (1+2n)^{-3}, \tag{1.11}$$

whose value is approximately 2.46. *Then*

$$\sqrt{\frac{(m+1)}{2}} \Big[e(1/\alpha) - 1/\alpha \Big] \xrightarrow{d} N\left(0, \frac{\sigma^2}{(\log c)^2}\right) \tag{1.12}$$

for $t \to \infty$ and then $m \to \infty$.

The proof is given in Section 4. Note that the variance of the limiting normal distribution in (1.12) does not depend on the parameter α, so that confidence intervals for $1/\alpha$ are easily constructed. The relation of (1.12) to the method of maximum likelihood is indicated in Section 5. It is also shown how the asymptotic variance in (1.12) may be reduced by the factor $2/\sigma^2 \approx 0.81$ by the well known correction based on root of the likelihood equation.

The main idea of this work, that of the use of the extreme values to estimate the tail of the spectral distribution is, in a general way, related to two other topics in statistical inference. The first is the use of level crossings to estimate the even order spectral moments when they are finite. This topic was introduced in (Lindgren, 1974), and then developed further in (Bjornham and Lindgren, 1976; Cuzick and Lindgren, 1980). The moments of a distribution furnish crude estimates of the order of its tail, and this represents the link to our study. The work on crossings is restricted to the case where, at the least, the second spectral moment is finite, which implies that the number of crossings in each bounded time interval is almost surely finite. However, the present work is based on the assumption (1.3) which implies that the second spectral moment is infinite, and that "most" levels are crossed infinitely often almost surely (see (Berman, 1970)). In the present work the crossings statistics are replaced by sets of sojourns above high levels. Processes whose covariances satisfy (1.3) have been of much interest in extreme value theory (see, for example, (Berman, 1980)).

The second topic in statistics to which this paper is related is that of the estimation of the tail of a distribution function on the basis of the extreme order statistics from a sample of independent observations. An early paper on this subject is (Hill, 1975); many others, too numerous to cite here, have appeared since then. Our problem obviously differs from the latter because of the observable in the former is the sample function of a stochastic process, while the observable in the latter is a sample of independent observations. But the analogy between the two models is enlightening. First of all, it is striking that the formula for the Hill estimator of $1/\alpha$, namely, an average of the logarithms of the ratios of m successive extreme order statistics, is very similar in structure to our estimator (1.10). Second, in the distribution tail estimation problem of Hill, it is obvious that the only relevant statistics are the extreme order statistics. These form at most a small fraction of the entire sample, thus, we are justified in discarding most of the other sample values. One question that is likely to arise about the present work is in the same issue of the discarding of information about the sample function at points where the values are not large. I believe that it is justified by the same principle as that for Hill's estimators: The tail of the spectrum reflects the distribution of the high frequency components of the sample function, and, as we pointed out at the beginning of this section, the high frequency components tend to be most discernible at the extreme values.

2. THE CENTRAL LIMIT THEOREM FOR THE EXTREME SOJOURNS

We assume in this work that the spectral distribution function F in (1.1) is absolutely continuous, and we denote the density as $f(\lambda)$. Then $(f(\lambda))^{1/2}$ belongs to $L_2(-\infty, \infty)$, and so its Fourier transform,

$$b(t) = \int_{-\infty}^{\infty} e^{i\lambda t}(f(\lambda))^{1/2} d\lambda, \tag{2.1}$$

is well defined almost everywhere and also belongs to $L_2(-\infty,\infty)$. Since r is real, dF is symmetric about 0, and so $(f(\lambda))^{1/2}$ is an even function; therefore, $b(t)$ is real valued. Furthermore, $r(t)$ is representable as

$$r(t) = \int_{-\infty}^{\infty} b(t+s)b(s)ds, \tag{2.2}$$

and X_t has the stochastic integral representation

$$X_t = \int_{-\infty}^{\infty} b(t+s)\xi(ds), \tag{2.3}$$

where $\xi(s)$, $s \geqslant 0$, is the standard Brownian motion.
The following theorem is the main result of (Berman, 1988):

THEOREM 2A. *Assume $b \in L_1(-\infty,\infty)$. If $u(t)$ increases to infinity sufficiently slowly for $t \to \infty$, then $L_{t,0} = L_t(u(t))$, suitably normalized, has, for $t \to \infty$, a limiting standard normal distribution. The maximum rate at which $u(t)$ may increase with t is described by the following condition:*
For some $\theta > 1$ and some $\delta > 0$

$$\lim_{t\to\infty} t\ e^{-u^2(t)\theta/2} = \infty \tag{2.4}$$

and

$$\lim_{t\to\infty} e^{\delta u(t)} \int_{|s|>\sqrt{t}\exp(-u^2(t)\theta/4)} b^2(s)ds = 0. \tag{2.5}$$

Let $\phi(z)$ be the standard normal density function, and define

$$\Psi(x) = \int_{x}^{\infty} \phi(z)dz. \tag{2.6}$$

Then, by Fubini's theorem and stationarity,

$$\mathbf{E}L_{t,0} = t\Psi(u(t)). \tag{2.7}$$

Let $\phi(x,y;\rho)$ be the standard bivariate normal density with correlation coefficient ρ:

$$\phi(x,y;\rho) = (2\pi)^{-1}(1-\rho^2)^{-1/2}\exp\left\{-\frac{x^2-2\rho xy+y^2}{2(1-\rho^2)}\right\}. \tag{2.8}$$

Then, there is a well-known formula for Var $L_t(u)$ (Cramer and Leadbetter, 1967, p. 214):

$$\text{Var } L_t(u) = 2\int_0^t (t-s)\int_0^{r(s)} \phi(u,u;y)\ dy\ ds. \tag{2.9}$$

It is shown in (Berman, 1988) that, under the conditions of Theorem 2A with $u = u(t)$,

$$2\int_0^t (t-s)\int_0^{r(s)} \phi(u,u;y)dy\ ds \sim 2t\int_0^{\varepsilon}\int_0^{r(s)} \phi(u,u;y)dy\ ds$$

for $t \to \infty$, for any $\varepsilon > 0$; thus,

$$\text{Var } L_t(u) \sim 2t\int_0^{\varepsilon}\int_0^{r(s)} \phi(u,u;y)\ dy\ ds. \tag{2.10}$$

Since the left hand member of (2.10) does not depend on ε, as is clear from (2.9), the right hand member of (2.10) is asymptotically the same for all $\varepsilon > 0$. Hence we define

$$B(u) = \int_0^{1}\int_0^{r(s)} \phi(u,u;y)\ dy\ ds, \tag{2.11}$$

and note that (2.10) implies

$$\text{Var } L_t(u) \sim 2t\ B(u), \tag{2.12}$$

for $t \to \infty$, $u \to \infty$.
The next result appears in ((Berman, 1988), Theorem 8.2):

THEOREM 2B. *Suppose, in addition to the assumptions of Theorem 2A, that (1.3) above holds. Let $v = v(u)$ be a function implicitly defined by the relation*

$$1 - r(1/v(u)) \sim u^{-2}, \quad for \quad u \to \infty. \tag{2.13}$$

Then,

$$B(u) \sim \frac{\Psi(u)}{v(u)}\frac{2^{\alpha/2}}{\sqrt{\pi}}\ \Gamma(1/2 + 1/\alpha), \tag{2.14}$$

for $u = u(t) \to \infty$.

Since the hypothesis of Theorem 2A is stated as a condition on $b(t)$, it is useful to express the assumption (1.2) or its equivalent (1.3) also in terms of $b(t)$. It follows from (2.2) that (1.3) is equivalent to the regular variation of index α of $\int_{-\infty}^{\infty} |b(t+s) - b(s)|^2 ds$ for $t \to 0$. Let $\hat{b}(\lambda)$ be the inverse Fourier transform of $b(t)$; then $\hat{b}(\lambda) = (f(\lambda))^{1/2}$ almost everywhere, and so the condition (1.2) is equivalent to the regular variation of index $-\alpha$ of $\int_{\lambda}^{\infty} |\hat{b}(x)|^2 dx$ for $\lambda \to \infty$.

3. ASYMPTOTIC INDEPENDENCE OF SOJOURNS AT LEVELS $c^j u(t)$

THEOREM 3.1. *For $0 < c < 1$, put*

$$L_t = L_t(u(t)), \quad L_t^c = L_t(cu(t)). \tag{3.1}$$

Then, under the conditions of Theorem 2A,

$$\lim_{t \to \infty} \text{Corr}\,(L_t, L_t^c) = 0. \tag{3.2}$$

Proof. By a direct extension of the proof of (2.9) given in the cited source, the formula for the variance can be extended to the covariance,

$$\text{Cov}(L_t, L_t^c) = 2 \int_0^t (t-s) \int_0^{r(s)} \phi(u, uc; y)\, dy\, ds. \tag{3.3}$$

We note that the integral in (3.3) cannot but increase if $r(s)$ is replaced by $|r(s)|$ because the integral over $\{s : r(s) < 0\}$ is nonpositive. Hence we will estimate the bound for the covariance (3.3), namely,

$$2 \int_0^t (t-s) \int_0^{|r(s)|} \phi(u, uc; y)\, dy\, ds. \tag{3.4}$$

For arbitrary $\varepsilon > 0$, we write (3.4) as the sum of the terms

$$2 \int_0^\varepsilon (t-s) \int_0^{|r(s)|} \phi(u, uc; y)\, dy\, ds \tag{3.5}$$

and

$$2 \int_\varepsilon^t (t-s) \int_0^{|r(s)|} \phi(u, uc; y)\, dy\, ds. \tag{3.6}$$

For the purpose of estimating the terms (3.5) and (3.6), we introduce the following identity, which follows from (2.8) by simple algebra:

$$\phi(u, uc; \rho) = [\phi(u, u; \rho)\phi(cu, cu; \rho)]^{1/2} \exp\left\{ -\frac{u^2 \rho (1-c)^2}{2\sqrt{1-\rho^2}} \right\}. \tag{3.7}$$

Let us apply (3.7) to the term (3.6). Then the latter is at most equal to

$$2 \int_\varepsilon^t (t-s) \int_0^{|r(s)|} \Big[\phi(u, u; y)\; \phi(uc, uc; y)\Big]^{1/2} dy\, ds.$$

By the Cauchy—Schwarz inequality, the latter is at most equal to the product

$$\left\{2\int_{\varepsilon}^{t}(t-s)\int_{0}^{|r(s)|}\phi(u,u;y)\,dy\,ds\right\}^{1/2}\left\{2\int_{\varepsilon}^{t}(t-s)\int_{0}^{|r(s)|}\phi(uc,uc;y)dy\,ds\right\}^{1/2}.$$

By the calculation in the proof of (Berman, 1988, Theorem 2.1), the product above is of order smaller than the product $(2t\ B(u))^{1/2}(2t\ B(uc))^{1/2}$, which, by (2.12), is asymptotically equal to $(\text{Var } L_t \cdot \text{Var } L_t^c)^{1/2}$. Thus the contribution of the term (3.6) to the covariance (3.3) is asymptotically negligible relative to the product of the standard deviations, so that (3.6) may be ignored in the determination of the limit of the correlation in (3.2). Hence, for the proof of (3.2), it suffices to show that the contribution of (3.5) to the covariance is similarly asymptotically negligible relative to the product of the standard deviations.
For arbitrary $\delta > 0$ there exists $\varepsilon > 0$ such that $r(s) \geqslant 1-\delta$ whenever $0 \leqslant s \leqslant \varepsilon$. For such a choice of δ and ε, we express (3.5) as the sum of

$$2\int_{0}^{\varepsilon}(t-s)\int_{0}^{1-\delta}\phi(u,uc;y)\,dy\,ds \tag{3.8}$$

and

$$2\int_{0}^{\varepsilon}(t-s)\int_{1-\delta}^{|r(s)|}\phi(u,uc;y)\,dy\,ds. \tag{3.9}$$

By an application of (3.7), the term (3.8) is at most equal to

$$2\varepsilon t(1-\delta)[\phi(u,u;1-\delta)\phi(uc,uc;1-\delta)]^{1/2},$$

which, by the definition (2.8), is equal to

$$\frac{\varepsilon t(1-\delta)}{\pi\sqrt{1-(1-\delta)^2}}\exp\left\{-\frac{u^2(1+c^2)}{2(2-\delta)}\right\}. \tag{3.10}$$

According to (Berman, 1988, Corollary 2.1) with $\theta = 2/(2-\delta)$, and the relation (2.12) above, the expression (3.10) is of order smaller than $(\text{Var } L_t \cdot \text{Var } L_t^c)^{1/2}$ for $t \to \infty$. Hence, the term (3.8) is asymptotically negligible relative to the product of the standard deviations, and so its contribution to the correlation of L_t and L_t^c is asymptotically equal to 0.
Finally we consider the contribution of (3.9). By (3.7), it is at most equal to the product

$$2t\exp\left\{-\frac{u^2(1-\delta)(1-c)^2}{2\sqrt{1-(1-\delta)^2}}\right\}\int_{0}^{\varepsilon}\int_{1-\delta}^{r(s)}\phi^{1/2}(u,u;y)\phi^{1/2}(uc,uc;y)\,dy\,ds,$$

which, by the Cauchy—Schwarz inequality and the definition (2.11) of $B(u)$, at most equal to

$$2t\ B^{1/2}(u)B^{1/2}(uc)\exp\left\{-\frac{u^2(1-\delta)(1-c)^2}{2\sqrt{1-(1-\delta)^2}}\right\}.$$

By (2.12), this is asymptotically equal to

$$(\mathrm{Var}\ L_t\cdot\mathrm{Var}\ L_t^c)^{1/2}\exp\left\{-\frac{u^2(1-\delta)(1-c)^2}{2\sqrt{1-(1-\delta)^2}}\right\},$$

which is, for $u\to\infty$, of order smaller than the product of the standard deviations. Thus, (3.9) has a negligible contribution to the correlation, and so we conclude from this and the previous estimates that (3.2) holds.
The following result is an almost immediate consequence of Theorem 3.1, but we state it separately because it requires an indication of proof.

THEOREM 3.2. *Under the conditions of Theorem 2A, and for a function $u(t)$ satisfying the specified conditions, the random variables $L^*_{t,j}$, $j=0,1,\ldots,m$, defined in (1.7), have, for $t\to\infty$, a limiting $(m+1)$–dimensional normal distribution which is a product of standard normal distributions.*

Proof. According to the proof of Theorem 2A, given in (Berman, 1988), the limiting joint distribution of the random variables is multivariate normal because the vector $(L^*_{t,j})$ is approximable by a sum of independent random vectors, and the same argument of (Berman, 1988) which proves the asymptotic normality of the components $L^*_{t,j}$, $j=0,1,\ldots,m$, also proves the asymptotic normality of the vector itself. Since, by Theorem 3.1, the limits of the correlations are all 0, the same is true of the correlations of the limiting distribution.

4. PROOF OF THEOREM 1.1

In several of the arguments below we will use the following well–known result: Let (ξ_n) be a sequence of random variables, (σ_n) a sequence of positive numbers converging to 0, and μ a constant. If $(\xi_n-\mu)/\sigma_n \xrightarrow{d} N(0,\sigma^2)$, for some $\sigma^2>0$, then for every function $f(y)$ which is continuously differentiable at $y=\mu$,

$$(f(\xi_n)-f(\mu))/\sigma_n \xrightarrow{d} N(0,\sigma^2(f'(\mu))^2). \tag{4.1}$$

LEMMA 5.1. *Let (Y_n) be an i.i.d. sequence such that $\mathbf{E}(\log|Y_1|)^2<\infty$. Put*

$$\mu=\mathbf{E}(\log|Y_1|),\qquad \sigma^2=\mathrm{Var}(\log|Y_1|). \tag{4.2}$$

Then

$$\sqrt{n}\left[\prod_{j=1}^{n}|Y_j|^{1/n}-e^{\mu}\right]\Big/\sigma e^{\mu} \tag{4.3}$$

has, for $n\to\infty$, a limiting standard normal distribution.

Proof. By the classical central limit theorem we have

$$\sqrt{n}\left(\frac{1}{n}\sum_{j=1}^{n}\log|Y_j| - \mu\right) \xrightarrow{d} N(0,\sigma^2). \tag{4.4}$$

Then the assertion of the lemma follows by applying (4.1) with $f(y) = e^y$, $\sigma_n = n^{-1/2}$, and $\xi_n = n^{-1}\sum_{j=1}^{n}\log|Y_j|$.

Suppose now that for some $\theta > 0$ we define $X_n = \theta Y_n$, $n \geqslant 1$, where (Y_n) is the sequence in Lemma 5.1; then $\mathbf{E}(\log X_1) = \mu + \log\theta$ and $\mathrm{Var}(\log X_1) = \mathrm{Var}(\log Y_1) = \sigma^2$. As a consequence of Lemma 5.1, we obtain

$$\sqrt{n}\left(\prod_{j=1}^{n}|X_j|^{1/n} - \theta\, e^{\mu}\right) \xrightarrow{d} N(0,\sigma^2\theta^2e^{2\mu}). \tag{4.5}$$

We apply this to the particular case where Y_1 has the standard Cauchy density, so that X_1 has the Cauchy density with scale parameter θ,

$$\frac{\theta}{\pi(\theta^2+x^2)}. \tag{4.6}$$

By a direct calculation, one sees that $Z = \log Y_1$ has the density function

$$\frac{1}{\pi\cosh z}. \tag{4.7}$$

It is an even function, and so $\mu = 0$. The variance σ^2 is calculated as follows:

$$\mathbf{E}(\log Y_1)^2 = \frac{2}{\pi}\int_0^\infty \frac{x^2}{\pi\cosh x}dx = \frac{4}{\pi}\int_0^\infty \frac{x^2e^{-x}}{1+e^{-2x}}dx. \tag{4.8}$$

By the relation

$$(1+e^{-2x})^{-1} = \sum_{n=0}^{\infty}(-1)^n e^{-2nx},$$

for $x > 0$, the right hand member of (4.8) is equal to

$$\frac{4}{\pi}\int_0^\infty x^2\left\{\sum_{n=0}^{\infty}(-1)^n e^{-x(1+2n)}\right\}dx.$$

An interchange of order of integration is justified, and so the expression above is equal to

$$\frac{4}{\pi}\sum_{n=0}^{\infty}(-1)^n\int_0^\infty x^2\, e^{-x(1+2n)}dx,$$

which, by a change of variable and elementary integration, is equal to

$$\sigma^2 = \frac{8}{\pi} \sum_{n=0}^{\infty} (-1)^n (1+2n)^{-3}. \tag{4.9}$$

For σ^2 simple calculation yields an approximate value

$$\sigma^2 = 2.46. \tag{4.10}$$

As an application of (4.5) to this case, we have

$$\sqrt{n}\left(\prod_{j=1}^{n} |X_j|^{1/n} - \theta \right) \xrightarrow{d} N(0, \sigma^2\theta^2). \tag{4.11}$$

Now we complete the proof of Theorem 1.1. As noted above in the discussion following (1.9), the random variables R_j, $j = 0, 1, \dots, \frac{1}{2}(m-1)$, which can be observed and calculated without the knowledge of α, are asymptotically independent and have limiting Cauchy distributions with common scale parameter $\theta = c^{1/\alpha}$, for $t \to \infty$, and fixed $m \geqslant 1$. Define

$$\tilde{\theta}_m = \prod_{j=0}^{(m-1)/2} |R_j|^{(m-1)/2}; \tag{4.12}$$

then, according to (4.11), with $X_j = R_j$, $n = \frac{1}{2}(m+1)$, and $\theta = c^{1/\alpha}$, we have

$$\sqrt{\frac{(m+1)}{2}}(\tilde{\theta}_m - c^{1/\alpha}) \xrightarrow{d} N(0, \sigma^2 c^{2/\alpha}). \tag{4.13}$$

Now apply (4.1) to this with $f(x) = (\log x)/\log c$. The estimator $e(1/\alpha)$ in (1.10) is clearly equal to $\log \tilde{\theta}_m / \log c$, and so it follows from (4.13) that (1.12) holds.

5. REMARKS ON THE ASYMPTOTIC EFFICIENCY OF THE ESTIMATOR

While the determination of the optimal estimator of a parameter is often possible in the case of a sample of independent observations, it is usually not possible in the case of inference based on a single realization of a stochastic process sample function. Indeed, the motivation for our suggested method of estimation is based on the fact that our estimator is actually proportional to an average of nearly independent random variables representing a sample of logarithms taken from a Cauchy population. Therefore it is of some interest to investigate the question of optimality in the restricted setting of estimating the scale parameter θ of a Cauchy density on the basis of a sample of n independent observations.

In this particular example, the likelihood equation has a unique solution $\hat{\theta}_n$ (Lehmann, 1983, p. 474); however, the solution is not explicitly representable, but must be calculated for each given set of observations. The Cauchy density satisfies the general regularity conditions under which these conclusions hold:

i) In the class of estimators $\bar{\theta}_n$ satisfying $\sqrt{n}(\bar{\theta}_n - \theta) \xrightarrow{d} N(0, V)$, for some $V > 0$, the estimator $\hat{\theta}_n$ has the minimum V, that is $\hat{\theta}_n$ is a best asymptotically normal estimator (Lehmann, 1983, p. 406).

ii) The value of V corresponding to $\hat{\theta}_n$ is $2\theta^2$, so that

$$\sqrt{n}(\hat{\theta}_n - \theta) \xrightarrow{d} N(0, 2\theta^2) \tag{5.1}$$

(Lehmann, 1983, p. 474).
It follows from a comparison of (4.11) and (5.1) that the asymptotic efficiency of our estimator of θ, namely, $\theta_n^* = |X_1 \dots X_n|^{1/n}$, relative to the best one, namely, $\hat{\theta}_n$, is $2/\sigma^2$, which has the approximate value 0.813 (see (4.10)). One advantage of our estimator over $\hat{\theta}_n$ is that the former has an explicit form which can be easily calculated. Furthermore, by a slight correction which is easy to carry out, θ_n^* can be modified to be asymptotically equal to $\hat{\theta}_n$. Let $X_1, \dots, X_n$ be observations, and let $L_n(\theta)$ be the logarithm of the likelihood function:

$$L_n(\theta) = -n \log \pi + n \log \theta - \sum_{j=1}^{n} \log(X_j^2 + \theta^2).$$

By verifying the conditions of (Lehmann, 1983, p. 422, Theorem 3.1), we infer that

$$\theta_n^{**} = \theta_n^* - \frac{L_n'(\theta_n^*)}{L_n''(\theta_n^*)} \tag{5.2}$$

is an estimator which is asymptotically as good as $\hat{\theta}_n$ in the sense that it can be substituted for the latter in the relation (5.1).
The advantage of first estimating $1/\alpha$ instead of α is that the variance in the limiting distribution, $\sigma^2/(\log c)^2$, does not depend on α. Suppose we want a $100(1-\varepsilon)\%$ confidence interval for α. Let $z = z(\varepsilon)$ be the upper $\varepsilon/2$–percentile of the standard normal distribution. By (1.12), an approximate confidence interval for $1/\alpha$ is

$$e(1/\alpha) \pm \frac{z\sigma}{|\log c|\sqrt{\frac{(m+1)}{2}}}.$$

Therefore, the reciprocals of these two values are the end points of the confidence interval for α itself. By contrast the asymptotic variance of $\hat{\alpha}$ depends on α; indeed, applying (4.1) to (1.12) we find for $f(x) = 1/x$,

$$\sqrt{\frac{(m+1)}{2}}\left[\frac{1}{e(1/\alpha)} - \alpha\right] \xrightarrow{d} N(0, \sigma^2\alpha^4/(\log c)^2).$$

A natural question is why the parameter α in (1.2) is restricted to $\alpha < 2$. The answer is that our estimates in (Berman, 1988) are based on the equivalence of (1.2) and (1.3), which holds only for $0 < \alpha < 2$. If (1.2) holds for $\alpha > 2$, then it can be shown that $1 - r(t) \sim \lambda_2 t^2/2$, for $t \to 0$, where λ_2 is the second spectral moment, and so our method is not applicable.

REFERENCES

Berman, S.M. (1970). Gaussian processes with stationary increments: local times and sample function properties. *Ann. Math. Statist.* **41**, 1260–1272.

Berman, S.M. (1980). A compound Poisson limit for stationary sums, and sojourns of Gaussian processes. *Ann. Probability* **8**, 511–538.

Berman, S.M. (1988). A central limit theorem for extreme sojourn times of stationary Gaussian processes. In: *Lect. Notes Statist.* **50**, Springer, Berlin–Heidelberg etc.

Bjornham, A. and Lindgren, G. (1976). Frequency estimation from crossings of an observed mean level. *Biometrika* **63**, 507–512.

Cramer, H. and Leadbetter, M. R. (1967). *Stationary and Related Stochastic Processes.* Wiley, New York.

Cuzick, J. and Lindgren, G. (1980). Frequency estimation from crossings of an unknown level. *Biometrika* **67**, 65–72.

Feller, W. (1971). *An Introduction to Probability Theory and its Applications.* **2**, Wiley, New York.

Hill, B. M. (1975). A simple general approach to inference about the tail of a distribution. *Ann. Statist.* **3**, 1163–1174.

Lindgren, G. (1974). Spectral moment estimation by means of level crossings. *Biometrika* **61**, 401–418.

Lehmann, E. (1983). *Theory of Point Estimation,* Wiley, New York.

Prob. Theory and Math. Stat., Vol. 1, pp. 141–151
B.Grigelionis *et al.* (Eds.)
1990 VSP/Mokslas

RENORMALIZATION OF DYSON'S VECTOR–VALUED HIERARCHICAL MODEL AT LOW TEMPERATURES

P.M. BLEHER and P. MAJOR
Keldysh Institute of Applied Mathematics of the Soviet Academy of Sciences, Moscow, USSR
Mathematical Institute of the Hungarian Academy of Sciences, Budapest, Hungary

ABSTRACT

We discuss the large–scale limit of Dyson's vector–valued hierarchical model at law temperatures. This is a limit theorem for sums of random variables, whose investigation requires essentially new methods. We discuss the most important analytical questions connected with this problem. A fairly complex picture arises, and it is shown that because of the continuous symmetry of the model some unexpected phenomena appear.

In this paper we discuss Dyson's vector–valued model and explain how its so–called large–scale limit can be calculated. Formally, this is a limit theorem for the distribution of sums of random variables. Hence one would expect that the classical methods of probability theory, worked out for the investigation of such problems, can be applied in this case too. However, this model has a much more complex behaviour than the analogous models in classical probability, and the usual methods of probability theory are not appropriate for its investigation. The aim of the present paper is to explain the main motivations (statistical physical and probabilistic) for studying Dyson's model and the method of of investigation. Here we restrict ourselves to vector–valued models. A detailed discussion of scalar–valued models is given in (Bleher and Major, 1987). On the other hand, vector–valued models have a peculiar behaviour which cannot be guessed from the behaviour of scalar models. The behaviour of these models strongly depends on a physical parameter, on the temperature T. In scalar–valued case there is a special, so–called critical value of the parameter, where the large–scale limit exists with a different scaling. This means a non–classical limit theorem with an unusual normalization. In vector–valued case a similar phenomenon appears not only at the critical, but at all low temperatures. On low temperatures there is a phase transition, hence first we have to construct the random field, called pure state in the literature, which we want to renormalize. This field has a spontaneous magnetization, and a different normalization is needed in the direction of the spontaneous magnetization and in the direction orthogonal to it. In the direction orthogonal to the spontanous magnetization one has to normalize similarly to the normalization at the critical temperature. In the direction of the magnetization the situation is even more complex. Here the right normalization and the large–scale limit depends on another parameter of the model which measures the strength of the interaction. Let us emphasize again that the above statements hold not only for a singular critical value of the parameter T, but for all sufficiently small temperatures T. A detailed description of Dyson's vector–valued hierarchical model together with a complete proof is contained in (Bleher and Major, 1984; 1988; 1989). Since these papers are burdened with many technical details we have found it useful to discuss separately the most important analytical problems one has to study during these investigations.

The peculiar behaviour of vector–valued models is related to their invariance with respect to rotations. This is a symmetry with respect to a continuous group. Actually, our main motivation for studying this problem is to understand the role of continuous symmetries.

The most important part of the problem we are dealing with can be translated to purely analytical questions. One has to investigate the action of large powers of a certain integral operator to a starting function. This operator is very similar to the convolution operator (i.e. to the convolution of a function with itself), which is a well–known object in classical probability. There is however one essential difference between these operators. The operator appearing in our problem has, unlike the convolution operator, the following instability property: When a large power of this operator is applied to a starting function then the result strongly depends on the starting function. This instability property is the main reason for the peculiar behaviour of models in statistical physics. (See paper (Bleher and Major, 1987) for a heuristic explanation of this question.) Actually this instability is behind the different behaviour of scalar and vector–valued models too.

Let us discuss our problem in some more detail. We are interested in the behaviour of equilibrium states. Equilibrium states are appropriately defined probability measures on the space of all configurations $\sigma = \{\sigma(j),\ j \in \mathbf{Z}\}$. The set $\mathbf{Z}$ is chosen generally in translation invariant models as the integer lattice of the d–dimensional Euclidean space $\mathbf{R}^d$ with some $d \geqslant 1$. In Dyson's model we choose $\mathbf{Z} = \{1, 2, \ldots\}$. The variables $\sigma(j)$, $j \in \mathbf{Z}$, which are called in the literature spin variables, take values in the Euclidean space $\mathbf{R}^p$ with some $p \geqslant 1$. We speak of scalar–valued models if $p = 1$ and vector–valued models if $p \geqslant 2$. The probability measure that we call equilibrium space is a Gibbs measure which is a well–known object in statistical physics. It depends on the following three quantities: A Hamiltonian function $\mathcal{H}(\sigma)$ defined on the configurations σ, a probability measure ν, called the free measure, on $\mathbf{R}^p$, a physical parameter T, called the temperature. Formally it is defined as the probability measure μ given by the formula

$$\mu(d\sigma) = \text{const.} \exp\left\{-\frac{1}{T}\mathcal{H}(\sigma)\right\} \prod_{j \in \mathbf{Z}} \nu(d\sigma(j))\,. \tag{1}$$

Formula (1) would require a more detailed explanation. We omit it in this discussion, because it is needed only to translate the problems we are interested in into analytical questions. A precise definition of equilibrium states can be found also in Appendix D of (Bleher and Major, 1988).

Let $\sigma = \{\sigma(j),\ j \in \mathbf{Z}\}$ be a μ distributed random field. We are interested in its large–scale limit. Let us explain the meaning of this problem. Since we restrict ourselves to Dyson's model we shall assume that $\mathbf{Z} = \{1,2,\ldots\}$. For all $n = 1,2,\ldots$ define the fields

$$Y_n(j) = \frac{1}{A_n} \sum_{k=(j-1)2^n+1}^{j2^n} \sigma(k) - \mathbf{E}\sigma(k), \quad j \in \mathbf{Z}. \tag{2}$$

Let us choose the norming constant A_n in formula (2) in such a way that the finite dimensional distributions of the fields $Y_n(j)$ have a nontrivial limit as $n \to \infty$. We are interested in the right choice of the norming factor A_n and the distribution of the limit field which is called the large–scale limit of the original field $\sigma(j)$ in the literature. In case of vector–valued models, i.e. when $\sigma(j) = (\sigma^{(1)}(j), \ldots, \sigma^{(p)}(j))$ with some $p \geqslant 2$ we normalize in each coordinate independently. This means that we define

$$Y_n^{(l)}(j) = \frac{1}{A_n^{(l)}} \sum_{k=(j-1)2^n+1}^{j2^n} \sigma^{(l)}(k) - E\sigma^{(l)}(k), \quad j \in \mathbf{Z}, \quad l = 1, \ldots, p. \tag{2'}$$

In formulas (2) and (2′) we have formulated a problem about the limit distribution of partial sums of random variables, and this is a very natural problem of classical probability theory. If the variables $\sigma(j)$ are independent or very weakly dependent then $A_n = 2^{n/2}$ is the right choice, and the large–scale limit consists of independent Gaussian random variables. But for certain choice of the Hamiltonian function $\mathcal{H}(\sigma)$ and the temperature T we have to apply a different normalization, and also the large–scale limit has a different structure. Our main goal is to obtain a possibly complete picture of different possibilities.

It is quite natural that the large–scale limit strongly depends on the Hamiltonian function of the equilibrium state. The surprising fact, which requires an explanation, is the dependence of the large–scale limit and of the norming constant on the temperature T. In interesting models the following picture holds true: There is a so–called critical temperature T_{cr}, where the model has a peculiar behaviour. For all $T \neq T_{cr}$ the large–scale limit is a field of independent Gaussian variables, and the norming constant is the classical one, i.e. one has to divide by the square–root of the number of summands in a bloc. The only exceptional case is when $T = T_{cr}$. In this case one has to divide by a different power, and the large–scale limit may be non–Gaussian. Moreover, the behaviour of translation invariant models at the critical temperature strongly depends on the dimension of the lattice where they are defined.

It is an outstanding problem of the statistical physics to find a satisfactory explanation for the above mentioned phenomena. Our paper (Bleher and Major, 1987) was devoted mainly to this problem. We have also understood that in vector–valued models some even more complex phenomena appear. Such models can show some phenomena similar to the critical behaviour for *all* low temperatures. The discussion of this question is the main topic of the present paper. Let us remark that since here a model living on the integer lattice is investigated, i.e. a model whose dimension is always one, it is not possible to investigate the dependence of the behaviour of different models on their dimension directly with the

help of this model. Nevertheless, this model can give some useful information also about this problem.
Our model has the Hamiltonian function

$$\mathcal{H}(\sigma) = -\sum_{i\in\mathbb{Z}}\sum_{\substack{j\in\mathbb{Z}\\ j>i}} d(i,j)^{-a}\sigma(i)\sigma(j), \tag{3}$$

where $1 < a < 2$, $d(i,j) = 2^{n(i,j)-1}$, and

$$n(i,j) = \min n, \ \exists \text{ some } k \text{ such that } (k-1)2^n < i,j \leqslant k2^n .$$

(In vector-valued models $\sigma(i)\sigma(j)$ denotes scalar product in formula (3).) For the sake of convenience we work with the number $c = 2^{2-a}$ instead of the parameter a. Here the function $d(i,j)$, called hierarchical distance, is a modification of the usual distance $|i-j|$. It behaves very similarly, and we have introduced it, because models defined with the help of such a distance are simpler to handle because of their symmetry. Let us observe that in formula (3) there is an interaction $d^{-a}(i,j)$ between all pairs (i,j). This interaction is power-like decreasing with respect to the hierarchical distance, and the exponent a of this power is a most important parameter of the model. It plays a role similar to the dimension in translation invariant models.
The description of the large-scale limit of an equilibrium state is a very natural probabilistic problem, but it needs some justification from the point of view of statistical physics. It seems very unlikely that some physical effect can be thought out which depends on what kind of limit distribution appears in the large-scale limit of an equilibrium state. In statistical physics one would be satisfied with the answer to such at the first sight much simpler questions like the order of decrease of the correlation function $\mathbf{E}\sigma(i)\sigma(j) - \mathbf{E}\sigma(i)\mathbf{E}\sigma(j)$ as $|i-j| \to \infty$. But, disregarding some special solvable models, there seems to be no way to determine the decrease of the covariance function without solving the more complex problem about the large-scale limit. This is a very important peculiarity of the problem we are dealing with, and it also indicates the special character of the probabilistic problem one has to solve when investigating the large-scale limit of an equilibrium state. In formula (2) the norming constant A_n has to be chosen in such a way that the variance of $Y_n(j)$ be separated both from zero and infinity. If one knows the right choice of A_n one can also determine the decrease of the correlation function. But we do not know the behaviour of the correlation function at the start. Moreover, the correlation function behaves essentially differently for $T = T_{cr}$ and $T \neq T_{cr}$, and no method is known to give an analytic formula for the critical temperature. Probably, it is principally impossible to give such an explicit formula. What we are able to do is to carry out a procedure which enables us to approximate the critical temperature at each step better and better. We can get the information we are interested in with the help of such a procedure. This fact also indicates an essential difference between the determination of the large-scale limit of equilibrium states and the usual limit theorems in classical probability. In problems of probability theory one generally knows at the start whether one has to deal with strongly or weakly dependent random variables. Here we learn it only after solving the limit problem.

To give a more detailed discussion we have to translate our problem to purely analytical questions. To do it first we have to define our model completely. We consider a model with the Hamiltonian function $\mathcal{H}(\sigma)$ defined in (3) and the free measure ν defined by the formula

$$p_0(x) = \frac{d\nu}{dx}(x) = C(t)\exp\left\{-\frac{x^2}{2} - \frac{t}{4}|x|^4\right\}, \tag{5}$$

where $t > 0$ is another parameter of the model, and $C(t)$ is an appropriate norming constant. We shall assume that $t > 0$ is sufficiently small. What is important for us is that $p_0(x)$ is a small perturbation of the normal density function, and it tends to zero very fast.

Our first problem is to construct an equilibrium state with the above defined Hamiltonian function and free measure at some temperature T and then to describe its large–scale limit. Both problems can be solved with the help of some limiting procedure if one solves first the following two problems:

Problem 1. Put $V_n = \{1, 2, \ldots, 2^n\}$, and

$$\mathcal{H}_{V_n}(x_1, \ldots, x_{2^n}) = -\sum_{i\in V_n}\sum_{\substack{j\in V_n\\ j>i}} d(i,j)^{-a} x_i x_j.$$

Define the probability measure $\mu_n = \mu_{n,T}$ on $\mathbf{R}^{V_n}$ (on $(\mathbf{R}^p)^{V_n}$ if we have a model with p–dimensional spins) with the density function $p_n(x_1, \ldots, x_{2^n})$ by the following formula:

$$p_n(x_1, \ldots, x_{2^n}) = \frac{d\mu_n(x_1, \ldots, x_{2^n})}{dx_1 \ldots dx_{2^n}} = C_n \exp\left\{-\frac{1}{T}\mathcal{H}_{V_n}(x_1, \ldots, x_{2^n})\right\}.$$

Let $(\sigma(1), \sigma(2), \ldots, \sigma(2^n))$ be a μ distributed random vector, and let $p_n(x)$ denote the density function of the average $2^{-n}\sum\limits_{i=1}^{2^n}\sigma(i)$. Give a good asymptotic formula for $p_n(x)$.

Problem 2. Let $N \geqslant n$, and $h > 0$. Define the probability measure μ_N^h on $\mathbf{R}^{V_N}$ (on $(\mathbf{R}^p)^{V_N}$ in p–dimensional models) with density function $p_N^h(x_1, \ldots, x_{2^N})$ by the formula

$$p_N^h(x_1, \ldots, x_{2^N}) = C_N^h \exp\left\{-\frac{1}{T}\mathcal{H}_{V_N}^h(x_1, \ldots, x_{2^N})\right\},$$

where

$$\mathcal{H}_{V_N}^h(x_1, \ldots, x_{2^N}) = \mathcal{H}_{V_N}(x_1, \ldots, x_{2^N}) - h\sum_{i=1}^{2^N} x_i,$$

and C_N^h is an appropriate norming constant. (In the vector–valued case we consider h as the vector $(h, 0, \ldots, 0)$ with some $h > 0$, and product means scalar product in the last formula.) Let $\mu^h_{n,N}$ denote the restriction of the above defined measure μ^h_N to the volume V_n, and let us consider the Radon–Nikodym derivative

$$f^h_{n,N}(x_1,\dots,x_{2^n}) = \frac{d\mu^h_{n,N}}{d\mu_n}(x_1,\dots,x_{2^n}).$$

Give a good asymptotic formula on the function $f^h_{n,N}(x_1,\dots,x_{2^n})$.
Both problems can be translated to purely analytical questions. It can be seen e.g. with the help of Appendix A of (Bleher and Major, 1988) that Problem 1 is equivalent to the following

Problem 1$'$. Define the sequence of density functions $p_n(x) = p_n(x,T)$ by the recursive formula

$$p_{n+1}(x) = C_n(T)\int \exp\left\{\frac{c^n}{T}(x^2-u^2)\right\} p_n(x-u)p_n(x+u)\,du\,,$$

and let $p_0(x)$ be defined by formula (5). Give a good asymptotic formula on $p_n(x)$.
Problem 2 can be translated with the help of the result in Appendix C of (Bleher and Major, 1988) to the following

Problem 2$'$. Define the functions $f^h_{n,N}(x)$, $N \geqslant n$, by the relations

$$f^h_{N,N}(x) = K(N,h)\exp\left(\frac{2^N h x^{(1)}}{T}\right), \tag{6}$$

$$f^h_{n,N}(x) = K(n,N,h)\mathbf{S}_n f^h_{n+1,N}(x) \tag{6'}$$

with

$$\mathbf{S}_n f(x) = \int \exp\left(\frac{c^n}{T}xy\right) f\left(\frac{x+y}{2}\right) p_n(y)\,dy\,, \tag{6''}$$

where $p_n(x)$ is the same as in Problems 1 and 1$'$, and $K(n,N,h)$ is an appropriate normig constant. Find a good asymptotic formula for the above defined functions $f^h_{n,N}(x)$.
It is proved in Appendix C of (Bleher and Major, 1988) that

$$\frac{d\mu^h_{n,N}}{d\mu_n}(x_1,\dots,x_{2^n}) = f^h_{n,N}\left(2^{-n}\sum_{j=1}^{2^n} x_j\right), \quad n \leqslant N,$$

hence Problems 2 and 2$'$ are equivalent. The main part of our investigation consists of solving Problems 1$'$ and 2$'$. In this paper the vector-valued case (i.e. the case when $p \geqslant 2$) is considered.
Formally the problems change very little when scalar-valued models are replaced with vector-valued ones. Thus in Problem 1$'$ the only change is that $|x|$ means the absolute value of a vector, and xy denotes scalar product. Nevertheless, and this is the most striking feature of the problem we are investigating, these seemingly unessential modifications radically change the behaviour of the model. Thus the functions $p_n(x)$ defined in Problem 1$'$ have the following behaviour for small T in the scalar-valued case: Since $p_n(x) = p_n(-x)$, it is enough to consider $p_n(x)$ for

$x \geqslant 0$. There is some sequence $M_n = M_n(T)$, $M_n > 0$, $M_n \to M$ with some $M = M(T) > 0$ such that $2^{-n/2}p(2^{-n/2}x+M_n)$ tends to a normal density function with expectation zero and some positive variance. (The number M_n is called the spontaneous magnetization in the literature.) This means a central limit theorem with the usual normalization.
The behaviour of the model in the vector-valued case is more complex. It is not difficult to see that $p_n(x)$ depends on x only through its absolute value $|x|$, i.e. there is a function $P_n(z) = P_n(z,T)$, $z \in \mathbf{R}^1$ and $z \geqslant 0$ such that $p_n(x) = P_n(|x|)$ for all $x \in \mathbf{R}^p$. Hence, it is natural to investigate the function $P_n(z)$ instead of $p_n(x)$. For small T the behaviour of the function $P_n(z,T)$ is essentially different for $1 < c < \sqrt{2}$ and $\sqrt{2} < c < 2$. For $\sqrt{2} < c < 2$ there is a sequence M_n such that $2^{-n/2}P_n(2^{-n/2}z+M_n)$ tends to a Gaussian density function with expectation zero and positive variance, i.e. the situation is similar to the scalar-valued case. On the other hand, for $1 < c < \sqrt{2}$ we have to normalize otherwise. In this case $c^{-n}P(c^{-n}z+M)$ has a non-trivial limit with some $M > 0$ which can be described as the solution of an integral equation. This means a non-central limit theorem with an unusual normalization. (In scalar-valued cases there is only at $T = T_{cr}$ such a big difference between the cases $1 < c < \sqrt{2}$ and $\sqrt{2} < c < 2$.)
The main goal of our investigations is to give an explanation for the above discussed phenomena and some related questions. For this aim let us rewrite Problem 1′ in a form more appropriate for us. Let us introduce the functions

$$\bar{q}_n(x) = \bar{q}_n(x,T) = B_n \exp\left\{\frac{a_0}{2a_1}c^n x^2\right\} p_n\left(\sqrt{\frac{T}{a_1}}x\right)$$

and

$$q_n(x) = c^{-n/2}\bar{q}_n\left(c^{-n/2}x\right)$$

with $a_0 = 2/(2-c)$ and $a_1 = a_0 + 1$, where the function $p_n(x)$ is the same as in Problems 1 and 1′. Simple calculation shows that

$$\bar{q}_{n+1}(x) = \int \exp\{-c^n u^2\}\bar{q}_n(x-u)\bar{q}_n(x+u)\,du \tag{7}$$

and

$$q_{n+1}(x) = \mathcal{R}q_n(x)$$

with

$$\mathcal{R}q(x) = \int \exp\{-u^2\} q\left(\frac{x}{\sqrt{c}} - u\right) q\left(\frac{x}{\sqrt{c}} + u\right) du \tag{7'}$$

with some B_n. (We are not really interested in the value of B_n, because it influences only the norming factor when $p_n(x)$ is expressed through $q_n(x)$. On the other hand, the norming constant in $p_n(x)$ is determined by the fact that $p_n(x)$ is a density function.) We also have

$$\bar{q}_0(x) = q_0(x) = B_0 \exp\left\{\frac{a_0 - T}{2a_1}x^2 - \frac{tT^2}{4a_1^2}|x|^4\right\}.$$

Since the function $p_n(x)$ can be simply expressed by $q_n(x)$ or $\bar{q}_n(x)$, their investigations are equivalent problems. It is more convenient to work with the function $q_n(x)$ or $\bar{q}_n(x)$ than directly with $p_n(x)$. Let us emphasize that the formulas expressing $q_{n+1}(x)$ and $\bar{q}_{n+1}(x)$ through $q_n(x)$ and $\bar{q}_n(x)$ do not depend on the parameter T. Moreover, in the case of the function $q_n(x)$ it depends neither on n. (This was our main reason for introducing the function $q_n(x)$.) All dependence on T is contained in the starting function $q_0(x)$. But this dependence is very essential, because the operator $\mathcal{R}$, unlike its classical probability counterpart the convolution operator, "remembers" of the starting function. The limit of $\mathcal{R}^n q_0(x)$, as $n \to \infty$, is a small perturbation of the starting function. In (Bleher and Major, 1987) we discussed this property and pointed out that this is the final cause of phase transitions and critical phenomena in statistical physics. Here we show that it has other far-reaching consequences. It also implies that vector-valued models with continuous symmetries have some properties which have no scalar-valued counterpart. Let us explain this in more detail.

The function $\bar{q}_n(x)$ depends on x only through its absolute value $|x|$, i.e. it remains invariant if we rotate x. This property, called $\mathbf{O}(p)$ symmetry in the literature, is a continuous symmetry, and it has far-reaching consequences. For the sake of simpler notations let us assume that $p = 2$. Because of this property it is natural to introduce the function $Q_n(x)$ defined by the relation $Q_n(x) = \bar{q}_n((x, 0))$ and to work with this function instead of the original function $\bar{q}_n(x)$. Let us rewrite formula (7) for the functions $Q_n(x)$. We get that

$$Q_{n+1}(x) = \tag{8}$$
$$= C_n \int \exp\left\{-c^n(u^2+v^2)\right\} Q_n\left(\sqrt{(x+u)^2+v^2}\right) Q_n\left(\sqrt{(x-u)^2+v^2}\right)\, du\, dv.$$

The argument of the function Q_n in formula (8) is rather complicated. Hence it is natural to substitute it by a simpler expression and to control the error caused by this substitution. This can be done in both cases $1 < c < \sqrt{2}$ and $\sqrt{2} < c < 2$, but the right substitutions are different in the two cases. This is the reason of their different behaviour. We give a short informal explanation for this.

In formula (8) we have to integrate in the whole space $\mathbf{R}^2$, but because of the kernel $\exp\{-c^n(u^2+v^2)\}$ in the integral an essential contribution for the integral is supplied only if u and v are small. (Actually this is the cause why the operator $\mathcal{R}$ "remembers" to the starting function.) The function $Q_n(x)$ has a maximum $M = M_n$ which depends very weakly on n, hence at this heuristical level we can simply disregard this dependence. Actually it is enough to give a good asymptotic formula for $Q_n(x)$ only in a small neighbourhood of M. When $x \sim M$ and u and v are small it is natural to try to make one of the following approximations in the argument of Q_n in formula (8):

$$\sqrt{(x \pm u)^2 + v^2} \sim x \pm u + \frac{v^2}{2(x \pm u)} \sim x \pm u + \frac{v^2}{2M},$$

or

$$\sqrt{(x \pm u)^2 + v^2} \sim x \pm u\,.$$

The first approximation would suggest the formula

$$\begin{aligned} Q_{n+1}(x) &= \\ &= C_n \int \exp\left\{-c^n(u^2+v^2)\right\} Q_n\left(x+u+\frac{v^2}{2M}\right) Q_n\left(x-u+\frac{v^2}{2M}\right) du\,dv + \\ &+ \varepsilon_n^1(x) = \\ &= \int \exp\left\{-c^n v^2\right\} Q_n\left(x+u+\frac{v^2}{2M}\right) Q_n\left(x-u+\frac{v^2}{2M}\right) du\,dv + \varepsilon_n(x), \quad (9) \end{aligned}$$

and the second one

$$\begin{aligned} Q_{n+1}(x) &= C_n \int \exp\left\{-c^n(u^2+v^2)\right\} Q_n(x+u)Q_n(x-u)\,du\,dv + \bar{\varepsilon}_n^1(x) = \\ &= \int Q_n(x+u)Q_n(x-u)\,du + \bar{\varepsilon}_n(x). \quad (9') \end{aligned}$$

(If the approximation in the first row is allowed in the latter formulas then the approximation in the second row is also permitted.) Whether the first approximation (9) has to be applied or the second rougher approximation (9′) is also permitted that depends on whether the error terms $\varepsilon_n(x)$ and $\bar{\varepsilon}_n(x)$ are negligible or not. A detailed analysis shows that in the case $1 < c < \sqrt{2}$ the first approximation (9) is the right choice and in the case $\sqrt{2} < c < 2$ the second one (9′).
For $1 < c < \sqrt{2}$ let us introduce the function $\bar{Q}_n(x) = c^{-3/2n}Q_n(c^{-n}x + M)$. Formula (9) can be rewritten as

$$\bar{Q}_{n+1}(x) = \mathbf{T}\bar{Q}_n(x) + \bar{\varepsilon}_n(x) \quad (10)$$

with

$$\mathbf{T}f(x) = \mathbf{T}_c f(x) = \int \exp(-v^2) f\left(\frac{x}{c} - u + \frac{v^2}{2M}\right) f\left(\frac{x}{c} + u + \frac{v^2}{2M}\right) du\,dv\,.$$

In this case $\bar{Q}_n(x)$ tends to the unique solution of the fixed point equation $Q = \mathbf{T}Q$, and this fact supplies the answer to Problem 1′. To prove it we have to show that the solution $Q(x)$ of the fixed point equation is sufficiently stable for our purposes, i.e. the relation $\lim_{n\to\infty} \mathbf{T}^n(Q(x)+\varepsilon(x)) = Q(x)$ holds for a small perturbation of the fixed point $Q(x)$, and this convergence is sufficiently fast. This stability property holds only in the case $1 < c < \sqrt{2}$, and this is the reason why the solution of Problem 1′ is different in the cases $1 < c < \sqrt{2}$ and $\sqrt{2} < c < 2$. To give a complete proof of the convergence $\bar{Q}_n(x) \to Q(x)$ we have to overcome several technical difficulties. The most important one among them is to show that the error term $\bar{\varepsilon}_n(x)$ is really negligibly small. This question is discussed in the Preface and in the first three Sections of Part I of paper (Bleher and Major, 1988) in more detail.
In the case $\sqrt{2} < c < 2$ it is more natural to work with the function $\tilde{Q}_n(x) = 2^{-n/2}Q_n(2^{-n/2}x + M)$. Relation (9′) implies that

$$\tilde{Q}_{n+1}(x) = \mathbf{U}\tilde{Q}_n(x) + \bar{\varepsilon}_n'(x) \quad (10')$$

with

$$\mathbf{U}f(x) = \int f\left(\frac{x}{\sqrt{2}} - u\right) f\left(\frac{x}{\sqrt{2}} + u\right) du .$$

The fixed point of the operator $\mathbf{U}$, i.e. of the convolution operator, are normal density functions with expectation zero. In the case $\sqrt{2} < c < 2$ the error term $\bar{\varepsilon}_n(x)$ is negligibly small for $\sqrt{2} < c < 2$. This implies that in this case $\sqrt{2} < c < 2$ $\tilde{Q}_n(x)$ tends to a normal density function, and this yields the solution of Problem 1′ in this case too.

We discuss the investigation of Problem 2 more briefly. We have introduced the dependence of the measure μ_N^h on h in order to investigate the influence of an external magnetic field to the model. We need such an approach in the investigation of the model at low temperatures. At low temperatures there is a phase transition, i.e. several different equilibrium states exist with the same Hamiltonian function, free measure and temperature. Hence first we must clarify which equilibrium state we are working with. We select out a pure state, i.e. an equilibrium state which cannot be written as the mixture of different equilibrium states. A natural way to construct pure states, and we choose this approach, is to introduce an external field h_N in the volume V_N and take the limit of the measures $\mu_N^{h_N}$ in V_N as $N \to \infty$ and $h_N \to 0$. To carry out this program we have to solve Problem 2. The pure state we construct in this way has a spontaneous magnetization in the direction $e^{(1)} = (1, 0)$, i.e. $\mathbf{E}\sigma^{(1)}(j) = M > 0$ and $\mathbf{E}\sigma^{(2)}(j) = 0$ for a random field $\sigma = \{\sigma(j),\, j \in \mathbf{Z}\}$ with the distribution of the pure state we have constructed. In the large–scale limit defined by formula (2′) we have to normalize differently in the direction of the magnetization and in the direction orthogonal to it. Our main interest in this work is the description of the large–scale limit in both directions. The operator S_n defined in formula (6″) depends on the function $p_n(x)$ appearing in Problem 1. This implies that the recursive formula expressing $f_{n,N}^h(x)$ through $f_{n+1,N}^h(x)$ has an essentially different form in the cases $1 < c < \sqrt{2}$ and $\sqrt{2} < c < 2$. Nevertheless, the asymptotic behaviour of the function $f_{n,N}^h(x)$ is the same in the two cases.

More precisely, we need a good asymptotic formula for the function $f_{n,N}^h(x)$ only in a typical region, and outside this region it is enough to give some upper bound on it. Actually we are interested in the product $p_n(x) f_{n,N}^h(x)$, and not the function $f_{n,N}^h(x)$ itself. Hence the typical region, where we need a good approximation, is a small neighbourhood of the maximum of the above mentioned product. In this domain the Radon–Nikodym derivative has the following form:

$$f_{n,N}^h(x) = C_n \exp\left\{g_n x^{(1)} + A_n x^{(2)2} + \varepsilon_n(x)\right\}, \tag{11}$$

where $\varepsilon_n(x)$ is a small error term, and the constants g_n and A_n are defined by a recursive formula. This formula holds both for $1 < c < \sqrt{2}$ and $\sqrt{2} < c < 2$, and even the recursive formulas on g_n and A_n have the same structure in the two cases. This means that in Problem 2′ there is no essential change between models with different parameters c. A heuristic explanation of this fact is contained in the last two formulas at page 466 of (Bleher and Major, 1987a), and they are the basis for our investigations of Problem 2. When giving an upper bound outside the typical

region we have to work differently if the absolute value $|x|$ is not typical and if it is typical, but the vector x is not in the typical region because of its direction. This question is discussed in Section 2 of Part II of (Bleher and Major, 1988) in more detail.
The solution of Problem 2 in the case $\sqrt{2} < c < 2$ is given in (Bleher and Major, 1984). Actually the greatest part of that work deals with this question. The solution of this problem (which also contains the investigation of the asymptotic behaviour of the sequences g_n and A_n) is the main ingredient in the description of the large-scale limit of the equilibrium state. For $\sqrt{2} < c < 2$ one has to normalize with $A_n^{(2)} = 2^n c^{-n/2}$ in the direction orthogonal to the direction of the spontaneous magnetization, and the limit is a field of dependent Gaussian random variables whose distribution we can describe explicitly. In the direction of the magnetization the classical norming $A_n^{(1)} = 2^{n/2}$ has to be applied, and the limit is a field of independent Gaussian random variables. (The boundary case $c = \sqrt{2}$ is similar to the above case $\sqrt{2} < c < 2$. (See (Bleher and Major, 1989).) The only difference is that in this case the normalization $A_n^{(1)} = 2^{n/2}\sqrt{n}$ has to be applied in the direction of the spontaneous magnetization.) This means that in the direction orthogonal to the spontaneous magnetization a "critical" normalization has to be applied for all low temperatures, i.e. the same normalization as at the critical temperature. This result has no equivalent in scalar-valued models.
The large-scale limit of Dyson's model in the case $1 < c < \sqrt{2}$ is similar to the case $\sqrt{2} < c < 2$ in the direction orthogonal to the spontaneous magnetization, and it is different in the direction of the magnetization. The reason for it lies in the fact that the solution of Problem 2 is similar in the two cases, and the solution of Problem 1 is different. In the direction orthogonal to the spontaneous magnetization one has to divide again by $A_n^{(2)} = 2^n c^{-n/2}$, and the limit is a field of dependent Gaussian random variables. In the direction of the spontaneous magnetization one has to divide by $A_n^{(1)} = 2^n c^{-n}$, and the limit field is non-Gaussian. The explicit form of the large-scale limit is given in Theorem 2 of Part II in (Bleher and Major, 1988). The proof of this result consists of a limit procedure which can be carried out if Problems 1 and 2 are already solved.

REFERENCES

Bleher, P.M. and Major, P. (1984). Renormalization of Dyson's hierarchical vector-valued φ^4 model at low temperature. *Comm. Math. Phys.* **95**, 487–532.

Bleher, P.M. and Major, P. (1987a). Critical phenomena and universal exponents in statistical physics. *Ann. Probab.* **15**, 431–477.

Bleher, P.M. and Major, P. (1987b). Limit Theorems in Statistical Physics. On Dyson's Hierarchical Model. In: *Proceedings of the First World Congress of the Bernoulli Society.* Ed. by Yu. V. Prohorov et al. V.1 VSP, Utrecht.

Bleher, P.M. and Major, P. (1988). The large-scale limit of Dyson's hierarchical vector-valued model at low temperature. *Ann. de l'Institute H. Poincaré, ser. Phys.* **49**, 1–143.

Bleher, P.M. and Major, P. (1989). The large-scale limit of Dyson's hierarchical vector-valued model at low temperature. The marginal case $c=\sqrt{2}$. *Comm. Math. Phys.* **125**, 43–69.

Prob. Theory and Math. Stat., Vol. 1, pp. 152–165
B.Grigelionis *et al.* (Eds.)
1990 VSP/Mokslas

ANALYTIC FUNCTIONALS OF STOCHASTIC PROCESSES AND INFINITE DIMENSIONAL OSCILLATORY INTEGRALS

V.I. BOGACHEV
Department of Mechanics and Mathematics,
Moscow State University, Moscow, USSR

ABSTRACT

We prove smoothness of joint distributions of several quadratic forms or integral–type functionals of diffusion and Gaussian processes.

In this paper we study quadratic forms and integral–type functionals of random processes (in particular, diffusion and Gaussian processes) and improve some results of (Bogachev, 1989). Actually, this paper continues (Bogachev, 1989), but can be read independently. Let (X, μ) be *a* measurable space, $F : X \to \mathbf{R}^m$ be μ–measurable. The measure $\mu \circ F^{-1}$ on $\mathbf{R}^m$ is defined by the formula $\mu \circ F^{-1}(B) = = \mu(F^{-1}B), B$ being Borel in $\mathbf{R}^m$. In recent years different methods of studying differential properties of $\mu \circ F^{-1}$ were developed. The method of (Bogachev, 1989) uses some ideas of the Malliavin calculus and the notion of differentiability along a vector field. Recall some basic constructions in the case when X is a locally convex space with a Radon measure μ. Let $v : X \to X$ be a Borel mapping (vector field). Differentiability of a function φ along v is defined in a usual way:

$$\partial_v \varphi(x) = \lim_{t \to 0} \frac{1}{t} \Big(\varphi\big(x + tv(x)\big) - \varphi(x) \Big).$$

A measure μ is said to be *differentiable along* v if there exists a measure ν with the following property:

$$\int \partial_v \varphi \mu = - \int \varphi \nu$$

for all bounded Borel φ with $\partial_v \varphi \in L^1(\mu)$. Such measure ν is called a *derivative* of μ along v and denoted by $d_v \mu$. If $d_v \mu = \rho_v \mu$ then ρ_v is called a logarithmic derivative of μ. Derivatives $d_v^2 \mu, \ldots,$ are defined by induction.

Example 1. a) For a constant vector field $v(x) \equiv h$ differentiability above coincides with Skorokhod differentiability (see (Bogachev, 1988a, 1989; Skorokhod, 1974)). If $d_h \mu \ll \mu$ we obtain Fomin differentiability; the set of all h with this property is denoted by $D(\mu)$ and called the subspace of differentiability of μ (see (Bogachev, 1986, 1988b)).

b) If X is a Hilbert space, γ is a centered Gaussian measure on X with the covariation operator R then γ is infinitely differentiable along $D(\gamma) = \sqrt{R}(X)$ and $\rho_h(x) = -(x, a)$ for every $h = \sqrt{R}a$.

c) Denote by $\{e_n\}$ and $\{r_n\}$ eigenvectors and eigenvalues of R in b). Consider a field $v(x) = Ru(x)$, where $u : X \to X$ is Borel, differentiable along $H = D(\gamma), u'_H(x)$ is a Hilbert - Schmidt operator γ-a.e., $\|u\|, \|u'_H\|_2 \in L^2(\gamma)$ $(\|u'_H(x)\|_2 = \left(\sum\limits_n \left(\partial_{\varphi_n} u_n(x)\right)^2\right)^{1/2}$, $\varphi_n = \sqrt{r_n} e_n$, is the Hilbert - Schmidt norm for operators on H). It is easy to prove (see (Daletsky, 1983)), that γ is differentiable along v and $d_v\gamma = -(\mathrm{tr}_H \sqrt{R} u'_H(x) - (\sqrt{R}u(x), x))\gamma, \mathrm{tr}_H \sqrt{R} u'_H(x) = = \sum\limits_n \sqrt{r_n} \partial_{\varphi_n} u_n(x)$. Instead of differentiability along H it is possible to require absolute continuity of functions $t \mapsto u_n(x + t\varphi_n)\gamma$-a.e.

Assume now that $m = 1$ and F, μ are infinitely differentiable along v with $\partial_v F \neq 0$ a.e., $\partial_v^k F, 1/\partial_v F \in L^p(d_v^n \mu)$ for all $k, n \geqslant 0, p \geqslant 1$. Then $\mu \circ F^{-1}$ admits a density $p \in C^\infty$. For the proof it suffices to obtain the following estimates by means of the integration by parts formula (see (Bogachev, 1989; Gaveau and Moulinier, 1983; Moulinier, 1985))

$$|\tau(t)| = \left|\int\limits_X \exp\left(itF(x)\right)\mu(dx)\right| \leqslant c_\ell(1 + |t|)^{-\ell}. \tag{1}$$

In the case $m > 1$ we shall use the following simple trick. Let $e \in \mathbf{R}^m, \|e\| = 1$, $F_e = (F, e), \tau_e(t) = \int \exp(itF_e(x))\mu(dx)$. If all $c_\ell(e)$ in the estimates (1) for τ_e can be taken locally bounded on the unit sphere in $\mathbf{R}^m$ then $\mu \circ F^{-1}$ also has a smooth density on $\mathbf{R}^m$. So the main problem we face is to estimate infinite dimensional oscillatory integrals, depending on parameters. This is very closely connected with estimations of $\mu(x : |\partial_v F(x)| \leqslant \varepsilon), \varepsilon \to 0$. Note that such problem appears in all approaches to the Malliavin calculus (in standard Malliavin's schemes we have $\langle DF, DF\rangle_H$ instead of $\partial_v F$), see (Bogachev, 1989; Bell, 1987; Ikeda and Watanabe, 1981; Smorodina, 1988; Moulinier, 1985). In (Bogachev, 1989) and here these estimates are obtained for some polynomials and integral-type functionals. An open problem: if f is a continuous polynomial on a Hilbert space X with a nondegenerate Gaussian measure γ and $f' \neq 0$ then must $\gamma \circ f^{-1}$ possess a smooth density ?

Let $L(\mathbf{R}^m)$ denote the normed space of operators in $\mathbf{R}^m$ and $S(\mathbf{R}^m)$ be the Schwartz space of rapidly decreasing smooth functions on $\mathbf{R}^m$.

Denote by ξ_t the solution of the stochastic differential equation

$$d\xi_t = A(t, \xi_t)\, dw_t + B(t, \xi_t)\, dt, \;\; \xi_0 = 0, \;\; t \in [0, T], \tag{2}$$

where w_t is m-dimensional standard Wiener process,

$$A : [0, T] \times \mathbf{R}^m \to L(\mathbf{R}^m), B : [0, T] \times \mathbf{R}^m \to \mathbf{R}^m$$

are Lipschitzian with Lipschitzian derivatives of all orders in the second argument. Consider the following quadratic form of this diffusion

$$Q(\omega) = \int\limits_0^T \int\limits_0^T \Big(K(s, t)\xi_s(\omega), \xi_t(\omega)\Big) ds dt, \tag{3}$$

are Lipschitzian with Lipschitzian derivatives of all orders in the second argument. Consider the following quadratic form of this diffusion

$$Q(\omega) = \int_0^T \int_0^T \Big(K(s,t)\xi_s(\omega), \xi_t(\omega) \Big) ds dt, \tag{3}$$

where $K : [0,T] \times [0,T] \to L(\mathbf{R}^m)$ is a Holder (of some order α) mapping, satisfying the condition $K(s,t) = K(t,s) = K(s,t)^*$. We say that K is nondegenerate if for every n there exist $t_1, \dots, t_n \in [0,T]$ such that the mappings $s \mapsto K(s,t_i)$ are linearly independent.

LEMMA 1. *If K satisfies the conditions above then for every n there exist $s_1, \dots, s_n \in [0,T]$ and $e \in \mathbf{R}^m$ such that the mappings $s \mapsto K(s,s_i)e$ are linearly independent. The equivalent condition: the integral operator $\widetilde{K}$ with the kernel K in $C([0,T], \mathbf{R}^m)$ has infinite dimensional range.*

Proof. If $\ell = \dim \operatorname{span}\{K(\cdot,s)h : s \in [0,T], h \in \mathbf{R}^m\} < \infty$ then we can take a base $v_1, \dots, v_\ell$ in this linear space of $\mathbf{R}^m$–valued functions, where $v_j = K(\cdot, s_j)h_j$. Let $v_j = (v_{1j}, \dots, v_{mj}), V = (v_{ij})$. For every $t \in [0,T]$ and $i \in \{1, \dots, m\}$ there exist $\lambda_{i1}(t), \dots, \lambda_{i\ell}(t)$ with $K(\cdot,t)e_i = \lambda_{i1}(t)v_1 + \dots + \lambda_{i\ell}(t)v_\ell$. Denote by $\Lambda(t)$ the matrix $(\lambda_{ij}(t))$. Among such matrices we can find the maximal number $k (k \leqslant \ell \times m)$ of linearly independent ones $\Lambda(t_1), \dots, \Lambda(t_k)$. Now we see that each map $s \mapsto K(s,t)$ is a linear combination of $V\Lambda(t_1), \dots, V\Lambda(t_k)$. Indeed $\Lambda(t) = \alpha_1(t)\Lambda(t_1) + \dots + \alpha_k(t)\Lambda(t_k)$ and $K(\cdot,t) = V\Lambda(t) = \alpha_1(t)V\Lambda(t_1) + \dots + \alpha_k(t)V\Lambda(t_k)$. This is a contradiction with a nondegeneracy of K. The second statement is obvious.

THEOREM 1. *If K is nondegenerate and for every n*

$$\mathbf{P}\left(\omega : \inf_t \Big| \det A(t, \xi_t(\omega)) \Big| \leqslant \varepsilon \right) = o(\varepsilon^n) \tag{4}$$

then Q has a density of distribution $p \in S(\mathbf{R}^1)$.

COROLLARY 1. *The conclusion of Theorem 1 is true if*

$$\inf_{t,x} \Big| \det\, A(t,x) \Big| > 0. \tag{5}$$

COROLLARY 2. *For each n*

$$\tau(t) = \int \exp\, \left(it\widetilde{Q}(x) \right) \mu^\xi(dx) = o(t^{-n}), \quad t \to \infty, \tag{6}$$

COROLLARY 3. *If nontrivial linear combinations of kernels $K_1,\dots,K_d$ satisfy the conditions above then the map $(Q_1,\dots,Q_d)$ induces a measure with the density $p \in S(\mathbf{R}^d)$.*

Proof. The proof is divided in four steps.

Step 1. Let $X = L^2([0,T],\mathbf{R}^m), \gamma$ be the Wiener measure on X, induced by the process w_t, starting from the origin, R be covariation operator of γ with the proper base $\{e_n\}$ and correspondent eigenvalues $\{r_n\}$. It is known (Ikeda and Watanabe, 1981) that there exists a Borel map $\Phi : X \to X$ with $\Phi(\omega)(t) = \xi_t(\omega)$ for almost all $\omega \in X$ (we consider (X,γ) being a probability space for w_t). Hence the distribution of Q coincides with the distribution of random variable $F = \widetilde{Q}\circ\Phi$ on (X,γ) and we can investigate

$$F(x) = \int_0^T\int_0^T \Big(K(s,t)\Phi(x)(s), \Phi(x)(t)\Big)dsdt. \tag{7}$$

Using differentiability of solutions of stochastic differential equations with respect to parameters we can show that almost sure Φ is Frechet differentiable along the subspace

$$H = D(\gamma) = \Big\{h \in AC\Big([0,T],\mathbf{R}^m\Big) : h' \in L^2, h(0) = 0\Big\}.$$

The mixed partial derivative $\partial^s_{a_1\dots a_s}\Phi$ along vectors $a_1,\dots,a_s \in H$ coincides with the solution of stochastic differential equation, which is obtained after differentiating with respect to $\lambda_1,\dots,\lambda_s$ at $\lambda_1 = \dots = \lambda_s = 0$ of the following equation

$$dx_t^{\lambda_1\dots\lambda_s} = A(t, x_t^{\lambda_1\dots\lambda_s})\Big[dw_t + (\lambda_1 a_1'(t) + \dots + \lambda_s a_s'(t))dt\Big] +$$

$$+B\left(t, x_t^{\lambda_1\dots\lambda_s}\right)dt, x_0^{\lambda_1\dots\lambda_s} = 0.$$

In particular, $y_t(\omega) = \partial_a\Phi(\omega)(t)$ satisfies the equation

$$dy_t = (A_x'(t,\xi_t)dw_t)y_t + \Big[B_x'(t,\xi_t)y_t + A(t,\xi_t)a'(t)\Big]dt.$$

Let $B_x'(t,\xi_t) = B(t), A_x'(t,\xi_t)dw_t = dA_t, A(t,\xi_t)a'(t) = \nu_t, Z_t$ be an operator–valued process, satisfying the equation

$$dZ_t = B(t)Z_t dt + dA_t Z_t,\ Z_0 = I. \tag{8}$$

Note that Z_t is almost sure invertible and Z_t^{-1} is also a solution of some linear stochastic equation. Indeed, (8) can be written in the Stratonovich's form

$$Z_t = I + \int_0^t A_x'(s,\xi_s)Z_s \circ dw_s + \int_0^t \widetilde{B}(s)Z_s ds$$

with some new drift term $\widetilde{B}$ (see (Ikeda and Watanabe, 1981, Chapter 5, § 2)). Let U_t be an operator-valued process, satisfying the following stochastic integral equation in the Stratonovich's form

$$U_t = I - \int_0^t U_s A'_x(s, \xi_s) \circ dw_s - \int_0^t U_s \widetilde{B}(s) ds.$$

Then

$$d(U_t Z_t) = U_t \circ dZ_t + dU_t \circ Z_t = 0.$$

Hence $U_t Z_t \equiv I$. As we have shown y_t is the solution of the equation

$$dy_t = dA_t y_t + B(t) y_t dt + \nu_t dt.$$

Hence y_t can be written in the following form

$$y_t = Z_t \int_0^t Z_s^{-1} A(s, \xi_s) a'(s) ds. \tag{9}$$

Hence for $a \in H$ the mapping $t \mapsto F(\omega + ta)$ is absolutely continuous for almost all ω and

$$\partial_a F(\omega) = 2 \int_0^T \int_0^T (K(s,t)\xi_s(\omega), y_t(\omega))\, ds dt. \tag{9'}$$

Step 2. Define the vector field v by the formula

$$v(x) = \sum_{n=1}^{\infty} r_n \partial_{e_n} F(x) e_n.$$

This series converges in X almost sure, since $\|a'\|_{L^2} \leqslant c\|a\|_H$ and therefore

$$\left|\partial_{e_n} F(\omega)\right| \leqslant \text{const} \sup\left\|\xi_t(\omega)\right\| \sup\left\|Z_t(\omega)\right\| \sup\left\|Z_t^{-1}\right\| \sup\left\|A(t, \xi_t(\omega))\right\| r_n^{-1/2}.$$

It is possible to prove (we omit rather hard details) that F and γ are infinitely differentiable along v,

$$\partial_v^n F \in \bigcap_p L^p(\gamma), \ d_v^n \gamma = \rho_v^n(\gamma)\gamma,$$

where

$$\rho_v^n(\gamma) \in \bigcap_p L^p(\gamma).$$

The proof of the analogous fact (for $F(\omega) = \Phi(\omega)(t_0)$) can be found in (Ikeda and Watanabe, 1981, Chapter 5, § 8), where it is shown that F belongs to all

Sobolev classes over (X,γ) $(F \in H_\infty$, see (Ikeda and Watanabe, 1981)). The random variable $\langle DF, DF\rangle_H$ which appears in the schemes based on the Malliavin–Ornstein–Uhlenbeck operator (see (Ikeda and Watanabe, 1981, Chapter 5, § 7; Bell, 1987; Gaveau and Moulinier, 1983; Moulinier, 1985)) coincides with $\partial_v F$. So in any case we have to prove the inclusion

$$1/\partial_v F = \langle DF, DF\rangle_H^{-1} \in \bigcap_p L^p(\gamma)$$

(this is the main difficulty).

Step 3. Now we are going to obtain estimates (the crucial point of our proof)

$$\gamma(\omega : |\partial_v F(\omega)| \leqslant \varepsilon) = o(\varepsilon^k),\ \varepsilon \to 0,\ k \in \mathbb{N}. \tag{10}$$

Suppose we have the estimate

$$\gamma\Big(\omega : \sup_{t\in[0,T]} \Big\| \int_0^T K(s,t)\xi_s(\omega)ds \Big\| \leqslant \varepsilon\Big) = o(\varepsilon^n). \tag{11}$$

We shall deduce (10) from (11). The proof of (11) is given below in Step 4. Let

$$\eta_t(\omega) = \int_0^T K(s,t)\xi_s(\omega)ds.$$

By the closed graph theorem two norms $a \mapsto \|R^{-1/2}a\|_X$ and $a \mapsto \|a'\|_{L^2}$ are equivalent on H and there exists c with $\|a'\|_{L^2} \geqslant c\|R^{-1/2}a\|_X, a \in H$. From (9), (9′) we obtain

$$|\partial_v F(\omega)| = \sum_n r_n \Big(\partial_{e_n} F(\omega)\Big)^2 = \sup\Big\{\Big|\sum_n \sqrt{r_n}x_n \partial_{e_n} F(\omega)\Big| : \sum_n x_n^2 \leqslant 1\Big\} =$$

$$= \sup\Big\{\Big|\partial_a F(\omega)\Big| : a = \sqrt{R}h, \|h\|_X \leqslant 1\Big\} =$$

$$= \sup\Big\{\Big|2\int_0^T (\eta_t(\omega), Z_t \int_0^t Z_s^{-1} A(s,\xi_s(\omega))a'(s)ds)dt\Big| : a = \sqrt{R}h, \|h\|_X \leqslant 1\Big\} \geqslant$$

$$\geqslant 2c \sup\Big\{\Big|\int_0^T (\eta_t(\omega), Z_t \int_0^t Z_s^{-1} A(s,\xi_s(\omega))h(s)ds)dt\Big| : \|h\|_X \leqslant 1\Big\}. \tag{12}$$

Now we shall estimate the distribution of the expression in the right-hand side. Without any loss of generality we can assume that $\alpha < 1/2$. Recall that supremums of norms of diffusion processes have all moments (Ikeda and Watanabe, 1981). Besides, the trajectories of these processes (under our assumptions) possess Holder properties of any orders less than 1/2. It is possible to verify that the correspondent Holder norms also have all moments. It means the following estimates are valid:

$$\gamma\Big(\omega: \sup_{t\in[0,T]} \Big\|Z_t(\omega)\Big\| \geqslant M\Big) = o(M^{-n}),\ M \to \infty,\ n \in \mathbb{N},$$

$$\gamma\Big(\omega: \sup_{t\in[0,T]} \Big\|Z_t^{-1}(\omega)\Big\| \geqslant M\Big) = o(M^{-n}),$$

$$\gamma\Big(\omega: \sup_{s,t} |t-s|^{-\alpha}\Big(\Big\|\xi_t(\omega)-\xi_s(\omega)\Big\| + \Big\|\eta_t(\omega)-\eta_s(\omega)\Big\| +$$

$$+\|Z_t - Z_s\| + \|Z_t^{-1} - Z_s^{-1}\|\Big) \geqslant M\Big) = o(M^{-n}). \quad (13)$$

The estimates above together with (4) imply

$$\gamma\Big(\omega: \int_0^T \Big\|A^{-1}(t,\xi_t(\omega))\Big\|^2 dt \geqslant M\Big) = o(M^{-n}),\ M \to \infty.$$

Fix $\sigma \in C^\infty(\mathbf{R})$ with supp, $\sigma \subset [0,1]$, $\int \sigma(s)ds = 1$.
Let

$$\psi(t) = Z_t^*\eta_t(\omega), 0 < \varepsilon \leqslant (4+2T^{1+\alpha})^{-1}, M = \varepsilon^{-1}, \beta = 3/\alpha, k = 5/\alpha + 3/\alpha^2.$$

Assume that ω is chosen to satisfy the conditions

i) $\|\psi(t_0)\| = \varepsilon$ for some $t_0 \in [0,T]$, $\|Z_t\| \leqslant M$;

ii) $\int_0^T \|A^{-1}(t,\xi_t(\omega))\|^2 dt \leqslant M$, $\|\psi(t)-\psi(s)\| \leqslant M|t-s|^\alpha$.

Extend ψ with the same Holder norm on $[0, T+1]$ and consider the approximation

$$\varphi(t) = \int_0^1 \psi(t+s\varepsilon^k)\sigma(s)ds.$$

Let

$$h(t) = A^{-1}(t,\xi_t)Z_t\varphi'(t).$$

Direct calculations show that the following inequalities are valid:

$$\|\varphi(t)-\psi(t)\| \leqslant M\varepsilon^{k\alpha},$$

$$\Big\|\varphi'(t)\Big\| \leqslant M\varepsilon^{-k}\int \Big|\sigma'(s)\Big| ds T^\alpha,$$

$$\|h\|_{L^2} \leqslant M^3\varepsilon^{-k}\int \Big|\sigma'(s)\Big| ds T^\alpha = \text{const } \varepsilon^{-k-3}.$$

$$\|h\|_{L^2} \leqslant M^3\varepsilon^{-k}\int\left|\sigma'(s)\right|dsT^\alpha = \text{const }\varepsilon^{-k-3}.$$

By the Holder condition $\|\psi(t)\| \geqslant \varepsilon - M\varepsilon^{\alpha\beta}$ for $t \in [t_0 - \varepsilon^\beta, t_0 + \varepsilon^\beta]$. Thus we obtain

$$\int_0^T\left(\eta_t(\omega), Z_t\int_0^t Z_s^{-1}A\left(s,\xi_s(\omega)\right)h(s)ds\right)dt =$$

$$= \int_0^T\left(\psi(t),\varphi(t)-\varphi(0)\right)dt \geqslant \int_0^T\left\|\psi(t)\right\|^2dt - \int_0^T\left\|\psi(t)\right\|\cdot\left\|\psi(t)-\varphi(t)+\varphi(0)\right\|dt \geqslant$$

$$\geqslant 2(\varepsilon - M\varepsilon^{\alpha\beta})^2\varepsilon^\beta - MT^\alpha(M\varepsilon^{k\alpha}+M\varepsilon^{k\alpha})T =$$
$$= \varepsilon^{2+3/\alpha}(2-4\varepsilon+2\varepsilon^2-2\varepsilon T^{\alpha+1}) \geqslant \varepsilon^{2+3/\alpha}.$$

Finally we have the following inequality:

$$\sup_{\|h\|_X\leqslant 1}\int_0^T\left(\eta_t(\omega), Z_t\int_0^t Z_s^{-1}A\left(s,\xi_s(\omega)\right)h(s)ds\right)dt \geqslant c\varepsilon^q, \tag{14}$$

where $q = 5+8/\alpha+3/\alpha^2$ and c is a constant, not depending on ε and ω. It is clear from (13) that for every $n \in \mathbf{N}$ we can choose C_n so that for any $\varepsilon > 0$ γ–measure of points ω, not satisfying conditions i) and ii) with $M = \varepsilon^{-1}$, is less than $C_n\varepsilon^n$. Together with (14) this implies (10).

Step 4. We complete the proof by demonstrating (11). Note that an analogous step is present in proofs of similar theorems, dealing with the functional $F(\omega) = \xi_{t_0}(\omega)$ (transition probabilities) or with integral functionals of the Brownian motion (see (Bell, 1987; Ikeda and Watanabe, 1981; Gaveau and Moulinier, 1983; Moulinier, 1985)), when the distribution of the maximum of the norm of some auxiliary process must be estimated. However, in these two cases such auxiliary processes turn out to be solutions of stochastic equations while in our case η_t need not be a Markov process or a martingale and so the traditional methods of estimating of it's supremum can not be used. For this aim we shall resort again to the Malliavin calculus. By Lemma 1 there exist $t_1,\ldots,t_n \in [0,T]$ and $e \in \mathbf{R}^m$ such that the mappings $s \mapsto K(s,t_i)e$ are linearly independent. The desired relationships (11) follow from the existence of bounded density of distribution of the random vector $f = ((\eta_{t_1}, e),\ldots,(\eta_{t_n}, e))$. Hence it suffices to establish the estimate

$$\left|\int\exp\left(i(t,f)\right)\gamma\right| \leqslant c\left(1+\|t\|\right)^{-n-2}, \quad t\in\mathbf{R}^n. \tag{15}$$

Fix $b \in \mathbf{R}^n$, $\|b\| = 1$. Let $g = (b,f)$. All our arguments which have been used in Steps 1, 2 for the functional F, are applicable for g. It means that the vector field

$$u(\omega) = \sum_n r_n\partial_{e_n}g(\omega)e_n$$

(in notations of (Ikeda and Watanabe, 1981) $g \in H_\infty$). Again we omit the details. Now the only thing to be done is to obtain the inclusion

$$\partial_u g^{-1} \in \bigcap_p L^p(\gamma)$$

which is equivalent to estimates

$$\gamma\Big(\omega : |\partial_u g(\omega)| \leqslant \varepsilon\Big) \leqslant k_\ell(b)\varepsilon^\ell,\ \ell \in \mathbf{N}. \tag{16}$$

From calculations above we have

$$\partial_a g(\omega) = \int_0^T \left(\sum_{j=1}^n b_j K(t,t_j)e, Z_t \int_0^t Z_s^{-1} A\Big(s, \xi_s(\omega)\Big) a'(s)ds\right) dt.$$

Let

$$\psi(t,b) = Z_t^* \sum_{j=1}^n b_j K(t,t_j)e.$$

Conditions of linear independence and continuity imply the following inequality:

$$\inf_{\|b\|=1} \sup_{[0,T]} \left\| \sum_{j=1}^n b_j K(t,t_j)e \right\| = T > 0.$$

Since for any $y \in \mathbf{R}^m$

$$\|y\| = \|Z^{-1}Zy\| \leqslant \|Z^{-1}\| \cdot \|Zy\|,$$

we come to the estimate $\|\psi(t,b)\| \geqslant T\|Z_t^{-1}\|^{-1}$. Our considerations show that all mappings $t \mapsto \psi(t,b)$, $\|b\| = 1$, satisfy the conditions i) and ii), indicated above for ψ, with probability more than $1 - d_\ell \varepsilon^\ell$, where d_ℓ can be chosen universal for all b with $\|b\| = 1$. Repeating the estimates of Step 3 we obtain inequalities (16) where constants k_ℓ can be taken not depending on $b \in \mathbf{R}^n$ with $\|b\| = 1$. As we know, estimates (16) imply inequalities

$$\left| \int \exp\Big(i\lambda(b,f)\Big)\gamma \right| \leqslant c_\ell(b)(1+|\lambda|)^{-\ell}, \quad \lambda \in \mathbf{R}^1. \tag{17}$$

It is not difficult to show that if the constants k_ℓ in (16) do not depend on $b \in S_n$ (unit sphere in $\mathbf{R}^n$) then the constants c_ℓ in (17) also can be chosen universal. This gives us the desired estimates (15). The theorem is proved.

Corollaries 1, 2 immediately follow from the theorem. For the proof of Corollary 3 note that by Theorem 1 for every $b \in \mathbf{R}^d$ with $\|b\| = 1$ there exist constants $\Lambda_\ell(b)$, satisfying

$$\left| \int \exp\left(it \sum_{j=1}^d b_j Q_j(x)\right) \mu^\xi(dx) \right| \leqslant \Lambda_\ell(b)\Big(1+|t|\Big)^{-\ell}. \tag{18}$$

Analysing the proof of Theorem 1 we see that the constants in (18) can be taken locally bounded on the unit sphere in $\mathbf{R}^d$. By compactness of the sphere we obtain estimates Λ_ℓ, uniform on the sphere in $\mathbf{R}^d$ (that is, with Λ_ℓ not depending on b with $\|b\| = 1$). This completes the proof.

Remark 1. If we are interested only in the boundedness of the density p of $\mu^\xi \circ \widetilde{Q}^{-1}$ or in the differentiability of some finite order r then, as we see from the proof, assumptions on A, B, K can be weakened. It suffices to require that A, B have $n(r) + 3$ Lipschitzian derivatives in the second argument, K is α–Holder and for some $t_1, \ldots, t_{n(r)} \in [0, T], e \in \mathbf{R}^m$ the mappings $s \mapsto K(s, t_i)e$ are linearly independent, where $n(r) = [2(r+2)(5 + 8/\alpha + 3/\alpha^2)] + 1$.
Note that in the case when A depends only on t much more general theorem is valid (see (Bogachev, 1989) and Theorem 4 below) since in this case the measure μ^ξ is differentiable along $D(\gamma)$ while in all other cases $D(\mu^\xi) = 0$ (Bogachev, 1989).
Now we consider the integral functionals on $C([0, T], \mathbf{R}^m)$ of the following form

$$F(x) = \int_0^T f\Big(t, x(t)\Big)\,dt, \tag{19}$$

where $f : \mathbf{R}^{m+1} \to \mathbf{R}$ is Holder, infinitely differentiable in x and satisfies the following conditions:

i) $\|\partial_x^s f(t, x)\| \leqslant c(s) \exp(b(s)\|x\|), c(s), b(s) > 0$,

ii) there exists a neighbourhood of zero in $V \subset \mathbf{R}^{m+1}$ such that the set $Z = \{(t, x) \in V : \partial_x f(t, x) = 0\}$ can be covered by the finite collection of smooth surfaces $\Pi_1, \ldots, \Pi_q$ and for some $k, \beta > 0$

$$\Big\|\partial_x f(t, x)\Big\| \geqslant k\Big(\operatorname{dist}\big((t, x), \bigcup \Pi_j\big)\Big)^\beta, \ (t, x) \in V. \tag{20}$$

Theorem 2. *Let μ^ξ be a measure on $C_0([0, T], \mathrm{R}^m)$ induced by the solution ξ_t of stochastic differential equation (2), where A and B satisfy the conditions of Theorem 1. If f has the properties listed above then $\mu^\xi \circ F^{-1}$ admits a density $p \in S(\mathbf{R}^1)$.*

Corollary 1. $\int \exp\Big(itF(x)\Big)\mu^\xi(dx) = o(t^{-n}),\ t \to \infty,\ n \in \mathbf{N}.$

Corollary 2. *If $f_1, \ldots, f_d$ are such that their nontrivial linear combinations satisfy the conditions* i), ii) *then the mapping $F = (F_1, \ldots, F_d)$ induces a measure $\mu^\xi \circ F^{-1}$ on $\mathbf{R}^d$ with the density $p \in S(\mathbf{R}^d)$.*

The proof is analogous to the proof of Theorem 1. The only difference there is connected with estimating the distribution of the random variable $Y = \sup\{\operatorname{dist}((t, \xi_t), \bigcup_j \Pi_j) : (t, \xi_t) \in V\}$.

Remark 2. a) More detailed analysis shows that the estimate ii) can be replaced by a weaker

$$\left\|\partial_x f(t,x)\right\| \geqslant k \exp\left(-1/\operatorname{dist}\left((t,x), \bigcup \Pi_j\right)^{\beta}\right), \ \beta < 2,$$

(see (Moulinier, 1985) for the case $\xi_t = w_t$).
b) If in the conditions of Theorem 2 f is only $r+2$ times differentiable in x then one can guarantee r–fold bounded differentiability of p.
Denote by H_α the space of α–Holder mappings $h : [0,T] \to \mathbf{R}^m$ with $h(0) = 0$ supplied with the natural Banach norm

$$\|h\|_\alpha = \sup \|h(t) - h(s)\| / |t-s|^\alpha.$$

Let W_n be the Sobolev space of functions $\varphi : [0,T] \to \mathbf{R}^m$ with $\varphi(0) = 0$, absolutely continuous derivative of the order $n-1$ and $\varphi^n \in L^2([0,T], \mathbf{R}^m)$. W_n is equipped with the Hilbert norm

$$\|\varphi\|_{W_n} = \left(\|\varphi\|_{L^2}^2 + \ldots + \|\varphi^{(n)}\|_{L^2}^2\right)^{1/2}.$$

THEOREM 3. *Let μ be a Gaussian measure on $C_0([0,T], \mathbf{R}^m)$ with $\mu(H_\alpha) = 1$ and $W_n \subset D(\mu)$ for some α and n or a Radon measure on H_α, infinitely differentiable along W_n and together with derivatives along vectors in W_n possessing all moments (with respect to the norm $\|\cdot\|_\alpha$). If f satisfies the conditions mentioned before Theorem 2, then $\mu \circ F^{-1}$ admits a density $p \in S(\mathbf{R}^1)$.*

The statements of corollaries of Theorem 2 are also valid.

Remark 3. The statement of Remark 2 b) in this case is true as well.

Example 2. If in some neighbourhood of the origin f is analytic and $\partial_x f \not\equiv 0$ then by Lojasiewicz's theorem f satisfies the condition ii), since $\|\partial_x f(t, x)\| \geqslant k \operatorname{dist}((t,x), Z)^\beta$ (see (Malgrange, 1966, Theorem 4.1)). In particular, a polynomial f with $\partial_x f \not\equiv 0$ satisfies all conditions indicated above (it follows also from the Seidenberg – Tarski theorem).
The proof of Theorem 3 uses the method described above. The analogous proof in the case $m = 1, f(t,x) \equiv f(x)$ can be found in (Bogachev, 1989).
Now discuss distributions of quadratic forms. We call Q a quadratic form on a linear space E if $Q(x) = V(x,x)$ where $V : E \times E \to \mathbf{R}^1$ is bilinear. Q can be written also in the form $Q(x) = (Ax)(x)$ where $A : E \to E'$ is linear and symmetric (that is $(Ax)(y) \equiv (Ay)(x)$). The dimension of $A(E)$ is called the rank of Q on E. A μ–measurable function on E, which coincides almost everywhere with some quadratic form, is called a μ–measurable quadratic form.

THEOREM 4. *Assume that a Radon measure μ on a locally convex space X is $r+2m+3$ times differentiable along directions of vectors from some linear subspace $D \subset X, Q_1, \ldots, Q_m$ are μ–measurable quadratic forms on X with*

$$\partial_a Q_j \in \bigcap_p L^p(d_{h_1 \cdots h_r} \mu)$$

for all $a, h_1, \ldots, h_r \in D$. If for every $(\alpha_1, \ldots, \alpha_m) \neq (0, \ldots, 0)$

$$\mathrm{rank}(\alpha_1 Q_1 + \ldots + \alpha_m Q_m)\Big|_D \geqslant 4r + 4m + 9$$

then the mapping $Q = (Q_1, \ldots, Q_m)$ induces a measure on $\mathbf{R}^m$ with the density possessing integrable partial derivatives of any order $s \leqslant r$.

COROLLARY. *If μ is Gaussian measure, $Q_1, \ldots, Q_m$ are measurable quadratic forms and their nontrivial linear combinations have infinite rank on $D(\mu)$ then $\mu \circ (Q_1, \ldots, Q_m)^{-1}$ has a density $p \in S(\mathbf{R}^m)$.*

Remark 4. The same is true for polynomials $Q_j + \ell_j + c_j, \ell_j$ being μ-measurable linear functionals on X.
The proof of Theorem 4 practically coincides with the proof of analogous theorem for continuous forms in

THEOREM 5. *Let μ be a Radon measure on a locally convex space X, a_1 and a_2 be independent vectors in X such that there exist $d_{a_1}\mu = \rho_1\mu$, $d_{a_2}\mu = = \rho_2\mu, d^2_{a_1a_2}\mu, and \rho_1, \rho_2 \in L^r(\mu)$ for some $r > 2$. If a μ-measurable quadratic form Q is positively or negatively definite on the plane* span(a_1, a_2) *then $\mu \circ Q^{-1}$ has a density p of bounded variation (in particular,* $\sup |p| < \infty$*).*

COROLLARY. *Let γ be a symmetric Gaussian measure on X and Q be γ-measurable quadratic form. The measure $\gamma \circ Q^{-1}$ has bounded density p if and only if there exists two-dimensional plane $L \subset D(\gamma)$ on which Q is positively or negatively definite (in this case $p \in BV$).*

The proof is based on the methods of (Bogachev, 1989). For double stochastic integrals the statement of the corollary was proved by Yu.A. Davydov. Note that it is the rank of Q on $D(\mu)$ but not on X to be of importance. It is clear from the following example.

Example 3. Let γ be the Wiener measure on $C_0[0, 1]$. Then

$$Q(\omega) = \lim_{n \to \infty} \sum_{j=1}^{n} \left(\omega\left(\frac{j}{n}\right) - \omega\left(\frac{j-1}{n}\right)\right)^2$$

exists γ-a.e. It is wellknown that $Q = 1$, γ-a.e. The set $E = \{Q \text{ exists}\}$ is a linear subspace with $\gamma(E) = 1$ and so Q is a quadratic form on E, and therefore a measurable quadratic form on X. It is not difficult to show that Q has an infinite rank on every linear subspace of a full measure in E. Degeneracy of $\gamma \circ Q^{-1}$ is explained by the equality $Q\,|_{D(\gamma)} = 0$.
Our last example illustrates theorems 4, 5.

Example 4. Let γ be again the Wiener measure on $C_0[0,T]$. Consider the double Ito's integral (see (Kallianpur, 1980))

$$Q(\omega) = \int_0^T \int_0^T q(s,t) dw_s dw_t,$$

$q \in L^2([0,T] \times [0,T])$ being symmetric. Q is a measurable quadratic form and it's rank on $D(\gamma)$ equals to the rank of the form

$$\widetilde{Q}(x) = \int_0^T \int_0^T q(s,t) x(s) x(t) ds dt$$

on $L^2[0,T]$ and coincides with the dimension of the integral operator with the kernel q (since $D(\gamma) = \{h \in AC[0,T] : h(0) = 0, h' \in L^2\}$). For instance, if q is non–constant, continuous aside of the diagonal and has a constant limit (finite or infinite) on approaching the diagonal, then Q admits a density of bounded variation. Note that Theorems 4, 5 are applicable to measures μ^ξ induced by solutions of (2) with $A(t,x) \equiv A(t)$ since in this case μ^ξ is infinitely differentiable along $D(\gamma)$ (this follows from the Girsanov's theorem). Distributions of continuous forms on Hilbert spaces were first studied in (Uglanov, 1982) by other methods. A shorter proof of the smoothness of distributions of nondegenerate quadratic mappings on Hilbert space was given in (Steblovskaya, 1989). Our method (Bogachev, 1989) differs from (Uglanov, 1982; Steblovskaya, 1989) and enables to deal with the general situation. It would be interesting to study distributions of arbitrary quadratic forms of diffusion processes, Markov processes, Ito processes, martingales.
Another perspective problem is application of described methods for investigating oscillatory Feynman's integrals (see (Maslov, 1970, 1976)). For example, if γ is a Gaussian measure with Fourier transform $e^{-Q(x)}$ and Φ is "Feynman measure" defined by it's Fourier transform $e^{iB(x)}$, B being a quadratic form, then by Parseval's equality

$$\int \exp\Big(itB(x)\Big)\gamma(dx) = \int \exp\Big(-tQ(x)\Big)\Phi(dx).$$

REFERENCES

Bell, D. (1987). *The Malliavin Calculus.* Wiley, New York.

Bogachev, V.I. (1986). Some results on differentiable measures. *Math. USSR Sb.* **55**, 335–349.

Bogachev, V.I. (1988a). On the differentiability of measures in the Skorokhod sense. *Theory Probab. Appl.* **33**, 349–354.

Bogachev, V.I. (1988b). Subspaces of differentiability of smooth measures on infinite dimentional spaces. *Soviet Math. Dokl.* **37**, 304–308.

Bogachev, V.I. (1989). Differential properties of measures on infinite dimentional spaces and the Malliavin calculus. *Acta Math. Univ. Carolina* **30**, 1–29.

Daletsky, Yu. L. (1983). Stochastic differential geometry. *Russian Math. Surveys.* **38**, 87–111.

Gaveau, B. and Moulinier, J. M. (1983). Integrales oscillantes stochastiques: estimation asymptotique de fonctionnelles caractèristiques. *J. Funct. Anal.* **54**, 161–176.

Ikeda, N. and Watanabe, S. (1981). *Stochastic differential equations and diffusion processes.* North Holland, Amsterdam.

Kallianpur, G. (1980). *Stochastic filtering theory.* Springer, Berlin–Heidelberg etc.

Malgrange, B. (1966). *Ideals of differentiable functions.* Oxford Univ, Press, Oxford.

Maslov, V.P. (1970). On the stationary phase method for Feynman continual integral. *Theor. Math. Phys.* **2**, 30–35.

Maslov, V. P. (1976). *Complex Markov chains and the continual Feynman integral.* Nauka, Moscow.

Moulinier, J. M. (1985). Fonctionnelles oscillantes stochastiques et hypoellipticité. *Bull. Sc. Math.* **109**, 37–60.

Skorokhod, A. V. (1974). *Integration in Hilbert space.* Springer, Berlin–Heidelberg etc,

Smorodina, N. V. (1988). Differential calculus in the space of configurations and stable measures. *Theory Probab. Appl.* **33**, 522–534.

Steblovskaya, V. R. (1989). Smoothness of finite dimentional images of measures. *Ukrainian Math. J.* **41**, 261–265.

Uglanov, A. V. (1979). Surface integrals in a Banach space. *Math. USSR Sb.* **110**, 189–217.

Uglanov, A. V. (1982). On division of generalized functions of infinite number of variables by polynomials. *Soviet Math. Dokl.* **264**, 1096–1099.

Prob. Theory and Math. Stat., Vol. 1, pp. 166–178
B.Grigelionis *et al.* (Eds.)
1990 VSP/Mokslas

EXPONENTIAL INEQUALITIES FOR THE DISTRIBUTIONS OF VON MISES AND U–STATISTICS

I.S. BORISOV
Institute of Mathematics, 630090 Novosibirsk, USSR

ABSTRACT

Bernstein type inequalities for the tail distributions of von Mises and U–statistics with Hilbert space valued kernels are obtained.

1. INTRODUCTION AND STATEMENT OF MAIN RESULT

Let $X_1, X_2, \ldots$ be a sequence of independent identically distributed random variables (r.v.'s) taking values in arbitrary measurable space $(Y, @)$ and with the distribution P. Let $h(\cdot)$ be a measurable function from $(Y^m, @^m)$ to a Hilbert space H. The functional

$$M_m(X_1, \ldots, X_n) = \sum_{i_1, \ldots, i_m \leqslant n} h(X_{i_1}, \ldots, X_{i_m})$$

is called m–variate von Mises statistic with kernel $h(\cdot)$. We shall suppose without any loss of generality that $h(\cdot)$ is a symmetric function. The kernel $h(\cdot)$ is called r–degenerate for P, $0 \leqslant r \leqslant m-1$, if

$$\max\Big(l \geqslant 0 : \mathbf{E}(h(X_1, \ldots, X_m) \mid G_l) = 0 \text{ a.s.}\Big) = r,$$

where G_l is the field generated by $X_1, \ldots, X_l$; G_0 is the trivial field. For $r = 0$ we deal with so–called centered non–degenerate kernel; $(m-1)$–degenerate kernel is called also degenerate for P.
In the paper we shall consider only such kernels $h(\cdot)$ which are estimated as follows

$$\|h(z_1, \ldots, z_m)\| \leqslant \prod_{i \leqslant m} g(z_i), \tag{1}$$

where $g : Y \to \mathbf{R}$ is a measurable function. For example, if there exists the function

$$f(z) = \operatorname*{ess\,sup}_{y_1, \ldots, y_{m-1} \in Y} \|h(z, y_1, \ldots, y_{m-1})\|$$

then we can set in (1) $g(z) = (f(z))^{1/m}$. In particular, if

$$\sup_{z_i \in Y} \|h(z_1, \ldots, z_m)\| \leqslant C$$

then (1) is fulfilled for $g(z) = C^{1/m}$.

THEOREM. *Let the function $g(\cdot)$ in (1) satisfy Bernstein's condition:*

$$\mathbf{E}g(X_1)^k \leqslant b^2 L^{k-2} k!/2 \tag{2}$$

for all $k \geqslant 2$. Let the kernel $h(\cdot)$ be r–degenerate for P, $0 \leqslant r \leqslant m-1$. Then for any $x \geqslant 0$

$$\mathbf{P}(\max_{k \leqslant n} \|M_m(X_1, \ldots, X_k)\| \geqslant n^m x) \leqslant$$

$$\leqslant C(m) \sum_{l=r+1}^{m} \exp\left(-\frac{C_0(m) n x^{2/l}}{b^{2m/l} + x^{1/l} L b^{m/l-1}}\right) \tag{3}$$

where positive constants $C(m), C_0(m)$ depend on m only.

Remark 1. The inequality (3) is unimprovable in some sence. Let, for example,

$$h(z_1, \ldots, z_m) = \prod_{i \leqslant m} g(z_i),$$

where $g(\cdot)$ is a real valued function with $\mathbf{E}g(X_1) = 0$. In this case $h(\cdot)$ is degenerate for P. Then $M_m(X_1, \ldots, X_k) = (\sum_{i \leqslant k} g(X_i))^m$ and we can obtain by classical Bernstein's inequality (see, for instance, (Yurinskii, 1976))

$$\mathbf{P}(\max_{k \leqslant n} |M_m(X_1, \ldots, X_k)| \geqslant n^m x) \leqslant 2\exp\left(-\frac{n x^{2/m}}{2(b^2 + L x^{1/m})}\right).$$

The right–hand side of (3) for such kernels is quite analogous to this estimate.

Remark 2. The inequality (3) is valid (with another constants $C(m), C_0(m)$) if we shall consider instead of $M_m(\cdot)$ m–variate U–statistic (without the normalizing factor)

$$U(X_1, \ldots, X_k) = \sum_{i_1 < \cdots < i_m \leqslant k} h(X_{i_1}, \ldots, X_{i_m})$$

because any U–statistic can be represented as linear combination (depending on m only) of the corresponding von Mises statistics. Therefore it is easy to obtain from the theorem the bounded law of the iterated logarithm for U–statistics under less restrictive moment conditions then in (Dehling et al., 1986).

2. PRELIMINARY RESULTS. POISSON APPROXIMATION

Denote by $P_k(\cdot)$ the empirical distribution based on the sample $X_1, \dots, X_k$. Then one can represent $M_m(X_1, \dots, X_k)$ in such a way:

$$k^{-m} M_m(X_1, \dots, X_k) = \int_{Y^m} \cdots \int h(z_1, \dots, z_m) P_k(dz_1) \dots P_k(dz_m).$$

It is easy to obtain from here (see, for instance, (Filippova, 1962))

$$M_m(X_1, \dots, X_k) - \mathbf{E} M_m(X_1, \dots, X_k) =$$

$$= \sum_{l=1}^{m} C_m^l k^{m-l} \int_{Y^m} \cdots \int h(z_1, \dots, z_m) \prod_{j=1}^{m-l} P(dz_j) \prod_{i=m-l+1}^{m} V_k(dz_i), \tag{4}$$

where $V_k(\cdot) = k(P_k(\cdot) - P(\cdot))$. If $h(\cdot)$ is r–degenerate for P then the first r summands of the sum in (4) vanish. It will be shown further that in fact the l–th summand of the sum in (3) is an upper bound for the tail distribution of the l–th summand of the sum in (4).

Introduce the following notation

$$J_{kr} = \int_{Y^r} \cdots \int f(z_1, \dots, z_r) V_{nk}(dz_1) \cdots V_{nk}(dz_r),$$

where $V_{nk}(\cdot) = n^{-1/2} V_k(\cdot)$. It is easy to see that the l–th summand of the sum in (4) coinsides with $k^{m-l} n^{1/2} J_{kl} C_m^l$ and

$$f(z_1, \dots, z_l) = \int_{Y^{m-l}} \cdots \int h(z_1, \dots, z_m) \prod_{j=l+1}^{m} P(dz_j). \tag{5}$$

Hence our goal is to obtain the corresponding estimate for the tail distribution of $\max_{k \leqslant n} \| J_{kl} \|$ for any l. We can suppose that $f(\cdot)$ in (5) satisfies (1) and (2) because one can set in (2) without loss of generality $b = 1$.

First we shall obtain the corresponding results for J_{nl} and real valued kernel $f(\cdot)$. Then we shall extend the results on r.v. $\max_{k \leqslant n} | J_{kl} |$ and Hilbert space valued kernels.

The main step of estimation of the tail distribution of J_{kl} is an transition from stochastic integral (multivariate) with respect to random measure $V_{nk}(\cdot)$ to the corresponding stochastic integral with respect to the Poisson random measure. Analogous idea is contained in (Major, 1988) where it was considered a problem close to that in present paper. But realization of the idea in our paper is much simpler than in (Major, 1988).

Denote by $Q_w(B)$, $B \in @$, the Poisson random measure with the intensity measure $w(\cdot)$, that is $Q_w(\cdot)$ is a measure on @ with probability 1 (random point process), r.v.'s $Q_w(B_1), \dots, Q_w(B_k)$ are independent for any pairwise disjoint subsets

$B_i \in @$ and r.v. $Q_w(B)$ has the Poisson distribution with parameter $w(B)$. Introduce also the normed Poisson random measure $Q^0_w(\cdot) = n^{-1/2}(Q_w(\cdot) - w(\cdot))$.
It is well-known that empirical measure $V_{nn}(\cdot)$ coinsides in distribution with conditional Poisson random measure $Q^0_w(\cdot)$ under condition $Q^0_w(Y) = 0$ (or $Q_w(Y) = n$) for $w(\cdot) = nP(\cdot)$. The following lemma is one of the main step of the proof. It can be considered as some transition from conditional random measure $Q^0_{nP}(\cdot) \mid Q^0_{nP}(Y) = 0$ to Poisson random measure $Q^0_{nP}(\cdot)$.

LEMMA 1. *Let $B_1, \dots, B_k$ be arbitrary pairwise disjoint measurable subsets of Y and $P(\bigcap_{i\leqslant k} \bar{B}_i) \neq 0$ where $\bar{B}_i = Y - B_i$. Then for any Borel subsets $A_1, \dots, A_k \subseteq \mathbf{R}$ the following inequality holds*

$$\mathbf{P}\Big(\bigcap_{i\leqslant k}\big(V_{nn}(B_i) \in A_i\big)\Big) \leqslant 2\Big(P\Big(\bigcap_{i\leqslant k}\bar{B}_i\Big)\Big)^{-1/2}\mathbf{P}\Big(\bigcap_{i\leqslant k}\big(Q^0_{nP}(B_i) \in A_i\big)\Big). \tag{6}$$

Proof. In fact this result is contained in (Borisov, 1983) (see also (Borisov, 1978)). But we shall reproduce here the proof for convenience of readers.
For any B_i, A_i defined above we have

$$\mathbf{P}\Big(\bigcap_{i\leqslant k}\big(V_{nn}(B_i) \in A_i\big)\Big) = \big(\mathbf{P}(Q_{nP}(Y) = n)\big)^{-1}\mathbf{P}\Big(\bigcap_{i\leqslant k}\big(Q^0_{nP}(B_i) \in A_i\big)\Big);$$

$$Q_{nP}\big((Y) = n\big) = \big(\mathbf{P}(Q_{nP}(Y) = n)\big)^{-1}\sum_{l=0}^{n}\mathbf{P}\Big(\bigcap_{i\leqslant k}\big(Q^0_{nP}(B_i) \in A_i\big);$$

$$Q_{nP}\Big(\bigcup_{i\leqslant k} B_i\Big) = l\Big)\mathbf{P}\Big(Q_{nP}\Big(\bigcap_{i\leqslant k}\bar{B}_i\Big) = n - l\Big) =$$

$$= \mathbf{P}\Big(\bigcap_{i\leqslant k}\big(Q^0_{nP}(B_i) \in A_i\big)\Big)\max_{l\leqslant n}\frac{\mathbf{P}\Big(Q_{nP}\Big(\bigcap_{i\leqslant k}\bar{B}_i\Big) = l\Big)}{\mathbf{P}(Q_{nP}(Y) = n)}.$$

By classical Stirling's formula it is easy to obtain

$$\max_{l\leqslant n}\frac{\mathbf{P}\Big(Q_{nP}\Big(\bigcap_{i\leqslant k}\bar{B}_i\Big) = l\Big)}{\mathbf{P}(Q_{nP}(Y) = n)} =$$

$$= \max_{l\leqslant n}\frac{n!}{l!n^n}\Big(P\Big(\bigcap_{i\leqslant k}\bar{B}_i\Big)n\Big)^l \exp\Big(n - nP\Big(\bigcap_{i\leqslant k}\bar{B}_i\Big)\Big) \leqslant 2\Big(P\Big(\bigcap_{i\leqslant k}\bar{B}_i\Big)\Big)^{-1/2}.$$

Lemma is proved.
We shall need further some special properties of Poisson distribution. The following two lemmas generalize and strengthen the corresponding results in (Major, 1988).

LEMMA 2. *Let $Q_1, \ldots, Q_s$ be a sequence of independent Poisson r.v.'s with parameters $v_1, \ldots, v_s$. Then for any $g_1, \ldots, g_s \in \mathbf{R}$ and integer $N \geqslant 2$*

$$m_g(N) = \sum_{l=0}^{N-2} C_{N-1}^l m_g(l) G(N-l) \tag{7}$$

where

$$m_g(k) = \mathbf{E}\Big(\sum_{i \leqslant s} g_i(Q_i - v_i)\Big)^k, \quad G(k) = \sum_{i \leqslant s} g_i^k v_i.$$

Proof. It is easy to see that

$$\mathbf{E}\Big(\sum_{j \leqslant s} g_i(Q_i - v_i)\Big)^{N-1} \sum_{i \leqslant s} g_i Q_i =$$

$$= \sum_{i \leqslant s} g_i v_i \sum_{\substack{k_r \geqslant 0, r \neq i \\ k_i \geqslant 1}} \Big(\sum_{j \leqslant s} g_j(k_j - v_j)\Big)^{N-1} \frac{v_i^{k_i - 1} e^{-v_i}}{(k_i - 1)!} \prod_{l \neq i} \frac{v_l^{k_l} e^{-v_l}}{k_l!} =$$

$$= \sum_{i \leqslant s} g_i v_i \mathbf{E}\Big(\sum_{j \leqslant s} g_j(Q_j - v_j) + g_i\Big)^{N-1}.$$

Hence it follows

$$m_g(N) = \sum_{i \leqslant s} g_i v_i \Big(\mathbf{E}\Big(\sum_{j \leqslant s} g_i(Q_j - v_j) + g_i\Big)^{N-1} - \mathbf{E}\Big(\sum_{j \leqslant s} g_j(Q_j - v_j)\Big)^{N-1}\Big) =$$

$$= \sum_{l=0}^{N-2} C_{N-1}^l m_g(l) G(N-l).$$

Lemma is proved.

COROLLARY 1. *For any Poisson r.v. Q with parameter v and $N \geqslant 2$*

$$\mathbf{E}(Q - v)^N = v \sum_{l=0}^{N-2} C_{N-1}^l \mathbf{E}(Q - v)^l.$$

It is follows from here

$$\mathbf{E}(Q - v)^N \geqslant 0, \qquad N = 0, 1, 2, \ldots; \tag{8}$$

$$\mathbf{E}(Q - v)^N \leqslant v(N-1)\mathbf{E}(Q - v)^{N-2} + 2^{-1}(N - l)\mathbf{E}(Q - v)^{N-1}. \tag{9}$$

COROLLARY 2. *If $v = np$ where $p \in (0,1), n \geqslant 1$, then from (9) it follows*

$$\mathbf{E}(Q - np)^N \leqslant n^{N/2} p N!;$$

in particular,

$$\mathbf{E}\big(Q^0_{nP}(B)\big)^N \leqslant N!\,P(B), \qquad N = 1, 2, \dots . \tag{10}$$

LEMMA 3. *Let $g_1, \dots, g_s \in \mathbf{R}$ satisfy Bernstein's type condition*

$$\sum_{i \leqslant s} |g_i|^k v_i \leqslant b^2 L^{k-2} k!/2, \qquad k = 2, 3, \dots . \tag{11}$$

Then

$$|m_g(N)| \leqslant 4^N (b^2 N)^{N/2}, \qquad \text{if} \qquad 0 \leqslant NL^2 \leqslant b^2; \tag{12}$$

$$|m_g(N)| \leqslant (6LN)^N, \qquad \text{if} \qquad L^2 N > b^2. \tag{13}$$

Proof. We shall prove (12) and (13) by the induction on N. If inequality (12) holds for all $N < N_0$, where $0 \leqslant N_0 L^2 \leqslant b^2$, then for $N = N_0$ we obtain from (7) and (11)

$$|m_g(N)| \leqslant \sum_{l=0}^{N-2} C^l_{N-1} (4^2 b^2 N)^{1/2} L^{N-2-l} b^2 (N-l)!/2 \leqslant b^2 N D(N),$$

where

$$D(N) = \sum_{l=0}^{N-2} C^l_{N-2} (4^2 b^2 N)^{1/2} L^{N-2-l} (N-l)!/2.$$

It is elementary deduced the following recurrent inequality

$$D(N) \leqslant (4^2 b^2 N)^{(N-2)/2} + 3LND(N-1).$$

From here and by estimate $N_0 L \leqslant b N_0^{1/2}$ it is easy to obtain the estimate

$$D(N) \leqslant 4^N b^{N-2} N^{(N-2)/2}. \tag{14}$$

Hence the relation (12) is proved.
To prove (13) we split the sum in (7) in two parts

$$m_g(N) = \sum_{l=0}^{[(b/L)^2]} + \sum_{l=1+[(b/L)^2]}^{N-2}, \tag{15}$$

where we suppose for the sake of simplicity that $(b/L)^2 < N-2$. The first sum in (15) is estimated by (12):

$$\left| \sum_{l=0}^{[(b/L)^2]} \right| \leqslant 4^N (b^2 N)^{N/2} \leqslant (4LN)^N.$$

For estimation of second sum in (15) we shall use the induction by $N \geqslant 3+(b/L)^2$ quite analogous to proving of (14) and (12):

$$\left|\sum_{l=1+[(b/L)^2]}^{N-2}\right| \leqslant \sum_{l=0}^{N-2} C_{N-1}^l (6LN)^l L^{N-2-l} b^2 (N-l)!/2 \leqslant (6LN)^N/24.$$

Hence it follows (13). Lemma is proved.

COROLLARY 3. *Let $B_1, \dots, B_s$ be an arbitrary measurable partition of Y. If*

$$\sum_{i\leqslant s} |g_i|^k P(B_i) \leqslant b^2 L^{k-2} k!/2, \qquad k = 2,3,\dots, \tag{16}$$

then

$$\left|\mathbf{E}\Big(\sum_{i\leqslant s} g_i Q_{nP}^0(B_i)\Big)^N\right| \leqslant 4^N (b^2 N)^{N/2}, \qquad \textit{if} \qquad 0 \leqslant NL^2 \leqslant nb^2; \tag{17}$$

$$\left|\mathbf{E}\Big(\sum_{i\leqslant s} g_i Q_{nP}^0(B_i)\Big)^N\right| \leqslant (6LN)^N n^{-N/2}, \qquad \textit{if} \qquad NL^2 > nb^2. \tag{18}$$

Let us consider a special Banach space L_2^* of measurable functions $f\colon\ Y^r \to \mathbf{R}$ with the norm

$$\|f\|_* = \sum_{q_i \geqslant 0} \mathbf{E}^{1/2}(f(\underbrace{X_1,\dots,X_1}_{q_1 \text{ times}}, \underbrace{X_2,\dots,X_2}_{q_2 \text{ times}}, \dots, \underbrace{X_s,\dots,X_s}_{q_s \text{ times}}))^2. \tag{19}$$

Such spaces arise in a natural way in the theory of von Mises and U–statistics (see, for instance (Dehling et al., 1986; Filippova, 1962)). Denote by T_s a finite partition ($B_i \in \mathcal{Q}$; $i \leqslant s$) of Y and by $f^*(\cdot) \in L_2^*$ a step function wich equals to $f_{i_1\dots i_r}$ on the set $B_{i_1} \times \dots \times B_{i_r}$. For any $f \in L_2^*$ there is a sequence ($f_j^*; j \geqslant 1$) such that

$$\lim_{j\to\infty} \|f - f_j^*\|_* = 0. \tag{20}$$

Suppose that for P and some $T_s, s \geqslant 3r$, there exists a partition $N_1, \dots, N_{3r}$ of the set $(1,2,\dots,s)$ such that for any $j \leqslant 3r$

$$P\Big(\bigcup_{i\in N_j} B_i\Big) \leqslant (2r)^{-1} \tag{21}$$

where $B_i \in T_s$. Further we shall drop this condition.
Now let us consider the stochastic integral

$$J_{kr}^* = \int \cdots \int f^*(z_1, \dots, z_r) V_{nk}(dz_1) \dots V_{nk}(dz_r) =$$

$$= \sum_{j_1,\dots,j_r \leqslant 3r} \quad \sum_{i_1\in N_{j_1},\dots,i_r \in N_{j_r}} f_{i_1\dots i_r} V_{nk}(B_{i_1}) \dots V_{nk}(B_{i_r}). \tag{22}$$

Note that for any $j_1, \ldots, j_r$ it follows from (21)

$$P\Big(\bigcup_{i\in N_{j_1}+\cdots+N_{j_r}} B_i\Big) \leqslant 1/2, \tag{23}$$

that is the conditions of Lemma 1 are fulfilled and we can use the Poisson approximation for distributions (or moments) of any inner sum in (22). Using this fact and (23) we can easy to obtain the estimate

$$\mathbf{E}(J^*_{kr})^2 \leqslant C(r)\|f^*\|^2_*.$$

It follows from here that J^*_{kr} converges in $\mathcal{L}_2$ to J_{kr} if (20) is fulfilled. Then we can write

$$\mathbf{P}\Big(\max_{k\leqslant n}|J_{kr}| \geqslant x\Big) \leqslant \lim_{f^*\to f}\sup \mathbf{P}\Big(\max_{k\leqslant n}|J^*_{kr}| \geqslant x/2\Big). \tag{24}$$

In other words we reduced the problem to that for a step kernel $f^*(\cdot)$. Now our goal is to obtain an uniform on T_s upper bound for the tail distribution of $\max\limits_{k\leqslant n}|J^*_{kr}|$.

3. ESTIMATION OF THE TAIL DISTRIBUTION OF J^*_{nr}

In this section we shall obtain an exponential bound for the right–hand side of the inequality

$$\mathbf{P}\Big(|J^*_{nr}| \geqslant x\Big) \leqslant \inf_{N\geqslant 1} x^{-2N}\mathbf{E}(J^*_{nr})^{2N}. \tag{25}$$

In fact this approach was elaborated in the classical theory for sums of r.v.'s (cf. (Major, 1988)). In the next section it will be shown that estimation of probability deviations of r. v. $\max\limits_{k\leqslant n}|J^*_{kr}|$ can be reduced to the considered one in this section. It follows from Lemma 1, (22) and (23) we need to estimate the moments

$$\mathbf{E}\Big(\sum_{i_q\in N_{j_q};q\leqslant r} f_{i_1\ldots i_r}Q^0_{nP}(B_{i_1})\ldots Q^0_{nP}(B_{i_r})\Big)^{2N}. \tag{26}$$

Without any loss of generality we shall assume $f^* \geqslant 0$. Then by (8) we can extend the summing in (26) on the domain $i_1, \ldots, i_r \leqslant s$. Denote

$$g^*(x) = \sum_{i\leqslant s} g_i I(x \in B_i),$$

where $B_i \in T_s$. We can suppose that $f^* \leqslant f$, $g^* \leqslant g$ and f^*, g^* are connected by relation (1). In this case the step function g^* satisfies (2). Then from (1) and (8) it follows the inequality (cf. (Major, 1988))

$$\mathbf{E}\Big(\sum_{i_1,\ldots,i_r} f_{i_1\ldots i_r}Q^0_{nP}(B_{i_1})\ldots Q^0_{nP}(B_{i_r})\Big)^{2N} \leqslant \mathbf{E}\Big(\sum_i g_iQ^0_{nP}(B_i)\Big)^{2Nr}. \tag{27}$$

First let us consider the case $Lx^{1/r} \leqslant 6eb^2n^{1/2}$. Then by (17) we can write (we change the factor 4^N in (17) by 10^N)

$$\inf_{N\leqslant nb^2/(2rL^2)} \left(\mathbf{E}\Big(\sum_i g_i Q^0_{nP}(B_i)\Big)^{2Nr} x^{-2N}\right) \leqslant$$

$$\leqslant \inf_{0<z\leqslant nb^2/L^2} \exp\left(z\log(10b) + 2^{-1}z\log z - r^{-1}z\log x\right) = I_1. \tag{28}$$

The extremal point of the expression in argument of the exponent in (28) is equal to

$$z_0 = e^{-1}(x^{1/r}/(10b))^2 \leqslant 6^2enb^2/(10L)^2 < nb^2L^{-2}.$$

Then

$$I_1 = \exp\left(-(2e)^{-1}\left(x^{1/r}/(10b)\right)^2\right). \tag{29}$$

Now we put $Lx^{1/r} > 6eb^2n^{1/2}$. We must estimate in this case

$$I_2 = \inf_{z>nb^2L^{-2}} \exp\left(z\Big(\log z - \log(n^{1/2}x^{1/r}/(6L))\Big)\right).$$

The corresponding extremal point equals to $z_1 = n^{1/2}x^{1/r}/(6eL)$. Hence it follows $(z_1 > nb^2L^{-2})$

$$I_2 = \exp\left(-n^{1/2}x^{1/r}/(6eL)\right). \tag{30}$$

Combining (29) and (30) we obtain finally the following result.

LEMMA 4. *For any $x > 0$*

$$\mathbf{P}\left(|J^*_{nr}| \geqslant x\right) \leqslant \exp\left(-\frac{C_0x^{2/r}}{b^2 + n^{-1/2}Lx^{1/r}}\right). \tag{31}$$

4. A MARTINGALE APPROACH TO ESTIMATION OF THE TAIL DISTRIBUTION OF $\max_{k\leqslant n}|J^*_{kr}|$

In this section we shall obtain a martingale representation for J^*_{kr} and reduce the problem for $\max|J^*_{kr}|$ to estimation of the right-hand side of (25).
Introduce the following recurrent sequence

$$S^{(r)}_k(i_1,\dots,i_r) = \prod_{q\leqslant r} V_{nk}(B_{i_q}) -$$

$$-\sum_{2\leqslant l\leqslant r} \sum_{\substack{j_1<\cdots<j_{r-l}\leqslant r \\ q_1<\cdots<q_l\leqslant r,\ j_d\neq q_v}}\sum S^{(r-l)}_k(i_{j_1},\dots,i_{j_{r-l}})\mathbf{E}\prod_{v=1}^{l} V_{nk}(B_{i_{q_v}}), \tag{32}$$

where $S^{(0)}_k(\cdot) = 1$, $S^{(1)}_k(i) = V_{nk}(B_i)$, $r = 2,3,\dots$.

Lemma 5. *For any natural r and $i_1, \dots, i_r$ the sequence $(S_k^{(r)}(\cdot); k \geqslant 1)$ is a martingale.*

We omit the proof which is not difficult and based on the well-known representation

$$\prod_{i \leqslant r}(a_i + d_i) = \prod_{i \leqslant r} a_i + \sum_{i \leqslant r} d_i \prod_{j \leqslant r,\ j \neq i} a_j + $$
$$+ \sum_{i_1 < i_2 \leqslant r} d_{i_1} d_{i_2} \prod_{\substack{j \leqslant r \\ j \neq i_1, i_2}} a_j + \dots + \prod_{i \leqslant r} d_i. \tag{33}$$

Now from (32) we can express $\prod V_{nk}(\cdot)$ and rewrite J_{kr}^*:

$$J_{kr}^* = \sum_{i_1, \dots, i_r \leqslant s} f_{i_1 \dots i_r} \sum_{l=0}^{m} \sum_{\dots} S_k^{(r-l)}(\cdot) \mathbf{E} \prod_{v \leqslant l} V_{nk}(B_{i_{q_v}}). \tag{34}$$

But this representation is not yet convenient because the factors $\mathbf{E} \prod V_{nk}(\cdot)$ depend on k too. We have for any fixed $j_1, \dots, j_l$

$$n^{l/2} \mathbf{E} \prod_{v \leqslant l} V_{nk}(B_{j_v}) = \sum_{q_1, \dots, q_l \leqslant k} \mathbf{E} \prod_{v \leqslant l} I_{q_v}(B_{j_v}) =$$
$$= kS_1^* + k(k-1)S_2^* + \dots + k(k-1)\dots(k - [l/2] + 1)S_{[l/2]}^*, \tag{35}$$

where $I_{q_v}(B) = I(X_{q_v} \in B) - P(B)$ is the centered indicator function; j-th summand of the right-hand side of (35) is obtained by the summing on vector indexes $(q_1, \dots, q_l)$ in (35) having only j different coordinates where $j \leqslant l/2$. It is very important that S_m^*, $m \leqslant l/2$, are not depend on k. Hence we obtain

$$J_{kr}^* = \sum_{i_1, \dots, i_r \leqslant s} f_{i_1 \dots i_r} S_k^{(r)}(i_1, \dots, i_r) +$$
$$+ \sum_{2 \leqslant l \leqslant r} kn^{-l/2} \sum_{i_1, \dots, i_r \leqslant s} f_{i_1 \dots i_r} \sum_{\dots} S_k^{(r-l)}(\cdot) S_1^* +$$
$$+ k(k-1)n^{-l/2} \sum_{i_1, \dots, i_r \leqslant s} f_{i_1 \dots i_r} \sum_{\dots} S_k^{(r-l)}(\cdot) S_2^* +$$
$$+ k(k-1)\dots(k - [l/2] + 1)n^{-l/2} \sum_{i_1, \dots, i_r \leqslant s} f_{i_1 \dots i_r} S_k^{(r-l)} S_{[l/2]}^*. \tag{36}$$

It follows from (36)

$$\max_{k \leqslant n} |J_{kr}^*| \leqslant \sum_{l=2}^{r} \sum_{m=1}^{[l/2]} \max_{k \leqslant n} |I_{krlm}^{(n)}| + \max_{k \leqslant n} |I_{kr00}^{(n)}|$$

where $S_0^* = 1$ and

$$I_{krlm}^{(n)} = n^{m-l/2} \sum_{i_1, \dots, i_r \leqslant s} f_{i_1 \dots i_r} \sum_{\dots} S_k^{(r-l)}(\cdot) S_m^*$$

is a martingale on k for any fixed n, r, l, m. Hence the problem of estimation of even moments of $\max_{k \leqslant n} |J_{kr}^*|$ is reduced to that for $I_{nrlm}^{(n)}$. We must note that $I_{nrlm}^{(n)}$ has the same structure as J_{nr}^*. For example consider $I_{n300}^{(n)}$. We have from (32)

$$S_n^{(3)}(i_1, i_2, i_3) = \prod_{v \leqslant 3} V_{nn}(B_{i_v}) - \sum_{\substack{v \leqslant 3, m < q \leqslant 3 \\ m, q \neq v}} S_n^{(1)}(i_v) \mathbf{E} I_1(B_{i_m}) I_1(B_{i_q}) -$$

$$-n^{-1/2} \mathbf{E} I_1(B_{i_1}) I_1(B_{i_2}) I_1(B_{i_3}). \tag{37}$$

One can represent the second summand of right-hand side of (37) as the sum

$$\sum S_n^{(1)}(i_v)\Big(P\Big(B_{i_m} \bigcap B_{i_q}\Big) - P(B_{i_m})P(B_{i_q})\Big).$$

Multiply this sum by $f_{i_1 i_2 i_3}$ and sum on all i_1, i_2, i_3. Then we obtain the expression

$$3 \sum_{i \leqslant s} f_i^0 S_n^{(1)}(i) = 3 \sum_{i \leqslant s} f_i^0 V_{nn}(B_i), \tag{38}$$

where

$$f_i^0 = \sum_{j \leqslant s} f_{ijj} P(B_j) - \sum_{j_1 < j_2 \leqslant s} f_{i j_1 j_2} P(B_{j_1}) P(B_{j_2})$$

satisfy the condition (16) if only $f^*(\cdot)$ satisfies the conditions (1) and (2) (we can put in (2) $b = 1$). In the other words the expression (38) has the same structure as J_{n1}^* (the estimate (33) for $r = 1$ is better than for $r = 3$). Moreover, since

$$\Big|\mathbf{E} I_1(B_{i_1}) I_1(B_{i_2}) I_1(B_{i_3})\Big| \leqslant C\Big(P\Big(B_{i_1} \bigcap B_{i_2} \bigcap B_{i_3}\Big) +$$

$$+ \sum_{v \neq q \neq m} P(B_{i_v}) P\Big(B_{i_q} \bigcap B_{i_m}\Big) + \sum_{v \neq q \neq m} P(B_{i_v}) P(B_{i_q}) P(B_{i_m})\Big)$$

then we can estimate under the conditions on $f^*(b = 1)$

$$n^{-1/2} \Big| \sum_{i_1, i_2, i_3} f_{i_1 i_2 i_3} \mathbf{E} I_1(B_{i_1}) I_1(B_{i_2}) I_1(B_{i_3}) \Big| \leqslant L C_1 n^{-1/2} \leqslant x/2.$$

The investigation of the general case is quite analogous to that above. Therefore we can reformulate Lemma 4 in such a way.

LEMMA 6. *If f^* satisfies (1) and (2) then for any $x \geqslant 0$*

$$\mathbf{P}\Big(\max_{k \leqslant n} |J_{kr}^*| \geqslant x\Big) \leqslant C(r) \exp\Big(-\frac{C_0(r) x^{2/r}}{b^2 + n^{-1/2} L x^{1/r}}\Big). \tag{39}$$

5. CONCLUSIVE REMARKS

Now we shall drop the restriction (21). It is obvious that for any partition T_s of Y there are less than $2r$ subsets B_i wich satisfy the condition $P(B_i) > (2r)^{-1}$. For the sake of simplisity we consider the case when there exists unique set B_{i_0} of such kind. Then the following inequality holds (see the notations above)

$$|J^*_{kr}| \leqslant \sum_{j=0}^{r} C^j_r R_k(i_0, r, j), \tag{40}$$

where

$$R_k(i_0, r, j) = |g_{i_0} V_{nk}(B_{i_0})|^{r-j} \Big| \sum_{i_1,\dots,i_j \neq i_0} g^{j-r}_{i_0} f_{i_0\dots i_0 i_1\dots i_j} \prod_{q\leqslant j} V_{nk}(B_{i_q}) \Big|$$

when $g_{i_0} \neq 0$.
Let us estimate the tail distribution of $R_k(i_0, r, j)$:

$$\mathbf{P}\Big(\max_{k\leqslant n} R_k(i_0, r, j) \geqslant x\Big) \leqslant \mathbf{P}\Big(\max_{k\leqslant n} |g_{i_0} V_{nk}(B_{i_0})| \geqslant x^{1/r}\Big) +$$

$$+\mathbf{P}\Big(\max_{k\leqslant n} \Big| \sum_{i_1,\dots,i_j \neq i_0} g^{j-r}_{i_0} f_{i_0\dots i_0 i_1\dots i_j} \prod_{q\leqslant j} V_{nk}(B_{i_q}) \Big| \geqslant x^{j/r}\Big). \tag{41}$$

The first probability of the right–hand side of (41) is estimated by Bernstein's inequality (see part 1 of the paper). The corresponding estimate for the second probability follows from Lemma 6 and inequality

$$|g^{j-r}_{i_0} f_{i_0\dots i_0 i_1\dots i_j}| \leqslant g_{i_1} \dots g_{i_j}.$$

Hence it follows from (39)—(41) that the theorem is valid for any P.
We proved the theorem for real valued kernels. The extension on the case of Hilbert space valued kernels follows from the monotony of the moments (26):

$$\mathbf{E}\| \sum_{i_1,\dots,i_r} f_{i_1\dots i_r} Q^0_{nP}(B_{i_1}) \dots Q^0_{nP}(B_{i_r}) \|^{2N} =$$

$$= \mathbf{E}\Big(\sum_{\substack{i_1,\dots,i_r \\ j_1,\dots,j_r}} (f_{i_1\dots i_r}, f_{j_1\dots j_r}) \prod_{q\leqslant r} Q^0_{nP}(B_{i_q}) \prod_{v\leqslant r} Q^0_{nP}(B_{j_v}) \Big)^N \leqslant$$

$$\leqslant \mathbf{E}\Big(\sum_{i_1,\dots,i_r} \|f_{i_1\dots i_r}\| Q^0_{nP}(B_{i_1}) \dots Q^0_{nP}(B_{i_r}) \Big)^{2N}.$$

The theorem is proved.

REFERENCES

Borisov, I.S. (1978). On the rate of convergence in the conditional invariance principle. *Theory Probab. Appl.* **23**, 67–79.

Borisov, I.S. (1983). A problem of accuracy of approximation in the central limit theorem for empirical measures. *Siberian Math. J.* **24**, 833–843.

Dehling, H., Denker, M. and Philipp, W. (1986). A bounded law of the iterated logarithm for Hilbert space valued martingales and its application to U-statistics. *Probab. Th. Rel. Fields* **72**, 111–131.

Filippova, A.A. (1962). Mises theorem on the asymptotic behaviour of functionals of empirical distribution functions and its statistical applications. *Theory Probab. Appl.* **7**, 24–57.

Major, P. (1988). On the tail behaviour of the distribution function of multiple stochastic integrals. *Probab. Th. Rel. Fields* **78**, 419–438.

Yurinskii, V.V. (1976). Exponential inequalities for sums of random variables. *J. Multivar. Anal.* **6**, 473–499.

Prob. Theory and Math. Stat., Vol. 1, pp. 179–188
B.Grigelionis *et al.* (Eds.)
1990 VSP/Mokslas

ERGODICITY AND STABILITY OF MARKOV CHAINS AND OF THEIR GENERALIZATIONS. MULTIDIMENSIONAL CHAINS

A.A. BOROVKOV
Institute of Mathematics, Novosibirsk, USSR

ABSTRACT

The paper is devoted to the following subjects:
1. Ergodicity (including rate of convergence) and stability of arbitrary Markov chains (MC).
2. Ergodicity of stochastically recursive sequences (SRS), which are stochastic processes of more general nature than MC.
3. Ergodicity of multidimensional MC (including rate of convergence).
4. Applications to stabilization problem for heat differential equation.

1. ERGODICITY AND STABILITY OF ARBITRARY MC

1. *Ergodicity*

A great number of papers is devoted to the investigation of MC. We have to mention first of all the fundamental works of Markov, Kolmogorov, Doeblin, Harris, Orey and others. About 10 years ago there appeared some new and very simple approaches to the proof of ergodicity of MC. I mean the ideas of (Atreya and Ney, 1978; Nummelin, 1978) concerning the existence of modified regenerations for a wide class of MC and the ideas of (Foster, 1953; Moustafa, 1957; Tweedie, 1976) concerning positivity of MC. The situation in the theory of MC after appearance of these papers became quite different and much better:

1. It turned out that now it is possible to present very general ergodicity theorems for arbitrary MC just in university text–books. It takes 5, may be 10 pages, no more.

2. The above–mentioned new approaches are not only of methodological value. They also help to understand much better the nature of things and to extend some ergodicity results to a much more general class of stochastic processes.
Let $(\mathcal{X}, \mathcal{B})$ be an arbitrary measurable space and let $X = \{X(n) = X(x, n),\ n \geqslant 0\}$ be a nonperiodic irreducible $\mathcal{X}$–valued MC with initial value $x = X(x, 0) = X(0) \in \mathcal{X}$. Let

$$\tau(x) = \min\{k \geqslant 1 :\ X(x, k) \in V\}$$

for some set $V \in \mathcal{B}$.

The simplest and fairly general conditions of ergodicity of X are the following ones.

Let there exist a set $V \in \mathcal{B}$, a probability measure φ on $(\mathcal{X}, \mathcal{B})$, a number $p \in (0,1)$, and a natural number $n_0 \geqslant 1$ such that

$$\text{(I)} \qquad \sup_{x \in V} \mathbf{E}\, \tau(x) < \infty$$

$$\text{(II)} \qquad P_{n_0}(x, B) \equiv \mathbf{P}\big(X(x, n_0) \in B\big) \geqslant p\, \varphi(B)$$

for all $x \in V, B \in \mathcal{B}$.

The condition (II) is a "mixing" condition. In all the preceeding papers it was supposed that $\varphi(V) > 0$. In this case it is a Doeblin—Doob type condition, weakened on the set V. It turned out that this property is unnecessary for ergodicity. Condition (II) is stated in terms of the local characteristics of X and is final in this sense.

The condition (I) is that of uniform positive recurrency of X. It requires further investigation.

THEOREM 1. *Let the conditions* (I), (II) *hold. Then there exists the unique probabilistic invariant measure* π *on* $(\mathcal{X}, \mathcal{B})$ *such that for any* x *and for* $n \to \infty$

$$\Delta(x, n) \equiv \| P_n(x, \cdot) - \pi(\cdot) \| = 2 \sup_{B \in \mathcal{B}} | P_n(x, B) - \pi(B) | \to 0,$$

where the norm $\| \mu(\cdot) \|$ *is the variation of measure* μ.

This assertion was proved in its main features in (Orey, 1971; Atreya and Ney, 1978; Nummelin, 1978, 1984).

2. *Rate of convergence*

Under additional conditions the assertion of Theorem 1 can be made more precise. Alongside with (I) let us consider stronger condition

$$\text{(Ia)} \qquad \sup_{x \in V} \mathbf{P}\,\big(\tau(x) > k\big) < \delta(k), \qquad \sum \delta(k) < \infty,$$

where $\delta(k)$ is the function of the form either $\delta(k) = c\, e^{-\beta k}$ or $\delta(k) = \varepsilon(k)\, k^{-\delta}$. Here $\varepsilon(k)$ is a slowly varying function, $c < \infty$, $\beta > 0$, $\delta \geqslant 1$. The constants c and β can be different in different formulas.

THEOREM 2. *Under conditions* (Ia), (II)

$$\sup_{x \in V} \Delta\,(x, n) \leqslant c\, n\, \delta(n). \tag{1}$$

These estimates are sharp. For $\delta(k) = c\, e^{-\beta k}$, $\beta > 0$, and $\delta(k) = c\, k^{-2}$ they were proved in (Nummelin, 1984).

3. *Large deviations*

The probabilities of "large deviations" of $X(n)$ from the set V in terms of some function $L\colon \mathcal{X} \to [0,\infty)$, $\sup_{x\in V} L(x) \leqslant c$, can be estimate as well.

THEOREM 3. *Under conditions* (Ia), (II) *and assumption*

$$\sup_x \mathbf{P}\Big(L(X(x,1)) - L(x) > t\Big) \leqslant \delta(t)$$

we have

$$\sup_{n,x\in V} \mathbf{P}\Big(L(X(x,n)) > t\Big) \leqslant c\,t\,\delta(t). \tag{2}$$

Instead of the set V in formulas (1), (2) one can put some broader set.

4. *Conditions of positivity*

Let us return to conditions (I), (Ia). For a wide class of MC of practical interest these conditions are very far from the final conditions (expressed in terms of local characteristics of MC). And still these conditions are the essential step in the investigations of positivity of MC. Theorem 2 shows us that behaviour of the distribution of $\tau(x)$ gives us the adequate terms for the description of convergence $\sup_{x\in V} \Delta(x,n) \to 0$.
But conditions (I), (Ia) themselves require further investigation. One of the possible ways here is to use the so-called method of Lyapunov functions.

THEOREM 4 (Foster–Moustafa–Tweedie criteria (Foster, 1953; Moustafa, 1957; Tweedie 1976)). *If for some nonnegative measurable function* $L\colon \mathcal{X} \to \mathbf{R}$

$$\mathbf{E}L(X(x,1)) - L(x) < -\varepsilon < 0, \qquad x \in V^c = \mathcal{X}\backslash V,$$

$$\mathbf{E}L(X(x,1)) < c < \infty, \qquad x \in V,$$

then (I) *holds.*

There are also other forms of sufficient conditions for (I) which provide (Ia) besides. The importance of the behaviour of the distribution of $\tau(x)$ can be illustrated as well by stability conditions.

5. *Stability*

Let ${}^rX$, $r = 1,2,\ldots$, be a sequence of MC. The left upper index r will denote the correspondence to MC ${}^rX$.
We present the simplest possible sufficient conditions for stability. They are quite

far from most general ones (which are also obtained).
Let us suppose that there exist the densities

$$ {}^{r}p(x,y) = \frac{{}^{r}P_1(x,dy)}{\mu(dy)}, \qquad p(x,y) = \frac{P_1(x,dy)}{\mu(dy)} $$

w.r. to some measure μ.

THEOREM 5. *If MC* X, ${}^{r}X$ *satisfy conditions* (Ia), (II) *uniformly with respect to* r,

$$ {}^{r}p \to p, \qquad {}^{r}p \leqslant \overline{p}, $$

where $\overline{p}$ *is integrable, then for any* $B \in \mathcal{B}$

$$ {}^{r}\pi(B) \to \pi(B). $$

2. ERGODICITY OF STOCHASTICALLY RECURSIVE SEQUENCIES

As it was said in the very beginning, some of the ideas and approaches mentioned above can be carried over to random sequences $X(n) \in \mathcal{X}$ of a more general nature than MC.
Let $(\mathcal{X}', \mathcal{B}(\mathcal{X}'))$ be some measurable space, $\{\xi_n\}_{-\infty}^{\infty}$ be $\mathcal{X}'$-valued stochastic sequence.

Definition. We call $\{X(n)\}_1^\infty$ *stochastically recursive sequence, governed by the sequence* $\{\xi_n\}$ (or simply SRS), if $\{X(n)\}$ and $\{\xi_n\}$ are defined on the same probability space and for $\mathcal{F}_n = \sigma(X(n), \ldots, X(1); \quad \xi_n, \xi_{n-1}, \ldots)$

$$ \mathbf{P}(X(n+1) \in B \mid \mathcal{F}_n) = \mathbf{P}(X(n+1) \in B \mid \sigma(X(n), \xi_n)) = P_{(n)}(X(n), \xi_n, B). $$

We will consider only homogeneous SRS, when

1. $P_{(n)} = P$ are independent of n.
2. The sequence $\{\xi_n\}$ is strictly stationary.

In this case we can call the sequence $\{X(n)\}$, satisfying

$$ \mathbf{P}(X(n+1) \in B \mid \mathcal{F}_n) = P(X(n), \xi_n, B), \tag{3} $$

a *MC, governed by stationary sequence* $\{\xi_n\}$ as well.
We can consider the function $P(x, x', B)$ again as "transition probability" to hit the set B from the point x when governing parameter is x'.
If σ-algebras $\sigma(\xi_n, \xi_{n-1}, \ldots)$ are trivial then according to (3) $\{X(n)\}$ forms the usual MC. If $\{\xi_n\}$ is the sequence of i.i.d.r.v. and hence ξ_n is independent of $\xi_{n-1}, \xi_{n-2}, \ldots;$ $X(n), X(n-1), \ldots$, then SRS $\{X(n)\}$ also forms the MC.
Obviously, in general case SRS is not a MC.
There exists a special important form of SRS, defined by the recurrent relations

$$ X(n+1) = f(X(n), \xi_n), \tag{4} $$

where the mapping $f\colon \mathcal{X} \times \mathcal{X}' \to \mathcal{X}$ is measurable. Obviously (4) satisfies (3). Vice versa: for a broad class of spaces $\mathcal{X}$ SRS (3) can be reduced to (4) (for some others $\mathcal{X}'$ and $\{\xi_n\}$).

SRS of types (3), (4) arise in many applications, in particular in queueing theory, in the theory of dynamical systems etc.

There are a lot of papers devoted to the study of SRS (I mean SRS of the form (4)). More general definition (3) is new as far as I know, and was suggested by myself very recently. From my point of view this definition is more convenient and natural in many aspects.

One of the main tools for the investigation of SRS is the so-called method of renovating events developed in (Borovkov, 1984).

We have the following interesting fact. Let us consider the method of modified regenerations, suggested in (Atreya and Ney, 1978; Nummelin, 1978) to prove ergodicity under conditions (I), (II) and the method of renovating events for the investigation of ergodicity of SRS. It turned out that it is possible to find such a generalization of these methods for SRS (3) that they practically coincide in all main points.

We can consider, for instance, the following version of conditions of type (I), (II) for SRS.

Let us suppose that there exists stationary sequence of so-called "renovating" events $A_n \in \sigma(\xi_n, \xi_{n-1}, \ldots)$ and families of numbers $p_{x'} > 0$ and measures $\varphi_{x'}(\cdot)$ on $\mathcal{B}$, $x' \in \mathcal{X}'$, such that

(I) $$\mathbf{P}(A_n) = \mathbf{P}(A_0) > 0\,,$$

(II) on the set A_n

$$\mathbf{P}\big(X(n+1) \in B \mid \mathcal{F}_n\big) \equiv \mathbf{P}\big(X(n), \xi_n, B\big) \geqslant p_{\xi_n}\varphi_{\xi_n}(B).$$

In some sense this more general version of conditions (I), (II) looks even simpler than the initial one.

THEOREM 1A. *If $\{\xi_n\}$ is metrically transitive and conditions (I), (II) are fulfilled then there exists stationary distribution π such that for any fixed $X(1)$*

$$\sup_B \;|\, \mathbf{P}\big(X(n) \in B\big) - \pi(B)\,| \to 0\,.$$

Condition of positivity (I) (about existence of the sequence A_n) requires some further investigation. Here we have again an interesting situation — method of Lyapunov function can also be extended to SRS case.

Quite often condition (II) can be fulfilled on the set of the form

$$\{L(X(n)) \leqslant N\} \cap \{\xi_n \in U\}$$

for some $U \subset \mathcal{X}'$, $\mathbf{P}(U) > 0$, $N \geqslant 0$ and for some suitable function $L\colon \mathcal{X} \to R$. Under this assumption the following result is very useful. I present it in the simplest and most transparant form.

THEOREM 4A (analogue of Theorem 4). *Let us suppose that there exists a function* $g: \mathcal{X}' \to R$ *such that* a.s.

$$L(X(n+1)) - L(X(n)) \leqslant \begin{cases} g(\xi_n), & \text{if } L(X(n)) > N, \\ g(\xi_n) + c, & \text{if } L(X(n)) \leqslant N, \end{cases}$$

where $\mathbf{E}g(\xi_n) < 0$.
Then there exists a stationary sequence A_n *satisfying* (I), (II).

This result can be generalized in many directions. In particular, a.s. inequality $\leqslant$ can be replaced by some much weaker inequalities for conditional distributions. Some of these generalizations were obtained by my colleague S.Foss.
As I have said above, the investigations of SRS are motivated mostly by applications. The presented results and the technique developed here proved to be very effective tools for the solution of many interesting applied problems, connected, for instance, with queueing systems and, in particular, with communication and computer networks.

3. ERGODICITY OF MULTIDIMENSIONAL MC

All the above-mentioned results are very useful for the investigation of ergodicity and stability of multidimensional random walks (r.w.) in the space $\mathcal{X} = \mathbf{R}^d$, $d \geqslant 1$, as well. It seems very difficult or well-nigh impossible to produce here a short description of ergodicity conditions for the general case. I restrict myself to the particular case of two-dimensional MC in the space

$$\mathcal{X} = \mathbf{R}_+^2 = \{x = (x_1, x_2): \quad x_1 \geqslant 0, \quad x_2 \geqslant 0\}.$$

It will be convenient for us to represent $\{X(n)\}$ in the recurrent form

$$X(n+1) = X(n) + \xi(X(n), n),$$

where the distribution of $\xi(x, n)$ depends only on x (in what follows $\xi(x)$ denotes a random variable with this distribution). The distribution of jumps $\xi(x)$ can be defined in terms of transition function $P(x, B)$; for instance we can put

$$\mathbf{P}(\xi(x) \in A) = P(x, x + A).$$

Another restriction which we need to get intelligible minimal conditions of ergodicity concerns regularity of $P(x, B)$.
First we present some results about one-dimensional case $\mathcal{X} = \mathbf{R}_+^1$. If we suppose that $\xi(x) \Rightarrow \xi$ as $x \to \infty$ then we can get quite acceptable almost minimal conditions of ergodicity: if $\xi^+(x)$ are uniformly integrable w.r. to x, $\mathbf{E}\xi < 0$ and if condition (II) of the general ergodic theorem is fulfilled for the set $V = \{x \leqslant N\}$ for some $N \geqslant 0$, then the MC X is ergodic.
In two-dimensional case we shall consider the so-called ***asymptotically homogeneous*** r.w., when the distributions of $\xi(x)$ converge to that of

$$\xi((\infty, x_2)),\ \xi((x_1, \infty)),\ \xi = \xi((\infty, \infty))$$

as $x_1 \to \infty$; $x_2 \to \infty$; $x_1, x_2 \to \infty$ respectively. For the sake of simplicity we suppose that we have convergence in variation.
We shall also suppose that our r.w. satisfies the condition (II) of the general ergodic theorem for the set $V = V_N = \{x : |x| \leqslant N\}$ for each $N \geqslant N_0$, $N_0 \geqslant 0$ (if there exists a density p then this condition is fulfilled always).
Now the question is: what will the ergodicity conditions be for the two-dimensional case? Can they be expressed in terms of expectations again? The answer is no. For the first time it was understood about 10 years ago by Malyshev and Menshikov for MC of very special form (Malyshev and Menshikov, 1979).
Now the ergodicity problem for MC in $\mathbf{R}_+^2$ and $\mathbf{R}_+^3$ can be considered as solved under almost minimal conditions and solved partly for $\mathbf{R}_+^d$, $d > 3$. It turned out that for the description of minimal ergodicity conditions for r.w. in $\mathbf{R}_+^d$ we need ergodicity conditions for the spaces $\mathbf{R}_+^1, \ldots, \mathbf{R}_+^{(d-1)}$ and the ergodic distributions themselves (if they exist) in these spaces.
We shall give the formulation of the ergodicity theorem for two-dimensional case $d = 2$. For the dimensions $d \geqslant 3$ conditions of ergodicity look more complicated.
We assume that

$$\sup_x \mathbf{P}(|\xi^+(x)| > t) \leqslant \delta(t)$$

(the function $\delta(\cdot)$ is defined in condition (Ia)) and that something else and not very essential is fulfilled, what is always fulfilled if

$$\sup_x \mathbf{P}(|\xi^-(x)| > t) \leqslant \alpha(t),$$

where $\alpha(\cdot)$ is some function, satisfying the same conditions as conditions for $\delta(\cdot)$.
Denote

$$a(x) = \mathbf{E}\,\xi(x), \quad a = (a_1, a_2) = \mathbf{E}\,\xi.$$

If $a_2 < 0$ then for large $X_1(n)$ the process $X_2(n)$ will be close to the one-dimensional r.w.

$$X^{(2)}(n+1) = X^{(2)}(n) + \xi_2\Big((\infty, X^{(2)}(n)), n\Big)$$

with negative drift for large $X^{(2)}(n)$. Hence this MC is ergodic. We denote π_2 its stationary distribution and set

$$\overline{a}_1 = \int a_1\big((\infty, x_2)\big)\pi_2(dx_2).$$

In a similar way we introduce the distribution π_1 (if $a_1 < 0$) and the quantity

$$\overline{a}_2 = \int a_2\big((x_1, \infty)\big)\pi_1(dx_1).$$

If $a_i > 0$ then we suppose that the following condition is fulfilled.
(III). For each $N > 0$ there exists $n_1 = n_1(N) > 0$ such that

$$\inf_{x_i \leqslant N,\ x_j > N} P_{n_1}(x, \{x_i > N\}) > 0; \ i \neq j; \ i, j = 1, 2.$$

THEOREM 1'. *Let all these assumptions be fulfilled. If one of the following three conditions*

$$
\begin{array}{ll}
1) & a_1 > 0, \quad a_2 < 0, \quad \bar{a}_1 < 0, \\
2) & a_1 < 0, \quad a_2 > 0, \quad \bar{a}_2 < 0, \\
3) & a_1 < 0, \quad a_2 < 0, \quad \bar{a}_1 < 0, \quad \bar{a}_2 < 0,
\end{array}
\tag{5}
$$

holds, then MC $X(n)$ is ergodic.
Moreover, for some $h > 0$

$$\sup_{|x|\leqslant hn} \Delta(x,n) < c\,n\,\delta(n),$$

$$\sup_{n,|x|\leqslant hn} \mathbf{P}(\mid X(x,n) \mid > t) < c\,t\,\delta(t).$$

The conditions (5) of ergodicity are minimal in the following sense: if either of the following conditions

$$
\begin{array}{ll}
4) & a_1 > 0, \quad a_2 > 0, \\
5) & a_2 < 0, \quad \bar{a}_1 > 0, \\
6) & a_1 < 0, \quad \bar{a}_2 > 0,
\end{array}
\tag{6}
$$

holds, then X is non–ergodic (transient). The inequalities (6) would be complementary to (5) if they were not strict.
The conditions of ergodicity can be obtained for MC in all the space $\mathbf{R}^2$ too but they look also more complicated as well as conditions for $\mathbf{R}^3_+$, $\mathbf{R}^3$.

4. STABILIZATION PROBLEM FOR HEAT EQUATION

Among others there is rather interesting application of results, obtained above, to the problems of classical analysis, in particular to the so–called ***stabilization problem*** for the second initial boundary value problem for parabolic partial differential heat equation

$$\frac{\partial u}{\partial t} = D(u) \equiv -\sum_{i=1}^{d} \frac{\partial}{\partial x_i}\Big(a_i(x)u\Big) + \frac{1}{2}\sum_{i,j=1}^{d} \frac{\partial^2}{\partial x_i \partial x_j}\Big(b_{ij}(x)u\Big) \tag{7}$$

with given initial data $u(x,t)\big|_{t=0} = u^{(y)}(x,t)\big|_{t=0} = \delta(x-y)$ (fundamental solution) and boundary conditions

$$\left.\frac{\partial u(x,t)}{\partial x_i}\right|_{x_i=0} = 0.$$

Non–zero stabilization problem for equation (7) is as follows: under which conditions there exist fundamental solution $u^{(y)}(x,t)$ and non-trivial solution $u(x) \not\equiv 0$ of equation $D(u) = 0$ such that $u(x,t) \to u(x)$ as $t \to \infty$.
The important points, which allow to get stabilization conditions for (7), are as follows.

1. The solution of the second initial–boundary value problem for the heat equation can be interpreted as a distribution of the Markov diffusion process

$w(t) = w(y,t)$, starting at the point $w(y,0) = y$, with local drift coefficients $a_i(x)$, diffusion coefficients $b_{ij}(x)$ and with reflection from the boundaries $x_i = 0$. So the stabilization problem for (7) is equivalent to the ergodic problem for Markov diffusion process, described by (7).

2. Ergodic results for MC in $\mathbf{R}^d$ allow to obtain similar results for Markov diffusion processes.
We restrict ourselves again to one– and two–dimensional cases. First we consider the case $d = 1$.

THEOREM 1″. *If for some $N > 0$*

$$a(x) \leqslant -a_0 < 0, \quad for \quad x > N, \tag{8}$$

$$b(x) + | a(x) | > b_0 > 0; \quad b(x) \geqslant 0, \quad b(x) \not\equiv 0, \tag{9}$$

then there exist solutions $u(x,t)$, $u(x) \not\equiv 0$ of equations (7) and $D(u) = 0$, respectively, such that $u(x,t) \to u(x)$ as $t \to \infty$.

The meaning of conditions (8), (9) is rather simple. The first one (negative drift) doesn't allow the trajectory to go to infinity. The second one keeps the trajectory moving.
For the class of equations satisfying (9) and conditions $a(x) \to a \neq 0$, $b(x) \to$ $\to b > 0$, as $x \to \infty$, the condition (8) is the necessary and sufficient one for ergodicity, that is for existence of $u(x) \not\equiv 0$ and stabilization $u(x,t) \to u(x)$. If $a > 0$ then $u(x,t) \to 0$ as $t \to \infty$.
Two–dimensional case ($d = 2$). To get here intelligible minimal stabilization conditions we need again some regularity of the coefficients $a_i(x)$, $b_{ij}(x)$. Let us assume that there exist functions $a_i(x_1,\cdot)$, $a_i(\cdot,x_2)$, $b_{ij}(x_1,\cdot)$, $b_{ij}(\cdot,x_2)$ and constants a_i, b_{ij}; $i,j = 1,2$, such that

$$\begin{aligned}
&a_i\big((x_1,x_2)\big) \to a_i(\cdot,x_2), && \text{if} \quad x_1 \to \infty,\\
&a_i\big((x_1,x_2)\big) \to a_i(x_1,\cdot), && \text{if} \quad x_2 \to \infty,\\
&a_i\big((x_1,x_2)\big) \to a_i, && \text{if} \quad x_1 \to \infty, x_2 \to \infty.
\end{aligned}$$

Similar relations hold for functions $b_{ij}\big((x_1,x_2)\big)$. For the sake of simplicity we suppose that

$$\text{Det} \parallel b_{ij}(x) \parallel \geqslant b_0 > 0.$$

What are the conditions of ergodicity for the equation (7) in this situation? The answer to this question will be related to the stabilization of the solutions of one–dimensional sub–equations

$$\frac{\partial v^{(1)}}{\partial t} = D_1(v^{(1)}) \equiv -\frac{\partial}{\partial x_1}\Big(a_1(x_1,\cdot)v^{(1)}\Big) + \frac{1}{2}\frac{\partial^2}{\partial x_1^2}\Big(b_{11}(x_1,\cdot)v^{(1)}\Big), \tag{10}$$

$$\frac{\partial v^{(2)}}{\partial t} = D_2(v^{(2)}) \equiv -\frac{\partial}{\partial x_2}\Big(a_2(\cdot,x_2)v^{(2)}\Big) + \frac{1}{2}\frac{\partial^2}{\partial x_2^2}\Big(b_{22}(\cdot,x_2)v^{(2)}\Big). \tag{11}$$

If $a_2 < 0$ then according to one–dimensional conditions (8), (9) there exist the solutions $v^{(2)}(x_2,t)$, $v^{(2)}(x_2) \not\equiv 0$ of equations (11), $D_2(v^{(2)}) = 0$, respectively, such that $v^{(2)}(x_2,t) \to v^{(2)}(x_2)$ (let $v^{(2)}(x,t)$ be the fundamental solution). Denote

$$\overline{a}_1 = \int a_1(\cdot, x_2)\, v^{(2)}(x_2)\, dx_2.$$

If $a_1 < 0$ we introduce in a similar way the value

$$\overline{a}_2 = \int a_2(x_1, \cdot)\, v^{(1)}(x_1)\, dx_1.$$

THEOREM 1‴. *If one of the three conditions* (5) *of the Theorem* 1′ *holds then equation* (7) *has non–zero stabilizing solution* $u(x,t) \to u(x) \not\equiv 0$; $D(u) = 0$.

These conditions of non–zero stabilization are minimal in the following sence: if one of three complimentary conditions (6) of Theorem 1′ holds then $u(x) \not\equiv 0$, $D(u) = 0$ such that $u(x,t) \to u(x)$ does not exist.
The results presented in this talk are fragments of the new book which will appear soon in the series of Proceedings of Institute of Mathematics, Novosibirsk.

REFERENCES

Atreya, K.B. and Ney, P. (1978). A new approach to the limit theory of recurrent Markov chains. *Trans.Amer.Math.Soc.* **245**, 493–501.

Borovkov, A.A. (1984). *Asymptotic Methods in Queueing Theory.* Wiley, New York.

Foster, F.G. (1953). On the stochastic matrices associated with certain queueing processes. *Ann.Math.Stat.* **24**, 355–360.

Malyshev, V.A. and Menshikov, V.M. (1979). Ergodicity, continuity and analyticity of countable Markov chains. *Proc. Moscow Math. Soc.* **39**, 3–48.

Moustafa, M.D. (1957). Imput–output Markov processes. *Proc. Kon. Netherlands Acad. Wet.* **60**, 112–118.

Nummelin, E.A. (1978). A splitting technique for Harris recurrent Markov chains. *Z. Wahr. verw. Geb.* **43**, 309–318.

Nummelin, E.A. (1984). *General Irreducible Markov Chains and Non–Negative Operators.* Cambridge Univ. Press, Cambridge.

Orey, S. (1971). *Limit Theorems for Markov Chain Transition Probabilities.* Van Nostrand, New York.

Tweedie, R.L. (1976). Criteria for classifying general Markov chains. *Adv. Appl. Probab.* **8**, 737–771.

Prob. Theory and Math. Stat., Vol. 1, pp. 189–196
B.Grigelionis *et al.* (Eds.)
1990 VSP/Mokslas

EVENT AND TIME AVERAGES: THE STATIONARY CASE AND THE NON–STATIONARY CASE

P. BRÉMAUD

Laboratoire des Signaux et Systémes, CNRS–ESE, Plateau du Moulon,
91190 Gif–sur–Yvette, Cedex, France

ABSTRACT

We present a calculus for point processes associating the theory of stochastic intensity and the theory of Palm measure via a representation lemma for predictable processes in a stationary context. We briefly discuss an application to Queueing Theory, the so–called PASTA result.

1. INTRODUCTION

We wish to briefly discuss the interactions of the Palm theory of point processes and of the stochastic intensity theory. The connection was first established in (Papangelou, 1972). Palm–martingale calculus for point processes is based on three ingredients: Matthes—Mecke formula (Matthes et al., 1974), the stochastic intensity integration formula (Brémaud, 1981) and the representation lemma for predictable processes in the stationary framework (Baccelli and Brémaud, 1987). We observe that Mecke's Palm characterization of Poisson processes (Mecke, 1967) becomes via Papangelou's theorem a version of Watanabe's martingale characterization of Poisson processes (Watanabe, 1964). Also, the celebrated PASTA results of Queueing Theory (König and Schmidt, 1980; Wolf, 1982) are shown to be just Papangelou's theorem. Finally we give a variant of Wolff's proof of PASTA (Mecke, 1967) in the non–stationary case, thus showing how the martingale strong law of large numbers (Liptser and Shiryaev, 1978) connects time averages, event averages and stochastic intensity.

2. KNOWN FACTS

2.1. *Matthes—Mecke formula*

Let be given a probability space (Ω, F, P) on which a measurable semi–group of automorphisms $\{\theta_t\}, t \in \mathbf{R}$, is defined (see (Baccelli and Brémaud, 1987) for the notations used in this note). Suppose that P is stationary in the sense that for all

$t \in \mathbf{R}$

$$P \circ \theta_t^{-1} = P. \tag{2.1.1}$$

A simple point process N is given, defined on $(\Omega, \mathcal{F}, P)$ and such for all $C \in B(\mathbf{R})$, $\omega \in \Omega$, $t \in \mathbf{R}$

$$N(\theta_t\omega, C) = N(\omega, C + t). \tag{2.1.2}$$

Suppose that its average intensity is finite and non null

$$0 < \lambda \triangleq \mathbf{E}[N(0,1]] < \infty. \tag{2.1.3}$$

The Palm probability P_0 is associated with (P, N) is defined by (Matthes et al., 1974)

$$P_0(A) = [\lambda(b-a)]^{-1}\mathbf{E}\left[\int_{(a,b]} (\mathbf{1}_A \circ \theta_s)N(ds)\right]. \tag{2.1.4}$$

It does not depend on $(a, b] \in \mathbf{R}$ and

$$\begin{aligned} P_0(N(\{0\}) = 1) &= 1, \\ P_0 \circ \theta_{T_n} &= P_0, \quad \forall n \in \mathbf{Z}, \\ \mathbf{E}_0[T_1] &= \lambda^{-1}, \end{aligned} \tag{2.1.5}$$

where $\{T_n\}, n \in \mathbf{Z}$, is the sequence of points of N with the convention $T_0 \leqslant 0 < T$. The Matthes—Mecke formula is an immediate consequence of (2.1.4)

$$\lambda\mathbf{E}_0\left[\int_{\mathbf{R}} v(s)ds\right] = \mathbf{E}\left[\int_{\mathbf{R}} (v(s) \circ \theta_s)N(ds)\right], \tag{2.1.6}$$

where$\{v(t)\}, t \in \mathbf{R}$, is an arbitrary non-negative measurable process.

2.2. *Stochastic intensity formula*

Let $\{\mathcal{F}_t\}, t \in \mathbf{R}$, be a filtration on $(\Omega, \mathcal{F})$ such that $(a, b] \subset (-\infty, t] \Rightarrow N(a, b]$ is $\mathcal{F}_t$-measurable. Let $\{\lambda(t)\}, t \in \mathbf{R}$, be a non-negative measurable process adapted to $\{\mathcal{F}_t\}$. If for all $(a, b] \subset \mathbf{R}$

$$\mathbf{E}[N(a,b] \mid \mathcal{F}_a] = \mathbf{E}\left[\int_{a,b} \lambda(s)ds \mid \mathcal{F}_a\right]. \tag{2.2.1}$$

$\{\lambda_a\}$ is called the *$\mathcal{F}_t$-stochastic intensity of N* (Brémaud, 1981). As an immediate consequence of (2.2.2) we have the stochastic intensity formula:

$$\mathbf{E}\left[\int_{\mathbf{R}} x(s)N(ds)\right] = \mathbf{E}\left[\int_{\mathbf{R}} x(s)\lambda(s)ds\right] \tag{2.2.2}$$

for all non-negative $\mathcal{F}_t$-predictable processes $\{X(t)\}$, $t \in \mathbf{R}$. Of course (2.2.2) is a particular case of Dellacherie—Meyer dual predictable projection theorem (see (Liptser and Shiryaev, 1978)).

3. THE PALM-MARTINGALE CALCULUS

3.1. *A representation lemma*

Suppose that for all $s,t \in \mathbf{R}$

$$\theta_t \mathcal{F}_s = \mathcal{F}_{s-t} \tag{3.1.1}$$

and that $\{T_t\}$ has a predictable structure adapted to $\{\theta_t\}$, that is a process $\{x(t)\}$, $t \in \mathbf{R}$, is $\mathcal{F}_t$-predictable if and only if for all $t \in \mathbf{R}$, $\omega \in \Omega$,

$$X(t,\omega) = v(t, \theta_t \omega) \tag{3.1.2}$$

for some measurable process $\{v(t)\}$, $t \in \mathbf{R}$, such that $v(t)$ is a $\mathcal{F}_{0-}$-measurable r. v. for all $t \in \mathbf{R}$. This concept is not empty as shows the representation lemma of (Baccelli and Brémaud, 1987):

LEMMA. *Let Z be a random variable on $(\Omega, \mathcal{F}, P)$ and define $\{U(t)\}$, $t \in \mathbf{R}$, by*

$$U(t) = Z \circ \theta_{T_n} \text{ on } [T_n, T_{n+1}). \tag{3.1.3}$$

Define $\mathcal{F}_t^U = \sigma(U(s),\ s \leqslant t)$. Then $\{\mathcal{F}_t^U\}, t \in \mathbf{R}$, satisfies (3.1.1) *and* (3.1.2).

$\{U(t)\}, T \in \mathbf{R}$, is a typical "jump process", such that $U(t,\omega) = U(0, \theta_t\omega)$, and filtrations of the type $\{\mathcal{F}_t^U\}$ cover all the cases of interest in Queueing Theory.

3.2. *A Palm inversion formula*

Here we assume that $\{\mathcal{F}_t\}$ has a predictable structure adapted to $\{\theta_t\}$ and that N admits the $\mathcal{F}_t$-intensity $\{\lambda(t)\}$. It is shown in (Brémaud, 1981) that $\{\lambda(t)\}$ can be chosen $\mathcal{F}_t$-predictable, and it can be shown that $\lambda(t,\omega) = \lambda(0, \theta_t\omega)$. By assumption on the predictable structure of $\{\mathcal{F}_t\}, \lambda(0)$ is $\mathcal{F}_{0-}$-measurable. Combining the stochastic intensity formula and the Matthes—Mecke formula one obtains, for all non-negative measurable stochastic process $\{v(t)\}$ such that $v(t)$ is $\mathcal{F}_{0-}$-measurable for all $t \in \mathbf{R}$:

$$\lambda \mathbf{E}_0 \left[\int_{\mathbf{R}} v(s) ds \right] = \mathbf{E} \left[\int_{\mathbf{R}} v(s)\lambda(0) ds \right]. \tag{3.2.1}$$

This last formula connects the stochastic intensity calculus and the Palm calculus. Together with Matthes—Mecke's formula and the stochastic intensity it constitutes the Palm-martingale calculus.

4. PAPANGELOU'S RND THEOREM

4.1. *Papangelou's formula*

In (Baccelli and Brémaud, 1987) the Palm–martingale calculus is used to provide a "calculus" proof of Papangelou's Randon—Nikodym Derivative theorem (Papangelou, 1972) which we quote here. Recall that the filtration $\{\mathcal{F}_t\}$ is supposed to have a structure adapted to $\{\theta_t\}$. Then $P \ll P_0$ on $\mathcal{F}_{0-}$ if and only if N admits a $\mathcal{F}_t$–intensity $\{\lambda(t)\}$. Moreover, in that case $\lambda(t,\omega) = \lambda(0,\theta_t\omega)$, where

$$\lambda(0) = \lambda \frac{dP}{dP_0}\Big|_{\mathcal{F}_{0-}} \tag{4.1.1}$$

Formula (4.1.1) is Papangelou's formula which we prefer to write in explicit form:

$$\lambda \mathbf{E}_0[X] = \mathbf{E}[\lambda(0)X] \tag{4.1.2}$$

for all non–negative $\mathcal{F}_{0-}$–measurable r.v. X.

4.2. *Mecke—Watanabe characterization of Poisson processes*

From Papangelou's result, one sees that $P \equiv P_0$ on $\mathcal{F}_{0-}$ if and only if N admits the $\mathcal{F}_t$–intensity $\lambda(t) \equiv \lambda$. But, by Watanabe's theorem of characterization of Poisson process (Watanabe, 1964) this is equivalent to: N is $\mathcal{F}_t$–Poisson, i.e. $N(a,b]$ is a Poisson r.v of mean $\lambda(b-a)$ for all $(a,b] \subset \mathbf{R}$, and moreover $N(a,b]$ is independent of $\mathcal{F}_a$.

5. PASTA

5.1. *Papangelou's PASTA*

If $(P,\{\theta_t\})$ is ergodic, then (see, for instance (Baccelli and Brémaud, 1987)). Papangelou's formula (4.1.2) becomes, calling $\{T_n\}$, $n \in \mathbf{Z}$, the sequence of points of N, with the convention $T_0 \leqslant 0 < T_1$:

$$\lambda \lim_{k\uparrow\infty} \frac{1}{k} \sum_{j=1}^{k} Z(T_j) = \lim_{t\uparrow\infty} \frac{1}{t} \int_{(0,t)} Z(s)\lambda(s)ds, \quad P\text{–a.s.}, \tag{5.1.1}$$

where $Z(t) = X \circ \theta_t$ for some $\mathcal{F}_{0-}$–measurable non–negative random variable X. In the Poisson case where $\lambda(t) \equiv \lambda$, this is the *PASTA* theorem of Queueing Theory: *If N is $\mathcal{F}_t$–Poisson, and $Z(t) = X \circ \theta_t$ for some $\mathcal{F}_{0-}$–measurable r.v. X, then the empirical "event average"*

$$\lim_{k\uparrow\infty} \frac{1}{k} \sum_{j=1}^{k} Z(T_j)$$

is equal to the empirical time average

$$\lim_{t\uparrow\infty} \frac{1}{t} \int_{(0,t)} Z(s)ds.$$

5.2. *Wolff's PASTA*

The *PASTA* theorem was first proven in (König and Schmidt, 1980) using Palm-calculus. Their result is not relative only to the Poisson case but is more general, not assuming the existence of stochastic intensity. In the intensity case, Papangelou's formula and its corollary (5.1.1) yield the most general result in the stationary case. Wolff in (Wolff, 1982) takes another point of view, rather interesting in that it gives a further connection between Palm probability and stochastic intensity. Wolff's result is following:

THEOREM. *Let N be a simple point process with the $\mathcal{F}_t$-intensity $\{\lambda(t)\}$, let $\{Z(t)\}$ be a $\mathcal{F}_t$-predictable process with values in the measurable space $(E, \mathcal{E})$ and let $C \in \mathcal{E}$. Define the following subsets of*

$$\Omega' = \{\omega | \lambda(t,\omega) \text{ is bounded}\}.$$

$$\Omega^2 = \{\omega | \exists \lim_{t\uparrow\infty} N((0,t],\omega)/t = \lambda(\omega) > 0\}.$$

Consider the quantities

$$A(t) = \frac{1}{N(0,t]} \int_{(0,t]} 1_C(Z(s))N(ds) \tag{5.2.1}$$

$$B(t) = \frac{1}{t} \int_{(0,t)} 1_C(Z(s))\lambda(s)ds. \tag{5.2.2}$$

Then

(i) *if $\omega \in \Omega^1 \cap \Omega^2$ and if the limit as $t \uparrow \infty$ of the quantity* (5.2.1) *exists, the limit as $t \uparrow \infty$ of the quantity in* (5.2.2) *exists and both limits are related by*

$$\lambda(\omega) \lim_{t\uparrow\infty} A(t,\omega) = \lim_{t\uparrow\infty} B(t,\omega), \tag{5.2.3}$$

(ii) *the same statement holds as in* (i), *only exchange* (5.2.1) *and* (5.2.2).

Proof. Preliminaries. We recall the following known facts

(α) the martingale strong law of large numbers (see (Liptser and Shiryaev, 1987)):

If $\{M_t\}$ is a square integrable $\mathcal{F}_t$ –martingale with associated Doob—Meyer compensator $\langle M \rangle_t$, then

$$\lim_{t \uparrow \infty} \frac{M_t}{\langle M \rangle_t} = 0 \text{ on } \{\langle M \rangle_\infty = \infty\}. \tag{5.2.4}$$

(β) Defining (see for instance (Brémaud, 1981))

$$M_t = \int_{(0,t]} \mathbf{1}_C(Z(s))N(ds) - \int_{(0,t)} \mathbf{1}_C(Z(s))\lambda(s)ds \tag{5.2.5}$$

we have

$$\langle M \rangle_t = \int_{(0,t)} \mathbf{1}_C(Z(s))\lambda(s)ds. \tag{5.2.6}$$

(γ) Also (see (Brémaud, 1981)):

$$\int_{(0,\infty)} \mathbf{1}_C(Z(s))N(ds) = \infty \Leftrightarrow \int_{(0,\infty)} \mathbf{1}_C(Z(s))\lambda(s)ds = \infty. \tag{5.2.7}$$

In particular

$$N(0,\infty) = \infty \Leftrightarrow \int_{(0,\infty)} \lambda(s)ds = \infty.$$

(Actually (5.2.7) is a particular case of the latter equivalence: just consider the point process N' defined by

$$N'(a,b] = \int_{(a,b]} \mathbf{1}_C(Z(s))N(ds)).$$

(δ) Finally observe that

$$A(t) - \frac{t}{N(0,t]}B(t) = \frac{M_t}{\langle M \rangle_t} \frac{t}{N(0,t]} \frac{1}{t} \int_{(0,t)} \mathbf{1}_C(Z(s))\lambda(s)ds. \tag{5.2.8}$$

We now turn to the proof of Wolff's result. We will not carry ω in the notations, but we recall that we work with $\omega \in \Omega^1 \cap \Omega^2$.

(i) Assume the existence of the limit in (5.2.1) and that this limit is positive. This implies (since $N(0,\infty) = \infty$ on Ω^2), that

$$\int_{(0,\infty)} \mathbf{1}_C(Z(s))N(ds) = \infty,$$

and therefore, by (γ), that

$$\langle M \rangle_\infty = \int_{(0,\infty)} \mathbf{1}_C(Z(s))\lambda(s)ds = \infty.$$

By the martingale SLLN

$$\lim_{t\uparrow\infty} \frac{M_t}{\langle M\rangle_t} = \infty,$$

and therefore since

$$t^{-1} \int\limits_{(0,t)} 1_C(Z(s))\lambda(s)ds$$

remains bounded ($\lambda(t)$ bounded), we have oquality of the event and time empirical averages, in view of (δ).
Next assume the existence of the limit (5.2.1), but suppose this limit is 0. If

$$\int\limits_{(0,\infty)} 1_C(Z(s))\lambda(s)ds < \infty,$$

then the result is proven. If not, then $\langle M\rangle_\infty = \infty$ and we catch up with the end of the proof in the case where the limit was positive.

(ii) The part of the proof when one starts from the existence of the limit in (5.2.2) is *mutatis mutandis* the same as in (i) when the limit is positive. When the limit in (5.2.2) is null, and $\langle M\rangle_\infty = \infty$, the arguments is also the same. And when $\langle M\rangle_\infty < \infty$, this implies

$$\int\limits_{(0,\infty)} 1_C(Z(s))N(ds) < \infty$$

by (γ), and therefore the limit in (5.2.1) is 0.

Remarks:

(1) Wolff's result does not assume stationarity.

(2) From the very conclusion of Wolff's result on $\Omega^1 \cap \Omega^2$,

$$\lim_{t\uparrow\infty} t^{-1} \int\limits_{(0,t)} \lambda(s,\omega)ds$$

exists and equals $\lambda(\omega)$. One can in fact show that if either limit: $\lim t^{-1} \int_0^t \lambda(s)ds$ or $\lim t^{-1}N(0,t]$ exists, then the other exists and they are equal. This remark is convenient in the doubly-stochastic situation, where $\lambda(t)$ is a functional of some process $\{Y(t)\}$ driving N and $\mathcal{F}_t = \mathcal{F}_t^N \vee \mathcal{F}_t^Y$.

(3) The proof of Wolff's result, given in this note, shows that if one does not wish to assume the existence of a limit for (5.2.1) or (5.2.2), still something can be said, namely that

$$\lim_{t\uparrow\infty} \left[\frac{\lambda(\omega)}{N((0,t],\omega)} \int\limits_{(0,t]} 1_C(Z(s,\omega))N(ds,\omega) - \frac{1}{t} \int\limits_{(0,t)} 1_C(Z(s,\omega))\lambda(s,\omega)ds \right] = 0.$$

A precise statement is left to the reader.

REFERENCES

Baccelli, F. and Brémaud, P. (1987). *Palm Probabilities and Stationary Queues.*. Springer, Berlin—Heidelberg etc.

Brémaud, P. (1981). *Point Processes and Queues: Martingale Dynamics.*. Springer, Berlin—Heidelberg etc.

Brémaud, P. (1989). Characteristics of queueing systems observed at events and the connection between stochastic intensity and Palm probability. In: *Queueing Systems, Theory and Applications,* **5**

Brémaud, P. (1990). Necessary and sufficient conditions for the equality of event and time averages. *J. Appl. Probab.* **27**

König, D. and Schmidt, D. (1980). Imbedded and non imbedded stationary characteristics of queueing systems with varying service rate and point processes. *J. Appl. Probab.* **17**, 753—767.

Liptser, R.S. and Shiryaev, A.N. (1978). *Statistics of Random Processes.*. Springer, Berlin—Heidelberg etc.

Matthes, K., Kersten, J. and Mecke, J. (1974). *Infinitely Divisible Point Processes.*. Wiley, New York.

Mecke, J. (1967). Stationäre Zufällige auf lokalkompakten Abelischen Gruppen. *Z. Wahr.verw. Geb.,* **8**, 39—56.

Papangelou, F. (1972). Integrability of expected increments of point processes and a related change of time. *Trans. Amer. Math. Soc.* **165**, 483—506.

Watanabe, S. (1964). On discontinuous additive functionals and Levy measures of a Markov process. *Japan J.Math.* **34**, 53—70.

Wolff, R.W. (1982). Poisson arrivals see time averages. *Oper.Res.* **30**, 223—231.

Prob. Theory and Math. Stat., Vol. 1, pp. 197–206
B.Grigelionis *et al.* (Eds.)
1990 VSP/Mokslas

ON THE CONVERGENCE OF SUMS OF INDEPENDENT RANDOM VECTORS NORMED BY MATRICES

V.V. BULDYGIN and S.A. SOLNTSEV

Kiev Polytechnical Institute, pr. Pobedy 37, 252056 Kiev, USSR

ABSTRACT

In this paper the general criterion for almost sure convergence of sums of independent symmetric random vectors normed by matrices is established. Using this criterion we state necessary and sufficient conditions for a.s. convergence of a multidimensional Gaussian Markov sequence. Particularly, an entropy conditions are obtained which are necessary and sufficient for a.s. continuity of a multidimensional Gaussian Markov process.

1. INTRODUCTION

Let $(X_k,\ k \geqslant 1)$ be a sequence of independent random vectors taking values in a m–dimensional Euclidean space $\mathbf{R}^m$ $(m \geqslant 1)$; $S_n = X_1 + X_2 + \ldots + X_n$, $n \geqslant 1$; $(A_n,\ n \geqslant 1)$ be a sequence of linear operators from $\mathbf{R}^m$ to $\mathbf{R}^d$. Consider normalized sums $(A_n S_n,\ n \geqslant 1)$. The purpose of the paper is to investigate the conditions under which the sequence $(A_n S_n,\ n \geqslant 1)$ converges almost surely (a.s.) with subsequent application tothe investigation of the sample behaviour of a multidimensional Gaussian Markov processes with discrete or with continuous time. Without loss of generality we assume the vectors X_k, $k \geqslant 1$, to be symmetric, i.e. X_k and $(-X_k)$ have the same distribution, $k \geqslant 1$, and $m = d$, i.e., $A_n : \mathbf{R}^m \to \mathbf{R}^m, n \geqslant 1$.

Weak and strong limit theorems for sums of independent random vectors with operator normalizations became the object of study not so long ago (Kandelaki and Sazonov,1964; Hahn, 1979, 1981; Buldygin and Solntsev, 1986, 1987a, 1987b, 1988). Strong laws of large numbers (SLLN) for sums of independent random vectors normed by operators were investigated in (Buldygin and Solntsev, 1986, 1987a). Here these results are generalized. Our interest in the study of limit theorems with operator normalization was caused by investigation of the asymptotic behaviour of the solutions of stochastic recurrent equations of high order and sample behaviour of multidimensional Gaussian processes. The corresponding applications are given in the final part of the present paper.

2. THE CONTRACTION PRINCIPLE IN THE SPACE OF CONVERGING SEQUENCES

Proofs of the statements on a.s. convergence of sums of independent random vectors with operator normalizations make use of the contraction principle in the space of converging sequences.

Let B be a separable Banach space with norm $\|\cdot\|$; $c(B)$ be the space of all sequences converging in B.

Consider the double array of B-valued random elements $(Y_{nk};\ n,k \geqslant 1)$ and assume the following conditions:

a) for each $n \geqslant 1$ the series $\sum\limits_{k=1}^{\infty} Y_{nk}$ converge a.s. in the B-norm;

b) sequences $(Y_{nk},\, n \geqslant 1)$, $k \geqslant 1$, are independent and symmetric as random elements in the space of all sequences in B.

Now consider the infinite real matrix $(b_{nk};\ n,\, k \geqslant 1)$ and assume the following conditions:

c) $\sup\limits_{n\geqslant 1} \sup\limits_{m\geqslant 1} \left(\sum\limits_{k=1}^{m-1} \mid b_{nk} - b_{n,(k+1)} \mid + \mid b_{nm} \mid \right) < \infty$;

d) for each $k \geqslant 1$

$$(b_{nk} Y_{nk},\ n \geqslant 1) \in c(B) \quad \text{a.s.}$$

Put

$$Z_n = \sum_{k=1}^{\infty} Y_{nk}, \quad \widetilde{Z}_n = \sum_{k=1}^{\infty} b_{nk} Y_{nk},\ n \geqslant 1.$$

THEOREM 1. *Under the above conditions*

$$(Z_n, n \geqslant 1) \in c(B) \quad a.s.$$

implies

$$(\widetilde{Z}_n, n \geqslant 1) \in c(B) \quad a.s.$$

In the proof the following auxiliary lemma is used.

LEMMA 1. *Let matrix (Y_{nk}) be the same as in Theorem 1 and*

$$(Z_n,\ n \geqslant 1) \in c(B) \quad a.s.$$

Then

1) *for each* $k \geqslant 1$

$$(Y_{nk},\ n \geqslant 1) \in c(B) \quad a.s.;$$

2) *the series* $\sum\limits_{k=1}^{\infty} (\lim\limits_{n\to\infty} Y_{nk})$ *converges* a.s. *in* B*-norm;*

3) $\lim\limits_{n\to\infty} Z_n = \sum\limits_{k=1}^{\infty} \left(\lim\limits_{n\to\infty} Y_{nk}\right)$ a.s.

3. NECESSARY CONDITIONS

Consider necessary conditions for a.s. convergence of sums of independent random vectors with operator normalizations in the general case.

Let $(X_k,\ k \geqslant 1)$ be a sequence of independent symmetric random elements with values in a separable Banach space B; $S_n = \sum\limits_{k=1}^{n} X_k,\ n \geqslant 1$; $L(B,F)$ be a class of all bounded linear operators from B into some separable Banach space F; $(A_n,\ n \geqslant 1) \subset L(B,F)$; $\mathbf{N}$ be the set of positive integers; $K(\mathbf{N})$ be a class of sets $J_n \subset \mathbf{N}$ satisfying the following condition:

$$J_n = \varnothing \quad \text{or} \quad J_n = [j_1(n),\ j_2(n)] \cap \mathbf{N},$$

where $j_1(n),\ j_2(n) \in \mathbf{N},\ \ j_1(n) \leqslant j_2(n),\ \ n \geqslant 1$. Denote by $K_c(\mathbf{N})$ the subclass of the class $K(\mathbf{N})$ satisfying the additional condition:

$$\exists \lim_{n\to\infty} j_1(n), \quad \exists \lim_{n\to\infty} j_2(n)$$

(finite or infinite). Let $M(\mathbf{N})$ be the class of all non-decreasing and unbounded sequences of positive integers. Assume that $\sum\limits_{k\in\varnothing} = 0$.

Theorem 2. *Let* $(J_n,\ n \geqslant 1) \in K_c(\mathbf{N})$ *and*

$$\left(A_n S_n,\ n \geqslant 1\right) \in c(F) \quad a.s.$$

Then

$$\Big(A_n \sum_{k\in J_n \cap [1,n]} X_k,\ n \geqslant 1\Big) \in c(F) \quad a.s.$$

Moreover

1) *if* $\lim\limits_{n\to\infty} j_1(n) = \infty$, *then*

$$\Big\| A_n \sum_{k\in J_n \cap [1,n]} X_k \Big\| \underset{n\to\infty}{\longrightarrow} 0 \quad a.s.$$

2) *if* $\lim\limits_{n\to\infty} j_1(n) = 0$ *and* $\lim\limits_{n\to\infty} j_2(n) = \infty$, *then*

$$\lim_{n\to\infty} A_n \sum_{k\in J_n \cap [1,n]} X_k = \lim_{n\to\infty} A_n S_n \quad a.s.$$

Example. Let $(k_n, n \geqslant 1) \subset \mathbf{N}$; $k_n \leqslant n,\quad n \geqslant 1$; $k_n \to \infty$ as $n \to \infty$. Set

$$j_1(n) = k_n, \quad j_2(n) = n, \quad n \geqslant 1.$$

Then by Theorem 2

$$\left(A_n S_n,\ n \geqslant 1\right) \in c\,(F) \quad \text{a.s.}$$

implies

$$\left(A_n S_{k_n},\ n \geqslant 1\right) \in c\,(F) \quad \text{a.s.}$$

and

$$\lim_{n\to\infty} A_n S_n = \lim_{n\to\infty} A_n S_{k_n} \quad \text{a.s.}$$

Moreover

$$\left\| A_n(S_n - S_{k_n}) \right\| \underset{n\to\infty}{\longrightarrow} 0 \quad a.s.$$

Note that above relations are rather unexpected because the sequence $(k_n,\, n \geqslant 1)$ can increase arbitrary slowly.

THEOREM 3. *If*

$$\left(A_n S_n,\ n \geqslant 1\right) \in c\,(B) \quad a.s.$$

then

1) *for each* $k \geqslant 1$

$$\left(A_n X_k,\ n \geqslant 1\right) \in c\,(B) \quad a.s.$$

2) *the series* $\sum_{k=1}^{\infty} Y_k$ *converges* a.s. *in* B *where* $Y_k = \lim_{n\to\infty} A_k X_k$, *and*

$$\lim_{n\to\infty} A_n S_n = \sum_{k=1}^{\infty} Y_k \quad a.s.$$

3) *for all sequences* $(n_j,\ j \geqslant 1) \in M(\mathbf{N})$

$$\lim_{j\to\infty} \left\| A_{n_{j+1}}(S_{n_{j+1}} - S_{n_j}) \right\| = 0 \quad a.s.$$

The proof of Theorem 2 is based on Theorem 1 and Lemma 1. Theorem 3 follows from Theorem 2.

4. A CRITERION FOR A.S. CONVERGENCE OF SUMS OF INDEPENDENT RANDOM VECTOR WITH OPERATOR NORMALIZATIONS IN $\mathbf{R}^m$

The necessary conditions for a.s. convergence in infinite-dimensional Banach space (Theorem 3) are not sufficient. However, if $B = F = \mathbf{R}^m$, $m \geqslant 1$, a complete solution of the problem can be obtained.

THEOREM 4. *Let*

1) *for each* $k \geqslant 1$

$$\left(A_n X_k,\ n \geqslant 1\right) \in c\,(\mathbf{R}^m) \quad a.s.;$$

2) *the series* $\sum_{k=1}^{\infty} Y_k$ *converges a.s. in* $\mathbf{R}^m$, *where*

$$Y_k = \lim_{n\to\infty} A_n X_k, \quad k \geqslant 1.$$

Then there is a finite set of sequences $M_0(\mathbf{N}) \subset M(\mathbf{N})$ *such that if*

3) *for all* $(n_j,\ j \geqslant 1) \in M_0(\mathbf{N})$

$$\lim_{j\to\infty} \left\| A_{n_{j+1}} \left(S_{n_{j+1}} - S_{n_j} \right) \right\| = 0 \quad a.s.$$

then

$$(A_n S_n,\ n \geqslant 1) \in c\,(\mathbf{R}^m) \quad a.s.;$$

and

$$\lim_{n\to\infty} A_n S_n = \sum_{k=1}^{\infty} Y_k \quad a.s.$$

The constructive description of the class $M_0(\mathbf{N})$ is rather complicated. So it is often convenient to pass from the class $M_0(\mathbf{N})$ to $M(\mathbf{N})$.

THEOREM 5.

$$(A_n S_n,\ n \geqslant 1) \in c\,(\mathbf{R}^m) \quad a.s.$$

iff

1) *for each* $k \geqslant 1$

$$(A_n X_k,\ n \geqslant 1) \in c\,(\mathbf{R}^m) \quad a.s.$$

2) *the series* $\sum_{k=1}^{\infty} Y_k$ *converges a.s. in* $\mathbf{R}^m$, *where*

$$Y_k = \lim_{n\to\infty} A_n X_k, \quad k \geqslant 1,$$

3) *for all sequences* $(n_j,\ j \geqslant 1) \in M(\mathbf{N})$

$$\lim_{j\to\infty} \left\| A_{n_{j+1}} \left(S_{n_{j+1}} - S_{n_j} \right) \right\| = 0 \quad a.s.$$

Moreover,

$$\lim_{n\to\infty} A_n S_n = \sum_{k=1}^{\infty} Y_k \quad a.s.$$

The proof of Theorems 4 and 5 makes use of SLLN for sums of independent random vectors normed by operators (Buldygin and Solntsev, 1987a).
Using Theorem 5 we can state conditions under which the a.s. convergence of
$(A_n S_n,\ n \geqslant 1)$ and convergence in probability are equivalent.

THEOREM 6.

$$(A_n S_n,\ n \geqslant 1) \in c(\mathbf{R}^m) \quad a.s.$$

iff

1) *the sequence* $(A_n S_n,\ n \geqslant 1)$ *converges in* $\mathbf{R}^m$ *in probability;*

2) *for all* $(n_j,\ j \geqslant 1) \in M(\mathbf{N})$

$$\lim_{j\to\infty} \left\| A_{n_{j+1}} \left(S_{n_{j+1}} - S_{n_j} \right) \right\| = 0 \quad a.s.$$

5. THE CASE OF GAUSSIAN SUMMANDS

Denote the trace of a matrix A by Tr A and let I be the unit matrix. If X is a centered Gaussian vector in $\mathbf{R}^m$, then

$$\mathbf{D}(X) = \mathbf{E}XX'$$

and

$$\operatorname{Tr} \mathbf{D}(X) = \mathbf{E} \parallel X \parallel^2 .$$

For jointly Gaussian centered random vectors X, Y let $\mathbf{D}(X|Y)$ be the conditional covariance matrix:

$$\mathbf{D}(X|Y) = \mathbf{E}\left(X - \mathbf{E}(X|Y)\right)\left(X - \mathbf{E}(X|Y)\right)'.$$

It is well-known that a sequence of independent centered Gaussian random vectors $(\Gamma_n,\ n \geqslant 1)$ converges a.s. in R^m iff for each $\varepsilon > 0$

$$\sum_{n=1}^{\infty} \exp\left(-\varepsilon \,/\, \operatorname{Tr} \mathbf{D}(\Gamma_n)\right) < \infty.$$

Theorem 6 implies

THEOREM 7. *Let* $(X_k,\ k \geqslant 1)$ *be a sequence of independent random vectors in* R^m; $\mathbf{E}X_k = 0$; $(A_n,\ n \geqslant 1) \subset L(\mathbf{R}^m, \mathbf{R}^m)$.
Then

$$\lim_{n\to\infty} A_n S_n = Y \quad a.s.$$

iff

1) *the sequence* $(A_n S_n,\ n \geqslant 1)$ *converges to* Y *in probability;*

2) *for all* $(n_j,\ j \geqslant 1) \in M(\mathbf{N})$ *and all* $\varepsilon > 0$

$$\sum_j \exp\left(-\varepsilon / \operatorname{Tr} \mathbf{D}\left(A_{n_{j+1}}(S_{n_{j+1}} - S_{n_j})\right)\right) < \infty.$$

Remark. In the particular case when $Y = 0$ from Theorem 7 we get SLLN for sums of independent random vectors with operator normalizations (Buldygin and Solntsev, 1987b).

6. GAUSSIAN MARKOV SEQUENCES

Let $(Y_n,\ n \geqslant 1)$ be a Gaussian Markov sequence in $\mathbf{R}^m$, $\mathbf{E}\,Y_n \equiv 0$. It is known that the Gaussian sequence $(Y_n,\ n \geqslant 1)$ is a Gaussian Markov sequence if and only if it satisfies the system of stochastic recurrent equations

$$Y_1 = Z_1, \qquad Y_n = C_n Y_{n-1} + Z_n \quad (n \geqslant 2),$$

where $(Z_n,\ n \geqslant 1)$ is a sequence of centered independent Gaussian random vectors in $\mathbf{R}^m$ and $(C_n,\ n \geqslant 2)$ is a sequence of non–stochastic matrices such that

$$\mathbf{E}(Y_n \mid Y_{n-1}) = C_n Y_{n-1},$$

$$\mathbf{D}(Y_n \mid Y_{n-1}) = \mathbf{D}(Z_n), \quad n \geqslant 2.$$

If matrices $C_n,\ n \geqslant 2,$ are nondegenerate one can write the sequence $(Y_n,\ n \geqslant 1)$ in the form

$$Y_n = A_n(X_1 + X_2 + \ldots + X_n), \quad n \geqslant 1,$$

where

$$A_1 = I, \quad A_n = C_n\, C_{n-1}\, \ldots\, C_2, \quad n \geqslant 2,$$

$$X_1 = Z_1, \quad X_k = (C_k\, \ldots\, C_2)^{-1} Z_k, \quad k \geqslant 2.$$

Theorem 7 leads to the following assertion.

THEOREM 8. *Let* $(Y_n,\ n \geqslant 1)$ *be a Gaussian Markov sequence in* $\mathbf{R}^m$; $\mathbf{E}\,Y_n \equiv 0,\ n \geqslant 1$. *Then*

$$\lim_{n\to\infty} Y_n = Y \quad a.s.$$

iff

1) *the sequence* $(Y_n,\ n \geqslant 1)$ *converges to* Y *in probability,*

2) *for all* $(n_j,\ j \geqslant 1) \in M(\mathbf{N})$ *and all* $\varepsilon > 0$

$$\sum_{j=1}^{\infty} \exp\Big(-\varepsilon \,/\, \mathrm{Tr}\, \mathbf{D}\,(Y_{n_{j+1}} \mid Y_{n_j})\Big) < \infty.$$

COROLLARY 1. *Let* $(Y_n,\ n \geqslant 1)$ *be a centered Gaussian Markov sequence. Then*

$$\lim_{n\to\infty} Y_n = 0 \quad a.s.$$

iff

$$\lim_{n\to\infty} \operatorname{Tr} \mathbf{D}(Y_n) = 0$$

and condition 2) *of Theorem* 7 *is satisfied.*

7. STOCHASTIC RECURRENT EQUATIONS

Let $(\zeta_n,\ n \geqslant 1)$ be a sequence of centered jointly Gaussian random variables satisfying the following system of stochastic recurrent equations:

$$\zeta_n = a_{n1}\,\zeta_{n-1} + a_{n2}\,\zeta_{n-2} + \ \dots\ + a_{nm}\,\zeta_{n-m} + b_n\,\gamma_n, \quad n \geqslant 1; \tag{1}$$

$$\zeta_0 = \zeta_{-1} = \ \dots\ = \zeta_{1-m} = 0,$$

where $(\gamma_n,\ \ n \geqslant 1)$ is a standard Gaussian sequence and $(a_{nk}, n \geqslant 1,\ k = 1, 2, \dots, m)$ and $(b_n,\ n \geqslant 1)$ are given arrays of real numbers. This sequence $(\zeta_n,\ n \geqslant 1)$ is called Gaussian m–Markovian. The question of necessary and sufficient conditions for a.s. convergence of $(\zeta_n,\ n \geqslant 1)$ can be reduced to the same question for a Gaussian Markov sequence in $\mathbf{R}^m$.

Let $(\zeta_n,\ n \geqslant 1)$ be a Gaussian m–Markovian sequence, $m > 1$. Set

$$Y_n' = (\zeta_n,\ \zeta_{n-1},\ \dots,\ \zeta_{n-m+1}),$$

$$Z_n' = (b_n\,\gamma_n,\ 0,\ \dots,\ 0), \quad n \geqslant 1.$$

By choosing suitable matrices $C_n,\ n \geqslant 2,$ one can write the equivalent vector form of system (1) as follows:

$$Y_1 = Z_1; \quad Y_n = C_n Y_{n-1} + Z_n, \quad n \geqslant 2.$$

Then Theorem 7 leads to the following assertion.

THEOREM 9. *Let* $(\zeta_n,\ n \geqslant 1)$ *be a Gaussian* m*–Markovian sequence. Then*

$$\lim_{n\to\infty} \zeta_n = \zeta \quad a.s.$$

iff

1) *the sequence* $(\zeta_n,\ n \geqslant 1)$ *converges to* ζ *in probability,*

2) *for all* $(n_j,\ j \geqslant 1) \in M(\mathbf{N})$ *and all* $\varepsilon > 0$

$$\sum_{j=1}^{\infty} \exp\left(- \varepsilon \,/\, d_j^2\right) < \infty,$$

where

$$d_j^2 = \mathbf{E}\left(\zeta_{n_j} - \mathbf{E}\left(\zeta_{n_j} \mid \zeta_{n_{j-1}},\ \zeta_{n_{j-1}-1},\ \dots,\ \zeta_{n_{j-1}-m+1}\right)\right)^2.$$

8. A CRITERION FOR A.S. CONTINUITY OF A MULTIDIMENSIONAL GAUSSIAN MARKOV PROCESS

Let $X = \big(X(t),\ t \in [a,b]\big)$ be a centered Gaussian process with values in a real separable Banach space B. Without loss of generality we assume that the process X is stochastically continuous. Set

$$M_t([a,b]) = \Big\{(t_k,\ k \geqslant 1) \subset [a,b]\ :\ t_k \uparrow t \ \text{ or } \ t_k \downarrow t\Big\}.$$

Let $C[a,b]$ be the space of all continuous on $[a,b]$ functions.

Lemma 2 (Buldygin and Solntsev, 1988).

$$\mathbf{P}(X \in C[a,b]) = 1$$

iff for all $t \in [a,b]$ *and all* $(t_k,\ k \geqslant 1) \in M_t([a,b])$

$$\mathbf{P}\Big(\lim_{k\to\infty} X(t_k) = X(t)\Big) = 1.$$

Let $Y = \big(Y(t),\ t \in [a,b]\big)$ be a centered stochastically continuous Gaussian Markov process in R^m. Note that for any $(t_k,\ k \geqslant 1) \in M_t([a,b])$ the sequence of random vectors $\big(Y(t_k),\ k \geqslant 1\big)$ is a Gaussian Markov sequence in R^m. Hence by Theorem 7 and Lemma 2 we get the following

Theorem 10.

$$\mathbf{P}\,(Y \in C[a,b]) = 1$$

iff for all $t \in [a,b]$, *all* $(t_k,\ k \geqslant 1) \in M_t([a,b])$ *and all* $\varepsilon > 0$

$$\sum_{k=1}^{\infty} \exp\left(-\frac{\varepsilon}{\operatorname{Tr}\mathbf{D}\left(Y(t_{k+1}) \mid Y(t_k)\right)}\right) < \infty.$$

Corollary 2 (Dynkin — Kinny criterion). *If*

$$\sup_{\substack{t,s\in[a,b]\\ |t-s|<h}} \operatorname{Tr}\mathbf{D}\Big(Y(t)-Y(s) \mid Y(s)\Big) = o(|\ln h|^{-1}),$$

as $h \downarrow 0$, *then*

$$\mathbf{P}\Big(Y \in C[a,b]\Big) = 1.$$

9. AN ENTROPY CRITERION FOR A.S. CONTINUITY OF A GAUSSIAN MARKOV PROCESS

Let $X = (X(t),\ t \in [a,b])$ be a centered stochastically continuous Gaussian process in R^m. Denote the ε–entropy of $[a,b]$ with respect to

$$\rho_X(t,s) = \left(\mathbf{E} \parallel X(t) - X(s) \parallel^2\right)^{1/2}, \quad t,s \in [a,b],$$

by $H_X(\varepsilon)$. According to (Sudakov, 1969) if

$$\mathbf{P}\left(X \in C[a,b]\right) = 1,$$

then

$$\lim_{\varepsilon \to 0} \varepsilon^2 H_X(\varepsilon) = 0.$$

This necessary condition of continuity in the general Gaussian case is not sufficient. However, for a Gaussian Markov process the Sudakov necessary condition is sufficient.

THEOREM 11. *Let $Y = (Y(t),\ t \in [a,b])$ be a separable stochastically continuous centered Gaussian Markov process in $\mathbf{R}^m$. Then*

$$\mathbf{P}\left(Y \in C[a,b]\right) = 1$$

iff

$$\lim_{\varepsilon \to 0} \varepsilon^2 H_Y(\varepsilon) = 0.$$

REFERENCES

Buldygin, V.V. and Solntsev, S.A. (1986). The contraction principle and strong laws of large numbers for weighted sums. *Theory Probab. Appl.* **31**, 452–462.

Buldygin, V.V. and Solntsev, S.A. (1987a). The contraction principle for c_0–summing operators and SLLN for weighted sums. In: *Probab. Theory and Math. Statistics. Proc. Fourth Vilnius Conf.* **1**, VSP, Utrecht, 269–290.

Buldygin, V.V. and Solntsev, S.A. (1987b). On the equivalence of sample and sequential continuity of Gaussian processes and continuity of Gaussian Markov processes. *Theory Probab. Appl.* **33**, 32–46.

Hahn, M. (1979). The generalized domain of attraction of Gaussian law on Hilbert space. *Lect. Notes Math.* **709**, 125–144.

Hahn, M. and Klass, M. (1981). The multidimensional central limit theorem for arrays normed by affine transformations. *Ann. Probab.* **9**, 611–623.

Kandelaki, N.P. and Sazonov, V.V. (1964). On a central limit theorem for random elements with values in Hilbert space. *Theory Probab. Appl.* **9**, 43–52.

Sudakov, V.N. (1969). Gaussian measures, Cauchy measures and ε–entropy. *Soviet Math. Dokl.* **10**, 310.

Prob. Theory and Math. Stat., Vol. 1, pp. 207–216
B.Grigelionis *et al.* (Eds.)
1990 VSP/Mokslas

CLT FOR FAMILIES OF INTEGRAL FUNCTIONALS ARISING IN SOLVING THE MULTIDIMENSIONAL BURGERS EQUATION

A.V. BULINSKII

Department of Mechanics and Mathematics, Moscow State University,
Moscow 119899, USSR

ABSTRACT

Asymptotic normality for the transformed solution $Z_t^\varepsilon(a)$ of the Cauchy problem for multidimensional Burgers equation with random initial data is established in the potential case. The initial potential is the generalized shot-noise field. The limiting behaviour of the random field $Z_t^\varepsilon(a)$ is different for "times" t with different dependence on the scale parameter ε.

Starting from some random fields $\zeta(y)$, $y \in \mathbf{R}^d$, we consider parametric families H_t of vector-valued functionals of the form

$$H_t(x,\zeta,K) = h\Big(\int\limits_{U_t} K(x,t,y,\zeta(y))dy\Big), x \in \mathbf{R}^d, t > 0, \tag{1}$$

where $U_t \subset \mathbf{R}^d$ denotes a Borel set "increasing" with t, K is a function with values in $\mathbf{R}^q, h$ maps $\mathbf{R}^q$ into $\mathbf{R}^l$. The main goal of the paper is to indicate the conditions implying the asymptotic normality of H_t as $t \to \infty$ for the family of such nonlinear functionals H_t related to the solution of the Cauchy problem for the Burgers equation with random initial data $\zeta(\cdot)$. Note that even for simple linear integral functionals of the field $\zeta(\cdot)$ belonging to the class (1) the normality may be sensitive both to the dimension d and to the "mode of growth" of the sets U_t as $t \to \infty$. Such effects are briefly illustrated for the random field $\zeta(\cdot)$ typical for physical models of disordered structures (see e.g. (Lifshitz et al., 1982)).
Namely, let $\{x_i\}$ be a $\mathbf{R}^d$-valued point Poisson field of intensity $\lambda > 0$ and $\{\xi_i\}, \{\theta_i\}, i \in \mathbf{N}$, be sequences of real r.v. on the same probability space $(\Omega, \mathcal{F}, \mathbf{P})$. Assuming that $\{x_i\}, \{\xi_i\}, \{\theta_i\}$ are mutually independent, every ξ_i is distributed as r.v. ξ with d.f. G (resp. θ_i, θ, F) and $\theta > 0$ a.s., introduce for $\varphi: \mathbf{R}^d \to \mathbf{R}$ the generalized shot-noise field

$$\zeta(x) = \sum_i \xi_i \varphi((x - x_i)/\theta_i), \quad x \in \mathbf{R}^d. \tag{2}$$

The trivial cases $\xi \equiv 0$ a.s. and $\varphi(x) \equiv 0$ for a.e. x are excluded.

PROPOSITION 1. *Let $\varphi \in L^1(\mathbf{R}^d)$, $I(\varphi) \neq 0$, where $I(\varphi) = \int_{\mathbf{R}^d} \varphi(x)dx$ and $\mathbf{E}|\xi|^s < \infty$, $\mathbf{E}\theta^{sd} < \infty$ for some $s \in (2,3]$. Then if $U_t \to \infty$ in the sense of Van Hove (see e.g. (Ruelle, 1969)) as $t \to \infty$ we have*

$$B_t^2 = \operatorname{var}\Big(\int_{U_t} \zeta(y)dy\Big) \sim \lambda \mathbf{E}\xi^2 \mathbf{E}\theta^{2d}(I(\varphi))^2|U_t|$$

and

$$B_t^{-1}\Big(\int_{U_t} \zeta(x)dx - \lambda \mathbf{E}\xi \mathbf{E}\theta^d I(\varphi)|U_t|\Big) \xrightarrow{D} \mathcal{N}(0,1), \tag{3}$$

where $|U_t|$ denotes the Lebesgue measure of U_t.

Remark 1. If one drops the condition $I(\varphi) \neq 0$ and retains all the other hypotheses of Proposition 1, then the CLT (relation (3)) may fail as the following example shows.

Let $U_t = \prod_{j=1}^{d} [a_j(t), b_j(t)]$, $Q(m) = ([-m,m])^d$, $d \geqslant 1$, $\xi = 1$, $\theta = 1$ a.s. Then $I(\varphi) = 0$ for $\varphi(x) = \mathbf{1}_{Q(m)}(x) \prod_{j=1}^{d} \operatorname{sign} x_j$, $x \in \mathbf{R}^d$, and moreover if $L_j(t) = b_j(t) - a_j(t) \geqslant 2m$, $j = 1, \ldots, d$, then $B_t^2 = \lambda(4m^3/3)^d$ and

$$\mathbf{E}\exp\Big\{i\tau \int_{U_t} \zeta(x)dx\Big\} =$$

$$= \exp\Big\{\lambda\Big(2^d \int_{Q(m)} \cos\Big(\tau \prod_{j=1}^{d} g(z_j)\Big)dz - (4m)^d\Big)\Big\}, \tag{4}$$

where $g(\tau) = (m - |\tau|)\mathbf{1}_{[-m,m]}(\tau)$, $\tau \in \mathbf{R}$.

Let now $\Lambda = \{\alpha = (\alpha_1, \ldots, \alpha_d): \alpha_j \in \{0,1\},\ j = 1, \ldots, d\}$ and let $T_\alpha = \{j \in \{1, \ldots, d\}: \alpha_j = 1\}$. For $z \in \mathbf{R}^d$, $\alpha \in \Lambda$, $\varphi \in L^1(\mathbf{R}^d)$ put

$$J_\alpha(z) = \int_{\Gamma_\alpha(z)} \varphi(x)dx,$$

where $\Gamma_\alpha(z) = \{x \in \mathbf{R}^d: x_j \leqslant z_j,\ j \in T_\alpha\}$, if $\alpha \neq 0$ and if $\alpha = 0$, then $\Gamma_0(z) = \mathbf{R}^d$. We define the rank of a function φ with $I(\varphi) = 0$ as the number $\operatorname{rg}\varphi \in \{1, \ldots, d\}$ such that

a) $J_\alpha(z) = 0$ for all $z \in \mathbf{R}^d$ and $\alpha \in \Lambda$ with $\sum_{j=1}^{d} \alpha_j \leqslant \operatorname{rg}\varphi - 1$,

b) $J_{\alpha_0}(z_0) \neq 0$ for some $z_0 \in \mathbf{R}^d$ and $\alpha_0 \in \Lambda$ with $\sum_{j=1}^{d} \alpha_j = \operatorname{rg}\varphi$.

Considering functions φ with the following properties $\varphi \in L^1(\mathbf{R}^d)$, $I(\varphi) = 0$, $\operatorname{supp}\varphi \subset \prod_{i=1}^{d}[\alpha_i, \beta_i]$, $l_i = \beta_i - \alpha_i \geqslant 1$, $i = 1, \ldots, d$, it is not difficult to construct such functions φ_p that $\operatorname{rg}\varphi_p = p$ for every $p \in \{1, \ldots, d\}$. If $\theta = 1$ a.s., $\mathbf{E}|\xi|^s < \infty$ for some $s \in (2, 3]$ and $\operatorname{rg}\varphi < d(s-2)/s$, then for the family of cubes $U_t = ([-t, t])^d$ we have

$$B_t^{-1} \int_{U_t} \zeta(x)dx \xrightarrow{D} \mathcal{N}(0, 1), \; t \to \infty.$$

For instance when $\operatorname{rg}\varphi = 1$, $s = 3$ the CLT holds for $d > 3$ (and may fail for $d = 1$ according to (4)). The details are given in (Bulinskii, 1989b). Some other aspects of the CLT for integral functionals of the (non–Poisson) shot-noise field can be found in (Heinrich and Schmidt, 1985).

A nontrivial example of nonlinear integral functionals with $U_t = \mathbf{R}^d$ is provided by the solution of the Cauchy problem for the multidimensional Burgers equation

$$\frac{\partial \bar{v}}{\partial t} + (\bar{v}, \nabla)\bar{v} = \nu \, \Delta \, \bar{v},$$

$$\bar{v}(0, x) = \bar{v}_0(x), \; x \in \mathbf{R}^d, \; t \geqslant 0$$

with random data.

For the solution belonging to the class of potential fields of velocities (i.e., $\bar{v}(t, x) = \nabla\Phi(t, x)$ and $\bar{v}_0(x) = \nabla\Phi_0(x)$) the Hopf–Cole substitution (see e. g. (Shandarin et al., 1988) leads to the formula

$$\bar{v}(t, x) = \int_{\mathbf{R}^d} dy \frac{x - y}{t} \exp\left\{ -\frac{1}{2\nu}\left(\Phi_0(y) + \frac{|x - y|^2}{2t}\right)\right\} \Big/$$

$$\Big/ \int_{\mathbf{R}^d} dy \exp\left\{ -\frac{1}{2\nu}\left(\Phi_0(y) + \frac{|x - y|^2}{2t}\right)\right\}.$$

In order to mark the dependence of the solution on Φ_0 and ν we write $\bar{v}(x, t; \Phi_0, \nu)$. After the transformation $x = a\sqrt{t}$, $a \in \mathbf{R}^d, t > 0$, we study the asymptotic behaviour of the vector field $Z_t(a; \Phi_0, \nu) = \bar{v}(a\sqrt{t}, t; \Phi_0, \nu)$ as $t \to \infty$. Note that M. Rosenblatt (1987) has investigated (in the case $d = 1$) a different problem of the behaviour of integrals with respect to x for the solution of Burgers equation with fixed t.

Since for any $\nu > 0$

$$Z_t(a; \Phi_0, \nu) = \sqrt{2\nu} Z_t\Big(a\sqrt{2\nu}; \Phi_0(\cdot\sqrt{2\nu})/(2\nu), 1/2\Big), \tag{5}$$

without loss of generality it is enough to consider the case $\nu = 1/2$.

S.A. Molchanov investigated the case of a homogeneous Gaussian initial potential $\Phi_0(x)$ with an isotropic correlation function. After that A.V. Bulinskii studied the case $\Phi_0(x) = \zeta(x)$ where $\zeta(x)$ is the shot–noise field (2). The physical interpretation of the both models is due to S.A. Molchanov. We mention only the important application in modelling the early Universe (see (Gurbatov et al., 1988; Shandarin and Zel'dovich, 1988)). The problem of asymptotic normality of the field Z_t was considered in the paper by Bulinskii and Molchanov, 1990. The report (Bulinskii, 1989) was also devoted to the following more general problem posed by S.A. Molchanov.

Let $\varepsilon > 0$ denote the scale parameter labelling the family of the fields

$$\zeta^{(\varepsilon)}(x) = \sum_i \xi_i \varphi((x - x_i^{(\varepsilon)})/\theta_i), \ x \in \mathbf{R}^d, \tag{6}$$

that is, now instead of $\{x_i\}$ in (2) we take (in general nonhomogeneous) Poisson point fields $\{x_i^{(\varepsilon)}\}$ with intensity function $\lambda_\varepsilon(x) = \lambda_0(\varepsilon x)$, $x \in \mathbf{R}^d$. The assumptions on $\xi, \theta, \varphi(\cdot), \lambda_0(\cdot)$ are given below. The behaviour of $Z_t^{(\varepsilon)}(a), a \in \mathbf{R}^d$ (i.e. the field (5) corresponding to $\Phi_0^{(\varepsilon)}(x) = -\zeta^{(\varepsilon)}(x)$, where minus is taken only for the sake of convenience) will be different for "times" with different dependence on the scale ε.

In this paper we deal with the case

$$\varepsilon \to 0, \ t \to \infty, \ \varepsilon\sqrt{t} \to c, \ 0 \leqslant c < \infty. \tag{7}$$

The case $c = \infty$ will be considered in a forthcoming paper.

From the physical point of view it is interesting to consider a periodical function $\lambda_0(\cdot)$ but here in the case (7) we do not use this assumption. Moreover, instead of $\zeta^{(\varepsilon)}(x)$ we can take here a more general field

$$\tilde{\zeta}^{(\varepsilon)}(x) = \zeta^{(\varepsilon)}(x) + A_0(\varepsilon x), \tag{8}$$

where $A_0(\cdot)$ is a nonrandom function whose behaviour is specified below.

So, putting $\Phi_0^{(\varepsilon)}(x) = -\tilde{\zeta}^{(\varepsilon)}(x)$, $\varepsilon > 0$, we study the vector–valued field

$$Z_t^{(\varepsilon)}(a) = V_t^{(\varepsilon)}(a)/J_t^{(\varepsilon)}(a), a \in \mathbf{R}^d, \ t > 0, \ \varepsilon > 0,$$

where

$$V_t^{(\varepsilon)}(a) = \int_{\mathbf{R}^d} dy \frac{a\sqrt{t} - y}{t} \exp\left\{\tilde{\zeta}^{(\varepsilon)}(y) - \frac{|a\sqrt{t} - y|^2}{2t}\right\}, \tag{9}$$

$$J_t^{(\varepsilon)}(a) = \int_{\mathbf{R}^d} dy \exp\left\{\tilde{\zeta}^{(\varepsilon)}(y) - \frac{|a\sqrt{t} - y|^2}{2t}\right\}; \tag{10}$$

and $|x|^2 = (x, x)$, $(\cdot, \cdot)$ being the inner product in $\mathbf{R}^d$.

The analysis of the asymptotic behaviour of the field $Z_t^{(\varepsilon)}$ is based on the study of the fields (9), (10), various approximation schemes and limit theorems for sums of dependent multiindexed random variables. The technique of characteristic functions and the technique of semiinvariants are used (see e.g., (Bulinskii, 1989; Saulis and Statulevičius, 1989)).

Let $\varphi_r(x) = \varphi(x)\mathbf{1}\{|x| \leqslant r\}, r > 0$. From now on the subscript r in the notation $J_{t,r}^{(\varepsilon)}(a)$ means that one takes the field $\tilde{\zeta}_r^{(\varepsilon)}(x)$ instead of $\tilde{\zeta}^{(\varepsilon)}(x)$, that is, $\varphi_r(\cdot)$ is taken instead of $\varphi(\cdot)$ in (6). We have to choose $r = r(t)$ in the course of the proofs.

The following lemmas are essential for the study of the auxiliary field $J_t^{(\varepsilon)}$ arising in the solution of the heat–equation.

LEMMA 1. *Assume* $\varphi \in L^1(\mathbf{R}^d)\bigcap L^\infty(\mathbf{R}^d), \mathbf{E}\exp\{\|\varphi\|_\infty|\xi|\} < \infty, \mathbf{E}\theta^d < \infty$. *Let the functions* $\lambda_0(\cdot), A_0(\cdot)$ *be continuous on* $\mathbf{R}^d$ *and such that for all* $x \in \mathbf{R}^d$ *and some positive constants* $c_i,\ i = 0,1,2$

$$\lambda_0(x) \leqslant c_0,\ |A_0(x)| \leqslant \max\{c_1, c_2|x|^2\}. \tag{11}$$

Assume condition (7) *to be satisfied and let moreover*

$$cc_2 < 1/2. \tag{12}$$

Then we have for all $a \in \mathbf{R}^d$

$$\mathbf{E}J_t^{(\varepsilon)}(a) = t^{d/2}\Big(\int\limits_{\mathbf{R}^d} \exp\Big\{-\frac{|y-a|^2}{2} + \lambda_0(cy)H + A_0(cy)\Big\}dy + g_t^{(\varepsilon)}(a)\Big).$$

Here $g_t^{(\varepsilon)}(a) \to 0, a \in \mathbf{R}^d$, *under* (7) *and*

$$H = \mathbf{E}\theta^d \int\limits_{\mathbf{R}^d} dx \int\limits_{\mathbf{R}} dG(v)\Big(e^{v\varphi(x)} - 1\Big).$$

This result can be obtained from the following representation

$$\mathbf{E}J_t^{(\varepsilon)}(a) = t^{d/2}\mathcal{E}_t^{(\varepsilon)}(a),$$

$$\mathcal{E}_t^{(\varepsilon)}(a) = \int\limits_{\mathbf{R}^d} dy \exp\Big\{-\frac{|y-a|^2}{2} + A_0(\varepsilon\sqrt{t}y) + \Lambda_t^{(\varepsilon)}(y)\Big\}, \tag{13}$$

$$\Lambda_t^{(\varepsilon)}(y) = \int\limits_{\mathbf{R}^d} dx \int\limits_{\mathbf{R}^+} dF(u) \int\limits_{\mathbf{R}} dG(v)\Big(\exp\Big\{v\varphi\Big(\frac{x}{u}\Big)\Big\} - 1\Big)\lambda_0\Big(\varepsilon\sqrt{t}y - \varepsilon x\Big).$$

Remark also that if one admits $cc_2 \geqslant 1/2$, while the other conditions of Lemma 1 remain unchanged then it is impossible to guarantee the existence of $\mathbf{E}J_t^{(\varepsilon)}(a)$.

LEMMA 2. *If the conditions of Lemma* 1 *are strengthened by the assumption* $\mathbf{E}\exp\{2\|\varphi\|_\infty|\xi|\} < \infty$, *then for all* $a \in \mathbf{R}^d$ *and* $t, \varepsilon > 0$

$$\mathbf{E}|J_t^{(\varepsilon)}(a) - J_{t,r}^{(\varepsilon)}(a)| \leqslant L_0(a)t^{d/2}R(\varphi, r), \tag{14}$$

where

$$R(\varphi, r) = \Bigg(\int\limits_{|x|>r} \varphi^2(x)dx + \Big(\int\limits_{|x|>r} \varphi(x)dx\Big)^2\Bigg)^{1/2} \tag{15}$$

and $L_0(a)$ does not depend on ε, t.

Let us note that if for some $k, \delta > 0$

$$|\varphi(x)| \leqslant k(1+|x|^{d+\delta})^{-1}, \ x \in \mathbf{R}^d, \tag{16}$$

then $R(\varphi, r) \leqslant Lr^{-\delta}$, $L = L(d, \delta)$.

LEMMA 3. *If the conditions of Lemma 1 are strengthened by the assumptions* $\mathbf{E}\exp\{2\|\varphi\|_\infty|\xi|\} < \infty$, $\mathbf{E}\theta^{2d} < \infty$, *then we have for all* $a, b \in \mathbf{R}^d$ *and* $r(t) \to \infty$

$$\operatorname{cov}\Big(J_t^{(\varepsilon)}(a), J_t^{(\varepsilon)}(b)\Big) = t^{d/2}\Big(\sigma(a,b) + g_t^{(\varepsilon)}(a,b)\Big), \tag{17}$$

where

$$\sigma(a,b) = \exp\Big\{-\frac{|a|^2+|b|^2}{2}\Big\}\int_{\mathbf{R}^d} dy M(y)\int_{\mathbf{R}^d} dw f_0(y,w), \tag{18}$$

$$M(y) = \exp\{-|y|^2 + (y, a+b) + 2A_0(cy) + 2\lambda_0(cy)H\},$$

$$f_0(y,w) = \exp\Big\{\lambda_0(cy)\int_{\mathbf{R}^d} dx \int_{\mathbf{R}_+} dF(u)\int_{\mathbf{R}} dG(v)\Big(e^{v\varphi(x/u)} - 1\Big)\Big(e^{v\varphi((x-w)/u)}-1\Big)\Big\} - 1 \tag{19}$$

and $g_t^{(\varepsilon)}(a,b) \to 0$ *under* (7) *for every* $a, b \in \mathbf{R}^d$.

The proof is based on the study of the following integrals

$$\operatorname{cov}(J_{t,r}^{(\varepsilon)},\ J_{t,r}^{(\varepsilon)}(b)) = t^{d/2}e^{-(|a|^2+|b|^2)/2}\times$$

$$\times\int_{\mathbf{R}^d} dy e^{-|y|^2+(y,a+b)+A_0(\varepsilon y\sqrt{t})}F_{\varepsilon,r}(y\sqrt{t})D_{\varepsilon,r}(y),$$

$$F_{\varepsilon,r}(x) = \exp\Bigg\{\int_{\mathbf{R}^d} dz\int_{\mathbf{R}_+} dF(u)\int_{\mathbf{R}} dG(v)\times(e^{v\varphi_r((x-z)/u)}-1)\lambda_0(\varepsilon z)\Bigg\},$$

$$D_{\varepsilon,r}(y) = \int_{\mathbf{R}^d} dw F_{\varepsilon,r}(y\sqrt{t}-w)\times K_{\varepsilon,r}(y\sqrt{t}, y\sqrt{t}-w)\exp\{h_t^{(\varepsilon)}(y,w)\},$$

$$K_{\varepsilon,r}(x,y) = \exp\Bigg\{\int_{\mathbf{R}^d} dz\int_{\mathbf{R}_+} dF(u)\int_{\mathbf{R}} dG(v)\times$$

$$\times(e^{v\varphi_r((x-z)/u)}-1)(e^{v\varphi_r((y-z)/u)}-1)\lambda_0(\varepsilon z)\Bigg\} - 1,$$

$$h_t^{(\varepsilon)}(y,w) = A_0(\varepsilon\sqrt{t}y - \varepsilon w) - \frac{|w|^2}{2t} + \frac{(y,w)}{\sqrt{t}} - \frac{(w,b)}{\sqrt{t}}.$$

Examining the behaviour of (20) we take into account (7), (12) and the following relation

$$\sup_{z\in\mathbf{R}^d}\Big\{-\Big(\frac{1}{2}-\tau\Big)|z|^2+(z,(1-2\tau)y-b)+\tau|y|^2\Big\}=\frac{1}{2}|y|^2-(b,y)+\frac{|b|^2}{2(1-2\tau)}$$

is valid for any $\tau\in(cc_2,1/2)$ and any $b,y\in\mathbf{R}^d$.
Other details are omitted.
Now let $Q(h)=([-h,h])^d$ be a family of cubes with $h=h(t)$ and introduce

$$J^{(\epsilon)}_{t,r,h}(a)=\int\limits_{Q(h)}dy\exp\Big\{\tilde{\zeta}^{(\epsilon)}_r(y)-\frac{|a\sqrt{t}-y|^2}{2t}\Big\}. \tag{20}$$

Lemma 4. *Suppose in addition to the conditions of Lemma 3 that $h(t)/\sqrt{t}\to\infty$ as $t\to\infty$. Then for any $a\in\mathbf{R}^d$*

$$\mathrm{var}\big(J^{(\epsilon)}_{t,r}(a)-J^{(\epsilon)}_{t,r,h}(a)\big)=t^{d/2}\tilde{g}^{(\epsilon)}_t(a),$$

where $\tilde{g}^{(\epsilon)}_t(a)\to 0$ for every $a\in\mathbf{R}^d$, provided that (7) holds.

Theorem 1. *Let $\varphi\in L^1(\mathbf{R}^d)\bigcap L^\infty(\mathbf{R}^d)$ and assume that for some $s\in(2,3]$ there exists a function $r(t)$ satisfying, as $t\to\infty$, the following conditions*

$$r(t)\to\infty,\ R(\varphi,r(t))=o(t^{-d/4}),\ r(t)=o(t^\beta),\ \beta=\frac{1}{4}(s-2)(s-1)^{-1}. \tag{21}$$

Let also $\mathbf{E}\exp\{s\|\varphi\|_\infty|\xi|\}<\infty$ and $\theta\leqslant\theta_0$ a.s. (θ_0 is a positive constant). Then under the other hypotheses of Lemma 1 all finite dimensional distributions of the field

$$\tilde{J}^{(\epsilon)}_t(a)=t^{-d/4}(J^{(\epsilon)}_t(a)-\mathbf{E}J^{(\epsilon)}_t(a))$$

weakly converge to the corresponding finite dimensional distributions of the centered Gaussian field $J(a)$ with

$$\mathrm{cov}(J(a),J(b))=\sigma(a,b);\quad a,b\in\mathbf{R}^d,$$

where $\sigma(a,b)$ is defined by (18), while $\mathbf{E}J^{(\epsilon)}_t(a)$ is given by (13), (14).

The proof of this theorem is based on the lemmas formulated above and on the CLT for the auxiliary fields

$$X^{(\epsilon,t,r)}_j=\int\limits_{K_j}dx\exp\Big\{\tilde{\zeta}^{(\epsilon)}_r(x)\Big\}f(x,t),$$

where $K_j = (j_1 - 1, j] \times \ldots \times (j_d - 1, j_d]$, $j \in \mathbf{Z}^d$,

$$f(x,t) = \sum_{1 \leqslant l \leqslant n} \tau_l \exp\Big\{ - \frac{|x - a_l\sqrt{t}|^2}{2t} \Big\}$$

and $\tau_l \in \mathbf{R}$, $a_l \in \mathbf{R}^d$, $l = 1, \ldots, n$, $n \in \mathbf{N}$, are arbitrary fixed. Thus

$$J^{(\varepsilon)}_{t,r,h}(a) = \sum_{j \in T(h)} X_j^{(\varepsilon,t,r)}, \quad Q(h) = \bigcup_{j \in T(h)} K_j, \quad T(h) \subset \mathbf{Z}^d.$$

Note that from Lemmas 3 and 4 we have

$$\operatorname{var} \sum_{j \in T(h)} X_j^{(\varepsilon,t,r)} = t^{d/2}\Big(\sum_{k,l=1}^{n} \tau_k \tau_l \sigma(a_k, a_l) +$$

$$+ g_t^{(\varepsilon)}(a_1, \ldots, a_n; \tau_1, \ldots, \tau_n)\Big),$$

where $g_t^{(\varepsilon)}(a_1, \ldots, a_n; \tau_1, \ldots, \tau_n) \to 0$ under (7). The case $\sum_{k,l=1}^{n} \tau_k \tau_l \sigma(a_k, a_l) = 0$ being trivial, further on we will assume that this sum is positive. It is essential to note that the variance of the sum $\sum_{j \in T(h)} X_j^{(\varepsilon,t,r)}$ has not the property of "regular" growth (meaning the quasilinear growth in the number of summands) because in our case $h(t)/\sqrt{t} \to \infty$ as $t \to \infty$.

For all $j \in \mathbf{Z}^d$, $\varepsilon, t > 0$

$$\mathbf{E}|X_j^{(\varepsilon,t,r)}|^s \leqslant \sup_{x \in K_j} \Big(\mathbf{E}\exp\{s\tilde{\zeta}_r^{(\varepsilon)}(x)\} \Big| \sum_{k=1}^{n} \tau_k \exp\Big\{ - \frac{|x - a_k\sqrt{t}|^2}{2t} \Big\} \Big|^s \Big) \leqslant c_s,$$

$$c_s = c_s(a_1, \ldots, a_n; \tau_1, \ldots, \tau_n).$$

The field $X_j^{(\varepsilon,t,r)}$ is $m(t) = 2\theta_0 r(t) + \sqrt{d}$ dependent on the set $T(h(t))$. So using the CLT for weakly dependent random fields (see (Bulinskii, 1987; 1989b)) we see that it is enough to establish the following relation as $(t \to \infty)$

$$h(t)^d r(t)^{d(s-1)} t^{-sd/4} + t^{-d/4} r(t)^d (\log h(t))^{(d-1)/2} + h(t)^{d/2} t^{-d/2} r(t)^{3d/2} \to 0.$$

It remains to take into account that for function $h(t)$ we have only the restriction $h(t)/\sqrt{t} \to \infty$ as $t \to \infty$. So we come to condition (21) on the behaviour of the function $r(t)$.

Remark 2. In particular, if $c = 0$ then

$$\mathbf{E}J_t^{(\varepsilon)}(a) = t^{d/2}((2\pi)^{d/2} e^{\lambda_0(0)H + A_0(0)} + g_t^{(\varepsilon)}(a)),$$

$$\sigma(a,b) = \sigma_0^2 e^{-|a-b|^2/4}, \sigma_0^2 = \text{const}$$

and $\mathbf{E}J_t^{(\varepsilon)}(a) = (2\pi t)^{d/2} e^{\lambda H}$ for all $a \in \mathbf{R}^d$ whenever $\lambda_0(\cdot) \equiv \lambda$, $A_0 \equiv 0$.

COROLLARY 1. *If φ satisfies condition (16) then instead of (21) in Theorem 1 one can assume $\delta > d(s-1)(s-2)^{-1}$.*

If the amplitudes ξ_i have all exponential moments it is possible to weaken the condition on the function φ. Using the technique of semiinvariants we have

THEOREM 2. *Let $\varphi \in L^1(\mathbf{R}^d)\bigcap L^\infty(\mathbf{R}^d)$ and $\mathbf{R}(\varphi, t^\beta) = o(t^{-d/4})$ as $t \to \infty$ for some $\beta < 1/4$. Assume that $\mathbf{E}\exp\{\mu|\xi|\} < \infty$ for all $\mu > 0$ and $\theta \leqslant \theta_0$ a.s. (θ_0 is a positive constant). Then under hypotheses (7), (12) and the conditions on the functions $\lambda_0(\cdot), A_0(\cdot)$ stated in Lemma 1 the conclusion of Theorem 1 is true.*

COROLLARY 2. *In the conditions of Theorem 2 instead of (21) it is possible to demand (16) with $\delta > d$.*

Remark 3. The conditions of Theorem 2 can be considered as the limit variant of the assumptions of Theorem 1 as $s \to \infty$.
In the general case of unbounded θ_i we introduce

$$q_0(s) = \inf\{k : k \geqslant s(s-2)^{-1},\ k \in \mathbf{N}\}$$

and define for $\gamma = \gamma(t) > 0$

$$\tilde{\zeta}^{(\varepsilon)}_{r,\gamma}(x) = \sum_{i:|x - x_i^{(\varepsilon)}| \leqslant r\gamma} \xi_i \varphi_r((x - x_i^{(\varepsilon)})/\theta_i) + A_0(\varepsilon x),\ x \in \mathbf{R}^d. \tag{22}$$

Using fields (22) it is possible to prove

THEOREM 3. *Let $\varphi \in L^1(\mathbf{R}^d)\bigcap L^\infty(\mathbf{R}^d)$, $\mathbf{E}\theta^{2d} < \infty$, $\mathbf{E}\exp\{s\|\varphi\|_\infty|\xi|\} < \infty$ for some $s \in (2,3]$ and let (7), (12) hold along with the assumptions on the functions $\lambda_0(\cdot), A_0(\cdot)$ made in Lemma 1. Assume that there exist functions $r(t), \gamma(t)$ satisfying the conditions*

$$\int_{\gamma(t)}^{\infty} u^d dF(u) = o\left(t^{-dq_0/4}\right),\quad q_0 = q_0(s),$$

$$r(t) \to \infty,\quad r(t)\gamma(t) = o(t^\beta),\quad \beta = \frac{1}{4}(s-2)(s-1)^{-1},$$

$$\mathbf{R}(\varphi, r(t)) = o(t^{-d/4}),$$

as $t \to \infty$. Then the conclusion of Theorem 1 is valid.

Here we have indicated the scheme of studying the functionals $J_t^{(\varepsilon)}(\cdot)$. The same approach permits to establish the asymptotic normality of the vector fields $V_t^{(\varepsilon)}(\cdot)$. Of course, appropriate modifications are necessary to take into account different integral representations of $J_t^{(\varepsilon)}(\cdot)$ and $V_t^{(\varepsilon)}(\cdot)$. After the investigation of the common asymptotic behaviour of $J_t^{(\varepsilon)}(\cdot)$ and $V_t^{(\varepsilon)}(\cdot)$ we can obtain the following result of asymptotic normality of the field $Z_t^{(\varepsilon)}(\cdot)$.

THEOREM 4. *Let the conditions of Theorem 3 be fulfilled. Then there exists a centered Gaussian field $Z(a), a \in \mathbf{R}^d$, which is the limit in the sense of weak convergence of all finite dimensional distributions for the family of fields*

$$\tilde{Z}_t^{(\varepsilon)}(a) = t^{(d+2)/4}\mathcal{E}_t^{(\varepsilon)}(a)Z_t^{(\varepsilon)}(a) + t^{d/4}\mathcal{M}_t^{(\varepsilon)}(a), \quad a \in \mathbf{R}^d,$$

where

$$\mathcal{M}_t^{(\varepsilon)}(a) = \int_{\mathbf{R}^d} dy(y-a)\exp\Big\{-\frac{|y-a|^2}{2} + A_0(\varepsilon\sqrt{t}y) + \Lambda_t^{(\varepsilon)}(y)\Big\}.$$

Remark 4. We do not give here the explicit formula for the matrix of covariances of the field $Z(\cdot)$. In the particular case $A_0 \equiv 0, \lambda_0 \equiv \lambda$ a constant, the function $\mathcal{M}_t^{(\varepsilon)}(\cdot) \equiv 0$ and the covariance matrix for the limiting field $Z(\cdot)$ can be found in (Bulinskii, 1989a; Bulinskii and Molchanov, 1990). Note also that analogously to Theorem 2 we can give the limit variants of Theorems 3 and 4 as $s \to \infty$ (then $q_0(s) \to 1$).

REFERENCES

Bulinskii, A. V. (1987). Limit theorems under weak dependence conditions. *In: Proc. Fourth Vilnius Conf.* **1**, VNU Sci. Press, Utrecht. 307—326.

Bulinskii, A. V. (1989a). Asymptotic normality for the families of integral functionals. *In: Abstr. Fifth Inter. Vilnius Conf. Probab. Theory and Math. Statist.* **1**, IMC, Vilnius. 83—84.

Bulinskii, A. V. (1989b). *Limit Theorems Under weak Dependence Conditions.* Moscow State Univ. Press, Moscow.

Bulinskii, A. V. and Molchanov, S. A. (1990). Asymptotic Gaussian behaviour of the solution of Burgers equation with random initial data. *Theory Probab. Appl.* (to appear).

Gurbatov, S. N. *et al.* (1988). The large-scale structure of the Universe in the frame of the model equation of nonlinear diffusion. MPA, **371**, 1—21.

Heinrich, L. and Schmidt, V. (1985). Normal convergence of multidimensional shot noise and rates of this convergence. *Adv. Appl. Probab.* **17**, 709—730.

Lifshitz, I. M. *et al.* (1982). *Introduction to the Theory of Disordered Systems.* Nauka, Moscow.

Rosenblatt, M. (1987). Scale renormalization and random solutions of the Burgers equation. *J. Appl. Probab.* **24**, 328—338.

Ruelle, D. (1969). *Statistical Mechanics. Rigorous Results.* Benjamin, New York–Amsterdam.

Saulis, L. and Statulevičius, V. (1989). *Limit Theorems for Large Deviations.* Mokslas, Vilnius.

Shandarin, S. F. and Zel'dovich, Ya. B. (1988). Turbulence, intermittency, structures in a self-gravitating medium: the large-scale structure of the Universe. MPA, **350**, 1—91.

Prob. Theory and Math. Stat., Vol. 1, pp. 217–227
B.Grigelionis *et al.* (Eds.)
1990 VSP/Mokslas

THE THEORY OF DYNAMICAL SEMIGROUPS AND ITS APPLICATIONS

A.M. CHEBOTAREV
Applied Mathematics Department, MIEM,
B.Wuzovski per. 3/12, 109028 Moscow, USSR

ABSTRACT

The problem of conservativity of a dynamical semigroup is considered for a class of unbounded Lindblad–type generators, including classical diffusions and Markov jump processes. In this class necessary and sufficient conditions of conservativity are obtained, which generalize the corresponding results of Feller, Skorokhod and Has'minsky for classical Markov processes. Efficient sufficient conditions are suggested, which imply conservativity for new examples of dynamical semigroups, including classical Markov semigroups.

1. INTRODUCTION

The transition probability of a Markov process can be regarded as a family of multiplication operators in a Hilbert space $\mathcal{H}$. If for example x is a point in finite dimensional space $\mathbf{R}^n$, then we can introduce the following definition:

$$\psi \in \mathcal{H} = L_2(\mathbf{R}^n), \qquad \dot{P}_t(\Gamma, s)\psi(x) = P(x, t|\Gamma, s)\psi(x),$$

where P_t in the left is the bounded operator in $\mathcal{H}$ and $P(x, t|\Gamma, s)$ in the right is the transition probability function.
The semigroup property for operators $P_t(\Gamma, s)$ follows from the same property for functions $P(x, t|\Gamma, s)$ and the Kolmogorov—Feller infinitesimal operator L acting originally on functions as follows

$$LP(x, t|\Gamma, s) = \int m(dq)\, |C(x, q)|^2 \Big\{ P(x + q, t|\Gamma, s) - P(x, t|\Gamma, s) \Big\}$$

has natural extension as the operator's map which acts on bounded operators in the Hilbert space $\mathcal{H}$:

$$LP_t(\Gamma, s) = \int m(dq)\, C^*(\cdot\,, q) \Big\{ D_q^* P_t(\Gamma, s) D_q - P_t(\Gamma, s) \Big\} C(\cdot\,, q), \qquad (1.1)$$

where D_q being the shift operator and D_q^* its conjugate:

$$D_q\psi(x) = \psi(x - q), \qquad D_q^*\psi(x) = \psi(x + q).$$

On the other hand if the transition probability is a smooth function, the operator of multiplication by its second derivative can be represented in the similar form:

$$\frac{1}{2}P''\psi = -(P\psi')' + \frac{1}{2}(P\psi)'' + \frac{1}{2}P\psi'' = \left\{-\partial_x P\partial_x + \frac{1}{2}\partial_x^2 P + \frac{1}{2}P\partial_x^2\right\}\psi. \quad (1.2)$$

So in both cases the infinitesimal maps associated with generators of Markov processes have the same structure:

$$L(P) = \Phi(P) - \Phi(I) \circ P,$$

where $A \circ B = (AB + BA)/2$ is the Jordan product of operators A and B, and the map Φ acts on bounded operators P as follows:

$$\Phi(P) = \int_{\mathbf{R}^n} m(dq)\, C^*(\cdot\,, q) D_q^* P D_q C(\cdot\,, q) \qquad \text{(for jump process)},$$

$$\Phi(P) = -\sum_{j,k,l=1}^{n} \partial_j a_{jk}^*(\cdot) P a_{kl}(\cdot)\partial_l \qquad \text{(for symmetric diffusion)}.$$

At last let us note that the vector field (b, ∇) acts on operators of multiplication by the transition probability as a commutator:

$$bP'\psi = \frac{1}{2}\left\{b(P\psi)' + (bP\psi)' - P(b\psi)' - Pb\psi'\right\} = i[H, P]\psi,$$

$$H = \frac{1}{2i}((b, \nabla) + (\nabla, b)),$$

where H is the symmetric operator of the first order.
So in quite general case generators of Markov processes have extensions as the bounded operator maps:

$$L(P) = \Phi(P) - \Phi(I) \circ P + i[H, P].$$

This form of L coinsides with the structure of dynamical semigroup generator discovered in (Lindblad, 1976) and independently in (Gorini et al., 1976). Since the equation

$$\frac{d}{dt}P_t = i[H, P_t], \qquad P|_{t=0} = P$$

is the ***Heisenberg evolution equation*** for quantum observable P, the ***Lindblad equation***

$$\frac{d}{dt}P_t = L(P_t) \quad (1.3)$$

generalizes both quantum and probability types of evolutions.

2. THE MINIMAL DYNAMICAL SEMIGROUP

The dynamical semigroup $P_t(\cdot)$ is a one parameter ultraweakly continuous semigroup of completely positive, and normal maps defined on some von Neumann *-algebra $\mathcal{A}$ of bounded operators in a Hilbert space $\mathcal{H}$ (Lindblad, 1976; Gorini, 1976). The Lindblad equation (1.3) describes its time evolution in the Heisenbag picture.

A dynamical semigroup is called *conservative* if it preserves the unit operator: $P_t(I) = I$ ($\forall t > 0$). The corresponding infinitesimal map $L(\cdot)$ is called *regular*.

Lindblad proved that all bounded maps $L(\cdot)$ of given structure are regular.

The first result on the dynamical semigroups with unbounded infinitesimal maps $L(\cdot)$ was obtained in (Davies, 1978). It was found that under some conditions on the domain of H and $\Phi(I)$ there exists a minimal solution of the Lindblad evolution equation and this solution is the least upper bound for the recurrent sequence of operators:

$$P_t^{(n+1)}(X) = W_t^* X W_t + \int\limits_{[0,t)} ds\, W_{t-s}^* \Phi(P_s^{(n)}(X)) W_{t-s},$$

where W_t is one parameter semigroup of contractions on the Hilbert space $\mathcal{H}$ with dissipative generator

$$G = -\left(iH + \frac{1}{2}\Phi(I)\right)$$

and W_t^* is the conjugate semigroup. This result is the natural operator-valued generalization of the minimal solution existance theorem for the Kolmogorov—Feller equation.

Under the conditions which provide the existence of the minimal solution for the Lindblad equation we describe here the necessary and sufficient conservativity conditions.

In order to describe the assumptions under which the strongly continuous minimal solution for the integral Lindblad equation exists we have to introduce a special class of unbounded completely positive maps.

Denote by $\mathcal{A}$ any von Neumann algebra of bounded operators in the Hilbert space $\mathcal{H}$ endowed with the strong operator topology and by $F^{ws}(\mathcal{A},\mathcal{H})$: $\mathcal{A} \to \mathcal{A}$ a class of linear maps of $\mathcal{A}$ such that image $P(X_n)$ of any strongly convergent sequence of operators $\{X_n\}$ is a weakly convergent sequence:

$$F^{ws}(\mathcal{A},\mathcal{H}) = \left\{P\colon \mathcal{A} \to \mathcal{A}, \quad P(X_n) \underset{w}{\to} P(X) \quad \text{if} \quad X_n \underset{s}{\to} X\right\}.$$

The space $F_*^s(\mathcal{A},\mathcal{H})$ contains all corresponding maps of A in the space of bilinear forms on $\mathcal{H}$:

$$F_*^s(\mathcal{A},\mathcal{H}) = \left\{P_*\colon P_*(X)[\varphi,\psi] = \langle\varphi, P(X)\psi\rangle, \quad P \in F^{ws}(\mathcal{A},\mathcal{H})\right\}.$$

Let $\|\cdot\|$ be the Hilbert norm on $\mathcal{H}$ and $\mathcal{B}$ — any Banach subspace in $\mathcal{H}$ such that $\|\cdot\|_{\mathcal{B}} \geqslant \|\cdot\|$, A and B — any absolutely convex compact sets in $\mathcal{A}$ and $\mathcal{B}$ and let

$\rho_{A,B}(\cdot)$ be the system of seminorms on $F_*^s(\mathcal{A},\mathcal{H})$:

$$\rho_{A,B}(\Phi_*) = \sup_{X\in A;\ \varphi,\psi\in B} \left|\Phi_*(X)[\varphi,\psi]\right|.$$

The completion of the vector space $F_*^s(\mathcal{A},\mathcal{H})$ in this separable locally convex topology will be denoted by $F_*^s(\mathcal{A},\mathcal{B})$ and by $CP_*^s(\mathcal{A},\mathcal{B})$ — the cone of all completely positive maps from $F_*^s(\mathcal{A},\mathcal{B})$:

$$CP_*^s(\mathcal{A},\mathcal{B}) = \Big\{\Phi_*\colon\ \Phi_* \in F_*^s(\mathcal{A},\mathcal{B}),$$

$$\sum_{k,j} \Phi_*(Y_k^* Y_j)[\psi_k,\psi_j] \geqslant 0 \quad (\forall\{Y_k\}\in\mathcal{A})\ (\forall\{\psi\}\in\mathcal{B})\Big\}.$$

The following proposition gives the main example of the Banach space $\mathcal{B}$ and of a set of completely positive maps from $CP_*^s(\mathcal{A},\mathcal{B})$.

PROPOSITION 2.1. *Let $\Phi(\cdot)$ be any operator map on $\mathcal{A}$. If $\operatorname{dom}\Phi(I)$ is dense in $\mathcal{H}$ and*

$$e^{-\lambda\Phi(I)}\Phi(\cdot)e^{-\lambda\Phi(I)} \in CP^{ws}(\mathcal{A},\mathcal{H}) \quad (\forall\lambda>0).$$

Then $\Phi_* \in CP_*^s(\mathcal{A},\mathcal{B})$, where $\|\psi\|_{\mathcal{B}}^2 = \|\Phi(I)^{1/2}\psi\|^2$.

Suppose that H and $\Phi(I)$ are densely defined selfadjoint operators in $\mathcal{H}$ and

$$\Phi(I)\geqslant I, \qquad \operatorname{dom}\Phi(I)\subseteq\operatorname{dom}H, \qquad \overline{\operatorname{dom}\Phi(I)} = \mathcal{H}. \tag{2.1}$$

In what follows we will suppose that there exists one-parameter strongly continuous semigroup of contractions

$$W_t = e^{-Gt}\colon \mathbf{R}_+ \to \mathcal{A}, \qquad G = iH + \frac{1}{2}\Phi(I) \tag{2.2}$$

and let W_t^* be its conjugate semigroup.
The important condition which prevents us to take an arbitrary small Banach space $\mathcal{B}$ and supplies the minimal semigroup P_t with strong continuity property is the condition of the boundedness and strong continuity of W_t in $\mathcal{B}$:

$$\|W_t\psi\|_{\mathcal{B}} \leqslant e^{ct}\|\psi\|_{\mathcal{B}}, \qquad \lim_{t\to 0}\|W_t\psi-\psi\|_{\mathcal{B}} = 0 \quad (\forall\psi\in\mathcal{B}). \tag{2.3}$$

The following proposition gives an example of the class of hamiltonians H such that the last conditions hold.

PROPOSITION 2.2. *Let $\|\psi\|_{\mathcal{B}} = \|\Phi(I)^{1/2}\psi\|$ and $i[\Phi(I),H] \geqslant -\gamma\Phi(I)$ on $\operatorname{dom}G$ for some $\gamma>1$. Then the conditions (2.3) hold with $c=\gamma-1$.*

Now we can formulate the theorem on the conditions sufficient for existence of a minimal dynamical semigroup.

THEOREM 2.3. *Let* $\Phi_*(\cdot) \in CP^s_*(\mathcal{A},\mathcal{B})$ *and the assumptions (2.1)—(2.3) be fulfilled. Then there exists the minimal strong continuous one parameter semigroup* $P_t(\cdot)$: $\mathcal{A} \to \mathcal{A}$ *of normal, completely positive maps. If this semigroup is conservative then there is no other conservative semigroups in this class.*

This semigroup satisfies the integral Lindblad equation

$$P_t(X) = W_t^* X W_t + \left\| \int\limits_{[0,t)} ds\, W_{t-s}^* \Phi_*(P_s(X)) W_{t-s} \right\|,$$

where $\|\mathcal{B}_*(X)\|$ means the operator, corresponding to the bilinear form $\mathcal{B}_*(\mathcal{A})$ (Kato, 1972). The bilinear form on $\mathcal{B}$

$$\Phi_*(X,t) = \int\limits_{[0,t)} ds\, W_{t-s}^* \Phi_*(P_s(X)) W_{t-s}$$

is bounded on $\mathcal{H}$:

$$\Phi_*(X,t)[\varphi,\psi] = \int\limits_{[0,t)} ds\, \Phi_*(P_s(X))[W_{t-s}\varphi, W_{t-s}\psi],$$

$$|\Phi_*(X,t)[\varphi,\psi]| \leqslant \|X\|\,\|\psi\|\,\|\varphi\|.$$

This form can be extended onto the whole Hilbert space $\mathcal{H}$: $\big\|[\Phi_*(X,t)]\big\| \leqslant \|X\|$, and $\|\Phi_*(\cdot\,,t)\|$ is a completely positive bounded map.
The necessary and sufficient conservativity conditions are given in the following theorem.

THEOREM 2.4. *Let the conditions of the theorem 2.3 be fulfilled. Then the following statements are equivalent.*

a) *The minimal solution of the Lindblad integral equation is conservative.*

b) *The sequence of contractions* $Q^n(I)$ *converges to* 0 *as* $n \to \infty$ *in the strong operator topology, where* $Q^n(I)$ *is the composition of* n *maps and*

$$Q(X) = \left\| \int\limits_{[0,\infty)} ds\, e^{-t} W_t^* \Phi_*(X) W_t \right\|.$$

c) *The operator equation* $X = Q(X)$ *has no bounded positive solutions in* $\mathcal{A}$.

d) *The operator equation* $\langle\psi, X\psi\rangle = \langle\psi, L(X)\psi\rangle$, $\forall\psi \in \operatorname{dom} G$, *has no bounded positive solutions in* $\mathcal{A}$.

The last form of the necessary and sufficient condition of conservativity is an operator form of the Feller—Has'minskii regularity condition for diffusion process (Has'minskii, 1960). On the other hand the second statement of the theorem generalizes the Gikhman and Skorokhod regularity condition for the Markov jump process (Gikhman and Skorokhod, 1977) which has completely another form in abelian case.

3. SUFFICIENT REGULARITY CONDITIONS

If we suppose that $m(dq)$ from the previous equation (1.1) is probability measure, then

$$\varkappa(x) = \int m(dq)\,|C(x,q)|^2$$

is the intensity of the jumps, and the conditional probability measure

$$m(x|B) = \frac{1}{\varkappa(x)} \int_B m\,(dq)\,|C(x,q)|^2$$

is the transition probability of the jump from the point x to the set $x + B$. The necessary and sufficient regularity condition from (Gikhman and Skorokhod, 1977) requires the divergence of the following series with probability 1:

$$\mathbf{P}_x\left\{ \sum \frac{1}{\varkappa(x_n)} = \infty \right\} = 1, \tag{3.1}$$

where $\{x_n\}_1^\infty$ is the corresponding Markov chain of the jump process:

$$x_0 = x, \qquad x_n = x_{n-1} + q_n, \qquad \mathbf{P}\{q_n \in B | x_{n-1}\} = m(x_n|B).$$

This result follows immediately from our second form of conservativity condition. In this case the operators

$$\Phi(I) = \int m\,(dq)\,|C(\cdot\,,q)|^2 = \varkappa(\,\cdot\,),$$

$$Q(I) = \int dt\, e^{-t(1+\varkappa(\cdot))}\varkappa(\,\cdot\,) = (1 + \varkappa(\,\cdot\,)^{-1})^{-1}$$

are the multiplication operators and $Q^n(I)$ is the mathematical expectation with respect to probability measure of the Markov chain:

$$Q^{n+1}(I)(x) = \mathbf{M}_x \prod_{k=0}^{n} 1 + 1/\varkappa(x_k) =$$

$$= \int \cdots \int m(x, dq_1) \cdots m(x_{n-1}, dq_n) \prod_{k=0}^{n} 1 + 1/\varkappa(x_k).$$

Since the product has the lower and the upper bounds of the additive type

$$\exp\left\{ -\sum \frac{1}{\varkappa(x_k)} \right\} \leqslant \prod \left(1 + \frac{1}{\varkappa(x_k)}\right)^{-1} \leqslant \left(1 + \sum \frac{1}{\varkappa(x_k)}\right)^{-1},$$

then $Q^{n+1}(I) \to 0$ if and only if the series $\sum 1/\varkappa(x_k)$ diverges with probability 1. Thus we came to the condition (3.1).
The same technics can be applied also for analysis of the Davies neutron diffusion model (Davies, 1977).

In his example there exist a separable Hilbert space $\mathcal{H}$ and an increasing family of projectors $\left\{\pi_k\right\}_1^\infty$ such that $\sum \pi_k = I$ and the infinitesimal operator of the dynamical semigroup is such that all subspaces $\mathcal{H}_n = \pi_n\mathcal{H}$ are invariant under the action of the semigroup W_t and another two conditions are fulfilled:

$$\pi_n\Phi(X)\pi_n = \pi_n\Phi(\pi_{n+1}X\pi_{n+1})\pi_n \qquad \text{(invariance equations)},$$

$$\pi_n\Phi(X)\pi_n \leqslant \varkappa_n\|X\| \qquad \text{(upper bounds conditions)},$$

where $\left\{\varkappa_n\right\}_1^\infty$ is some increasing sequence of positive numbers. Then for any $\psi \in \mathcal{H}_k$ we have the invariance equation

$$\langle\psi, Q^{n+1}(I)\psi\rangle = \langle\psi, \pi_k Q(\dots\pi_{k+n}Q(I)\pi_{k+n}\dots)\pi_k\psi\rangle.$$

On the other hand from the upper bounds conditions we obtain the following inequalities

$$\pi_m Q(I)\pi_m \leqslant \int dt\, e^{-t} e^{-(iH+\varkappa_m/2)t}\varkappa_m e^{(iH-\varkappa_m/2)t} = \left(1+\frac{1}{\varkappa_m}\right)^{-1},$$

$$\pi_k Q(\dots\pi_{k+n}Q(I)\pi_{k+n}\dots)\pi_k \leqslant$$
$$\leqslant \prod_{m=1}^{n}\left(1+\frac{1}{\varkappa_{k+m}}\right)^{-1} \leqslant \left(1+\sum_{m+1}^{n}\frac{1}{\varkappa_{k+m}}\right)^{-1}.$$

So the conservativeness condition $s-\lim Q^n(I) = 0$ is fulfilled if the series $\sum_{n=1}^{\infty} 1/\varkappa_n$ diverges.

In the model of the neutron diffusion all assumptions on the map $\Phi(\cdot)$, the semigroup W_t in the symmetric Fock space $\mathcal{H} = \bigoplus_{k=0}^{\infty}\mathcal{H}_k(L_2(\mathbf{R}^n))$ and the sequence $\{\varkappa_k\}$ are fulfilled for the map

$$\Phi(\cdot) = \sum_{m=1}^{3}\Phi_m(\cdot), \qquad \Phi_m(Y) = \left(B_p^{(m)*}YB_p^{(m)}\right),$$

where the first term with $B_p^{(1)} = a(f_1(p))$ describes absorbtion of a neutron ($a(\cdot)$ — annihilation operator), the second with

$$B_p^{(2)} = a^*(f_2(p))a(f_3(p))$$

describes scattering ($a^*(\cdot)$ — creation operator) and the third with

$$B_p^{(3)} = a^*(f_4(p))a^*(f_5(p))a(f_6(p))$$

describes creation of two neutrons from another one. The Høegh—Krohn lemma, used by Davies gives the upper bounds for projection of $\Phi(I)$:

$$\pi_k\Phi(I)\pi_k \leqslant ckI$$

if $\{f_k(p)\}$ are the functions from the Schwartz space. So the series $\sum 1/\varkappa(x_k)$ diverges and the Lindblad infinitesimal operator

$$L(X) = \sum_{m=1}^{3} \Phi_m(X) + i[H, X], \qquad H = -\sum \pi_k |p_k|^2 \pi_k$$

is regular. In this example operators $\Phi(I)$ and H commute and this symplifies the problem of conservativity.
More complicated examples of regular dynamical semigroups will be obtained by using an operator generalization of the Jensen inequality which in Abelian case has the following form:

$$\mathbf{M}_x \left(1 + \sum \frac{1}{\varkappa(x_n)}\right)^{-1} \leqslant \left(1 + \sum \frac{1}{\mathbf{M}_x \varkappa(x_n)}\right)^{-1}$$

and for completely positive map $Q(\cdot)$ and positive operators $\{X_n\}$ its form is almost the same:

$$Q\left(I + \sum X_n^{-1}\right)^{-1} \leqslant \left(Q(I)^{-1} + \sum Q(X_n)^{-1}\right)^{-1} \tag{3.2}$$

if the inverse operators X_n^{-1} and $Q(X_n)^{-1}$ exist (Chebotarev, 1989).
Note that the Lindblad generator is invariant under additive transform of the map Φ:

$$\Phi(X) \to \Phi(X) + \lambda X, \qquad \lambda \geqslant 0.$$

So in what follows we will suppose that $\Phi(I) \geqslant I$.
In this case one can prove that there exists a positive selfadjoint operator S such that $Q(I)$ has the following form

$$Q(I) = (I + S^{-1})^{-1}, \qquad S \geqslant I. \tag{3.3}$$

For example when we consider operator S corresponding to the Markov jump process then $S = \varkappa(\cdot) + I$ and in case of the symmetrical diffusion $S = -(\nabla, A(\cdot)\nabla) + I$ is the second order operator.
If we apply operator-valued Jensen inequality (3.2) and expansion (3.3) for $Q(I)$ we obtain the upper bounds for $Q^n(I)$:

$$\langle \psi, Q^n(I), \psi \rangle \leqslant \left(1 + \sum_{k=1}^{n} \frac{1}{\langle \psi, Q^k(S)\psi \rangle}\right)^{-1} \qquad (\forall \psi,\ \|\psi\| \leqslant 1).$$

So the sufficient regularity condition is fulfilled if the following series diverges

$$\sum_{k=1}^{\infty} \frac{1}{\langle \psi, Q^k(S)\psi \rangle} = \infty \quad (\forall \psi \in \operatorname{dom}\Phi(I)).$$

For example if the assumptions (2.1) hold and

$$i\Big\{\langle \Phi(I)\psi, H\psi \rangle - \langle H\psi, \Phi(I)\psi \rangle\Big\} \geqslant -c\langle \psi, \Phi(I)\psi \rangle, \tag{3.4}$$

$$-\langle\psi,\Phi(\Phi_\lambda(I))\psi\rangle+\langle\Phi(I)\psi,\Phi(I)\psi\rangle\geqslant -c\langle\psi,\Phi(I)\psi\rangle, \tag{3.5}$$

where

$$\Phi_\lambda(I)=\Phi(I)(I+\lambda\Phi(I))^{-1},\qquad \lambda>0,$$

for fixed positive constant c and arbitrary $\psi\in\operatorname{dom}\Phi(I)$, then there exist three constants c_1,c_2,c_3 such that the operators S an $Q^n(S)$ are bounded with respect to $\Phi(I)$:

$$S\leqslant c_1\Phi(I),\qquad Q^n(S)\leqslant c_2(nI+c_3\Phi(I)),$$

$$\sum_{n=1}^{\infty}\frac{1}{\langle\psi,Q^n(S)\psi\rangle}\geqslant\sum_{n=1}^{\infty}\frac{1}{nc_2\|\psi\|^2+c_3\langle\psi,\Phi(I)\psi\rangle}=\infty\quad(\forall\psi\in\operatorname{dom}\Phi(I)).$$

Hence $\langle\psi,Q^n(I)\psi\rangle\to 0$ as $n\to\infty$ $(\forall\psi\in\mathcal{H})$ and following theorem take place.

THEOREM 3.1. *Let the assumptions* (2.1)—(2.3) *be fulfilled and* $\Phi_*(\cdot)\in CP_*^s(\mathcal{A},\mathcal{B})$. *If inequalities* (3.4)—(3.5) *take place for any* $\psi\in\operatorname{dom}\Phi(I)$, *then the minimal dynamical semigroup is conservative.*

Example 3.1. Consider a completely positive map on von Neumann algebra $\mathcal{A}$ of all bounded operators:

$$\Phi(Y)=\int_{\mathbf{R}^n} m(dq)\,C^*(\cdot,q)D_q^*YD_qC(\cdot,q)$$

for Hilbert space $\mathcal{H}=L_2(\mathbf{R}^n)$ and $\mathcal{B}=\left\{\psi\colon\ \|\psi\|_{\mathcal{B}}=\|\Phi(I)^{1/2}\psi\|<\infty\right\}$, where m is a probability measure and $C(x,q)$ is square integrable function for every fixed x. Then $\Phi_*(\cdot)\in CP_*^s(\mathcal{A},\mathcal{B})$ and

$$-\Phi(\Phi_\lambda(I))+\Phi(I)^2\geqslant-\int m(dq)|C(x,q)|^2(\varkappa(x+q)-\varkappa(x)),$$

where $\varkappa(x)=\int m(dq)\,|C(x,q)|^2$.
Hence the sufficient conservativity condition in terms of the conditional probability measure $m(X|B)$ and intensity of the jumps $\varkappa$ looks as follows:

$$\sup_x\int m(x|dq)\,(\varkappa(x+q)-\varkappa(x))<\infty.$$

In such form this condition was obtained earlier in (Chebotarev, 1988), as the regularity condition for the Markov jump processes.

Example 3.2. If we take the simplest one-dimensional symmetric diffusion, then the corresponding map Φ acts as follows:

$$\Phi[Y]=-\frac{\partial}{\partial x}a(x)Ya(x)\frac{\partial}{\partial x},\qquad \Phi(I)=-\frac{\partial}{\partial x}a^2(x)\frac{\partial}{\partial x}$$

and the following regularity condition holds:

$$-\Phi\Big[\Phi_\lambda(I)\Big]+\Phi^2(I)\geqslant-\Phi\Big[\Phi(I)\Big]+\Phi^2(I)=$$

$$= \frac{\partial}{\partial x} a \left[-\frac{\partial}{\partial x} a^2 \frac{\partial}{\partial x} + a \frac{\partial^2}{\partial x^2} a \right] a \frac{\partial}{\partial x} \geqslant c \frac{\partial}{\partial x} a^2 \frac{\partial}{\partial x}.$$

Since the operator $\frac{\partial}{\partial x} a^2 \frac{\partial}{\partial x}$ is negative, then the sufficient regularity condition has another equivalent form:

$$-\frac{\partial}{\partial x} a^2 \frac{\partial}{\partial x} + a \frac{\partial^2}{\partial x^2} a \leqslant cI, \quad \text{or} \quad \sup_x a(x) a''(x) < \infty.$$

Such inequality implies the well-known Feller's regularity condition for one-dimensional diffusion and admits the grows of the coefficient $a^2(x)$ as $c(1+|x|)^2 \ln(2+|x|)$.

Example 3.3. Let the operator H and the map $\Phi(\,\cdot\,)$ act as follows:

$$H\psi(x) = \frac{1}{2} \sum_{j=1}^{n} \left\{ b^j(x) \frac{\partial}{\partial x_j} \psi(x) + \frac{\partial}{\partial x_j} (b^j(x) \psi(x)) \right\},$$

$$\Phi(P)\psi(x) = \sum_{j,k,l=1}^{n} \frac{\partial}{\partial x_j} a_{jk}(x) P a_{kl}(x) \frac{\partial}{\partial x_l} \psi(x) +$$

$$+ \int_{\mathbf{R}^n} m(dq)\, c^*(x,q) D_q^* P D_q c(x,q) \psi(x).$$

If P is a multiplication operator by the function $P(x)$ from $C_0^\infty(\mathbf{R}^n)$, then the Lindblad infinitesimal operator

$$L(P) = \Phi(P) - \Phi(I) \circ P - i[H, P] \tag{3.6}$$

is also the multiplication operator by the function

$$L(P)(x) = \left\{ \frac{1}{2} (\nabla, A\nabla) + (b, \nabla) \right\} P(x) + \int m(dq)\, |c(x,q)|^2 (P(x+q) - P(x)),$$

where $\nabla = \text{grad}$ and $A = a^* a$.

In this case the Lindblad evolution equation coinsides with the Kolmogorov—Feller equation, and the conservativity condition gives the conservation of total probability for transition probability function.

In this example the operators $\Phi(I)$, $\Lambda = i[H, \Phi(I)]$ and $D = \Phi(\Phi(I)) - \Phi^2(I)$ are the symmetrical operators of the second order:

$$\Phi(I) = -(\nabla, A\nabla) + V + I, \qquad \Lambda = \frac{1}{2}(\hat{\Lambda} + \hat{\Lambda}^*), \qquad D = \frac{1}{2}(\hat{D} + \hat{D}^*),$$

$$\hat{\Lambda} = \Lambda_0 + (\Lambda_1, \nabla) + (\nabla, \Lambda_2 \nabla), \qquad \hat{D} = D_0 + (D_1, \nabla) + (\nabla; D_2 \nabla),$$

where Λ_0 and D_0 are the scalar functions, Λ_1 and D_1 ars vectors from $\mathbf{R}^n$ and Λ_2 and D_2 are matrices:

$$\Lambda_0 = \left(\frac{1}{2} (\nabla, A\nabla)(\nabla, b) + (b, \nabla) V \right) 1(x), \qquad V(x) = \int m(dq)\, |c(x,q)|^2,$$

$$\Lambda_1 = (\nabla A(\nabla, b) - (\nabla, A\nabla)b)1(x), \qquad \Lambda_2(x) = (2(A\nabla)\otimes b - (b,\nabla)A)1(x),$$

$$D_0 = \int m(dq)\,|c(x,q)|^2(V(x+q) - V(x)) - \mathrm{div}(A\,\mathrm{grad}|c(x,q)|^2),$$

$$D_1 = 2\int m(dq)\,|c(x,q)|^2 A(x+q)\,\mathrm{grad}\,(\arg c(x,q)),$$

$$D_2 = \Big((\nabla, A\nabla)a - (a,\nabla)\otimes(\nabla A)\Big)1(x) + \int m(dq)\,|c(x,q)|^2(A(x+q) - A(x)).$$

PROPOSITION 3.2. *Let there exist the one-parameter semigroup of the constractions* $W_t = \exp(-iH + \frac{1}{2}\Phi(I))t$ *in* $\mathcal{H} = L_2(\mathbf{R}^n)$. *The minimal dynamical semigroup with the infinitesimal generator* (3.6) *is conservative if the matrix* $A^{(x)}_{jl} = a^*_{jk}(x)a_{kl}(x)$ *is positive and there exists a positive constant* $\lambda > 0$ *such that*

$$D_2(x) + D_2^*(x) \leqslant \lambda A(x), \qquad D_0(x) + D_1(x)A^{-1}(x)D_1(x) \leqslant \lambda(V(x)+1),$$

$$\Lambda_2(x) + \Lambda_2^*(x) \leqslant \lambda A(x), \qquad \Lambda_0(x) + \Lambda_1(x)A^{-1}(x)\Lambda_1(x) \leqslant \lambda(V(x)+1).$$

The examples 3.1 and 3.2 follow from this result.

REFERENCES

Chebotarev, A.M. (1988). Sufficient conditions of regularity of Markov point processes. *Theory Probab. Appl.* **33**, 25—39.

Chebotarev, A.M. (1989). Sufficient conditions of conservativity of dynamic semigroups. *Theor. and Math. Phys.* **80**, 192—211.

Davies, E.B. (1977). Quantum dynamical semigroup and the neutron diffusion equation. *Reports Math. Phys.* **11**, 169—188.

Gikhman, I.I. and Skorokhod, A.V. (1977). *Introduction to the Theory of Random Processes.* Nauka, Moscow.

Gorini, V. *et al.* (1976). Completely positive dynamical semigroups of n-level systems. *J. Math. Phys.* **17**, 821—825.

Has'minskii, R.Z. (1960). Ergodic properties of the recurrent diffusion processes. *Theory Probab. Appl.* **5**, 196—214.

Kato, T. (1966). *Perturbation Theory for Linear Operators.* Springer, Berlin—Heidelberg etc.

Lindblad, G. (1976). On the generators of quantum dynamical semigroup. *Commun. Math. Phys.* **48**, 119—130.

Prob. Theory and Math. Stat., Vol. 1, pp. 228–237
B.Grigelionis *et al.* (Eds.)
1990 VSP/Mokslas

AN APPROXIMATION OF INTEGER–VALUED MEASURES BY GENERALIZED POISSON MEASURES

V. ČEKANAVIČIUS
Department of Mathematics, Vilnius University,
Naugarduko 24, 232006 Vilnius, Lithuania

ABSTRACT

Let F be an integer–valued distribution with non–negative characteristic function $\hat{F}(t)$, and let F^n be n–fold convolution of F. We approximate F^n by infinite divisible signed measure D with Fourier—Stieltjes transform

$$\hat{D}(t) = \exp\Big\{\sum_{i=1}^{l}(-1)^{i+1}(\hat{F}(t)-1)^i/i\Big\}.$$

Let $\mathcal{F}$ be the set of all one dimensional probability measures, $\mathcal{F}_+ \subset \mathcal{F}$ be the set of distributions with non–negative for all t characteristic function, $\mathcal{F}_\mathbf{Z} \subset \mathcal{F}$ be the set of integer–valued distributions, $\mathbf{D}$ be the set of infinitely divisible distributions. Let E_a be the distribution concentrated at a point $a, E \equiv E_0$. The notation $C(m)$ will be used for positive constants depending only on the argument m.
Let W be a measure of finite variation, $\|W\|$ be it's total variation,

$$|W| = \sup_x |W\{(-\infty, x)\}|, \quad \widehat{W}(t) = \int_{-\infty}^{\infty} e^{itx} W\{dx\},$$

$$\exp\{W\} = \sum_{j=0}^{\infty} W^j/j!.$$

(In this paper all products and powers of measures will be understood in the convolution sense, $W^0 = E, W^2 = W * W$.)
For $F \in \mathcal{F}$ let σ^2 be the dispersion of F,

$$\beta_k = \int_{-\infty}^{\infty} |x|^k F\{dx\}, \quad k = 1, 2, \ldots$$

In early fifties A.N. Kolmogorov proposed to approximate n–fold convolution F^n by the class $\mathbf{D}$. Various aspects of this problem were considered by

A.N. Kolmogorov, Yu.V. Prohorov, L. Le Cam, L.D. Meshalkin, I.A. Ibragimov and E.L. Presman, A.Yu. Zaitsev, T.V. Arak and others (complete history of the problem is given in (Arak and Zaitsev, 1986)).
The subset of $\mathbf{D}$ given by

$$\mathcal{A} = \{E_a \exp\{\lambda(W - E)\} \quad | a \in \mathbf{R}, \lambda \geqslant 0, W \in \mathcal{F}\} \tag{1}$$

plays an important role in the Kolmogorov's problem.
It was proved that if F is suitabily centered then the accompanying distribution $\exp\{n(F-E)\}$ approximates F^n with accuracy of order $C_1 n^{-1/3}$ (the upper bound was proved by L. Le Cam and the lower one by I. Ibragimov and E. Presman). The definitive solution of Kolmogorov's problem was given by T. V. Arak and required more complicated distributions. This result may be written in the following form

$$C_2 n^{-2/3} \leqslant \sup_{F \in \mathcal{F}} \inf_{D \in \mathbf{D}} |F^n - D| \leqslant \sup_{F \in \mathcal{F}} \inf_{A \in \mathcal{A}} |F^n - A| \leqslant C_3 n^{-2/3} \tag{2}$$

(see (Arak and Zaitsev, 1986)).
Let us consider some possible ways to improve (2).
1. *$\mathcal{F}$ subsets*
Approximation of F^n when F belongs to some subset of $\mathcal{F}$. For example T. V. Arak proved that

$$\sup_{F \in \mathcal{F}_+} |F^n - \exp\{n(F - E)\}| \leqslant C_4 n^{-1} \tag{3}$$

(Arak and Zaitsev, 1986, p. 116).
2. *Other metrics*
For example E. L. Presman approximated binomial distribution in the metric $\|\cdot\|$ (Arak and Zaitsev, 1986).
3. *Asymptotic expansions*
The form of any concrete asymptotic expansion depends on the quantity of available information. For example from general results proved in (Čekanavičius, 1989) directly folows that

$$\sup_{F \in \mathcal{F}_+} |F^n - \exp\{n(F - E)\}(E - n(F - E)^2/2)| \leqslant C_5 n^{-2}. \tag{4}$$

4. *Extension of the set of approximating measures*
We will consider infinitely divisible signed measures. By a generalised Poisson measure (P–measure) we mean the signed measure $\exp\{\lambda(W - E)\}$, where $\lambda \in \mathbf{R}$, $W \in \mathcal{F}$ and exponential of measure is understood in the sense as defined above. Let $\mathcal{V}$ be a set of all possible finite convolutions of P-measures, i. e.

$$\mathcal{V} = \{E_a \exp\{\lambda_1(W_1 - E) + \lambda_2(W_2 - E)\} | a \in \mathbf{R}, \lambda_1 \geqslant 0, \lambda_2 \leqslant 0, W_1, W_2 \in \mathcal{F}\}. \tag{5}$$

It is obvious that $\mathcal{V} \supset \mathcal{A}$ and $\mathcal{V} \setminus \mathcal{F} \neq \varnothing$. For some subsets of $\mathcal{F}$ it is possible to approximate F^n by the class $\mathcal{V}$ quite sharp, but W_1 and W_2 are more complicated then F^n. So it is natural to consider only W_1, W_2, which are comparable with F^k, where k is a constant.

Let

$$\mathcal{G} = \exp\{(F - E) - (F - E)^2/2\}. \tag{6}$$

Obviously $\mathcal{G} \in \mathcal{V}$. E.L.Presman used $\mathcal{G}$ to approximate binomial and symmetrized binomial distributions (see e. g.(Arak and Zaitsev, 1986, p. 92)). Similar results were proved in (Kruopis, 1986) for integer-valued distributions under assumptions on factorial moments (it must be mentioned that in (Kruopis, 1986) the dispersion of F was smaller then 4 as a rule). Some other results connected with the use of $\mathcal{G}$ may be found in (Čekanavičius, 1988b).

In this paper we approximate an integer valued distribution with non-negative characteristic function F^n by P-measures under assumptions which differ from those implied on factorial moments in (Kruopis, 1986; Šiaulys and Čekanavičius, 1988). Note that for $m = 0$, $r = 1$, $l = 2$, $F = (p^2 + (1-p)^2)E + p(1-p)E_1 + p(1-p)E_{-1}$, the result of E. L. Presman (Arak and Zaitsev, 1986, p. 92) follows directly from (9).

For given integer-valued measure W we define the quantities:

$$|W|_r = \left(\sum_{l=-\infty}^{\infty} |W\{l\}|^r\right)^{1/r}, \quad \text{for} \quad r \geqslant 1,$$

$$|W|_\infty = \sup_l |W\{l\}|.$$

Note that

$$|W|_1 = \|W\|.$$

At first we consider Bergström type (see (Bergström, 1951)) asymptotic expansion. Let

$$B_m(H) = \sum_{j=0}^{m} \binom{n}{j} H^{n-j}(F - H)^j, \tag{7}$$

$$H = \exp\left\{\sum_{i=1}^{l}(-1)^{i+1}(F - E)^i/i\right\}.$$

THEOREM 1. *Let $F \in \mathcal{F}_+ \cap \mathcal{F}_{\mathbf{Z}}$, $n, m, l \in \mathbf{N}$, $0 \leqslant m < n$, $p_0 = F\{0\}$, $q_0 = 1 - p_0$, then*

1) *for $r \geqslant 2$*

$$|F^n - B_m(H)|_r \leqslant$$

$$\leqslant C_6(l, m, r) \min\left\{{q_0}^{(1-r)/2r} n^{-l(m+1)+(1-r)/2r},\ n^{m+1}\sigma^{2(l+1)(m+1)}\right\}, \tag{8}$$

2) *for $1 \leqslant r < 2$*

$$|F^n - B_m(H)|_r \leqslant C_7(l, m, r) \times \tag{9}$$

$$\times \min\left\{{q_0}^{-1/4}(1 + \sigma\sqrt{n})^{(2-r)/2r} n^{-l(m+1)-1/4},\ n^{m+1}\sigma^{2(l+1)(m+1)}(1 + n\sigma^2)\right\},$$

3)

$$|F^n - B_m(H)|_\infty \leqslant$$

$$\leqslant C_8(l,m)\min\left\{q_0{}^{-1/2}n^{-l(m+1)-1/2},\ n^{m+1}\sigma^{2(l+1)(m+1)}\right\}. \tag{10}$$

Here $C_6(l,m,r), C_7(l,m,r)$ depends on r as expresed below:

$$C_6(l,m,r) = C_9(l,m)C_{10}^{(r-1)/r},$$

$$C_7(l,m,r) = C_{11}(l,m)\left(1+\pi\sqrt{r/(2-r)}\right)^{(2-r)/(2r)}.$$

Taking $m=0$, $l=1,2$, $r=1$ we get following corollary.

COROLLARY 1. *For $F\in\mathcal{F}_+\cap\mathcal{F}_{\mathbf{Z}}, n\in\mathbf{N}$*

$$\|F^n - \exp\{n(F-E)\}\| \leqslant C_{12}q_0{}^{-1/4}n^{-1-1/4}(1+\sigma\sqrt{n})^{1/2},$$

$$\|F^n - \exp\{n(F-E) - n(F-E)^2/2\}\| \leqslant C_{13}q_0{}^{-1/4}n^{-2}(n^{-1/2}+\sigma)^{1/2}. \tag{11}$$

For the uniform metric the estimate (11) can be improved.

THEOREM 2. *For $F\in\mathcal{F}_+\cap\mathcal{F}_{\mathbf{Z}},\ n,l\in\mathbf{N}$*

$$|F^n - \exp\{n(F-E) - n(F-E)^2/2\}| \leqslant C_{14}(l)(1+\beta_1 n^{-l})n^{-2}. \tag{12}$$

Now we consider some lower estimations. $B_m(H)$ and H are from (7).

THEOREM 3. *Let $F\in\mathcal{F}_+\cap\mathcal{F}_{\mathbf{Z}},\ \ n,m\in\mathbf{N},\ \ 0\leqslant m<n$, and let $\beta_3/(\sigma^3\sqrt{n})\leqslant$ $\leqslant C_{15}$, for all n. Then*

1) for $r\geqslant 1$

$$|F^n - B_m(H)|_r \geqslant C_{16}(l,m)n^{-l(m+1)}\left(1+C_{17}(l,m)\sigma\sqrt{n}\sqrt{(r-1)/r}\right)^{(1-r)/r},$$

2) $|F^n - B_m(H)|_\infty \geqslant C_{18}(l,m)n^{-l(m+1)-1/2}\sigma^{-1}$,
3) $|F^n - B_m(H)| \geqslant C_{19}(l,m)n^{-l(m+1)}$.

COROLLARY 2. *Under assumptions of Theorem 3*

$$|F^n - \exp\{n(F-E)\}| \geqslant C_{20}n^{-1},$$

$$|F^n - \exp\{n(F-E) - n(F-E)^2/2\}| \geqslant C_{21}n^{-2}.$$

Now we will prove our propositions. We need some auxilary results.

LEMMA 1. *Let $F \in \mathcal{F}_+ \cap \mathcal{F}_{\mathbf{Z}}$, $a > 0$. Then*

$$\int_{-\pi}^{\pi} \exp\{a(\widehat{F}(t) - 1)\}dt \leqslant C_{22} {q_0}^{-1/2} a^{-1/2}. \tag{13}$$

Proof. From inequalities for Levy's concentration function Q (see e.g., (Arak and Zaitsev, 1986, p.48–51)) we have

$$\int_{-\pi}^{\pi} \exp\{a(\widehat{F}(t) - 1)\}dt \leqslant C_{23} Q(\exp\{a(F - E)\}, 1/\pi) \leqslant$$

$$\leqslant C_{22}\{aF\{x : |x| > 1/\pi\}\}^{-1/2} = C_{22} {q_0}^{-1/2} a^{-1/2}.$$

LEMMA 2. *Let $F \in \mathcal{F}_+$, $m \in \mathbf{N}$, $a > 0$. Then*

$$|(F - E)^m \exp\{a(F - E)\}| \leqslant C_{24}(m) a^{-m}. \tag{14}$$

Proof. (14) follows directly from (7) in (Čekanavičius, 1989).

LEMMA 3 (Šiaulys and Čekanavičius, 1988). *Let W be an integer-valued measure of finite variation, $W\{\mathbf{R}\} = 0$. Then*

1) *for $r \geqslant 2$*

$$|W|_r \leqslant \left\{ \frac{1}{2\pi} \int_{-\pi}^{\pi} |\widehat{W}(t)|^{r/(r-1)} dt \right\}^{(r-1)/r},$$

2) $|W|_\infty \leqslant \frac{1}{2\pi} \int_{-\pi}^{\pi} |\widehat{W}(t)|dt$,

3) $|W| \leqslant \frac{1}{4\pi} \int_{-\pi}^{\pi} |\widehat{W}(t)|/|\sin \frac{t}{2}|dt$,

4) *if also*

$$\sum_{k=-\infty}^{\infty} |k| \, |W\{k\}| < \infty,$$

then for $a \in \mathbf{R}$, $b > 0$, $1 \leqslant r < 2$,

$$|W|_r \leqslant \left\{ 1 + b\pi \sqrt{\frac{r}{2 - r}} \right\}^{(2-r)/(2r)} \times$$

$$\times\left\{\frac{1}{2\pi}\int_{-\pi}^{\pi}(|\widehat{W}(t)|^2+\frac{2-r}{rb^2}|(e^{-ita}\widehat{W}(t))'_t|^2)dt\right\}^{1/2}.$$

LEMMA 4. *Let W be an integer-valued measure of finite variation,*

$$W\{\mathbf{R}\}=0,\ h\geqslant 1,\ T(h)=\int_{-\infty}^{\infty}e^{-t^2/2}\widehat{W}(t/h)dt.$$

Then

1) *for* $r\geqslant 1$

$$|W|_r\geqslant C_{25}|T(h)|\left(1+h\sqrt{\frac{2\pi(r-1)}{r}}\right)^{(1-r)/r}, \tag{15}$$

2)

$$|W|_\infty\geqslant C_{26}|T(h)|\left(1+h\sqrt{2\pi}\right)^{-1}, \tag{16}$$

3)

$$|W|\geqslant C_{27}|T(h)|. \tag{17}$$

Proof. (15)—(17) (and even more general results) were proved in (Šiaulys and Čekanavičius, 1988).

Proof of Theorem 1. Let

$$\beta(t)=\exp\{\widehat{F}(t)-1\}.$$

Using standard methods it is easy to prove that

$$|\widehat{F}(t)-\widehat{H}(t)|\leqslant C_{28}(l)|\widehat{F}(t)-1|^{l+1}, \tag{18}$$

$$\max(\widehat{F}(t),|\widehat{H}(t)|)\leqslant \beta(t). \tag{19}$$

We have

$$\widehat{H}'(t)=\widehat{H}(t)\left(\sum_{j=1}^{l}(-1)^{j+1}(\widehat{F}(t)-1)^j/j\right)'_t=$$

$$=\widehat{F}'(t)\sum_{j=0}^{l-1}(1-\widehat{F}(t))^j\left(1-\left(1-\widehat{F}(t)\right)+\theta_{l1}|\widehat{F}(t)-1|^{l+1}\right)=$$

$$= \left(1-(1-\widehat{F}(t))\right)\widehat{F}'(t)\sum_{j=0}^{l-1}\left(1-\widehat{F}(t)\right)^j + \theta_{l2}|\widehat{F}'(t)||\widehat{F}(t)-1|^{l+1} =$$

$$= \widehat{F}'(t)\left(1-\left(1-\widehat{F}(t)\right)^l\right) + \theta_{l2}|\widehat{F}'(t)||\widehat{F}(t)-1|^{l+1} =$$

$$= \widehat{F}'(t) + \theta_{l3}|\widehat{F}'(t)||\widehat{F}(t)-1|^l. \tag{20}$$

Here

$$|\theta_{li}| \leqslant C_{29}(l), \quad i = 1,2,3.$$

From (20) it follows that

$$|(\widehat{F}(t)-\widehat{H}(t))'_t| \leqslant C_{30}(l)|\widehat{F}(t)-1|^l, \tag{21}$$

$$|\widehat{H}'(t)| \leqslant C_{31}(l)|\widehat{F}'(t)|. \tag{22}$$

Using Bergström identity (Bergström, 1951) and (18), (19) we have

$$|\widehat{F}^n(t) - B_m(\widehat{H}(t))| =$$

$$= \left|\sum_{j=m+1}^{n}\binom{j-1}{m}\widehat{F}^{n-j}(t)\left(\widehat{F}(t)-\widehat{H}(t)\right)^{m+1}\widehat{H}^{j-m-1}(t)\right| \leqslant$$

$$\leqslant \sum_{j=m+1}^{n}\binom{j-1}{m} C_{32}(l,m)\beta^n(t)|\widehat{F}(t)-1|^{(l+1)(m+1)} \leqslant$$

$$\leqslant C_{33}(l,m)\binom{n}{m+1}\beta^{n/2}(t)\min\left\{n^{-(l+1)(m+1)},(\sigma t)^{2(l+1)(m+1)}\right\}. \tag{23}$$

Analoguously from (18), (21)—(22) it follows

$$\left|\left(\widehat{F}^n(t) - B_m(\widehat{H}(t))\right)'_t\right| \leqslant$$

$$\leqslant \sum_{j=m+1}^{n}\binom{j-1}{m}\left\{\left|\left(\widehat{F}^{n-j}(t)\widehat{H}^{j-m-1}(t)\right)'\right||\widehat{F}(t)-\widehat{H}(t)|^{m+1}+\right.$$

$$\left.+|\widehat{F}^{n-j}(t)\widehat{H}^{j-m-1}(t)|(m+1)|\widehat{F}(t)-\widehat{H}(t)|^m|\widehat{F}'(t)-\widehat{H}'(t)|\right\} \leqslant$$

$$\leqslant C_{34}(l,m)\binom{n}{m+1}|\widehat{F}'(t)|\beta^n(t)\left\{n|\widehat{F}(t)-1|^{(l+1)(m+1)}+\right.$$

$$\left.+|\widehat{F}(t)-1|^{m(l+1)+l}\right\} \leqslant C_{35}(l,m)|\widehat{F}'(t)|\beta^{2n/3}(t)\times$$

$$\times\min\left\{n^{-l(m+1)-m},\ (\sigma t)^{2(m(l+1)+l)}(1+n\sigma^2t^2)\right\}n^{m+1}. \tag{24}$$

We have $F \in \mathcal{F}_+ \cap \mathcal{F}_{\mathbf{Z}}$. Hence

$$\widehat{F}(t) = \sum_j p_j \cos tj, \ \widehat{F}(t) - 1 = -2\sum_j p_j \sin^2 \frac{tj}{2}, \tag{25}$$

$$|\widehat{F}'(t)| = \Big|\sum_j jp_j \sin tj\Big| \leqslant 2\Big(\sum_j j^2 p_j\Big)^{1/2}\Big(\sum_k p_k \sin^2\frac{tk}{2}\Big)^{1/2} \leqslant$$

$$\leqslant 2\sigma\left(\sum_j p_j \sin^2 \frac{tj}{2}\right)^{1/2}, \tag{26}$$

where $p_j = F\{j\}$.
From (25)—(26) it follows that

$$|\widehat{F}'(t)\beta^{n/6}(t)| \leqslant C_{36} \min\{\sigma/\sqrt{n},\ (t\sigma)^2\}. \tag{27}$$

From (24), (27) it follows that

$$\Big|\Big\{\widehat{F}^n(t) - B_m(\widehat{H}(t))\Big\}'_t\Big| \leqslant$$

$$\leqslant C_{37}(l,m)\min\Big\{\sigma\beta^{n/2}(t)n^{-l(m+1)+1/2},\ n^{m+1}(\sigma t)^{2(l+1)(m+1)}(1+n\sigma^2t^2)\Big\}. \tag{28}$$

Estimates (23), (28), (13) and Lemma 3 imply (8)—(10). Note that for $1 \leqslant r < 2$ in Lemma 3 we take $a = 0$, $b = \max(1, \sqrt{n}\sigma)$.

Proof of Theorem 2. We have

$$\Big|F^n - \exp\Big\{n(F-E) - n(F-E)^2/2\Big\}\Big| \leqslant$$

$$\leqslant \Big|F^n - \exp\Big\{n(F-E)\Big\}\Big(E - n(F-E)^2/2\Big)\Big| +$$

$$+\Big|\exp\Big\{n(F-E)\Big\}\Big(\exp\{-n(F-E)^2/2\} - -\sum_{j=0}^{l+2}(-n/2)^j(F-E)^{2j}/j!\Big)\Big| +$$

$$+\sum_{j=2}^{l+2}(n/2)^j\Big|\exp\{n(F-E)\}(F-E)^{2j}\Big|/j! = I_1 + I_2 + I_3. \tag{29}$$

From (4) and (14) we have

$$I_1 + I_3 \leqslant C_{38}(l)n^{-2}. \tag{30}$$

From Lemma 3 it follows

$$I_2 \leqslant C_{39}(l)\int_{-\pi}^{\pi} \beta^n(t)n^{l+3}|\widehat{F}(t) - 1|^{2(l+3)}/|t|dt \leqslant$$

$$\leqslant C_{40}(l)n^{-(l+2)}\int_{-\pi}^{\pi}|\widehat{F}(t)-1|/|t|dt \leqslant C_{41}(l)n^{-(l+2)}\beta_1. \tag{31}$$

Now follows (12) from (29)—(31).

Proof of Theorem 3. We use the same idies as in (Čekanavičius, 1988a; Šiaulys and Čekanavičius, 1988) only our estimations are more accurate.
From H definition it follows

$$\widehat{H}(t)-\widehat{F}(t)=-K(\widehat{F}(t)-1)^{l+1}+C_{42}(l)\theta|\widehat{F}(t)-1|^{l+2}, \tag{32}$$

where

$$K=\sum\left\{(-1)^l\prod_{i=1}^{l}\left(-\frac{1}{i}\right)^{k_i}\frac{1}{k_i!}\,\middle|\,\sum_{i=1}^{l}ik_i=l+1\right\},\ |\theta|\leqslant 1.$$

We have

$$\widehat{F}^n(t)-B_m(\widehat{H}(t))=\binom{n}{m+1}K^{m+1}\Big(-\sigma^2t^2/2\Big)^{(l+1)(m+1)}+$$

$$+\binom{n}{m+1}K^{m+1}\Big(-\sigma^2t^2/2\Big)^{(l+1)(m+1)}\Big(\widehat{F}^{n-(m+1)}(t)-1\Big)+$$

$$+\binom{n}{m+1}\widehat{F}^{n-(m+1)}(t)K^{m+1}\Big\{(\widehat{F}(t)-1)^{(l+1)(m+1)}-(-\sigma^2t^2/2)^{(l+1)(m+1)}\Big\}+$$

$$+\binom{n}{m+1}\widehat{F}^{n-(m+1)}(t)\Big\{\Big(\widehat{F}(t)-\widehat{H}(t)\Big)^{m+1}-\Big(K(\widehat{F}(t)-1)^{l+1}\Big)^{m+1}\Big\}+$$

$$+\binom{n}{m+1}\Big(\widehat{H}^{n-(m+1)}(t)-\widehat{F}^{n-(m+1)}(t)\Big)\Big(\widehat{F}(t)-\widehat{H}(t)\Big)^{m+1}+$$

$$+\Big(\widehat{F}^n(t)-B_{m+1}(\widehat{H}(t))\Big)=A(t)+\sum_{i=1}^{5}R_i(t).$$

Thus

$$|T(h)|=\left|\int_{-\infty}^{\infty}\Big(\widehat{F}^n(t/h)-B_m(\widehat{H}(t/h))\Big)e^{-t^2/2}dt\right|\geqslant$$

$$\geqslant\left|\int_{-\infty}^{\infty}A(t/h)e^{-t^2/2}dt\right|-\sum_{i=1}^{5}\int_{-\infty}^{\infty}|R_i(t/h)|e^{-t^2/2}dt. \tag{33}$$

Using (23), (18), (32) and expression

$$\widehat{F}(t)=-\sigma^2t^2/2+\hat{\theta}\beta_3|t|^3/6,\ |\hat{\theta}|\leqslant 1,$$

after some calculations from (33) it follows that

$$|T(h)|\geqslant\binom{n}{m+1}\Big\{C_{42}(l,m)(\sigma^2/h^2)^{(m+1)(l+1)}\Big(1-C_{43}(l,m)n\sigma^2/h^2\Big)-$$

$$-C_{44}(l,m)\sum_{j=1}^{(l+1)(m+1)}(\sigma^2/h^2)^{(m+1)(l+1)-j}(\beta_3/h^3)^j-$$

$$-C_{45}(l,m)(\sigma^2/h^2)^{(m+1)(l+1)+1}-C_{46}(l,m)n(\sigma^2/h^2)^{(l+1)(m+2)}\Big\}. \qquad (34)$$

We have

$$\beta_3 < C_{15}\sigma^3\sqrt{n},\ \beta_3=\sum_k |k|^3p_k \geqslant \sum_k k^2p_k=\sigma^2,$$

so

$$\sigma\sqrt{n}\geqslant 1/C_{15}. \qquad (35)$$

Now let

$$h=\alpha\sigma\sqrt{n},$$

where

$$\alpha\geqslant \max(1,C_{15}).$$

From (34), (35) and the inequality

$$\binom{n}{m+1}\geqslant C_{47}(m)n^m,\ n\geqslant m+1,$$

it follows that

$$\big|T(h)\big|\geqslant C_{48}(l,m)n^{-l(m+1)}\alpha^{-2(m+1)(l+1)}\Big\{\Big(1-C_{49}(l,m)\alpha^{-2}\Big)-$$

$$-C_{50}(l,m)\alpha^{-1}-C_{51}(l,m)(n\alpha^2)^{-1}-C_{52}(l,m)\Big(n^l\alpha^{2(l+1)}\Big)^{-1}\Big\}. \qquad (36)$$

It is obvious that for $\alpha\geqslant C_{53}(l,m)$, where $C_{53}(l,m)$ is large enough we have

$$\big|T(h)\big|\geqslant C_{54}(l,m)n^{-l(m+1)}.$$

An application of Lemma 4 completes the proof.

REFERENCES

Arak, T.V. and Zaitsev, A.Yu. (1986). Uniform limit theorems for sums of independent random variables. *Proc. Steklov Inst. Math.* **174**, Nauka, Leningrad

Bergström, H. (1951). On asymptotic expansions of probability functions. *Skandinavisk Aktuar.* **1–2**, 1–34.

Čekanavičius, V. (1988a). On lower bounds in Bergström identity. *Liet. Mat. Rink.* **28**, 153–169.

Čekanavičius, V. (1988b). On approximation by generalized measures of Poisson type. I. *Liet. Mat. Rink.* **28**, 582–588.

Čekanavičius, V. (1989). Approximation by accompanying distributions and asymptotic expansions. I. *Liet. Mat. Rink.* **29**, 171–178.

Kruopis, Yu. (1986). Approximation of distributions of sums of lattice random variables. I. *Liet. Mat. Rink.* **26**, 455–467.

Šiaulys, Y. and Čekanavičius, V. (1988). Approximation of distributions of integer-valued additive functions by discrete charges. I. *Liet. Mat. Rink.* **28**, 795–810.

Prob. Theory and Math. Stat., Vol. 1, pp. 238–244
B.Grigelionis *et al.* (Eds.)
1990 VSP/Mokslas

LIM INF RESULTS FOR THE STANDARDIZED EMPIRICAL DISTRIBUTION FUNCTION

E. CSÁKI*
Math. Inst. Hung. Acad. Sci. Budapest,
Reáltanoda u. 13–15 P.O.B. 127, H–1364

ABSTRACT

We give certain lim inf results for the supremum of the standardized empirical distribution function over sub-intervals. First we prove analogue results for the Kiefer process and then use strong approximation to derive the required results for the empirical process.

1. INTRODUCTION

Let $X_1, X_2, \ldots$ be a sequence of i.i.d. random variables, each uniformly distributed on the interval $(0,1)$. The empirical distribution function $F_n(x)$ is defined by

$$F_n(x) = \frac{1}{n} \sum_{i=1}^{n} I(X_i \leqslant x), \tag{1.1}$$

where $I(\cdot)$ denotes the indicator function. Let $\{\delta_n\}_n n = 1, 2, \ldots$ be a decreasing sequence of numbers such that $0 \leqslant \delta_n < 1/2$ and for $0 \leqslant \delta < 1/2$ put

$$T_n^+(\delta) = \sup_{\delta < x < 1-\delta} \frac{F_n(x) - x}{(x(1-x))^{1/2}} \tag{1.2}$$

and

$$T_n(\delta) = \sup_{\delta < x < 1-\delta} \frac{|F_n(x) - x|}{(x(1-x))^{1/2}}. \tag{1.3}$$

In (Csáki, 1977) we investigated the lim sup behaviour of $T_n^+(\delta_n)$ and $T_n(\delta_n)$ under various conditions on δ_n. In the present paper we derive certain lim inf results for the.

It was shown in (Csáki, 1981, 1982) that

$$\liminf_{n \to \infty} \frac{n^{1/2} T_n(\delta_n)}{(2 \log \log n)^{1/2}} = 1 \qquad \text{a.s.} \tag{1.4}$$

provided δ_n satisfies

$$\lim_{n \to \infty} \frac{\log \log(1/\delta_n)}{\log \log \log n} = \infty. \tag{1.5}$$

---

* Research supported by Hungarian National Foundation for Scientific Research Grant No. 1808

The assertion (1.4) is in contrast with the result of (Mogul'skii, 1979) and (Kuelbs, 1979) who showed that

$$\liminf_{n\to\infty}(n\log\log n)^{1/2}\sup_{0\leqslant x\leqslant 1}|F_n(x)-x|=\pi/\sqrt{8}\qquad \text{a.s.} \tag{1.6}$$

The limiting distributions of T_n^+, T_n were derived in (Jaeschke, 1979). For these and related results we may also refer to (Csörgő and Révész, 1981) as well as to (Shorack and Wellner, 1986).
Our aim is to prove the following result.

THEOREM 1. *Assume that $\delta_n, n = 1,2,\ldots,$ is a non-increasing sequence of numbers such that $0 \leqslant \delta_n < 1/2$ and*

$$\lim_{n\to\infty}\frac{\log(1/\delta_n)}{\log\log n}=\infty. \tag{1.7}$$

Then

$$\liminf_{n\to\infty}\frac{n^{1/2}T_n^+(\delta_n)}{(2\log\rho_n')^{1/2}}=\liminf_{n\to\infty}\frac{n^{1/2}T_n(\delta_n)}{(2\log\rho_n')^{1/2}}=1\qquad a.s., \tag{1.8}$$

where

$$\rho_n'=\min(\rho_n,\log n), \tag{1.9}$$

$$\rho_n=\frac{\log(1/\delta_n)}{\log\log n}. \tag{1.10}$$

Here $\delta_n = 0$ is possible. In this case (1.7) is satisfied, $\rho_n = \infty$, $\rho_n' = \log n$ and we obtain (1.4).
The proof of Theorem 1 will be based on the invariance principle (Komlós et al., 1975) and on analogous results for the Kiefer process.
The Kiefer process $K(x, y)$ $(0 \leqslant x \leqslant 1,\ 0 \leqslant y)$ is a two-parameter Gaussian process with

$$\mathbf{E}K(x,y)=0$$

and

$$\mathbf{E}K(x_1,y_1)K(x_2,y_2)=y_1\wedge y_2(x_1\wedge x_2-x_1x_2).$$

The corresponding result for the Kiefer process reads as follows.

THEOREM 2. *Let δ_n be a non-increasing sequence of numbers such that $0 < \delta_n < 1/2$ and (1.7) holds true. Put*

$$K_n^+(\delta)=\sup_{\delta<x<1-\delta}\frac{K(x,n)}{(x(1-x))^{1/2}}, \tag{1.11}$$

$$K_n(\delta)=\sup_{\delta<x<1-\delta}\frac{|K(x,n)|}{(x(1-x))^{1/2}}. \tag{1.12}$$

$$K_n(\delta) = \sup_{\delta<x<1-\delta} \frac{|K(x,n)|}{(x(1-x))^{1/2}}. \tag{1.12}$$

Then

$$\liminf_{n\to\infty} \frac{K_n^+(\delta_n)}{(2n\log\rho_n)^{1/2}} = \liminf_{n\to\infty} \frac{K_n(\delta_n)}{(2n\log\rho_n)^{1/2}} = 1 \qquad a.s., \tag{1.13}$$

where ρ_n *is defined by* (1.10).

The difference between the formulation of Theorems 1 and 2 is that $\delta_n = 0$ is meaningless in the latter case and we have ρ_n in (1.13) instead of ρ_n'. First we prove Theorem 2 and then show how Theorem 1 follows from Theorem 2.

2. PROOF OF THEOREM 2

We prove the result only for $K_n(\delta_n)$, the proof for $K_n^+(\delta_n)$ is similar.
It is easy to see that

$$\mathbf{P}\Big(K_n(\delta) \leqslant zn^{1/2}\Big) = \mathbf{P}\Big(\sup_{0<s<\log(1-\delta)/\delta} |U(s)| \leqslant z\Big), \ z \geqslant 0, \tag{2.1}$$

where $U(s)$ is the Ornstein—Uhlenbeck process, i.e., the Gaussian process with

$$\mathbf{E}U(s) = 0, \quad s \in \mathbf{R},$$

$$\mathbf{E}U(s)U(t) = e^{-|t-s|}, \ s,t \in \mathbf{R}.$$

It follows from extreme value theory (cf. e.g., (Newell, 1962)) that the asymptotic relation

$$\mathbf{P}\Big(\sup_{0\leqslant s\leqslant t} |U(s)| \leqslant z\Big) \sim \exp\Big\{-\Big(\frac{2}{\pi}\Big)^{1/2} tze^{-z^2/2}\Big\}, z \to \infty, t \to \infty, \tag{2.2}$$

holds true.
In order to show that

$$\liminf_{n\to\infty} \frac{K_n(\delta_n)}{(2n\log\rho_n)^{1/2}} \geqslant 1 \qquad \text{a.s.} \tag{2.3}$$

define

$$n_k = \Big[e^{k/(\log k)^2}\Big] \tag{2.4}$$

and

$$A_k = \Big\{\min_{n_k\leqslant n<n_{k+1}} K_n(\delta_{n_k}) \leqslant (1-2\varepsilon)(2n_{k+1}\log\rho_{n_k})^{1/2}\Big\}. \tag{2.5}$$

We apply an inequality of Mogul'skii (1979):

$$\mathbf{P}\Big(\min_{m\leqslant k<n} \|S_k\| \leqslant \lambda_1\Big) \leqslant \frac{\mathbf{P}\Big(\|S_n\| \leqslant \lambda_1+\lambda_2\Big)}{\min_{m\leqslant k<n} \mathbf{P}(\|S_n - S_k\| \leqslant \lambda_2)}, \tag{2.6}$$

Since $K(x,n) = B_1(x) + \ldots + B_n(x)$, where $B_i(x), i = 1,\ldots,n$, are independent Brownian bridges, we can apply (2.6) with $m = n_k, n = n_{k+1}, \|S_n\| = K_n(\delta_{n_k}), \lambda_1 = (1-2\varepsilon)(2n_{k+1}\log\rho_{n_k})^{1/2}, \lambda_2 = \varepsilon(2n_{k+1}\log\rho_{n_k})^{1/2}$ and get from (2.2) and (2.6) for large enough k

$$\mathbf{P}(A_k) \leqslant C\mathbf{P}\Big(K_{n_{k+1}}(\delta_{n_k}) \leqslant (1-\varepsilon)n_{k+1}^{1/2}(2\log\rho_{n_k})^{1/2}\Big) \leqslant$$

$$\leqslant Ce^{-2\log\log n_k} \leqslant C(\log k)^4/k^2,$$

which by Borel—Cantelli lemma yields (2.3).
Next we show that

$$\liminf_{n\to\infty} \frac{K_n(\delta_n)}{(2n\log\rho_n)^{1/2}} \leqslant 1 \qquad \text{a.s.} \tag{2.7}$$

Put $n_k = \left[e^{k\log k}\right]$ and

$$B_k = \left\{\sup_{\delta_{n_k}<x<1-\delta_{n_k}} \frac{|K(x,n_k) - K(x,n_{k-1})|}{(x(1-x))^{1/2}} \leqslant (1+\varepsilon)(2n_k\log\rho_{n_k})^{1/2}\right\}.$$

Then

$$\mathbf{P}(B_k) = \mathbf{P}\left(K_{n_k} - n_{k-1}(\delta_{n_k}) \leqslant (1+\varepsilon)(2n_k\log\rho_{n_k})^{1/2}\right)$$

and from (2.1) and (2.2) one easily gets $\sum_k \mathbf{P}(B_k) = \infty$. Hence since B_k, $k = 1,2,\ldots$, are independent, we have

$$\mathbf{P}(B_k \text{ i.o.}) = 1. \tag{2.8}$$

But

$$K_{n_k}(\delta_{n_k}) \leqslant K_{n_{k-1}}(\delta_{n_k}) + \sup_{\delta_{n_k}<x<1-\delta_{n_k}} \frac{|K(x,n_k) - K(x,n_{k-1})|}{(x(1-x))^{1/2}}, \tag{2.9}$$

hence

$$B_k \subseteq \left\{\frac{K_{n_k}(\delta_{n_k})}{(2n_k\log\rho_{n_k})^{1/2}} \leqslant 1 + \varepsilon + \frac{K_{n_{k-1}}(\delta_{n_k})}{(2n_k\log\rho_{n_k})^{1/2}}\right\}.$$

In (Csörgő and Révész, 1981) it is shown that

$$\limsup_{n\to\infty} \sup_{0<x<1} \frac{|K(x,n)|}{(4x(1-x)n\log\log(n/x(1-x)))^{1/2}} = 1 \qquad \text{a.s.,} \tag{2.10}$$

consequently

$$K_{n_{k-1}}(\delta_{n_k}) \leqslant \left(\log\log\frac{n_{k-1}}{\delta_{n_k}(1-\delta_{n_k})}\right)^{1/2} \times$$

$$\times \sup_{0<x<1} \frac{|K(x,n_{k-1})|}{(x(1-x)\log\log(n_{k-1}/x(1-x)))^{1/2}} =$$

$$= O\left(\left(n_{k-1}\log\log\frac{n_{k-1}}{\delta_{n_k}(1-\delta_{n_k})}\right)^{1/2}\right) \qquad \text{a.s.}$$

and it is easy to verify that

$$\lim_{k\to\infty}\frac{K_{n_{k-1}}(\delta_{n_k})}{(2n_k\log\rho_{n_k})^{1/2}} = 0 \qquad \text{a.s.} \tag{2.11}$$

Now (2.8), (2.9) and (2.11) imply (2.7). Relations (2.3) and (2.7) together yield (1.13), and Theorem 2 is proved.

3. PROOF OF THEOREM 1

The invariance principle due to (Komlós et al., 1975) says that on a suitably rich probability space one can define a sequence $X_1, X_2, \ldots$ of i.i.d. uniform on $(0,1)$ random variables together with the Kiefer process $K(x,y)$ such that as $n\to\infty$,

$$\sup_{0\leqslant x\leqslant 1}\left|n\big(F_n(x)-x\big)-K(x,n)\right| = O(\log^2 n) \qquad \text{a.s.} \tag{3.1}$$

An easy consequence of this result is

$$\sup_{\delta_n<x<1-\delta_n}\left|n^{1/2}\frac{F_n(x)-x}{(x(1-x))^{1/2}}-\frac{K(x,n)}{(nx(1-x))^{1/2}}\right| = O\left(\frac{\log^2 n}{(n\delta_n)^{1/2}}\right) \qquad \text{a.s.} \tag{3.2}$$

Hence Theorem 2 implies Theorem 1 under the additional condition

$$\lim_{n\to\infty}\frac{\log^2 n}{(n\delta_n\log\log n)^{1/2}} = 0. \tag{3.3}$$

Let

$$\delta_n^* = \max\left(\frac{\log^4 n}{n}, \delta_n\right). \tag{3.4}$$

Since δ_n^* satisfies (3.3), we have

$$\liminf_{n\to\infty}\frac{n^{1/2}T_n(\delta_n^*)}{(2\log\rho_n^*)^{1/2}} = 1 \qquad \text{a.s.,} \tag{3.5}$$

where

$$\rho_n^* = \frac{\log(1/\delta_n^*)}{\log\log n}. \tag{3.6}$$

But $T_n(\delta_n) \geqslant T_n(\delta_n^*)$ and $\log\rho_n' \sim \log\rho_n^*$, hence we have

$$\liminf_{n\to\infty}\frac{n^{1/2}T_n(\delta_n)}{(2\log\rho_n')^{1/2}} \geqslant \liminf_{n\to\infty}\frac{n^{1/2}T_n(\delta_n^*)}{(2\log\rho_n^*)^{1/2}} = 1 \qquad \text{a.s.} \tag{3.7}$$

On the other hand, if $\delta_n = \delta_n^*$ for all n large enough, then δ_n satisfies (3.3) and we have (1.8). Now assume that

$$\delta_{n_i} < \frac{\log^4 n_i}{n_i}, \quad i = 1, 2, \dots, \tag{3.8}$$

for $n_1 < n_2 < \dots$ and $n_i \to \infty$ as $i \to \infty$. Since $T_n(\delta_n) \leqslant T_n(0)$, we have

$$\mathbf{P}(n_i^{1/2} T_{n_i}(\delta_{n_i}) \leqslant (1+\varepsilon)(2 \log\log n_i)^{1/2}) \geqslant$$

$$\geqslant \mathbf{P}(n_i^{1/2} T_{n_i}(0) \leqslant (1+\varepsilon)(2 \log\log n_i)^{1/2}).$$

It follows from results of (Jaeschke, 1979) that

$$\lim_{n\to\infty} \mathbf{P}\left(n^{1/2} T_n(0) \leqslant (1+\varepsilon)(2\log\log n)^{1/2}\right) = 1, \tag{3.9}$$

hence

$$\lim_{i\to\infty} \mathbf{P}(n_i^{1/2} T_{n_i}(\delta_{n_i}) \leqslant (1+\varepsilon)(2\log\log n_i)^{1/2}) = 1, \tag{3.10}$$

therefore

$$\liminf_{i\to\infty} \frac{n_i^{1/2} T_{n_i}(\delta_{n_i})}{(2\log\log n_i)^{1/2}} \leqslant 1 \qquad \text{a.s.} \tag{3.11}$$

From (3.8) we obtain $\log\log n_i \sim \log \rho'_{n_i}$ as $i \to \infty$; consequently

$$\liminf_{i\to\infty} \frac{n_i^{1/2} T_{n_i}(\delta_{n_i})}{(2\log \rho'_{n_i})^{1/2}} \leqslant 1 \qquad \text{a.s.} \tag{3.12}$$

or

$$\liminf_{n\to\infty} \frac{n^{1/2} T_n(\delta_n)}{(2\log \rho'_n)^{1/2}} \leqslant 1 \qquad \text{a.s.} \tag{3.13}$$

This completes the proof of Theorem 1.
Finally we make some remarks.

Remark 1. It is obvious that Theorem 1 remains true if $T_n^+(\delta_n)$ is replaced by

$$T_n^-(\delta_n) = \sup_{\delta_n < x < 1-\delta_n} \frac{x - F_n(x)}{(x(1-x))^{1/2}} \tag{3.14}$$

or when the definitions of T_n, T_n^+ and T_n^- are modified in such a way that the sup is taken on intervals of the form (δ_n, a) ar $(a, 1-\delta_n)$ with $0 < a < 1$. We have in particular

$$\liminf_{n\to\infty} \left(\frac{n}{2\log\log n}\right)^{1/2} \sup_{0<x<1/2} \frac{x - F_n(x)}{(x(1-x))^{1/2}} = 1 \qquad \text{a.s.} \tag{3.15}$$

This answers an open question raised by (Shorack and Wellner, 1986, p. 605).

Remark 2. It is worthwile to compare some results of the present paper with those of (Csáki, 1977). The assertion (3.15) may be compared to

$$\limsup_{n\to\infty}\left(\frac{n}{2\log\log n}\right)^{1/2}\sup_{0<x<1/2}\frac{x-F_n(x)}{(x(1-x))^{1/2}}=\sqrt{2}\qquad\text{a.s.}$$

In the case when $\delta_n=\exp\{-(\log n)^c\}$ $(0<c<1)$ Theorem 3.1 in (Csáki, 1977) gives

$$\limsup_{n\to\infty}\frac{n^{1/2}T_n(\delta_n)}{(2\log\log n)^{1/2}}=\sqrt{c+1}\qquad\text{a.s.}$$

while our present Theorem 1 gives

$$\liminf_{n\to\infty}\frac{n^{1/2}T_n(\delta_n)}{(2\log\log n)^{1/2}}=\sqrt{c}\qquad\text{a.s.}$$

In general, however, the lim sup and lim inf differ even in the norming factor.

Remark 3. In the case when (1.5) holds, we have $\log\rho_n'\sim\log\log n$ as $n\to\infty$, hence Theorem 1 gives (1.4).

Remark 4. The present results can be extended to other cases when a.s. invariance of type (3.1) holds. Similar results can be given e.g. for the standardized quantile process.

Remark 5. General lim inf results for the weighted empirical distributions $\sup_x|F_n(x)-x|\psi(x)$ or $\sup_x(F_n(x)-x)\psi(x)$ are open problems. In particular lim inf results for $\sup_{0\leqslant x\leqslant 1}(F_n(x)-x)$ are unknown.

REFERENCES

Csáki, E. (1977). The law of the iterated logarithm for normalized empirical distribution function. *Z. Wahr. verw. Geb.* **38**, 147–167.

Csáki, E. (1981). A note on the lower limit of the standardized empirical distribution function *Mimeograph Ser.* **81-19**, Depart. Statist., Purdue University.

Csáki, E. (1982). On the standardized empirical distribution function. *Coll. Math. Soc. J. Bolyai* **32**, North–Holland, Amsterdam, pp. 123–138.

Csörgő, M. and Révész, P. (1981). *Strong Approximations in Probability and Statistics.* Academic Press, New York.

Jaeschke, D. (1979). The asymptotic distribution of the supremum of the standardized empirical distribution function on sub–intervals. *Ann. Statist.* **7**, 108–115.

Komlós, J., Major, P. and Tusnády, A. (1975). An approximation of partial sums of independent r.v.'s and the sample df.I. *Z. Wahr.verw.Geb.* **32**, pp. 111–131.

Kuelbs, J. (1979). Rates of growth for Banach space valued independent increment processes, *Lect. Notes in Math.* **709**, 151–169.

Mogul'skii, A. A. (1979). On the law of the iterated logarithm in Chung's form for functional spaces. *Theor.Probab.Appl.* **24**, pp. 405–413.

Newell,G.F. (1962). Asymptotic extreme value distributions for one dimensional diffusion processes. *J.Math. Mech.* **11**, pp. 481–496.

Shorack, G. and Wellner, J. (1986). *Empirical Processes with Applications to Statistics.* Wiley, New York.

Prob. Theory and Math. Stat., Vol. 1, pp. 245–257
B.Grigelionis *et al.* (Eds.)
1990 VSP/Mokslas

THE SMOOTH MEASURES IN BANACH SPACES AND THEIR SMOOTH MAPPINGS

Yu.L. DALETSKY and V.R. STEBLOVSKAYA
Polytechnic Institute, Kiev, USSR

ABSTRACT

The concept of admissible operator in $L_2(X,\mu)$ with respect to measure μ on the space X is introduced. This concept generalizes the concepts of the admissible shift of measure. The various transformations preserving the μ–admissibility are considered. The μ–admissibility of differential operators and operators connected with them is considered. The smoothness of the finite dimensional images of measures is shown.

1. THE ADMISSIBLE TRANSFORMATIONS OF MEASURE

Let $(X,\mathcal{B},\mu)$ be a space with a measure, $K = L_2(X,\mu)$ be the space of the functions $f : X \to \mathbf{C}$, $\int\limits_X |f(x)|^2\mu(dx) < \infty$. Consider the linear operator A in K with a domain D_A which is dense in K. Suppose that there exists the measure μ' on $(X,\mathcal{B})$, defined by the relation

$$\langle A\varphi,\mu\rangle = \int\limits_X (A\varphi)(x)\mu(dx) = \int\limits_X \varphi(x)\mu'(dx) = \langle\varphi, A'\mu\rangle, \tag{1}$$

where brackets $\langle,\rangle$ mean the pairing between the function and the measure; A' is a dual operator on measures with respect to A.
Let the measure $\mu' = A'\mu$ be absolutely continuous with respect to $\mu : \mu' \prec \mu$ and the density $\rho(A,\mu,x) = \mu'(dx)/\mu(dx) \in K$. In this case the operator A is said to be μ–admissible.
Denote by A^* the adjoint operator in the space K:

$$(A\varphi,\psi) = \int\limits_X (A\varphi)(x)\psi(x)\mu(dx) = \int\limits_X \varphi(x)(A^*\psi)(x)\mu(dx) =$$
$$= (\varphi, A^*\psi); (\varphi \in D_A, \psi \in D_{A^*}).$$

The following result is obvious.

PROPOSITION 1. *An operator A is μ–admissible if and only if $1 \in D_{A^*}$. In this case $\rho(A,\mu,x) = (A^*\cdot 1)(x)$.*

Consider the simplest properties of μ–admissible operators.

a) Let A_1, A_2 be μ–admissible, $D_{A_1} = D_{A_2}$, $\lambda_1, \lambda_2 \in \mathbb{C}$. Then the operator $A = \lambda_1 A_1 + \lambda_2 A_2$ is μ–admissible and

$$\rho(A, \mu, x) = \lambda_1 \rho(A_1, \mu, x) + \lambda_2 \rho(A_2, \mu, x).$$

b) Let the operator $A = A_1 \cdot A_2$ be defined, A_1 be μ–admissible and $(A_1^* \cdot 1) \in D_{A_2^*}$. Then A is μ–admissible and $\rho(A, \mu, x) = A_2^* \cdot A_1^* \cdot 1$.

c) Let A_n be the sequence of μ–admissible operators with common domain D, converging to the operator A on D. Let the corresponding sequence of densities $\rho(A_n, \mu, x)$ converges in K to the function $\lambda(x)$. Then the operator A is μ–admissible and $\rho(A, \mu, x) = \lambda(x)$.

d) Let the operator $A(t)$ possesses the domain D_A independent on t and let for each $t \in [a; b]$ $A(t)$ be μ–admissible; let the density $\rho(A(t), \mu, x)$ be integrable with respect to t as the element of K with respect to parameter x. Then the operator

$$B = \int_a^b A(t)dt$$

(in the sense of convergence on the dense set) is μ–admissible and

$$\rho(B, \mu, x) = \int_a^b \rho(A(t), \mu, x)dt.$$

e) Let $A(t)$ be the operator described above, and the density $\rho(A(t), \mu, x)$ be differentiable with respect to t in the sense of convergence in K. Then the operator $B(t) = dA(t)/dt$ (in the sense of differentiation on the dense set D_A) is also μ–admissible for each t and

$$\rho(B(t), \mu, x) = \frac{d}{dt}\rho(A(t), \mu, x).$$

2. THE ADMISSIBLE OPERATORS, DEFINED BY THE DIFFERENTIAL EQUATIONS

THEOREM 1. *Let the operator A be μ–admissible and the operator $V(t) = \exp(At) : D \to D$ $(t \geqslant 0)$ be defined. Let the function $v(t, x)$ solves the Cauchy problem*

$$\frac{\partial v}{\partial t} = A^* v; \quad v(0, x) = 1. \tag{2}$$

Then the operator $V(t)$ is μ–admissible for each $t \geqslant 0$ and

$$\rho(V(t), \mu, x) = V(t)^* \cdot 1 = v(t, x).$$

Proof. It is sufficient to prove the relation

$$\int_X V(t)\varphi(x)\mu(dx) = \int_X \varphi(x)v(t,x)\mu(dx) \qquad (\varphi \in D). \tag{3}$$

Consider the function

$$\gamma(t) = \int_X \{\exp(At)\varphi(x) - \varphi(x)v(t,x)\}\mu(dx).$$

Obviously $\gamma(0) = 0$. Then find

$$\gamma'(t) = \int_X A \exp(At)\varphi(x)\mu(dx) - \int_X \varphi(x)\frac{\partial v(t,x)}{\partial t}\mu(dx).$$

Because of μ–admissibility of the operator A we obtain

$$\int_X A \exp(At)\varphi(x)\mu(dx) = \int_X \exp(At)\varphi(x)(A^* \cdot 1)(x)\mu(dx) =$$

$$= \int_X \varphi(x)(\exp At)^*(A^* \cdot 1)(x)\mu(dx).$$

It is easy to show that $(\exp At)^* = \exp(A^* t)$. So

$$\gamma'(t) = \int_X \varphi(x)\left\{A^* \exp(A^* t) \cdot 1 - \frac{\partial v(t,x)}{\partial t}\right\}\mu(dx) \equiv 0.$$

Therefore $\gamma(t) \equiv 0$, thus the theorem is proved.

Remark 1. Let the operator A be μ–admissible and the operator $U(t) = \exp(iAt) : D \to D$ be defined. Let the function $u(t,x)$ solves the Schrödinger equation

$$\frac{\partial u}{\partial t} = -iA^* u; \; u(0,x) = 1. \tag{4}$$

Then the operator $U(t)$ is μ–admissible for $t \in \mathbf{R}^1$ and

$$\rho(U(t),\mu,x) = U(t)^* \cdot 1 = u(t,x).$$

Remark 2. Let the operator A be μ–admissible and the operator $\cos(At)$ be defined. Let the function $u(t,x)$ solves the problem

$$\frac{\partial^2 u}{\partial t^2} = -(A^*)^2 u; \; u(0,x) = 1, \; u_t'(0,x) = 0. \tag{5}$$

Then the operator $\cos(At)$ is μ–admissible and $\rho(\cos At, \mu, x) = u(t, x)$. Analogously for the operator $\sin(At)\rho(\sin(At), \mu, x) = v(t, x)$, where $v(t, x)$ is the solution of the Cauchy problem

$$\frac{\partial^2 v}{\partial t^2} = -(A^*)^2 v;\ v(0, x) = 0,\ v_t'(0, x) = A^* \cdot 1 = \rho(A, \mu, x). \tag{6}$$

THEOREM 2. *Let the operator $A(t)$ be μ–admissible for each $t \in [a; b]$ and there exists the function $v_t(\tau, x)$ solving the Cauchy problem*

$$\frac{\partial v}{\partial \tau} = -A(\tau)^* v; \quad v(t, x) = 1 \quad (0 \leqslant \tau \leqslant t) \tag{7}$$

representable in the form of the multiplicative integral

$$v_t(\tau, x) = \overleftarrow{\exp} \int_t^\tau (-A(\theta)^*) d\theta \cdot 1 =$$

$$= \lim_{\max|\Delta_k t| \to 0} \prod_{k=1}^n \exp(A(t_{k-1})^* \Delta_k t) \cdot 1$$

$$(\tau = t_0 < t_1 < \ldots < t_n = t; |\Delta_k t| = t_k - t_{k-1}).$$

Then the operator $V(t, \tau) = \overleftarrow{\exp} \int_\tau^t A(\theta) d\theta$ is μ–admissible and

$$\rho(V(t, \tau), \mu, x) = v_t(\tau, x).$$

To prove the theorem we remark that the operator $V(t, \tau)$ is the limit of the sequence of μ–admissible operators

$$V(t, \tau) = \lim_{|q| \to 0} B_q;\ B_q = \prod_{k=1}^n \exp(A(t_{k-1}) \Delta_k t),$$

where $q = (t_0, t_1, \ldots, t_n)$ is the subdivision of the interval $[\tau; t], |q| = \max|\Delta_k t|$. In addition the sequence of densities $\rho(B_q, \mu, x) = B_q^* \cdot 1$ converges in K to the function $v_t(\tau, x)$. Therefore the limit operator $V(t, \tau)$ is μ–admissible $(t, \tau \in [a; b])$ by virtue of the property c).

Remark 3. Let the operator $A(t)$ be μ–admissible for each $t \in [a; b]$. Let in addition there exists the function $z_t(\tau, x)$ solving the Cauchy problem

$$\frac{\partial^2 z}{\partial \tau^2} = -(A(\tau)^*)^2 z;\ z(t, x) = 1, z'(t, x) = 0 \tag{8}$$

(or $z(t, x) = 0,\ z'(t, x) = 1$).

Then the operator $V(t,\tau) : u(\tau,x) \mapsto u(t,x)$ giving the solution of the Cauchy problem

$$\frac{\partial^2 u}{\partial t^2} = -A(t)^2 u; \; u(\tau,x) = u_0, \; u'(\tau,x) = 0 \tag{9}$$

(or $u(\tau,x) = 0, u'(\tau,x) = u_0'$) is μ-admissible and

$$\rho(V(t,\tau),\mu,x) = -\frac{\partial}{\partial\tau} z_t(\tau,x) \quad (\text{or } \rho(V(t,\tau),\mu,x) = z_t(\tau,x)).$$

It is possible to obtain the proof writing the second order equation (9) as a system of first-order equations and then applying the results of Theorem 2.

3. FUNCTIONS OF ADMISSIBLE OPERATORS

Using the representations of operator functions in the form of Fourier and Laplace transforms and the previous results it is possible to prove the μ-admissibility for the sufficiently large class of operators.

Let be possible for the function $z(t)$ to write the Laplace transformation formula

$$f(p) = \int_0^\infty e^{-pt} z(t)dt$$

and the inversion formula

$$z(t) = \frac{1}{2\pi i} \int_{\sigma-i\infty}^{\sigma+i\infty} f(p)e^{pt}dp \quad (\sigma \geqslant \gamma_0 > 0).$$

Let A be the generator of semi-group e^{-tA} and the spectrum $\sigma(A) \subset @ = \{\mathrm{Re}\, p \geqslant \gamma_0 > 0\}$ be the domain of regularity of function f. Define the function of operator

$$f(A) = \int_0^\infty z(t)e^{-tA}dt. \tag{10}$$

THEOREM 3. *Assume that the operator $(-A)$ satisfies the conditions of Theorem 1. Then the operator $f(A)$ is μ-admissible and the density can be represented in the form*

$$\rho(f(A),\mu,x) = \int_0^\infty v(t,x)z(t)dt, \tag{11}$$

where $v(t,x)$ is the solution of the Cauchy problem (2) for the operator $(-A^*)$ (if the integral converges in K).

Proof. It follows from Theorem 1 that the operator e^{-tA} is μ-admissible and $\rho(\exp(-tA),\mu,x) = v(t,x)$. Then for $\varphi \in D_A$

$$\int_X f(A)\varphi(x)\mu(dx) = \int_X \left\{ \int_0^\infty z(t)e^{-tA}dt \right\} \varphi(x)\mu(dx) =$$

$$= \int_0^\infty z(t) \left\{ \int_X e^{-tA}\varphi(x)\mu(dx) \right\} dt = \int_0^\infty z(t) \left\{ \int_X \varphi(x)v(t,x)\mu(dx) \right\} dt =$$

$$= \int_X \varphi(x) \left\{ \int_0^\infty z(t)v(t,x)dt \right\} \mu(dx) = \int_X \varphi(x)\rho(f(A),\mu,x)\mu(dx).$$

Remark 4. Consider the function $f \in L_2(\mathbf{R}^1)$. Denote

$$g(t) = \int_{-\infty}^{\infty} f(y)e^{-ity}dy$$

its Fourier transformation. Then

$$f(y) = \frac{1}{2\pi} \int_{-\infty}^{\infty} g(t)e^{ity}dt.$$

Define the function f of the operator A connected with the group e^{itA}

$$f(A) = \frac{1}{2\pi} \int_{-\infty}^{\infty} g(t)e^{itA}dt. \tag{12}$$

Let the operator iA satisfies the conditions of the Remark 1 and the integral (12) converges in K.
It is possible by analogy with the Theorem 3 to show the μ–admissibility of operator (12) and to obtain the expression for the density

$$\rho(f(A),\mu,x) = \frac{1}{2\pi} \int_{-\infty}^{\infty} g(t)u(t,x)dt, \tag{13}$$

where $u(t,x)$ is the solution of the Cauchy problem (4) if the integral (13) converges in K.

Example 1. Let $(-A)$ be the generator of strongly continuous semi–group e^{-tA}, satisfying the conditions of Theorem 3. Using the Laplace transform of function $f(y) = (y^2+a^2)^{-1/2}$ it is possible to define the function

$$f(A) = (A^2 + a^2I)^{-1/2} = \int_0^\infty e^{-tA}J_0(at)dt.$$

The μ–admissibility of $f(A)$ follows from Theorem 3. The density can be calculated like this

$$\rho((A^2 + a^2 I)^{-1/2}, \mu, x) = \int_0^\infty v(t,x) J_0(at)dt,$$

where $v(t,x)$ is the solution of the problem (2) for the operator $(-A^*)$.

Example 2. Having chosen the proper pairs of functions connected by the Fourier transforms it is possible to obtain the examples of μ–admissible functions of operator A which was defined in Remark 4. For example

$$\rho(e^{-1/2A^2}, \mu, x) = \frac{1}{2\pi} \int_{-\infty}^{\infty} e^{-t^2/2} u(t,x)dt,$$

$$\rho(e^{-|A|}, \mu, x) = \frac{1}{2\pi} \int_{-\infty}^{\infty} \frac{u(t,x)}{1+t^2} dt,$$

where $u(t,x)$ is the solution of Schrödinger equation (4).

4. THE ADMISSIBILITY OF DIFFERENTIAL OPERATORS AND THEIR FUNCTIONS

Below we shall illustrate the properties of the μ–admissible operators by the examples related to the differential operators. In addition we shall obtain in some general way the results which were deduced earlier from the special constructions (see (Daletsky and Belopolskaya, 1989; Daletsky and Sokhadze, 1988)).

Let $X = H_- \supset H_0 \supset H_+$ be the equipped Hilbert space. Introduce the operation $\langle h, D\rangle \varphi(x) = \langle h, \varphi'(x)\rangle$ for the differentiation of function along the direction $h \in H_+$. The μ–admissibility of this operation means the differentiability of the measure μ along the direction h and the existence of logarithmic derivative

$$\rho_\mu(h; x) = -\rho(\langle h, D\rangle, \mu, x) = \langle \lambda(x), h\rangle \in K$$

satisfying the formula of integration by parts

$$\int_X \langle h, D\rangle \varphi(x)\mu(dx) = -\int_X \varphi(x)\langle \lambda(x), h\rangle \mu(dx). \tag{14}$$

The measure μ will be said to be sufficiently smooth if the function $\lambda(x)$ (its vector logarithmic derivative) is sufficiently smooth.

Consider the first–order differential operator $A_1 = \langle z(x), D\rangle$, where $z(x)$ is a smooth vector field ($z : H_- \to H_+$). The μ–admissibility of this operator means the existence of logarithmic derivative of measure μ along the vector field z (Daletsky and Belopolskaya, 1989)

$$\rho_\mu(z(x); x) = -\rho(\langle z(x), D\rangle, \mu, x) = \langle \lambda(x) + D, z(x)\rangle \cdot 1$$

satisfying the formula of integration by parts

$$\int_X \langle z(x), D\rangle \varphi(x)\mu(dx) = -\int_X \varphi(x)\langle \lambda(x) + D, z(x)\rangle \cdot 1\mu(dx). \tag{15}$$

Analogously the μ-admissibility of the operator

$$A_n = \langle z_1(x), D\rangle \dots \langle z_n(x), D\rangle,$$

where $z_1, z_2, \dots, z_n$ is the set of smooth vector fields, means the existence of logarithmic derivative of measure μ along this set of vector fields

$$\rho_n(z_1, \dots, z_n; x) = (-1)^n \rho(A, \mu, x) =$$
$$= \langle \lambda(x) + D, z_n(x)\rangle \dots \langle \lambda(x) + D, z_1(x)\rangle \cdot 1 \tag{16}$$

THEOREM 4. (see (Daletsky and Belopolskaya, 1989)). *If μ is sufficiently smooth measure,*

$$@_c = \langle C(x), D \otimes \dots \otimes D\rangle$$

is the differential operator of order n with smooth coefficient $C(x)$ then $@_c$ is μ-admissible and

$$\rho(@_c, \mu, x) = \langle (\lambda(x) + D) \otimes \dots \otimes (\lambda(x) + D), C(x)\rangle, \tag{17}$$

where the differentiation D operates on all functions followed it.

Consider the differential operator $A = \langle z(x), D\rangle$, determined by the vector field z. Let the family of mappings

$$S_t : x \mapsto x(t), \quad \frac{dx}{dt} = z(x), x(0) = x$$

correspond to it. Consider the shift operator on functions

$$V(t) = e^{tA} : \quad V(t)f(x) = f(S_t x).$$

The dual operator $V(t)'$ realizes the corresponding shift on measures:

$$\mu_t = V(t)'\mu = \mu \circ S_t^{(-1)}.$$

THEOREM 5. (see (Daletsky and Sokhadze, 1988)). *Let the measure μ possesses the logarithmic derivative along the vector field z. Then the operator $V(t)$ is μ-admissible and the density $\rho(V(t), \mu, x) = \mu_t(dx)/\mu(dx)$ is given by the formula*

$$\rho(V(t), \mu, x) = \exp\left\{ -\int_0^t \rho_\mu(z(S_\tau^{-1}x);\ S_\tau^{-1}x)d\tau \right\}. \tag{18}$$

Proof. The operator A is μ–admissible. Consider the Cauchy problem (2) for given operator

$$\frac{\partial v(t,x)}{\partial t} = -\langle \lambda(x) + D, z(x)\rangle v(t,x); \qquad v(0,x) = 1$$

or after the obvious transformations

$$\frac{\partial v(t,x)}{\partial t} + \langle z(x), D\rangle v = -\rho_\mu(z;x)v. \tag{19}$$

Write the characteristic system for the equation (19)

$$\frac{dx}{dt} = z(x), \frac{dv}{dt} = -\rho_\mu(z;x)v.$$

Solving this system we obtain

$$v(t,x) = V(t)^* \cdot 1 = \exp\left\{ -\int_0^t \rho_\mu(z(S_\tau^{-1}x);\ S_\tau^{-1}x)d\tau \right\}.$$

It follows from Theorem 1 that $\rho(V(t),\mu,x) = v(t,x)$.
Then consider the second–order elliptic differential operator

$$A = \langle C(x), D \otimes D\rangle + \langle a(x), D\rangle + b(x),$$

where $C(x)$ is a smooth tensor coefficient, $a(x)$ is a smooth vector field.
The pair of Kolmogorov equations is connected with this operator: the equation in the space of functions

$$\frac{\partial u}{\partial t} = Au \tag{20}$$

and the equation in the space of measures

$$\frac{\partial \mu_t}{\partial t} = A'\mu_t. \tag{21}$$

One can analyse the Cauchy problems for these equations by means of methods of the stochastic equations theory (see (Daletsky and Fomin, 1983)). If the operator e^{tA} is defined then $\mu_t = e^{tA'}\mu_0$ is the weak solution of Kolmogorov equation in measures, and μ–admissibility of the operator e^{tA} implies the absolute continuity $\mu_t \prec \mu_0$.
The condition of μ–admissibility of this operator follows from the solvability of the Cauchy problem

$$\frac{\partial v}{\partial t} = \{\langle(\lambda(x) + D) \otimes (\lambda(x) + D),\ C(x)\rangle +$$
$$+\langle \lambda(x) + D,\ a(x)\rangle + b(x)\}v;\ v(0,x) = 1. \tag{22}$$

Not considering the conditions in details formulate the result.

THEOREM 6. *If the Cauchy problems* (20), (22) *are correct then the solution of Kolmogorov equation in measures* (21) *satisfies the absolute continuity relation* $\mu_t \prec \mu_0$, *and the following formula for the density holds* $\mu_t(dx)/\mu(dx) = v(t,x)$ *where* $v(t,x)$ *is the solution of the Cauchy problem* (22).

5. THE MAPPINGS OF ADMISSIBLE OPERATORS

Let $f : X \to Y$ be a measurable mapping, μ be a measure on X and μ^f be the image of the measure μ. The operators A in $L_2(X,\mu)$ and B in $L_2(Y,\mu^f)$ are said to be f-compatible if $(\varphi \in D_B) \Rightarrow (\varphi \circ f \in D_A)$ and $(B\varphi)(f(x)) = A(\varphi \circ f)(x)$.

PROPOSITION 2. *If operators* A *and* B *are* f*-compatible,* A *is* μ*-admissible then operator* B *is* μ^f*-admissible and*

$$||\rho(B,\mu^f,y)||_{L_2(Y,\mu^f)} \leqslant ||\rho(A,\mu,x)||_{L_2(X,\mu)}. \tag{23}$$

Proof. Indeed one can obtain for any $\varphi \in D_B$

$$\left|\int_Y (B\varphi)(y)\mu^f(dy)\right| = \left|\int_X A(\varphi\circ f)(x)\mu(dx)\right| =$$

$$= \left|\int_X (\varphi\circ f)(x)\rho(A,\mu,x)\mu(dx)\right| \leqslant$$

$$\leqslant \left\{\int_Y |\varphi(y)|^2\mu^f(dy)\right\}^{1/2} \left\{\int_X |\rho(A,\mu,x)|^2\mu(dx)\right\}^{1/2} =$$

$$= ||\rho(A,\mu,x)||_{L_2(X,\mu)}||\varphi||_{L_2(Y,\mu^f)}.$$

It follows from this estimate that $\int_Y (B\varphi)(y)\mu^f(dy)$ is a linear continuous functional on $L_2(Y,\mu^f)$. So there exists the function $\beta \in L_2(Y,\mu^f)$ such that

$$\int_Y (B\varphi)(y)\mu^f(dy) = \int_Y \varphi(y)\beta(y)\mu^f(dy).$$

Obviously $\beta(y) = \rho(B,\mu^f,y)$ and the estimate (23) holds.
Then let X, Y be the Hilbert spaces, $X = H_- \supset H_0 \supset H_+$, f be sufficiently smooth mapping.
Consider the differential operators $@_x = \langle\xi_1, D\rangle \ldots \langle\xi_n, D\rangle$ and $@_y = \langle\eta_1, D\rangle \ldots \ldots \langle\eta_n, D\rangle$, where ξ_j, η_j $(j = 1,2,\ldots,n)$ are f-compatible pairs of sufficiently smooth vector fields. In addition $@_x$ and $@_y$ are also f-compatible and if the measure μ is sufficiently smooth then $@_y$ is μ^f-admissible.

Using formula (23) and estimates for the logarithmic derivatives of the measure μ (Daletsky and Belopolskaya, 1989; Steblovskaya, 1989) one can obtain the inequality

$$\int_Y |\rho^f(\eta_1,\ldots,\eta_n;y)|^2\mu^f(dy) \leqslant \int_X |\rho(\xi_1,\ldots,\xi_n;x)|^2\mu(dx) \leqslant$$

$$\leqslant \text{const}\sum_{k=1}^{n}\sigma_{n,2n}^{2n}(\xi_k) \tag{24}$$

where

$$\sigma_{n,m}(\xi_k) = \sum_{j=0}^{n}\left\{\int_X \sigma_2^m(\xi_k^{(j)}(x))\mu(dx)\right\}^{1/m}$$

(here $\sigma_2(\xi_k^{(j)}(x))$ is a Hilbert—Schmidt norm of j-linear mapping $\xi_k^{(j)}(x)$ – the derivative of order j of vector field ξ_k).
Now consider the finite-dimensional space $Y = \mathbf{R}^m$. It is known (see (Averbuch et al., 1971)) that in this case the smoothness of measure means the existens of smooth density of this measure with respect to Lebesque measure.
Let $f = (f_1, f_2, \ldots, f_m)$ be the set of measurable square mappings $f_k(x) = (A_k x, x)$, where $A_k : H_- \to H_+$ is the operator possessing the symmetric trace class restriction on H_0. Then we consider the condition of the existence of the smooth density of measure μ^f (see (Steblovskaya, 1989)). The analogous results were obtained in another way (see (Uglanov, 1982)); for $m = 1$ see (Smorodina, 1987; Bogachev, 1989).
Let at first $m = 1$.

PROPOSITION 3. *If the measure μ is infinitely smooth and the operator A is not finite-dimensional then the measure μ^f possesses the infinitely differentiable density.*

Proof. One can show that it is sufficient to inspect the following estimate (see (Steblovskaya, 1989)):

$$\int_{H_-} \frac{\mu(dx)}{||Ax||^{2q}} < \infty \quad (q = 1, 2, \ldots). \tag{25}$$

Using the spectral decomposition of the operator A it is possible for each q to find the invariant finite-dimensional subspace Z_q of operator A ($\dim Z_q \geqslant 2q + 1$) on which $||Ax|| \geqslant \varepsilon(q)||x||, \varepsilon(q) > 0$ ($x \in Z_q$) and, therefore

$$\int_{H_-} \frac{\mu(dx)}{||Ax||^{2q}} \leqslant \int_{Z_q} \frac{\mu_{Z_q}(dx)}{\varepsilon^{2q}||x||_{Z_q}^{2q}} < \infty.$$

Now let $m > 1$.

THEOREM 7. *If the operator* $A_{\mathbf{h}} = \sum_{k=1}^{m} h_k A_k$ *is not finite–dimensional for all* $\mathbf{h} = (h_1, \ldots, h_m) \in \mathbf{R}^m$ *and the finite–dimensional projections of infinitely smooth measure* μ *possesses the moments of any order, then the measure* μ^f *possesses the infinitely differentiable density with respect to Lebesque measure in* $\mathbf{R}^m$.

Proof. It is sufficient to establish for the characteristic function $\chi(\theta)$ of measure μ^f the following estimates

$$||\theta||^k |\chi(\theta)| \leqslant c_k \quad (k = 1, 2, \ldots;\ \theta \in \mathbf{R}^m).$$

Note that

$$||\theta||^k \chi(\theta) = \int_{\mathbf{R}^m} ||\theta||^k e^{i(\theta, y)} \mu^f(dy) =$$

$$= \int_{-\infty}^{\infty} ||\theta||^k e^{iz||\theta||} \nu(dz) = \frac{1}{i^k} \int_{-\infty}^{\infty} \frac{\partial^k}{\partial z^k} e^{iz||\theta||} \nu(dz),$$

where ν is the image of measure μ^f with respect to mapping $\varphi : y \mapsto (\mathbf{h}, y) = z$, $\mathbf{h} = \theta / ||\theta||$.
Consider the superposition of mappings $\varphi \circ f : H_- \to \mathbf{R}^1$ and vector field $v(x)$ on H_-, $\varphi \circ f$–compatible with vector field $w \equiv 1$ in $\mathbf{R}^1$. Then, as it follows from (24) and the results, relating to the case $m = 1$

$$\left| \int_{-\infty}^{\infty} \frac{\partial^k}{\partial z^k} e^{iz||\theta||} \nu(dz) \right| \leqslant \left\{ \int_{H_-} |\rho_k(v, v, \ldots, v; x)|^2 \mu(dx) \right\}^{1/2} \leqslant \text{const}\ \sigma^k_{k,2k}(v(x)).$$

The estimate of last expression can be reduced to the estimation of the expression (25) for the operator $A_{\mathbf{h}}$

$$\sigma^k_{k,2k}(v(x)) \leqslant \text{const} \left\{ \int_{H_-} \left\| \sum_{j=1}^{m} h_j A_j x \right\|^{-4k(k+1)} \mu(dx) \right\}^{1/2}.$$

Then it is necessary to obtain for the right–hand side of last expression the uniform with respect to $\mathbf{h}(||\mathbf{h}|| = 1)$ estimate. Let $\mathbf{h}_0$ be a fixed element of $\mathbf{R}^m$ on a unit sphere ($||\mathbf{h}_0|| = 1$). Using the spectral decomposition of the operator $A_{\mathbf{h}_0} = \sum_{k=1}^{m} h_{0j} A_j$ and methods of spectrum perturbations theory the existence of the invariant subspace $Z_{\mathbf{h}}$ of the operator $A_{\mathbf{h}}(\dim Z_{\mathbf{h}} \geqslant 4k(k+1) + 1)$ can be shown. Therefore the inequality

$$||A_{\mathbf{h}}x|| \geqslant \varepsilon||x|| \quad (x \in Z_{\mathbf{h}}) \tag{26}$$

holds for fixed $\varepsilon > 0$ and for all $\mathbf{h} \in U_{\mathbf{h}_0}$, where $U_{\mathbf{h}_0}$ is some neighborhood of $\mathbf{h}_0$ on the sphere. Covering the sphere by the finite set of such neighborhoods (it is possible due to its compactness) we find the parameter $\varepsilon > 0$ such that the relation (26) holds for all $\mathbf{h}$ on the sphere.

REFERENCES

Averbuch, V.I., Smolyanov, O.G. and Fomin, S.V. (1971). Distributions and Differential Equations in Linear Spaces. *Proc. Moscow Math. Soc.* **24**, 133–174.

Bogachev, V.I. (1989). Distributions of smooth functionals of random processes and infinite dimentional oscillating integrals. In: *Abstr. Fifth Intern. Vilnius Conf. Probab. Theory and Math. Statist.* MCI, Vilnius, pp. 76–77.

Daletsky, Yu.L. and Belopolskaya, Ya.I. (1989). *Stochastic Equations and Differential Geometry.* Visha Shkola, Kiev.

Daletskŷ, Yu.L. and Fomin, S.V. (1983). *Measures and Differential Equations in Infinite-Dimensional Spaces.* Nauka, Moscow.

Daletsky, Yu.L. and Sokhadze, G.A. (1988). Absolute continuity of smooth measures. *Funct. Analys. and Appl.* **22**, 173–174.

Smorodina, N.V. (1987). Differential calculus on measurable spaces and conditions of smoothness of densities of random variables. *Soviet. Math. Dokl.* **292**, 1053–1058.

Steblovskaya, V.R. (1989). Smoothness of finite dimentional images of measures. *Ukrain. Math. J.* **41**, 261–265.

Uglanov, A.V. (1982). Feynman measures with alternating sign correlation operator. *Soviet. Math. Dokl.* **264**, 1096–1099.

Prob. Theory and Math. Stat., Vol. 1, pp. 258–260
B.Grigelionis *et al.* (Eds.)
1990 VSP/Mokslas

LOCAL INVARIANCE PRINCIPLE FOR I.I.D. RANDOM VARIABLES

Yu.A. DAVYDOV

Faculty of Mathematics and Mechanics, Leningrad State University, Bibliotechnaya pl.2, Stary Peterhof, 198904 Leningrad, USSR

ABSTRACT

Let P_n be the distribution of random broken line $\xi_n(t)$, $t \in [0,1]$, $n \in \mathbf{N}$; W be the Wiener measure on the space $X = \mathbf{C}[0,1]$ and f be the measurable mapping of $(X, \mathbf{B}_X)$ with values in $(\mathbf{R}, \mathbf{B}_{\mathbf{R}})$ some condition under which we can obtain the convergence of $P_n f^{-1}$ to $W f^{-1}$ in variation are given.

Let $(\xi_n, n \in N)$ be a sequence of i.i.d. random variables with common distribution F. We suppose that $\mathbf{E}\xi_n = 0, \mathbf{D}\xi_n = \sigma^2 < \infty$. Let $S_n = \sum_{k=1}^{n} \xi_k$. We consider the sequence of continuous random broken lines

$$\xi_n(t) = \sigma^{-1} n^{-1/2} (S_{[nt]} + (nt - [nt])\xi_{[nt]+1}), \ t \in [0,1].$$

We denote by P_n the distribution of the process ξ_n on the space $X = \mathbf{C}[0,1]$ of continuous functions and by W the Wiener measure on X.
The measurable mapping f of $(X, \mathbf{B}_X)$ with values in $(\mathbf{R}, \mathbf{B}_{\mathbf{R}})$ will be called functional. The main question considered here is:
Under what conditions we can obtain the convergence in variation:

$$P_n f^{-1} \to W f^{-1} \tag{1}$$

This fact is particularly interesting in the case when the distribution Wf^{-1} is absolutely continuous, because the strong convergence (1) in this case is equivalent to the convergence of densities in $\mathbf{L}^1(\mathbf{R})$, in another words to the local limit theorem for functional f. We shall use two classes of functionals M_W and $M_W^{(1)}$. For their description we need some auxiliary concepts.
The set $\Delta \subset X, \Delta = \{x + cl, c \in [0,a]\}$ will be called the segment parallel to l. We say that $\Delta_n = \{x_n + cl_n, c \in [0, a_n]\}$ converges to Δ if $x_n \to x, l_n \to l, a_n \to a$. We denote λ_Δ the Lebesgue measure λ transfered from $[0,a]$ to Δ by the mapping $c \to x + cl$. The restriction of a functional f on Δ will be denoted by f_Δ.
Let

$$H_W = \{l \in X | l \text{ is abs. continuous}, \ l(0) = 0, \ l' \in \mathbf{L}^1[0,1]\}.$$

It is well known that H_W is the set of admissible shifts for the Wiener measure W.

Let $D_l f(x)$ be the derivative of f in the direction l at the point x:

$$D_l f(x) = \lim_{c \to 0} \frac{f(x+cl) - f(x)}{c}.$$

Now we can define the classes M_W and $M_W^{(1)}$.

Definition 1. The class M_W is the family of all functionals f for which the following property is fulfilled: for W–almost all $x \in X$ there exist an open ball $V \ni x$ and $l \in H_W$ such that for any segment $\Delta \subset V$ parallel to l the convergence $\Delta_n \to \Delta$ implies that in variation:

$$\lambda_{\Delta_n} f_{\Delta_n}^{-1} \longrightarrow \lambda_\Delta f_\Delta^{-1}.$$

Some properties of this class and examples of particular functionals from M_W may be found in (Davydov, 1984).

Definition 2. The class $M_W^{(1)}$ is the family of all functionals f having the following property: for W–almost all $x \in X$ there exist $l \in H_W$ and an open ball $V \ni x$ such that f is Lipchitzian in $V, D_l f(x) \neq 0$ and mapping $(h, y) \to D_h f(y)$ is continuous in the point (l, x).

It is not difficult to see that

1)$M_W^{(1)} \subset M_W$.

2) For all $f \in M_W^{(1)}$ the distribution Wf^{-1} is absolutely continuous.

Examples. 1. Let f be a W–almost surely differentiable in the sense of Frechet functional having W–a.s. non zero derivative. Then $f \in M_W^{(1)}$.

2. Let

$$f(x) = \int_0^1 h(x|t|)dt, \quad x \in X,$$

where h is locally bounded and such that for any ε there exists an interval $(\alpha, \beta) \subset (-\varepsilon, \varepsilon)$ on which h is continuously differentiable and $h' \neq 0$. Then $f \in M_W^{(1)}$. In (Davydov, 1980,1983) (see also (Davydov, 1984,1988)) the result that gives practically a local limit theorem for functionals from M_W has been proved. It can be named by local invariance principle (by analogy with the case of weak convergence).

THEOREM 1 .(Davydov, 1980,1983). *Suppose that F has a density $\varphi = dF/d\lambda$ which is absolutely continuous and which has the finite informational Fisher's quantity*

$$\int_{\mathcal{R}} \frac{(\varphi')^2}{\varphi} d\lambda < \infty.$$

Then for all $f \in M_W$ we have the convergence in variation (1).

In this report we suggest another form of local invariance principle:

THEOREM 2. *Suppose that*

1) *F has a density $\varphi = dF/d\lambda$,*

2) *$f \in M_W^{(1)}$. Then the strong convergence (1) is fullfiled.*

Let us consider the linear functional $f(x) = x(1)$. It is obvious that $f \in M_W^{(1)}$. In this case the distribution of $P_n f^{-1}$ is exactly the distribution of the normalized sum $S_n/(\sigma\sqrt{n})$ and we see that the classical Prokhorov local limit theorem (see Ibragimov and Linnik, 1965, p.159) for i.i.d. random variables is almost covered by the theorem 2. The necessary part of the Prokhorov's theorem shows that the condition 1) is very close to complete form.
The proof of theorem 2 is sufficiently long and we have no possibility to give it here. We note only three essential moments which allow us to pass from theorem 1 to theorem 2.

1.The proof is based on the method of suspension (detailed exposition of this method is represented in (Davydov, 1988)).

2.We actively use the representation of variables ξ_n by the function of orthogaussian random variables $\eta_n : \xi_n = G(\eta_n)$.

3. We use the following nontrivial one–dimensional result about convergence in variation:

LEMMA . *Suppose that absolutely continuous functions $\{h_n, n \in \mathbb{Z}\}, h$ defined on $[0,1]$ have the properties:*

1) $h_n(0) \longrightarrow h(0), n \to \infty;$

2) $\|h'_n - h'\|_{L^1[0,1]} \to 0, \ n \to \infty;$

3) *$h'(s) \neq 0$ for almost all $s \in [0,1]$.*
Then $\lambda h_n^{-1} \to \lambda h^{-1}$ in variation.

REFERENCES

Davydov, Yu.A. (1980). Local invariance principle. I. *Notes Sci. Seminars Leningrad Branch of Math. Inst.* **97**, 45–50.

Davydov, Yu.A. (1983). Local invariance principle. II. *Notes Sci. Seminars Leningrad Branch of Math. Inst.* **130**, 69–77.

Davydov, Yu.A. (1988). Local limit theorems for functionals of random processes. *Theory Probab. Appl.* **33**, 788–793.

Davydov, Yu.A. and Lifshits, M.A. (1984). *The method of stratification in some probabilistic problems. Itogi Nauki i tekhniki. Ser.: Probab. Theory, Math. Statist. and Theor. Cibern.* **22**, 61–157.

Ibragimov, I.A. and Linnik, Yu.V. (1965). *Independent and Stationary Connected Variables.* Nauka, Moscow.

Prob. Theory and Math. Stat., Vol. 1, pp. 261–270
B.Grigelionis *et al.* (Eds.)
1990 VSP/Mokslas

CONDITIONALLY POSITIVE DEFINITE FUNCTIONS ON THE COEFFICIENT ALGEBRA OF A NONABELIAN GROUP

L.V. DENISOV and A.S. HOLEVO
Steklov Mathematical Institute, Vavilova 42, 117966 Moscow GSP–1, USSR

ABSTRACT

Generalizations of the Shoenberg theorem and the Levy–Khintchine representation in quantum probability theory are obtained, giving a full description of convolution semigroups of instruments on a Moore group.

1. INTRODUCTION

In the last years there was a considerable interest in the theory of continuous quantum measurement processes which started in (Davies, 1976; Barchielli et al., 1982). In (Holevo, 1987, 1988) an analogue of the method of characteristic functions was developed, making it possible to describe such processes, taking values in an Abelian locally compact group. The description is based on an analogue of the Levy—Khintchine representation for a convolution semigroup of instruments (operation–valued measures), which is a non–commutative generalization of the usual convolution semigroup of probability measures. A natural problem arises to generalize this approach to the case of non–Abelian groups.

In (Barchielli and Lupieri, 1988) a wide class of convolution semigroups of instruments on the Moore groups was constructed by combining methods of non–Abelian harmonic analysis with methods of quantum stochastic calculus. In this article we present new results, which can be considered as noncommutative generalizations of Schoenberg's theorem and Levy—Khintchine representation, giving a full description for an arbitrary convolution semigroup of instruments on a Moore group.

2. POSITIVE DEFINITE AND CONDITIONALLY POSITIVE DEFINITE FUNCTIONS ON THE COEFFICIENT ALGEBRA OF AN ALMOST PERIODIC GROUP

Let $(G, \mathcal{B})$ be a locally compact group with the σ–algebra $\mathcal{B}$ of the Borel subsets and the unit element e, and let us denote by $\mathcal{H}$ a separable Hilbert space, by $\mathcal{A} \subset \mathfrak{B}(\mathcal{H})$ some von Neumann algebra of bounded operators in $\mathcal{H}$, and by $\mathfrak{F}_\sigma = \mathfrak{F}_\sigma(\mathcal{A})$ the Banach algebra of ultraweakly continuous maps of $\mathcal{A}$ into itself. Two topologies will be considered in $\mathfrak{F}_\sigma$: the first one is the usual norm topology, and

topologies will be considered in $\mathfrak{F}_\sigma$: the first one is the usual norm topology, and the second one is the locally convex topology τ_2 (cf. (Holevo, 1988)) defined by the family of seminorms

$$\Phi \in \mathfrak{F}_\sigma(\mathcal{A}) \to \sup_{\|X\|\leqslant 1} \|\Phi[X]\Psi\|, \quad \Psi \in \mathcal{H}.$$

Let us remind some facts and notions from the harmonic analysis on non–Abelian groups (Heyer, 1977). Let D be some finite dimensional unitary representation of G in the space $\mathbf{C}^{n(D)}$, where $n(D)$ is the dimension of D; $\mathrm{Rep}_n G$ be the set of all such representations for fixed $n(D) = n$; $\mathrm{Rep}G = \bigcup_{n\geqslant 1} \mathrm{Rep}_n G$ and let $\mathfrak{K}(G)$ be the coefficient algebra of G, i. e. the algebra of all functions f on G of the form $f(x) = \langle D(x)\xi, \eta\rangle \equiv d_{\xi,\eta}(D)$ where $D \in \mathrm{Rep}G$, $\xi, \eta \to \mathbf{C}^{n(D)}$.
In the coefficient algebra $\mathfrak{K}(G)$ we consider the Fourier topology τ_F (Heyer, 1977, Def. 1.3.10), and in Rep$G$ the compact–open topology τ_{co} which corresponds to the following convergence: a net $\{D_\alpha\} \subset \mathrm{Rep}G$ τ_{co}–converges to the element $D \in \mathrm{Rep}_n G$ if $\{D_\alpha\} \subset \mathrm{Rep}_n G$ and $\|D_\alpha(x) - D(x)\| \to 0$ uniformly on every compact set $K \subset G$. A net $\{f_\alpha\} \subset \mathfrak{K}(G)$ is called τ_F–convergent, if it converges uniformly on every compact $K \subset G$, and is a norm–bounded set in $\mathfrak{K}(G)$. We also denote $f^*(x) = f(x^{-1})$, $\forall f \in \mathfrak{K}(G)$, and use the bar for the complex–conjugation. One should remember that the algebra $\mathfrak{K}(G)$ separates the points of G iff G belongs to the class $\mathbf{A}$ of almost periodic groups (Heyer, 1977).

Definition 1. A linear function $\Phi(f)$, $f \in \mathfrak{K}(G)$, with values in $\mathfrak{F}_\sigma(\mathcal{A})$ is called
i) hermitean (h.) if

$$\Phi(\bar{f})[X^*] = \Phi(f)[X]^* \quad \forall f \in \mathfrak{K}(G), \quad X \in \mathcal{A};$$

ii) positive definite (p. d.) if the inequality

$$\sum_{jk} (\Psi_j \mid \Phi(\bar{f}_j f_k)[X_j^* X_k]\Psi_k) \geqslant 0 \tag{1}$$

holds for all finite sets $\{f_j\} \subset \mathfrak{K}(G), \{\Psi_j\} \subset \mathcal{H}, \{X_j\} \subset \mathcal{A}$;
iii) conditionally positive definite (c. p. d.), if the inequality (1) holds under the additional condition $\sum_j f_j(e) X_j \Psi_j = 0$.

Let us mention that our definition of p.d. function describes the positivity properties of quantum characteristic function of instrument (cf. (Holevo, 1988) for the abelian case), while the c.p.d. functions appear as logarithms of characteristic functions. Moreover, there is a Bochner–type theorem which states that the necessary and sufficient conditions for $\mathfrak{F}_\sigma(\mathcal{A})$–valued function $\Phi(f)$ on $\mathfrak{K}(G)$ to be the characteristic function of a unique completely positive instrument are: 1) $\Phi(1)[I] = I$; 2) Φ is a p.d. function; 3) $\Phi(\cdot)$ is $\tau_F \div \tau_2$–continuous. Assume $G \to \mathbf{A}$. Using Proposition 1.6 of (Barchielli and Lupieri, 1988), which is a generalization of Theorem 1.3.3 of (Heyer, 1977), one can establish a one–to–one correspondence between $\mathfrak{F}_\sigma(\mathcal{A})$–valued linear functions Φ on $\mathfrak{K}(G)$ and linear maps $F_\Phi : \mathrm{Rep}G \to \bigcup_{n\geqslant 1} \mathbf{M}(n, \mathfrak{F}_\sigma(\mathcal{A}))$, where $\mathbf{M}(n, \mathfrak{F}_\sigma(\mathcal{A}))$ denotes the algebra of $(n \times n)$–matrices with entries from $\mathfrak{F}_\sigma(\mathcal{A})$, by the relation

$$\Phi(d_{\xi,\eta}(D)) = \langle \xi \mid F_\Phi(D)\eta\rangle, \quad \forall d_{\xi,\eta}(D) \in \mathfrak{K}(G). \tag{2}$$

We remark also that $\Phi(g)$, $g \in \mathfrak{K}(G)$ is called τ_{co}–continuous if the corresponding map $F_\Phi(D)$, $D \in \mathrm{Rep}G$, in (2) is τ_{co}–continuous.
Let us introduce a noncommutative product of $\mathfrak{F}_\sigma(\mathcal{A})$–valued functions Φ, Ψ on $\mathfrak{K}(G)$ by the relation

$$\Phi \odot \Psi(d_{\xi,\eta}(D))[X] = \sum_{\alpha=1}^{n(D)} \langle \xi \mid F_\Phi(D) e_\alpha \rangle \circ \langle e_\alpha \mid F_\Psi(D)\eta \rangle, \tag{3}$$

where $d_{\xi,\eta}(D) \in \mathfrak{K}(G)$, $X \in \mathcal{A}$ and $\{e_\alpha\}$, $\alpha = \overline{1, n(D)}$, is an arbitrary orthonormal set in $\mathbf{C}^{n(D)}$. It is possible also to define the exponent $\mathrm{Exp}\mathcal{L}(g)$ corresponding to the product $\odot$ of a bounded $\mathfrak{F}_\sigma(\mathcal{A})$–valued linear function $\mathcal{L}(g)$, in the following way

$$\exp \mathcal{L}(d_{\xi,\eta}(D)) = \langle \xi | \exp F_{\mathcal{L}}(D)\eta \rangle, \quad \forall d_{\xi,\eta}(D) \in \mathfrak{K}(G). \tag{4}$$

The following is a noncommutative generalization of Schoenberg's theorem.

THEOREM 1. *Let $\mathcal{L}(g)$, $g \in \mathfrak{K}(G)$ be an $\mathfrak{F}_\sigma(\mathcal{A})$–valued function. Then the following conditions are equivalent:*

1) $\mathrm{Exp}t\mathcal{L}(g)$ *is a p.d. function for all* $t \geqslant 0$;

2) $\mathcal{L}(g)$ *is a h.c.p.d. function;*

3) $\mathcal{L}(g)$ *is hermitean and the kernel*

$$D\mathcal{L}(g, h; X, Y) = \mathcal{L}(gh)[X^*Y] - \bar{g}(e)X^*\mathcal{L}(h)[Y] -$$

$$-h(e)\mathcal{L}(\bar{g})[X^*]Y + \bar{g}h(e)X^*\mathcal{L}(1)[I]Y \tag{5}$$

is positive definite.

Proof. 1) $\Rightarrow$ 2) follows by differentiation with respect to t at $t = 0$ of the inequality (1) for $\Phi(f) = \mathrm{Exp}t\mathcal{L}(f)$ and the equivalence 2) $\Leftrightarrow$ 3) is rather standard (cf. (Holevo, 1988)). To prove 2) $\Rightarrow$ 1) we use the following theorem (Evans and Hanche–Olsen, 1979): if $\mathcal{A}$ is a C^*–algebra in Hilbert space $\mathcal{H}, \mathcal{L} \in \mathfrak{F}(\mathcal{A}), \mathcal{L}^* = \mathcal{L}$ then $\exp t\mathcal{L}$ is positive for all $t \geqslant 0$ if $X\Psi = 0$, where $X \in \mathcal{A}, \Psi \in \mathcal{H}$, implies $(\Psi | \mathcal{L}(X^*X)\Psi) \geqslant 0$. We are going to prove inequality (5) for $\Phi(f) = \mathrm{Exp}t\mathcal{L}(f)$ and arbitrary finite sets $\{\Psi_j\} \subset \mathcal{H}$, $\{f_j\} \subset \mathfrak{K}(G)$, $\{X_j\} \subset \mathcal{A}$. Since

$$f_j(x) = \langle D_j(x)\xi_j, \eta_j \rangle, \quad \xi_j, \eta_j \in \mathbf{C}^{n(D_j)}$$

then

$$\Phi(\bar{f}_j f_k) = \langle \eta_j \otimes \xi_k | \Phi(D_j^* \otimes D_k)\xi_j \otimes \xi_k \rangle.$$

Introducing orthonormal bases $\{e_\alpha^j\}$, $\alpha = \overline{1, n(D_j)}$, $\{e_\beta^k\}$, $\beta = \overline{1, n(D_k)}$, in $\mathbf{C}^{n(D_j)}$, $\mathbf{C}^{n(D_k)}$, respectively, we denote

$$\Phi_{(\alpha\gamma)(\beta\delta)}(D_j^* \otimes D_k) = \langle e_\alpha^j \otimes e_\gamma^k | \Phi(D_j^* \otimes D_k) e_\beta^j \otimes e_\delta^k \rangle.$$

Then (5) holds iff for all $N = 1, 2, \ldots$ and all sets $\{\Psi_{j\alpha}\} \subset \mathcal{H}$, $\{X_{j\alpha}\} \subset \mathcal{A}$

$$\sum_{j,k=1}^{N} \sum_{\alpha,\beta=1}^{n(D_j)} \sum_{\gamma,\delta=1}^{n(D_k)} (\Psi_{j\alpha}|\Phi_{(\alpha\gamma)(\beta\delta)}(D_j^* \otimes D_k)[X_{j\beta}^* X_{k\gamma}]\Psi_{k\delta}) \geqslant 0. \tag{6}$$

On the other hand the condition 2) is equivalent to the following: $\sum_{k=1}^{N} \sum_{\gamma=1}^{n(D_k)} X_{k\gamma}\Psi_{k\gamma} = 0$ implies a similar inequality for $\mathcal{L}$. Given N and $\{D_j\}$, $j = \overline{1, N}$, consider the Hilbert space consisting of vectors $\underline{\Psi} = [\Psi_k^\gamma]$; $k = \overline{1, N}, \gamma = \overline{1, n(D_k)}$ with the direct sum inner product. In $\mathcal{H}$ we consider the $\mathbf{C}^*$–algebra $\mathcal{A}$ consisting of matrices $\underline{X} = \left[X_{\alpha,\gamma}^{j,k}\right]$, $j, k = \overline{1, N}$, $\alpha = \overline{1, n(D_j)}$, $\gamma = \overline{1, n(D_k)}$, with elements from $\mathcal{A}$, so that

$$(\underline{X}\,\underline{\Psi})_j^\alpha = \sum_{k=1}^{N} \sum_{\gamma=1}^{n(D_k)} X_{\alpha,\gamma}^{j,k} \Psi_k^\gamma.$$

The product in $\underline{\mathcal{A}}$ is then

$$(\underline{X}\,\underline{Y})_{\alpha,\lambda}^{j,l} = \sum_{k=1}^{N} \sum_{\gamma=1}^{n(D_k)} X_{\alpha,\gamma}^{j,k} Y_{\gamma,\lambda}^{k,l} \tag{7}$$

and the involution

$$\underline{X}^* = \left[(X_{\gamma,\alpha}^{k,j})^*\right].$$

Consider the map $\underline{\mathcal{L}} : \underline{\mathcal{A}} \to \underline{\mathcal{A}}$ defined as

$$(\underline{\mathcal{L}}\,[\underline{X}])_{\alpha,\gamma}^{j,k} = \sum_{\beta=1}^{n(D_j)} \sum_{\delta=1}^{n(D_k)} \mathcal{L}_{(\alpha\gamma)(\beta\delta)}(D_j^* \otimes D_k)\left[X_{\beta,\delta}^{j,k}\right]. \tag{8}$$

Let $\underline{X}\,\underline{\Psi} = 0$ that is

$$\sum_{k=1}^{N} \sum_{\gamma=1}^{n(D_k)} X_{\alpha,\gamma}^{j,k} \Psi_k^\gamma = 0$$

for all $j = \overline{1, N}$, $\quad \alpha = \overline{1, n(D_j)}$. Then

$$(\underline{\Psi}|\underline{\mathcal{L}}(\underline{X}^*\underline{X})\underline{\Psi}) = \sum_{j,k=1}^{N} \sum_{\alpha,\beta=1}^{n(D_j)} \sum_{\gamma,\delta=1}^{n(D_k)} \Bigg(\Psi_j^\alpha|\mathcal{L}_{(\alpha\gamma)(\beta\delta)}(D_j^* \otimes D_k)$$

$$\left[\sum_{l=1}^{N} \sum_{\lambda=1}^{n(D_l)} (X_{\lambda,\beta}^{l,j})^* X_{\lambda,\gamma}^{l,k}\right] \Psi_k^\delta\Bigg), \tag{9}$$

which is non-negative by the condition 2). Therefore $\exp t\underline{\mathcal{L}}$ is positive for all $t \geqslant 0$. Due to the definition (7) of the product in $\underline{\mathcal{A}}$ and to (2) we have

$$(\exp t\underline{\mathcal{L}}[\underline{X}])^{j,k}_{\alpha,\gamma} = \sum_{\beta=1}^{n(D_j)} \sum_{\delta=1}^{n(D_k)} (\Phi)_{(\alpha\gamma)(\beta\delta)}(D_j^* \otimes D_k)\left[X^{j,k}_{\beta,\delta}\right],$$

where $\Phi = \exp t\mathcal{L}$, hence from positivity of $\exp t\underline{\mathcal{L}}$ as in (9), we deduce (6). The theorem is proved.

The next theorem is a generalization to the non-Abelian case of the corresponding result from (Holevo, 1988).

THEOREM 2. *Let $\mathcal{L}(g)$ be a h.c.p.d. function on $\mathfrak{K}(G)$ with values in $\mathfrak{K}(\mathcal{A})$. There exists the following set of objects:*

*a) a Hilbert space $\mathcal{K}$, *–representations $\rho(g)$ and $\pi[X]$ of algebras $\mathfrak{K}(G)$ and $\mathcal{A}$, respectively; such that*

$$[\rho(g), \pi[X]] = 0 \quad \forall g \in \mathfrak{K}(G), \quad X \in \mathcal{A};$$

b) an $\mathfrak{B}(\mathcal{H},\mathcal{K})$–valued function $B(g), g \in \mathfrak{K}(G)$, satisfying the 1–cocycle equation

$$B(gh) = \rho(g)B(h) + h(e)B(g) \tag{10}$$

and such that $B(\bar{g})^*B(h) = B(\bar{h})^*B(g)$, $\pi[X]B(g) = B(g)X$ $\forall X \in \mathcal{A}$, $g,h \in \mathfrak{K}(G)$;

c) an $\mathfrak{Z} = \mathcal{A}' \cap \mathcal{A}''$–valued function $\mathcal{I}(g), g \in \mathfrak{K}(G)$, satisfying the equation

$$\mathcal{I}(gh) - g(e)\mathcal{I}(h) - h(e)\mathcal{I}(g) = \frac{1}{2i}\{B(\bar{g})^*B(h) - B(\overline{g^*})^*B(h^*)\} \tag{11}$$

and having the properties

$$\mathcal{I}(g^*) = -\mathcal{I}(g), \quad \mathcal{I}(\bar{g}) = -\mathcal{I}(g)^* \quad \forall g,h \in \mathfrak{K}(G);$$

d) operators $C \in \mathcal{A}''$, $A \in \mathfrak{B}(\mathcal{H},\mathcal{K})$, such that

$$\mathcal{L}(g)[X] = A^*\rho(g)\pi[X]A + A^*B(g)X + XB(\bar{g})^*A +$$

$$+ X[\theta(g) + i\mathcal{I}(g)] + g(e)(C^*X + XC), \tag{12}$$

where $\theta(d_{\xi,\eta}(D)) = -\frac{1}{2}\sum_{\nu=1}^{n(D)} B(d_{e_\nu,\xi}(D)^*)^*B(d_{e_\nu,\eta}(D)^*)$, for every orthonormal set $\{e_\nu\}$, $\nu = \overline{1,n(D)}$, $\forall d_{\xi,\eta}(D) \in \mathfrak{K}(G)$; and vice versa, if there is a set of objects satisfying the conditions a)– d), then the formula (12) determines a h.c. p.d. function.

Sketch of the proof. We first formulate several lemmas, which are more or less direct generalizations of the corresponding results in the case of Abelian groups.

LEMMA 1. *Let $\Phi(f)$ be a $\mathfrak{F}_\sigma(\mathcal{A})$–valued p.d. function on $\mathfrak{K}(G)$. Then there exists a Hilbert space $\mathcal{K}$, an operator $F \in \mathfrak{B}(\mathcal{H},\mathcal{K})$, a*–representation ρ : $\mathfrak{K}(G) \to \mathfrak{B}(\mathcal{K})$ a*–representation $\pi : \mathcal{A} \to \mathfrak{B}(\mathcal{K})$ such that $[\pi[X], \rho(f)] = 0 \quad \forall X \in \mathcal{A}, \quad f \in \mathfrak{K}(G)$. Moreover there is a representation*

$$\Phi(f)[X] = F^*\pi[X]\rho(f)F \tag{13}$$

and vice versa, if there exists a set of objects $\{F, \pi, \rho\}$, which fits the above conditions, then the formula (13) determines a p.d. function.

LEMMA 2. *Let $\mathcal{L}(g), g \in \mathfrak{K}(G)$ be a h.c.p.d. function with values in $\mathfrak{F}_\sigma(\mathcal{A})$. Then there exists a Hilbert space $\mathcal{K}$, a*–representation $\rho : \mathfrak{K}(G) \to \mathfrak{B}(\mathcal{K})$, a *–representation $\pi : \mathcal{A} \to \mathfrak{B}(\mathcal{K})$ such that*

$$[\rho(g), \pi[X]] = 0 \quad \forall g \in \mathfrak{K}(G), \quad X \in \mathcal{A};$$

and a family of bilinear functions $\mathcal{B}(g)[X], g \in \mathfrak{K}(G), X \in \mathcal{A}$ with values in $\mathfrak{B}(\mathcal{H},\mathcal{K})$ which satisfy the 1–cocycle equation

$$\mathcal{B}(hg)[YX] = \rho(h)\pi[Y]\mathcal{B}(g)[X] + g(e)\mathcal{B}(h)[Y]X \tag{14}$$

$$\forall g, h \in \mathfrak{K}(G); \quad X, Y \in \mathcal{A}.$$

Moreover we can write

$$D\mathcal{L}(g, h; X, Y) = \mathcal{B}(g)[X]^*\mathcal{B}(h)[Y]. \tag{15}$$

Now we introduce in $B(\mathcal{H},\mathcal{K})$ an ultraweakly closed subspace $\mathfrak{M} = \mathrm{cl}\{\mathcal{B}(g)[X]Y, g \in \mathfrak{K}(G), X, Y \in \mathcal{A}\}$. It is clear that if $A_1, A_2 \in \mathfrak{M}$ then $A_1^* A_2 \in \mathcal{A}''$.

LEMMA 3. *Every solution of the 1–cocycle equation (14) is of the form*

$$\mathcal{B}(g)[X] = \rho(g)\pi[X]A - g(e)AX + B(g)X \tag{16}$$

where $A \in \mathfrak{M}$, and $B(g)$ is an $\mathfrak{M}$–valued 1–cocycle such that $B(g)X = \pi[X]B(g)$, $B(\bar{g})^*B(h) = B(\bar{h})^*B(g)$ and

$$B(hg) = \rho(h)B(g) + g(e)B(h)$$

$$\forall h, g \in \mathfrak{K}(G), \quad X \in \mathcal{A}. \tag{17}$$

If we now define

$$\mathcal{L}_0(g)[X] = \mathcal{L}(g)[X] - A^*\rho(g)\pi[X]A - A^*B(g)X - XB(\bar{g})^*A \tag{18}$$

which is always possible because of Lemmas 2 and 3, then we can prove the following identity

$$D\mathcal{L}_0(g, h; X, Y) = X^*YB(g)^*B(h) \tag{19}$$

$\forall g, h \in \mathfrak{K}(G)$; $X, Y \in \mathcal{A}$. The next step is to introduce a function $\Delta(g) = \mathcal{L}_0(g)[I]$.

LEMMA 4. *The function $\Delta(g)$, $g \in \mathfrak{K}(G)$ is linear and posesses the following properties*

1) $\Delta(g) \in \mathcal{A}''$, $g \in \mathfrak{K}(G)$;
2) $\Delta(\bar{g}) = \Delta(g)^*$;
3) $\Delta(gh) - h(e)\Delta(g) - g(e)\Delta(h) + hg(e)\Delta(1) = B(\bar{g})^* B(h)$ $\forall g, h \in \mathfrak{K}(G)$.

Let us denote by $\mathcal{E}$ a conditional expectation from $\mathcal{A}''$ onto its centre $\mathfrak{Z}$. Then we can introduce a pair of functions on $\mathfrak{K}(G)$ with values in $\mathfrak{Z}$:

$$\theta_0(g) = \mathcal{E}\left(\frac{\Delta(g) + \Delta(g^*)}{2}\right), \tag{20}$$

$$\mathcal{I}_0(g) = \mathcal{E}\left(\frac{\Delta(g) - \Delta(g^*)}{2i}\right). \tag{21}$$

They are symmetric and antisymmetric with respect to the *–operation in $\mathfrak{K}(G)$, respectively. There is an obvious decomposition $\mathcal{E}(\Delta(g)) = \theta_0(g) + i\mathcal{I}_0(g)$. Having noticed that $f^*(e) = f(e)$ for every $f \in \mathfrak{K}(G)$, we can describe the properties of θ_0 and $\mathcal{I}_0$ in the same way as in Lemma 4.

For the antisymmetric part $\mathcal{I}_0$ we have:

1) $\mathcal{I}_0(g) \in \mathfrak{Z}$, $\mathcal{I}_0(1) = 0$;
2) $\mathcal{I}_0(g^*) = -\mathcal{I}_0(g)$, $\mathcal{I}_0(\bar{g}) = -\mathcal{I}(g)^*$;
3) $\mathcal{I}_0(gh) - g(e)\mathcal{I}_0(h) - h(e)\mathcal{I}_0(g) = \{B(\bar{g})^* B(h) - B(\overline{g^*})^* B(h^*)\}/2i$. (22)

And for the symmetric part θ_0 we can write:

1) $\theta_0(g) \in \mathfrak{Z}$;
2) $\theta_0(g^*) = \theta_0(g)$, $\theta_0(\bar{g}) = \theta_0(g)^*$;
3) $\theta_0(gh) - h(e)\theta_0(g) - g(e)\theta_0(h) + hg(e)\theta_0(1) = \frac{1}{2}\{B(\bar{g})^* B(h) + B(\overline{g^*})^* B(h^*)\}$. (23)

If we consider now the function $\mathcal{E}(\Delta(g))$, $g \in \mathfrak{K}(G)$, then we see that it posesses the same algebraic properties as $\Delta(g)$ in Lemma 4, but takes its values in $\mathfrak{Z}$. So, we can subtract one more term from $\mathcal{L}_0(g)$ by defining

$$\Lambda(g)[X] = \mathcal{L}_0(g)[X] - \mathcal{E}(\Delta(g))X \tag{24}$$

and prove by calculation that

$$D\Lambda(g, h; X, Y) = 0 \quad \forall g, h \in \mathfrak{K}(G), \quad X, Y \in \mathcal{A}. \tag{25}$$

The map $\Lambda(g)$ is evidently an $\mathcal{A}''$–valued one.

LEMMA 5. *Let $\Lambda(g)$, $g \in \mathfrak{K}(G)$ be a hermitean linear function with values in $\mathfrak{F}_\sigma$, such that $\Lambda(g)[X] \in \mathcal{A}''$ for all $X \in \mathcal{A}, g \in \mathfrak{K}(G)$ and satisfying the equation (25). Then it has the structure.*

$$\Lambda(g)[X] = g(e)(C_1^* X + XC_1) + \lambda(g)X, \tag{26}$$

where $C_1 \in \mathcal{A}''$ and $\lambda(g)$ is a $\mathfrak{Z}$-valued function on $\mathfrak{K}(G)$, such that

$$\lambda(gh) = h(e)\lambda(g) + g(e)\lambda(h). \tag{27}$$

Let us prove that $\lambda(g)$ is an antisymmetric function on $\mathfrak{K}$ (G), i.e.

$$\lambda(g^*) = -\lambda(g) \quad \forall g \in \mathfrak{K}(G). \tag{28}$$

From (27) it follows that $\lambda(1) = 0$. We use the following identity for the functions from the coefficient algebra $\mathfrak{K}(G)$ (cf. (Barchielli and Lupieri, 1988)). If we take $g(x) = \langle D(x)\xi, \eta\rangle$, then

$$g + g^* - 2g(e) = -\sum_{\nu=1}^{n(D)} \left(\langle D\xi|e_\nu\rangle - \langle\xi|e_\nu\rangle\right)\left(\langle e_\nu|D\eta\rangle - \langle e_\nu|\eta\rangle\right), \tag{29}$$

where $\{e_\nu\}, \nu = \overline{1, n(D)}$, is an arbitrary o.n.b. in $\mathbf{C}^{n(D)}$.
Taking the function λ from both sides of (29), and using (27), we have

$$\lambda(g^* + g - 2g(e)1) = 0$$

whence (28) follows.
It is useful to introduce the operator $\Theta = \theta_0(1)$, where θ_0 is defined by (21), and to consider the function

$$\theta(g) = \theta_0(g) - g(e)\Theta. \tag{30}$$

This function has the same properties 1), 2) as θ_0, but the equation for it is of the form

$$\theta(gh) - h(e)\theta(g) - g(e)\theta(h) = \frac{1}{2}\{B(\bar{g})^* B(h) + B(\overline{g^*})^* B(h^*)\}. \tag{31}$$

One can also write the final definitions

$$\mathcal{I}(g) = \mathcal{I}_0(g) + \lambda(g), \quad C = C_1 + \frac{1}{2}\Theta. \tag{32}$$

If we summarize the formulas (18), (24), (26), (30) and (32), then we obtain the desired representation (12), and to complete the proof we need only to find the explicit form of $\theta(g)$.
The unique symmetric solution of equation (31) can be written with the help of the identity (29). It has the form (cf. (Barchielli and Lupieri, 1988) for a similar structure):

$$\theta(d_{\xi,\eta}(D)) = -\frac{1}{2}\sum_{i=1}^{n(D)} B\left(d_{e_i,\xi}(D)^*\right)^* B\left(d_{e_i,\eta}(D)^*\right) \tag{33}$$

$$\forall d_{\xi,\eta}(D) \in \mathfrak{K}(G),$$

where $\{e_i\}$, $i = \overline{1, n(D)}$, is an arbitrary o.n.b. in $\mathbf{C}^{n(D)}$.

3. APPLICATION TO THE CASE OF LIE—MOORE GROUPS

The algebraic representation formula (12) can be considered as a very general noncommutative analogue of the Levy—Khintchine formula, which gives a full characterization of logarithms of infinitely divisible characteristic functions. To learn more details about the structure of the $\tau_{co} \div \tau_2$–continuous h.c.p.d. function $\mathcal{L}(g), g \in \mathfrak{K}(G)$ with values in $\mathfrak{F}_\sigma(\mathfrak{B}(\mathcal{H}))$, or to separate in (12) the "Gaussian" part from the "Poissonian" integral part, we must restrict the class of groups under consideration. So, it is enough for our purposes to investigate the structure of $\tau_{co} \div \div \tau_2$–continuous h.c.p.d. functions for the class of Bochner groups **B** (Heyer, 1977). If $G \in \mathbf{B}$ then the representation $\rho(g)$ in (12) becomes τ_F–strongly continuous, and it is possible then to extend it to the *–representation of the C^*–algebra $C^b_{\mathbb{C}}(G)$ of bounded continuous functions on G. After that, using the general theory of representations (Dixmier, 1967) we can decompose $\rho(g)$ into the direct integral of one dimensional representations. This decomposition leads naturally to the separation of "Gaussian" part, which corresponds to the subspace in $\mathcal{K}$, where $\rho(g)$ acts as a multiplication by $g(e)$, from the "Poissonian" integral part, where the integration is carried over the set $G\backslash\{e\}$. It is crucial that all the 1–cocycles $B(g)$ are subjected to the same direct integral decomposition into the one–dimensional 1–cocycles. The key point here is in the solution of these 1–cocycle equations by a method, which is similar to the one used on (Parthasarathy and Schmidt, 1972). Let us impose now another restriction on the class of groups under consideration. We shall consider only Lie groups, which are at the same time Moore groups (Heyer, 1967), i.e. they are Lie groups, which posess only finite–dimensional irreducible representations. Then theorem 2 implies the following partial noncommutative analogue of Hunt's theorem.

COROLLARY. *Let G be an n–dimensional real Lie—Moore group and $\mathcal{L}(g), g \in \in \mathfrak{K}(G)$ be a $\tau_{co} \div \tau_2$–continuous h.c.p.d. function with values in $\mathfrak{F}_\sigma(\mathfrak{B}(\mathcal{H}))$. Then there is a decomposition*

$$\mathcal{L}(g) = \mathcal{L}_1(g) + \mathcal{L}_2(g) + g(e)(C^*X + XC), \tag{34}$$

where

$$\mathcal{L}_1(g)[Y] = g(e)\sum_{j=1}^{\infty} R_j^* Y R_j + \sum_{j=1}^{n}(\widetilde{R}_j^* Y + Y\widetilde{R}_j)X_j(g)(e) + \\ +\frac{1}{2}\sum_{j,k=1}^{n} a_{jk}X_jX_k(g)(e)Y, \tag{35}$$

$$g \in \mathfrak{K}(G), \quad Y \in \mathfrak{B}(\mathcal{H}),$$

and $\{X_j\}_{j=1}^n$ is the canonical set of right derivatives for G, where $\{R_j\} \in \mathfrak{B}(\mathcal{H})$ are such that the series $\sum_{j=1}^{\infty} R_j^* R_j$ converges ultraweakly, $\widetilde{R}_k = \sum_{j=1}^{\infty} R_j\bar{C}_{jk} + b_k I$, $\{c_{jk}\}_{\substack{j=\overline{1,\infty}\\ k=\overline{1,n}}}$ are such that $\sum_{\nu=1}^{\infty} |c_{\nu k}|^2 < \infty, a_{kl} = \sum_{\nu=1}^{\infty} \bar{c}_{\nu k}c_{\nu l}, k,l = \overline{1,n}$ are real,

$\{b_k\}_{k=\overline{1,n}} \subset \mathbf{R}$.

The second term $\mathcal{L}_2(g)$ is given by the formula

$$\mathcal{L}_2(g)[Y] = \sum_{j,k=1}^{\infty} V_j^* Y V_k \int_{G\setminus e} g(x)\varphi^{jk}(x)\mu(dx) +$$

$$+\sum_{j=1}^{\infty} \int_{G\setminus e} (g(x)-g(e))\left[V_j^* Y \varphi^j(x) + Y V_j \overline{\varphi^j(x)}\right]\mu(dx) +$$

$$+Y \int_{G\setminus e} \left[g(x) - g(e) - \sum_{i=1}^{n} x_i(x) X_i(g)(e)\right]\varphi(x)\mu(dx), \tag{36}$$

where μ is a finite measure on $G\setminus e$, $\{V_j\}_{j=\overline{1,\infty}} \subset \mathfrak{B}(\mathcal{H})$, φ is a non-negative real function, such that

$$\int_{G\setminus e} |f(x) - f(e)|^2 \varphi(x)\mu(dx) < \infty \quad \forall f \in \mathfrak{K}(G)$$

and the matrix of functions

$$\begin{pmatrix} \varphi^{jk}(x) & \vdots & \overline{\varphi^j}(x) \\ \dots & \vdots & \dots \\ \varphi^k(x) & \vdots & \varphi(x) \end{pmatrix}_{j,k=1,\infty}$$

is positive definite μ–almost everywhere. The series

$$\sum_{j,k=1}^{\infty} V_j^* V_k \int_{G\setminus e} \varphi^{jk}(x)\mu(dx)$$

in (36) is ultraweakly convergent.

REFERENCES

Barchielli, A., Lanz, L. and Prosperi, G.M. (1982). A model for macroscopic description and continuous observations in quantum mechanics. *Nuovo Cimento.* **B72**, 79–121.

Barchielli, A. and Lupieri, G. (1988). Convolution semigroups in quantum probability and quantum stochastic calculus. *Preprint* IFUM 432/FT.

Davies, E.B. (1976). *Quantum Theory of Open Systems.* Academic Press, New York.

Dixmier, J. (1967). *C^*-algebres et leurs representations.* Gautier–Villars, Paris.

Evans, E. and Hanche–Olsen, H. (1979). The generators of positive semigroups. *J.Funct.Anal.* **32**, 207–212.

Heyer, H. (1977). *Probability Measures on Locally Compact Groups.* Springer, Berlin.

Holevo, A.S. (1987). Representations of Levy—Khintchine type in quantum probability theory. *Theory Probab. Appl.* **32**, 142–146.

Holevo, A. S. (1988). A noncommutative generalization of conditionally positive definite functions. *Lect. Notes in Math.* **1303**, 128–148.

Parthasarathy, K. R. and Schmidt, K. (1972). Positive definite kernels, continuous tensor products and central limit theorems of probability theory. *Lect.Notes in Math.* **272**, 1–107.

Prob. Theory and Math. Stat., Vol. 1, pp. 271–284
B.Grigelionis *et al.* (Eds.)
1990 VSP/Mokslas

ON THE INTEGRABILITY OF THE MAXIMUM AND THE LOCAL PROPERTIES OF GAUSSIAN FIELDS

V.A. DMITROVSKII*

Kaliningrad Technical Institute for the Fishing Industry and Economy,
Sovetskij Prospect 1, Kaliningrad, 236000, USSR

ABSTRACT

In the represented paper two classical problem of the theory of Gaussian fields are considered. The first one is the integrability of some functions on the maximum of Gaussian field. The second one is the formulation of the constructive conditions, which imply fields continuity or boundedness a.s. The results on the first problem have a complete apearence. The results on the second one represent the weakest constructive conditions known up today.

1. INTRODUCTION

Let $(\Omega, \mathfrak{A}, \mathbf{P})$ be a fundamental complete probability space, R be the set of real numbers, T be an arbitrary set, $\xi : (\Omega \times \mathsf{T}) \to \mathsf{R}$ be a centered Gaussian field. For $\forall t \in \mathsf{T}$, $s \in \mathsf{T}$ let us denote $r(t,s) = \mathbf{E}\xi(\cdot\,,t)\xi(\cdot\,,s)$, $\sigma(t) = [r(t,t)]^{1/2}$, $\sigma = \sup\limits_{t\in\mathsf{T}} \sigma(t)$.

In this paper two classical problems of the theory of Gaussian fields are considered. The first one is the evaluation of the greatest lower bound γ such that

$$\mathbf{E}\exp\Big((2\sigma^2)^{-1}\big(\sup_{t\in\mathsf{T}}|\xi(\cdot\,,t)| - \gamma\big)^2\Big) < \infty. \tag{1}$$

In (Talagrand, 1984) it had been shown, that (1) took place when

$$\gamma > \tau = \inf\Big\{u : \mathbf{P}\{\omega : \sup_{t\in\mathsf{T}}|\xi(\omega,t)| \leqslant u\} > 0\Big\}.$$

But the question of the strictness of this condition has been left open. In Theorem 3 we give the strict condition on γ and determine the value of τ. Moreover, the analogous problem is solved for the functional $\sup\limits_{t\in\mathsf{T}}\xi(\cdot\,,t)$ (Theorems 3, 4).

The second problem is the formulation of the constructive conditions, which guarantee continuity or boundedness almost surely (a.s.) of ξ. The well-known

* The represented results have been developed while postdoctoral fellowship in the Moscow State University in autumn, 1988.

Dudley's condition (Dudley, 1967) guarantees continuity and boundedness of ξ simultaneously. The necessary and sufficient majorising measure condition (Fernique, 1975; Talangrand, 1987) is non-constructive. In (Samorodnitsky, 1988) the condition of continuity in terms of the local Dudley's characteristics has been given, which had not yet demanding the convergence of Dudley's integral on the whole parametric set $\mathbf{T}$. In Theorem 5 this condition is weakened. Moreover, Theorem 5 gives sufficient condition for boundedness a.s. of ξ on $\mathbf{T}$.

2. THE BASIC NOTIONS AND NOTATIONS

For $\forall t \in \mathbf{T}$, $s \in \mathbf{T}$ let us denote $\rho(t,s) = [\sigma^2(t) - 2r(t,s) + \sigma^2(s)]^{1/2}$, ρ is a pseudometric on $\mathbf{T}$, $\mathsf{B}(t,\varepsilon) = \{s \in \mathbf{T} : \rho(t,s) < \varepsilon\}$, $t \in \mathbf{T}$, $\varepsilon > 0$, is the ball on $\mathbf{T}$ of the radius ε with the center t, $(\mathbf{T},\rho)$ is the topological space generated by the set of the balls $\mathsf{B}(t,\varepsilon)$, $t \in \mathbf{T}$, $\varepsilon > 0$. In this paper we shall always assume completness of $(\mathbf{T},\rho)$. Let $\bar{\mathbf{R}} = \mathbf{R} \cup \{+\infty\} \cup \{-\infty\}$, $\mathbf{R}_+ = \{u \in \mathbf{R} : u > 0\}$, $\bar{\mathbf{R}}_+ = \mathbf{R}_+ \cup \{+\infty\}$. $\mathfrak{B}, \bar{\mathfrak{B}}, \mathfrak{B}_+, \bar{\mathfrak{B}}_+$ are σ-algebras of Borel subsets on $\mathbf{R}, \bar{\mathbf{R}}, \mathbf{R}_+, \bar{\mathbf{R}}_+$ respectively. $\mathfrak{C}$ is the minimal σ-algebra, containing all $\mathsf{B}(t,\varepsilon)$, $t \in \mathbf{T}$, $\varepsilon > 0$. $\mathfrak{A} \otimes \mathfrak{C}$ is the minimal σ-algebra, containing all Cartesian products $A \times \mathsf{B}(t,\varepsilon)$, $A \in \mathfrak{A}$, $t \in \mathbf{T}$, $\varepsilon > 0$.
Let $\mathbf{M}$ be the set of all probability measures, defined on $\mathfrak{C}$. For $\forall m \in \mathbf{M}$, $n \in \mathbf{M}$ there exists integral

$$\mathcal{R}(m,n) = \int\int r(t,s)m(dt)n(ds)$$

(from there and up to the end of the paper $\mathbf{T}$ is the set of integration). Let $\mathcal{P}(m,n) = [\mathcal{R}(m,m) - 2\mathcal{R}(m,n) + \mathcal{R}(n,n)]^{1/2}$. The balls $\mathcal{B}(m,\varepsilon)$, topological space $(\mathbf{M},\mathcal{P})$ and σ-algebras $\mathfrak{M}$, $\mathfrak{A} \otimes \mathfrak{M}$, are defined by analogy with $\mathsf{B}(t,\varepsilon)$, $(\mathbf{T},\rho)$, $\mathfrak{C}$, $\mathfrak{A} \otimes \mathfrak{C}$ respectively.
The next statements are the refinement of the well-known results (Fernique, 1975; Ito and Nisio, 1968). They play an auxiliary role and are evident enough.

THEOREM 1. *Let $(\mathbf{T},\rho)$ be a separable topological space. Then there exist a measurable Gaussian field $\xi^* : (\Omega \times \mathbf{T}, \mathfrak{A} \otimes \mathfrak{C}) \to (\bar{\mathbf{R}}, \bar{\mathfrak{B}})$, an event $\Omega^* : \mathbf{P}(\Omega^*) = 1$, and a measurable function $\alpha : (\mathbf{T},\mathfrak{C}) \to (\bar{\mathbf{R}}_+, \bar{\mathfrak{B}}_+)$ such that*
1) For $\forall t \in \mathbf{T}$

$$\mathbf{P}\{\omega : \xi(\omega,t) = \xi^*(\omega,t)\} = 1.$$

2) ξ^* is separable, i.e. for any open $\mathbf{T}' \subseteq \mathbf{T}$, for any countable dense on $(\mathbf{T},\rho)$ set S and for $\forall \omega \in \Omega^*$

$$\sup_{t\in\mathbf{T}'} \xi^*(\omega,t) = \sup_{s\in\mathbf{T}'\cap S} \xi^*(\omega,t), \qquad \inf_{t\in\mathbf{T}'} \xi^*(\omega,t) = \inf_{s\in\mathbf{T}'\cap S} \xi^*(\omega,t).$$

3) For $\forall \omega \in \Omega^*$, $t \in \mathbf{T}$, $s \in \mathbf{T}$: $\rho(t,s) = 0$

$$\xi^*(\omega,t) = \xi^*(\omega,s).$$

4) For $\forall \omega \in \Omega^*$, $t \in \mathbf{T}$ $\lim_{\varepsilon\to 0} \sup_{s\in\mathsf{B}(t,\varepsilon)} \xi^*(\omega,s) = \xi^*(\omega,t) + \alpha(t)/2,$

$$\lim_{\varepsilon\to 0} \inf_{s\in\mathsf{B}(t,\varepsilon)} \xi^*(\omega,s) = \xi^*(\omega,t) - \alpha(t)/2.$$

But in the following these properties will be too poor to grant the validity of our statements. So we have to construct some another version of ξ.

THEOREM 2. *Let* $(\mathbf{T},\rho)$ *be a separable topological space and* ξ^* *(defined in Theorem 1) be bounded on* $\mathbf{T}$ *a.s. Then there exist an event* $\Omega^{**} : \mathbf{P}(\Omega^{**}) = 1$, *a version* $\xi^{**} : (\Omega \times \mathbf{T}, \mathfrak{A} \otimes \mathfrak{C}) \to (\bar{\mathbf{R}}, \bar{\mathfrak{B}})$ *of the field* ξ, *a field* $\Xi : (\Omega \times \mathbf{M}, \mathfrak{A} \otimes \mathfrak{M}) \to (\bar{\mathbf{R}}, \bar{\mathfrak{B}})$ *and the function* $\mathcal{A} : \mathbf{M} \to \mathbf{R}_+$, *that satisfy all statements of Theorem 1 (after replacing in notation every single star by the pair), and*
5) *For* $\forall \omega \in \Omega^{**}$ *the function* $\xi^{**}(\omega,\cdot) : (\mathbf{T},\mathfrak{C}) \to (\mathbf{R},\mathfrak{B})$ *is measurable.*
6) *For* $\forall \omega \in \Omega^{**}$, $m \in \mathbf{M}$

$$\Xi(\omega,m) = \int \xi^{**}(\omega,t)m(dt).$$

7) *For* $\forall \omega \in \Omega^{**}$, $m \in \mathbf{M}$, $n \in \mathbf{M}$: $\quad \mathcal{P}(m,n) = 0$

$$\Xi(\omega,m) = \Xi(\omega,n).$$

8) *For* $\forall \omega \in \Omega^{**}$, $m \in \mathbf{M}$

$$\lim_{\varepsilon\to 0} \sup_{n\in\mathcal{B}(m,\varepsilon)} \Xi(\omega,n) = \Xi(\omega,m) + \mathcal{A}(m)/2,$$

$$\lim_{\varepsilon\to 0} \inf_{n\in\mathcal{B}(m,\varepsilon)} \Xi(\omega,n) = \Xi(\omega,m) - \mathcal{A}(m)/2.$$

9) *For* $\forall m \in \mathbf{M}$

$$\mathcal{A}(m) = \sup_{n\in\mathbf{K}(m)} \int \alpha(t)m(dt),$$

where

$$\mathbf{K}(m) = \{n \in \mathbf{M} : \mathcal{P}(m,n) = 0\}.$$

Note that α and $\mathcal{A}$ are the Ito—Nisio's oscillation functions for ξ^* and Ξ respectively.
In this paper we shall always assume (without any remarks) a separability of $(\mathbf{T},\rho)$ and a validity of all properties, posed in Theorems 1 and 2, for any field under consideration. Moreover, for the simplicity of notation the stars in ξ^*, ξ^{**} will be omited (except for the proofs of Theorems 1, 2).
Let $\Phi(u) = (2\pi)^{-1/2} \int_{-\infty}^{u} \exp(-x^2/2)dx$, $\bar{\Phi}(u) = 1 - \Phi(u)$.
Let Ψ, $\bar{\Psi}$ be the inverse functions to Φ and $\bar{\Phi}$ respectively (formaly $\Psi(0) = \bar{\Psi}(1) = -\infty$, $\Psi(1) = \bar{\Psi}(0) = +\infty$),

$$F(u) = \mathbf{P}\{\omega : \sup_{t\in\mathbf{T}} \xi(\omega,t) \leqslant u\}, \quad g(u) = \Psi\Big(\mathbf{P}\{\omega : \sup_{t\in\mathbf{T}} |\xi(\omega,t)| \leqslant u\}\Big),$$

$$h(u) = \Psi\big(F(u)\big), \quad \tau = \inf\{u : g(u) > -\infty\}, \quad \theta = \inf\{u : h(u) > -\infty\},$$

(formaly $\theta = -\infty$ if $h(u) > -\infty$ for $\forall u \in \mathbf{R}$; $h(u) = -\infty$ for $\forall u < \theta$; $g(u) = -\infty$ for $\forall u < \tau$),

$$\underline{\sigma} = \big[\inf_{m\in\mathbf{M}} \mathcal{R}(m,m)\big]^{1/2},$$

$$\mathbf{M}_* = \{m \in \mathbf{M} : \mathcal{R}(m,m) = \underline{\sigma}^2\}, \mathbf{T}^* = \{t \in \bar{\mathbf{T}} : \sigma(t) = \sigma\},$$

$$\alpha_* = \sup_{m\in\mathbf{M}_*} \int \alpha(t)m(dt), \quad \alpha^* = \sup_{t\in\mathbf{T}^*} \alpha(t), \quad \bar{\alpha} = \sup_{t\in\mathbf{T}} \alpha(t).$$

For any subset $\mathbf{T}' \subseteq \mathbf{T}$, for $\forall\varepsilon > 0$ we set $N(\mathbf{T}',\varepsilon)$ for the minimal number of the balls of radius ε with the centeres in $\mathbf{T}'$, covering whole $\mathbf{T}'$. $H(\mathbf{T}';\varepsilon) = \ln N(\mathbf{T}';\varepsilon)$ will be called ε-entropy of $\mathbf{T}'$, $D(\mathbf{T}';\varepsilon) = \int_0^\varepsilon H^{1/2}(\mathbf{T}';x)dx$ will be called Dudley's integral for $\mathbf{T}'$.

3. THE MAIN RESULTS

THEOREM 3. *Let ξ be bounded a.s. on $\mathbf{T}$. Then $\tau = \bar{\alpha}/2$ and for $\forall\gamma > \alpha^*/2$*

$$\mathbf{E}\exp\left((2\sigma^2)^{-1}(\sup_{t\in\mathbf{T}}|\xi(\cdot,t)| - \gamma)^2\right) < \infty, \tag{2}$$

$$\mathbf{E}\exp\left((2\sigma^2)^{-1}(\sup_{t\in\mathbf{T}}\xi(\cdot,t) - \gamma)^2\right) < \infty. \tag{3}$$

For $\forall\gamma \leqslant \alpha^*/2$ (2) and (3) are invalid.

THEOREM 4. *Let ξ be bounded a.s. on $\mathbf{T}$. If $\underline{\sigma} = 0$, then $\theta = \alpha_*/2$. If $\underline{\sigma} > 0$, then $\theta = -\infty$ and for $\forall\gamma > \alpha_*/2$*

$$\mathbf{E}\exp\left((2\underline{\sigma}^2)^{-1}\left(\left(\inf_{t\in\mathbf{T}}\xi(\cdot,t)\right)_+ - \gamma\right)^2\right) =$$

$$= \mathbf{E}\exp\left((2\underline{\sigma}^2)^{-1}\left(\left(\sup_{t\in\mathbf{T}}\xi(\cdot,t)\right)_- + \gamma\right)^2\right) < \infty, \tag{4}$$

where $(x)_+ = \max\{0,x\}$, $(x)_- = \min\{0,x\}$. *For $\forall\gamma \leqslant \alpha_*/2$ (4) is invalid.*

THEOREM 5. *Let some constants $a > 0$ and $b \geqslant 0$ be such that for $\forall t \in \mathbf{T}$*

$$\overline{\lim_{i\to\infty}} D\Big(\mathsf{B}(t, a(\ln i)^{-1/2})\backslash\mathsf{B}\big(t, a\big(\ln(i+1)\big)^{-1/2}\big); a(4\ln i)^{-1/2}\Big) \leqslant b. \tag{5}$$

Then ξ is bounded a.s. on $\mathbf{T}$, the set of points of discontinuity (if they exist) is finite and the maximal value of oscillation $\bar{\alpha}$ satisfies the inequality: $\bar{\alpha} \leqslant 2^{3/2}(4b + a)$. If (5) is valid for $\forall a > 0$, $b = 0$, then ξ is sample continuous a.s. on $(\mathbf{T},\rho)$.

Remark. Theorem 5 guarantees boundedness a.s. of the process

$$\xi(\omega,t) = w(\omega,t)\left(t\ln\ln(3t^{-1})\right)^{-1/2},$$

where w is the standard Brownian motion process, $t \in [0,1]$. This example shows that in spite of generality of the problem under consideration the conditions of Theorem 5 are close to the strict ones.

4. PROOF OF THE AUXILIARY RESULTS

Proof of Theorem 1. Separability of $(\mathbf{T},\rho)$ implies (Fernique, 1975, Theorem 3.2.2) an existence of Gaussian field $\xi' :(\Omega \times \mathbf{T}, \mathfrak{A} \otimes \mathfrak{C}) \to (\bar{\mathbf{R}}, \bar{\mathfrak{B}})$ and of an event Ω': $\mathbf{P}(\Omega') = 1$, which satisfy the statements 1)—3) of Theorem 1 (after the corresponding changes in notation $\xi^* \to \xi'$, $\Omega^* \to \Omega'$). Moreover (Fernique, 1975, Theorem 3.3.2), there exists an event $\Omega^* \subseteq \Omega'$: $\mathbf{P}(\Omega^*) = 1$ and a function $\alpha : \mathbf{T} \to \bar{\mathbf{R}}_+$ such that for $\forall \omega \in \Omega^*$, $t \in \mathbf{T}$

$$\lim_{\varepsilon\to 0} \sup_{s\in \mathbf{B}(t,\varepsilon)} \xi'(\omega,s) - \lim_{\varepsilon\to 0} \inf_{s\in \mathbf{B}(t,\varepsilon)} \xi'(\omega,s) = \alpha(t),$$

and for $\forall t \in \mathbf{T}$ there exists an event $\Omega_t : \mathbf{P}(\Omega_t) = 1$ such that for $\forall \omega \in \Omega_t$

$$\lim_{\varepsilon\to 0} \sup_{s\in \mathbf{B}(t,\varepsilon)} \xi'(\omega,s) = \xi'(\omega,t) + \alpha(t)/2,$$

$$\lim_{\varepsilon\to 0} \inf_{s\in \mathbf{B}(t,\varepsilon)} \xi'(\omega,s) = \xi'(\omega,t) - \alpha(t)/2.$$

Let us set $\mathbf{T}_\infty = \{t \in \mathbf{T} : \alpha(t) = \infty\}$ and

$$\xi^*(\omega,t) = \begin{cases} (\lim\limits_{\varepsilon\to 0} \sup\limits_{s\in \mathbf{B}(t,\varepsilon)} \xi'(\omega,s) + \lim\limits_{\varepsilon\to 0} \inf\limits_{s\in \mathbf{B}(t,\varepsilon)} \xi'(\omega,s))/2, & \text{if } t \in \mathbf{T}\backslash\mathbf{T}_\infty, \\ \xi'(\omega,t), & \text{if } t \in \mathbf{T}_\infty, \end{cases}$$

for $\forall \omega \in \Omega^*$. Statements 1)—4) are evident now. The proof of the measurability of α and ξ^* is not hard, but it is too long to be included in the restricted space of this paper.

Proof of Theorem 2. Since the complete proof of Theorem 2 is very long, we draft its scheme only.

Let us define a sequence $\{t_i\}_1^\infty$ by the next recurrent procedure: $t_1 \in \mathbf{T}^*$ is arbitrary and

$$t_i \in \arg\max_{t\in\mathbf{T}} \mathbf{E}\xi_i^2(\cdot\,,t), \qquad i = 2,3,\ldots,$$

where $\xi_i(\omega,t) = \xi(\omega,t) - \mathbf{E}\left(\xi(\cdot\,,t)|\mathfrak{F}_i\right)$, $\mathfrak{F}_i \subseteq \mathfrak{A}$ are the σ–algebras, generated by the random variables $\xi(\cdot\,,t_j)$, $j = 1,2,\ldots,i-1$. Let us consider the new version of ξ:

$$\xi^{**}(\omega,t) = \sum_{i=1}^{\infty} \zeta_i(\omega) h_i(t),$$

and its continuation onto $\mathbf{M}$:

$$\Xi(\omega,m) = \sum_{i=1}^{\infty} \zeta_i(\omega) \int h_i(t) m(dt),$$

where $\zeta_i(\omega) = \xi_i(\omega, t_i)$, $\xi_1(\omega, t) = \xi(\omega, t)$, $h_i(t) = \mathbf{E}\xi(\cdot, t)\xi_i(\cdot, t_i)/\mathbf{E}\xi_i^2(\cdot, t_i)$. These versions satisfy all statements of Theorem 2, but its proof and construction of Ω^{**} is very long.

LEMMA 1. *Let $\mathbf{T}'$ be an open set on $(\mathbf{T}, \rho)$. Then for $\forall \omega \in \Omega^*$ (Ω^* being defined in Theorem 1) $\Rightarrow$*

$$\sup_{t \in \mathbf{T}'} \xi(\omega, t) = \sup_{t \in \mathbf{T}'} \big(\xi(\omega, t) + \alpha(t)/2\big), \tag{6}$$

$$\inf_{t \in \mathbf{T}'} \xi(\omega, t) = \inf_{t \in \mathbf{T}'} \big(\xi(\omega, t) - \alpha(t)/2\big), \tag{7}$$

$$\sup_{t \in \mathbf{T}'} |\xi(\omega, t)| = \sup_{t \in \mathbf{T}'} \big(|\xi(\omega, t)| + \alpha(t)/2\big). \tag{8}$$

Proof. Let (6) is not true, i.e. for some $\omega_0 \in \Omega^*$

$$\sup_{t \in \mathbf{T}'} \xi(\omega_0, t) < \sup_{t \in \mathbf{T}'} \big(\xi(\omega_0, t) + \alpha(t)/2\big) = a.$$

Then there exists a sequence $\{t_i\}_1^\infty$ such that $\xi(\omega_0, t_i) + \alpha(t_i)/2 > a - i^{-1}$. But according to the statement 4) of Theorem 1 we may choose for $\forall i$ such sequences $\{s_{i,j}\}_{j=1}^\infty$, that

$$\xi(\omega_0, s_{i,j}) > \xi(\omega_0, t_i) + \alpha(t_i)/2 - j^{-1}, \qquad s_{i,j} \in \mathbf{T}'.$$

Then

$$\sup_{t \in \mathbf{T}} \xi(\omega_0, t) \geqslant \overline{\lim_{i \to \infty}}\, \xi(\omega_0, s_{i,i}) \geqslant \lim_{i \to \infty} (a - 2i^{-1}) = a.$$

This contradiction affirms (6). We may prove (7) by analogy, (8) is an immediate consequence of (6), (7).

LEMMA 2. *Let ξ be bounded on $\mathbf{T}$ a.s. Then for $\forall u \in \mathbf{R}$*

$$g(u) < (u - \alpha^*/2)/\sigma, \qquad h(u) \leqslant (u - \alpha^*/2)/\sigma.$$

Moreover, if $\underline{\sigma} > 0$, then for $\forall u \in \mathbf{R}$

$$h(u) \leqslant (u - \alpha_*/2)/\underline{\sigma}.$$

Proof. Let us take a point $t_0 \in \mathbf{T}^*$ such that $\alpha(t_0) = \alpha^*$. Then due to Lemma 1 for $\forall \omega \in \Omega^*$

$$\sup_{t \in \mathbf{T}} |\xi(\omega, t)| = \sup_{t \in \mathbf{T}} \big(|\xi(\omega, t)| + \alpha(t)/2\big) \geqslant |\xi(\omega, t_0)| + \alpha^*/2,$$

$$\sup_{t \in \mathbf{T}} \xi(\omega, t) = \sup_{t \in \mathbf{T}} \big(\xi(\omega, t) + \alpha(t)/2\big) \geqslant \xi(\omega, t_0) + \alpha^*/2,$$

and

$$g(u) \leqslant \Psi\Big(\mathbf{P}\{\omega : |\xi(\omega, t_0)| + \alpha^*/2 \leqslant u\}\Big) =$$

$$= \Psi\big(2\Phi((u - \alpha^*/2)/\sigma) - 1\big) < (u - \alpha^*/2)/\sigma,$$

$$h(u) \leqslant \Psi\Big(\mathbf{P}\{\omega : \xi(\omega, t_0) + \alpha^*/2 \leqslant u\}\Big) \leqslant (u - \alpha^*/2)/\sigma.$$

Let now $\underline{\sigma} > 0$. Let us take a measure $m_0 \in \mathbf{M}_*$ such that $\int \alpha(t) m_0(dt) = \alpha_*$. The random variable $\Xi(\cdot, m_0)$ is the centered normal one with the variance $\underline{\sigma}^2$. Then as a consequence of Lemma 1 for $\forall \omega \in \Omega^*$ we have

$$\sup_{t\in\mathbf{T}} \xi(\omega, t) = \sup_{t\in\mathbf{T}} \big(\xi(\omega, t) + \alpha(t)/2\big) \geqslant \int \big(\xi(\omega, t) + \alpha(t)/2\big) m_0(dt) =$$

$$= \Xi(\omega, m_0) + \alpha_*/2,$$

and $h(u) \leqslant \Psi\Big(\mathbf{P}\{\omega : \Xi(\omega, m_0) + \alpha_*/2 \leqslant u\}\Big) = (u - \alpha_*/2)/\underline{\sigma}$. Lemma 2 is proved.

Let us set for $\forall \mathbf{T}' \in \mathfrak{C}$

$$h(\mathbf{T}'; u) = \Psi\Big(\mathbf{P}\{\omega : \sup_{t\in\mathbf{T}'} \xi(\omega, t) \leqslant u\}\Big),$$

$$g(\mathbf{T}'; u) = \Psi\Big(\mathbf{P}\{\omega : \sup_{t\in\mathbf{T}'} |\xi(\omega, t)| \leqslant u\}\Big),$$

$$\varkappa_1(\mathbf{T}') = \inf\big\{u : h(\mathbf{T}'; u) \geqslant 0\big\},$$

$$\varkappa_2(\mathbf{T}') = \inf\big\{u : g(\mathbf{T}'; u) \geqslant 0\big\}, \qquad \sigma(\mathbf{T}') = \sup_{t\in\mathbf{T}'} \sigma(t).$$

Lemma 3. *Let $\mathbf{T}' \subseteq \mathbf{T}$ be open. Then for $\forall u \geqslant \varkappa_1(\mathbf{T}')$*

$$h(\mathbf{T}'; u) \geqslant \big(u - \varkappa_1(\mathbf{T}')\big)/\sigma(\mathbf{T}'), \tag{9}$$

for $\forall u \geqslant \varkappa_2(\mathbf{T}')$

$$g(\mathbf{T}'; u) \geqslant \big(u - \varkappa_2(\mathbf{T}')\big)/\sigma(\mathbf{T}'), \tag{10}$$

and the functions $h(\mathbf{T}'; \cdot)$, $g(\mathbf{T}'; \cdot)$ are concave on $\mathbf{R}$.

Proof. It was shown independently in (Sudakov, 1974; Borell, 1975), that for $\forall v_i \in \mathbf{R}$, $w_i \in \mathbf{R}$, $i = 1, 2$,

$$h(\mathbf{T}' : v_1 + w_1\sigma(\mathbf{T}')) \geqslant h(\mathbf{T}'; v_1) + w_1, \;\; g(\mathbf{T}'; v_2 + w_2 \cdot \sigma(\mathbf{T}')) \geqslant g(\mathbf{T}'; v_2) + w_2.$$

Let us take $v_1 = \varkappa_1(\mathbf{T}')$, $v_2 = \varkappa_2(\mathbf{T}')$, $u_1 = v_1 + w_1\sigma(\mathbf{T}')$, $u_2 = v_2 + w_2\sigma(\mathbf{T}')$. Then (9), (10) follow from the evident inequalities

$$h(\mathbf{T}'; \varkappa_1(\mathbf{T}')) \geqslant 0, \qquad g(\mathbf{T}'; \varkappa_2(\mathbf{T}')) \geqslant 0.$$

Concavity of $h(\mathsf{T}';\cdot)$, $g(\mathsf{T}';\cdot)$ was proved in (Ehrhard, 1983).

LEMMA 4. *Under assumptions of Lemma 3*

$$\varkappa_1(\mathsf{T}') \leqslant \mathbf{E}\sup_{t\in\mathsf{T}'} \xi(\cdot,t), \qquad \varkappa_2(\mathsf{T}') \leqslant \mathbf{E}\sup_{t\in\mathsf{T}'} |\xi(\omega,t)|. \tag{11}$$

Proof. Let $h^{-1}(\mathsf{T}';\cdot)$ be the inverse function for $h(\mathsf{T}';\cdot)$, i.e. $h^{-1}(\mathsf{T}';x) = u$ for $x = h(\mathsf{T}';u)$, $u > \theta$, and $h^{-1}(\mathsf{T}';x) = \theta$ for $\forall x \leqslant h(\mathsf{T}';\theta)$. It is evident, that $h^{-1}(\mathsf{T}';\cdot)$ is continuous convex function on $\mathbf{R}$. Using Jensen's inequality, we have

$$\mathbf{E}\sup_{t\in\mathsf{T}'} \xi(\omega,t) = \int_{-\infty}^{\infty} u\, d\Phi\big(h(\mathsf{T}';u)\big) = \int_{-\infty}^{\infty} h^{-1}(\mathsf{T}';x)\, d\Phi(x) \geqslant$$

$$\geqslant h^{-1}\left(\mathsf{T}'; \int_{-\infty}^{\infty} x d\Phi(x)\right) = h^{-1}(\mathsf{T}';0) = \varkappa_1(\mathsf{T}').$$

The proof of the second estimation is analoguous.

LEMMA 5. *Let ξ be bounded a.s. on T. Then for $\forall u \geqslant \varkappa_1(\mathsf{T}) + \sigma(2\ln 2)^{1/2}$*

$$h(u) \geqslant \left[\sigma^{-2}(u-\alpha^*/2)^2 - 2\chi_1(u)\right]^{1/2}, \tag{12}$$

for $\forall u \geqslant \varkappa_2(\mathsf{T}) + \sigma(2\ln 2)^{1/2}$

$$g(u) \geqslant \left[\sigma^{-2}(u-\alpha^*/2)^2 - 2\chi_2(u)\right]^{1/2}, \tag{13}$$

where $\chi_i(u) = \inf_{\varepsilon\geqslant w_i(u)} \chi_i(u,\varepsilon)$, $\quad w_i(u) = v\big(\sigma(\varkappa_i(\mathsf{T}) - \alpha^*/2)/(u-\alpha^*/2)\big)$,

$$v(\delta) = \sup_{t\in\mathsf{T}^*_\delta} \inf_{s\in\mathsf{T}^*} \rho(t,s),$$

$$\mathsf{T}^*_\delta = \{t\in\mathsf{T} : \sigma(t) > \sigma - \delta\},$$

$$\chi_i(u,\varepsilon) = \ln\left[N(\mathsf{T}^*;\varepsilon)+1\right] + (u-\alpha^*/2)\big(\sup_{t\in\mathsf{T}^*} \varkappa_i(\mathsf{B}(t,2\varepsilon)) - \alpha^*/2\big)/\sigma^2 -$$

$$-\big(\sup_{t\in\mathsf{T}^*} \varkappa_i(\mathsf{B}(t,2\varepsilon)) - \alpha^*/2\big)^2/(2\sigma^2), \qquad i = 1,2.$$

Proof. For an arbitrary $u \geqslant \varkappa_1(\mathsf{T}) + \sigma(2\ln 2)^{1/2}$ let us take $\delta = \sigma(\varkappa_1(\mathsf{T}) - -\alpha^*/2)/(u-\alpha^*/2)$, $\varepsilon \geqslant v(\delta)$, and fix on T^* some minimal ε-net $S = \{s_i\}_1^N$, $N = N(\mathsf{T}^*;\varepsilon)$. Then

$$\mathsf{T}^*_\delta \subseteq \bigcup_{i=1}^{N} \mathsf{B}(s_i, 2\varepsilon). \tag{14}$$

Indeed, if (14) is not true, we may take such $t_1 \in \mathbf{T}^*_\delta$, that $\max_{i \leqslant N} \rho(t_1, s_i) \geqslant 2\varepsilon$. But the definitions of v and S imply an existence of $t_2 \in \mathbf{T}^*$ and $j \in \{1, \ldots, N\}$ such that $\rho(t_1, t_2) < v(\delta) \leqslant \varepsilon$, $\rho(t_2, s_j) < \varepsilon$. Consequently $\rho(t_1, s_j) \leqslant \rho(t_1, t_2) + \rho(t_2, s_j) < 2\varepsilon$. This contradiction proves the inclusion (14).
Due to the choice of δ and Lemma 3 we have

$$\mathbf{P}\{\omega : \sup_{t \in \mathbf{T}} \xi(\omega, t) \geqslant u\} \leqslant \sum_{i=1}^{N} \mathbf{P}\{\omega : \sup_{t \in \mathbf{B}(s_i, 2\varepsilon)} \xi(\omega, t) \geqslant u\} +$$

$$+\mathbf{P}\{\omega : \sup_{t \in \mathbf{T} \setminus \mathbf{T}^*_\delta} \xi(\omega, t) \geqslant u\} \leqslant N \bar{\Phi}\Big(\Big(u - \sup_{t \in \mathbf{T}^*} \varkappa_1(\mathbf{B}(t, 2\varepsilon))\Big)/\sigma\Big) +$$

$$+\bar{\Phi}\Big((u - \varkappa_1(\mathbf{T}))/(\sigma - \delta)\Big) \leqslant \bar{\Phi}\Big(\Big(\sigma^{-2}\Big(u - \sup_{t \in \mathbf{T}^*} \varkappa_1(\mathbf{B}(t, 2\varepsilon))\Big)^2 -$$

$$-2\ln(N+1)\Big)^{1/2}\Big) = \bar{\Phi}\Big(\Big(\sigma^{-2}(u - \alpha^*/2)^2 - \chi_1(u, \varepsilon)\Big)^{1/2}\Big).$$

Here we have used the simple inequality: for $\forall u \geqslant (2a)^{1/2}$

$$\bar{\Phi}(u)\exp(a) \leqslant \bar{\Phi}((u^2 - 2a)^{1/2}).$$

A validity of (12) follows from an arbitrary choice of ε. The proof of (13) is analogous.

LEMMA 6. *Let $\underline{\sigma} > 0$. Then for $\forall m \in \mathbf{M}_*$, $n \in \mathbf{M}$*

$$\mathcal{R}(m, n) \geqslant \underline{\sigma}^2.$$

Proof. Let us suppose, that the statement is not valid, i.e. there exist measures $m_0 \in \mathbf{M}_*$, $m_1 \in \mathbf{M}$ such that $\mathcal{R}(m_0, m_1) = r^2 < \underline{\sigma}^2$. Let us take the new measure $m_\lambda = (1 - \lambda)m_0 + \lambda m_1$, where $\lambda = (\underline{\sigma}^2 - r^2)/(\underline{\sigma}^2 - 2r^2 + \mathcal{R}(m_1, m_1))$. Since $\lambda \in]0, 1[$, then $m_\lambda \in \mathbf{M}$, and

$$\mathcal{R}(m_\lambda, m_\lambda) = (1 - \lambda)^2 \underline{\sigma}^2 + 2\lambda(1 - \lambda) r^2 + \lambda^2 \mathcal{R}(m_1, m_1) = (1 - \lambda)\underline{\sigma}^2 + \lambda r^2 < \underline{\sigma}^2.$$

This contradiction with the definition of $\underline{\sigma}$ proves our lemma.

LEMMA 7. *Let $\mathbf{T}' \subseteq \mathbf{T}$ be open and $d = D(\mathbf{T}'; \sigma(\mathbf{T}')/2) < \infty$. Then*

$$\mathbf{E} \sup_{t \in \mathbf{T}'} \xi(\cdot, t) \leqslant 2^{5/2} d.$$

Proof. Let us fix on $\mathbf{T}'$ some minimal ε_k–nets

$$S_k = \{s_k^i\}_{i=1}^{N(k)},\ N(k) = N(\mathbf{T}'; \varepsilon_k),\ \varepsilon_k = \sigma(\mathbf{T}')2^{-k}, \quad k = 1, 2, \ldots .$$

For $\forall s \in S_k$ we set

$$\mathsf{U}_k(s) = \mathsf{B}(s, \varepsilon_k) \qquad \text{if} \qquad s = s_k^1,$$

$$\mathsf{U}_k(s) = \mathsf{B}(s, \varepsilon_k) \setminus \bigcup_{j=1}^{i-1} \mathsf{U}_k(s_k^j) \qquad \text{if} \qquad s = s_k^i,\ i = 2, \ldots, N(k),$$

and define the mappings $f_k : \mathsf{T}' \to S_k$, $\quad k = 1, 2, \ldots$: for $\forall t \in \mathsf{T}'$.

$$f_k(t) = s \in S_k \quad \Leftrightarrow \quad t \in \mathsf{U}_k(s).$$

Note that $\rho(t,s) < \varepsilon_k$ if $f_k(t) = s$. As far as $S = \bigcup_{k=1}^{\infty} S_k$ is countable dense set on (T', ρ), we have a.s.

$$\sup_{t \in \mathsf{T}'} \xi(\omega, t) = \sup_{t \in S} \xi(\omega, t) \leqslant \sup_{t \in S_1} \xi(\omega, t) + \sum_{k=1}^{\infty} \sup_{s \in S_{k+1}} \Big(\xi(\omega, s) - \xi(\omega, f_k(s))\Big).$$

Using Jensen's inequality for $\lambda_1 = \sigma^{-1}(\mathsf{T}')\big(2 \ln N(1)\big)^{1/2}$, we deduce

$$\mathbf{E} \sup_{s \in S_1} \xi(\cdot, s) \leqslant \lambda_1^{-1} \ln \mathbf{E} \exp\big(\lambda_1 \sup_{s \in S_1} \xi(\cdot, s)\big) \leqslant$$

$$\leqslant \lambda_1^{-1} \ln \left[\sum_{i=1}^{N(1)} \mathbf{E} \exp\big(\lambda_1 \xi(\cdot, s_1^i)\big)\right] \leqslant \lambda_1^{-1} \ln \Big[N(1) \exp\big(\lambda_1^2 \sigma^2(\mathsf{T}')/2\big)\Big] =$$

$$= \lambda_1^{-1} \ln N_1 + \lambda_1 \sigma^2(\mathsf{T}')/2 = \sigma(\mathsf{T}')\big(2 \ln N(1)\big)^{1/2} = 2^{3/2} \varepsilon_1 \big(\ln N(1)\big)^{1/2},$$

and, by analogy, for $\lambda_k = \varepsilon_{k-1}^{-1}\big(2 \ln N(k)\big)^{1/2}$, $\quad k = 2, 3 \ldots$:

$$\mathbf{E} \sup_{s \in S_k} \Big(\xi(\cdot, s) - \xi(\cdot, f_{k-1}(S))\Big) \leqslant$$

$$\leqslant \lambda_k^{-1} \ln \left[\sum_{i=1}^{N(k)} \mathbf{E} \exp\Big(\lambda_k \Big(\xi(\cdot, s_k^i) - \xi(\cdot, f_{k-1}(s_k^i))\Big)\Big)\right] \leqslant$$

$$\leqslant \lambda_k^{-1} \ln \big[N(k) \exp(\lambda_k^2 \varepsilon_{k-1}^2/2)\big] = \varepsilon_{k-1}\big(2 \ln N(k)\big)^{1/2} \leqslant$$

$$\leqslant 2^{3/2} \varepsilon_k \big(\ln N(k)\big)^{1/2}.$$

Consequently due to monotonicity of $N(\mathsf{T}'; \cdot)$ we have

$$\mathbf{E} \sup_{t \in \mathsf{T}'} \xi(\cdot, t) \leqslant 2^{3/2} \sum_{k=1}^{\infty} \varepsilon_k \big(\ln N(k)\big)^{1/2} \leqslant 2^{5/2} \int_0^{\varepsilon_1} H^{1/2}(\mathsf{T}'; x)\, dx.$$

Lemma 7 is proved.

5. PROOF OF THE MAIN RESULTS

Proof of Theorem 4. Let us take into consideration Hilbert space $\mathbf{H} = \mathcal{L}^2(\Omega, \mathfrak{A}, \mathbf{P})$ of the square–integrable random variables with the inner product $\langle h, g\rangle = \int_\Omega h(\omega)g(\omega)\mathbf{P}(d\omega)$ for $\forall h \in \mathbf{H}$, $g \in \mathbf{H}$. The set $\mathbf{G} = \{\Xi(\cdot\,, m) : m \in \mathbf{M}\}$ is a closed convex subset of $\mathbf{H}$. The condition $\underline{\sigma} = 0$ imply that $\mathbf{G}$ contains zero of $\mathbf{H}$. Let us denote

$$\theta_{\mathbf{G}} = \inf \{u : \mathbf{P}\{\omega : \sup_{g\in\mathbf{G}} g(\omega) \leqslant u\} > 0\},$$

$$IN(\mathbf{G}; h) = \lim_{\varepsilon\to 0} \sup_{\|g-h\|<\varepsilon, g\in\mathbf{G}} g(\cdot) - \lim_{\varepsilon\to 0} \inf_{\|g-h\|<\varepsilon, g\in\mathbf{G}} g(\cdot).$$

In (Lifschitz, 1989, Theorem 2) it was shown, that $\theta_{\mathbf{G}} = IN(\mathbf{G}, 0)/2$. But it is evident, that $\theta_{\mathbf{G}} = \theta$ and $IN(\mathbf{G}; 0) = \alpha_*$ (following Theorem 2). Therefore $\theta = \alpha_*/2$.

Let now $\underline{\sigma} > 0$. Let us take a measure $m_0 \in \mathbf{M}_*$ such that $\int \alpha(t)m_0(dt) = \alpha_*$. Then $\Xi(\cdot\,, m_0)$ is a centered normal random variable with the variance $\underline{\sigma}^2$. Let us consider the new field $\eta(\omega, t) = \xi(\omega, t) - \Xi(\omega, m_0)$. For (and only for) $m \in \mathbf{M}_*$ we have

$$\int\!\!\int (\mathbf{E}\eta(\cdot\,, t)\eta(\cdot\,, s))m(dt)m(ds) = 0.$$

Therefore η satisfies the condition $\underline{\sigma}_\eta = 0$. Since for $\forall t \in \mathbf{T}$, $\alpha_\eta(t) = \alpha(t)$ and $\mathbf{E}\eta(\cdot\,, t)\Xi(\cdot\,, m_0) \geqslant 0$ (Lemma 6) then, using Slepian's theorem, we obtain for $\forall u \geqslant -\alpha_*/2$, $\varepsilon > 0$

$$\mathbf{P}\{\omega : \sup_{t\in\mathbf{T}} \xi(\omega, t) \leqslant -u\} = \mathbf{P}\{\omega : \sup_{t\in\mathbf{T}} \eta(\omega, t) + \Xi(\omega, m_0) \leqslant -u\} \geqslant$$

$$\geqslant \mathbf{P}\Big(\{\omega : \sup_{t\in\mathbf{T}} \eta(\omega, t) \leqslant \alpha_*/2 + \varepsilon\} \cap \{\omega : \Xi(\omega, m_0) \leqslant -u - \alpha_*/2 - \varepsilon\}\Big) \geqslant$$

$$\geqslant \mathbf{P}\{\omega : \sup_{t\in\mathbf{T}} \eta(\omega, t) \leqslant \alpha_*/2 + \varepsilon\}\bar{\Phi}((u + \alpha_*/2 + \varepsilon)/\underline{\sigma}) \geqslant$$

$$\geqslant \bar{\Phi}\Big(\big(\underline{\sigma}^{-2}(u + \alpha_*/2)^2 + 2\chi_3(u, \varepsilon)\big)^{1/2}\Big), \tag{15}$$

where

$$\chi_3(u, \varepsilon) = \varepsilon(u + \alpha_*/2)/\underline{\sigma}^2 + \varepsilon^2/(2\underline{\sigma}^2) - \ln \mathbf{P}\{\omega : \sup_{t\in\mathbf{T}} \eta(\omega, t) \leqslant \alpha_*/2 + \varepsilon\}. \tag{16}$$

Here we have used the simple inequality for $\forall a > 0,\ u > 0$

$$\bar{\Phi}(u)\exp(-a) \geqslant \bar{\Phi}((u^2 + 2a)^{1/2}).$$

Estimates (15), (16) imply $\theta = -\infty$. Let

$$\varepsilon(u) = \inf \Big\{\varepsilon > 0 : -\ln \mathbf{P}(\{\omega : \sup_{t\in\mathbf{T}} \eta(\omega, t) \leqslant \alpha_*/2 + \varepsilon\}) \leqslant (u + \alpha_*/2)^{1/2}\Big\}.$$

Then $\chi_3(u,\varepsilon(u)) = \varepsilon(u)(u+\alpha_*/2)/\underline{\sigma}^2 + \varepsilon^2(u)/(2\underline{\sigma}^2) + (u+\alpha_*/2)^{1/2} = u\varphi(u)$, where $\varphi(u)\to 0$ as $u\to\infty$. The last observation proves (4) since

$$\mathbf{P}\{\omega : \sup_{t\in\mathbf{T}} \xi(\omega,t) \leqslant -u\} = \mathbf{P}\{\omega : \inf_{t\in\mathbf{T}} \xi(\omega,t) \geqslant u\}, \quad u \geqslant 0.$$

Proof of Theorem 3. In order to prove (3) we have to estimate separately the summands in $\chi_2(u,\varepsilon)$ (Lemma 5). On the one hand for $\forall t\in\mathbf{T}$, $\varepsilon>0$

$$\alpha(t)/2 \leqslant \varkappa_2(\mathbf{B}(t,\varepsilon)),$$

on the other hand for $\forall t\in\mathbf{T}$ we have

$$\alpha(t)/2 = \lim_{\varepsilon\to 0}\mathbf{E}\sup_{t\in\mathbf{B}(t,\varepsilon)}\xi(\omega,t) \geqslant \lim_{\varepsilon\to 0}\varkappa_1(\mathbf{B}(t,\varepsilon))$$

(see Lemma 4 and (Fernique, 1975, p. 46)). Since $\mathbf{T}^*$ is compact, the term $\left(\sup_{t\in\mathbf{T}^*}\varkappa_2(\mathbf{B}(t,2\varepsilon)) - \alpha^*/2\right)$ tends to zero as $\varepsilon\to 0$.
Boundedness of ξ implies (Fernique, 1975, Theorem 2.3.3) that

$$\overline{\lim_{\varepsilon\to 0}}\ln\ln N(\mathbf{T}^*;\varepsilon)/\ln\varepsilon^{-1} \leqslant 2.$$

For this reason the inequality $\ln\left(N(\mathbf{T}^*;\varepsilon)+1\right) \leqslant \varepsilon^{-3}$ is valid for sufficiently small $\varepsilon>0$. Let us take

$$\varepsilon(u) = \max\left\{(u-\alpha^*/2)^{-1/4},\quad v\left(\sigma(\varkappa_2(\mathbf{T})-\alpha^*/2)/(u-\alpha^*/2)\right)\right\}$$

(v is defined in Lemma 5). The function $v(\delta)$ tends to zero as $\delta\to 0$ because of continuity of function σ on $(\mathbf{T},\rho)$. Therefore $\varepsilon(u)\to 0$ as $u\to\infty$. Thus we have

$$0 \leqslant (u-\alpha^*/2)/\sigma - \left[\sigma^{-2}(u-\alpha^*/2)^2 - 2\chi_2(u)\right]^{1/2} \leqslant 2\sigma\chi_2(u)(u-\alpha^*/2)^{-1} \leqslant$$

$$\leqslant 2\sigma^{-1}\left(\sup_{t\in\mathbf{T}^*}\varkappa_2\left(\mathbf{B}(t,2\varepsilon(u))\right) - \alpha^*/2\right) +$$

$$+2\sigma(u-\alpha^*/2)^{-1}\ln\left(N\left(\mathbf{T}^*,(u-\alpha^*/2)^{-1/4}\right)+1\right) \leqslant$$

$$\leqslant 2\sigma^{-1}\left(\sup_{t\in\mathbf{T}^*}\varkappa_2\left(\mathbf{B}(t,2\varepsilon(u))\right) - \alpha^*/2\right) + 2\sigma(u-\alpha^*/2)^{-1/4}$$

for all u large enough. Together with Lemma 2 this implies immediately (2), (3). To prove the equality $\tau = \overline{\alpha}/2$, we have to consider the continuation of ξ on the set $\mathbf{Q}$ of δ-measures, any of which is concentrated in one of the points $t\in\mathbf{T}$, and takes value $+1$ or -1. Let us denote $\eta(\omega,t) = \int\xi(\omega,t)q(dt)$ for $\forall\omega\in\Omega$, $q\in\mathbf{Q}$. It is plain that $\sup_{t\in\mathbf{T}}|\xi(\omega,t)| = \sup_{q\in\mathbf{Q}}\eta(\omega,q)$, $\underline{\sigma}_\eta = 0$, $\alpha_{*\eta} = \bar{\alpha}$. Indeed, let $t_0\in\mathbf{T}$, $q_1\in\mathbf{Q}$, $q_2\in\mathbf{Q}$ be such that $\alpha(t_0)=\bar{\alpha}$, $q_1(\{t_0\})=1$, $q_2(\{t_0\})=-1$. Let measure m_0 on $\mathbf{Q}$ be such that $m_0(\mathbf{Q})=1$, $m_0(\{q_1\}) = m_0(\{q_2\}) = 1/2$. Then

$$\underline{\sigma}_\eta^2 \stackrel{!}{=} \int_{\mathbf{Q}}\int_{\mathbf{Q}}\mathbf{E}\eta(\cdot,p_1)\eta(\cdot,p_2)m_0(dp_1)m_0(dp_2) = 0, \quad \alpha_{*\eta} = \int\alpha(p)m_0(dp) = \bar{\alpha}.$$

We finish our proof by use of Theorem 4.

Proof of Theorem 5. Let us set for some $t_0 \in \mathsf{T}$, $c > 0$, $i = 2, 3, \dots$

$$\mathsf{T}_i = \mathsf{B}(t_0, a(\ln i)^{-1/2}) \backslash \mathsf{B}\big(t_0, a\big(\ln(i+1)\big)^{-1/2}\big),$$

$$A_i = \Big\{\omega : \sup_{t \in \mathsf{T}_i} \xi(\omega, t) - \xi(\omega, t_0) \geqslant 2^{1/2}\Big(a + c + 4D\big(\mathsf{T}_i; a(4\ln i)^{-1/2}\big)\Big)\Big\}.$$

Then

$$\mathbf{P}(A_i) \leqslant \bar{\Phi}\Big(h\Big(\mathsf{T}_i; 2^{1/2}(a + c + 4D(\mathsf{T}_i; a(4\ln i)^{1/2}))\Big)\Big) \leqslant$$

$$\leqslant \bar{\Phi}\Big(a^{-1}(\ln i)^{1/2}\big(2^{1/2}(a + c + 4D(\mathsf{T}_i; a(4\ln i)^{1/2})) - \varkappa_1(\mathsf{T}_i)\big)\Big) \leqslant$$

$$\leqslant \bar{\Phi}\big(a^{-1}(a+c)(2\ln i)^{1/2}\big) < i^{-(a+c)^2/a^2}.$$

The convergence of the set $\sum_{k=2}^{\infty} \mathbf{P}(A_i)$ imply, that a.s. only finite number of A_i takes place, i.e. due to supposition we have a.s.

$$\lim_{i \to \infty} \sup_{t \in \mathsf{T}_i} \xi(\omega, t) - \xi(\omega, t_0) \leqslant$$

$$\leqslant \overline{\lim_{i \to \infty}}\, 2^{1/2}\Big(a + c + 4D\big(\mathsf{T}_i; a(4\ln i)^{-1/2}\big)\Big) \leqslant 2^{1/2}(a + 4b + c).$$

The inequality $\bar{\alpha} \leqslant 2^{3/2}(a + 4b)$ follows now from an arbitrary choice of $t_0 \in \mathsf{T}$ and $c > 0$.

Suppose now, that number of points of discontinuity of ξ is not finite. Then there exists a sequence $\{t_j\}_1^\infty$ such that $\alpha(t_j) > 0$, $j = 1, 2, \dots$. Let t_0 be some of its limit point. In this case T_i with arbitrary large i contain points of discontinuity. Since Dudley's integral is not convergent for such T_i, we have

$$\overline{\lim_{i \to \infty}}\, D\big(\mathsf{T}_i; a(4\ln i)^{-1/2}\big) = \infty,$$

that contradicts the conditions of Theorem 5.

Suppose now (5) is valid for $b = 0$ and for $\forall a > 0$. In this case $\bar{\alpha} \leqslant 2^{3/2}(a + c)$, where a and c may be arbitrary small, i.e. $\bar{\alpha} = 0$. Together with the statement 4) of Theorem 1 this gives us the proof of Theorem 5.

The proof of boundedness of the process in Remark demands of accurate evaluation of the expression

$$d_i = D\Big(\mathsf{B}(0, a(\ln i)^{-1/2}) \backslash \mathsf{B}\big(0, a\big(\ln(i+1)\big)^{-1/2}\big); a(4\ln i)^{-1/2}\Big) =$$

$$= D\Big(\big[3\exp\big(-(i+1)^{a^{-2}}\big), 3\exp(-i^{a^{-2}})\big[; a(4\ln i)^{-1/2}\Big),$$

which shows, that $\overline{\lim_{i \to \infty}}\, d_i < \infty$ for $\forall a > 1$.

Acknowledgements
Author thanks V.I.Piterbarg, Ju.K.Beljaev, Ju.A.Davydov and particularly M.A.Lifschitz for their attention to the represented paper and useful remarks.

REFERENCES

Borell, C. (1975). The Brunn—Minkovsky inequality in Gauss space. *Invent. Math.* **30**, 207–216.

Dudley, R.M. (1967). The sizes of compact subsets of Hilbert spaces and continuity of Gaussian processes. *J. Funct. Analysis* **1**, 290–330.

Ehrhard, A. (1983). Symetrisation dans l'espace de Gauss. *Math. Scand.* **53**, 281–301.

Fernique, X. (1975). Regularite des trajectoires des fonctions aleatoires gaussiennes. *Lect. Notes in Math.* **480**, 1–96.

Ito, K. and Nisio, M. (1968). On the oscillation function of Gaussian processes. *Math. Scand.* **22**, 209–223.

Lifschitz, M.A. (1989). An oscillation and a lower·bound for the distribution of maximum of Gaussian field. *Notes Sci. Seminars Leningrad Branch of Math. Inst.*

Samorodnitsky, G. (1988). Continuity of Gaussian processes. *Ann. Probab.* **16**, 1019–1033.

Sudakov, V.N. and Tsyrelson, B.S. (1974). An extremal properties of the half–spaces for the spherical invariant measures. *Notes Sci. Seminars Leningrad Branch of Math. Inst.* **41**, 14–24.

Talagrand, M. (1984). Sur l'integrabilite des vecteurs gaussienne. *Z. Wahr. verw. Geb.* **68**, 1–8.

Talagrand, M. (1987). Regularity of Gaussian processes. *Acta Math.* **159**, 99–149.

Prob. Theory and Math. Stat., Vol. 1, pp. 285–292
B.Grigelionis *et al.* (Eds.)
1990 VSP/Mokslas

ON LARGE DEVIATION PROBABILITIES FOR THE MAXIMUM LIKELIHOOD ESTIMATORS

K. DZHAPARIDZE and E. VALKEILA*
Centre for Mathematics and Computer Science,
P.O. Box 4079, 1009 AB Amsterdam, the Netherlands
Computer Centre, University of Helsinki,
Teollisuuskatu 23, SF-00510 Helsinki, Finland

ABSTRACT

This paper is devoted to estimating large deviation probabilities of the maximum likelihood estimator from the "true" value of an unknown parameter. In contrast with (Ibragimov and Has'minskii, 1981, Theorem I.5.1) we consider a sequence of parametric families of filtered experiments (cf. (Valkeila and Vostrikova, 1987)). This allows us to verify the conditions of this theorem in so–called "predictable" terms — in terms of the Hellinger processes $h(\alpha; P, P')$ of order $\alpha \in (0,1)$, associated with two probability measures P and P' in a filtered space as, e.g., in (Jacod and Shiryaev, 1987).

1. INTRODUCTION

1.1. Let

$$(\Omega^n, F^n, \mathcal{F}^n, \{P_\theta^n, \theta \in \Theta\}, Q^n), \qquad n = 1, 2, \ldots$$

be a sequence of parametric families of filtered experiments with a measure Q^n dominating a family $\{P_\theta^n, \theta \in \Theta\}$ which is indexed by a certain set Θ in $\mathbf{R}^d$. A filtration $\mathcal{F}^n = (F_t^n)_{t\geqslant 0}$, satisfying the usual assumptions with respect to Q^n, is supposed to be such that $\vee_t F_t^n = F_\infty^n = F^n$ and $F_0^n = \{\emptyset, \Omega^n\}$, for simplicity.

1.2. For a $\mathcal{F}^n$ stopping time T restrict the measures P_θ^n and Q^n to the sub–σ–field F_T^n of F^n. Consider the density process $z_t^n(\theta) = dP_\theta^n/dQ^n|F_t^n$, and for each θ take a version with paths in the space $\mathbf{D}$ of right–continuous functions with left–hand limits. The density process $z^n(\theta)$ is supposed to be continuous in θ.
For a fixed time t define the maximum likelihood estimator θ_t^n as the value of θ in Θ that renders $z_t^n(\theta)$ as large as possible: $z_t^n(\theta_t^n) = \sup_{\theta\in\Theta} z_t^n(\theta)$.
It is supposed that a solution to this equation exists P_θ^n – a.s., for all θ (this is certainly true if Θ is compact), and therefore, as is shown in (Sieders and Dzhaparidze, 1987), it is measurable.
Our result on properties of the maximum likelihood estimator θ_t^n is of asymptotical nature, i.e. it is valid for n and R large enough and t fixed, where $n \to \infty$ describes

* We are grateful to the Finnish Academy and CWI for financial support.

the approach to the "limiting experiment", and R the normalized deviation from the "true" value θ (see 3.2. below).

1.3. Following (Ibragimov and Has'minskii, 1981, Theorem I.5.1) to characterize the rate of convergence (cf. also (Sieders and Dzhaparidze, 1987)), for each fixed n we consider a function g_n on $[0, \infty[$, monotonically increasing to infinity and such that $R^L \exp\{-g_n(R)\} \to 0$ as n and R tend to infinity, for each L.

1.4. As in the works mentioned above, conditions are sought for establishing the following inequality concerning large deviation probabilities:

$$P_\theta^n\{|\varphi_n^{-1}(\theta_t^n - \theta)| > N\} \leqslant B\exp\{-bg_n(N)\}$$

valid uniformly in $\theta \in \Theta$, for n and N large enough and t fixed, with certain positive constants b and B and a certain normalizing sequence of positive definite matrices φ_n, $n = 1, 2, \ldots$, which usually depends on θ, i.e. $\varphi_n = \varphi_n(\theta)$, $n = 1, 2, \ldots$

Unlike these works, however, our conditions are expressed in so-called "predictable" terms, namely in terms of the Hellinger processes of order $\alpha \in (0, 1)$, associated to our sequence of parametric families of experiments in the same fashion as, e. g., in (Jacod and Shiryaev, 1987, Ch. IV) (see Section 2 below for more details). In this sense the results presented here can be viewed as an extention of the results in (Valkeila and Vostrikova, 1987), as well as the ideas and results in the works mentioned above, especially (Ibragimov and Has'minskii, 1981).

1.5. The present paper is organized as follows. In Section 2 we restrict our attention to the case of binary experiments involving two probability measures P and P' in order to associate to it the Hellinger processes and related, so-called "divergency" process, and to present in 2.7 a necessary result (Dzhaparidze and Valkeila, 1989), concerning the Hellinger-type distances between P and P'. Next, following (Valkeila and Vostrikova, 1987), we present in 2.6 a simple, useful estimate for the Hellinger integral associated to P and P'. We end this Section by giving more upper bounds for the Hellinger integrals.

In Section 3 we formulate and prove our main theorem concerning large deviation probabilities mentioned in Subsection 1.4 above. Based on the results mentioned in the preceding section, the proof is essentially reduced to verifying the conditions of a version by (Sieders and Dzhaparidze, 1987) of Theorem I.5.1 in (Ibragimov and Has'minskii, 1981).

We refer to (Kutoyants, 1984) and (Ibragimov and Has'minskii, 1981) for various applications. As noted above, we shall formulate our results for a general filtered experiments. We hope to give applications of our results elsewhere.

2. BINARY EXPERIMENTS

2.1. Consider a binary experiment $(\Omega, F, \mathcal{F}, P, P')$ and let a measure Q dominate both P and P'. As in 1.2 define the density processes z and z' by

$$z_T = dP_T/dQ_T \text{ and } z'_T = dP'_T/dQ_T.$$

2.2. As usual, z^c and Δz will denote the continuous part and jumps of the process z. We associate to them the quadratic characteristic $\langle z^c \rangle$ and the jump measure μ^z with the $(Q, \mathcal{F})$–compensator $\nu^{z,Q}$. According to (Jacod and Shiryaev, 1987, Theorem IV.1.33), μ^z and its compensator only charges the set

$$\{(\omega, t, x) : z_{t-}(\omega) > 0, z'_{t-}(\omega) > 0, -z_{t-}(\omega) \leqslant x \leqslant z'_{t-}(\omega)\}.$$

For bravity, we use the following notations $X = zz'$ and $Y(\alpha) = z^\alpha z'^{1-\alpha}$ with $\alpha \in (0,1)$. Put also $\lambda = \lambda(x) = 1 + x/z_-$ and $\lambda' = \lambda'(x) = 1 - x/z'_-$.

2.3. We present here briefly certain known facts concerning the Hellinger integrals and Hellinger processes; for more details see (Jacod and Shiryaev, 1987, Section IV.1).

Recall that

$$H(\alpha; P, P') = \mathbf{E}_Q(z^\alpha z'^{1-\alpha}) = \mathbf{E}_Q Y(\alpha)$$

is called the Hellinger integral of order $\alpha \in (0,1)$. Obviously, it is free of a particular choise of a dominating measure Q.

By definition, the Hellinger process $h(\alpha) = h(\alpha; P, P')$ is a predictable increasing process for which $Y(\alpha) + Y(\alpha)_- \circ h(\alpha)$ is a $(Q, \mathcal{F})$–martingale. Recall that not only $H(\alpha; P, P')$ is free of a particular choise of a measure Q, but also the process $h(\alpha; P, P')$. Hence, for a $\mathcal{F}$–stopping time T we have $H(\alpha; P_T, P'_T) = 1 - \mathbf{E}_Q(Y(\alpha)_- \circ h(\alpha; P_T, P'_T))$. Besides, the following multiplicative decomposition takes place (Jacod and Shiryaev, 1987, Section V.4):

$$Y(\alpha) = N(\alpha)\mathcal{E}(-h(\alpha)) \tag{2.1}$$

with a certain $(Q, \mathcal{F})$–supermartingale $N(\alpha)$. As usual, $\mathcal{E}$ denotes the Dolean's exponential. Hence

$$H(\alpha; P_T, P'_T) = \mathbf{E}_Q Y(\alpha) = \mathbf{E}_Q\{N_T(\alpha)\mathcal{E}_T(-h(\alpha))\}. \tag{2.2}$$

2.4. Recall also the following explicite expression for $h(\alpha)$: if $Q = (P + P')/2$, then by (Jacod and Shiryaev, 1987, Section IV.1, Corollary 1.35)

$$h(\alpha) = 2\alpha(1-\alpha)X_-^{-2} \circ \langle z^c \rangle + \psi_\alpha(\lambda, \lambda') \star \nu^{z,Q},$$

where $\psi_\alpha(x, y) = \alpha x + (1-\alpha)y - x^\alpha y^{1-\alpha}$ (see 2.2 for notations).

2.5. As in (Dzhaparidze and Valkeila, 1989) (where more details can be found), for $p \geqslant 2$ we introduce the following divergency process $k(p) = k(p; P, P')$ related to the discontinuous part of z only: $k(p) = |(\lambda)^{1/p} - (\lambda')^{1/p}|^p \star \nu^{z,Q}$. It exists since

$k(p) \leqslant 2h(1/2)$, and it is free of a measure Q. Note that in case of an even $p > 2$ the process $k(p)$ is related to the Hellinger processes $h(\alpha)$ as follows:

$$k(p) = -\sum_{j=1}^{p-1} \binom{p}{j} (-1)^j h(j/p).$$

2.6. Here and in the next section we will evaluate from above the Hellinger integral $H(\alpha; P_T, P'_T)$ in terms of the Hellinger process $h(\alpha; P, P')$. First we present an inequality due to (Valkeila and Vostrikova, 1987), and give a simple alternative proof for an arbitrary $\alpha \in (0,1)$.

LEMMA 2.1. *For any $\mathcal{F}$–stopping time T and constant $c > 0$, and each $\alpha \in (0,1)$*

$$H(\alpha; P_T, P'_T) \leqslant e^{-c} + P^{\alpha}\{(h_T(\alpha; P, P') < c)\}. \tag{2.3}$$

Proof. Clearly, $\mathcal{E}(-h(\alpha)) \leqslant e^{-c} + \mathcal{E}(-h(\alpha))I\{\mathcal{E}(-h(\alpha)) > e^{-c}\}$ since $x - 1 \leqslant \leqslant xI(x \geqslant 1)$ for $x \geqslant 0$. Hence, by (2.1) and (2.2)

$$H(\alpha; P_T, P'_T) \leqslant e^{-c}\mathbf{E}_Q N_T(\alpha) + \mathbf{E}_Q(Y_T(\alpha)I\{\mathcal{E}(-h_T(\alpha; P, P')) > e^{-c}\}) \leqslant$$
$$\leqslant e^{-c} + \mathbf{E}_Q(Y_T(\alpha)I\{h_T(\alpha; P, P') < c\}),$$

as $\mathbf{E}_Q N_T(\alpha) \leqslant 1$ and $\mathcal{E}(-h(\alpha)) \leqslant \exp\{-h(\alpha)\}$; see (Jacod and Shiryaev, 1987, Section V.4). Finally, apply Hölder's inequality with exponents $1/\alpha$ and $1/(1-\alpha)$ to the second term on the right–hand side of the last inequality: this gives the desired estimate, since $\mathbf{E}_Q z'_T \leqslant 1$ and

$$\mathbf{E}_Q(Y_T(\alpha)I\{h_T(\alpha; P, P') < c\}) = \mathbf{E}_P((z'_T/z_T)^{1-\alpha}I\{h_T(\alpha; P, P') < c\}) \leqslant$$
$$\leqslant \mathbf{E}_Q(z'_T)^{1-\alpha}P^{\alpha}\{h(\alpha; P_T, P'_T) < c\}.$$

2.7. As an alternative to Lemma 2.1 we give below Corollary 2.1 to the following

LEMMA 2.2. *For any $\alpha \in]0,1[$ and stopping time T we have*

$$H(\alpha; P_T, P'_T) \leqslant \left(\mathbf{E}_P \mathcal{E}^{1/\alpha}(-h_T(\alpha; P, P'))\right)^{\alpha/2}. \tag{2.4}$$

Remark 2.1. Compare with a result from (Kabanov et al., 1980): for any stopping time T and any pair $\alpha < \beta$ we have

$$H(\beta; P_T, P'_T) \leqslant \{\mathbf{E}_P(\mathcal{E}_T(-h(\alpha; P, P')))^{(1-\beta)/(\beta-\alpha)}\}^{(\beta-\alpha)/(1-\alpha)}$$

(see (Jacod and Shiryaev, 1987), p. 278).

Proof. By (2.1)

$$Y(\alpha) = z^{\alpha/2}\sqrt{\mathcal{E}(-h(\alpha))}z'^{(1-\alpha)/2}\sqrt{N(\alpha)}.$$

Apply now Hölder's inequality to the right-hand side of the above inequality with exponents $p = 2/\alpha$ and $q = 2/(2-\alpha)$ to get

$$H(\alpha; P_T, P'_T) \leqslant (\mathbf{E}_P(\mathcal{E}_T(-h(\alpha)))^{1/\alpha})^{\alpha/2}(\mathbf{E}_Q(N_T(\alpha)^{1/(2-\alpha)} z'^{(1-\alpha)/(2-\alpha)}_T))^{(2-\alpha)/\alpha}.$$

It remains to verify that

$$\mathbf{E}_Q(N_T(\alpha)^{1/(2-\alpha)} z'^{(1-\alpha)/(2-\alpha)}_T) \leqslant 1. \tag{2.5}$$

To see this apply again Hölder's inequality to the left-hand side of the above inequality with exponents $p = (2-\alpha)/(1-\alpha)$ and $q = 2-\alpha$. Since $\mathbf{E}_Q z'_T = 1$ and $\mathbf{E}_Q N_T(\alpha) \leqslant 1$, this gives (2.5).

COROLLARY 2.1. *For any stopping time T and constant c we have*

$$H(\alpha; P_T, P'_T) \leqslant e^{-c/2} + P^{\alpha/2}(h_T(\alpha; P, P') < c). \tag{2.6}$$

Proof. Apply the inequality $\mathcal{E}^{1/\alpha}(-h(\alpha)) \leqslant e^{-h(\alpha)/\alpha}$, along with

$$\mathbf{E}e^{-X} \leqslant e^{-c} + \mathbf{P}(X < c) \tag{2.7}$$

and the inequality $(a+b)^\alpha \leqslant a^\alpha + b^\alpha$ to get (2.6) from (2.4).

2.8. In (Dzhaparidze and Valkeila, 1989) the following Hellinger-type distances between stopped measures P_T and P'_T are studied:

$$\rho_p(P_T, P'_T) = \{\mathbf{E}_Q|(z_T)^{1/p} - (z'_T)^{1/p}|^p\}^{1/p},$$

where $p \geqslant 2$. It has been shown that in the simplest case of the usual Hellinger distance ρ_2 we have

$$\rho_2(P_T, P'_T) \leqslant \sqrt{8\mathbf{E}_P\{h_T(P, P')\}},$$

with $h(P, P') = h(1/2; P, P')$, and, generally, the following assertion holds:

LEMMA 2.3. *For any even integer $p \geqslant 2$ there is a constant $C_p > 0$ such that*

$$\rho_p^p(P_T, P'_T) \leqslant C_p \mathbf{E}_P\{h_T(P, P')^{p/2} + k_T(p; P, P')\}.$$

Proof. See (Dzhaparidze and Valkeila, 1989).
As it is shown in the same paper, for deterministic processes h and $k(p)$ the following inequalities hold:
for any $p > 2$ and any fixed t we have that

$$\rho_p^p(P_t, P'_t) \leqslant C_p\{h_t^{p/2} + k_t(p)\} \tag{2.8}$$

and

$$\rho_p^p(P_t, P'_t) \geqslant c(p; h(t))\{h_t^{p/2} + k(p)\}, \tag{2.9}$$

where the constant in (2.9) is $c(p;h^n(t)) = c_p(\mathcal{E}_t(-h^n))^{2p}$, and so $c(p;h^n(t)) \to$ $\to c_p > 0$ as $h^n(t) \to 0$.

2.9. Finally, we mention the following relation between the Hellinger–type distances:

$$\rho_2(P,P') \leqslant C(p)\rho_p(P,P') \tag{2.10}$$

valid for $p \geqslant 2$ (see (Dzhaparidze and Valkeila, 1989)).

3. LARGE DEVIATION PROBABILITIES

3.1. Consider a sequence of parametric families of experiments, indexed by a set Θ in $\mathbf{R}^d$ as in 1.1, and relate to it the density process as in 1.2.
As is said in 1.4, our statement will be uniform in $\theta \in \Theta$. It will be valid for a certain normalizing sequence of positive definite matrices depending usually on θ, i.e. $\varphi_n = \varphi_n(\theta)$, $n = 1, 2, \ldots$.
Denote $U_n(\theta) = \varphi_n^{-1}(\theta)(\Theta - \theta)$ and $\Gamma_{n,R}(\theta) = \overline{U_n(\theta)} \cap \{u : R \leqslant |u| \leqslant R+1\}$.

3.2. For all n and R large enough, and t fixed, consider the following three conditions concerning the Hellinger processes $h_t(P_\theta^n, P_{\theta'}^n)$ and the divergency processes $k_t(P_\theta^n, P_{\theta'}^n)$:
Condition I. Given Θ in $\mathbf{R}^d$, there corresponds to it a function g_n on $[0,\infty[$, introduced in 1.3, such that

$$P_\theta^n\{h_t^n(\theta, \theta + \varphi_n u) < Cg_n(|u|)\} \leqslant \exp\{-cg_n(R)\} \tag{3.1}$$

for each $\theta \in \Theta$ and $u \in \Gamma_{n,R}(\theta)$, where c and C are certain positive constants.
The following is an alternative to Condition I.
Condition I$'$. Given Θ in $\mathbf{R}^d$, there corresponds to it a function g_n on $[0,\infty[$, introduced in 1.3, such that

$$\mathbf{E}_\theta^n \exp\{-h_t^n(\theta, \theta + \varphi_n u)\} \leqslant \exp\{-cg_n(R)\}, \tag{3.2}$$

for each $\theta \in \Theta$ and $u \in \Gamma_{n,R}(\theta)$, where c is a certain positive constant.

LEMMA 3.1. *We have*

Condition I $\Rightarrow$ *Condition* I$'$

and if the function g_n in (3.2) is such that $g_n(R+1)/g_n(R) \leqslant \gamma$ for some $\gamma > 0$ for all R large enough, then

Condition I$'$ $\Rightarrow$ *Condition* I.

Proof. Assume first that (3.1) holds. Then apply (2.7) with $c = Cg_n(|u|)$ to guarantee Condition I with $g_n'(R) = (c \wedge C)g_n(R) - \log 2$.
Assume that (3.2) holds with some c'. For a non-negative random variable X and a non-negative constant a we have

$$\mathbf{P}(X < a) \leqslant e^a \mathbf{E} e^{-X}.$$

Taking here $a = Cg_n(|u|)$ with some $C < c'/2\gamma$ we get (3.1) with $c = c'/2$.
Condition II. Given Θ in $\mathbf{R}^d$, there corresponds to it numbers p and γ, $p \geqslant \gamma > d$, p is even, and a polynomial $\mathrm{pol}_\Theta(R)$ in R such that for $|u| < \delta$

$$\mathbf{E}_\theta^n\{h_t^n(\theta, \theta + \varphi_n u)^{p/2} + k_t^n(p; \theta, \theta + \varphi_n u)\} \leqslant C(\theta)|u|^\gamma \tag{3.3}$$

for each $\theta \in \Theta$, where $C(\theta)$ is a certain positive function of θ admitting the following estimate for each $\theta \in \Theta$ and $u \in \Gamma_{n,R}(\theta)$: $C(\theta + \varphi_n u) \leqslant \mathrm{pol}_\Theta(R)$.

3.3. THEOREM 3.1. *Under Conditions* I *and* II *or* I$'$ *and* II, *there exist positive constants b and B such that for all n and N large enough*

$$\sup_{\theta \in \Theta} P_\theta^n\{|\varphi_n^{-1}(\theta_t^n - \theta)| > N\} \leqslant B\exp\{-bg_n(N)\}. \tag{3.4}$$

Moreover, the constant b can be made arbitrary close from below to $(\gamma - d)/(\gamma - d + pd)$ by choosing B large enough.

Proof. Denote $Z_t^n(\theta, \theta+\varphi_n u) = z_t^n(\theta+\varphi_n u)/z_t^n(\theta)$ with $\theta \in \Theta$ and $\theta+\varphi_n u \in \Theta$. To prove (3.4) it suffices, according to (Sieders and Dzhaparidze, 1987), to check the following two conditions for n and R large enough:

(i) $$P_\theta^n\{Z_t^n(\theta, \theta + \varphi_n u) \geqslant \varepsilon\} \leqslant \eta \exp\{-g_n(R)\}$$

for each $\theta \in \Theta$ and each $u \in \Gamma_{n,R}(\theta)$, where $0 < \varepsilon < 1$ and $\eta > 0$ are some constants, and

(ii) $$\rho_p^p(P_{t,\theta+\varphi_n u}^n, P_{t,\theta+\varphi_n v}^n) \leqslant |u - v|^\gamma \mathrm{pol}_\Theta(R)$$

for each $\theta \in \Theta$ and each $u, v \in \Gamma_{n,R}(\theta)$ such that $|u - v| < \delta$.
First we will check condition (i). Assume that Condition I holds. Fix an $\varepsilon \in (0,1)$ and note that by Markov's inequality and the definition of the Hellinger integral $(H(1/2; P, P') = H(P, P'))$ we have

$$P_\theta^n\{Z_t^n(\theta, \theta + \varphi_n u) \geqslant \varepsilon\} \leqslant \varepsilon^{-1/2} H(P_{t,\theta}^n, P_{t,\theta+\varphi_n u}^n).$$

Apply now Lemma 2.1 with $\alpha = 1/2$ and $c = g_n(u)$. Then by the last inequality and Condition I we get (i) with $\eta = 2\varepsilon^{-1/2}$.
To deduce from the same inequality that Condition I$'$ implies (i), we extend it by using first (2.4) with $\alpha = 1/2$ then the inequality $\mathcal{E}(-h) \leqslant \exp(-h)$.
We show now that Condition II implies (ii). By Lemma 2.3 we have that

$$\rho_p^p(P_{t,\theta+\varphi_n u}^n, P_{t,\theta+\varphi_n v}^n) \leqslant C_p \mathbf{E}_{\theta+\varphi_n u}^n\{h_t(\theta + \varphi_n u, \theta + \varphi_n v)^{p/2} +$$
$$+ k_t(p; \theta + \varphi_n u, \theta + \varphi_n v)\}.$$

Due to Condition II the right-hand side of this inequality is bounded by

$$C(\theta + \varphi_n u)|u - v|^\gamma \leqslant \mathrm{pol}_\Theta(R)|u - v|^\gamma.$$

Hence (ii) is satisfied. The assertion conserning the constant b is proved in (Sieders and Dzhaparidze, 1987).

3.4. We specify the conditions I, I′ and II in the case of deterministic processes h^n and $k^n(p)$. Note meanwhile that by the inequality $\log(EX) \geqslant E\log(X)$ it is easily seen that (3.2) implies

$$\mathbf{E}_\theta^n(h_t^n(\theta, \theta + \varphi_n u)) \geqslant c g_n(R) \tag{3.5}$$

for each $\theta \in \Theta$ and $u \in \Gamma_{n,R}(\theta)$. The inverse implication is not true in general, except for deterministic processes h^n. Therefore introduce the following condition:
Condition I(d). Let the Hellinger processes h^n be deterministic satisfying

$$h_t^n(\theta, \theta + \varphi_n u) \geqslant g_n(R)$$

for each $\theta \in \Theta$ and $u \in \Gamma_{n,R}(\theta)$, with c and $g_n(R)$ as in Condition I′.
Condition II is specified as follows.
Condition II(d). Let h^n and k^n be deterministic processes. Given Θ in $\mathbf{R}^d$, there corresponds to it numbers p and γ, $p \geqslant \gamma > d$ and a polynomial $\mathrm{pol}_\Theta(R)$ in R such that for $|u| < \delta$

$$h_t^n(\theta, \theta + \varphi_n u) \leqslant C(\theta)|u|^{2\gamma/p}$$

and

$$k_t^n(p; \theta, \theta + \varphi_n u) \leqslant C(\theta)|u|^\gamma$$

for each $\theta \in \Theta$, where $C(\theta)$ is as in Condition II.

COROLLARY 3.1. *Suppose that the processes $h_t^n(\theta, \theta + \varphi_n u)$ and $k_t^n(p; \theta, \theta + +\varphi_n u)$ are deterministic, satisfying Conditions* I(d) *and* II(d). *Then the conclusion of Theorem* 3.1 *holds.*

REFERENCES

Dzhaparidze, K. and Valkeila, E. (1989). On Hellinger type distances for filtered experiments. *Probab. Theory Rel. Fields* (*to appear*).

Ibragimov, I.A. and Has'minskii, R.Z. (1981). *Statistical Estimation: Asymptotic Theory.* Springer, Berlin–Heidelberg etc.

Jacod, J. and Shiryaev, A.N. (1987). *Limit Theorems for Stochastic Processes.* Springer, Berlin–Heidelberg etc.

Kabanov, Yu.M., Liptser, R.Sh. and Shiryaev, A.N. (1980). Some limit theorems for simple point processes (a martingale approach). *Stochastics* **3**, 203–216.

Kutoyants, Yu.A. (1984). *Parameter Estimation for Stochastic Processes.* Heldermann, Berlin.

Sieders, A. and Dzhaparidze, K. (1987). A large deviation result for parameter estimators and its application to nonlinear regression analysis. *Ann. Statist.* **15**, 1031–1049.

Valkeila, E. and Vostrikova, L. (1987). On predictable criteria of (c_n)-consistancy of estimates. *Theory Probab. Appl.* **32**, 477–489.

Prob. Theory and Math. Stat., Vol. 1, pp. 293–303
B.Grigelionis *et al.* (Eds.)
1990 VSP/Mokslas

FUNCTIONAL INTEGRALS, VARIATIONAL METHOD AND SOME PROBLEMS OF STOCHASTICAL PHYSICS

G.V. EFIMOV
Joint Institute for Nuclear Research, Dubna, USSR

ABSTRACT

At the present time it is very popular to represent different physical characteristics in the form of functional integrals (see, for example (Glimm and Jaffe, 1981; Feynman, 1979)). However, calculations of functional integrals with the exception of the Gaussian integrals and a few integrals of a special form are of a serious difficulty. The main computing methods of functional integrals are, first, the quasi–classical approach or the method of stationary phase when the main contribution to the integral comes from a function which minimized an integrand action (see, for example (Feynman, 1979)) and, second, the variational calculations (see, (Feynman and Hibbs, 1965)).
In this paper we propose a variational method improving the Feynman method (Feynman and Hibbs, 1965). The idea of this method was formulated in (Efimov, 1985).
We apply this method for investigating the asymptotic behaviour of the Green functions in stochastic fields and studying the effective potential in the two–dimensional ϕ^4 field theory.

1.VARIATIONAL METHOD

Here we formulate our variational method which will be used in what follows. Let the functional integral be given

$$I(g) = \int d\sigma_\phi e^{-gW[\phi]}, \tag{1}$$

$$d\sigma_\phi = \frac{1}{N_\phi}\delta\phi \exp\Big\{-\frac{1}{2}\int\limits_V\int\limits_V dx_1 dx_2 \phi(x_1) D^{-1}(x_1,x_2)\phi(x_2)\Big\}. \tag{2}$$

The notation is the following: $D^{-1}(x_1,x_2)$ is a distribution or a differential operator. The Green function $D(x_1,x_2)$ is defined by the equation

$$\int\limits_V dy D^{-1}(x_1,y)D(y,x_2) = \delta(x_1 - x_2)$$

and it satisfies some given boundary conditions. The volume $V \subset \mathbf{R}^d$ over which the integration is performed in (2) can be both finite and infinite ($V \to \mathbf{R}^d$). The functional differential in a lattice approximation is defined as

$$\delta\phi = \prod_{x \in V} d\phi(x).$$

The normalization constant N_ϕ is determined from the condition $\int d\sigma_\phi = 1$. $W[\phi]$ is a real functional, and g is a "coupling constant".
It is assumed that the functional integral (1) is defined on the Gaussian measure (2), i.e. at least there exists a perturbative series in the coupling constant g.
Let us formulate our variational method. The succession of our actions is following. First of all, let us diagonalize the quadratic form in (2). We introduce the function $\Delta(x_1, x_2)$ satisfying the condition

$$\int_V dy \Delta(x_1, y)\Delta(y, x_2) = D(x_1, x_2).$$

In the case under consideration this function can easily be found but it is enough for us to suppose its existence. Let us introduce the functional variable

$$\phi(x) = \int_V dy \Delta(x, y)\Phi(y) = (\Delta, \Phi)(x).$$

The functional integral (1) can be written

$$I(g) = \frac{1}{N_\Phi} \int \delta\Phi \exp\Big\{ -\frac{1}{2} \int_V dx \Phi^2 - gW[(\Delta, \Phi)]\Big\}, \tag{3}$$

where the new constant N_Φ is defined by the condition $I(0) = 1$.
Let us choose in the volume $V \subset \mathbf{R}^d$ some orthonormal system of functions $\{g_{\{n\}}(x)\}$, where $\{n\} = (n_1, \ldots, n_d)$, $n_j = 0, 1, \ldots$ $(j = 1, \ldots, d)$, satisfies the conditions

$$\int_V d^d g_{\{n\}}(x) g_{\{n'\}}(x) = \delta_{\{n,n'\}} = \delta_{n_1 n'_1} \ldots \delta_{n_d n'_d},$$

$$\sum_{\{n\}} g_{\{n\}}(x) g_{\{n\}}(x') = \delta^{(d)}(x - x') = \delta(x - x').$$

The choice of this system is arbitrary enough. The unique condition imposed on this system is that the functions $D(x_1, x_2)$ and $\Delta(x_1, x_2)$ can be developed over the functions of this system.
Let us represent the function $\Phi(x)$ over which the integration is performed in (3) in the form

$$\Phi(x) = \sum_{\{n\}} u_{\{n\}} g_{\{n\}}(x),$$

where the coefficients $u_{\{n\}}$ are independent variables. Then,

$$\int_V dx \Phi^2(x) = \sum_{\{n\}} u^2_{\{n\}},$$

$$(\Delta, \Phi)(x) = \sum_{\{n\}} \Delta_{\{n\}}(x) u_{\{n\}}, \Delta_{\{n\}}(x) = \int_V dy \Delta(x,y) g_{\{n\}}(x).$$

The functional integral (3) can be written in the form of infinitely multiple integral

$$I(g) = \int d\sigma_u \exp\{-gW[(\Delta, \Phi)]\},$$

$$d\sigma_u = \prod_{\{n\}} \frac{du_{\{n\}}}{(2\pi)^{d/2}} \exp\left(-\frac{1}{2} \sum_{\{n\}} u^2_{\{n\}} \right), \tag{4}$$

where the normalization constant is written in the explicit form.
We want to stress that the representation (4) is equivalent to (1).
Let us proceed to the variational estimation of the integral (4). We introduce the new variables

$$u_{\{n\}} = \frac{u'_{\{n\}}}{\sqrt{1 + q_{\{n\}}}} + s_{\{n\}}, \tag{5}$$

where the quantities $q_{\{n\}}$ and $s_{\{n\}}$ will be variational parameters. They satisfy the conditions

$$\left| \sum_{\{n\}} q_{\{n\}} \right| < \infty, \qquad \left| \sum_{\{n\}} s_{\{n\}} \right| < \infty.$$

We would like to make the following remarks. Instead of (5) it is possible to do the substitution

$$u_{\{n\}} = \sum_{\{l\}} U_{\{n,l\}} \frac{u'_{\{l\}}}{\sqrt{1 + q_{\{l\}}}} + s_{\{n\}},$$

where U is an orthogonal real matrix: $\det U = 1, UU^T = 1$. This matrix defines some rotation in the space of variables $u_{\{n\}}$. However, it means transition to another orthonormal basis. In other words, the basis enters into the set of variational parameters.
Let us substitute (5) into (4). One can get

$$I(g) = \prod_{\{n\}} \frac{1}{\sqrt{1 + q_{\{n\}}}} \int d\sigma_u \exp\left\{ \frac{1}{2} \sum_{\{n\}} \frac{q_{\{n\}}}{1 + q_{\{n\}}} u^2_{\{n\}} - \right.$$

$$\left. - \sum_{\{n\}} \frac{s_{\{n\}}}{1 + q_{\{n\}}} u_{\{n\}} - \frac{1}{2} \sum_{\{n\}} s^2_{\{n\}} - gW\left[(\Delta_q, \Phi) + (\Delta, s) \right] \right\},$$

where

$$(\Delta_q, \Phi) = \sum_{\{n\}} \frac{\Delta_{\{n\}}(x)}{\sqrt{1 + q_{\{n\}}}} u_{\{n\}};$$

$$(\Delta, s)(x) = \sum_{\{n\}} \Delta_{\{n\}}(x) s_{\{n\}}.$$

The measure $d\sigma_u$ is the same as in (4).
Let us use the unequality

$$\int d\sigma \exp\{-W\} \geqslant \exp\left\{ - \int d\sigma W \right\}$$

which is valid for any positive definite measures and any real functionals W. We obtain

$$I(g) \geqslant \exp\left\{ - L[q] - \frac{1}{2}(s,s) - \int d\sigma_u W[(\Delta_q, \Phi) + (\Delta, s)] \right\},$$

$$L[q] = \frac{1}{2} \sum_{\{n\}} \left[\ln(1 + q_{\{n\}}) - \frac{q_{\{n\}}}{1 + q_{\{n\}}} \right], \quad (s,s) = \sum_{\{n\}} s^2_{\{n\}}. \tag{6}$$

Representing our integral $I(g)$ in the form

$$I(g) = \exp(-E(g))$$

one can obtain from (6) for $E(g)$ the upper estimation

$$E(g) \leqslant E_+(g),$$

$$E_+(g) = \min_{\{q,s\}} \left\{ L[q] + \frac{1}{2}(s,s) + \int d\sigma_u W[(\Delta_q, \Phi) + (\Delta, s)] \right\}. \tag{7}$$

This formula is the desired unequality.
Thus, the variational parameters are, first, the orthonormal system and, second, the parameters $\{q_{\{n\}}, s_{\{n\}}\}$ over which we have to compute the minimum in (7). It should be noted that this variational estimation (7) gives the exact result for the quadratic functional $W[\phi]$.
In conclusion, we want to remark that this variational method differs from the Feynman method in that additional parameters $s_{\{n\}}$ are introduced and the parameters $q_{\{n\}}$ are connected with the pure Gaussian measure just as the specific properties of the differential operator $D^{-1}(x_1, x_2)$ enter into the interaction functional W. Therefore, the variational equation obtained from (7) connect directly the parameters $q_{\{n\}}$ and $s_{\{n\}}$ with the behaviour of the Green function $D(x_1, x_2)$ so that a more precise estimation can be achieved.

2. GREEN FUNCTION IN RANDOM FIELDS

Let us consider the Green function satisfying the following equation

$$\left[\left(i\frac{\partial}{\partial x_\mu}+V_\mu(x)\right)^2+W(x)+m^2\right]G(x,y|V,W)=\delta(x-y), \tag{8}$$

where $W(x)\geqslant 0$. This equation is defined in the Euclidean space $\mathbf{R}^d$,

$$x^2=x_1^2+\cdots+x_d^2,\quad \left(i\frac{\partial}{\partial x_\mu}\right)^2=-\left(\frac{\partial^2}{\partial x_1^2}+\cdots+\frac{\partial^2}{\partial x_d^2}\right).$$

The solution of equation (8) can be written in the form according to the Feynman functional integral representation (see (Feynman, 1951))

$$G(x,y|V,W)=\int\limits_0^\infty d\alpha\, e^{-\alpha m^2}\int\delta\Phi(\beta)\exp\Bigg\{-\int\limits_0^1 d\beta\Phi_\mu^2(\beta)+ \tag{9}$$

$$+2i\sqrt{\alpha}\int\limits_0^1 d\beta\Phi_\mu(\beta)V_\mu\left(x-2\sqrt{\alpha}\int\limits_\beta^1 d\beta'\Phi(\beta')\right)-$$

$$-\alpha\int\limits_0^1 d\beta W\left(x-2\sqrt{\alpha}\int\limits_\beta^1 d\beta'\Phi(\beta')\right)\Bigg\}\delta\left(x-y-2\sqrt{\alpha}\int\limits_0^1 d\beta\Phi(\beta)\right).$$

The normalization of the functional integral in (9) is chosen in the following way:

$$\int\delta\Phi\exp\left\{-\int\limits_0^1 d\beta\Phi^2(\beta)\right\}\delta\left(x-2\sqrt{\alpha}\int\limits_0^1 d\beta\Phi(\beta)\right)=$$

$$=\int\left(\frac{dk}{2\pi}\right)^d e^{-ikx-\alpha k^2}=\frac{1}{(4\pi\alpha)^{d/2}}e^{-x^2/(4\alpha)}.$$

Let us perform the following transformations in (9). We will calculate the functional integral in the representation of basic vectors. The orthonormal basis in the interval $0<\beta<1$ will be taken in the form

$$g_{\{n\}}(\beta)=\{1,\sqrt{2}\cos(2\pi n\beta),\sqrt{2}\sin(2\pi n\beta)\quad (n=1,2,\ldots)\}$$

We introduce the new variables of integration in (9):

$$\Phi_\mu(\beta)=\Phi_{0\mu}+a_\mu(\beta),$$

$$a_\mu(\beta)=\sum_{n=1}^\infty\left(u_{n\mu}\cos(2\pi n\beta)+v_{n\mu}\sin(2\pi n\beta)\right),$$

$$A_\mu(\beta)=\int\limits_0^\beta d\beta' a_\mu(\beta')=\sum_{n=1}^\infty\frac{1}{2\pi n}\left(u_{n\mu}\sin(2\pi n\beta)+v_{n\mu}(1-\cos(2\pi n\beta))\right).$$

After introducing the new variables and performing the integration over $\Phi_{0\mu}$, one can obtain for the functional integral (9) .

$$G(x,y|V,W) = \int_0^\infty \frac{d\alpha}{4\pi\alpha^{d/2}} \exp\Big\{-\alpha m^2 - \frac{(x-y)^2}{4\alpha}\Big\} R(x,y|V,W), \tag{10}$$

$$R(x,y|V,W) = \int d\sigma_a I_V(x,y|V) I_S(x,y|W),$$

$$I_V(x,y|V) = \exp\Big\{i\int_0^1 d\beta (x-y-2\sqrt{\alpha}a(\beta))_\mu V_\mu(x\beta + y(1-\beta) + 2\sqrt{\alpha}A(\beta))\Big\},$$

$$I_S(x,y|W) = \exp\Big\{-\alpha\int_0^1 d\beta W(x\beta + y(1-\beta) + 2\sqrt{\alpha}A(\beta))\Big\},$$

$$d\sigma_a = \prod_{n=1}^\infty \Big(\frac{dudv}{2\pi}\Big)^d \exp\Big\{-\frac{1}{2}\sum_{n=1}^\infty (u^2_{n\mu} + v^2_{n\mu})\Big\}.$$

The representation (10) is a basis of our further calculations.
As the first example of the application of our variational method we consider the asymptotical behaviour of the Green function of scalar particles in the stochastic or random Gaussian scalar and vector fields. Our results can be formulated in the form of the following two statements.

STATEMENT 1. *Let the equation*

$$\Big(-\frac{\partial^2}{\partial x^2} + g\phi^2(x)\Big) G(x,y|\phi) = \delta(x-y)$$

be given in the Euclidean space $\mathbf{R}^4$. *The field* $\phi(x)$ *is a random Gaussian field with the correlation function*

$$\langle\phi(x_1)\phi(x_2)\rangle_\phi = D(x_1-x_2) = \int \Big(\frac{dk}{2\pi}\Big)^4 \widetilde{D}(k^2) e^{-ik(x_1-x_2)}.$$

The function $\widetilde{D}(k^2)$ *decreases rapidly enough so that*

$$D_m = \int \Big(\frac{dk}{2\pi}\Big)^4 \widetilde{D}(k^2)(k^2)^m < \infty \qquad (m=0,1).$$

Then, the following inequality is valid for the Green function averaging over the random field

$$G(x-y) = \langle G(x,y|\phi)\rangle_\phi \geqslant \frac{\text{const}}{\sqrt{(x-y)^2}} \exp\Big\{-M_+\sqrt{(x-y)^2}\Big\}.$$

Here

$$M_+ = \min_{\xi>0,\sigma>0,\lambda>0} \left\{ \frac{1}{4\xi} + 2\sigma + \frac{\lambda}{2} + \right.$$

$$\left. +g\xi \int_0^\infty ds e^{-s} \int \left(\frac{dk}{2\pi}\right)^4 \widetilde{D}(k^2) \left[1 - \exp\left\{i(kn)\frac{s}{2\lambda} - \frac{\xi}{\sigma}k^2(1-e^{-\sigma s/\lambda})\right\}\right] \right\},$$

where n is a Euclidean vector with $n^2 = 1$.

For the weak and strong coupling we have

$$M_+ = \begin{cases} \sqrt{gD_0} & (g \ll 1), \\ 1.09\sqrt[4]{gD_1} & (g \gg 1). \end{cases}$$

The second example is relevant to (Simonov, 1988a, 1988b, 1988c) where the author claims that stochastic gauge fields (electromagnetic fields, for example) can lead to the confinement. The confinement is considered to be reached if the Green function decreases at large distances more rapidly than any linear exponent, i.e. for any $N > 0$

$$\lim_{|x|\to\infty} |G(x)| \exp\left\{N\sqrt{x^2}\right\} = 0.$$

This condition means that these Green functions cannot describe asymptotically free states of particles.
Here, we show that it is not true. Our conclusion is based on the following statement.

STATEMENT 2. *Let the equation*

$$\left[\left(i\frac{\partial}{\partial x_\mu} + V_\mu(x,y)\right)^2 + m^2\right] G(x,y|F) = \delta(x-y),$$

be given in the Euclidean space $\mathbf{R}^4$. The vector field V_μ is defined

$$V_\mu(x,y) = \int_0^1 ds s F_{\mu\nu}(xs + y(1-s))(x-y)_\nu.$$

Here $F_{\mu\nu}(x)$ is a random Gaussian field with the correlation function

$$\langle F_{\mu\nu}(x) F_{\rho\sigma}(y)\rangle_F = (\delta_{\mu\rho}\delta_{\nu\sigma} - \delta_{\mu\sigma}\delta_{\nu\rho}) D(x-y),$$

$$D(x) = \int \left(\frac{dk}{2\pi}\right)^4 \widetilde{D}(k^2) e^{-ikx},$$

where the function $\widetilde{D}(k^2)$ decreases rapidly enough.
Then, the following inequalities are valid for the Green function averaging over the random field $F_{\mu\nu}$ in the limit $x^2 \to \infty$

$$G(x) = \langle G(x,0|F)\rangle_F \geqslant \frac{\text{const}}{\sqrt{x^2}} \exp\{-m\sqrt{x^2}\} \sim G_0(x),$$

$$G(x,0;-x,0) = \langle G(x,0|F)G(-x,0|F)\rangle_F \geqslant \frac{\text{const}}{x^2}\exp\{-2m\sqrt{x^2}\} \sim$$
$$\sim G_0(x)G_0(-x).$$

These inequalities mean that the scalar particles are free ones as $x^2 \to \infty$. Detailes of proving these statements can be found in (Dinejchan et al., 1989). Here, we want to say that the variational parameters q_n in these cases should be chosen in the form

$$q_n = \left(\frac{\sigma\sqrt{x^2}}{\pi n}\right)^2,$$

where σ is a variational parameter.

3. ON THE NATURE OF PHASE TRANSITION IN A TWO–DIMENSIONAL ϕ^4 THEORY

Here, we apply our variational method to investigate the character of the phase transition in the quantum field model described by the Lagrangian density

$$L(x) = \frac{1}{2}\phi(x)(\partial^2 - m^2)\phi(x) - \frac{g}{4}\phi^4(x) \tag{11}$$

in the two–dimensional space–time $\mathbf{R}^2$. There exist theorems proving that the second order phase transition takes place in this model for a coupling constant $g = g_c$ (see, for example (Simon, 1974; Chang, 1975; Stevenson, 1984, 1985)). Investigations of the effective potential are a standard way of studying the phase structure. In (Palley and Ritschel, 1989) it has been shown that the loop approximation of the effective potential leads to the first order phase transition. It has been shown also that a variational method can give the second order phase transition. Our method (Efimov, 1985; Dinejchan et al., 1989) gives a more precise estimation. The effective potential is defined by the following functional integral in the Euclidean space:

$$V_{eff}(\phi_c, g) = -\lim_{V\to R^2}\frac{1}{V}\ln I_V(\phi_c),$$

$$I_V(\phi_c) = C_m\int\delta\phi(x)\delta\Big(\phi_c - \frac{1}{V}\int\limits_V dx\phi(x)\Big)\exp\left\{\int\limits_V d^2x\left[\frac{1}{2}\phi(x)(\partial^2 - m^2)\phi(x) - \right.\right.$$
$$\left.\left. - \frac{g}{4}N_m\phi^4(x)\right]\right\}, \tag{12}$$

where $C_m = \sqrt{\det(\partial^2 + m^2)}$ and N_m denotes the normal ordering

$$N_m\phi^2(x) = \phi^2(x) - D_m,$$

$$N_m\phi^3(x) = \phi^3(x) - 3D_m\phi(x),$$

$$N_m\phi^4(x) = \phi^4(x) - 6D_m\phi^2(x) + 3D_m^2,$$

$$D_m = \int \frac{d^2k}{(2\pi)^2}\frac{1}{m^2+k^2}.$$

The phase transition takes place if there exists such g_c that

$$\min_{\phi_c} V_{eff}(\phi_c, g) = V_{eff}(\phi_{min}, g) < V_{eff}(0, g)$$

for $g > g_c$ and $\phi_{min} \neq 0$. The character of the phase transition in this model is defined by the behaviour of the coefficient $\alpha(g)$ in expantion the effective potential $V_{eff}(\phi_c)$ for small ϕ_c:

$$V_{eff}(\phi_c, g) = E_0(g) + \alpha(g)\phi_c^2 + O(\phi_c^4). \tag{13}$$

The second order phase transition arises when the function $\alpha(g)$ in (13) is equal to zero $\alpha(g_c) = 0$ for some g_c and is negative for $G > G_c$. The first order phase transition takes place if $\alpha(g) > 0$ for any g.
Let us introduce the new field variable

$$\phi(x) = \phi_0 + \Phi(x), \tag{14}$$

$$\int\limits_V dx\Phi(x) = 0.$$

Substituting (14) into (12) and performing integration over the constant field ϕ_c one can get

$$I_V(\phi_c) = C_m \int \delta\Phi(x)\exp\left\{\int d^2x\left[\frac{1}{2}\Phi(x)(\partial^2 - m^2)\Phi(x) - \right.\right.$$

$$\left.\left. -\frac{m^2}{2}\phi_c^2 - \frac{g}{4}N_m(\Phi(x) + \phi_c)^4\right]\right\}, \tag{15}$$

where the integration function $\Phi(x)$ satisfies (14). Now we make the following transformation. We introduce in (15) a new mass μ, go over to the normal-ordering with respect to this new mass μ and demand that the coefficient before ϕ^2 in the Lagrangian density should be equal to μ^2. After some transformations we obtain

$$I_V(\phi_c) = \exp\{-VV_{loop}(\phi_c)\}J_V(\phi_c),$$

$$V_{loop}(\phi_c) = \frac{m^2}{2}\phi_c^2 + \frac{g}{4}\phi_c^4 - \frac{3}{8\pi}g\phi_c^2\ln\frac{\mu^2}{m^2} + \tag{16}$$

$$+\frac{1}{8\pi}(\mu^2 - m^2 - m^2\ln\frac{\mu^2}{m^2}) + \frac{3}{2}g\left(\frac{1}{4\pi}\ln\frac{\mu^2}{m^2}\right)^2,$$

$$\mu^2 - m^2 + \frac{3g}{4\pi}\ln\frac{\mu^2}{m^2} = 3g\phi_c^2,$$

$$J_V(\phi_c) = C_\mu \int \delta\Phi \exp\left\{ \int d^2x \left[\frac{1}{2}\phi(x)(\partial^2 - \mu^2)\phi(x) - \right.\right.$$

$$\left.\left. -g\phi_c N_\mu \Phi^3(x) + \frac{g}{4} N_\mu \Phi^4(x)\right]\right\}.$$

Thus, the effective potential $V(\phi_c)$ consists of two parts

$$V_{eff} = V_{loop}(\phi_c) + V_{sc}(\phi_c).$$

The function $V_{loop}(\phi_c)$ is called the "all–loop" potential because it contains all "cactus–type" loop diagrams on the language of the Feynman diagrams. The function $V_{sc}(\phi_c)$ is called the "strong–connected" potential because it contains the strong–connected Feynman diagrams.
The behaviour of $V_{loop}(\phi_c)$ as $\phi_c^2 \to 0$ is the following:

$$V_{loop}(\phi_c) = \frac{m^2}{2}\phi_c^2 + O(\phi_c^4).$$

The behaviour of $V_{sc}(\phi_c)$ as $\phi_c \to 0$ can be obtained by our variational method. As $\phi_c \to 0$ the functional integral in (16) can be written

$$J_V(\phi_c) = C_\mu \int \delta\Phi(x) \exp\left\{ \int_V dx \left[\frac{1}{2}\Phi(x)(\partial^2 - \mu^2)\Phi(x) - \right.\right.$$

$$\left.\left. -\frac{g}{4} N_\mu \Phi^4(x)\right] + \frac{1}{2} g^2 \phi_c^2 \left(\int_V dx N_\mu \Phi^3(x)\right)^2 \right\}. \tag{17}$$

Applying our variational method to (17) (details see (Efimov, 1985)), one can get

$$V_{sc}(\phi_c) = -\lim_{V\to R^2} \frac{1}{V} \ln\{J_V(\phi_c)\} \leqslant V_{sc}^+(\phi_c),$$

$$V_{sc}^+(\phi_c) = \min_{q,A} \left\{ \frac{1}{2} \int \frac{d^2k}{(2\pi)^2} \left[\ln(1 + q(k^2)) - \frac{q(k^2)}{1 + q(k^2)}\right] + \right.$$

$$\left. + \frac{\mu^2}{2} A^2 + \frac{g}{4}\left[A^4 - 6A^2\Delta_q + 3\Delta_q^2\right] - \frac{g^2\phi_c^2}{2}\left[6R_3 + 18A^2 R_2\right] \right\},$$

where

$$\Delta_q = \int \frac{d^2k}{(2\pi)^2} \frac{1}{\mu^2 + k^2} \frac{q(k^2)}{1 + q(k^2)},$$

$$R_n = \int d^2x (D_q(x))^n,$$

$$D_q(x) = \int \frac{d^2k}{(2\pi)^2} \frac{e^{-ikx}}{(\mu^2+k^2)(1+q(k^2))}.$$

The function $q(k^2)$ and the constant A^2 are our variational parameters. It turns out that

$$q(k^2) = \frac{\mu^2 f}{\mu^2+k^2},$$

where f is a variational parameter. Then, all integrations and calculations can be performed explicitly.
Thus, we have for the effective potential upper estimation

$$V_{eff}(\phi_c) \leqslant V^+_{eff}(\phi_c) = V_{loop}(\phi_c) + V^+_{sc}(\phi_c),$$

for $\phi_c \ll 1$

$$V^+_{eff}(\phi_c) = E^+_0(g) + \alpha^+(g)\phi_c^2.$$

Our estimations are undoubtedly true as $g \to \infty$ true. In this limit one can get

$$V^+_{eff}(\phi_c) = -\frac{3g}{2}\left(\frac{\ln(g/m^2)}{4\pi}\right)^2 + 3g\left(\frac{\ln(g/m^2)}{4\pi}\right)\phi_c^2,$$

i.e. $\alpha(g) > 0$. In other words the first order phase transition takes place and this phase transition is defined by the kinetic part of the effective potential. Thus our estimation confirms the result of the loop approximation.

REFERENCES

Chang, S.J. (1975). *Phys. Rev.* **12**, 1071.

Dinejchan, M., Efimov, G.V. and Namsrai, K. (1989). *Preprint.* JINR, Dubna, EE2–89–120.

Efimov, G.V. (1985). *Problems of Quantum Theory of Non-Local Interactions.* Nauka, Moscow.

Feynman, R.P. (1951). *Phys. Rev.* **84**, 106.

Feynman, R.P. (1979). Path Integral. *Lect. Notes in Phys.* **106**

Feynman, R.P. and Hibbs, A.R. (1965). *Quantum Mechanics and Path Integrals.* McGraw–Hill, New York.

Glimm, J. and Jaffe, A. (1981). *Quantum Physics. A Functional Integral Point of View.* Springer, Berlin—Heidelberg etc.

Palley, L. and Ritchel,V. (1989). *Phys. Lett.* **221**, 44.

Simon, B. (1974). *The $P(\Phi)_2$ Euclidean (Quantum) Field Theory.* Princeton Univ. Press, Princeton.

Simonov, Yu.A. (1988a). *Sov. Nucl. Phys.* **47**, 1165.

Simonov, Yu.A. (1988b). *Sov. Nucl. Phys.* **48**, 1381.

Simonov, Yu.A. (1988c). *Nucl. Phys.* **307**, 512.

Stevenson, P.M. (1984). *Phys. Rev.* **30**, 1712.

Stevenson, P.M. (1985). *Phys. Rev.* **32**, 1389.

Prob. Theory and Math. Stat., Vol. 1, pp. 304–313
B.Grigelionis *et al.* (Eds.)
1990 VSP/Mokslas

THE LAW OF THE ITERATED LOGARITHM AND ORDER STATISTICS

V.A. EGOROV

Leningrad Ulyanov (Lenin) Electrotechnical Institute, Chair of Mathematics, Professor Popov Street, 5, 197022 Leningrad, USSR

ABSTRACT

The upper estimates in the iterated logarithm type law for censored sums of dependent random variables are obtained.

There are two types of censoring. For positive random variables $(Y_j)_{j=1}^n$ for the first type of censoring we replace these variables by the sequence $(\bar{Y}_j)_{j=1}^n$, where $\bar{Y}_j = 0$, if $Y_j > a(n), \bar{Y}_j = Y_j$, if $Y_j \leqslant a(n)$, $j \leqslant n$, $a(n)$ is some level of censoring. For the second type of censoring we replace the sequence $(Y_j)_{j=1}^n$ by the sequence $(Y_{j,n})_{j=1}^{n-k}$, where $Y_{1,n} \leqslant Y_{2,n} \leqslant \cdots \leqslant Y_{n,n}$ is the variation row, $0 \leqslant k \leqslant n$, $k = k(n)$ (see (Kendall and Stuart, 1966)). If we increase $k(n)$ or decrease $a(n)$, then we decrease the speed of the increasing of the corresponding sums:

$$T_{k,n} = \sum_{j=1}^{n-k+1} Y_{j,n}, \qquad L_n = \sum_{j=1}^{n} \bar{Y}_j.$$

Finally, for $a(n) = o(\bar{B}_n/\ln\ln\bar{B}_n)$ or $k(n)/\ln\ln B_n \to \infty (n \to \infty)$ correspondingly, we obtain the upper estimates in the strong law of the large numbers

$$\limsup_{n\to\infty} \frac{T_{k,n}}{B_n} \leqslant 1, \qquad \limsup_{n\to\infty} \frac{L_n}{B_n} \leqslant 1 \quad \text{a.s.,} \tag{1}$$

where $\bar{B}_n = \sum_{j=1}^n \mathbf{E}\bar{Y}_j$, $B_n = \sum_{j=1}^n \mathbf{E}Y_j$. Obviously, the further essential decreasing of these upper estimates is impossible. Similary, we can construct other upper estimates, for example, for independent symmetric random variables $(X_j)_{j=1}^n$. Denoting $Y_j = X_j^2, j \leqslant n$, and using the same schemes of censoring, finally, we receive, that the above mentioned conditions on $a(n)$ or $k(n)$ imply the upper estimates in the law of the iterated logarithm

$$\limsup_{n\to\infty} \frac{S_{k,n}}{\sqrt{B_n \ln\ln B_n}} \leqslant 1, \qquad \limsup_{n\to\infty} \frac{U_{k,n}}{\sqrt{2B_n \ln\ln B_n}} \leqslant 1 \quad \text{a.s.,}$$

where

$$S_{k,n} = \sum_{j=1}^{n-k+1} X_{j,n}, \qquad U_{k,n} = \sum_{j=1}^{n} \bar{X}_j,\ X_j^2 = \bar{Y}_j,\ j \leqslant n.$$

Here we study these effects for the second type of censoring. Note, that such results are known mainly for i.i.d. case (Feller, 1968; Mory, 1976; Häusler and Mason, 1987). Here we consider the general case and, moreover, we replace the independence condition by the condition (2) defined later.
Let $(Y_n)_{n=1}^{\infty}$ be positive random variables,

$$T_n = \sum_{j=1}^{n} Y_j, \qquad B_n = \mathbf{E}T_n \to \infty \quad (n \to \infty),$$

$k(x)$ be an integer valued positive function,

$$T_{k,n} = \sum_{j=1}^{n-k+1} Y_{j,n},$$

where $k = k(B_n)$. With the help of the Tchebycheff inequality $\mathbf{P}(T_n > xB_n) \leqslant 1/x$ and the Borel—Cantelli lemma we obtain a.s. that $T_n = o(B_n(\ln B_n)^{1+\delta})$ $(n \to \infty)$ a.s. for any $\delta > 0$. Moreover, following (Petrov, 1975), we denote by Ψ_c the class of functions $f(x)$ such that $f(x) \geqslant 0$, $f(x)$ is non-decreasing for $x > x_0$ for some $x_0 > 0$ and

$$\int_{x_0}^{\infty} \frac{dx}{xf(x)} < \infty;$$

and such function $f \in \Psi_d$ if this integral diverges. Then, for any function $f \in \Psi_c$ the following condition is fulfilled:

$$T_n = o(B_n f(B_n)) \quad (n \to \infty) \quad \text{a.s.}$$

On the other hand from the example, constructed by the author (Egorov, 1972) we obtain, that for any function $f \in \Psi_d$, there is a sequence of independent positive random variables Y_j such that

$$\limsup_{n \to \infty} \frac{T_n}{B_n f(B_n)} = \infty \quad \text{a.s.}$$

We show, that the results of this type are fulfilled for censorized sums $T_{k,n}$. Instead of the independence of the sequence Y_n we suppose, that the following condition is fulfilled

$$\mathbf{E}\Big(\prod_{i=1}^{m} Y_{j_i}\Big) \leqslant c\Big(\prod_{i=1}^{m} \mathbf{E}Y_{j_i}\Big) \tag{2}$$

for any set of indexes $(j_1, j_2, \ldots, j_m : 1 \leqslant j_1 < j_2 < \cdots < j_m \leqslant n)$ and for any $m, m \leqslant n, 1 \leqslant m \leqslant \lim_{x \to \infty} k(x) \leqslant \infty$. In this paper $c, c_1, c_2, \ldots$ mean positive constants.

Note, that, if $\mathbf{E}Y_n < \infty$, $n = 1, 2, \ldots$, then the condition (2) is fulfilled for independent random variables, for multiplicative systems, for such random variables $Y_n > 0$, that the sequence $(T_n - \mathbf{E}T_n)$, $n = 1, 2, \ldots$, is a supermartingale.
We denote by F_c^k (where $k = k(x)$ is a non-decreasing function) the set of functions $f(x) \geqslant 1$ such that $f(x)$ is non-decreasing and

$$\int_{x_0}^{\infty} \frac{dx}{x(f(x))^{k(x)}} < \infty$$

for some $x_0 > 0$, and such function $f \in F_d^k$, if this integral diverges. For example, $f(x) = (\ln x)^\delta \in F_c^k$, if $\delta > 1/k$, but $f(x) = (\ln x)^{1/k} \in F_d^k$, where k is constant.

THEOREM 1. *Let (2) be fulfilled.*
1) *If k is a constant, then*

$$T_{k,n} = o(B_n f(B_n)) \quad (n \to \infty) \qquad \text{a.s.} \tag{3}$$

for any $f \in F_c^k$.
2) *If $k \uparrow \infty$, then, for any $f \in F_c^k$,*

$$\limsup_{n\to\infty} \frac{T_{k,n}}{B_n f(B_n)} \leqslant 2e \qquad \text{a.s.}, \tag{4}$$

$$Y_{n-k+1,n} = O\left(\frac{B_n f(B_n)}{k}\right) \qquad \text{a.s.} \tag{5}$$

3) *If in addition to 2)*

$$Y_{n-k+1,n} = o\left(\frac{B_n f(B_n)}{k}\right) \quad (n \to \infty) \qquad \text{a.s.} \tag{6}$$

(compare with (5)), then

$$\limsup_{n\to\infty} \frac{T_{k,n}}{B_n f(B_n)} \leqslant 1 \qquad \text{a.s.} \tag{7}$$

Moreover, if $f(x) \to \infty$ $(x \to \infty)$, then

$$\lim_{n\to\infty} \frac{T_{k,n}}{B_n f(B_n)} = 0 \qquad \text{a.s.}, \tag{8}$$

and, if $\lim_{x\to\infty} f(x) < \infty$, then

$$\limsup_{n\to\infty} \frac{T_{k,n}}{B_n} \leqslant 1 \qquad \text{a.s.} \tag{9}$$

4) *In 3) we can replace (6) by the condition*

$$Y_{n-k+1,n} = o\left(\frac{B_n f(B_n)}{k_1}\right) \quad (n \to \infty) \qquad \text{a.s.}, \tag{10}$$

where $k_1 = k_1(B_n)$ is a positive non-decreasing integervalued function, $f \in F_c^{k_1}$. Moreover, if $f(x) \to \infty$, then it is sufficiently to suppose, that $c_2 f \in F_c^{[c_1 k_1]}$ for some $c_1 > 0, c_2 > 0$.
5) Let $x/k(x) \sim \alpha(x)$, where $\alpha(x)$ is non-decreasing function, $f(2x) \leqslant Cf(x)$, $x > x_0$, for some $x_0 > 0$. If $c_2 f \in F_d^{[c_1 k]}$ for any constants $c_1 > 0, c_2 > 0$, then we can construct such sequence of independent positive random variables Y_n, that (6) is fulfilled, but

$$\limsup_{n\to\infty} \frac{T_{k,n}}{B_n f(B_n)} = \infty \qquad \text{a.s.} \tag{11}$$

This theorem is new even for $k = 1$, when $T_{k,n} = T_n$.

We mark important consequences of this theorem.
If $k(x) > c \ln\ln x, x > x_0$, for some $x_0 > 0$ and $c > 0$, then there is such a constant $\theta > 1$, that $f(x) \equiv \theta \in F_c^k$. That is why in this case we have a.s. $T_{k,n} = O(B_n)$. If $Y_{n-k+1,n} = o(B_n/\ln\ln B_n)$ a.s., then we can construct such a function $k_1(x)$, that $k_1(x)/\ln\ln x \to \infty$ and $Y_{n-k+1,n} = o(B_n/k_1(B_n))$ a.s. Then, $f(x) \equiv \theta \in F_c^{k_1}$ for any $\theta > 1$. That is why, using Theorem 1, we receive

$$\limsup_{n\to\infty} \frac{T_{k,n}}{B_n} \leqslant 1 \qquad \text{a.s.}$$

COROLLARY 1. *Let $Y_n, n = 1, 2, \ldots$, be a sequence of positive independent random variables and for any $\varepsilon > 0$*

$$\sum_{j=1}^{n} \mathbf{E}(Y_j; Y_j > \varepsilon B_n/\ln\ln B_n) = o(B_n). \tag{12}$$

Then for $k(x) = [c \ln\ln x]$

$$\lim_{n\to\infty} \frac{T_{k,n}}{B_n} = 1 \qquad \text{a.s.} \tag{13}$$

Proof of Corollary 1. Using Lemmas 3, 4 from (Egorov, 1970) and (12), we obtain that for some sequence $\varepsilon_j \downarrow 0$

$$\sum_{j=1}^{n} \mathbf{E}(Y_j; Y_j > \varepsilon_j M_j) = o(B_n), \tag{14}$$

where $M_j = B_j/\ln\ln B_j$. Let $\bar{Y}_n = Y_n$, if $Y_n \leqslant \varepsilon_n M_n$, and $Y_n = 0$, if $Y_n > \varepsilon_n M_n, \underline{Y}_n = Y_n - \bar{Y}_n$, $n = 1, 2, \ldots$. Using Prohorov criterion about almost sure stability (Loeve, 1960), we get

$$\frac{1}{B_n} \sum_{j=1}^{n} \bar{Y}_j \to 1 \quad (n \to \infty) \qquad \text{a.s.} \tag{15}$$

Let $\underline{B}_n = \sum_{j=1}^{n} \mathbf{E}\underline{Y}_j$. If $\underline{B}_n < C < \infty$, then with the probability one $\sum_{j=1}^{\infty} \underline{Y}_j < \infty$, that is why

$$\sum_{j=1}^{n-k+1} \underline{Y}_{j,n} = o(B_n), Y_{n-k+1,n} = o(M_n) \qquad \text{a.s.} \tag{16}$$

Here $\underline{Y}_{j,n}$ is the corresponding order statistic for the sequence $(\underline{Y}_j)_{j=1}^n$. If $\underline{B}_n \to \infty$, then (16) is followed from (14) and Theorem 1, applied to the sequence $\underline{Y}_j$, $j = 1, 2, \ldots$. Corollary 1 is a consequence of (15), (16) and of the following inequality

$$\left| \sum_{j=1}^{n-k+1} Y_{j,n} - \sum_{j=1}^{n} \bar{Y}_j \right| \leqslant 2k\varepsilon_n M_n + \sum_{j=1}^{n-k+1} \underline{Y}_{j,n}. \tag{17}$$

For the proof of (17), we may assume that $\varepsilon_n M_n \uparrow \infty$. Let $s(j)$ be the rank of the order statistic $Y_{j,n}, j_m = \max(j : Y_{s(j)} = \bar{Y}_{s(j)}, j \leqslant n)$. If $j_m \geqslant n - k + 1$, then

$$\left| T_{k,n} - \sum_{j=1}^{n} \bar{Y}_j \right| = \left| T_n - \sum_{j=n-k+2}^{n} Y_{s(j)} - \sum_{j=1}^{n} \bar{Y}_{s(j)} \right| = \left| \sum_{j=1}^{n} \underline{Y}_{s(j)} - \sum_{j=n-k+2}^{n} Y_{s(j)} \right| =$$

$$= \left| \sum_{j=1}^{j_m} \underline{Y}_{j,n} - \sum_{j=n-k+2}^{j_m} Y_{s(j)} \right| \leqslant \sum_{j=1}^{n-k+1} \underline{Y}_{j,n} + 2 \sum_{j=n-k+2}^{j_m} Y_{j,n} \leqslant \sum_{j=1}^{n-k+1} \underline{Y}_{j,n} + 2k\varepsilon_n M_n.$$

If $j_m \leqslant n - k + 1$, then

$$T_{k,n} - \sum_{j=1}^{n} \bar{Y}_j = \sum_{j=1}^{n} \underline{Y}_{s(j)} - \sum_{j=n-k+2}^{n} Y_{s(j)} =$$

$$= \sum_{j=1}^{n} \underline{Y}_{j,n} - \sum_{j=n-k+2}^{n} \underline{Y}_{j,n} = \sum_{j=1}^{n-k+1} \underline{Y}_{j,n}.$$

Corollary 1 is proved.

The proof of Theorem 1 is based on some combinatorial lemmas.

LEMMA 1.(Egorov, 1970). *Let* $(Y_j)_{j=1}^n$ *be positive real numbers,* $Q_{k,n} = \sum_{J_{k,n}} Y_{j_1} Y_{j_2}, \ldots, Y_{j_k}$, *where* $J_{k,n} = (j_1, j_2, \ldots, j_k : 1 \leqslant j_1 < j_2 < \cdots < j_k \leqslant n)$. *Then, for any positive integer* n *and* k *and for any* $A > 0$, *the following statement is fulfilled: if* $Y_j \leqslant A, j \leqslant n$, *and* $Q_{k,n} \leqslant A^k$, *then*

$$T_n = \sum_{j=1}^{n} Y_j \leqslant (2k-1)A.$$

LEMMA 2. *Let* $\max_{j \leqslant n} Y_j \leqslant \varepsilon A, Q_{k,n} \leqslant A^k$, *where* ε, A, k *depend on* n, *so that* $\varepsilon \to 0, k \to \infty$, *if* $n \to \infty$. *Then* $\limsup_{n \to \infty} (T_n / Ak) \leqslant 1/e$.

Proof of Lemma 2. Let $T_n = T_n(Y_1, Y_2, \ldots, Y_n)$,

$$T = \min_K T_n(Y_1, Y_2, \dots, Y_n) = T_n(Y_1^0, \dots, Y_n^0),$$

where $K = (Y_1, \dots, Y_n : Q_{k,n} \leqslant A^k, \max_{j \leqslant n} Y_j \leqslant \varepsilon A)$. It is possible, that some components of vector $Y^0 = (Y_1^0, \dots, Y_n^0)$ are equal to zero, some components are equal to εA, for the other the inequalities $0 < Y_j < \varepsilon A$ are satisfied. We suppose, that $0 < Y_j < \varepsilon A$ for $j = 1, 2, \dots, p$, $Y_j = \varepsilon A$ for $j = p+1, \dots, p+m, p+m \leqslant n$. Let

$$t(Y_1, Y_2, \dots, Y_p) = T_n(Y_1, Y_2, \dots, Y_p, Y_{p+1}^0, \dots, Y_n^0).$$

The function t reaches its minimum on the boundary of its domain. That is why for $Y = Y^0$ we have

$$Q_{k,n} = Q_{k,p+m} = A^k. \tag{18}$$

Using Lagrange multiplier λ, we obtain for $Y = Y^0$ the equalities

$$1 - \lambda \partial Q_{k,n} / \partial Y_j = 0, \quad j = 1, 2, \dots, p.$$

After differentiation and simple transfors we obtain $Y_j^0 = Y_1^0$, $j = 1, 2, \dots, p$. Therefore, denoting $Y_1^0 = \gamma \varepsilon A$, where $\gamma < 1$, we have

$$\frac{T}{\varepsilon A} = \max_{m,\gamma,p} (m + p\gamma), \tag{19}$$

provided $m, p \geqslant 0$, $k = m + p \leqslant n, \gamma < 1$ and

$$\varepsilon^k \sum_r C_m^r C_p^{k-r} \gamma^{k-r} = 1. \tag{20}$$

If $[\gamma p] > k - r$, then $C_p^{k-r} \gamma^{k-r} \geqslant C_{[\gamma p]}^{k-r}$, therefore, (20) involves

$$C_{m+[\gamma p]}^k = \sum_r C_m^r C_{[\gamma p]}^{k-r} \leqslant \frac{1}{\varepsilon^k}. \tag{21}$$

If Lemma 2 is wrong, then $T > (1 + 2\beta) Ak/e$ for some $\beta > 0$ and some sequence $n = n_s$. Therefore, in this case

$$t = m + [\gamma p] > \frac{(1+\beta)k}{\varepsilon e}, \quad k = o(t). \tag{22}$$

The inequality (22) contradicts (21). Really, from (21) we obtain, that for any $\alpha > 0$ and for $n_s \to \infty$, the following inequalities are fulfilled

$$1 \geqslant \varepsilon^k C_t^k \geqslant \frac{t^k \varepsilon^k}{k!} \exp\left(- \frac{k(k-1)}{2t}(1+\alpha)\right) \geqslant \left((1-2\alpha)\frac{t\varepsilon e}{k}\right)^k. \tag{23}$$

Here we use the Stirling formula, the conditions (22) and the inequality

$$\frac{t(t-1)\dots(t-k+1)}{t^k} \geqslant \exp\Big(-(1+\alpha)\sum_{i=1}^{k}\frac{i-1}{t}\Big) = \exp\Big(-(1+\alpha)\frac{k(k-1)}{2t}\Big).$$

As $\alpha > 0$ is arbitrary, (23) contradicts (22). Lemma 2 is proved.
These Lemmas involve corresponding inequalities in the theorems on the large deviations.

THEOREM 2.
1)

$$\mathbf{P}(T_n > x) \leqslant \mathbf{P}\Big(Y_{n,n} > \frac{x}{2k-1}\Big) + \frac{c_1}{\sqrt{k}}\Big(\frac{2eB_n}{x}\Big)^k, \tag{24}$$

$$\mathbf{P}(T_{k,n} > x) \leqslant \frac{c_1}{\sqrt{k}}\Big(\frac{2eB_n}{x}\Big)^k. \tag{25}$$

2) *If $k \to \infty$ $(n \to \infty)$, then for any sequence $\varepsilon_n \to 0$, there is a sequence $\beta_n \to 1$ such that*

$$\mathbf{P}(T_{k,n} > x) \leqslant \mathbf{P}\Big(Y_{n-k+1,n} > \varepsilon_n \frac{x}{k_1}\Big) + c_1\Big(\beta_n\frac{B_n}{x}\Big)^{k_1} \tag{26}$$

for any $k_1, 1 \leqslant k_1 \leqslant n$.

Proof. By Lemma 1

$$\{T_n > x\} \subset \Big\{Y_{n,n} > \frac{x}{2k-1}\Big\} \cup \Big\{Q_{k,n} > \Big(\frac{x}{2k-1}\Big)^k\Big\}. \tag{27}$$

Therefore, using Tchebycheff inequality and (2), we obtain

$$\mathbf{P}(T_n > x) \leqslant \mathbf{P}\Big(Y_{n,n} > \frac{x}{2k-1}\Big) + \mathbf{E}Q_{k,n}\Big(\frac{2k-1}{x}\Big)^k \leqslant \mathbf{P}\Big(Y_{n,n} > \frac{x}{2k-1}\Big) +$$

$$+\frac{B_n^k}{k!}\Big(\frac{2k-1}{x}\Big)^k \leqslant \mathbf{P}\Big(Y_{n,n} > \frac{x}{2k-1}\Big) + \frac{c_1}{\sqrt{k}}\Big(\frac{2eB_n}{x}\Big)^k.$$

The inequality (24) is proved. To obtain (25) it is sufficient to note that the inequality $Q_{k,n} < (x/(2k-1))^k$ implies the inequality $Y_{n,n}Y_{n-1,n}\dots Y_{n-k+1,n} \leqslant$ $\leqslant (x/(2k-1))^k$, which gives us the inequality $Y_{n-k+1,n} \leqslant x/(2k-1)$.
To prove (26), note that by Lemma 2 the inequality $T_{k,n} > x$ implies one of the inequalities $Y_{n-k+1,n} > \varepsilon_n x/k_1$ or $Q_{k,n} > (\beta_n x e/k_1)^{k_1}$. Therefore

$$\mathbf{P}(T_{k,n} > x) \leqslant P\Big(Y_{n-k+1,n} > \frac{\varepsilon_n x}{k_1}\Big) + \Big(\frac{k_1}{\beta_n x e}\Big)^{k_1}\mathbf{E}Q_{k_1,n} \leqslant$$

$$\leqslant \mathbf{P}\left(Y_{n-k+1,r} > \varepsilon_n \frac{x}{k_1}\right) + c_1 \left(\frac{B_n}{\beta_n x}\right)^{k_1}.$$

Theorem 2 is proved.
Let us prove Theorem 1. Let $n_p = \min(n : n > n_{p-1}, B_{n+1} > \alpha^p)$, where $\alpha > 1$. Then,

$$\frac{B_{n_p}}{B_{n_{p-1}+1}} \leqslant \alpha, \qquad B_{n_p} \geqslant \alpha^{p-1}. \tag{28}$$

Denote $k_p = k(B_{n_{p-1}}), x = x_p = 2eB_{n_p} f(B_{n_{p-1}})$. Then, (25) implies the inequalities

$$\mathbf{P}(T_{k_p,n_p} > x_p) \leqslant C_1 \left(\frac{1}{f(B_{n_{p-1}})}\right)^{k_p} \leqslant C_1 \left(\frac{1}{f(\alpha^{p-1})}\right)^{k(\alpha^{p-1})} \leqslant$$

$$\leqslant C_1 \int\limits_{\alpha^{p-2}}^{\alpha^{p-1}} \frac{dx}{x(f(x))^{k(x)}} \frac{\alpha}{\alpha - 1}. \tag{29}$$

Using (29) and the Borel—Cantelli lemma we have for $f \in F_c^k$

$$\limsup_{p\to\infty} \frac{T_{k_p,n_p}}{2eB_{n_p} f(B_{n_{p-1}})} \leqslant 1 \quad \text{a.s.} \tag{30}$$

If $n_{p-1} < n \leqslant n_p$, then by (30) for all sufficiently large n

$$\frac{T_{k,n}}{B_n f(B_n)} \leqslant \frac{T_{k_p,n_p}}{B_{n_{p-1}+1} f(B_{n_{p-1}})} \leqslant \frac{\alpha T_{k_p,n_p}}{B_{n_p} f(B_{n_{p-1}})} \leqslant \alpha^2 \cdot 2e.$$

This inequality for $\alpha \to 1$ gives us (4). The proposition (3) follows from (4), if we note, that in this case the statement $f \in F_c^k$ implies the statement $\alpha f \in F_c^k$ for any $\alpha > 0$. Now, by analysing the proof of Theorem 2, notice, that the right-hand side of the inequality (25) estimates the probability $\mathbf{P}(Q_{k,n} > (x/(2k-1))^k)$. Therefore, using (29) and the Borel—Cantelli lemma, we obtain, that a.s. $Q_{k_p,n_p} \leqslant (x_p/(2k-1))^{k_p}$ for all sufficiently large p. Hence, a.s., $Y_{n_p-k_p+1,n_p} \leqslant x_p/(2k-1)$ for all sufficiently large p. These inequalities prove (5).
Let (10) be fulfilled. Then by (26) we have for some sequence $\beta_n \to 1$ $(n \to \infty)$

$$\mathbf{P}(T_{k_p,n_p} > x) \leqslant C_1 \left(\beta_{n_p} \frac{B_{n_p}}{x}\right)^{k_1(B_{n_{p-1}})}.$$

Denoting here $x = \beta_{n_p} B_{n_p} f(B_{n_{p-1}})$ and repeating the above estimations, beginning from (29), we get (7).
For proving (8), (9) denote

$$\theta = \lim f(x)\ (x \to \infty),\ 0 < \theta \leqslant \infty,\ k_2 = [\beta k_1],$$

where $\beta > 0$ is a constant. Then, for $\alpha > \theta^{(k_1/k_2)-1}$, the inequality $(\alpha\theta)^{k_2} > \theta^{k_1}$ is fulfilled. Therefore, $\alpha f \in F_c^{k_2}$, if $f \in F_c^{k_1}$. Notice, that if we replace f by αf and k_1 by k_2, then the condition (10) remains true. Hence, using (7), we have

$$\limsup_{n\to\infty} \frac{T_{k_2,n}}{B_n f(B_n)} \leqslant \alpha \quad \text{a.s.} \tag{31}$$

The condition (10) implies, that a.s. $T_{k_1,n} - T_{k_2,n} = o(B_n f(B_n))$ Therefore, we obtain from (31)

$$\limsup_{n\to\infty} \frac{T_{k_1,n}}{B_n f(B_n)} \leqslant \alpha \quad \text{a.s.} \tag{32}$$

for any $\alpha > \theta^{(k_1/k_2)-1}$ and for any $\beta > 0$. If $\beta \to \infty$, then we obtain that (32) is fulfilled for any $\alpha > \theta^{-1}$. If $\theta = \infty$, then (32) is true for any $\alpha > 0$, and we receive (8). If $\theta < \infty$, then we may replace $f(B_n)$ by θ and α replace by $1/\theta$ in (32). It gives us (9). In order to prove the second part of the item 4), note, that, if $f(x) \to \infty$ $(x \to \infty)$, then the conditions (10) and (8) are invariant under replacement f by αf and k_1 by $k_2 = [\beta k_1]$, where $\alpha > 0, \beta > 0$ are constant.
It remains to prove the item 5). In this case we can construct two functions k_1 and $f_1 \in F_d^{k_1}$ such that $x/k_1(x) \uparrow, f_1 \uparrow, f(2x) \leqslant cf(x)$ for $x > x_0$ for some $x_0 > 0$, and $f(x) = o(f_1(x)), k(x) = o(k_1(x))$ $(x \to \infty)$. Now, let Y_n be the sequence of independent random variables such that

$$\mathbf{P}(Y_n = B_n f_1(B_n)/k_1(B_n)) = p_n = 1/f_1(B_n), \mathbf{P}(Y_n = 0) = 1 - p_n.$$

We propose, that $f_1(x)k(x) = o(f(x)k_1(x))$ $(x \to \infty)$. Then, $Y_n = o(B_n f(B_n)/k)$ and the condition (6) is fulfilled. For $B_n = \mathbf{E}T_n$ we have

$$B_n = \sum_{j=1}^{n} \frac{B_j}{4k_1(B_j)}, \quad B_n - B_{n-1} = \frac{B_n}{4k_1(B_n)} = o(B_n).$$

Therefore, denoting $n_k = \min(n : B_{n+1} > 4^k)$, we note, that $B_{n_p} \sim 4^p (p \to \infty)$, and for all sufficiently large k

$$n_{k+1} - n_k \geqslant \frac{4(B_{n_{k+1}} - B_{n_k})}{B_{n_{k+1}}/k_1(B_{n_k})} \geqslant 2k_1(B_{n_k}). \tag{33}$$

Let $\eta_k = \#(Y_n : Y_n > 0, n_k < n \leqslant n_{k+1}), x = k_1(B_{n_k})$. Then, after simple transformations we obtain

$$\mathbf{P}(\eta_k > x) \geqslant C^x_{n_{k+1}-n_k} p^x_{n_{k+1}} (1 - p_{n_k})^{n_{k+1}-n_k-x} \geqslant c\left(2\frac{n_{k+1} - n_k}{x} p_{n_{k+1}}\right)^x \geqslant$$

$$\geqslant (f_1(B_{n_k}))^{-k_1(B_{n_k})} \geqslant (f_1(4^k))^{-k_1(4^k)} \geqslant \frac{1}{3} \int_{4^k}^{4^{k+1}} \frac{dx}{x(f_1(x))^{k_1(x)}}. \tag{34}$$

Using (34) and the Borel—Cantelli lemma, we obtain, that, for $f_1 \in F_d^{k_1}$, with the probability one infinitely often (i.o.) $\eta_k > k_1(B_{n_k})$. Therefore, with the probability one i.o.

$$T_{n_{k+1}} > \frac{B_{n_k} f_1(B_{n_k}) k_1(B_{n_k})}{k_1(B_{n_k})} \geqslant c B_{n_{k+1}} f_1(B_{n_{k+1}}).$$

Hence,

$$\mathbf{P}\left(\limsup_{n\to\infty} \frac{T_n}{B_n f(B_n)} = \infty \right) = 1.$$

The theorem is proved.
With the help of these results we can obtain estimates like the upper estimations in the law of the iterated logarithm. Here we formulate one of such results.

COROLLARY 2. *Let* $\{X_n\}$ *be a sequence of independent symmetric random variables,* $Y_n = X_n^2, n = 1, 2, \ldots, \{X_{j,n}\}_{j=1}^n$ *be the corresponding absolute order statistics, that is* $Y_{j,n} = X_{j,n}^2, j \leqslant n$, *or* $|X_{1,n}| \leqslant |X_{2,n}| \leqslant \cdots \leqslant |X_{n,n}|$. *Denote*

$$S_{k,n} = \sum_{j=1}^{n-k+1} X_{j,n},$$

where $k = k(B_n)$. *If* $B_{n+1}/B_n \to 1, \ln k(x) = o(\ln \ln x), \ln \ln x = o(k(x))$ $(x \to \infty)$, *then the following inequality is fulfilled*

$$\limsup_{n\to\infty} \frac{S_{k,n}}{\sqrt{2B_n \ln \ln B_n}} \leqslant 1 \qquad \text{a.s.}$$

REFERENCES

Egorov, V.A. (1970). On the strong law of large numbers and the law of the iterated logarithm for a sequence of independent random variables. *Theory Probab. Appl.* **15**, 520–527.

Egorov, V.A. (1972). Some theorems on the strong law of large numbers and the law of the iterated logarithm. *Theory Probab. Appl.* **17**, 84–89.

Feller, W. (1968). An extension of the law of the iterated logarithm to variables without variances. *J. Math. and Mechanics* **18**, 343–355.

Hausler, E. and Mason, D.M. (1987). Laws of the iterated logarithm for sums of the middle portion of the sample. *Math. Proc. Cambridge Phil. Soc.* **101**, 301–311.

Kendall, M. and Stuart, A. (1966). *The Advance Theory of Statistics.* **2**, Charles Griffin, London.

Loeve, M. (1960). *Probability Theory. D.* Van Nostrand, Princeton.

Mory, T. (1976). The strong law of large numbers when extreme terms are excluded from sums. *Z. Wahrsch. verw. Geb.* **36**, 189–194.

Petrov, V.V. (1975). *Sums of Independent Random Variables.* Springer, Berlin–Heidelberg etc.

Prob. Theory and Math. Stat., Vol. 1, pp. 314-322
B.Grigelionis *et al.* (Eds.)
1990 VSP/Mokslas

MARTINGALES IN NON-LIFE INSURANCE

P. EMBRECHTS
Department Mathematik ETH-Z, CH-8092, Zürich, Switzerland

ABSTRACT

The present paper should be looked at as a summary of ideas on the subject at the time of the Vilnius meeting and is strongly based on (Haezendonck, 1988) talk. I very much hope that it will stimulate further research in this rapidly expanding field. I have deliberately stayed away from the theorem-proof format. Interested readers will find the precise formulations in the referenced papers.

1. THE MODEL FOR THE LIABILITIES

Consider a counting process $\{N_t : t \in \mathbf{R}_+\}$ that count the number of claims of an insurance portfolio, i.e. $N_0 \equiv 0$.
N_t takes values in $\{0, 1, 2, \ldots\}$ and its sample paths are right-continuous and increasing and the jumps of the sample paths are of size one.
Let $T_1, T_2, \ldots, T_n, \ldots$ be the moments of successive claims, $T_0 \equiv 0$. At these random times *claim amounts* have to be paid out. They are represented by a sequence of random variables $X_1, X_2, \ldots, X_n, \ldots$ which are positive. All these random variables are defined on some fixed probability space $(\Omega, \mathcal{A}, \mathbf{P})$. The amount to be paid out by the insurance company up to time t is described by the random process

$$S_t = \sum_{n=1}^{N_t} X_n \quad (t \in \mathbf{R}_+) \quad \text{(classical risk process)}.$$

This model can be generalized by introducing macroeconomic elements:

$I(t)$ is the inflation index figure at time t, $t \in \mathbf{R}_+$,

$i(t)$ is the force of interest at time t, $t \in \mathbf{R}_+$.

The amount of the nth claim is then equal to $I(T_n)X_n$ and the present value of this amount is equal to

$$\exp\left\{ -\int_0^{T_n} i(s)ds \right\} I(T_n)X_n.$$

Let us introduce the discounting factor

$$f(t) = \exp\left\{ -\int_0^{t} i(s)ds \right\} I(t),$$

then f has the folowing properties:

1. $f(0) = 1$;
2. f is decreasing (but not necessarily strictly);
3. f is continuous;
4. $f > 0$.

The random process

$$\bar{S}_t = \sum_{n=1}^{N_t} f(T_n)X_n \quad (t \in \mathbf{R}_+)$$

is the present value of the aggregate claims up to time t. The case $f \equiv 1$ corresponds to the classical one where income from riskless securities (interest) is cancelled by inflation.

2. THE MAIN PROBLEM

In order to compensate its liabilities described by the risk process $\{\bar{S}_t : t \in \mathbf{R}_+\}$, the company has to fix the *premium income rate*. The general idea is: the net premium income flow should be a deterministic function $p(t)$ $(t \in \mathbf{R}_+)$ such that:

$$p(t) = \mathbf{E}[\bar{S}_t] \quad \text{for all} \quad t \in \mathbf{R}_+.$$

However this idea is too narrow and does not represent the real insurance situation. A much more deeper idea, which is essentially due to B. de Finetti, can be summarized as follows:

— Since premiums have to be paid *in advance* by the insured, the net premium income flow should be a "predictable" random process $\{P_t : t \in \mathbf{R}_+\}$, which is defined in order to make the difference

$$M_t = P_t - \bar{S}_t \quad (t \in \mathbf{R}_+)$$

a *fair game* between insured and insurer.

The assertion that $\{M_t : t \in \mathbf{R}_+\}$ represents a fair game is mathematically translated by stating that $\{M_t : t \in \mathbf{R}_+\}$ is a *martingale*. But what is the mathematical meaning of "predictability". If time t is discrete, i.e. $t = 0, 1, \ldots, n, \ldots,$ then we can define

$$P_{t+1} = \mathbf{E}\Big[\bar{S}_{t+1} \mid \bar{S}_1, \ldots, \bar{S}_t\Big].$$

If time t is continuous, i.e. $t \in \mathbf{R}_+$, then the notion of "predictability" is much more intricate. Roughly speaking it is defined as some generalization of "having left continuous sample paths". See for instance (Dellacherie and Meyer, 1982). The following theorem (due to Meyer and Doob) gives the formal mathematical frame for the introduction of martingales in insurance and formalizes the basic idea of B. de Finetti (see (Dellacherie and Meyer, 1982)):

THEOREM. *If a random process $\{Y_t : t \in \mathbf{R}_+\}$ is a submartingale, then there exists under quite general assumptions a unique predictable random process $\{P_t : t \in \mathbf{R}_+\}$ such that*

$$M_t = P_t - Y_t \quad (t \in \mathbf{R}_+)$$

is a martingale.

This theorem is known as the *Doob—Meyer decomposition.* Since the aggregate claims $\{S_t : t \in \mathbf{R}_+\}$ are increasing, this random process is trivially a submartingale, and the Doob—Meyer decomposition can be applied! Usually $\{P_t : t \in \mathbf{R}_+\}$ is called the *compensator* or the *predictable term* of this decomposition.
Hence martingales appear in risk theory in a very natural way. Therefore it is surprising that the first paper dealing with martingales in risk theory was only published in 1973 (see (Gerber, 1973)).
However, with the Doob—Meyer decomposition not all problems are solved. Indeed how can one find and compute the predictable term $\{P_t : t \in \mathbf{R}_+\}$? In general there is no easy technique at our disposal. Only when the submartingale $\{Y_t : t \in \mathbf{R}_+\}$ has independent increments the problem is easily solved when $t \to \mathbf{E}[Y_t]$ is left continuous, since $\{Y_t - \mathbf{E}[Y_t] : t \in \mathbf{R}_+\}$ is a martingale! Hence $P_t = \mathbf{E}[Y_t]$ is the predictable term.

3. APPLICATIONS

3.1. If the classical risk process $\{S_t : t \in \mathbf{R}_+\}$ is a *compound Poisson process*, it has independent increments. Since the present value of the aggregate claims can be written as

$$\bar{S}_t = \int_0^t f(u)dS_u$$

it can be shown that $\{S_t : t \in \mathbf{R}_+\}$ also has independent increments. Hence

$$M_t = \mathbf{E}[\bar{S}_t] - \bar{S}_t =$$

$$= \lambda\mu \int_0^t f(u)du - \sum_{n=1}^{N_t} f(T_n)X_n$$

(with $\lambda = \mathbf{E}[N_1]$ and $\mu = \mathbf{E}[X_1]$) is a martingale. If $\int_0^\infty f(u)du < \infty$ it can *even* be proved that this martingale is *equally integrable* and so the limit $\lim_{t\to\infty} M_t$ exists a.s.
This martingale $\{M_t : t \in \mathbf{R}_+\}$ can be used for the computation of bounds on the probability of crossing a dividend barrier. Let $\Delta(t)$ be a dividend barrier, i.e. $\Delta : \mathbf{R}_+ \to \mathbf{R}_+$ such that

(i) $\Delta(0) > k$ (k = initial reserve or surplus), and

(ii) Δ is increasing and continuous.

If $\bar{Z}_t^{(k)} = k + (1+\theta)\lambda\mu \int_0^t f(u)du - \bar{S}_t$ is the present value of the surplus at time t (θ = safety loading) then we define

$$\bar{D} = \inf\{t \geqslant 0 \mid \bar{Z}_t > \Delta_t f(t)\}.$$

$\bar{D}$ represents the first time the present value of the surplus $\bar{Z}_t^{(k)}$ crosses the (present value) of the dividend barrier Δ_t. By the optional sampling theorem of Doob for martingales, one has

$$\mathbf{E}[M_0] = \mathbf{E}[M_{\bar{D}\wedge t}] \quad \text{for all } t > 0$$

and this yields upper bounds of the type

$$\mathbf{P}(\bar{D} > t) \leqslant C \quad \text{and} \quad \mathbf{P}(\bar{D} = \infty) \leqslant C.$$

3.2. Suppose that we have the same situation as in 3.1. Let us define the first time ruin occurs:

$$\bar{R} = \inf\left\{t \geqslant 0 \mid \bar{Z}_t^{(k)} < 0\right\}.$$

It is well-known that if $\theta = 0$ and $f \equiv 1$ (i.e. no economic environment) then $\bar{R} < \infty$ a.s.
However, using the same martingale $\{M_t : t \in \mathbf{R}_+\}$ as in 3.1, one can show that

$$\mathbf{P}(\bar{R} < \infty) < 1$$

if $\theta = 0$ and if $\int_0^\infty f^2(u)du < \infty$.

3.3. If $\{Y_t : t \in \mathbf{R}_+\}$ is a random process with independent increments then for each $r \in \mathbf{R}_+$:

$$M_t = \exp\left\{rY_t - \log \mathbf{E}[e^{rY_t}]\right\} \quad (t \in \mathbf{R}_+)$$

is a martingale (provided $\mathbf{E}[e^{rY_t}] < \infty$). This is a *multiplicative version of the Doob—Meyer decomposition.*
Again if $\{S_t : t \in \mathbf{R}_+\}$ is a compound Poisson process and if we put $Y_t = \bar{S}_t$, one obtains that

$$M_t = \exp\left\{r\bar{S}_t - \lambda \int_0^t \mathbf{E}\left[e^{rf(u)X_1} - 1\right] du\right\} \quad (t \in \mathbf{R}_+)$$

is a martingale (provided the integral in the preceding equality is finite). If $f \equiv 1$ (no economic environment) then $\{M_t : t \in \mathbf{R}_+\}$ is the martingale introduced by H.Gerber in his paper of 1973.
The martingale $\{M_t : t \in \mathbf{R}_+\}$ allows to compute upperbounds on the probability of ruin of the type

$$\mathbf{P}(\bar{R} \leqslant t) \leqslant C \text{ and } \mathbf{P}(\bar{R} \leqslant \infty) \leqslant C$$

which generalize and improve other known upperbounds. It should be mentioned that for a negative safety loading $\theta < 0$ and a strictly decreasing discounting factor f it is possible to find useful upperbounds on $\mathbf{P}(\bar{R} \leqslant t)$ and on $\mathbf{P}(\bar{R} = \infty)$. More details on this are to be found in (Delbaen and Haezendonck, 1987; Boogaert and Crijns, 1987).

4. GENERAL MARKOV PROCESSES

Let $Z_t^{(0)} = (1+\theta)\lambda\mu t - S_t$, $t \in \mathbf{R}_+$, be the surplus in the classical case ($\{S_t : t \in \mathbf{R}_+\}$ is a compound Poisson process and $f \equiv 1$). Since $\left\{Z_t^{(0)} : t \in \mathbf{R}_+\right\}$ has independent increments, it is a Markov process and

$$Q_t h(z) = \mathbf{E}\left[h\left(z + Z_t^{(0)}\right)\right]$$

(where h belongs to a suitable function space) defines a semi-group $\{Q_t : t \in \mathbf{R}_+\}$ of operators on that function space.
If

$$\mathcal{A}h(z) = \lim_{t \to 0} \frac{1}{t}(Q_t h(z) - h(z))$$

represents the infinitesimal generator of this operator semi-group, then martingales can be constructed by integrating this infinitesimal generator along sample paths of the Markov process. These sophisticated techniques lead to martingales of the type

$$M_t = e^{-\alpha t + \beta S_t} + \left(\alpha + \lambda - \lambda \mathbf{E}\left[e^{\beta X_1}\right]\right) \int_0^t e^{-\alpha u + \beta S_u} du \quad (t \in \mathbf{R}_+)$$

(with $\alpha, \beta \in \mathbf{R}_+$ and $\mathbf{E}\left[e^{\beta X_1}\right] < \infty$).
This is again a Doob—Meyer decomposition. Here the predictable term (or compensator)

$$\left(\alpha + \lambda - \lambda \mathbf{E}\left[e^{\beta X_1}\right]\right) \int_0^t e^{-\alpha u + \beta S_u} du$$

is random with continuous sample paths.
By these methods one can obtain information on the distribution of the number of claims before ruin occurs, or explicit estimates in the case of "absolute ruin", i.e. ruin in models were borrowing is included. See (Dassios and Embrechts, 1989). The key methodology used is piecewise–deterministic Markov process theory initially developed by (Davis, 1984).In this theory one recognises the following fundamental feature of many insurance related models:

i) The process consists of a jump–structure and, inbetween jumps, the paths are deterministic.

ii) With probability one, there are only finitely many jumps in any compact interval.

iii) The process has a Markovian structure.
Rather than going to diffusion approximations, in the above papers the rather specific "stochastic–jumps deterministic–interjump" evolution is explicitely exploited. As an example (Dassios and Embrechts, 1989) consider the process $Z_t^{(k)} = k +$ $+(1+\theta)\lambda\mu t - S_t$, $t \in \mathbf{R}_+$, where $\{S_t : t \in \mathbf{R}_+\}$ is compound Poisson with exponential jumps with mean μ. Whenever $Z_t^{(k)}$ becomes negative, money is borrowed with

a force of interest δ. It is shown that the natural (called ***absolute***) ruin probability for this process can be calculated as follows:

$$\mathbf{P}(\bar{R}_{abs} < \infty) =$$

$$= e^{-(\mu^{-1}-\lambda c^{-1})k} \left\{ 1 + \left(\mu^{-1} - \lambda c^{-1}\right) \int_{-c/s}^{0} e^{-\mu^{-1}x} \left(1 + \delta c^{-1} x\right)^{\lambda\delta^{-1}-1} dx \right\}^{-1},$$

where $c = (1+\theta)\lambda\mu$. Here $\bar{R}_{abs} = \inf\left\{t > 0 : Z_t^{(k)} \leqslant -c/\delta\right\}$.
As a further application, optimal dividend barriers are obtained for this model.

5. BACKWARD MARTINGALES

5.1. Up to now we have introduced martingales $\{M_t : t \in \mathbf{R}_+\}$ with respect to their internal history $\mathcal{H}_t (t \in \mathbf{R}_+)$, i.e. if $\mathcal{H}_t = \sigma\{M_s : s \leqslant t\}$ then M_t fulfils the property $\mathbf{E}[M_t \mid \mathcal{H}_s] = M_s$ whenever $s \leqslant t$.
Let us now reverse the time. We introduce the "inversed" internal history, which is defined as the future of the classical risk process $\{S_t : t \in \mathbf{R}_+\}$:

$$\mathcal{F}_t = \sigma\{S_u : u \geqslant t\}.$$

We have the following result:

$$M_t^{(k)} = \frac{S_t}{k+ct} + \int_t^{\infty} \frac{S_u}{u} \frac{k}{(k+cu)^2} du \quad (\text{with} \quad c = (1+\theta)\lambda\mu)$$

is a backward martingale with respect to $\{\mathcal{F}_t : t \in \mathbf{R}_+\}$, i.e. if $s \leqslant t$ then

$$\mathbf{E}\left[M_s^{(k)} \mid \mathcal{F}_t\right] = M_t^{(k)}.$$

Here we have a "backward" Doob—Meyer decomposition: indeed $S_t/(k+ct)$ is a backward supermartingale and

$$\int_t^{\infty} \frac{S_u}{u} \frac{k}{(k+cu)^2} du$$

is continuous in $t \in \mathbf{R}_+$ and predictable w.r.t. the future $\mathcal{F}_t$.

5.2. *An application*

Let T_t be the last time before $t > 0$ that the surplus $k + cu - S_u$ was negative, i.e.

$$T_t = \sup\{u \leqslant t \mid k + cu - S_u < 0\}.$$

T_t is dependent on the future (up to time t) and so it is a stopping time w.r.t. the future $\mathcal{F}_t$. Now apply the optional sampling theorem of Doob on the backward martingale $\{M_t^{(k)} : t \in \mathbf{R}_+\}$. This leads to the following equality (Delbaen and Haezendonck, 1987, 1988):

$$\mathbf{P}(\bar{R} > t) = \mathbf{E}\left[\left(1 - \frac{S_t}{k + ct}\right)^+\right] + \mathbf{E}\left[\mathbf{1}_{(S_t \leqslant k+ct)} \int_{T_t}^{t} \frac{S_u}{u} \frac{k}{(k + cu)^2} du\right]$$

($\bar{R}$—ruin time)

$$\mathbf{P}(\bar{R} = \infty) = \frac{\theta}{1+\theta} + \mathbf{E}\left[\int_{T_\infty}^{\infty} \frac{S_u}{u} \frac{k}{(k + cu)^2} du\right].$$

For $k = 0$ one obtains two well-known results

$$\mathbf{P}(\bar{R} > t) = \mathbf{E}\left[\left(1 - \frac{S_t}{ct}\right)^+\right] \quad \text{(H.Cramér)},$$

$$\mathbf{P}(\bar{R} = \infty) = \frac{\theta}{1+\theta}.$$

If $k > 0$ one obtains the following upperbound on the probability of ruin

$$\mathbf{P}(\bar{R} \leqslant t) \leqslant 1 - \mathbf{E}\left[\left(1 - \frac{S_t}{k + ct}\right)^+\right] - \int_0^t \frac{k}{(k + cu)^2} \mathbf{E}\left[\mathbf{1}_{(S_t \leqslant k+cu)} \frac{S_t}{t}\right] du.$$

This upperbound can be numerically computed and gives (especially for small t) much better results then other existing upper bounds.

6. MARTINGALE–EQUIVALENCES AND PREMIUM CALCULATION PRINCIPLES

Let $S_t = \sum_{n=1}^{N_t} X_n$ be a compound Poisson process on the probability space $(\Omega, \mathcal{H}_\infty, P)$, where $\mathcal{H}_\infty = \sigma\{S_t : t \in \mathbf{R}_+\}$. It is well-known that $\mathbf{E}_P[S_1] = \lambda \mathbf{E}_P[X_1]$ defines the pure premium density, i.e. risk parameter $\lambda\times$ net premium of a claim $\mathbf{E}_P[X_1]$.

Problem: characterize the other probability distributions Q on $(\Omega, \mathcal{H}_\infty)$ such that

1) $\{S_t : t \in \mathbf{R}_+\}$ is still a compound Poisson process on $(\Omega, \mathcal{H}_\infty, Q)$,

2) by taking $\mathbf{E}_Q[S_1]$ one obtains the other premium calculation principles.

For the solution of this problem martingales play again a crucial role.

Let β be a Borel measurable mapping from $\mathbf{R}_+$ into $\mathbf{R}$ such that

$$\mathbf{E}_P\left[e^{\beta(X_1)}\right] < \infty$$

then one obtains the martingale:

$$M_t^{(\beta)} = \exp\left\{\sum_{k=1}^{N_t} \beta(X_k) - \lambda t \mathbf{E}_P\left[e^{\beta(X_1)} - 1\right]\right\}.$$

Remark. $\beta(x) = rx$ gives the Gerber–martingale.

MAIN RESULT

Let Q be a probability distribution on $(\Omega, \mathcal{H}_\infty)$ such that

1) $\{S_t : t \in \mathbf{R}_+\}$ is a compound Poisson process on $(\Omega, \mathcal{H}_\infty, Q)$;

2) P and Q are equivalent on each $\mathcal{H}_t = \sigma\{S_u : u \leqslant t\}$, i.e. if $t \in \mathbf{R}_+$ and $A \in \mathcal{H}_t$ then

$$P(A) = 0 \Longleftrightarrow Q(A) = 0.$$

Then there exist $\beta : \mathbf{R}_+ \longrightarrow \mathbf{R}$ with $\mathbf{E}_P\left[e^{\beta(X_1)}\right] < \infty$ such that

$$Q(A) = \int_A M_t^{(\beta)} dP \text{ for each } A \in \mathcal{H}_t. \tag{$*$}$$

Conversely: let $\beta : \mathbf{R}_+ \longrightarrow \mathbf{R}$ with $\mathbf{E}_P\left[e^{\beta(X_1)}\right] < \infty$, then equality $(*)$ defines a unique probability distribution on $(\Omega, \mathcal{H}_\infty)$ such that 1) and 2) are fulfilled.
We say that P and Q are *martingale-equivalent* !
One has a one to one correspondence between

$$\left\{\beta \in L^0(\mathbf{R}_+, \mathcal{R}_+, P) \mid \mathbf{E}_P\left[e^{\beta(X_1)}\right] < \infty\right\}$$

and

$$\{Q \text{ is a probability distribution on } (\Omega, \mathcal{H}_\infty) \mid Q \text{ and } P \text{ are martingale–equivalent}\}.$$

Let us denote by $Q^{(\beta)}$ the probability distribution corresponding to the function β. One can show that

$$\mathbf{E}_{Q^{(\beta)}}[X_1] = \frac{\mathbf{E}_P\left[X_1 e^{\beta(X_1)}\right]}{\mathbf{E}_P\left[e^{\beta(X_1)}\right]}; \quad \mathbf{E}_{Q^{(\beta)}}[N_1] = \lambda \mathbf{E}_P\left[X_1 e^{\beta(X_1)}\right].$$

Relation to premium calculation principles:

6.1. Take $\beta \equiv a > 0$ (constant), then:

$$\mathbf{E}_Q[X_1] = \mathbf{E}_P[X_1]; \quad \mathbf{E}_Q[N_1] = \lambda e^a.$$

We obtain the ***net premium calculation principle*** for the claim amount X_1 and a Poisson process with the risk parameter $\lambda' = \lambda e^a$.

6.2. Take $\beta(x) = \log(a + bx)$ with $a = 1 - b\mathbf{E}_P[X_1]$, then

$$\mathbf{E}_Q[X_1] = \mathbf{E}_P[X_1] + b\mathrm{Var}_P[X_1] \quad \text{(variance principle)},$$

$$\mathbf{E}_Q[N_1] = \lambda.$$

6.3. Take $\beta(x) = \alpha x$, then:

$$\mathbf{E}_{Q^{(\beta)}}[X_1] = \frac{\mathbf{E}_P\left[X_1 e^{\alpha X_1}\right]}{\mathbf{E}_P\left[e^{\alpha X_1}\right]} \quad \text{(Esscher principle)},$$

$$\mathbf{E}_{Q^{(\beta)}}[N_1] = \lambda \mathbf{E}_P\left[X_1 e^{\alpha X_1}\right].$$

Observe that for each of these functions β, the surplus process

$$k + \mathbf{E}_{Q^{(\beta)}}[N_1]\mathbf{E}_{Q^{(\beta)}}[X_1] - S_t \quad (t \in \mathbf{R}_+)$$

is a martingale under $Q^{(\beta)}$, since $\{S_t : t \in \mathbf{R}_+\}$ is a compound Poisson process under $Q^{(\beta)}$. For more details see (Delbaen and Haezendonck, 1988).

Dedication

This paper is dedicated to the memory of Prof. Dr. Jean M.Haezendonck (University of Antwerp) who is died on 12/4/1989. Professor Haezendonck was a leading scientist in the field of mathematical models in insurance. His fundamental work on the stochastic foundation of risk theory leads our subject into the reality modern stochastic calculus. His sad death is a great loss personally as well as scientifically. On my request, Prof. Haezendonck delivered a talk on the subject of *Martingale Methods in Non-Life Insurance* at the 18th European Meeting of Statisticians, August 22–26, 1988, Berlin, GDR (see Haezendonck, 1988). My talk at the Vilnius Conference basically combined Haezendonck's work with results from (Dassios and Embrechts, 1989) on piecewise deterministic Markov processes, a theory in which Haezendonck took great interest in the months before his death.

REFERENCES

Boogaert, P. and Crijns, V. (1987). Upperbounds on ruin probabilities in case of negative loadings and positive interest rates. *Insurance: Math. Econ.* **6**, 221–232.

Dassios, A. and Embrechts, P. (1989). Martingales and insurance risk. *Commun. Statist. Stochastic Models* **5**, 181–217.

Davis, M.H.A. (1984). Piecewise–deterministic Markov processes. A general class of non–diffusion stochastic models. *J.Roy. Statist. Soc. B* **46**, 353–388.

Delbaen, F. and Haezendonck, J.M. (1987). Classical risk theory in an economic environment. *Insurance: Math. Econ.* **6**, 85–116.

Delbaen, F. and Haezendonck J.M. (1988). A martingale approach to premium calculation principles in an arbitrage free market. *Preprint.* University of Antwerp., Antwerpen.

Dellacherie, C. and Meyer, P.A. (1982). *Probabilities and Potential.* North–Holland, Amsterdam.

Gerber, H.U. (1973). Martingales in risk theory. *Mitteilungen der Vereinigung Schweizerischer Versicherungsmathematiker* **73**, 205–216.

Haezendonck, J.M. (1988). Martingale methods in non–life insurance. In: *Abstracts–18th EMS,* Berlin (GDR), pp. 26.

Prob. Theory and Math. Stat., Vol. 1, pp. 323–331
B.Grigelionis et al. (Eds.)
1990 VSP/Mokslas

ASYMPTOTIC MINIMAXITY OF USUAL GOODNESS OF FIT TESTS

M.S. ERMAKOV
Leningrad University, 198904 Leningrad, USSR

ABSTRACT

Let T_n be functionals on the set $\mathbf{F}$ of all d.f.'s $F(x)$, $x \in (0,1)$. Let $\hat{F}_n$ be empirical distribution functions. We show that the sequence of tests with test statistics $T_n(\hat{F}_n)$ is asymptotically minimax for the problems of hypothesis testing $T_n(F) = 0$ against the sets of alternatives $\{F : T_n(F) > b_n,\ F \in \mathbf{F}\}$, $b_n > 0$. Test statistics $T_n(\hat{F}_n)$ are test statistics of Kolmogorov, ω^2, χ^2 or rank tests.

1. INTRODUCTION

Let $X_1, \ldots, X_n$ be i.i.d.r.v.'s with d.f. $F(x), x \in [0,1]$. Let $\hat{F}_n$ be the empirical d.f. of $X_1, \ldots, X_n$. Denote by $\mathbf{F}$ the set of all continuous d.f.'s. The problem is to test the hypothesis $F(x) = F_0(x) = x, x \in [0,1]$, against the set of alternatives

$$\mathbf{F}(T_n, b_n) = \{F : T_n(F) > b_n, F \in \mathbf{F}\},$$

where T_n are functionals on $\mathbf{F}$ corresponding to Kolmogorov, ω^2, χ^2 or rank tests and $\{b_n\}$ is some sequence of real numbers.

Example. The case of Kolmogorov and ω^2 tests:

$$T_n(F) = T(F) = \max|F(x) - F_0(x)|,$$

$$T_n(F) = T(F) = \mathrm{int}\,(F(x) - F_0(x))^2\, dF_0(x).$$

The main result of the paper is the following. For the sets of alternatives $\mathbf{F}(T_n, b_n)$ we show that the sequences of tests with test statistics $T_n(\hat{F}_n)$ are asymptotically minimax on the sets $\mathbf{F}(T_n, b_n)$ (definitions of asymptotic minimaxity are given below).

The sets $\mathbf{F}(T_n, b_n)$ are nonparametric. We may mention only papers (Mann and Wald, 1942) and (Ingster, 1982, 1989) where the problem of nonparametric testing of hypothesis were considered against nonparametric sets of alternatives.

2. RANK STATISTICS

Let $X_1, \ldots, X_n$ be r.v.'s with continuous d.f.'s $F_1(x), \ldots, F_n(x), x \in [0,1]$.

Put

$$D(t) = n^{-1} \sum_{i=1}^{n} F_i(t), \qquad G(t) = n^{-1} \sum_{i=1}^{n} c_i F_i(t),$$

$$\hat{D}_n = n^{-1} \sum_{i=1}^{n} \chi \{X_i < t\}, \quad \hat{G}_n(t) = n^{-1} \sum_{i=1}^{n} c_i \chi \{X_i < t\},$$

where c_i are regression constants, $i = 1, \ldots, n$, and $\chi(A)$ is the indicator of A. Define

$$T_n(D, G) = n \int_0^1 h \left(\frac{n}{n+1} D(t) \right) dG(t)$$

where h is a function of bounded total variation on each compact subset of (0,1). Then $T_n(\hat{D}_n, \hat{G}_n)$ is a rank statistic. Denote

$$\mathbf{F}(b_n) = \{F : T_n(D, G) > b_n, F = (F_1, \ldots, F_n), F_i \in \mathbf{F}, i = 1, \ldots, n\}.$$

Throughout this paper $\beta(K_n, F)$ is the second kind error of the test K_n under the alternative F and

$$\beta(K_n) = \beta(K_n, b_n) = \sup \{\beta(K_n, F) : F \in \mathbf{F}(b_n)\}.$$

Definition 2.1. We say that a sequence of tests K_n with level $\alpha, \alpha \in (0,1)$, is asymptotically minimax if for each sequence of tests S_n with the same level α

$$\lim_{n \to \infty} \beta(K_n)/\beta(S_n) \leqslant 1.$$

Let K_n be an arbitrary asymptotically minimax sequence of tests. We denote

$$\beta(\alpha) = \beta(\alpha, b) = \lim_{n \to \infty} \beta(K_n, b_n),$$

where $\mathbf{b} = \{b_n\}_1^\infty$.
We shall assume that the function h satisfies the conditions of Theorem 3.4.3 (Denker, 1985). According to (Denker, 1985) such h is called a proper function. In particular h is a proper function if

$$\int_0^1 (z(1-z))^{-1/2} |h(z)| dz < \infty$$

and there exists $\delta > 0$ such that h is monotone in intervals $(0, \delta), (1 - \delta, 1)$.

THEOREM 2.1. *Let h be a proper function, $b_n = b/\sqrt{n}, b > 0$, and we should test the hypothesis $T_n(D, G) = 0$ against the sets of alternative $\mathbf{F}(b_n)$. Then the sequence of tests with the test statistics $T_n(\hat{D}_n, \hat{G}_n)$ is asimptotically minimax.*

For signed rank tests a similar result is valid. Let $X_1, \ldots, X_n$ be random variables with common d.f. $F(x), x \in [0, 1]$.
Put $g(t) = h(t)$ if $t > 1/2$ and $g(t) = -h(1 - t)$ if $t < 1/2$.

Denote

$$T_n(F) = \frac{1}{2}\int_0^1 g\left(\frac{n}{n+1}H(s)\right) dF(s)$$

where $H(s) = (F(s) + 1 - F(-s))/2$.
Then $T_n(\hat{F}_n)$ is a signed rank statistics.

THEOREM 2.2. *Let h be a proper function and $b_n = b/\sqrt{n}$, $b > 0$. Then the sequence of tests with test statistics $T_n(\hat{F}_n)$ is asymptotically minimax for the problem of testing the hypotheses $T_n(F) = 0$ against the sets of alternatives $\mathbf{F}(b_n)$.*

Now we give a brief outline of the proofs of Theorems 2.1, 2.2. We consider the problem of asymptotically minimax non–parametric hypothesis testing about functional values and find the asymptotically minimax bounds for the second kind errors. The statement of the question and results are similar to those obtained in (Koshevnik and Levit, 1976) for the problem of asymptotically minimax estimation of functional values. Hence, using Theorems 3.4.3, 3.5.3 (Denker, 1985), we prove that the asymptotically minimax bounds are achieved at the sequences of rank tests.

Remark. For the proof of asymptotic minimaxity of $T(\hat{F}_n)$ we construct a family of distributions $\{F_\theta\}$ on which the sequence of statistics $T_n(\hat{F}_n)$ is asymptotically efficient both for the problem of parameter estimation and that of hypothesis testing. In this way the sequence of statistics $T_n(\hat{F}_n)$ turns out to be simultaneously asymptotically optimal for statistical problems about functional values and values of parameter.

3. χ^2–TESTS

Let $p_{1n}, \ldots, p_{kn}$ be probabilities of cells for d.f. $F_0(x)$ and let the number k depend on n. Let $g_{1n}, \ldots, g_{kn}$ be constants such that $0 < c_1 < g_{jn} < c_2 < \infty$ for $j = 1, \ldots, k$.
Denote $r_{jn} = F(l_{jn}) - F(l_{j-1,n})$, where $l_{jn} = p_{1n} + \ldots + p_{jn}, l_{0n} = 0$, $j = 1, \ldots, k$.
Define functionals

$$T_n(F) = n \sum_{j=1}^{k} g_{jn}(r_{jn} - p_{jn})^2 / p_{jn}.$$

Then $T_n(\hat{F}_n)$ is the test statistics of χ^2–test. The problem is to test the hypothesis $T_n(F) = 0$ against the sets of alternatives $\mathbf{F}(T_n, b_n) = \{F : T_n(F) > b_n, F \in \mathbf{F}\}$.
Put

$$\sigma_n^2 = 2k^{-1} \sum_{j=1}^{k} g_j^2$$

and define numbers x_{α_n} by the equation

$$\alpha_n = \frac{1}{\sqrt{2\pi}\sigma_n} \int_{-\infty}^{x_{\alpha_n}} \exp\left\{-x^2/(2\sigma_n^2)\right\} dx.$$

THEOREM 3.1. *Assume that* $n \min\{p_{in} : 1 \leqslant i \leqslant k\} \to \infty, \max\{p_{in} : 1 \leqslant i \leqslant k\} \to 0$ *as* $n \to \infty$. *Assume also that*

$$0 < \liminf_{n\to\infty} b_n/\sqrt{k} \leqslant \limsup_{n\to\infty} b_n/\sqrt{k} < \infty.$$

Then the sequence of tests K_n *with test statistics* $T_n(\hat{F}_n)$ *is asymptotically minimax. If* K_n *have the level* $\alpha, 0 < \alpha < 1$, *then* $K_n = \chi\{k^{-1/2}T_n(\hat{F}_n) > x_{\alpha_n}(1 + o(1))\}$ *and*

$$\beta(K_n, b_n) = \frac{1}{\sqrt{2\pi}\sigma_n} \int_{-\infty}^{x_{\alpha_n}} \exp\{-(x + b_n/\sqrt{k})^2/(2\sigma_n^2)\}dx(1 + o(1))$$

as $n \to \infty$.

The proof of Theorem 3.1 is based on some asymptotic variant of the Theorem about the least favourable distribution (Lehman, 1986). For the proof of Theorem 3.1 we constructed the Bayesian prior distributions similar to (Efroimovich and Pinsker, 1982) and (Ermakov, 1988).

4. KOLMOGOROV AND ω^2 TESTS

For this problem we prove the asymptotic minimaxity of tests in terms of large deviations of the test statistics. In other words we obtain that these sequences of tests are asymptotically minimax if their first and second kind errors are small. Let

$$T(F) = \max_{0<x\leqslant 1} |F(x) - F_0(x)|q(F_0(x))$$

or

$$T(F) = \left(\int_0^1 (F(x) - F_0(x))^2 q(F_0(x))dF_0(x)\right)^{1/2},$$

where $q(t)$ is a continuous function, $t \in [0, 1]$.

Definition 4.1. Let K_n be a sequence of tests with levels $\alpha_n = \alpha_n(K_n) > 0$. We say that the sequence of tests K_n is *LD* asymptotically minimax if one of the following two statements is fulfilled.

i. If

$$\limsup_{n\to\infty} \ln \alpha_n(K_n)/\ln \beta(K_n) < \infty$$

then for each sequence of tests L_n with levels $\gamma_n < \alpha_n$

$$\limsup_{n\to\infty} \ln \beta(K_n)/\ln \beta(L_n) \leqslant 1.$$

ii. If

$$\limsup_{n\to\infty} \ln \beta(K_n)/\ln \beta(L_n) \leqslant \infty$$

then for each sequence of tests L_n such that $\beta(L_n) < \beta(K_n)$

$$\limsup_{n\to\infty} \ln \alpha_n(K_n)/\ln \alpha_n(L_n) \leqslant 1.$$

If the alternatives are simple then we shall say that the sequence of tests K_n is LD asymptotically efficient.

Remark 4.1. The study of the LD efficiency is of interest if the first kind errors of tests or the second kind errors of tests tend to zero as $n \to \infty$. In other cases the use of Pitman efficiency is more natural.

Remark 4.2. If a set of alternatives is fixed and $\alpha_n(K_n) = \alpha, 0 < \alpha < 1$, $\beta_n(K_n) = \beta, 0 < \beta < 1$, respectively then the LD-asymptotic efficiency coincides with the Hodges—Lehman efficiency (the Bahadur efficiency respectively).

THEOREM 4.1. *Let $b_n \to 0, nb_n^2 \to \infty$ as $n \to \infty$. Then the sequence of tests with test statistics $T(\hat{F}_n)$ are LD-asymptotically minimax.*
If $b_n = b = \text{const} > 0$, $\alpha(K_n) = \alpha = \text{const}$, $0 < \alpha < 1$, or $b_n = b = \text{const}$, $\beta(K_n) = \beta = \text{const}, 0 < \beta < 1$, then the Kolmogorov and ω^2-tests are LD-asymptotically minimax (such types of asymptotic minimaxity are minimax analogies of the Hodges—Lehman and Bahadur efficiencies).

The proof is based on new variants of the theorem about large deviations of empirical measures. This results are awkward and we omit them. In section 6 one interesting version of these theorems will be given.

THEOREM 4.2. *Suppose that D_n is a sequence of ω^2-tests such that $\ln\beta(\alpha_n, D_n)/\ln\alpha_n \to 0$, $\beta_n = \beta(\alpha_n, D_n) \to 0$ as $n \to \infty$. Let $nb_n^2 \to \infty$, $b_n n^{1/3} \to 0$ as $n \to \infty$. Then for each sequence of tests S_n such that $\alpha(S_n) < \alpha(D_n)$*

$$\lim_{n\to\infty} \sup \beta(D_n)/\beta(S_n) \leqslant 1.$$

We now indicate the asymptotics of large deviations of the first and second kind errors of Kolmogorov tests. Let $d = 1/2$, $q(t) \equiv 1$. Then

$$\alpha(D_n) = \exp\left\{-\frac{1}{2}r_n^2 nd^{-2}\right\}(1 + o(1)) \quad \text{as } n \to \infty, \tag{4.1}$$

$$\beta(D_n) = d^{-1}(2\pi n^{1/2}(b_n - r_n))^{-1} \exp\left\{-\frac{1}{2}(a_n - r_n)^2 nd^{-2}\right\}(1 + o(1)) \quad \text{as } n \to \infty, \tag{4.2}$$

where $D_n = \chi(T(\hat{F}_n) > r_n)$.
For the ω^2-test with $q(t) = 1$ we have

$$\beta(D_n) = \sup_{T(F) > a_n n^{-1/2}\pi^{-1}} P(T^2(\hat{F}_n) \geqslant r_n^2 \pi^{-2} n^{-1}) =$$

$$= (\exp\{\pi\sqrt{l_n}\} - \exp\{-\pi\sqrt{l_n}\}) l_n^{-1/2}(a_n - r_n)^{-1} \times$$

$$\times \exp\left\{-\frac{1}{2}(a_n - (r_n^2 - \gamma)^{1/2})^2 + \gamma l_n\right\}(1 + o(1)) \quad \text{as } n \to \infty, \tag{4.3}$$

$$\alpha(D_n) = (2\pi)^{-1/2} r_n^{-1} \exp\left\{-\frac{1}{2}r_n^2\right\}(1 + o(1)) \quad \text{as } n \to \infty, \tag{4.4}$$

where $\gamma = \pi^2/3 - 2$, $l_n = a_n(r_n^2 - \gamma)^{-1/2} - 1, a_n \to \infty, a_n n^{-1/6} \to 0$ as $n \to \infty$.
The computation of asymptotics (4.1)—(4.4) is based on some generalization results of the (Bentkus, 1987) about large deviations of the norms of sums of i.i.d.r.v.'s with values in a Banach space.
In conclusion of this section we indicate the following important properties of the LD-efficiency.

THEOREM 4.3. *Let for all sequences b_n, such that $b_n \to 0, b_n^2 n \to \infty$ as $n \to \infty$, the sequences of tests with the test statistics S_n be LD-efficient.*
Let K_{nb} be sequences of tests with the test statistics S_n for the problem of testing $T_n(F) = 0$ against the alternatives $\mathbf{F}(b/\sqrt{n}), b > 0$. Then

d. *If*

$$\limsup_{b\to\infty} \limsup_{n\to\infty} \ln \alpha(K_{nb})/ \ln \beta(K_{nb}) < \infty$$

then for any sequences of tests L_{nb} with levels $\gamma_{nb} \leqslant \alpha(K_{nb})$

$$\limsup_{b\to\infty} \limsup_{n\to\infty} \ln \beta(K_{nb})/ \ln \beta(L_{nb}) \leqslant 1.$$

dd. *If*

$$\limsup_{b\to\infty} \limsup_{n\to\infty} \ln \beta(K_{nb})/ \ln \alpha(K_{nb}) < \infty$$

then for any sequences of tests L_{nb} such that $\beta(L_{nb}) < \beta(K_{nb})$

$$\limsup_{b\to\infty} \limsup_{n\to\infty} \ln \alpha(K_{nb})/ \ln \alpha(L_{nb}) \leqslant 1.$$

If M_{nb} are sequences of test with the test statistics S_n for the problem of testing $T_n(F) = 0$ against the alternatives $\mathbf{F}(b), b > 0$. Then

c. *If*

$$\limsup_{b\to 0} \limsup_{n\to\infty} \ln \alpha(M_{nb})/ \ln \beta(M_{nb}) \leqslant \infty$$

then for any sequences of tests L_{nb} with levels $\gamma_{nb} < \alpha(M_{nb})$

$$\limsup_{b\to 0} \limsup_{n\to\infty} \ln \beta(M_{nb})/ \ln \beta(L_{nb}) \leqslant 1.$$

cc. *If*

$$\limsup_{b\to 0} \limsup_{n\to\infty} \ln \beta(M_{nb})/ \ln \alpha(M_{nb}) < \infty$$

then for any sequences of tests L_{nb} such that $\beta(L_{nb}) < \beta(K_{nb})$

$$\limsup_{b\to 0} \limsup_{n\to\infty} \ln \alpha(M_{nb})/ \ln \alpha(L_{nb}) \leqslant 1.$$

The proof is rather simple. If the statements of theorems 4.3 do not hold then it is easy to construct a sequence b_n, such that $b_n \to 0, b_n^2 n \to \infty$ as $n \to \infty$ and the sequence of tests K_{nb_n} or M_{nb_n} is not LD-minimax.

5. ASYMPTOTICAL BOUNDS FOR *LD*–EFFICIENCY

All types of efficiency (Bahadur, Pitman, Hodges—Lehman and Chernoff) have asymptotical bounds. We now indicate such bounds for the *LD*–efficiency. Let $X_1, \ldots, X_n$ be i.i.d.r.v.'s with density $f(x,\theta), \theta \in \mathbf{R}^1$, with respect to the Lebesque measure, $x \in \mathbf{R}^1$.
Suppose that the family $f(x,\theta)$ satisfies the conditions of local asymptotic normality at $\theta = 0$ (see (Ibragimov and Hasminskii 1979)).
The problem is to test the hypothesis $\theta = 0$ against the alternatives $\theta_n = b_n$, where $b_n^2 n \to \infty, b_n \to 0$ as $n \to \infty$.

THEOREM 5.1. *Let K_n be a sequence of tests. Let there exist $\tau < 1$ such that $\alpha(K_n) < \tau, \beta(K_n) < \tau$ for all n. Then*

$$\lim_{n\to\infty} \sup (nb_n^2 I)^{-1/2}(|2\ln\alpha(K_n)|^{1/2} + |2\ln\beta(K_n)|^{1/2}) \leqslant 1, \tag{5.1}$$

where I is the Fisher information at $\theta = 0$.

For the likelihood ratio tests the left hand–side in (5.1) equals one.
We now give the interpretation of this inequality. Suppose that $X_1, \ldots, X_n$ are i.i.d. normal r.v.'s $\mathcal{D}X_1 = 1$. Let $K_n = \chi(\bar{X} > c_n)$, where $\bar{X} = n^{-1}(X_1 + \ldots + X_n)$ Then

$$J_{1n} = \sqrt{2\ln\alpha(K_n)} = c_n n^{1/2}(1 + o(1)) \quad \text{as} \quad n \to \infty,$$

$$J_{2n} = \sqrt{2\ln\beta(K_n)} = (b_n - c_n) n^{1/2}(1 + o(1)) \quad \text{as} \quad n \to \infty.$$

Therefore $(J_{1n} + J_{2n})/(b_n^2 n)^{1/2} \to 1$ as $n \to \infty$.
The next two corollaries show the connection between the asymptotical bounds of the Bahadur, Hodges—Lehman, Chernoff, Pitman and *LD*–efficiencies.

COROLLARY 5.1. *Let K_{nb} be sequencies of tests of hypothesis testing $\theta = 0$ against the alternatives $\theta_n = b/\sqrt{n}, b > 0$. Let there exist $\tau < 1$ such that $\alpha(K_{nb}) < < \tau$ and $\beta(K_{nb}) < \tau$ for all n.*
Then

$$\limsup_{b\to\infty} b^{-1} \limsup_{n\to\infty} I^{-1/2}(|2\ln\alpha(K_{nb})|^{1/2} + |2\ln\beta(K_{nb})|^{1/2}) \leqslant 1. \tag{5.2}$$

COROLLARY 5.2. *Let L_{nb} be sequencies of tests of hypothesis testing $\theta = 0$ against the alternatives $\theta = b, b > 0$. Let $\alpha(L_{nb}) < \tau, \beta(L_{nb}) < \tau$ for some $\tau < 1$ and all n. Then*

$$\limsup_{b\to 0} b^{-1} \limsup_{n\to\infty} n^{-1} I^{-1/2}(|2\ln\alpha(L_{nb})|^{1/2} + |2\ln\beta(L_{nb})|^{1/2}) \leqslant 1. \tag{5.3}$$

Note that if K_{nb}, L_{nb} are the likelihood ratio tests then the left hand–sides in (5.2) equal one.
The proofs of corollaries 5.1, 5.2 are similar to the proof of Theorem 4.3.

Remark 5.1. Using (5.1) it is easy to find the asymptotical bounds for the local Bahadur, Hodges—Lehman and Chernoff efficiencies. One can take $\beta(L_{nb}) = = \beta = \text{const}, \alpha(L_{nb}) = \alpha = \text{const}$ and $\beta(L_{nb}) = \alpha(L_{nb})$ respectively.

6. TWO THEOREMS ABOUT LARGE DEVIATIONS OF EMPIRICAL MEASURES

For the proof of Theorem 4.1 new methods have been developed. Their application allowed us to prove two interesting theorems about large deviation of empirical measures. The first of them is a version of the theorem used for the proof of Theorem 4.1. In the second theorem a convenient technique is proposed for the study of moderate deviations of test statistics.
Let S be a Hausdorff space and let $\mathbf{B}$ be the σ-field of Borel sets in S. Let Λ be the set of all probability measures on $\mathbf{B}$. For $P, Q \in \Lambda$ the Kulback—Leibler information number is defined by

$$K(Q, P) = \underset{S}{\text{int}} \frac{dQ}{dP} \log \frac{dQ}{dP} dP \quad \text{if } Q \ll P$$

and $K(Q, P) = \infty$ otherwise.
For $\Omega, \Phi \subset \Lambda$ and $P \in \Lambda$ we denote

$$K(\Omega, P) = \inf\{K(Q, P) : Q \in \Omega\}, K(\Omega, \Phi) = \inf\{K(Q, P) : Q \in \Omega, P \in \Phi\}.$$

Let τ be the topology of weak convergence on Λ. The sequence Q_m in Λ converges to $Q \in \Lambda$ in topology τ, sumbdically $Q_m \xrightarrow{\tau} Q$ as $m \to \infty$, if for each set $B \in \mathbf{B}$ which boundary having Q-measure zero $Q_m(B) \to Q(B)$ as $m \to \infty$. The closure and the interior of a set $\Omega \subset \Lambda$ in the topology τ will be denoted by $\text{cl}_\tau \Omega$ and $\text{int}_\tau \Omega$.
Let $X_1, \ldots, X_n$ be i.i.d.r.v.'s taking values in S with the same probability distribution $P \in \Lambda$. Let $\hat{P}_n$ be the empirical probability measure based on $X_1, \ldots, X_n$.

THEOREM 6.1. *Let $\Omega \subset \Lambda$ and let $\Phi \subset \Lambda$ be a compact set in the topology τ. If $K(\text{int}_\tau \Omega, P) = K(\text{cl}_\tau \Omega, P)$ for all $P \in \Phi$ then*

$$\lim_{n\to\infty} (\sup_{P \in \Phi} |n^{-1} \ln P(\hat{P}_n \in \Omega) + K(\Omega, P)|) = 0.$$

If $K(\underset{\tau}{\text{int}}\, \Omega, \Phi) = K(\text{cl}_\tau \Omega, \Phi)$ then

$$\lim_{n\to\infty} (\sup_{P \in \Phi} n^{-1} \ln P(\hat{P}_n \in \Omega)) = -K(\Omega, \Phi)$$

Let P belong to Λ and let f be the measurable function on $(S, \mathbf{B})$, $Ef^2 < \infty$. For any $c > 0$ denote

$$U(c) = \Big\{Q : |\text{int}\, f(dQ - dP)| < c, \quad Q \in \Lambda\Big\}.$$

Let $T(P)$ be the functional on Λ satisfied the following Assumption.
A. For any $c > 0$ there exists the function $w(h) > 0, h \in \mathbf{R}^1$ such that $w(h) \to 0$ as $h \to 0$ and

$$\left|T(P + h(Q - P)) - T(P) - h \operatorname{int} f(dQ - dP)\right| < w(h)h$$

for all $Q \in U(c)$ and $h > 0$.

THEOREM 6.2. *Assume* A. *Then for any sequence b_n such that $b_n \to 0, nb_n^2 \to \infty$ as $n \to \infty$ it holds*

$$\lim_{n\to\infty} n^{-1} b_n^{-2} \ln P(T(\hat{P}_n) - T(P) > b_n) = -\frac{1}{2}(Ef^2(X))^{-1}$$

REFERENCES

Bentkus, B.Yu. (1986). On large deviations in Banach spaces. *Theory Probab. Appl.* **31**, 710–716.

Denker, M. (1985). *Asymptotic distribution theory in nonparametric statistics.* Vieweg, Wiesbaden.

Efroimovich, S.J. and Pinsker, M.S. (1982). Estimation of square-integrable density distribution of random variable. *Probl. Inform. Transm.* **18**, 19–38.

Ermakov, M.S. (1988). Asymptotically minimax tests for testing of non-parametric hypothesis about density of distribution. *Notes Sci. Seminars Leningrad Branch of Math. Inst.* **166**, 44–53.

Ibràgimov, I.A. and Hasminskii, R.Z. (1979). *Asymptotic estimation theory.* Nauka, Moscow.

Ingster, Yu.I (1982). On minimax distinction of families of non-parametric hypothesis. *Soviet Math. Dokl.* **267**, 536–540.

Ingster, Yu.I. (1987). On comparison of minimax properties of Kolmogorov, ω^2 and χ^2-tests. *Theory Probab. Appl.* **32**, 374–378.

Koshevnik, Ya.A. and Levit, B.Ya. (1976). On a non-parametric analogue of the information matrix. *Theory Probab. Appl.* **21**, 738–753.

Lehman, E.L. (1986). *Testing statistical hypothesis.* Wiley, New York.

Mann, H.B. and Wald, A. (1972). On the choice of the number of intervals in the application of the chi-square test. *Ann. Math. Statist.* **13**, 306–318.

Prob. Theory and Math. Stat., Vol. 1, pp. 332-349
B.Grigelionis *et al.* (Eds.)
1990 VSP/Mokslas

LES FONCTIONS ALÉATOIRES À VALEURS DANS LES ESPACES LUSINIENS ET LEURS MODIFICATIONS RÉGULIÈRES

X. FERNIQUE
Institut de Recherche Mathématique Avancée,
Unité associée au C.N.R.S., Université Louis Pasteur,
7, rue René Descartes, 67084 Strasbourg Cédex (France)

ABSTRACT

La construction de modifications régulières des fonctions aléatoires à valeurs dans les espaces lusiniens est entreprise: modifications mesurables, à trajectoires continues ou cadlag, mesurables et bornées. Les propriétés de convergence en loi sont étudiées pour le cadre des trajectoires continues, puis pour celui des trajectoires cadlag.

0.1. Les progrès du Calcul Stochastique conduisent spécialistes et utilisateurs à manier des fonctions aléatoires à valeurs dans certains espaces fonctionnels; la résolution des équations différentielles en dimension infinie introduit par exemple de manière naturelle les fonctions aléatoires à valeurs distributions. Leur maniement est quelque fois difficile du fait que les Probabilistes privilégient les seules fonctions aléatoires à valeurs dans les espaces de Banach séparables; sans doute depuis une dizaine d'années, utilise-t-on aussi le cadre efficace des "countable hilbertian spaces", mais ce cadre reste trop particulier ou imprécis pour de nombreuses études. Les espaces de Lusin constituent un cadre plus large et nous voudrions montrer ici qu'il est excellent pour l'étude des différents modes de régularité des fonctions aléatoires; on présentera dans ce papier quelques énoncés fondamentaux pour le maniement des fonctions aléatoires à valeurs dans les espaces de Lusin; le lecteur trouvera d'ailleurs dans (Fernique, 1989) une présentation plus complète.

0.2. Le premier paragraphe rappelle les principales propriétés des espaces de Lusin; sur ces espaces, on pourra d'ailleurs consulter (Fernique, 1967) où ils sont présentés sous le nom d'espaces standards et (Bourbaki, 1974) où l'étude utilise malheureusement des hypothèses superflues. Le deuxième paragraphe est consacré à la construction de modifications régulières des fonctions aléatoires à valeurs dans les espaces lusiniens: modifications mesurables, modifications à trajectoires continues ou càdlàg, modifications à trajectoires mesurables et bornées. Le troisième paragraphe étudie des problèmes de topologie pour les lois de ces fonctions aléatoires. Le quatrième paragraphe présente des exemples.

1. SUR LES ESPACES LUSINIENS

1.1. Un espace topologique $E = (E, \mathcal{T})$ est lusinien si sa topologie $\mathcal{T}$ est séparée et s'il existe sur E une topologie polonaise $\mathcal{P}$ plus fine que $\mathcal{T}$. Les espaces lusiniens généralisent donc les espaces polonais et conservent pourtant les propriétés essentielles qui permettent un bon calcul des probabilités. L'argument fondamental pour le montrer est le suivant:

1.2. THÉORÈME ((Fernique, 1967, proposition 1.2.2); cf. (Bourbaki, 1974, IX.66, théorème 3)). *Soit E un espace lusinien; pour qu'un sous-espace de E soit lusinien, il faut et il suffit qu'il soit borélien dans E.*

Cet énoncé résulte des deux lemmes suivants (cf. (Bourbaki, 1974, IX.66, lemmes 6 et 7)).

1.2.1. LEMME. *Dans un espace lusinien E, tout ensemble borélien est un sous-espace lusinien de E.*

1.2.2. LEMME. *Tout sous-espace lusinien d'un espace séparé E est un ensemble borélien dans E.*

Pour les démonstrations de ces lemmes, nous renvoyons à (Bourbaki, 1974, IX.66—67); l'hypothèse inutile de régularité doit y être supprimée; dans la démonstration du deuxième lemme où cette hypothèse semble intervenir, reprendre la preuve de (Bourbaki, 1974) à l'identique jusqu'à ses quatre dernières lignes, y utiliser ensuite pour conclure et à la place de l'argument de régularité le fait que dans un espace séparé tout élément est l'intersection de ses voisinages fermés.
De ce théorème, on déduit alors:

1.2.3. COROLLAIRE. *Si f est une application continue injective d'un espace lusinien E dans un espace séparé F, l'image $f(E)$ est un ensemble borélien dans F.*

On constate en effet que $f(E)$ est un sous-espace lusinien de F.

1.2.4. En reprenant alors les termes de la définition d'un espace lusinien, le corollaire implique que la tribu borélienne $\mathcal{B} = \mathcal{B}(E, \mathcal{T})$ est identique à la tribu polonaise $\mathcal{B}(E, \mathcal{P})$; il en résulte donc que les probabilités sur l'espace lusinien E sont identiques aux probabilités sur l'espace polonais $(E, \mathcal{P})$.

1.3. Nous citons maintenant quelques propriétés des espaces lusiniens (cf. (Bourbaki, 1974, IX.68—69)) qui résultent de la définition 1.1 et du théorème 1.2 ci-dessus ou des propriétés analogues des espaces polonais.

THÉORÈME. (1) *Le produit d'une suite d'espaces lusiniens est un espace lusinien et sa tribu borélienne est la tribu produit des tribus boréliennes.* (2) *La réunion et l'intersection d'une suite de sous-espaces lusiniens d'un espace séparé sont des sous-espaces lusiniens.*

1.3.3. COROLLAIRE. *Soient E et F deux espaces lusiniens et f une application borélienne injective de E dans F, alors pour tout ensemble borélien A dans E, $f(A)$ est un ensemble borélien dans F.*

La diagonale de $A \times F$ est en effet un sous-espace fermé lusinien de l'espace lusinien produit $E \times F$; or $f(A)$ est sa projection continue et injective sur F.

1.3.4. COROLLAIRE. *Soient E un espace lusinien, alors sa tribu borélienne $\mathcal{B}(E)$ est engendrée par toute suite (g_n) de fonctions boréliennes réelles séparant les points:*

L'application g: $t \to \{g_n(t), n \in \mathbb{N}\}$ est en effet une application borélienne de E dans l'espace lusinien $\mathbb{R}^{\mathbb{N}}$ muni de sa tribu borélienne produit; la tribu $\mathcal{B}(E)$ contient donc la tribu engendrée par g. Comme g est injective, le corollaire 1.3.3 montre alors l'inclusion inverse et l' égalité.

1.4. THÉORÈME (cf (Fernique, 1967, 1.6.1); (Bourbaki, 1974, IX.75—77)). *Soit E un espace lusinien.* (1) *De tout recouvrement ouvert de E, on peut extraire un recouvrement dénombrable.* (2) *De tout ensemble $(f_i, i \in I)$ de fonctions réelles et continues sur E séparant ses points, on peut extraire une suite séparante.*

La première affirmation résulte en effet de la propriété analogue pour les espaces polonais; pour justifier la seconde, il suffit d'extraire de la famille $\{(x, y) \in E \times E : f_i(x) - f_i(y) > 0\}$, $i \in I$, un recouvrement ouvert dénombrable du complémentaire lusinien de la diagonale fermée dans $E \times E$.

1.4.3. COROLLAIRE. *Soit E un espace lusinien, tout ensemble de fonctions continues sur E séparant ses points engendre la tribu $\mathcal{B}(E)$.*

Ceci résulte immédiatement en effet de 1.3.4 et 1.4.2.

1.5. Nous donnons maintenant les exemples les plus utiles d'espaces lusiniens:

THÉORÈME (Fernique, 1967, 1.5.1). (a) *Tout espace de Fréchet séparable est lusinien.*

(b) *La limite inductive stricte de toute suite d'espaces de Fréchet séparables est lusinienne.*

(c) *Le dual faible de tout espace de Fréchet séparable est lusinien.*

(d) *Le dual faible de la limite inductive stricte de toute suite d'espaces de Fréchet séparables est lusinien.*

(e) *Le dual fort de tout espace de Fréchet—Montel est lusinien.*

(f) *Le dual fort de la limite inductive stricte de toute suite d'espaces de Fréchet—Montel est lusinien.*

1.6. Dans ce papier, nous utiliserons en particulier des suites bornées et des suites relativement compactes dans des espaces vectoriels topologiques lusiniens. Dans certains de ces espaces vectoriels, ces suites seront facilement maniables:

THÉORÈME (Fernique, 1989, théorème 1.8). *Soit E un espace vectoriel topologique de l'un des types* (a),(b),(c),(d),(e) *ou* (f) *du théorème* 1.5; *alors dans l'espace produit $E^{\mathbf{N}}$, l'ensemble $B(E)$ des suites bornées dans E et l'ensemble $K(E)$ des suites relativement compactes dans E sont tous mesurables. De plus pour toute probabilité μ sur $E^{\mathbf{N}}$, on a:*

$$\mu(B(E)) = \sup\left\{\mu\left[x = (x_n) \in E^{\mathbf{N}}: \forall\, n \in \mathbf{N}, x_n \in b\right], b \subset E,\ b \text{ borné dans } E\right\},$$

$$\mu(K(E)) = \sup\left\{\mu\left[x = (x_n) \in E^{\mathbf{N}}: \forall\, n \in \mathbf{N}, x_n \in k\right], k \subset E,\ k \text{ compact dans } E\right\}.$$

Démonstration. Elle résultera de constructions effectives des ensembles indiqués; nous distinguerons successivement suivant le type de E.

(a) Dans ce cas (a), nous notons (s_k) une suite dense dans E, d une distance définissant la topologie de E et (N_k) une suite de semi-normes définissant cette topologie; on a alors:

$$B(E) = \bigcap_{p \in \mathbf{N}} \bigcup_{q \in \mathbf{N}} \bigcap_{n \in \mathbf{N}} \{x \in E^{\mathbf{N}}:\ N_p(x_n) \leqslant q\},$$

$$K(E) = \bigcap_{p \in \mathbf{N}} \bigcup_{q \in \mathbf{N}} \bigcap_{n \in \mathbf{N}} \bigcup_{k=1,\dots,q} \left\{x \in E^{\mathbf{N}}:\ d(x_n, s_k) \leqslant \frac{1}{p} + 1\right\}.$$

(b) Nous notons (E_p) une suite d'espaces de Fréchet séparables définissant E, alors:

$$B(E) = \bigcup_{p \in \mathbf{N}} B(E_p), \quad K(E) = \bigcup_{p \in \mathbf{N}} K(E_p).$$

(c) ou (e) Dans ces deux cas, nous notons (K_p) une suite de parties compactes de E le recouvrant telle que toute autre partie compacte de E soit contenue dans l'un des K_p; on a alors:

$$B(E) = K(E) = \bigcup_{p \in \mathbf{N}} \bigcap_{n \in \mathbf{N}} \{x \in E^{\mathbf{N}}:\ x_n \in K_p\}.$$

(d) ou (f) Dans ces deux cas, soit (F_p) une suite d'espaces de Fréchet séparables (ou de Fréchet—Montel) dont la limite inductive stricte F ait pour dual faible (ou fort) l'espace E; pour tout $p \in \mathbb{N}$, nous notons h_p l'application canonique de E dans le dual faible (ou fort) E_p de F_p, on a alors:

$$B(E) = K(E) = \bigcap_{p \in \mathbb{N}} \{x \in E^{\mathbb{N}}\colon \ (h_p\, x_n)_{n \in \mathbb{N}} \in B(E_p)\}.$$

Dans tous les cas, les mesurabilités indiquées résultent des formules ci–dessus. Pour établir ensuite les égalités proposées, on remarque d'abord que les seconds membres sont inférieurs ou égaux aux premiers et il suffit donc d'étudier le cas où le premier membre étudié est positif; les propriétés de mesurabilité établies pour $B(E)$ et $K(E)$ permettent ensuite de se réduire au cas où le premier membre est égal à 1. Nous détaillons les preuves concernant $B(E)$; celles qui concernent $K(E)$ seraient analogues et nous les laissons au lecteur. Nous fixons un nombre $\varepsilon > 0$. Dans le cas (a) et si $\mu\{B(E)\} = 1$, alors pour tout entier positif p, il existe un entier positif $q(p)$ tel que:

$$\mu\Big\{ \bigcap_{n \in \mathbb{N}} \big[N_p(x_n) \leqslant q(p)\big]\Big\} > 1 - \varepsilon 2^{-(p+1)};$$

l'ensemble $b = \bigcap_{p \in \mathbb{N}} \big[N_p(x) \leqslant q(p)\big]$ est alors borné dans E et on a:

$$\mu\{x \in E^{\mathbb{N}}\colon \ \forall n \in \mathbb{N}, x_n \in b\} > 1 - \varepsilon,$$

c'est le résultat dans ce cas.

Dans le cas (b) et sous la même hypothèse, la suite $\{B(E_p), p \in \mathbb{N}\}$ étant croissante, il existe un entier positif p_0 tel que $\mu\{B(E_{p_0})\} > 1 - \varepsilon$. Au vu du résultat (a), il existe donc un ensemble b borné dans E_{p_0} et par suite dans E tel que:

$$\mu\{x \in E^{\mathbb{N}}\colon \ \forall n \in \mathbb{N}, x_n \in b\} > 1 - \varepsilon,$$

c'est le résultat dans ce cas.

Dans les cas (c) ou (e) et sous la même hypotèse, il existe un entier p_0 tel que $\mu\Big\{\bigcap_{n \in \mathbb{N}} [x_n \in K_{p_0}]\Big\}$ soit supérieur à $1 - \varepsilon$; l'ensemble $b = K_{p_0}$ est compact et donc borné dans E et on a:

$$\mu\{x \in E^{\mathbb{N}}\colon \ \forall n \in \mathbb{N}, x_n \in b\} > 1 - \varepsilon,$$

c'est le résultat dans ce cas.

Dans les cas (d) ou (f) et sous la même hypothèse, pour tout entier positif p, il existe au vu du résultat (c) ou (e), un ensemble borné b_p dans E_p tel que:

$$\mu\{x \in E^{\mathbb{N}}\colon \ \forall n \in \mathbb{N}, h_p x_n \in b_p\} > 1 - \varepsilon 2^{-(p+1)};$$

l'ensemble $b = \bigcap_{p \in \mathbb{N}} \big[h_p^{-1}(b_p)\big]$ est alors borné dans E et on a:

$$\mu\{x \in E^{\mathbb{N}}\colon \ \forall n \in \mathbb{N}, x_n \in b\} > 1 - \varepsilon,$$

c'est le résultat dans ce cas. Le théorème est donc établi.

2. MODIFICATIONS RÉGULIÈRES

2.0. Les fonctions aléatoires et leurs modifications

Dans toute la suite de l'article $(\Omega, \mathcal{A}, \mathbf{P})$ est un espace d'épreuves complet, E est un espace lusinien muni de sa tribu borélienne $\mathcal{B}(E)$, T est un ensemble. *Une fonction aléatoire* $X = \{X(\omega,t),\ \omega \in \Omega,\ t \in T\}$ *sur* T *à valeurs dans* E est une famille $\{X(t),\ t \in T\}$ d'applications mesurables de $(\Omega, \mathcal{A}, \mathbf{P})$ dans $(E, \mathcal{B}(E))$ indexées par T; *ses trajectoires* sont les applications $\{X(\omega)\colon t \to X(\omega,t)\}$ indexées par Ω. Soient X et Y deux fonctions aléatoires sur T à valeurs dans E; pour tout $t \in T$, l'ensemble $\Big\{X(t) = Y(t)\Big\} = \Big\{\big(X(\omega,t), Y(\omega,t)\big) \in \Delta\Big\}$ où Δ est la diagonale fermée de $T \times T$ est alors mesurable dans Ω; on dira que X *et* Y *sont modifications l'une de l'autre* si pour tout $t \in T$, $\mathbf{P}\{X(t) = Y(t)\} = 1$.

2.1. Modifications mesurables

Dans ce paragraphe, l'ensemble T est muni d'une tribu $\mathcal{T}$; une fonction aléatoire X sur T à valeurs dans E est mesurable si l'application: $(\omega, t) \to X(\omega,t)$ de $(\Omega \times \times T, \mathcal{A} \otimes \mathcal{T})$ dans $(E, \mathcal{B}(E))$ est mesurable. Les propriétés des espaces lusiniens réduisent la recherche des modifications mesurables d'une fonction aléatoire sur T à valeurs dans E à celle de modifications mesurables pour des suites de fonctions aléatoires réelles:

THÉORÈME (Fernique,1989, théorème 4.1.5). *Soit X une fonction aléatoire sur un espace mesurable $(T, \mathcal{T})$ à valeurs dans un espace lusinien E, soit de plus F un ensemble dénombrable de fonctions boréliennes réelles sur E séparant ses points; alors les deux propriétés suivantes sont équivalentes:*

(a) X a une modification mesurable Y,

(b) Pour tout élément de F, $f \circ X$ a une modification mesurable y_f.

Démonstration. Puisque F est dénombrable, l'espace $\mathbf{R}^F$ muni de sa topologie et de sa tribu produit est lusinien; l'application F de E dans $\mathbf{R}^F$: $x \to \{f(x), f \in \in F\}$ est injective et borélienne; notons $G = F(E)$ son image, c'est une partie borélienne de $\mathbf{R}^F$ (corollaire 1.3.3) et F est une application borélienne bijective de E sur G; l'application réciproque F^{-1} est donc une application borélienne de G sur E. L'hypothèse (b) exprime que $F \circ X$ a une modification mesurable $y = \{y_f, f \in F\}$ sur $\Omega \times T$ à valeurs dans $\mathbf{R}^F$. On pose alors:

$$Y = F^{-1}(y) \text{ si } y \in G, \quad Y = e_0 \in E \text{ si } y \notin G;$$

puisque G est une partie mesurable dans $\mathbf{R}^F, Y$ est une application mesurable de $(\Omega \times T, \mathcal{A} \otimes \mathcal{T})$ dans $(E, \mathcal{B}(E))$ et pour tout t appartenant à T, on a:

$$\mathbf{P}\{Y(t) \neq X(t)\} \leqslant \mathbf{P}\{F \circ Y(t) \neq F \circ X(t)\} \leqslant \sum_{f \in F} \mathbf{P}\{y_f(t) \neq f \circ X(t)\} = 0,$$

de sorte que Y est une modification mesurable de X; ceci prouve (b) $\Rightarrow$ (a); l'implication inverse est immédiate et le théorème est établi.
Le corollaire suivant sera particulièrement utile:

COROLLAIRE 2.1.1. *Soit E un espace vectoriel topologique localement convexe lusinien et T un espace métrique séparable; soit de plus F un ensemble de formes linéaires continues sur E séparant ses points, soit enfin X une fonction aléatoire sur T à valeurs dans E; on suppose que pour tout y appartenant à F, $\langle X,y\rangle$ est continu en probabilité. Dans ces conditions, X a une modification mesurable.*

On sait en effet que pour tout y appartenant à F, la fonction aléatoire réelle $\langle X,y\rangle$ continue en probabilité a une modification réelle mesurable; on peut d'ailleurs extraire de F (corollaire 1.4.3) une suite séparant les points de E et le théorème 2.1 fournit donc le résultat.

2.2. Modifications à trajectoires continues ou cadlag
Les propriétés des espaces lusiniens réduisent de même sous des hypothèses adéquates de compacité, la recherche des modifications à trajectoires continues d'une fonction aléatoire sur T à valeurs dans E à celle de modifications à trajectoires continues pour des suites de fonctions aléatoires réelles:

THÉORÈME. *Soit X une fonction aléatoire sur un espace séparable T à valeurs dans un espace lusinien E. On suppose qu'il existe une partie dénombrable dense S de T et une suite de parties compactes (K_n) de E telles que:*

(a) $\mathbf{P}\{\exists n \in \mathbb{N}\colon \forall s \in S,\ X(s) \in K_n\} = 1;$

on suppose aussi qu'il existe une suite F de fonctions réelles continues sur E séparant ses points tel que:

(b) *Pour tout $f \in F$, $f \circ X$ possède une modification x_f ayant ses trajectoires continues.*

Dans ces conditions, X possède aussi une modification ayant ses trajectoires continues.

Démonstration. Les hypothèses impliquent qu'il existe une partie négligeable Ω_0 de Ω et aussi pour tout $t \in T$, une partie négligeable Ω_t de Ω telles que:

(1) $\forall\omega \notin \Omega_0,\ \exists n \in \mathbb{N}\colon \forall s \in S,\ X(\omega,s) \in K_n,$

(2) $\forall\omega \notin \Omega_0,\ \forall f \in F,\ \forall s \in S,\ f \circ X(\omega,s) = x_f(\omega,s),$

(3) $\forall t \in T,\ \forall f \in F,\ \forall\omega \notin \Omega_t,\ f \circ X(\omega,t) = x_f(\omega,t).$

Pour tout $\omega \notin \Omega_0$, il existe donc (cf. (1)) un entier n tel que pour tout $t \in T$, le filtre des $\{X(\omega,s),\ s \in S,\ s \to t\}$ soit un filtre à base compacte dans K_n, n'ayant qu'un point adhérent (cf. (b) et (4)) et convergeant donc vers une limite $Z(\omega,t)$ appartenant à K_n; comme les filtres $\{f\circ X(\omega,s) = x_f(\omega,s),\ s \in S,\ s \to t\}$ convergent respectivement vers $x_f(\omega,t)$, on a $f \circ Z(\omega,t) = x_f(\omega,t)$ de sorte que les fonctions $f \in F$ étant continues sur le compact K_n et séparant ses points, les applications $x_f(\omega)$, $f \in F$ étant toutes continues sur $T, Z(\omega)$ est une application continue de T dans K_n et donc dans E. Pour tout $\omega \in \Omega_0$ et tout $t \in T$, on donne à $Z(\omega,t)$ une valeur fixe $z_0 \in E$. Pour tout $t \in T$ et tout $\omega \notin \Omega_0 \bigcup \Omega_t$, $Z(\omega,t)$

est égal (cf. (3)) à $X(\omega,t)$; puisque l'espace d'épreuves est complet, ceci suffit à montrer que Z est une fonction aléatoire à valeurs dans E et une modification de X; par construction, toutes ses trajectoires sont continues. Le théorème est donc établi.
De la même manière on montre:

THÉORÈME 2.2′ (Fernique, 1989, théorème 4.2.5). *Soit X une fonction aléatoire sur un intervalle $T \subset \mathbf{R}$ à valeurs dans un espace lusinien E. On suppose qu'il existe une partie dénombrable et dense S de T et une suite de parties compactes (K_n) de E telles que:*

(*a*) $\mathbf{P}\{\exists n \in \mathbf{N}\colon \forall s \in S,\ X(s) \in K_n\} = 1;$

on suppose aussi qu'il existe un ensemble dénombrable F de fonctions réelles continues sur E séparant ses points tel que:

(*b*) *Pour tout $f \in F$, $f \circ X$ possède une modification x_f ayant ses trajectoires cadlag.*

Dans ces conditions, X possède aussi une modification ayant ses trajectoires cadlag.

Démonstration. Les hypothèses impliquent qu'il existe une partie négligeable Ω_0 de Ω et pour tout $t \in T$, une partie négligeable Ω_t de Ω telles que:

(1) $\forall \omega \notin \Omega_0,\ \exists n \in \mathbf{N}\colon \forall s \in S,\ X(\omega,s) \in K_n,$

(2) $\forall \omega \notin \Omega_0,\ \forall f \in F,\ \forall s \in S,\ f \circ X(\omega,s) = x_f(\omega,s),$

(3) $\forall t \in T,\ \forall f \in F,\ \forall \omega \notin \Omega_t,\ f \circ X(\omega,t) = x_f(\omega,t).$

Pour tout $\omega \notin \Omega_0$, il existe donc (cf. (1)) un entier n tel que pour tout $t \in T$, le filtre des $\{X(\omega,s),\ s \in S,\ s > t,\ s \to t\}$ soit un filtre à base compacte dans K_n, n'ayant qu'un point adhérent (cf. (b) et (4)) et convergeant donc vers une limite $Z(\omega,t)$ appartenant à K_n; pour tout $\omega \in \Omega_0$ et tout $t \in T$, on donne à $Z(\omega,t)$ une valeur fixe $z_0 \in E$. Pour tout $t \in T$ et tout $\omega \notin \Omega_0 \bigcup \Omega_t$, $Z(\omega,t)$ est égal (cf. (3)) à $X(\omega,t)$; puisque l'espace d'épreuves est complet, ceci suffit à montrer que Z est une fonction aléatoire à valeurs dans E et une modification de X; par construction, elle est continue à droite; par construction aussi, pour tout $\omega \notin \Omega_0$ et tout $t \in T$, le filtre des $\{Z(\omega,s),\ s \in S,\ s \to t,\ s < t\}$ est un filtre à base compacte n'ayant qu'un point adhérent de sorte que $Z(\omega)$ a des limites à gauche. Le théorème est donc établi.

Remarque. Dans les deux théorèmes ci-dessus, les conditions (a) pourraient apparaître comme des conditions purement techniques. Ce n'est pas le cas: supposons en effet que l'espace ou l'intervalle T soit compact et que la fonction aléatoire X sur T à valeurs dans E ait une modification à trajectoires continues ou cadlag, alors cette modification est à fortiori à trajectoires relativement compactes dans E et la suite aléatoire $\{X(s),\ s \in S\}$ à valeurs dans $E^{\mathbf{N}}$ est presque sûrement à valeurs dans $K(E)$ (cf. théorème 1.6); si de plus E a l'un des types (a), (b), (c), (d), (e) ou (f), alors le même théorème montre que les conditions (a) des théorèmes précédents sont nécessairement vérifiées.

2.3. Modifications à trajectoires bornées

Nous supposons que l'espace E est un espace vectoriel topologique localement convexe et lusinien et nous désirons construire des modifications à trajectoires bornées. Pour la construction de modifications à trajectoires continues, nous avions été amenés à introduire des hypothèses de compacité relativement naturelles; elles seraient moins adaptées ici au moins dans les espaces où les parties bornées ne sont pas relativement compactes. On utilisera alors le théorème suivant:

THÉORÈME (Fernique, 1989, théorème 4.1.9). *Soit X une fonction aléatoire sur un espace métrique séparable T à valeurs dans un espace vectoriel topologique localement convexe et lusinien E ayant un espace d'épreuves complet; on suppose que pour tout y appartenant à E', $\langle X, y\rangle$ est continu en probabilité. Soient de plus $S = \{(s_n),\ n \in \mathbf{N}\}$ une partie dénombrable et dense de T et (B_k) une suite de parties convexes symétriques et fermées de E. Alors X a une modification mesurable X' telle que pour toute partie ouverte U de T,*

$$\forall k \in \mathbf{N},\ \mathbf{P}\{\forall t \in T \cap U,\ X'(t) \in B_k\} = \mathbf{P}\{\forall s \in S \cap U,\ X(s) \in B_k\}.$$

Démonstration. Le théorème 1.4 permet de construire une suite $F = (y_m) \subset E'$ séparant les points de E entre eux et les séparant aussi de chacun des B_k de sorte que:

$$\forall k \in \mathbf{N},\ B_k = \{x \in E\colon \forall m \in \mathbf{N},\ \langle x, y_m\rangle \leqslant \sup_{u \in B_k} \langle u, y_m\rangle\}.$$

Notons δ la distance sur T et $\mathbf{d}$ la distance en probabilité des v.a. réelles; pour tout t appartenant à T et tout k appartenant à $\mathbf{N}$, nous posons:

$$n(t,k) = \inf\left\{n \in \mathbf{N}\colon \forall m \leqslant k,\ \delta(t; s_n) + \mathbf{d}\Big(\langle X(s_n), y_m\rangle, \langle X(t), y_m\rangle\Big) \leqslant 2^{-k}\right\},$$

$$s_k(t) = s_n(t,k);$$

alors l'application: $t \to \{s_n(t), n \in \mathbf{N}\}$ de T dans $T^{\mathbf{N}}$ est mesurable, la suite $(s_n(t))$ converge vers t et pour tout y appartenant à F, $(\langle X(s_n(t), y\rangle)$ converge presque sûrement vers $\langle X(t), y\rangle$. Pour tout (ω, t, y) appartenant à $\Omega \times T \times F$, nous posons:

$$x(\omega, t, y) = \lim \Big\langle X\big(\omega, s_n(t)\big), y\Big\rangle \text{ si cette limite existe dans } \mathbf{R},$$

$$x(\omega, t, y) = 0 \text{ sinon;}$$

pour tout y appartenant à F, l'application: $(\omega, t) \to x(\omega, t, y)$ est mesurable sur $\Omega \times T$. Nous posons enfin pour tout $(\omega, t) \in \Omega \times T$, $X'(\omega, t) = x(\omega, t)$ s'il existe un élément $x(\omega, t)$ tel que pour tout y appartenant à F, $x(\omega, t, y) = \langle x(\omega, t), y\rangle$, $X'(\omega, t) = 0$ sinon. Alors l'argumentation du théorème 2.1 montre que X' est une fonction aléatoire mesurable et une modification de X; de plus pour tout (ω, t, y) appartenant à $\Omega \times T \times F$, on a l'alternative suivante: ou bien $\langle X'(\omega, t), y\rangle$ est nul ou bien il est égal à $\lim \Big\langle X\big(\omega, s_n(t)\big), y\Big\rangle$ où $\big(s_n(t)\big)$ appartient à $S^{\mathbf{N}}$ et $\big(s_n(t)\big)$ converge vers t; il en résulte donc que:

$$\forall m \in \mathbf{N},\ \forall t \in T,\ \langle X'(\omega, t), y_m\rangle \leqslant \sup\left\{0, \limsup_{s \to t\, s \in S} \langle X(\omega, s), y_m\rangle\right\}.$$

Les propriétés de F relativement aux parties B_k impliquent donc pour tout ouvert U de T:

$$\forall k \in \mathbf{N},\ \mathbf{P}\{\forall t \in T \cap U,\ X'(t) \in B_k\} \geqslant \mathbf{P}\{\forall s \in S \cap U,\ X(s) \in B_k\};$$

l'inégalité inverse est triviale, d'où le résultat annoncé.

3. SUR LA CONVERGENCE ÉTROITE DES LOIS

3.0. Nous étudions dans ce paragraphe les propriétés de convergence en loi des fonctions aléatoires à valeurs dans les espaces lusiniens; nous présenterons successivement deux situations, celle des fonctions aléatoires continues sur un espace métrique compact T et celle des fonctions aléatoires cadlag sur $T = [0, 1]$, toutes à valeurs dans un espace lusinien régulier E. On commencera par définir sur $\mathcal{C}(T, E)$ noté $\mathcal{C}(E)$ et sur $\mathcal{D}([0, 1], E)$ noté $\mathcal{D}(E)$ des topologies raisonnables; sur $\mathcal{C}(E)$, l'emploi de la topologie de convergence uniforme s'impose; sur $\mathcal{D}(E)$, la définition d'une topologie de Skorokhod canonique exige quelques précautions; si E est compact ou si E est un espace $\mathbf{R}^n$, ces difficultés s'évanouissent; on les rencontre peut-être déjà si E est polonais et si plusieurs distances différentes sont compatibles avec sa topologie. Par ailleurs les tribus à utiliser sur $\mathcal{C}(E)$ et sur $\mathcal{D}(E)$ nous sont imposées indépendamment du choix des topologies sur ces ensembles: les fonctions aléatoires sur T à valeurs dans E sont liées à la tribu produit $\mathcal{B}(E^T)$; nous devons donc utiliser les tribus induites par cette tribu produit. Ces tribus induites ne seront pas en général les tribus boréliennes sur les espaces topologiques $\mathcal{C}(E)$ et $\mathcal{D}(E)$ de sorte que les fonctions continues sur ces espaces ne seront pas toutes mesurables, les fonctions continues et bornées ne seront pas toutes intégrables. Dans ces conditions, la définition générale usuelle (Dellacherie and Meyer, 1975, III, 54) de la topologie de la convergence étroite des mesures devra être adaptée avec soin. Nous détaillerons ci-dessous les éléments liés à la convergence étroite des mesures sur $\mathcal{C}(T, E)$; nous avons choisi une rédaction adaptable intégralement et mot pour mot pour l'étude des mesures sur $\mathcal{D}(E)$.

3.1. La topologie de la convergence étroite des lois des fonctions aléatoires continues

Dans tout ce paragraphe, T désigne un espace métrique *compact*, S désignera un sous-ensemble dénombrable et dense de T; E désigne un espace lusinien *régulier*. On sait alors (Fernique, 1967, I.6.1; Bourbaki, 1974, IX.76) que E est paracompact et donc normal. Nous notons $\mathcal{C}(E)$ l'ensemble des fonctions continues sur T à valeurs dans E; à toute fonction continue réelle g sur E, nous associons l'application $\hat{g}$: $x \to g \circ x$ de $\mathcal{C}(E)$ dans $\mathcal{C}(\mathbf{R})$; $\mathcal{C}(\mathbf{R})$ étant naturellement muni de la topologie de la convergence uniforme, nous munissons $\mathcal{C}(E)$ de la topologie $\mathcal{U}(E)$ la moins fine pour laquelle les applications $\hat{g}$ soient continues, cette topologie est séparée; nous munissons aussi $\mathcal{C}(E)$ de la tribu $\mathcal{T}(E)$ induite par la tribu produit $\mathcal{B}(E^T)$. Remarquons que puisque $\mathcal{C}(E)$ ne semble pas nécessairement lusinien, la tribu $\mathcal{T}(E)$ n'est peut-être pas la tribu borélienne associée à $\mathcal{U}(E)$; c'est une sous-tribu de cette tribu borélienne et elle est engendrée par les fonctions π_t: $x \to x(t)$, $t \in S$, mesurables sur $\mathcal{C}(E)$ et séparant ses points. Sur tout

partie lusinienne de $\mathcal{C}(E)$, $\mathcal{T}(E)$ induit donc la tribu borélienne (théorème 1.4.2 et corollaire 1.3.4). Puisque T est compact, la topologie $\mathcal{U}(E)$ est identique à la topologie de la convergence uniforme et ceci pour toute structure uniforme ou toute distance compatible avec la topologie de E; si E est polonais, $\mathcal{C}(E)$ est donc lusinien de sorte que la tribu induite par $\mathcal{B}(E^T)$ est la tribu borélienne de $\mathcal{C}(E)$. Pour tout espace lusinien régulier E et pour tout $g \in \mathcal{C}(E; \mathbf{R})$, l'application $\hat{g}$ est une application mesurable de $\mathcal{C}(E)$ dans l'espace polonais $\mathcal{C}(\mathbf{R})$.
Nous notons $\mathcal{M}(E)$ l'ensemble des mesures positives bornées sur $(\mathcal{C}(E), \mathcal{T}(E))$; nous munissons $\mathcal{M}(E)$ de *la topologie de la convergence étroite* définie par les fonctions réelles continues bornées **mesurables** sur $(\mathcal{C}(E), \mathcal{T}(E))$ et que nous notons $\mathcal{E}(E)$. L'espace topologique $(\mathcal{M}(E), \mathcal{E}(E))$ sera muni de sa tribu borélienne.

3.1.1. PROPOSITION. *La topologie $\mathcal{E}(E)$ est séparée.*

Démonstration. Nous fixons un entier n et une famille $g = (g_1, \ldots, g_n)$ d'applications continues de E dans $\mathbf{R}$. L'application $\hat{g} = (\hat{g}_1, \ldots, \hat{g}_n)$ est alors une application mesurable de $(\mathcal{C}(E), \mathcal{T}(E))$ dans $\mathcal{C}(\mathbf{R}^n)$; on peut donc lui associer l'application $\tilde{g}$ de $\mathcal{M}(E)$ dans $\mathcal{M}(\mathbf{R}^n)$ qui à toute mesure positive bornée μ sur $\mathcal{C}(E)$ associe la mesure image par l'application $\hat{g}$. L'application $\hat{g}$ est mesurable et continue. Soient alors μ et ν deux mesures positives bornées sur $\mathcal{C}(E)$; si elles ne sont pas séparées par $\mathcal{E}(E)$, alors pour tout g de la forme ci-dessus, les mesures images $\hat{g}(\mu)$ et $\hat{g}(\nu)$ ne sont pas non plus séparées par la topologie séparée $\mathcal{E}(\mathbf{R})$ et sont donc égales; puisque E est un espace normal, pour toute partie fermée F de E, il existe une application continue g^F de E dans [0, 1] dont le noyau soit F; pour toute famille finie $(F_1, \ldots, F_n)$ de parties fermées de E et toute suite $(t_1, \ldots, t_n)$ d'éléments de T, on aura:

$$\mu\{x \in \mathcal{C}(E)\colon \forall i \in [1,n],\ x(t_i) \in F_i\} = \nu\{x \in \mathcal{C}(E)\colon \forall i \in [1,n],\ x(t_i) \in F_i\};$$

μ et ν coincident donc sur une classe de parties de $\mathcal{C}(E)$ stable par intersection finie et engendrant la tribu $\mathcal{T}(E)$ de sorte qu'elles sont égales, d'où le résultat.
Si E est un espace métrique compact, l'espace topologique $\mathcal{M}(E)$ est un espace polonais; dans le cas général, rien n'indique que la topologie $\mathcal{E}(E)$ soit aussi simple. Nous allons pourtant mettre en évidence, à partir de propriétés de compacité, des parties de $\mathcal{M}(E)$ sur lesquelles $\mathcal{E}(E)$ induit une topologie simple; nous énonçons d'abord quelques remarques introductives.

3.1.2. Soit F une partie fermée de E; puisque E est normal, toute fonction continue sur F est prolongeable en une fonction continue sur E; il en résulte que $\mathcal{U}(E)$ induit $\mathcal{U}(F)$ sur $\mathcal{C}(F)$; de même $\mathcal{T}(E)$ induit $\mathcal{T}(F)$ sur $\mathcal{C}(F)$.

3.1.3. Soit K une partie compacte de E; soit de plus (g_m) une suite de fonctions continues réelles sur E séparant les points de K; puisque K est compact, la suite (g_m) suffit pour définir la topologie de K; il en résulte que sur $\mathcal{C}(K)$, les applications $(\hat{g}_m)$ définissent la topologie $\mathcal{U}(K)$.

3.1.4. Soient F et G deux parties fermées de E, supposons $G \subset F$; alors l'ensemble $\{x \in \mathcal{C}(F)\colon \forall t \in T,\ x(t) \in G\}$ est égal à $\{x \in \mathcal{C}(F)\colon \forall t \in S,\ x(t) \in G\}$ et est mesurable dans $\mathcal{C}(E)$; à toute mesure positive bornée μ sur $\mathcal{C}(F)$, on peut donc associer $I\{\forall t \in T,\ x(t) \in G\}\mu$, c'est une mesure positive bornée sur $\mathcal{C}(G)$; on définit ainsi une application $\theta_{F,G}$ de $\mathcal{M}(F)$ dans $\mathcal{M}(G)$. Si F et G sont compacts, l'application $\theta_{F,G}$ est mesurable.

3.2. Convergence étroite et compacité

THÉORÈME (Fernique, 1989, théorème 4.3.2). *Soient (K_n) une suite croissante de parties compactes de E et (g_m) une suite de fonctions réelles continues sur E séparant ses points. On note M l'ensemble des probabilités μ sur $\mathcal{C}(E)$ telles que:*

(1) $\forall n \in \mathbf{N},\ \mu\{x \in \mathcal{C}(E)\colon\ \forall t \in T,\ x(t) \in K_n\} > 1 - 2^{-n}$;

dans ces conditions, M est un sous-espace lusinien de $\mathcal{M}(E)$; de plus pour qu'une partie M' de M soit relativement compacte, il faut et il suffit que:

(2) *pour tout $m \in \mathbf{N}$, l'ensemble $\hat{g}_m(M')$ des mesures images $\hat{g}_m(\mu)$, $m \in M'$, soit relativement compact dans l'espace polonais $\mathcal{M}(\mathbf{R})$.*

Démonstration. (a) Utilisant la suite (K_n) de l'énoncé, nous notons Π l'espace produit des $\mathcal{M}(K_n)$, $n \in \mathbf{N}$, et nous posons:

$$\theta_n = \theta_{K_{n+1},K_n},\ A = \big\{\mu = (\mu_n) \in \Pi\colon\ \theta_n\mu_{n+1} = \mu_n,\ \|\mu_n\| \in [1-2^{-n},1]\big\}.$$

A est alors un sous-ensemble mesurable (cf. remarque 3.1.4) d'un produit dénombrable d'espaces polonais et est donc lusinien. Notons dans ces conditions u l'application de A dans $\mathcal{M}(E)$ définie par $u(\mu) = \sup_{n\in\mathbf{N}} \mu_n$; alors u est une application injective de l'espace lusinien A dans l'espace séparé $\mathcal{M}(E)$ et $u(A) = M$ sera lusinien si u est continue.

(b) Montrons que u est continue: soit en effet Φ un filtre sur A convergeant dans A vers $m = (m_n)$; soit de plus f une fonction continue sur $\mathcal{C}(E)$ à valeurs dans $[0, 1]$; pour tout entier n et puisque A a la topologie produit, on a $\lim_{\mu\in\Phi} \int f\, d\mu_n = $ $= \int f\, dm_n$; on en déduit:

$$\int f\, d(u(m)) = \sup_{n\in\mathbf{N}} \int f\, dm_n = \sup_{n\in\mathbf{N}} \lim_{\mu\in\Phi} \int f\, d\mu_n \leqslant \varliminf_{\mu\in\Phi} \int f\, d(u(\mu));$$

appliquant le même calcul à $(1-f)$ et utilisant le fait que $u(m)$ et les $u(\mu)$ sont des probabilités, on en déduit $\int f\, d\big(u(m)\big) = \lim_{\mu\in\Phi} \int f\, d\big(u(\mu)\big)$, c'est la continuité de u. Le caractère lusinien de M est donc établi.

(c) Si M' est relativement compact, la continuité des applications $\widetilde{g}_m$ montre que pour tout entier m, $\widetilde{g}_m(M')$ est relativement compact dans $\mathcal{M}(\mathbf{R})$; supposons inversement que M' soit inclus dans M et que pour tout entier m, $\widetilde{g}_m(M')$ soit relativement compact dans $\mathcal{M}(\mathbf{R})$. Puisque $\mathcal{C}(\mathbf{R})$ est polonais, ceci implique que pour tout couple (m,p) d'entiers, il existe une partie compacte $C_{m,p}$ dans $\mathcal{C}(\mathbf{R})$ telle que pour tout $\mu \in M'$:

$$\mu\Big\{\big[\widetilde{g}_m(\mu)\big]^{-1}(C_{m,p})^c\Big\} < 2^{-(m+p)},\ \mu\bigg\{\Big\{\bigcap_m \big[\widetilde{g}_m(\mu)\big]^{-1}(C_{m,p})\Big\}^c\bigg\} < 2^{-p};$$

pour tout entier n, notons θ'_n l'application θ_{E,K_n}, alors l'ensemble $\theta'(M')$ des suites $\{(\theta'_n\mu,\ n \in \mathbf{N},\ \mu \in M'\}$, est un sous–ensemble de A contenu dans l'ensemble:

$$\left\{\mu = (\mu_n) \in \Pi\colon \forall n \in \mathbf{N},\ \forall p \in \mathbf{N},\ \|\mu_n\| \leqslant 1,\ \mu_n\left\{\left\{\bigcap_m \left[\widetilde{g}_m(\mu)\right]^{-1}(C_{m,p})\right\}^c\right\} < 2^{-p}\right\};$$

il sera donc relativement compact et son image M' par l'application continue u sera aussi relativement compacte si pour tout couple (n,p) d'entiers, $\bigcap_m\left[\widetilde{g}_m(\mu)\right]^{-1}(C_{m,p})$ a une trace compacte sur $\mathcal{C}(K_n)$; or c'est le cas puisque les applications $\widetilde{g}_m$, $m \in \mathbf{N}$, engendrent la topologie de $\mathcal{C}(K_n)$. Le théorème est donc établi.

Remarque. Soit M un ensemble de probabilités sur $\mathcal{C}(E)$ possédant la propriété 3.2.1; le schéma de preuve du théorème montre que tous les éléments μ de M sont des mesures de Radon sur $\mathcal{C}(E)$; il montre aussi que toutes les fonctions continues réelles et bornées sur $\big(\mathcal{C}(E), \mathcal{T}(E)\big)$ sont intégrables pour tout élément μ de M et que la topologie induite par $\mathcal{E}(E)$ sur M peut être définie par ces fonctions, qu'elles soient mesurables ou non. Ceci signifie que sur un tel ensemble M, toutes les anomalies que la convergence étroite pourrait présenter s'évanouissent.

3.3. Sur la convergence étroite des lois des fonctions aléatoires à trajectoires cadlag

Dans cet alinéa, où nous utiliserons les notations de l'alinéa 3.1, T désigne l'intervalle $[0, 1]$; S désigne un sous–ensemble dénombrable et dense de T contenant 1; E est un espace lusinien régulier. Nous notons $\mathcal{D}(E)$ l'ensemble des fonctions cadlag sur T à valeurs dans E; $\mathcal{D}(\mathbf{R})$ étant naturellement muni de la topologie de Skorokhod, nous munissons $\mathcal{D}(E)$ de la topologie $\mathcal{S}(E)$ la moins fine pour laquelle les applications $\hat{g}$ soient continues. En recopiant mot pour mot l'exposé et les preuves précédents, on peut énoncer

THÉORÈME (Fernique, 1989, théorème 4.3.3). *Soient T un intervalle borné de $\mathbf{R}$ et $\mathcal{D}(T,E)$ l'ensemble des fonctions cadlag sur T à valeurs dans E muni de la topologie de Skorokhod; on munit $\mathcal{D}(T,E)$ de la tribu induite par $\mathcal{B}(E^T)$ et $\mathcal{M}\left[\mathcal{D}(T,E)\right]$ de la topologie définie par les fonctions continues bornées et mesurables sur $\mathcal{D}(T,E)$. Soient de plus (K_n) une suite croissante de parties compactes de E et (g_m) une suite de fonctions continues sur E séparant ses points. On note M l'ensemble des probabilités sur $\mathcal{D}(T,E)$ telles que:*

$$\forall n \in \mathbf{N},\ \mu\{x \in \mathcal{D}(T,E)\colon \forall t \in T,\ x(t) \in K_n\} > 1 - 2^{-n};$$

dans ces conditions, M est un sous–espace lusinien de $\mathcal{M}\left[\mathcal{D}(T,E)\right]$; de plus pour qu'une partie M' de M soit relativement compacte, il faut et il suffit que pour tout $m \in \mathbf{N}$, l'ensemble $\widetilde{g}_m(M')$ des mesures images $\widetilde{g}_m(\mu)$, $m \in M'$, soit relativement compact dans l'espace polonais $\mathcal{M}\left[\mathcal{D}(T,\mathbf{R})\right]$.

4. EXEMPLES D'UTILISATION

4.1. Processus de Wiener non homogènes

Soient E un espace vectoriel topologique localement convexe quasi-complet lusinien et μ une probabilité gaussienne sur E; pour tout t appartenant à $T = [0,\ 1]$, soit de plus R_t une forme quadratique semi-positive sur E' telle que:

$$0 \leqslant s \leqslant t \leqslant 1 \Rightarrow \forall y \in E',\ R_s(y) \leqslant R_t(y) \leqslant R_1(y) = \int \langle x, y\rangle^2 d\mu(x);$$

il existe alors une fonction aléatoire gaussienne W (dite "fonction aléatoire de Wiener non homogène") sur T à valeurs dans E telle que:

$$\forall (t, y; s, z) \in (T \times E')^2,\ \mathbf{E}\big\{\langle W(t), y\rangle\langle W(s), z\rangle\big\} = R_{t\wedge s}(y, z).$$

Pour toute partie symétrique convexe et compacte K de E et toute partie finie ou dénombrable S de T, les inégalités de Lévy impliquent:

$$\mathbf{P}\big\{\exists t \in S,\ W(t) \notin K\big\} \leqslant 2\mathbf{P}\big\{W(1) \notin K\big\} = 2\mu(E - K).$$

Le théorème 2.3 implique donc que W a une modification à trajectoires compactes. Par ailleurs, pour tout y appartenant à E', l'application: $t \to \big\langle W(t), y\big\rangle$ a même loi que $w \circ R_{(\cdot)}(y)$ où w est une fonction aléatoire de Wiener réelle usuelle. Pour que W ait une modification à trajectoires continues, il faut et il suffit donc (théorème 2.2) que pour tout y appartenant à E', la fonction réelle non aléatoire: $t \to R_t(y)$ soit continue; dans le cas contraire, W a une modification à trajectoires cadlag dont les discontinuités sont celles des fonctions $R_{(\cdot)}(y),\ y \in E'$ (Fernique, 1989, exemple 4.2.6).
Soit $\{R_t^n,\ t \in T\}$, $n \in \mathbf{N}$, une suite de données du type précédent; le théorème 3.2 montre que pour que les lois des fonctions aléatoires de Wiener correspondantes W_n, $n \in \mathbf{N}$, soient relativement compactes dans l'espace des probabilités sur $\mathcal{C}(T, E)$ pour la topologie de la convergence étroite, il faut et il suffit que les deux conditions suivantes soient vérifiées:

(1) Les lois des $\big\{W_n(1), n \in \mathbf{N}\big\}$ sont relativement compactes dans l'espace des probabilités sur E.

(2) Pour tout y appartenant à E', les lois des $\big\langle W_n(\cdot\,), y\big\rangle, n \in \mathbf{N}$, sont relativement compactes dans l'espace des probabilités sur $\mathcal{C}\big([0,\ 1]; \mathbf{R}\big)$.
La condition (2) peut être explicitée à partir des critères classiques de compacité (Billingsley, 1968, théorème 8.2); en utilisant les propriétés d'homogénéité de w, on constate que l'ensemble des conditions $\{(1), (2)\}$ est équivalent à $\{(1), (2')\}$ où on a:

(2′) Pour tout y appartenant à E', les applications non aléatoires: $t \to R_t^n(y)$, $n \in \mathbf{N}$, sont équicontinues sur $[0,\ 1]$.
En utilisant les propriétés des suites convergentes de fonctions croissantes, on en déduit:

THÉORÈME. *Pour que la suite (W_n), $n \in \mathbb{N}$, converge en loi vers W_0 dans l'espace des fonctions continues sur $[0, 1]$ à valeurs dans E, il faut et il suffit que la suite $\{W_n(1),\ n \in \mathbb{N}\}$ converge en loi vers $W_0(1)$ dans E et que de plus pour tout y appartenant à E', la suite $\{R^n_{(\cdot)}(y), n \in \mathbb{N}\}$ converge vers $R^0_{(\cdot)}(y)$ pour la topologie de la convergence simple.*

4.2. Dans l'exemple précédent, on constate à l'évidence comment les théorèmes des paragraphes précédents permettent de décomposer les études de régularité des fonctions aléatoires à valeurs vectorielles suivant deux étapes correspondant respectivement aux deux types d'hypothèses de ces théorèmes. On retrouvera cette dualité dans la démonstration du théorème:

THÉORÈME. *Soit (X_n) une suite de fonctions aléatoires gaussiennes centrées sur un espace métrique compact T à valeurs dans un espace vectoriel topologique localement convexe lusinien quasi-complet E; on suppose que les conditions suivantes sont vérifiées:*

(1) *X_0 a presque toutes ses trajectoires continues sur T dans E,*

(1′) *Il existe une partie compacte K de E telle que $\mathbf{P}\{\forall t \in T,\ X_0(t) \in K\} > 0$,*

(2) *Pour tout couple $(t, y), (s, z)$ d'éléments de $T \times E'$ et tout entier n, on a:*

$$\mathbf{E}\Big(\big|\langle X_n(t), y\rangle - \langle X_n(s), z\rangle\big|^2\Big) \leqslant \mathbf{E}\Big(\big|\langle X_0(t), y\rangle - \langle X_0(s), z\rangle\big|^2\Big),$$

(3) *Pour tout couple $(t, y), (s, z)$ d'éléments de $T \times E'$ on a:*

$$\lim_{n\to\infty} \mathbf{E}\langle X_n(t), y\rangle\langle X_n(s), z\rangle = \mathbf{E}\langle X_0(t), y\rangle\langle X_0(s), z\rangle.$$

Dans ces conditions, pour tout entier n, X_n possède une modification X'_n ayant toutes ses trajectoires continues sur T dans E et les lois $\mathbf{P}_{X'_n}, n \in \mathbb{N}$, convergent étroitement vers $\mathbf{P}_{X'_0}$ dans $\mathcal{M}[\mathcal{C}(T, E)]$.

Démonstration. Attention au piège, l'hypothèse (1) n'implique peut-être pas la propriété (1′) malgré la compacité de T sauf pour certains espaces lusiniens E (cf. théorème 1.6). Sous l'hypothèse (1′), les évaluations gaussiennes montrent en substituant éventuellement à K son enveloppe symétrique convexe fermée et en notant N_K sa jauge, que $\mathbf{E} = \mathbf{E}\sup_{t\in T} N_K[X_0(t)]$ est fini; les propriétés de comparaison gaussiennes et la relation (2) montre alors que pour toute partie finie S de T et pour tout entier n, on a aussi:

4.2.1. $\quad \mathbf{E}\sup_{t\in S} N_K[X_n(t)] \leqslant \mathbf{E}.$

Dans ces conditions, chaque X_n vérifie la propriété 2.2 (a) et a une modification à trajectoires compactes (théorème 2.3). Par ailleurs, l'hypothèse (2) pour $y = z$, l'hypothèse (1) et les propriétés de comparaison gaussiennes montrent que pour tout y appartenant à E' et tout n, $\langle X_n, y\rangle$ a une modification continue. Le théorème 2.2 fournit donc la première conclusion du théorème. De ce fait, la propriété 4.2.1 montre que l'ensemble M des lois des $\mathbf{P}_{X'_n}$ vérifie la propriété 3.2.1; l'hypothèse (2) pour $y = z$, l'hypothèse (1) et les propriétés de comparaison gaussiennes montrent par ailleurs que M vérifie aussi 3.2.2; le théorème 3.2 montre donc que M est relativement compact dans $\mathcal{M}\left[\mathcal{C}(T,E)\right]$. La propriété (3) montre finalement que $\mathbf{P}_{X'_0}$ est le seul point adhérent de M; c'est le théorème.

4.3. Les exemples suivants vont utiliser le théorème de Talagrand caractérisant la régularité des fonctions aléatoires gaussiennes réelles à partir de mesures majorantes (Talagrand, 1987). Ce théorème permet en effet dans des situations assez générales d'évaluer les fonctions aléatoires gaussiennes vectorielles. L'outil technique est le lemme suivant:

4.3.1. LEMME. (Fernique, 1990). *Soient T un ensemble dénombrable, E un espace de Banach séparable, v un vecteur aléatoire gaussien à valeurs dans E et X une fonction aléatoire gaussienne sur T à valeurs dans E; on suppose que:*

$$\forall (t,y) \in T \times E', \ \mathbf{E}\langle X(t), y\rangle^2 \leqslant \mathbf{E}\langle v, y\rangle^2;$$

dans ces conditions, on a:

$$\mathbf{E}\sup_{t\in T}\|X(t)\| \leqslant c\left\{\mathbf{E}\|x\| + \sup_{\|y\|\leqslant 1}\int_0^\infty \left[\log N_{\langle X,y\rangle}(u)\right]^{1/2} du\right\},$$

où c est une constante absolue finie.

Ce lemme permet en effet de démontrer:

4.3.2. COROLLAIRE. *Soient E un espace vectoriel topologique localement convexe lusinien et B une partie symétrique convexe fermée bornée de E; soient de plus X une fonction aléatoire gaussienne sur un ensemble dénombrable S à valeurs dans E et x un vecteur gaussien à valeurs dans E; on suppose que:*

$$\forall (t,y) \in S \times E', \ \mathbf{E}\langle X(t), y\rangle^2 \leqslant \mathbf{E}\langle x, y\rangle^2,$$

on a alors:

$$\mathbf{E}\sup_{t\in S} N_B X(t) \leqslant c\left\{\mathbf{E}N_B(x) + \sup_{y\in B^0}\int \left[\log N_{\langle X,y\rangle}(u)\right]^{1/2} du\right\}.$$

Démonstration. Puisque E est lusinien et régulier, il existe (Corollaire 1.4.3) une suite d'hyperplans fermés séparant B de son complémentaire et donc une suite croissante (N_k) de semi-normes continues sur E convergeant simplement vers N_B; l'inégalité énoncée sera établie si on établit son analogue pour chaque semi-norme N_k pour laquelle son second membre est fini. Or l'espace E muni de la seule semi-norme N_k définit par passage au quotient et complétion un espace de Banach séparable $E_k = (E, N_k)$ dans lequel l'image canonique X_k de X vérifie

les hypothèses du lemme 4.3.1. La conclusion de ce lemme fournit le résultat du corollaire.
On en déduit alors:

4.3.3. THÉORÈME (Fernique, 1990). *Soit X une fonction aléatoire gaussienne sur un espace métrique séparable T à valeurs dans un espace vectoriel topologique localement convexe et lusinien E. On suppose qu'il existe un vecteur gaussien x à valeurs dans E tel que:*

$$\forall (t,y) \in T \times E', \ \mathbf{E}\big\langle X(t), y\big\rangle^2 \leqslant \mathbf{E}\langle x, y\rangle^2.$$

On suppose aussi que pour tout y appartenant à E', $\langle X, y\rangle$ est continu en probabilité sur T. Dans ces conditions:

(a) *Pour que X ait une modification ayant toutes ses trajectoires bornées sur T dans E, il suffit que l'application:*

$$y \to \int_0^\infty \big[\log N_{\langle X,y\rangle}(u)\big]^{1/2} du$$

soit continue sur E' pour la topologie forte.

(b) *On suppose que E est quasi-complet; pour que X ait une modification ayant toutes ses trajectoires continues sur T dans E, il suffit que la même application soit continue sur E' pour la topologie de la convergence compacte.*

Démonstration du théorème. Nous notons S une partie dénombrable et dense de T.

(a) Sous les hypothèses (a), il existe une partie symétrique convexe fermée et bornée B de E telle que:

$$\mathbf{E}\left\{N_B(x) + \sup_{y\in B^0} \int \big[\log N_{\langle X,y\rangle}(u)\big]^{1/2} du\right\} \leqslant 1.$$

Le corollaire 4.3.2 montre donc que $\mathbf{E}\sup_{t\in S} N_B\big(X(t)\big)$ est fini; de plus X est continu en probabilité. Le théorème 2.3 permet donc de construire une modification X' de X telle que $\mathbf{E}\sup_{t\in T} N_B\big(X'(t)\big)$ soit égal à $\mathbf{E}\sup_{t\in S} N_B\big(X(t)\big)$ de sorte que X' a presque toutes ses trajectoires sur T bornées dans E.

(b) Sous les hypothèses (b), il existe une partie symétrique convexe compacte K de E telle que

$$\mathbf{E}\left\{N_K(x) + \sup_{y\in K^0} \int \big[\log N_{\langle X,y\rangle}(u)\big]^{1/2} du\right\} \leqslant 1.$$

Le corollaire 4.3.2 montre donc que $\mathbf{E}\sup_{t\in S} N_K\big(X(t)\big)$ est fini; de plus, pour tout y appartenant à E', $\langle X, y\rangle$ a une modification à trajectoires continues sur T dans

R. Le théorème 2.2 permet donc de construire une modification X' de X ayant toutes ses trajectoires sur T continues dans E. Le théorème est démontré.

RÉFÉRENCES

Billingsley, P. (1968). *Convergence of Probability Measures.* Wiley, NewYork.

Bourbaki, N. (1974). *Eléments de Mathématique. Topologie générale.* Hermann, Paris.

Dellacherie, C. and Meyer, P.A. (1975). *Probabilités et potentiel.* Hermann, Paris.

Fernique, X. (1967). Processus linéaires, processus généralisés. *Ann. Inst. Fourier. Grenoble.* **17**, 1—92.

Fernique, X. (1989). Fonctions aléatoires à valeurs dans les espaces lusiniens. *Exposit. Math.* (to appear).

Fernique, X. (1990). Régularité de fonctions aléatoires gaussiennes à valeurs vectorielles. *Ann. Probab.* (to appear).

Talagrand, M. (1987). Regularity of Gaussian processes. *Acta Math.* **159**, 99—150.

Prob. Theory and Math. Stat., Vol. 1, pp. 350-359
B.Grigelionis *et al.* (Eds.)
1990 VSP/Mokslas

CRITICAL BRANCHING CAUSED BY INTERACTION WITH POINT CATALYSTS

K. FLEISCHMANN
Karl Weierstrass Institute of Mathematics, Academy of Sciences of the GDR, Box 1304, Berlin, DDR-1086

ABSTRACT

The model under consideration is a measure-valued branching process in $\mathbf{R}$ where the coefficient ρ of the branching term depends on time and space variables and is random. For fixed time t, typically $\rho(t)$ is a weighted sum of Dirac's delta functions on the real line, each interpreted as a catalyst and its weight as its action intensity. We are interested mainly in the large scale clumping effects of the system in the case when the random medium ρ has infinite overall density.

1. INTRODUCTION

The *growth of rare clumps* in space-time branching models in small dimensions has recently been exhibited in terms of a scaling limit theorem, see (Dawson and Fleischmann, 1988). We are interested in the effects if additionally a *space-time random medium* ρ enters into the model. More precisely, we do not change the motion component and also not the (local) criticality of the splitting law, but we assume that the *branching coefficient* $\rho \geqslant 0$ in the underlying equation may depend on time and space variables $t \geqslant 0$ and $x \in \mathbf{R}$, respectively (varying medium). Moreover, we allow that, for t fixed, $\rho(t)$ is a generalized function, in fact, some measure $\rho(t, dx)$ supported by a countable set and interpreted as the ensemble of *point catalysts*. Finally, we suppose that $\rho \not\equiv 0$ is random, and we restrict our consideration to two intrinsic examples of measure-valued *catalyst processes* ρ which are homogeneous and ergodic with respect to the time and space shift.
It turns out that the behavior of the resulting system essentially depends on the fact whether in the *expectation* $\langle \rho(t, dx) \rangle = \overline{\rho}\, dx$ of $\rho(t)$ the factor $\overline{\rho}$ in front of the Lebesgue measure dx is finite or not. In the former case the usual *averaging principle* works, i.e., the normalization as in the scaling theorem quoted above leads to the same limit as in the "classical" model with the constant branching coefficient (branching rate) $\overline{\rho}$. Hence, at a macroscopic level, here we also have a *Poissonian field of clumps* of a random size (evolving according to some i.i.d. critical continuous state Galton-Watson processes).
On the other hand, if $\overline{\rho}$ is infinite, the normalization which leaves invariant the expectation yields a degeneration to zero: the clumps are too rare to catch them in that way. However, we found a stronger (also expectation invariant) scaling

which catches those random clumps. In this case the catalyst process ρ is assumed to have the form

$$\rho(t) = \sum_{i=1}^{\infty} r_i \delta_{[w^{x(i)}(t)]}, \quad t \geqslant 0,$$

where the initial system of catalysts

$$\rho(0) = \sum_{i=1}^{\infty} r_i \delta_{x(i)}, \quad x(i) \neq x(j) \quad \text{for} \quad i \neq j,$$

is a *stable* random measure with index $\gamma \in (0,1)$, and the $w^{x(i)}$ are independent symmetric *stable* processes (of the catalysts) with some index $\alpha > 1$ (the same as the spatial transport of mass in the branching model has) and starting at time 0 at the sites $x(i)$. For instance, imagine $\alpha = 2$, i.e., that the mass transport is given by the heat flow and the catalysts' motion is governed by independent standard Brownian motions.

Roughly speaking, if at time t the i-th catalyst's position $w^{x(i)}(t)$ is close to x so that $|w^{x(i)}(t) - x| < \varepsilon$, then this catalyst causes branching at site x with rate r_i/ε.

That also in the limit $\varepsilon \to 0$ the model makes sense at all is based on the fact that in dimension one for the symmetric stable processes with index $\alpha > 1$ a *continuous local time* exists.

A basic tool in defining the model as well as the limiting field of random clumps in the case $\gamma < 1$ is a *nonlinear integral equation in the random medium* ρ which can formally be written as the differential equation:

$$\frac{\partial}{\partial t} v(t,x) = \kappa \Delta_\alpha v(t,x) - \rho(t,dx)|v|^{1+\beta}(t,x), \tag{1.1}$$

$t \geqslant 0$, $x \in \mathbf{R}$. Here the constant $\kappa > 0$ is the *mass transport coefficient* ("diffusion" constant), the fractional power $\Delta_\alpha = -(-\partial^2/\partial x^2)^{\alpha/2}$ is the generator of the symmetric stable semigroup with index $\alpha > 1$, and $\beta \in (0,1]$ is a constant which enters the so-called *branching index* $1 + \beta$.

Given a realization ρ of the random medium, the *measure-valued Markov branching process* $X^\rho = \{X^\rho(t); t \geqslant 0\}$ is defined in such a way that for some test functions $\varphi \geqslant 0$ the distribution of the random quantity $\int X^\rho(t,dx)\, \varphi(x)$ given $X^\rho(s)$ for some $s < t$ is expressed in terms of $\int X^\rho(s,dx)\, v(t,x)$ where v is a *mild* solution to (1.1) with initial condition $v(s+,\cdot) = \varphi$.

The model we are dealing with has essentially been introduced in (Dawson and Fleischmann, 1989a), based on (Dawson and Fleischmann, 1989b), and its study has been continued in (Dawson et al., 1989).

2. PRELIMINARIES

Let $S = \{S_t;\, t \geqslant 0\}$ denote the contraction semigroup of the symmetric stable Markov process on the real line $\mathbf{R}$ with index α corresponding to the generator $\kappa\Delta_\alpha$, $\kappa > 0, 1 < \alpha \leqslant 2$. It possesses continuous transition probability density functions

$$p(t,x,y) = p(t,y-x), \qquad t>0, \quad x,y \in \mathbf{R}$$

with the characteristic function (Fourier transform)

$$\int dy\ p(t,y)e^{iyx} = \exp[-\kappa t|x|^\alpha], \qquad t>0, \quad x\in\mathbf{R}.$$

In particular, in the case $\alpha = 2$,

$$p(t,x) = (4\pi\kappa t)^{-1/2}\exp[-x^2/4\kappa t], \qquad t>0, \quad x\in\mathbf{R}.$$

We denote by $\mathbf{F}$ the linear space of all continuous real-valued functions f on $\mathbf{R}$ with the property that $f(x)\exp(c_f x^2)$ is bounded in $x \in \mathbf{R}$, where c_f is some positive constant, provided that $\alpha = 2$ holds. Otherwise, i.e., if $\alpha < 2$, we require that $f(x)|x|^{1+\alpha}$ is bounded in x. Note that $\mathbf{F}$ is just defined in such a way that it contains all continuous density functions $p(t)$. We equip $\mathbf{F}$ with the supremum norm $\|\cdot\|_\infty$.

Fix a number $\beta \in (0,1]$. Let $\mathcal{M}^{\mathbf{F},\beta}$ denote the set of all those (non-negative) measures μ on $\mathbf{R}$ for which (μ, f^β) is finite for all $f \in \mathbf{F}_+$. Here the lower index $+$ at a set A refers to the subset of all non-negative members of A, and (m,h) is defined by $\int m(dx)\,h(x)$ (provided that the integral makes sense). Note that each μ in $\mathcal{M}^{\mathbf{F},\beta}$ is locally finite: $\mu(B) < \infty$ if B belongs to the ring $\mathfrak{B}$ of all bounded Borel subsets of $\mathbf{R}$. We equip $\mathcal{M}^{\mathbf{F},\beta}$ with the coarsest topology such that for each $f \in \mathbf{F}_+$ the map $\mu \mapsto (\mu, f^\beta)$ of $\mathcal{M}^{\mathbf{F},\beta}$ into $\mathbf{R}_+$ is continuous. Put $\mathcal{M}^{\mathbf{F}} = \mathcal{M}^{\mathbf{F},\infty}$.

3. THE CATALYST PROCESS

Let be given a constant $\kappa' \geqslant 0$. We introduce a family of independent symmetric stable Markov processes $w^x = \{w^x(t);\ t \geqslant 0\}$, $x \in \mathbf{R}$, with the generator $\kappa'\Delta_\alpha$, starting at $w^x(0) = x$, and trajectories in $\mathbf{D}[\mathbf{R}_+, \mathbf{R}]$. Here $\mathbf{D}[\mathbf{R}_+, \mathsf{E}]$ denotes the space of all functions from $\mathbf{R}_+$ into a topological space E which are right continuous and have left limits.

In addition to the indices $\alpha \in (1,2]$ and $\beta \in (0,1]$, let be given a number $\gamma \in (0,1]$. In the case $\alpha < 2$, additionally $\beta\gamma(1+\alpha) > 1$ is required. The latter condition guarantees that $f^{\beta\gamma}$ is integrable with respect to the Lebesgue measure dx, for all $f \in \mathbf{F}_+$. If $\gamma < 1$ holds, we define the *stable random measure* $\rho(0)$ on $\mathbf{R}$ with index γ by means of its Laplace functional

$$\langle\exp[-(\rho(0), f)]\rangle = \exp[-(dx, f^\gamma)], \qquad f\in\mathbf{F}_+. \tag{3.1}$$

(We remark here the brackets $\langle\cdot\rangle$ refer to the expectation with respect to the random medium; otherwise the expectation will be denoted by $\mathbf{E}$.) Note that with probability one the random measure $\rho(0)$ has *infinite asymptotic density* (i.e., infinite overall density) and is supported by a countable and dense set. Moreover, $\rho(0)$ has independent increments, i.e., it is "independent at each point".

On the other hand, if $\gamma = 1$, let $\rho(0)$ be a *compound Poisson point measure* defined by

$$\langle \exp[-(\rho(0), f)] \rangle = \exp[-(dx, 1 - L(f(\cdot)))], \quad f \in \mathbf{F}_+,$$

where

$$L(r) = \langle \exp(-r\xi) \rangle, \quad r \geqslant 0,$$

is the Laplace function of a non-negative random variable ξ with finite expectation $\langle \xi \rangle = \overline{\rho}$. In other words, here the random measure $\rho(0)$ consists of a Poissonian field of points with uniform density but all points carry masses the amount of which are independent copies of ξ. Of course, this $\rho(0)$ also has independent increments. In both cases,

$$\rho(0) = \sum_{i=1}^{\infty} r_i \delta_{x(i)}, \quad x(i) \neq x(j) \quad \text{for} \quad i \neq j,$$

is the initial state of the *catalyst process*

$$\rho(t) = \sum_{i=1}^{\infty} r_i \delta_{[w^{x(i)}(t)]}, \quad t \geqslant 0. \tag{3.2}$$

Here $\rho(0)$ is assumed to be independent of the family $\{w^x; x \in \mathbf{R}\}$. Consequently, the catalysts are initially situated at the sites $x(i)$, move then along $w^{x(i)}$ and carry their "*action intensity*" r_i with them.
Note that $\rho = \rho(t)$ (3.2) is a Markov process which can be realized in $\mathbf{D}[\mathbf{R}_+, \mathcal{M}^{\mathbf{F},\beta}]$. The distribution of ρ is *invariant* with respect to the time and space shift and, in the case $\rho < 1$, is also *self-similar* with respect to a time–space–mass transformation.
The catalyst process ρ will serve as a random medium for our branching process and also for the underlying nonlinear integral equation.

4. THE UNDERLYING INTEGRAL EQUATION IN A RANDOM MEDIUM

We fix some $L \in \mathbf{R}_+$ and $T \in (0, \infty)$ and consider the half-open interval $J = (L, L+T]$. By $\mathbf{F}^J$ we denote the set of all continuous mappings u of J into $\mathbf{F}$ such that

$$\|u\|_J = \int_J dt\ \|u(t)\|_\infty < \infty.$$

Then $[\mathbf{F}^J, \|\cdot\|_J]$ is a normed space.
We introduce the set Θ of all finite measures ϑ defined on $\mathbf{R}$ which are either concentrated at a point (i.e. degenerate) or absolutely continuous with a density which is bounded from above by a function in $\mathbf{F}_+$. An example is the *uniform distribution* on the interval $[-\varepsilon^{1/\alpha}, \varepsilon^{1/\alpha}]$, $0 < \varepsilon < 1$, with density function denoted by $g(\varepsilon)$. For $\vartheta \in \Theta$ and $f \in \mathbf{F}$, the convolution $\vartheta * f$ is defined by $\vartheta * f(x) =$

$= \int \vartheta(dy)\, f(y-x),\ x \in \mathbf{R}$. We will now write equation (1.1) in its integral form for $t \in J,\ x \in \mathbf{R}$

$$v(t,x) = \vartheta * p(t-L)(x) - \int_L^t ds \int \rho(s,dy)p(t-s,y-x)|v|^{1+\beta}(s,y). \qquad (4.1)$$

THEOREM 4.1. *For almost all $\rho^J = \{\rho(t);\ t \in J\}$, the following statement is true. To all $\vartheta \in \Theta$ with the exception of those have an atom in some set (depending on ρ^J) of the Lebesgue measure zero, there is a unique element $v = V^J[\vartheta, \rho^J]$ in $\mathbf{F}_+^J$ which solves the equation (4.1). Given such ρ^J and ϑ, we can find a sequence $\{\varepsilon_n;\ n = 1,2,\ldots\}$ in $(0,1)$ converging to zero as $n \to \infty$, such that in $\mathbf{F}_+^J$ the convergence*

$$V^J[\vartheta * g(\varepsilon_n)(x)\,dx, \rho^J] \longrightarrow V^J[\vartheta, \rho^J] \quad (n \to \infty)$$

holds. The sequence ε_n can be chosen independently of the weight of the possible atom of ϑ.

Proof. It can be found in (Dawson et al., 1989) for $\gamma < 1$ and carries over to the simpler case $\gamma = 1$ with obvious changes.

THEOREM 4.2. *Let $\varphi_n \in \mathbf{F}_+$, $n \to \infty$ pointwise monotonically increase to some constant c. Then, for almost all ρ^J, the solutions $V^J[\varphi_n(x)\,dx,\ \rho^J]$ of (4.1) defined in Theorem 4.1 also increase to some function denoted by $v_c = V^J[c\,dx, \rho^J]$, bounded on $J \times \mathbf{R}$, and which satisfies the equation (4.1) with ϑ replaced by the Lebesgue measure $c\,dx$. This solution does not depend on the choice of the sequence φ_n. As a random field on $J \times \mathbf{R}$, v_c is invariant and ergodic with respect to the space shift.*

Proof. See (Dawson and Fleischmann, 1989b) with modifications for $\gamma = 1$.

5. THE BRANCHING PROCESS

First we have to establish the *existence* of the *critical measure-valued branching process controlled by point catalysts.*

THEOREM 5.1. *For almost all realizations ρ of the catalyst process, there is a time-inhomogeneous family of $\mathcal{M}^{\mathbf{F}}$-valued Markov processes X^ρ with the Laplace transition functional*

$$\mathbf{E}\{\exp[-(X^\rho(t),\varphi)]\,|\,X^\rho(s)\} = \exp[-(X^\rho(s), v(t))],$$

$\varphi \in \mathbf{F}_+$, $0 \leqslant s < t$, where $v = V^J[\varphi(x)dx,\ \rho^J]$ is the solution of (4.1) according to Theorem 4.2 with $J = (s,t]$. We have

$$\mathbf{E}\{(X^\rho(t),\varphi)\,|\,X^\rho(s)\} = (X^\rho(s), S_{t-s}\varphi).$$

Proof. We refer to (Dawson and Fleischmann, 1989a) with obvious changes for $\gamma = 1$.
Despite the high singularity of the random medium we have the following *regularity* property of the states of the branching process.

THEOREM 5.2. *Assume that $X^\rho(0) = \delta_a$ holds for some fixed point $a \in \mathbf{R}$. Then for almost all realizations ρ of the catalyst process and each $t > 0$ the random measure $X^\rho(t)$ is a.s. absolutely continuous with respect to the Lebesgue measure.*

Proof. Using Theorem 4.2, this can be proved as in (Dawson et al., 1989) with simple modifications in the case $\gamma = 1$.

6. SCALING LIMITS

From now on we assume that our branching process X^ρ starts with a Lebesgue measure $X^\rho(0, dx) = \ell\, dx$, $\ell \geqslant 0$. We fix a constant $\zeta \geqslant 0$ and introduce the following *scaling*:

$$X_K^\rho(t, B) = K^{-\zeta} X^\rho(Kt, K^\zeta B), \quad t \geqslant 0, \quad K > 0, \quad B \in \mathfrak{B}. \tag{6.1}$$

Note that this scaling leaves invariant the expectation

$$\mathbf{E} X_K^\rho(t, dx) = \ell\, dx \tag{6.2}$$

for a given ρ. Set $\zeta_c = [\gamma(\alpha - 1) + 1]/\alpha\beta\gamma$.
Recall our assumptions $1 < \alpha \leqslant 2$, $0 < \beta \leqslant 1$, $0 < \gamma \leqslant 1$, and the requirement $\beta\gamma(1 + \alpha) > 1$ in the case $\alpha < 2$. Now we can formulate the *main result* (the case $\gamma < 1$ is taken from (Dawson and Fleischmann, 1989a)).

THEOREM 6.1. *Fix $t > 0$. As $K \to \infty$, the random distribution of the random measure $X_K^\rho(t)$ converges (in distribution) to the non-random distribution of some random measure $X_\infty(t)$. For $\zeta > \zeta_c$ a law of large numbers holds: $X_\infty(t, dx)$ equals the Lebesgue measure $\ell\, dx$. On the other hand, if $\zeta < \zeta_c$ then (even at macroscopic scale) we get a local extinction: $X_\infty(t)$ coincides with the zero measure. However, for the critical scaling exponent $\zeta = \zeta_c$ the limiting random measure $X_\infty(t)$ is non-degenerate. It is independent at each point, infinitely divisible, homogeneous, and has uniform density ℓ. In the case $\gamma = 1$, the limit measure is a compound Poisson point measure with the Laplace functional*

$$\mathbf{E} \exp[-(X_\infty(t), \varphi)] = \exp[-((\bar\rho\beta t)^{-1/\beta} \ell\, dx, 1 - L_t(\varphi(\cdot)))], \quad \varphi \in \mathbf{F}_+,$$

where

$$L_t(r) = 1 - r[(\bar\rho\beta t)^{-1} + r^\beta]^{-1/\beta}, \qquad r \geqslant 0,$$

is the Laplace function of some positive random variable ω with the expectation $\mathbf{E}\omega = (\bar\rho\beta t)^{1/\beta}$. In the case $\gamma < 1$,

$$\mathbf{E} \exp[-(X_\infty(t), \varphi)] = \exp\left[-\ell \int dz \langle v_{\varphi(z)}(t, 0)\rangle\right], \quad \varphi \in \mathbf{F}_+,$$

where for fixed z, $v_{\varphi(z)} = V^J[\varphi(z)\,dx, \rho^J]$ *is the "homogeneous" solution to* (4.1) *according to Theorem* 4.2 *with the Lebesgue measure* $\varphi(z)\,dx$ *as the initial measure and* $J = (0, t]$.

(Of course, $\langle v_{\varphi(z)}(t, 0)\rangle$ is the expectation of the random solution $v_{\varphi(z)}$ at time t.) $X_\infty(t)$ describes the random field of "*clumps*" on a macroscopic level. For $\gamma = 1$ they are situated in Poissonian points with density $(\overline{\rho}\beta t)^{-1/\beta_\ell}$ and their size is given by independent copies of ω. Note that ω is exponentially distributed in the finite variance case $\beta = 1$. As t increases, the *intermittency* is growing in the sense that, roughly speaking, less Poissonian points will be present and they will have a larger size. Actually one can expect that macroscopically the size of a clump will obey a critical continuous state Galton–Watson process with branching index $1+\beta$, as in the "classical" case of a constant medium $\overline{\rho}$, see (Dawson and Fleischmann, 1988). In other words, for $\gamma = 1$ the limiting *process* should be a measure–valued Markov branching process with vanishing transport coefficient, branching index $1+\beta$, and constant branching coefficient $\overline{\rho}$, i.e. related to the ordinary differential equation

$$\frac{dv}{dt} = -\overline{\rho}|v|^{1+\beta}. \tag{6.3}$$

Consequently, in the limit as $K \to \infty$, the scaling (6.1) finally transfers equation (1.1) into (6.3).
On the other hand, in the case $\gamma < 1$ the "limiting equation" is again given by equation (1.1) in the random medium ρ, but only solutions with constant initial density function and only their expectation enter into the limiting Laplace functional. Moreover, here it is not yet clear, whether (for fixed t) the limiting clumps are isolated or not, and which structure the limiting *process* (i.e. in dependence on t) should have.
A further essential difference of the two γ cases consists in their behavior as $\kappa \to 0$. For $\gamma = 1$ we have *continuity*, whereas in the other case we have *discontinuity*.
Finally we mention that if $\beta = 1$ and $\gamma < 1$, the Laplace transform of the *random variance* of $(X^\rho_K(t), \varphi)$ is given by

$$\langle \exp[-r\,\mathrm{Var}(X^\rho_K(t), \varphi)]\rangle =$$
$$= \exp\left[-(2r\ell)^\gamma K^{(1-2\zeta)\gamma+\zeta} \int dx \left\langle \left[\int_0^t ds\ (S_{K^{1-\alpha\zeta}s}\varphi)^2(w^x(K^{1-\alpha\zeta}s)) \right]^\gamma \right\rangle \right],$$

$r \geqslant 0$. Hence, for $\gamma > 1/2$ and $\zeta_v = \gamma/(2\gamma - 1)$ we get the interesting fact that the random variance of $(X^\rho_K(t), \varphi)$ has a non-degenerate stable limit distribution as $K \to \infty$ whereas in that case $(X^\rho_K(t), \varphi)$ degenerates to $\ell\|\varphi\|_1$ in probability since $\zeta_v > \zeta_c$, where $\|\cdot\|_1$ is the norm in the Lebesgue space $\mathbf{L}^1(\mathbf{R}, dx)$.

7. SKETCH OF PROOF OF THEOREM 6.1

By the definition (6.1) and Theorem 5.1, for $\varphi \in \mathbf{F}_+$,

$$\mathbf{E}\exp[-(X_K^\rho(t),\varphi)] = \exp[-(X^\rho(Kt),\, K^{-\zeta}\varphi(K^{-\zeta}\cdot))] =$$

$$= \exp[-\ell\|v(Kt)\|_1],$$

where v equals $V^J[K^{-\zeta}\varphi(K^{-\zeta}x)\,dx, \rho^J]$ according to Theorem 4.2 with $J = (0,t]$. Introduce

$$u(s,x) = K^\zeta v(Ks, K^{1/\alpha}x), \qquad K, s > 0, \quad x \in \mathbf{R},$$

then

$$\|v(Kt)\|_1 = \|u(t, K^\eta\cdot)\|_1, \quad \text{where} \quad \eta = \zeta - 1/\alpha.$$

Moreover, using the self-similarity

$$p(s,x) = K^{1/\alpha}p(Ks, K^{1/\alpha}x), \qquad K, s > 0, \quad x \in \mathbf{R}, \tag{7.1}$$

from (4.1) we have that

$$u(t,x) = S_t\varphi(K^{-\eta}\cdot)(x) - K^{\beta(\zeta_c-\zeta)}\int_0^t ds \int K^{-1/\alpha\gamma}\rho(Ks, K^{1/\alpha}dy)\times$$

$$\times p(t-s, y-x)u^{1+\beta}(s,y). \tag{7.2}$$

By the self-similarity of the catalysts' motion law, see (7.1), we obtain

$$\rho(Ks, K^{1/\alpha}\cdot) \overset{D}{=} \int \rho(0, K^{1/\alpha}dx)\delta_{w^x(s)}.$$

Suppose now that $\gamma < 1$. Then by (3.1)

$$K^{-1/\alpha\gamma}\rho(0, K^{1/\alpha}dy) \overset{D}{=} \rho(0,dy), \tag{7.3}$$

and we finally get

$$K^{-1/\alpha\gamma}\rho(Ks, K^{1/\alpha}dy) \overset{D}{=} \rho(s,dy). \tag{7.4}$$

The solution u of (7.2) depends on ρ. But if ρ is transformed in such a way that its distribution does not change, then the distribution of u does not change as well. Because we are only interested in the convergence in distribution, we may therefore apply (7.4) to get

$$u(t,x) = S_t\varphi(K^{-\eta}\cdot)(x) - K^{\beta(\zeta_c-\zeta)}\int_0^t ds \int \rho(s,dy)\times$$

$$\times p(t-s, y-x)u^{1+\beta}(s,y). \tag{7.5}$$

Under the "supercritical" scaling $\zeta > \zeta_c$, a detailed inspection shows that the nonlinear term in (7.5) can be neglected, more precisely,

$$\|u(t, K^\eta \cdot)\|_1 \xrightarrow{D} \|\varphi\|_1 \quad \text{as} \quad K \to \infty.$$

This yields the law of large numbers.
If $\zeta < \zeta_c$, again after integration, the bounded integral expression in (7.5) may be estimated from below to conclude that

$$\|u(t, K^\eta \cdot)\|_1 \xrightarrow{D} 0, \quad \text{as} \quad K \to \infty,$$

i.e., the local extinction follows.
Finally, if $\zeta = \zeta_c$, one can show that

$$\int dz\ u(s, K^\eta z + y) \stackrel{D}{\approx} \int dz\ v_{\varphi(z)}(s, K^\eta z + y), \quad \text{as} \quad K \to \infty,$$

uniformly in $s \in J$ and $y \in \mathbf{R}$, where (as in the formulation of the theorem) for *fixed* z, $v_{\varphi(z)} = V^J[\varphi(z)dx, \rho^J]$ is the "homogeneous" solution to (4.1) according to Theorem 4.2 with the *Lebesgue measure* $\varphi(z)\,dx$ as the initial measure. If now φ has the form $c1_{[a,b]}$ where $1_{[a,b]}$ is the indicator function of some interval $[a, b]$, $a < b$, then

$$\int dz\ v_{\varphi(z)}(t, Kz) = \int_a^b dz\ v_c(t, Kz) = K^{-1} \int_{Ka}^{Kb} dz\ v_c(t, z)$$

(we replaced K^η by K). Using the ergodicity of v_c according to Theorem 4.3, Birkhoff's individual ergodic theorem shows that, as $K \to \infty$, the latter expression a.s. converges to

$$(b - a)\langle v_c(t, 0)\rangle = \int dz\ \langle v_{\varphi(z)}(t, 0)\rangle$$

as desired. The step to general φ can be done by an obvious approximation procedure.
If $\gamma = 1$, then instead of (7.3) additionally the strong law of large numbers

$$K^{-1}\rho(0, K dy) \longrightarrow \overline{\rho}\, dy \qquad (K \to \infty)$$

has to be used.
Because of (6.2), to show that also in the case $\gamma < 1$ the scaling limit has intensity ℓ , we have only to prove that

$$\mathbf{E}(X_\infty(t), \varphi) \geqslant \ell \|\varphi\|_1. \tag{7.6}$$

By Jensen's inequality, from (6.4) we get

$$r\mathbf{E}(X_\infty(t), \varphi) \geqslant \ell \int dz \langle v_{r\varphi(z)}(t, 0)\rangle$$

for $r > 0$. By (4.1), we have

$$v_{r\varphi(z)}(t,0) = r\varphi(z) - \int_0^t ds \int \rho(s,dy)p(t-s,y)v^{1+\beta}_{r\varphi(z)}(s,y).$$

But obviously

$$v_{r\varphi(z)}(s,y) \leqslant r\varphi(z) \leqslant r\|\varphi\|_\infty, \quad s \in J, \quad y \in \mathbf{R}, \tag{7.7}$$

hence the latter integral may be estimated from above by

$$r^{1+\beta}\varphi(z)\|\varphi\|^\beta_\infty \int_0^t ds \int \rho(s,dy)p(t-s,y)$$

which is finite a.s. Hence, with probability one

$$r^{-1}\int dz\, v_{r\varphi(z)}(t,0) \longrightarrow \|\varphi\|_1, \quad \text{as} \quad r \to 0,$$

and again with the help of (7.7) by the dominated convergence theorem we get

$$r^{-1}\int dz \langle v_{r\varphi(z)}(t,0)\rangle \longrightarrow \|\varphi\|_1, \quad \text{as} \quad r \to 0.$$

Thus (7.6) holds.

REFERENCES

Dawson, D.A. and Fleischmann, K. (1988). Strong clumping of critical space-time branching models in subcritical dimensions. *Stoch. Proc. Appl.* **30**, 193–208.

Dawson, D.A. and Fleischmann, K. (1989a). Critical branching in a highly fluctuating random medium. *Carleton Univ., Lab. Research Stat. Probab., Techn. Report* **123**, 1–35.

Dawson, D.A. and Fleischmann, K. (1989b). Diffusion and reaction caused by point catalysts. *Carleton Univ., Lab. Research Stat. Probab., Techn. Report* **123**, 36–66.

Dawson, D.A., Fleischmann, K. and Roelly-Coppoletta, S. (1989). Absolute continuity of the measure states in a branching model with catalysts. *Manuscript,* 45 p.

Prob. Theory and Math. Stat., Vol. 1, pp. 360–374
B.Grigelionis *et al.* (Eds.)
1990 VSP/Mokslas

STATISTICAL ANALYSIS AND DATING OF THE OBSERVATIONS ON WHICH PTOLEMY'S "ALMAGEST" STAR CATALOGUE IS BASED

A.T. FOMENKO, V.V. KALASHNIKOV and G.V. NOSOVSKII
Moscow State University, Department of Mathematics, Len.gory, Moscow
VNIISI, prospect 60–let Oktyabrya, 9, Moscow, USSR

ABSTRACT

This work is devoted to a describing new statistical and geometrical procedures of dating ancient star catalogues considering numerical data contained in these catalogues and known real configurations of stars on the celestial sphere. The method was tested on several star catalogues whose dates are well–known (Tycho Brahe etc.) and on several artificially generated star catalogues. Then the same method was applied to the Almagest. The results obtained do not confirm the traditional dating of the Almagest (2–nd century A. D. or 2–nd century B. C.) but shift its dating to Arab's epoch, namely: 600–1300 A. D.

1. HISTORY OF THE PROBLEM AND SUBJECT OF THE WORK

The purpose of the work is to obtain the "intrinsic" date of Ptolemy's observations, using only data from his catalogue (contained in "Almagest") and methods of modern statistics and geometry. Our paper deals with the star catalogue only (not Almagest as a whole) and hence all our conclusions must be referred to the star catalogue. In fact, we have obtained not the precise date itself (which would be impossible), but the interval of possible dates of Ptolemy's observations, namely: from 600 A. D. to 1300 A. D. It comes to contradiction to the traditional date of Ptolemy's observations, accepted in traditional chronology and history of science (138 A. D.), as well as to the traditional date, suggested by most of the astronomers — investigators of Almagest (namely: about II century B. C. — the epoch of Hipparchus), see (Peters and Knobel, 1915; Newton, 1975; Baily, 1843).
In this paper we have not room to discuss the origin of this contradiction and interpretations of our statistical result. We should only mention that the date of observations really pointed out in Almagest –"the second year of the reign of Antonius" (Ptolemy's Almagest, 1984) — can be interpreted in many different ways. For more detailes see (Morozov, 1928; Fomenko, 1981a, 1981b, 1984, 1986; Fedorov and Fomenko, 1986). Interest in dating of the Almagest is not new. Detailes and further references see in (Peters and Knobel, 1915; Newton, 1975;

Morozov, 1982; Fomenko et al., 1988, 1989a,b). Complete information about first printed editions of Almagest one can find in Baily's work (Baily, 1843).
Besides the printed editions of XVI c., there exist various manuscripts of Almagest. Global and particular investigation of all main Almagest's manuscripts was carried out in (Peters and Knobel, 1915).
We have used in our work the Ptolemy's star catalogue from (Peters and Knobel, 1915), but the results we obtained is based on the data which are identical in all originally printed editions of Almagest (Baily, 1843) and in almost all manuscripts, investigated in (Peters and Knobel, 1915).

2. SOME NOTIONS FROM ASTRONOMY

Suppose that stars belong to a celestial sphere with center in the "eye of an observer". To fix positions of stars we need a spherical coordinate system. Following the "Almagest" and other Middle Ages catalogues we shall use *ecliptic* coordinates. The intersection of the celestial sphere with the plane of the Earth's orbit is called the ecliptic. Zodiac constellations are placed along the ecliptic. Ecliptical latitude b is measured in arc degrees ($-90° < b < 90°$), ecliptical longitude l — also in arc degrees ($0° < l < 360°$). The point of spring equinox is taken as "zero meridian" for ecliptical longitudes. This coordinate system is not fixed. It evolves in time because of the different reasons. We shall not discuss different astronomical aspects. It is important for us that there exists an astronomical and mathematical theory (by Newcomb) permitting to calculate stars positions in past and future. We used Newcomb's theory in the form of (Kinoshita, 1975) for computer calculating real positions of stars in the past.
Proper motions of stars have been taken into account too. All data about directions and velocities of stars were taken from catalogues (Hoffit, 1982) and (Fricke and Kopff, 1963).
We measure time t using centures as units. So the value $t = 0$ corresponds to 1900 A. D. as coordinates of "modern" stars are referred namely to this year. The value $t = 1$ corresponds to 1800 A. D., the value $t = 3.75$ corresponds to 1525 A. D. and so on. The parameter t we changed in some a'priori fixed time. For the Almagest we choose this interval from 600 B. C. till 1900 A. D., i.e., $0 < t < 25$.

3. SOME CHARACTERISTICS OF ANCIENT STAR CATALOGUES AND PRELIMINARY ANALYSIS OF THE ALMAGEST

We study star catalogues by Ptolemy, Tycho Brahe, Ulugbek, Al-Sufi, Hevelius. They all were compiled without telescope. Each catalogue contains about 1000 stars.
Some stars from ancient catalogues have proper names - the named stars. A set of named stars in the Almagest which consists of 12 stars designated "vocatur" in the text. They are (in bracets we show their modern name and Baily's number): Arcturus (α Boo, 110), Sirius (α CMa, 818), Aquila (α Aql, 288), Previndemiatrix (ε Vir, 509), Antares (α Sco, 553), Aselli (γ Cnc, 452), Procyon (α CMi, 848), Regulus (α Leo, 469), Spica (α Vir, 510), Lyra (α Lyr, 149), Capella (α Aur, 222), Canopus (α Car, 892). There are some stars with considerable proper motion

(more than $0,5''$/year) among them, e.g., Arcturus, Procyon, Sirius. It is natural to suggest that proper names were given to important famous stars and their coordinates were measured with special care. The list of named stars can depend on the underlying catalogue in general.

Numerous investigations (see (Newton, 1975)) force us to conclude that the real accuracy of ancient catalogues can be worse than the claimed one (their scales).

At the starting point of our investigation we had to identify stars from the Almagest catalogue with modern ones.

For this purpose we chose from the modern star catalogue a set of about 30 named stars and 50 "fast" stars (with the velocity of proper motion $v > 0.5''/1$ year). For solving "Who is who?" problem Newcomb's theory was used. Namely, we calculated (using a computer) ecliptical coordinates for all mentioned stars varying t from 0 till 25 (i.e., from 600 B. C. till 1900 A. D.). Then these coordinates were compared with those given in the Almagest.

The most of traditional identifications were confirmed. But we discovered several modern stars (e.g., o^2 Eridanus) which can be identified with different Almagest's stars (778, 779, 780 in Baily's numeration) in different epochs.

After having completed the identification procedure we obtained some list of all stars having reliable one-to-one identification with Almagest's stars.

4. ERRORS IN THE ALMAGEST STAR CATALOGUE

4.1. *Types of errors occurring in the catalogues*

We saw that it is necessary to analyse all possible types of errors in the catalogue. We divide these errors into 3 types: systematic, random and spikes.

Systematic errors appear as a result of global measurements or recalculations which imply the global rigid rotation of some sets of stars on the celestial sphere. Such systematic errors do occur in the Almagest (see below).

Random errors are resulted from mistakes in individual measurements. It is reasonable to assume that these errors have random distribution with zero mean value.

Spikes are caused by circumstances beyond the control of the observer and unknown to him — e.g., the errors of later copyists, refraction, etc. These errors usually change coordinates much larger than the unit of the scale and occur rarely.

4.2. *Systematic errors*

In order to compile a catalogue in terms of ecliptical coordinates the astronomer must know the position of the ecliptic and the position of the equinox axis in the epoch to which the catalogue is reduced. This position is known only with some error τ_1 in general.

Next, the astronomer can make a mistake τ_2 in definition of the longitudes of all stars simultaneously. These two errors are to be summarized to obtain the systematic error in longitude $\tau = \tau_1 + \tau_2$.

The next possible systematic error is implied by translation of the equinox point C along the meridian.

This error is denoted by β in Fig.1. Instead of β we can introduce a parameter ϕ which is an angle between the real axis of equinox and intersection line of the equatorial plane with the "catalogue's" ecliptical plane (see Fig.1).

These two errors β and τ (or ϕ and τ) totally describe all possible shifts of the point C on the celestial sphere: any error is some combination of β and τ.
The third error (which we denote γ) can appear due to a calculation of the angle ε between equator and ecliptic.
In reality parameters γ and β (or γ and ϕ) completely define a position P' of the pole of the ecliptic on the celestial sphere, see Fig.1.

Fig. 1

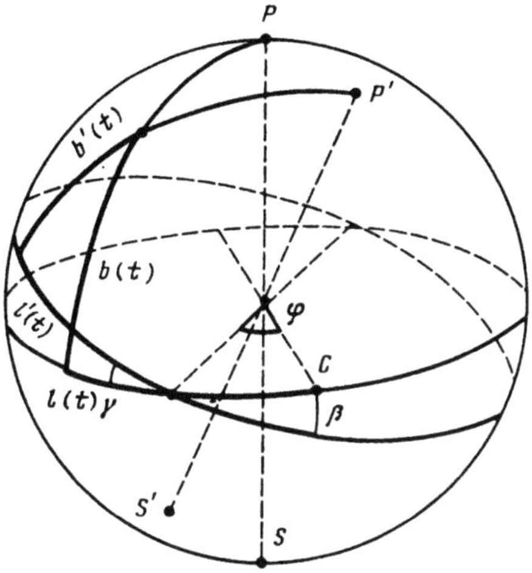

It is clear that any rotation of the celestial sphere can be decomposed into the composition of three orthogonal rotations defined by the parameters τ, β and γ. Thus, they reflect all possible systematic errors (if any).

4.3. ***Random errors and spikes***

Let us consider some star with Baily's number i, l_i and b_i being its ecliptical coordinates. We denote by $L_i\ (t, \tau, \beta, \gamma)$ and $B_i\ (t, \tau, \beta, \gamma)$ the ecliptical longitude and latitude of this star at time t given that systematic errors are equal to τ, β and γ.
Here and below we always can substitute β by ϕ if it is necessary without any explanation. Now it is possible to compare coordinates l_i, b_i from the Almagest with the above ones. Let us consider the following latitude and longitude deviations:

$$\Delta_b(i, t, \beta, \gamma) = B_i(t, \beta, \gamma) - b_i,$$

$$\Delta_l(i, t, \beta, \gamma, \tau) = L_i(t, \tau, \beta, \gamma) - l_i.$$

Here we used the obvious fact that the latitude (and hence the latitude deviation) does not depend on τ, i.e.,

$$B_i(t, \tau, \beta, \gamma) = B_i(t, \beta, \gamma).$$

This is a reason why latitudes are more informative than longitudes.
If measurements for the i-th star do not contain some unforseen errors (copyist's mistake, refraction etc.) then deviations Δ_b and Δ_l must be within the scale of

the underlying catalogue. The real accuracy of a catalogue can be unknown. As the author can choose as the catalogue scale unit the "record" accuracy that is the accuracy of observations of the most famous stars. In order to find and eliminate "spikes" we can use the following (rough) method (given that the values β and γ are fixed):
(1) Find a mean–square deviation

$$\sigma = \left[\left(\sum \Delta^2(i,t,\beta,\gamma)/N\right)\right]^{1/2}$$

where N is a number of considered stars. In fact the value σ does not depend on t since the most of stars have a negligible proper motion. Thus we can take the resulting value σ or even $\sigma/2$ as a "record" accuracy Δ of the given catalogue. The "real" accuracy is equal to $2\sigma \div 3\sigma$. We note too that about 40% of stars are within the record accuracy interval.
(2) Stars whose coordinates are not within the real accuracy must be excluded from the further analysis. Either these stars are "spikes" or they received large errors in the measurements of their coordinates.
Since there are few such "spikes" they do not affect on σ. We excluded from the list also all stars whose coordinates were considered to be doubtful by previous researchers, see (Peters and Knobel, 1915).
It is interesting to note that 2 stars (Canopus and Previndemiatrix) were removed from the catalogue – they turned to be spikes, see (Fomenko et al., 1988) for detailes. Thoroughful analysis of latitude errors is provided in Sect. 6.

5. STATISTICAL ANALYSIS OF THE ALMAGEST STAR CATALOGUE

5.1. *Preliminary remarks*
Let again l_i and b_i be ecliptical longitude and latitude of the i–th star from the clean catalogue. Let $L_i(t)$ and $B_i(t)$ be real corresponding values for time t. The detailed and careful statistical analysis shows (see (Newton, 1975)) that longitudes in the Almagest can not be considered reliable numerical data. R. Newton showed that these data were the result of some complicated recalculations of the initial ones. But all specialists agree that latitudes are the initial observed data.
We demonstrate in our paper that star catalogues (not only the Almagest but many others!) can be dated with help of latitudes behaviour only.
5.2. *A classification of latitude errors*
Let t^* be the real (but unknown to us) year of the observation of the stars. We started with a decomposition of the real latitude deviation $\Delta b_i(t^*) = b_i - B_i(t^*)$ on two components:

$$\Delta b_i(t^*) = \xi_i + r_i(t^*). \tag{1}$$

Let us call value ξ_i error of the observation (random error). It could be inspired by many various causes and we have not any reason to discuss here them. It is natural to suggest that ξ_i is a Gaussian random variable with zero mean value $\mathbf{E}\ \xi_i = 0$, and with finite variation $d = \mathbf{E}\ \xi_i^2 > 0$. The component $r_i(t^*)$ we can call an error due to a wrong determination of the ecliptic pole. The position of the ecliptic was known to the ancient astronomer with some error which can be characterized by two parameters γ and ϕ (see Fig. 1.5). It is easy to obtain from the definition that

$$r_i(t^*) = \gamma_i \ \sin(L_i(t^*) + \phi_i) + \delta_i, \tag{2}$$

where $|\delta_i| < 1''$ if $|B_i(t^*)| < 80°$. Consequently the value δ_i can be neglected in our calculations.

The idea of the proposed method is a determination of γ_i and ϕ_i by mathematical statistics and compensation these errors in order to deal with the real observation error only. Such approach leads us to some dating method. The realization of the method is based on the fact that the parameters γ_i and ϕ_i have "group nature", i.e., they are the same for some groups of stars (e.g., for constellations). This is really true in many cases because γ_i and ϕ_i do not depend on individual measurements but on preliminary determination of the ecliptic position for mentioned groups.

We assume that each constellation G in the ancient catalogue has an individual group error (i.e., this error is common for all stars from the constellation) in the determination of the position of the ecliptic pole. Let us parametrize it by values γ_G and ϕ_G. I.e., for each star $i \in G$ we assume that equalities $\gamma_i = \gamma_G$ and $\phi_i = \phi_G$ are true. Our aim is to estimate γ_G and ϕ_G for each group G from the catalogue. Note that the Almagest star catalogue contains 48 constellations.

5.3. ***Analysis of errors.7 regions on the Almagest star atlas***

Let us suppose that t is the year of observation. Determine the value

$$\Delta b_i(t, \gamma, \phi) = b_i - B_i(t) - \gamma_G \sin(L_i(t) + \phi_G) \tag{3}$$

for the i-th star and consider a constellation G containing N_G stars. Then we calculate values $\widehat{\gamma}_G$ and $\widehat{\phi}_G$ from the condition of minimization of the function

$$\sigma_G^2(t, \gamma, \phi) = \left[\sum_{i=1}^{N_G} \Delta b_i^2(t, \gamma, \phi)\right] \Big/ N_G \rightarrow \ \min \tag{4}$$

varying γ and ϕ. This problem can be solved easily analytically.

Let us call the value

$$\sigma_G^{\min}(t) = \sigma_G\left(t, \widehat{\gamma}_G, \widehat{\phi}_G\right)$$

a minimal mean-square error in the constellation G. We calculate additionally the percent $p_G^{\min}(t)$ of stars from G which meet the inequality $|\Delta b_i(t, \widehat{\gamma}_G, \widehat{\phi}_G)| <$ $< 10'$, i.e.,

$$p_G^{\min}(t) = [N\{i : |\Delta b_i(t, \widehat{\gamma}_G, \widehat{\phi}_G)| < 10'\}]/N_G. \tag{5}$$

(We write $N(A)$ for the number of elements in A.)

The concrete values $\sigma_G^{\min}$ and $p_G^{\min}$ for different constellations G see below. The calculated values $\widehat{\gamma}_G$ and $\widehat{\phi}_G$ are some estimates of the real parameters γ_G and ϕ_G determining the group error. Though it is possible to prove some asymptotic properties of these estimates (see Theorem 1 below) we can not consider $\widehat{\gamma}_G$ and $\widehat{\phi}_G$ to be close to real values γ_G and ϕ_G because we have not firm statistical reasons for such closeness as the total number of stars in constellations does not exceed 20–30. Consequently the values $\widehat{\gamma}_G$ and $\widehat{\phi}_G$ can only to calculate a lower boundary σ_G^{min} for the mean-square latitude error in the constellation G. The value p_G^{min} gives us some additional useful information about group errors. We need in considerable large groups of stars to estimate group error reliable. It occurs that *there are 7 regions on the Almagest star atlas which differ one from another from the point of view of measurements accuracy of latitudes. Each of these 7 regions is "homogeneous", i.e., the measurement accuracy in this region is more or less the same for the most of stars. This fact is very important.* It was discovered in our computer experiments with the data from the Almagest star catalogue. We should like to note that the same division of the star atlas follows from systematization of the results of preccedings researchers but it is beyond of the scope of this paper. List of these 7 regions see on Fig.2. A number of stars in each region varies from 94 to 262.

Fig. 2

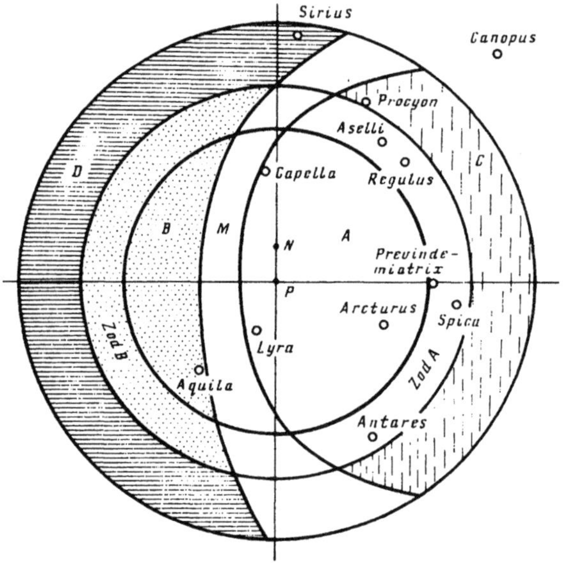

Region A contains all stars ($N_A = 249$) of the northern part of the sky and of the zodiac which are located to one hand of the Milk Way containing the spring equinox point.

Region B is the similar region ($N_B = 262$) located on the other hand of the Milk Way.

Region Zod A contains all zodiacal stars ($N_{ZodA} = 124$) from the region A and consists of 6 constellations: Gemini, Cancer, Leo, Virgo, Libra, Scorpius.

Region Zod B contains all zodiacal stars ($N_{ZodB} = 168$) from the region B.

Region C contains all southern stars ($N_C = 116$) located from the same side of the Milk Way as the region A.

Region D contains all southern stars ($N_D = 143$) located from the same side of the Milk Way as the region B.

Region M is the Milk Way ($N_M = 94$).
Let us consider a "large" group of stars R and determine parameters $\widehat{\gamma}_R$ and $\widehat{\phi}_R$ using the above relation (4) where one should replace G by R.

THEOREM 1. *Let us suppose that for all stars $i \in R$ parameters γ_i and ϕ_i are equal for all i (see (1), (2)) and coincide with γ_R and ϕ_R respectively. Then the values $\widehat{\gamma}_R$ and $\widehat{\phi}_R$ have the following properties:*

1) *$\widehat{\gamma}_R$ is a non–biased estimate of the value γ_R having a normal distribution with a variation*

$$\sigma^2(\widehat{\gamma}_R) = d[N_R(s_{20}\cos^2\phi_R + 2d_0\cos\phi_R\sin\phi_R + c_{20}\sin^2\phi_R)]^{-1},$$

where

$$s_{20} = \left[\sum_{i=1}^{N_R} \sin^2 L_i(t)\right]/N_R, \quad c_{20} = \left[\sum_{i=1}^{N_R} \cos^2 L_i(t)\right]/N_R,$$

$$d_0 = \left[\sum_{i=1}^{N_R} \sin L_i(t)\cos L_i(t)\right]/N_R.$$

2) *An estimate $\widehat{\phi}_R$ is asymptotically (when $N_R \to \infty$) non–biased for the value ϕ_R, its distribution function being able to be calculated in terms of a normal distribution (we do not need here in concrete formula).*

3) *The value $\sigma_R^{\min}(t^*)$ is asymptotically non–biased estimate for the real mean–square error $d^{1/2} = (\mathbf{E}\xi_i^2)^{1/2}$ of measurements.*

We shall call parameters ϕ_R and γ_R systematic errors in the group R. The value $\sigma_R^{\min}$ characterizes the exactness of measurements in the region R. Thus, in order to discover groups of well–measured stars we can use the following algorithm.
The algorithm of the choice of well–measured groups of stars

(1) Calculate values $\widehat{\gamma}_R$, $\widehat{\phi}_R$, $\sigma_R^{\min}$ for each "large" group R of stars;

(2) Choose the group $R_0 = \text{argmin}\sigma_R^{\min}$;

(3) Test that the calculated values $\widehat{\gamma}_{R_0}$ and $\widehat{\phi}_{R_0}$ are really parameters of the group error for all individual constellation from R_0. Consequently, Theorem 1 is valid. All such constellations G form the set of well–measured stars. Of course this set can be empty.

(4) Delete the set R_0 from the initial catalogue and repeat the algorithm beginning with step 1, etc.

As a result we obtain the hierarchy of well–measured collections of stars according with the accuracy of latitudes measurements.
Step 3 in the above algorithm is needed in some comments which we shall give below.
Let us note that the epoch t of real observations is unknown to us. Hence, all conclusions made above have a conventional character (given that the catalogue

was compiled in the epoch t). Consequently we need to test all values t from our a'priory time–interval. As we know the trajectory of the real ecliptic pole from Newcomb's theory it is sufficient to obtain $\widehat{\gamma}_R$ and $\widehat{\phi}_R$ only for some fixed $t = t_0$. These two parameters determine the pole of the "catalogue ecliptic" and it gives us the possibility to calculate $\widehat{\gamma}_R$ and $\widehat{\phi}_R$ for all t.

5.4. ***Errors in the Almagest star catalogue***

Computer calculations discovered the following values of minimal mean–square errors (practically they do not depend on t):

$$\sigma_{\mathrm{A}}^{\min} = 16.5'; \sigma_{\mathrm{B}}^{\min} = 19.2'; \sigma_{\mathrm{ZodA}}^{\min} = 12.8'; \sigma_{\mathrm{ZodB}}^{\min} = 19.3';$$

$$\sigma_{\mathrm{C}}^{\min} = 22.5'; \sigma_{\mathrm{D}}^{\min} = 24.4'; \sigma_{\mathrm{M}}^{\min} = 20.5'.$$

It follows that the region ZodA is the most well–measured one in the star atlas. These data confirmed that the claimed accuracy (10′) is a real accuracy of the catalogue (more than 50% stars have such deviations).

One can see the curve $\widehat{\gamma}_{\mathrm{ZodA}}(t)$ (which is a line in fact) in Fig.3. This curve is embedded in tolerant set corresponding to a confidence level $\varepsilon = 0.05$. Similar curves were obtained for all other regions. We calculated all functions $\widehat{\phi}_R(t)$ too. An example one can see in Fig.3.

Fig. 3

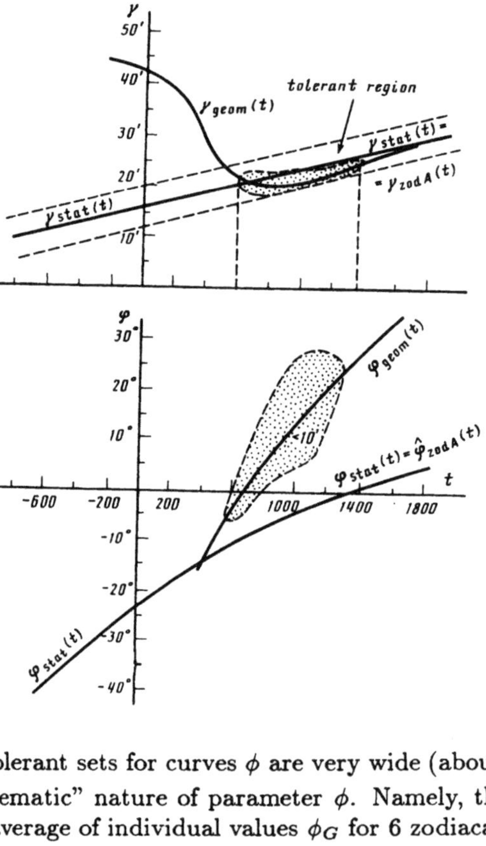

But the tolerant sets for curves ϕ are very wide (about 40°). This fact shows the "non–systematic" nature of parameter ϕ. Namely, the calculated value $\widehat{\phi}_{\mathrm{ZodA}}$ is only the average of individual values ϕ_G for 6 zodiacal constellations.

This fact can be considered as some indirect confirmation of the hypothesis that measurements were made by some instrument fixing an angle between the ecliptic and equator (of course with some error in the value of this angle). It is probable also that the axis of the rotation was fixed each time when a measurement occurs. One of such ancient instruments is well-known "astrolabe" or "armillary sphere" described by Ptolemy. Now let us turn to the procedure of the hypothesis testing that the value $\widehat{\gamma}_{\mathrm{ZodA}}$ determined by our calculations is common for all constellations from ZodA, i.e., this value really represents the group error. For each constellation G from ZodA we calculate and compare an "initial" error $\sigma_G^{\mathrm{in}} = \sigma_G(t,0,0)$, "minimal" error $\sigma_G^{\mathrm{min}}(t)$ and error σ_G^{gr} which is resulted after rotation on angles $\widehat{\gamma}_{\mathrm{ZodA}}(t)$ and $\widehat{\phi}_{\mathrm{ZodA}}(t)$, i.e.,

$$\sigma_G^{\mathrm{gr}} = \sigma_G\left(t, \widehat{\gamma}_{\mathrm{ZodA}}(t), \widehat{\phi}_{\mathrm{ZodA}}(t)\right).$$

The result is shown in Fig. 4 for $t = 100$ A.D. Similar calculations were made for all t. We can see from Fig. 4 that the resulting effect induced by the "optimal" individual rotation for each individual constellation practically coincides with the effect induced by the "common" rotation calculated for the total ZodA. We can see also the additional confirmation of the group nature of the error $\gamma_{\mathrm{ZodA}}(t)$ in Fig. 5 where we demonstrate graphs of the percents of the stars with latitude deviations not exceeding $10'$ after corresponding "optimal" rotation and without rotation (initial percent).

Fig. 4

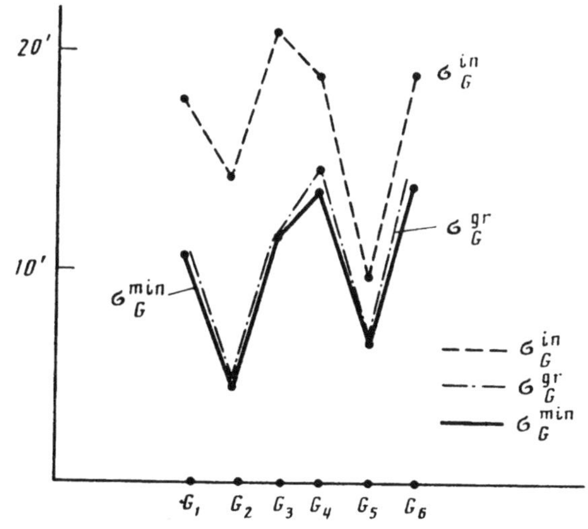

We investigated also neighbourhoods of the mentioned named stars: Antares, Arcturus, Aselli, Lyra, Capella, Procyon, Regulus, Spica. Two of these stars (Arcturus, Procyon) have large velocity of proper motion.

It turned out that their group errors for all these stars are the same (or very close) with stars from ZodA. Numerical data contained in the star catalogue are not sufficient to determine reliable group errors for neighbourhoods of only two stars: Aquila and Sirius. It was the reason that we excluded them from further consideration.

Fig. 5

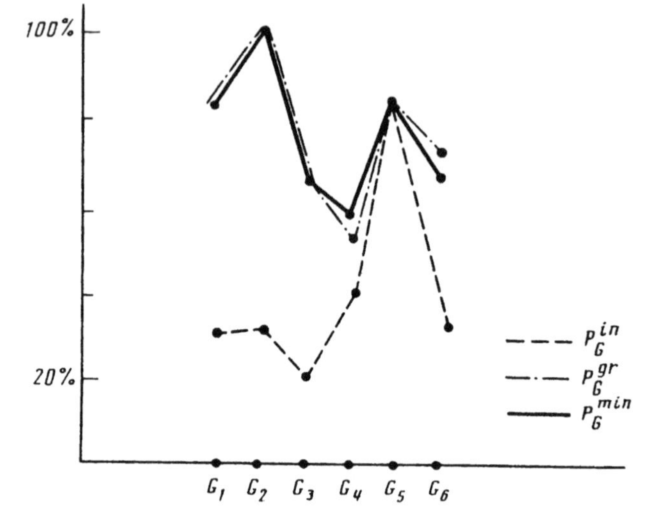

6. THE DATING OF THE ALMAGEST STAR CATALOGUE

6.1. *Statistical dating procedure*
Let I be the set of 8 named stars (see above) and

$$\Delta(t, \gamma, \phi) = \max_{i \in I} |\Delta b_i(t, \gamma, \phi)|.$$

We base our dating procedure on a hypothesis that the latitudes of *all* named stars from I must have *individual* errors not more than $10'$ in the year t^* of observations. In other words

$$\Delta(t^*, \gamma, \phi) < 10'$$

and the value γ is to belong to the statistical tolerant interval (see Fig. 3). Corresponding adoptable values of γ are "marked" with points in Fig. 3. Consequently *we declare that the time-interval* 600 *A. D.* $< t^* <$ 1300 *A. D. can be considered as dating interval.* Of course, this interval depends on different parameters in general: claiming accuracy ($10'$), confidence probability ε and some others. Stability of the method will be analysed in Section 7 below.
6.2. *Geometrical dating procedure*
Though we determined a dating interval some doubts about it can appear due to statistical nature of some assertions. Really we based our assertions on the fact that group errors for neighbourhoods of 8 named stars are the same. This fact was proved with help of statistics. Hence there is some positive probability (though it is very small) that this fact is wrong.
Let us consider again the value $\Delta(t, \gamma, \phi)$ and find for every t quantities:

$$(\gamma_{\text{geom}}(t), \phi_{\text{geom}}(t)) = \arg\min_{\gamma, \phi} \Delta(t, \gamma, \phi),$$

$$\Delta_{\min}(t) = \Delta(t, \gamma_{\text{geom}}(t), \phi_{\text{geom}}(t)).$$

These quantities depend on only positions of 8 named stars whereas $\widehat{\gamma}_{\text{ZodA}}(t)$ and $\widehat{\phi}_{\text{ZodA}}(t)$ do not depend on them (they depend on positions of all stars from Zod A). It is clear that $\Delta_{\min}(t) < 10'$ if 600 A. D. $\leqslant t \leqslant$ 1300 A. D. But it turned out that $\Delta_{\min}(t) < 10'$ *if and only if* 600 A. D. $\leqslant t \leqslant$ 1300 A. D., see Fig. 6. Besides $\gamma_{\text{geom}}(t) \cong \widehat{\gamma}_{\text{ZodA}}(t)$ for these t, see Fig. 3. Hence, *we confirmed without any statistical arguments that the above interval is a dating one. There do not exist such γ and ϕ leading to the inequality*

$$\Delta(t, \gamma, \phi) < 10',$$

when $t <$ 600 A. D. or $t >$ 1300 A. D. We confirmed also that the systematic error calculated with help of statistics (using coordinates of all stars from ZodA) is in fact "geometrically optimal" for 8 named stars. Hence:

(1) We confirmed the accuracy claimed by a compiler of the Almagest star catalogue;

(2) We calculated the time–interval containing the actual date of observation. We proved also that the catalogue could not have been compiled (on the basis of actual observations) outside this time–interval;

(3) We proved that the compiler made an error in the determination of the position of the ecliptic pole and calculated it ($\gamma \cong 20'$); besides he made an error in the determination of the position of the equator ($\beta < 5'$); it is important to note that the systematic error γ explains the existence of a strange "Peters' sinus" in latitudes deviations for zodiacal stars (Peters and Knobel, 1915, p. 6).

(4) We defined the information kernel (8 named stars) in accordance with the accuracy of measurements of coordinates.

Fig. 6

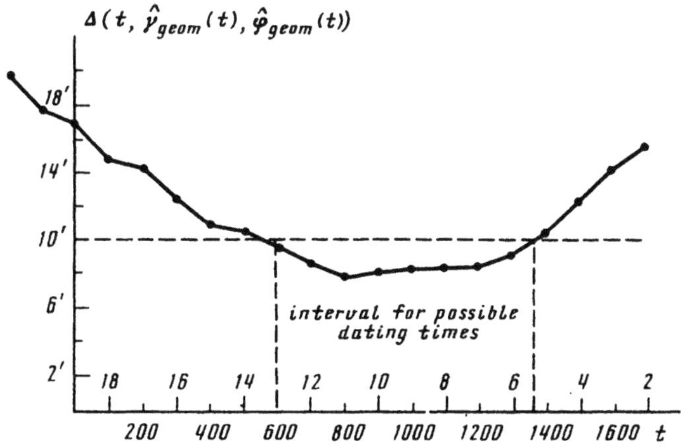

7. STABILITY OF THE METHOD

7.1. Our calculations showed that the decreasing of the confidence probability ε (beginning from $\varepsilon = 0.02$) does not shift the time–interval of probable dating. We have obtained also that this interval does not depend on the assumption about normality of the distribution of random variables ξ_i (a kind of robastness).
7.2. Let us show now how the final results depend on the content of the group of named stars (information kernel). E. g. if we remove Arcturus (the most fast star in the group) then the left boundary of the dating time–interval shifts to approximately 350 A. D. but it does not touch traditional Ptolemy's epoch. Some useful information about the dependence on the content of the group of named stars is contained in Fig. 7. There for some fixed t empirical distribution functions are drawn (these graphs refer to all 12 named stars):

$$F^t(x) = N\{i : |\Delta(i,t,\gamma,\phi)| < x\}/12.$$

We see that the "best" distribution function corresponds to $t = 10$ and $\gamma = 21'$. It confirms the above assertions.

Fig. 7

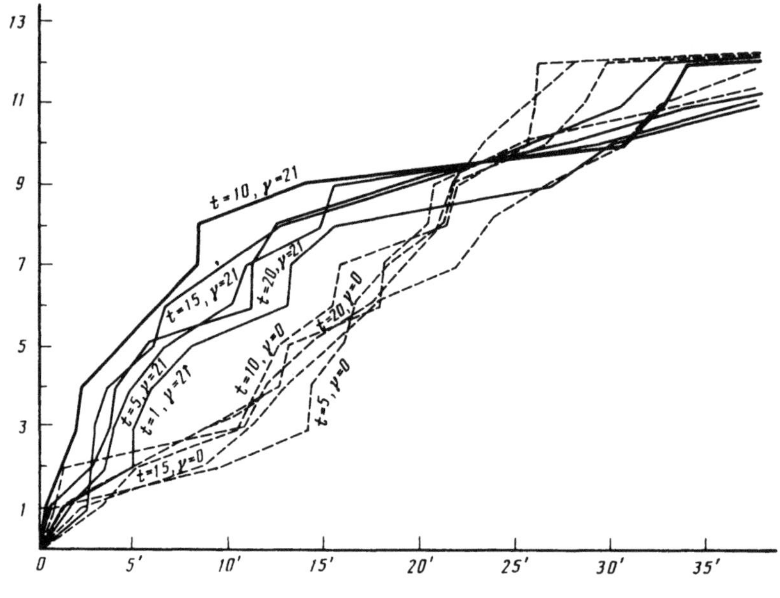

7.3. Let us change the claimed accuracy level Δ. Remind that we started with $\Delta = 10'$. Then the "Ptolemy's epoch" will be captured only when $\Delta = 25'$.
7.4. Now consider not only the "rigid rotation" of the celestial sphere (as group errors) but an arbitrary diffeomorphism of the coordinates (which is indeed close to an identity mapping) reflecting possible distortions (deformations) of astronomical instruments. Then it occures that we can reach "Ptolemy's epoch" only for such deformations which are implied by about 4% deviations of real instruments (e.g., armillary sphere) from ideal ones. This is quite impossible even for usual "customary" instruments (let alone scientific ones).

Consequently our results are stable in respect to different deviations both of numerical data and proposed suppositions.

8. DATING OF OTHER CATALOGUES

8.1. *Tycho Brahe's catalogue*

The observations of T. Brahe were made at the end of the 16–th century but the catalogue was reduced to 1600 A. D.

Applying the above method to his catalogue we discovered that T. Brahe did not make any significant systematic error in his determination of the position of the ecliptic pole ($\gamma \cong 1'$). On the other hand, the $1'$ accuracy claimed by T. Brahe is not achieved even for named bright famous stars. In fact his accuracy is about $2'$ for the set of his based stars. For this catalogue statistical and geometrical procedures lead to the same dating interval from 1510 A. D. till 1620 A. D. containing the real observation epoch.

8.2. Similar dating interval we received for Hevelius' catalogue (Baily, 1843). It was shown that the actual accuracy of latitudes in this catalogue is about $2'$ but not $1''$ claimed by Hevelius. It is possible also that Hevelius used not only personal observations but some other ones, perhaps that of T. Brahe.

8.3. The application of the proposed statistical and geometrical procedures to Ulugbeck's catalogue (Baily, 1843) gave a dating interval 700 A. D.–1400 A.D. which contains a real date of the compilation. Besides we received that a significant part of Ulugbeck's catalogue has a systematic error coinciding with that of Almagest's one. It is very probable that this part was taken from the Almagest.

8.4. As for al–Sufi's catalogue (Fomenko, 1988) it is in fact the Almagest star catalogue: latitudes for almost all stars are the same (even for fast stars) and longitudes differ on $12°\ 42'$ exactly! So no need to date this catalogue.

REFERENCES

Baily, F. (1843). The catalogues of Ptolemy, UlughBeigh, Tycho Brahe, Halley and Hevelius, deduced from the best authorities. *Roy. Astr. Soc. Memoirs* **13**, 1–248.

Fedorov, V. V. and Fomenko, A. T. (1986). Statistical estimation of chronological nearness of historical texts. *J. Sov. Math.* **32**, 668–675.

Fomenko, A. T. (1981a). The jump of the second derivative of the moon's elongation. *Celestian Mech.* **29**, 33–40.

Fomenko, A. T. (1981b). A method of duplicate recognition and some applications. *Soviet Math. Dokl.* **258**, 1326–1333.

Fomenko, A. T. (1984). New empirical and statistical methods of discovering parallels and dating duplicates. In: *Stability Problems for stochastic Models.* VNIISI, Moscow, pp. 154–177.

Fomenko, A. T. (1986). New empirico–statistical dating methods and statistics of certain astronomical data. In: *Abstr. 1–st World Congress of Bernoulli Soc.* VNU Science Press, Utrecht. **2**, pp. 892.

Fomenko, A. T. (1987). New empirical and statistical methods for dating events of ancient history and applications to global chronology of ancient world and middle ages. *British Library, Dept. of Printed Books.* Cup. 818–870.

Fomenko, A. T. (1988). Empirico-statistical methods of ordering narrative texts. *Intern. Statist. Review,* **56, 3**, 279–301.

Fomenko, A. T., Kalashnikov, V. V. and Nosovsky, G. V. (1988). Geometry of moving stars configurations and dating of the Almagest star catalogue. In: *Stability Problems of Stochastic Models.* VNIISI, Moscow, pp. 59–78.

Fomenko, A. T., Kalashnikov, V. V. and Nosovsky, G. V. (1989a). When was Ptolemy's star catalogue in Almagest compiled in reality? *Preprint* No 1989–04.. Dept. Math. Chalmers Univ. of Technology, Univ. Gøteborg, Gøteborg.

Fomenko, A. T., Kalashnikov, V. V. and Nosovsky, G. V. (1989b). Dating of Almagest basing on the variable configurations of stars. *Soviet. Math. Dokl.* **307.**

Hoffit, D. (1982). *The Bright Stars Catalogue.* Yale Univ. Obs., New Haven.

Fricke, W. and Kopff, A. (1963). *FK4. Verof. Astr. Inst.* **10**, Heildelberg.

Kinoshita, H. (1975). *Formulas for Precession.* Astrophys. Obs., Smithsonian Inst., Cambridge.

Morozov, N. A. (1928). *Christ.* Gosizdat, Moscow.

Peters, C. H. F. and Knobel, E. B. (1915). *Ptolemy's catalogue of stars. A revision of the Almagest.* Carnegie Inst., Washington.

Newton, R. R. (1975). *The Crime of Claudious Ptolemy.* J. Hopkins Press, London.

Toomer, G. J. (1984). *Ptolemy's Almagest.* J. Hopkins Press, London.

Prob. Theory and Math. Stat., Vol. 1, pp. 375–381
B.Grigelionis *et al.* (Eds.)
1990 VSP/Mokslas

GLOBAL ASYMPTOTICS OF SOLUTIONS OF THE FIRST AND SECOND BOUNDARY VALUE PROBLEMS FOR KOLMOGOROV EQUATION WITH SMALL DIFFUSION

S.M. FROLOVICHEV
MIEM, B. Vuzovski 3/12, 109028 Moscow, USSR

ABSTRACT

Global exponential asymptotics for solutions of the first and second boundary value problems are constructed for Kolmogorov's equation with small diffusion by using the method of tunnel canonical operator.

We construct the global exponential asymptotics of solutions of the following boundary value problems

$$\partial u/\partial t = L_h u, \quad x \in D \subset \mathbf{R}^n, t > 0;$$

$$u|_{t=0} = 0, x \in D; \quad u|_{x\in\partial D} = \Psi_0(x,t), t > 0; \tag{1}$$

$$\partial u/\partial t = L_h u, \quad x \in D \subset \mathbf{R}^n, \ t > 0;$$

$$u|_{t=0} = 0, x \in D; \quad u|_{x\in\partial D} = \Psi_1(x,t), t > 0; \tag{2}$$

where $u = u(x,t,h), L_h = h\langle A(x,t)\nabla_x, \nabla_x\rangle + v(x,t)$ is an operator which is uniformly parabolic in $\Omega = (\bar{D}_0 \times [0,T]), A(x,t) \in C^\infty(\Omega)$ is a symmetric $(n \times n)$–matrix; $v(x,t) \in C^\infty(\Omega)$; D_0 is a bounded domain such that $(D \bigcup \partial D) \subset D_0 \subset \mathbf{R}^n$, $\bar{D}_0$ is the closure of the domain $D_0, \Psi_i(x,t) \neq 0$ for $(x,t) \in (\partial D \times [0,T])$ and $\Psi_i(x,t) \in C(\partial D \times ([0,T]), i = 0,1$. The boundary ∂D of the domain D is of the class C^2, $h \to +0$ is a small parameter, $\langle\cdot,\cdot\rangle$ is the scalar product in $\mathbf{R}^n, \nabla_x = (\partial/\partial x_1, \dots, \partial/\partial x_n)$ is the gradient operator, $\partial/\partial\nu$ is the conormal derivative.

The global asymptotics of solution of the following boundary value problem

$$\partial u/\partial t = L'_h u, \quad x \in (x_0, x_1) \subset \mathbf{R}^1, t > 0;$$

$$u|_{t=0} = 0, \quad x \in (x_0, x_1); \tag{3}$$

$$u|_{x=x_0} = \mu_0(t), u|_{x=x_1} = \mu_1(t), t > 0;$$

where $L'_h = ha(x,t)\partial^2/\partial x^2 - b(x,t)\partial/\partial x + v(x,t)$, is also constructed. The functions $a(x,t) \geqslant C > 0$, $b(x,t)$, $v(x,t)$ are infinitely differentiable in $(x,t) \in ([x_0,x_1] \times [0,T]), \mu_i(t) \in C([0,T]), \mu_i(t) \neq 0$ for $t \in [0,T], i = 0,1$.

Numerous papers have been devoted to the study of logarithmic asymptotics of the solutions of the problems (1) – (3), i.e. to the construction of the function

$$S(x,t) = -\lim_{h\to+0} h\ln u(x,t,h)$$

(see (Ventsel and Freidlin, 1984) for a detailed bibliography). In (Maslov et al., 1988) the global asymptotics of the problem (3) is constructed in the multi–dimensional case when the equation coefficients are independent of the variable t. The global asymptotic of the problem (2) is given in this paper for the first time. It is known (Ventsel and Freidlin, 1984) that the solution of problems (1) – (3) has a probabilistic interpretation. Below we carry out some necessary constructions. We denote by D^δ the δ–neighbourhood of the boundary ∂D of the domain D in $\mathbf{R}^n_x$, i.e.

$$D^\delta = \left\{x : \min_{y\in\partial D}|x-y| < \delta\right\},$$

where $|\cdot|$ is the Euclidean norm in $\mathbf{R}^n_x$. In the space $\mathbf{R}^n_x \times \mathbf{R}^n_p$ we construct the Lagrangian manifold (Maslov, 1985) of dimension n:

$$\Lambda^0_n = \{p = \rho n(x), x \in \partial D\},$$

where $\rho \in \mathbf{R}^1$ and $n(x)$ is the unit outward normal vector to ∂D at the point x. We introduce the notation $\Lambda^t_n = g^t_H\Lambda^0_n$ for the translation of the manifold Λ^0_n along the trajectories of the Hamiltonian system

$$\dot{x} = H_p(x,p,t), \quad \dot{p} = -H_x(x,p,t) \tag{4}$$

with the Hamiltonian $H(x,p,t) = \langle A(x,t)p,p\rangle$ in the case of the problems (1) and (2), and $H(x,p,t) = a(x,t)p^2 + b(x,t)p$ in the case of the problem (3) for fixed time t, r_t is a point on the manifold Λ^t_n, $\{x(r_t),p(r_t)\}$ is its coordinates in $\mathbf{R}^n_x\times\mathbf{R}^n_p$, and $K_{\Lambda^t_n}$ is the tunnel canonical operator constructed on Λ^t_n:

$$K_{\Lambda^t_n} : \varphi_1(r_t) \to \varphi_2(x,t,h),$$

where $\varphi_1(r_t) \in C^\infty_0(\Lambda^t_n)$ and $\varphi_2(x,t,h) \in C^\infty(\mathbf{R}^n_x \times (0,T] \times (0,1]), 0 < T < \infty$. A detailed construction of the tunnel operator is given in (Maslov, 1985, p. 158).

THEOREM 1. *For any numbers $\varepsilon > 0, \delta > 0, T > 0$ which are independent of h, the solution $u = u(x,t,h)$ of the boundary value problem* (1) *can be represented in $(D \setminus D^\delta) \times [\varepsilon, T]$ in the form:*

$$u = 2(h/2\pi)^{1/2}K_{\Lambda^t_n}\Big(b_0(r_t)e(r_t)(1+o(1))\Big),$$

where

$$b_0(r_t) = \big(\rho(r_t)\big)^{-1}\exp\left\{\frac{1}{2}\int_0^t\Big(\mathrm{Sp}H_{p,x}(X,P,\tau)+v(X,\tau)\Big)d\tau\right\}\Psi_0(y,0),$$

$$\rho(r_t) = |p(g_H^{-t} r_t)|,$$

$$y = x(g_H^{-t} r_t), e(r_t) \in C_0^\infty(\Lambda_n^t);$$

$$e(r_t) = 1 \text{ if } x(r_t) \in D \setminus D^\delta, e(r_t) = 0 \text{ if } x(r_t) \notin D \setminus D^{\delta/2} \tag{5}$$

and X, P is the solution of Hamiltonian system (4) with the initial data

$$X|_{\tau=0} = y, \quad P|_{\tau=0} = -\rho(r_t) n(y).$$

Remark 1. There exists $\delta_1 > 0$ independent of h such that the solution of the problem (1) can be represented in the form of a single integral in the domain $D \cap D^{\delta_1}$ when the function $\Psi_0(x,t) = \Psi_0(x) \in C^1(\partial D)$.

THEOREM 2. *For any numbers $\varepsilon > 0, \delta > 0, T > 0$ which are independent of h, the solution $u = u(x,t,h)$ of the boundary value problem (2) can be represented in $(D \setminus D^\delta) \times [\varepsilon, T]$ in the form:*

$$u = 2h^{3/2}(2\pi)^{-1/2} K_{\Lambda_n^t}\Big(b_1(r_t)e(r_t)(1+o(1))\Big),$$

where

$$b_1(r_t) = \Big(\rho(r_t)\Big)^{-2} \exp\left\{\frac{1}{2}\int_0^t \Big(\mathrm{Sp} H_{p,x}(X,P,\tau) + v(X,\tau)\Big) d\tau\right\} \times$$

$$\times \Psi_1(y,0)\Big(\langle A(y,0)n(y), n(y)\rangle\Big)^{-1}.$$

There X, P is the solution of Hamiltonian system (4) with the initial data (5) and $e(r_t)$ is defined in Theorem 1.

Remark 2. There exists $\delta_1 > 0$ independent of h such that the solution of the problem (2) can be represented in the form of a single integral in the domain $D \setminus D^{\delta_1}$.

Remark 3. A point $x(r_t^*) \in D$ is called focal if any neighbourhood of the point r_t^* in Λ_n^t projects onto $\mathbf{R}_x^n$ in a non–diffeomorphic way. In a neighbourhood of focal points (provided r_t^* belongs to the essential domain of the manifold Λ_n^t (Maslov, 1985)) the asymptotics of the solution of the problem (1) cannot be represented in the form $h^{1/2}\exp\{-S(x,t)/h + O(1)\}$ but is given by the tunnel canonical operator and can be expressed in the form of an integral whose multiplicity equals the defect of the matrix of the projection of the manifold Λ_n^t onto $\mathbf{R}_x^n$ at the point r_t^* (Maslov, 1985). In boundary value problems, focal points appear even in very simple examples. Let $x \in \mathbf{R}^2$ and $L_h = h\Delta$ (Δ is the Laplacian), and let the domain D be an ellipse. Then for each $t \geqslant \varepsilon > 0$ two points of the evolute of D lying on its major axis are essential focal points.

Let us consider the one–dimensional case. We denote by x_0^t the translation of a point x_0 for the fixed time t along the trajectory of the problem $\dot{x} = b(x,t), x|_{t=0} = x_0$. Let x_1^t be defined similarly. We shall consider the following three cases:

(i) $b(x_0,t) > 0, \ b(x_1,t) < 0;$

(ii) $b(x_0,t)>0,\ b(x_1,t)>0$;

(iii) $b(x_0,t)<0,\ b(x_1,t)>0$ for $t\in[0,T]$.

We give the statement of Theorem in the case (i).

THEOREM 3. *For any numbers $\varepsilon>0,\delta>0,T>0$, independent of h the solution $u=u(x,t,h)$ of the problem (3) in the case (i) can be represented as:*
I. For $x\in[x_0,x_0^t-\delta),t\in[\varepsilon,T]$

$$u=\mu_0(\tau_0)\exp\left\{\int_\tau^t v(y,\lambda)d\lambda\Big|_{\tau=\tau_0}\right\}+o(1),$$

where $y=y(x_0,\tau,\lambda)$ is the solution of the problem $\dot y=b(y,\lambda),y|_{\lambda=\tau}=x_0$ and $\tau_0=\tau_0(t,x)$ is defined by $y(x_0,\tau_0,t)=x$.
II. For $x\in(x_1^t+\delta,x_1],t\in[\varepsilon,T]$

$$u=\mu_1(\tau_1)\exp\left\{\int_\tau^t v(y,\lambda)d\lambda\Big|_{\tau=\tau_1}\right\}+o(1),$$

where $y=y(x_1,\tau,\lambda)$ is the solution of the problem $\dot y=b(y,\lambda),y|_{\lambda=\tau}=x_1$ and the function $\tau_1=\tau_1(t,x)$ is defined by $y(x_1,\tau_1,t)=x$.
III. There exists $\delta_1>0$ independent of h, such that for $x\in(x_0^t-\delta_1,x_0^t],t\in[\varepsilon,T]$

$$u=\mu_0(\tau_0)\Phi_h\Big(-(S(x,t))^{1/2}\Big)\exp\left\{\int_\tau^t v(y,\lambda)d\lambda\Big|_{\tau=\tau_0}\right\}+o(1),$$

where

$$\Phi_h(\eta)=(2\pi h)^{-1/2}\int_0^\infty\exp\Big\{-(z-\eta)^2/h\Big\}dz,$$

$$S(x,t)=\int_0^t\Big(PH_p(X,P,\tau)-H(X,P,\tau)\Big)d\tau\Big|_{p_0=p_0(x,t)}.$$

There X,P is the solution of the system (4) with the initial data $X|_{\tau=0}=x_0,P|_{\tau=0}=p_0$, $p_0(x,t)$ is the solution of $x=X(p_0,x_0,t)$; finally, y,τ_0 are defined in I.
IV. There exists $\delta_1>0$ independent of h such that for $x\in[x_1^t,x_1^t+\delta_1),t\in[\varepsilon,T]$

$$u=\mu_1(\tau_1)\Phi_h\Big((S(x,t))^{1/2}\Big)\exp\left\{\int_\tau^t v(y,\lambda)d\lambda\Big|_{\tau=\tau_1}\right\}+o(1),$$

where the function $S(x,t)$ is defined in III, and X,P is the solution of the system (4) with the initial data: $X|_{\tau=0}=x_1$, $P|_{\tau=0}=p_0$; $p_0(x,t)$ is solution of the

equation $X(x_1, p_0, t) = x$ and $\tau_1 = \tau_1(t, x), y = y(x_1, \lambda)$ are defined in II.
V. For $x \in (x_0^t + \delta, x_1^t - \delta) = d^\delta, t \in [\varepsilon, T]$

$$u = \Big(h/(2\pi)\Big)^{1/2} K_{\Lambda^t}\Big(e_1(r_t)g(r_t)f(r_t)\Big(1 + o(1)\Big)\Big),$$

where

$$g(r_t) = \frac{2a(y,0)\rho(r_t) + b(y,0)}{a(y,0)\rho(r_t) + b(y,0)},$$

$$f(r_t) = \Big(\rho(r_t)\Big)^{-1} \exp\left\{\frac{1}{2}\int_0^t H_{p,x}\Big(X, P, \tau\Big)d\tau\right\} \mu^y(0) \exp\left\{\int_0^t v(X,\tau)d\tau\right\},$$

$e_1(r_t) \in C_0^\infty(\Lambda^t), e_1(r_t) = 1$ if $x(r_t) \in d^\delta, e_1(r_t) = 0$ if $x(r_t) \notin d^{\delta/2}andX$, P is the solution of the Hamiltonian system (4) with the initial data

$$X|_{\tau=0} = y, \quad P|_{\tau=0} = \rho, \quad \rho \in \mathbf{R}^1;$$

$$y = x(g_H^{-t} r_t), \quad \rho(r_t) = p(g_H^{-t} r_t),$$

$$\mu^{x_0}(0) = \mu_0(0), \quad \mu^{x_1}(0) = \mu_1(0),$$

$$\Lambda^0 = \{p = \beta_0, \ x = x_0\} \cup \{p = \beta_1, x = x_1\},$$

$$\beta_0, \beta_1 \in \mathbf{R}^1.$$

VI. There exists $\delta_1 > 0$ independent of h such that for $x \in [x_0^t, x_0^t + \delta_1), t \in [\varepsilon, T]$

$$u = (2\pi h)^{-1/2} g_0(x,t)\mu_0(0)\big|I(x,t)\big|^{-1/2}\widetilde{f}(x,t)\times$$

$$\times \int_0^t \exp\Big\{-S(x,t,\tau)/h\Big\}d\tau \exp\left\{\int_0^t v(X,\tau)d\tau\right\}\Big(1 + o(1)\Big),$$

where

$$g_0(x,t) = 2a(x_0,0)p_0(x,t) + b(x_0,0),$$

$$\widetilde{f}(x,t) = \exp\left\{\frac{1}{2}\int_0^t H_{p,x}(X,P,\tau)d\tau\right\}\Big|_{p_0=p_0(x,t)},$$

$$I(x,t) = \left(\frac{dX}{dp_0}\right)\Big|_{p_0=p_0(x,t)}$$

and X, P is the solution of the Hamiltonian system (4) with the initial data $X|_{\tau=0} = x_0, P|_{\tau=0} = p_0$, $p_0(x,t)$ is the solution of the equation $X(p_0, x_0, t) = x$. Finally,

$$S(x,t,\tau) = \int_\tau^t \Big(P^* H_p(X^*, P, \lambda) - H(X^*, P^*, \lambda)\Big)d\lambda\Big|_{p_0=p_0(x,t,\tau)},$$

where X^*, P^* is the solution of the system (4) with the initial data , $P^*|_{\lambda=\tau} = p_0 X^*|_{\lambda=\tau} = x_0$; $p_0(x,t,\tau)$ *is the solution of the equation* $X^*(p_0, x_0, t, \tau) = x$.
VII. *There exists* $\delta_2 > 0$ *independent of* h *such that for* $x \in (x_1^t - \delta_2, x_1^t], t \in [\varepsilon, T]$

$$u = -(2\pi h)^{-1/2} g_1(x,t)\mu_1(0)\big|I(x,t)\big|^{-1/2}\, \widetilde{f}(x,t)\times$$

$$\times \int_0^t \exp\Big\{-S(x,t,\tau)/h\Big\}d\tau \exp\Big\{\int_0^t v(X,\tau)d\tau\Big\}(1+o(1)),$$

where

$$g_1(x,t) = 2a(x_1,0)p_0(x,t) + b(x_1,0),$$

$I(x,t)$ *and* $\widetilde{f}(x,t)$ *are defined similarly as in* VI, *with the difference that here* X, P *is the solution of the system (4) with the initial data* $X|_{\tau=0} = x_1, P|_{\tau=0} = p_0$; *the function* $S(x,t)$ *is defined in* VI, *only here* X, P *is the solution of the system (4) with the initial data* $X|_{\lambda=\tau} = x_1$, $P|_{\lambda=\tau} = p_0, p_0(x,t,\tau)$ *is the solution of the equation* $X(p_0, x_1, t, \tau) = x$.

We sketch below the proof of Theorem 1.
We look for the solution of the problem (1) in the form of the double field potential (Ladyzenskaya et al., 1968):

$$u(x,t,h) = \int_0^t d\tau \int_{\partial D} 2h\langle A(\xi,\tau)\nabla_\xi \Gamma(x,t,\xi,\tau,h), n(\xi)\rangle \mu^h(\xi,\tau)d\gamma_\xi, \tag{6}$$

where $\mu^h(\xi,\tau)$ is an unknown function and $\Gamma(x,t,\xi,\tau,h)$ is the fundamental solution of the equation:

$$\frac{\partial}{\partial t}\Gamma\Big(x,t,\xi,\tau,h\Big) = L_h\Gamma(x,t,\xi,\tau,h), \quad t > \tau; \tag{7}$$

$\Gamma(x,\tau,\xi,\tau,h) = \delta(x-\xi)$. It is obvious that for $x \in D$ the function $u(x,t,h)$ satisfies (7) and $u(x,0,h) = 0$. It is known (Ladyzenskaya et al., 1968) that the limit value of $u(x,t,h)$ from (6) as $x \to x' \in \partial D$ equals

$$\lim_{x \to x' \in \partial D} u(x,t,h) =$$

$$= \int_0^t d\tau \int 2h\langle A(\xi,\tau)\nabla_\xi \Gamma(x',t,\xi,\tau,h), n(\xi)\rangle \mu^h(\xi,\tau)d\gamma_\xi - \mu^h(x',t). \tag{8}$$

Relation (8) allows us to reduce the solution of the problem (1) to the solution of the following integral equation:

$$\mu^h(x',t) - \int_0^t d\tau \int_{\partial D} 2h < A(\xi,\tau)\nabla_\xi \Gamma(x',t,\xi,\tau,h), n(\xi) > \mu^h(\xi,\tau)d\gamma_\xi =$$

$$= \Psi_0(x', t), x' \in \partial D, \quad t > 0. \tag{9}$$

This is an integral equation of Volterra type which can be solved by the method of succesive approximations. The global asymptotics of the fundamental solution of the equation (7) and of a more general equation was constructed in (Maslov, 1985) and is given by the tunnel canonical operator. By applying the Laplace method (Fedoryuk, 1977) we can show that the solution of (9) has the form $\mu^h(x,t) = -\Psi_0(x,t) + o(1), x \in \partial D, t > 0$. The proof of Theorem 1 consists in application of the Laplace method to the integral (6) with the function $\mu^h(\xi,\tau) = -\Psi_0(\xi,\tau)$. The solution of the problem (2) is sought in the form of the potential of layer of charge (Ladyzenskaya et al., 1968), and the solution of the problem (3) we seek already in the form of a linear combination of potential of double layer and potential of layer of charge. The scheme of the proof in the case of the problems (2) and (3) coincides with the scheme of the proof of the problem (1).

REFERENCES

Fedoryuk, M. V. (1977). *The Saddle-Point Method.* Nauka, Moscow.

Ladyzenskaya, O. A., Solovnikov, V. A. and Uraltseva, N. N. (1968). *Linear and Quasilinear Equations of Parabolic Type.* Amer. Math. Soc., Providence.

Maslov, V. P. (1972). *Perturbation Theory and Asymptotic Methods.* Dunod, Paris.

Maslov, V. P. (1985). *Global exponential asymptotics of solutions of tunnel equations.* Proc. Steklov Inst. Math., **163**,

Maslov, V. P., Frolovichev, S. M. and Chernykh, S. I. (1988). Precise asymptotics of large deviations in boundary value problems for diffusion processes. *Soviet Math. Dokl.* **36**, 255–259.

Ventsel, A. D. and Freidlin, M. I. (1984). *Random Perturbations of Dynamical Systems* Springer, Berlin–Heidelberg, etc.

Prob. Theory and Math. Stat., Vol. 1, pp. 382–390
B.Grigelionis *et al.* (Eds.)
1990 VSP/Mokslas

HYDRODYNAMIC LIMIT FOR GINZBURG—LANDAU TYPE CONTINUUM MODEL

T. FUNAKI
Mathematical Institute, HAS, H-1364, Budapest, Pf.127, Hungary, and
Department of Mathematics, Nagoya University, Nagoya, 464-01, Japan

ABSTRACT

The hydrodynamic limit of the time dependent Ginzburg—Landau (TDGL) equation is obtained. The discussion is based on the characterization of reversible measures of the TDGL equation in terms of invariant measures of an auxiliary SPDE of non-conservative type.

1. INTRODUCTION

We shall discuss the hydrodynamic behavior of a certain stochastic dynamics of continuum field $S : \mathbf{R}^d \to \mathbf{R}$ over $\mathbf{R}^d$. The evolution law of our model is prescribed by the following stochastic partial differential equation (SPDE):

$$dS_t(x) = \frac{1}{2}\Delta\{\gamma + (-\Delta)^m\}S_t(x)dt + \frac{1}{2}\Delta\{V'(S_t(x))\}dt + \\ + d\{\operatorname{div} \mathbf{w}_t(x)\}, \qquad x \in \mathbf{R}^d,\ t > 0. \tag{1.1}$$

Here in this equation, $\gamma > 0$, m is an integer satisfying $m > d/2$, Δ is the Laplacian on $\mathbf{R}^d$, $V \in C^3(\mathbf{R})$ is a function such that $V' \in C_b^2(\mathbf{R})$ and $\mathbf{w}_t(x) = \{w_t^i(x)\}_{i=1}^d$ is a system of d independent cylindrical Brownian motions (c.B.m.'s) on $L^2(\mathbf{R}^d)$; we mean by the c.B.m. on $L^2(\mathbf{R}^d)$ an $\mathcal{S}'(\mathbf{R}^d)$-valued continuous process $w_t(x)$ such as $\langle w_t, \varphi\rangle / \ \|\varphi\|_{L^2(\mathbf{R}^d)}$ is a 1-dimensional standard Brownian motion for every non-zero $\varphi \in \mathcal{S}(\mathbf{R}^d)$. The equation (1.1) is called the time-dependent Ginzburg—Landau equation (TDGL eq.) of conservative type.
The model has the conservation law in the following sense: If we consider (1.1) restricting in a finite but arbitrary large box Λ in $\mathbf{R}^d$ under a proper, for example, the periodic boundary condition, then the total density (or charge) $\int_\Lambda S_t(x)\,dx$ is preserved in time. As a result, the equation (1.1) has a family of non-unique stationary probability measures (equilibrium states). The investigation of reversible measures is actually one of the main subjects of the present paper; see Sections 3 and 4. However, instead of (1.1), our study is mainly concentrated on the following auxiliary SPDE which is called the TDGL equation of non-conservative type:

$$dS_t(x) = -\frac{1}{2}\{\gamma + (-\Delta)^m\}S_t(x)dt - \frac{1}{2}V'(S_t(x))dt + dw_t(x), \tag{1.2}$$

where $w_t(x)$ is a c.B.m. on $L^2(\mathbf{R}^d)$. This is because reversible measures of these two equations are closely related.
In order to explain these circumstances, we discuss our model from the heuristic point of view briefly. Let us introduce the Ginzburg—Landau type formal Hamiltonian (total energy) of the configuration $S : \mathbf{R}^d \to \mathbf{R}$ by

$$\mathcal{H}(S) = \int_{\mathbf{R}^d} \left\{ \frac{1}{2}(-\Delta)^m S(x) \cdot S(x) + U(S(x)) \right\} dx,$$

where

$$U(\sigma) = \frac{\gamma}{2}\sigma^2 + V(\sigma), \ \sigma \in \mathbf{R}, \tag{1.3}$$

is the self–potential of the model. Then the equations (1.1) and (1.2) are rewritten simultaneously into one equation

$$dS_t(x) = -\frac{1}{2}(-\Delta)^\alpha D\mathcal{H}(x, S_t)dt + (-\Delta)^{\alpha/2}\, dw_t(x), \tag{1.4}$$

where $D\mathcal{H}(x, S)$ denotes the functional derivative of $\mathcal{H}(S)$. Actually (1.1) and (1.2) correspond to (1.4) with $\alpha = 1$ and $\alpha = 0$, respectively. Note that $(-\Delta)^{1/2} w_t(x)$ has the same distribution as div $\mathbf{w}_t(x)$. The rewritten form (1.4) suggests that the reversible measures of (1.1) and (1.2) are closely connected and moreover they might be the *Gibbs states* corresponding to the Hamiltonian $\mathcal{H}$.
Since the equilibrium states of the equation (1.1) is not unique, it is natural to expect that this equation exhibits the hydrodynamic behavior. Indeed, under appropriate space–time scaling, a non–random limit is obtained and it is shown that the limit is described by a certain nonlinear diffusion equation; see Sect. 5.
The results presented here are the multi–dimensional extension of those obtained in the one–dimensional case (Funaki, 1989). The lattice version of the model was studied in (Fritz, 1989). The paper containing full proofs will be published elsewhere .

2. EXISTENCE AND UNIQUENESS OF SOLUTIONS OF THE SPDE's

Before discussing the problems stated in Sect.1, we formulate the existence and uniqueness results for the SPDE's (1.1) and (1.2). We call $S_t = \{S_t(x);\ x \in \mathbf{R}^d\}$ the solution of one of these SPDE's if it satisfies the stochastic integral equation obtained by rewriting them formally; e.g. S_t is called the solution of (1.1) with initial data S_0 if it satisfies

$$S_t(x) = \int_{\mathbf{R}^d} q(t,x,y)S_0(y)\,dy - \sum_{i=1}^{d} \int_0^t \int_{\mathbf{R}^d} \frac{\partial}{\partial y_i} q(t-u,x,y)\,dw_u^i(y)dy +$$

$$+\frac{1}{2}\int_0^{t} \int_{\mathbf{R}^d} \Delta_y q(t-u,x,y)V'(S_u(y))\,dudy,$$

where $q(t,x,y), t > 0, x, y \in \mathbf{R}^d$, is the fundamental solution of the parabolic operator $\frac{\partial}{\partial t} - \frac{1}{2}\Delta\{\gamma + (-\Delta)^m\}$. A similar definition can be given for (1.2).
Let $\chi \in C^\infty(\mathbf{R}^d)$ be a positive function satisfying $\chi(x) = |x|$ for $|x| \geqslant 1$. We introduce Hilbert spaces $\mathbf{H}_r = L^2(\mathbf{R}^d, e^{-r\chi(x)}dx)$, $r > 0$, having norms

$$|S|_r = \left\{ \int_{\mathbf{R}^d} S^2(x) e^{-r\chi(x)}\, dx \right\}^{1/2}, \quad S \in \mathbf{H}_r,$$

and set

$$\mathbf{H}_e = \bigcap_{r>0} \mathbf{H}_r,$$

a countably Hilbertian space. Using Eidel'man's estimate on q, we obtain the following.

THEOREM 2.1. *For every $S_0 \in \mathbf{H}_r$, $r > 0$, the solution of the SPDE* (1.1) *exists uniquely and satisfies*

$$S_t \in C([0,\infty), \mathbf{H}_r) \cap C((0,\infty), C(\mathbf{R}^d)) \quad a.s.$$

Especially, if $S_0 \in \mathbf{H}_e$, then $S_t \in C([0,\infty), \mathbf{H}_e)$ a.s. The similar result holds also for the SPDE (1.2).

Remark. (i) It is convenient to investigate these two equations simultaneously in more general setting .

(ii) The well-posedness of the martingale problems corresponding to both (1.1) and (1.2) can be established.

3. CHARACTERIZATION OF REVERSIBLE MEASURES AS GIBBS STATES

A probability measure μ on $C(\mathbf{R}^d) \cap \mathbf{H}_e$ is called a reversible measure of the SPDE's (1.1) or (1.2) if, for every $0 \leqslant t_1 < t_2 < \cdots < t_n \leqslant t$, $(S_{t_1}, S_{t_2}, \cdots, S_{t_n})$ has the same distribution as $(S_{t-t_n}, S_{t-t_{n-1}}, \cdots, S_{t-t_1})$ on the space $\{C(\mathbf{R}^d) \cap \mathbf{H}_e\}^n$, where $\{S_u, 0 \leqslant u < \infty\}$ is the solution of the SPDE's with initial distribution μ. In this section we discuss only the equation (1.2).

3.1. *Regularity condition*

Let B be the class of all bounded open sets in $\mathbf{R}^d$ having C^∞–boundaries. For a real valued random field (r.f.) $X = \{X(x), x \in \mathbf{R}^d\}$ and $\Gamma = \partial G$ with $G \in B$, we set

$$F_X(h,\psi) \equiv F_X(h,\psi;\Gamma) = \\ = \int_\Gamma \psi(x) X(x + h \cdot n(x))\, d\sigma(x),$$

for every $|h| \leqslant h_0$, $h_0 > 0$, and $\psi \in L^2(\Gamma) \equiv L^2(\Gamma, d\sigma)$, where $n(x)$ is the inner normal unit vector at $x \in \Gamma$ and $d\sigma$ is the volume element on Γ. Suppose a family of m generalized random fields (g.r.f.'s) $Y = \{Y_i(\psi),\ \psi \in L^2(\Gamma)\}_{i=0}^{m-1}$ on Γ is given. The Sobolev space of order $s \in \mathbf{R}$ on Γ is denoted by $H^s(\Gamma)$ as usual.

Definition 3.1. (i) The r.f. X is called *weakly* C^{m-1} *at* Γ if there exists $h_0 > 0$ such that $F_X(\cdot, \psi; \Gamma) \in C^{m-1}(-h_0, h_0)$ a.s. for every $\psi \in L^2(\Gamma)$.

(ii) We say Y satisfies *the regularity condition* $(RC)_\Gamma^s$, $s > 0$, if

$$E[Y_i(\psi)^2] \leqslant \text{const}\, \|\psi\|^2_{H^{-s+i+1/2}(\Gamma)},\quad \psi \in L^2(\Gamma),\ 0 \leqslant i \leqslant m-1.$$

By using Eidel'man's estimate on the fundamental solutions, we can prove the following assertion for the solution S_t of (1.2) satisfying $E[|S_0|_r^2] < \infty$ with some $r > 0$.

PROPOSITION 3.1. (i) *For every $t > 0$ and $G \in \mathcal{V}$, S_t is weakly C^{m-1} at $\Gamma = \partial G$ and therefore*

$$Y_{i,t}(\psi) = \frac{d^i}{dh^i} F_{S_t}(h, \psi; \Gamma)|_{h=0},\quad 0 \leqslant i \leqslant m-1,$$

exists a.s.

(ii) $Y_t = \{Y_{i,t}\}_{i=0}^{m-1}$ *satisfies* $(RC)_\Gamma^s$ *for every* s; $0 < s < m$.

3.2. *Stochastic Dirichlet problem*

Let I_G be a centered Gaussian g.r.f. on $G \in \mathcal{V}$ with covariance

$$E[I_G(\varphi)I_G(\varphi')] = \int_G \varphi \mathcal{A} \varphi' \, dx,\qquad \varphi, \varphi' \in C_0^\infty(G),$$

where $\mathcal{A} = \frac{1}{2}\{\gamma + (-\Delta)^m\}$. Assume that a family of m g.r.f.'s $Y = \{Y_i\}_{i=0}^{m-1}$ on $\Gamma = \partial G$ is given and satisfies the regularity condition $(RC)_\Gamma^s$ for some $s : (d/2) \vee (m - 1/2) < s < m$. With these given data let us consider the following two stochastic Dirichlet problems:

$$\begin{cases} \mathcal{A}X = I_G & \text{in } G, \\ \partial_i^+ X = 0 & \text{on } \Gamma,\quad 0 \leqslant i \leqslant m-1, \end{cases} \tag{3.1}$$

$$\begin{cases} \mathcal{A}X' = 0 & \text{in } G, \\ \partial_i^+ X' = Y_i & \text{on } \Gamma,\quad 0 \leqslant i \leqslant m-1, \end{cases} \tag{3.2}$$

where $\partial_i^+ = \partial^i / \partial n(x)^i$. The mathematical meaning to these equations can be specified in terms of their weak forms and it is possible to verify that the *solutions* of (3.1) and (3.2) satisfy $X \in C(G)$ a.s. and $X' \in C(G)$ a.s. We need the following technical condition (3.3) for the equation (3.2):

$$m > d \quad \text{or} \quad m = d = 1. \tag{3.3}$$

3.3. *Definition of the Gibbs states and the result*
Set $\Omega = C(\mathbf{R}^d)$ and denote $X(x,\omega) = \omega(x)$ for $x \in \mathbf{R}^d$ and $\omega \in \Omega$. Let B be the Borel field of Ω, $B(G) = \sigma\{X(x);\ x \in G\}$ if G is open in $\mathbf{R}^d$ and $B(C) = = \cap\{B(G);\ G \text{ is open}, G \supset C\}$ if C is closed. Assume the condition (3.3).

Definition 3.2. We call a probability measure μ on Ω *an $\mathcal{H}$-Gibbs state* if it satisfies the following two conditions for every $G \in \mathcal{V}$:

(1) The r.f. X distributed by μ is weakly C^{m-1} at $\Gamma = \partial G$ and

$$Y_i(\psi;\Gamma) = \frac{d^i}{dh^i} F_X(h,\psi;\Gamma)|_{h=0}, \quad 0 \leqslant i \leqslant m-1,$$

satisfies the regularity condition $(RC)^s_\Gamma$ for every s; $0 < s < m$.

(2) For every $B(G)$-measurable bounded function Ψ on Ω and $B((G^c)^0)$--measurable bounded function Φ on Ω, μ satisfies the so-called DLR equation:

$$\mathbf{E}^{\mu}[\Phi \cdot \Psi] = \mathbf{E}^{\mu}[\Phi(\omega)\mathbf{E}^{\mu^G_\omega}[\Psi]].$$

In this definition the finite-volume Gibbs state μ^G_ω appears and it is defined as follows. Let ν^G_0 be the probability distribution on the space $C(G)$ of the solution $\{X(x),\ x \in G\}$ of the stochastic Dirichlet problem (3.1) and let $X'(\omega) = = \{X'(x,\omega),\ x \in G\}$ be the solution of another stochastic Dirichlet problem (3.2) with $Y_i = Y_i(\cdot;\Gamma)$ which is determined by the condition (1). Note that X' is $B(\Gamma)$-measurable. Define $\nu^G_\omega(\cdot) = \nu^G_0(\cdot - X'(\omega))$ and then

$$d\mu^G_\omega(X(\cdot)) = Z^{-1}_{\omega,G} \exp\left\{ -\int_G V(X(x))\,dx \right\} d\nu^G_\omega(X(\cdot)),$$

where $Z_{\omega,G}$ is a normalization constant.
Now we can formulate one of the main results in this paper:

THEOREM 3.1. *Assume the condition* (3.3). *Then the following two statements for a probability measure μ on $C(\mathbf{R}^d) \cap \mathbf{H}_e$ are mutually equivalent.*

(1) *μ is a reversible measure of the SPDE* (1.2).

(2) *μ is an $\mathcal{H}$-Gibbs state satisfying $\mathbf{E}^{\mu}[|S|^2_r] < \infty$ for some $r > 0$.*

Concerning the proof we only notice Proposition 3.1 implies that every reversible measure μ fulfills the condition (1) of Definition 3.2.

4. CONSTRUCTION OF REVERSIBLE MEASURES

4.1. *Energy inequalities*
The results in this section are based on some coupling argument and energy inequalities. Let us consider the following SPDE which is a slight generalization of (1.2):

$$dS_t(x) = -\frac{1}{2}\{\gamma + (-\Delta)^m\}S_t(x)\,dt + b(x,S_t)\,dt + dw_t(x). \tag{4.1}$$

Here $b = b(x, S)$ is a bounded function being Lipschitz continuous as a map of $\mathbf{H}_r \to \mathbf{H}_r$ for every $r > 0$. We assume two functions b and $\bar{b}$ of such kind are given and the following bound is fulfilled:

$$\{b(x,S) - \bar{b}(x,\bar{S})\}\{S(x) - \bar{S}(x)\} \leqslant \frac{c_1}{2}\{S(x) - \bar{S}(x)\}^2 + c_2(x)^2, \text{a.e. } x \in \mathbf{R}^d, \quad (4.2)$$

for every $S, \bar{S} \in C(\mathbf{R}^d)$, where c_1; $c_1 < \gamma$ is a constant and $c_2(x)$ is a measurable function on $\mathbf{R}^d$. Let two initial data S and $\bar{S} \in \mathbf{H}_e$ be given. The solutions of the SPDE (4.1) with $S_0 = S$ and (4.1) with $S_0 = \bar{S}$ and $\bar{b}$ instead of b are denoted by S_t and $\bar{S}_t$, respectively. The c.B.m.'s $w_t(x)$ appearing in these two SPDE's are common.

PROPOSITION 4.1. *For every c such that $0 < c < \gamma - c_1$,*

$$|S_t - \bar{S}_t|_r^2 \leqslant e^{-ct}|S_0 - \bar{S}_0|_r^2 + \frac{2}{c}|c_2(\cdot)|_r^2, \quad t \geqslant 0, \text{ a.s.}$$

if $r > 0$ is sufficiently small.

4.2. *Stationary measures*

A probability measure μ on $C(\mathbf{R}^d) \cap \mathbf{H}_e$ is called *a stationary measure of* (4.1) if, for every $0 \leqslant t_1 < t_2 < \cdots < t_n$ and $t > 0$, $(S_{t_1+t}, S_{t_2+t}, \cdots, S_{t_n+t})$ has the same distribution as $(S_{t_1}, S_{t_2}, \cdots, S_{t_n})$, where $\{S_u; 0 \leqslant u < \infty\}$ is the solution of (4.1) with initial distribution μ. Take $\bar{b} \equiv 0$ in Proposition 4.1. Then, since $\bar{S}_t$ is a Gaussian process with the unique reversible probability measure ν having the characteristic functional

$$\mathbf{E}^{\nu}[e^{\sqrt{-1}<S,\varphi>}] = \exp\left\{-\frac{1}{2}\|\{\gamma + (-\Delta)^m\}^{-1/2}\varphi\|_{L^2(\mathbf{R}^d)}^2\right\}, \quad \varphi \in \mathcal{S}(\mathbf{R}^d), \quad (4.3)$$

we obtain

$$\mathbf{E}[|S_t|_r^2] \leqslant \text{const}\{1 + |S_0|_r^2\}, \ t \geqslant 0,$$

for the solution S_t of (4.1) starting from $S_0 \in \mathbf{H}_\epsilon$, where const is independent of t and S_0. The following result is an easy consequence of this estimate.

THEOREM 4.1. *The SPDE* (4.1) *and therefore* (1.2) *has at least one stationary probability measure.*

Now we introduce another condition on b:

$$\{b(x,S) - b(x,\bar{S})\}\{S(x) - \bar{S}(x)\} \leqslant \frac{c_1}{2}\{S(x) - \bar{S}(x)\}^2, \text{ a.e. } x \in \mathbf{R}^d, \ c_1 < \gamma. \quad (4.4)$$

Apply Proposition 4.1 by taking $b = \bar{b}$ to obtain

THEOREM 4.2. *Under the assumption* (4.4), *the stationary probability measure of the SPDE* (4.1) *is unique.*

Let $B_b(\mathbf{R}^d)$ and $B_1(\mathbf{R}^d)$ be the class of all bounded measurable functions on $\mathbf{R}^d$ and its subclass consisting of all g satisfying $0 \leqslant g(x) \leqslant 1$, a.e. $x \in \mathbf{R}^d$, respectively. For $g \in B_1(\mathbf{R}^d)$ and $\lambda \in B_b(\mathbf{R}^d)$, the function $b_{g,\lambda}$ is defined by

$$b_{g,\lambda}(x,S) = -\frac{1}{2}g(x)V'(S(x)) + \frac{1}{2}\lambda(x),\ x \in \mathbf{R}^d,\ S \in \mathbf{H}_e.$$

If the self-potential U is strictly convex, i.e.

$$\gamma_0 \equiv \inf_{\sigma \in \mathbf{R}} U''(\sigma) > 0, \tag{4.5}$$

then the condition (4.4) is fulfilled by $b = b_{g,\lambda}$ for every $g \in B_1(\mathbf{R}^d)$, $\lambda \in B_b(\mathbf{R}^d)$ and c_1; $\gamma - \gamma_0 < c_1 \cdot < \gamma$, so that the SPDE (4.1) with $b = b_{g,\lambda}$ has a unique stationary measure which will be denoted by $\mu_{g,\lambda}$. Note that the SPDE (1.2) is just the SPDE (4.1) with $b = b_{1,0}$. Now we define the so-called Vasershtein metric between two probability measures μ_1 and μ_2 on $\mathbf{H}_r$, $r > 0$, satisfying $\mathbf{E}^{\mu_i}[|S|_r^2] < \infty$, $i = 1,2$, by

$$d_r(\mu_1,\mu_2) = \inf_m \sqrt{\mathbf{E}^m[|S_1 - S_2|_r^2]},$$

where the infimum is taken over all probability measures m on $\mathbf{H}_r \times \mathbf{H}_r$ such that $m \circ \pi_i^{-1} = \mu_i$, $i = 1,2$; the maps π_i are projections defined by $\pi_i : \mathbf{H}_r \times \mathbf{H}_r \ni$ $\ni (S_1,S_2) \mapsto S_i \in \mathbf{H}_r$.
Another consequence of Proposition 4.1 is the following.

PROPOSITION 4.2. *Assume* (4.5). *Then we have*

$$d_r(\mu_{g,\lambda},\mu_{h,\xi}) \leqslant \text{const}\sqrt{\int_{\{x;g\neq h\}} e^{-r\chi(x)}\,dx + |\lambda - \xi|_r^2}$$

for every $g,h \in B_1(\mathbf{R}^d)$, $\lambda,\xi \in B_b(\mathbf{R}^d)$ *and sufficiently small* r. *The const can be taken independently of* g,λ,h,ξ *and* r.

This proposition implies that, if U satisfies (4.5), $\mu_{g,\lambda}$ can be defined uniquely for every $g \in B_1(\mathbf{R}^d)$ and $\lambda \in \mathbf{H}_e$ by taking the limit.

4.3. *Reversibility of* $\mu_{g,\lambda}$

Now we discuss the TDGL eq. of both conservative type (1.1) and non-conservative type (1.2). We consider the following SPDE as a generalization of the former type of equation for $g \in B_1(\mathbf{R}^d)$ and $\lambda \in B_b(\mathbf{R}^d)$:

$$\begin{aligned} dS_t(x) &= \frac{1}{2}\Delta\{\gamma + (-\Delta)^m\}S_t(x)\,dt + \\ &+ \frac{1}{2}\Delta\{g(x)V'(S(x)) - \lambda(x)\}dt + d\{\operatorname{div}\mathbf{w}_t(x)\}. \end{aligned} \tag{4.6}$$

Let $B_{1,0}(\mathbf{R}^d)$ and $B_{b,0}(\mathbf{R}^d)$ be the subclasses of $B_1(\mathbf{R}^d)$ respectively $B_b(\mathbf{R}^d)$ consisting of such functions that have compact supports. For $g \in B_{1,0}(\mathbf{R}^d)$, $\lambda \in B_{b,0}(\mathbf{R}^d)$ and $\theta \in \mathbf{H}_e$, we define a probability measure $\mu_{g,\lambda;\theta}$ on $C(\mathbf{R}^d) \cap \mathbf{H}_e$ by

$$d\mu_{g,\lambda;\theta}(S) = Z_{g,\lambda;\theta}^{-1} \exp\left[-\int_{\mathbf{R}^d} \{g(x)V(S(x)) - \lambda(x)S(x)\}\, dx \right] d\nu_\theta(S),$$

where $\nu_\theta(\cdot) = \nu(\cdot - \theta)$ and $Z_{g,\lambda;\theta}$ is a normalization constant. Then the following assertion can be verified. The proof begins with the investigation of the case $g = \lambda = 0$ and afterwards we use Cameron—Martin—Girsanov's formula for the SPDE's.

PROPOSITION 4.3. *Assume $g \in B_{1,0}(\mathbf{R}^d)$ and $\lambda \in B_{b,0}(\mathbf{R}^d)$.*

(i) *$\mu_{g,\lambda;0}$ is a reversible measure of the SPDE (4.1) with $b = b_{g,\lambda}$.*

(ii) *$\{\mu_{g,\lambda;\theta};\ \theta \in \mathbf{H}_e \cap C^\infty(\mathbf{R}^d),\ \Delta\theta = 0\}$ is a family of reversible measures of the SPDE (4.6).*

Especially, if (4.5) is fulfilled, then $\mu_{g,\lambda} = \mu_{g,\lambda;0}$ for $g \in B_{1,0}(\mathbf{R}^d)$ and $\lambda \in B_{b,0}(\mathbf{R}^d)$ and moreover it is not difficult to see $\mu_{g,\lambda;\theta} = \mu_{g,\lambda+2A\theta}$ if θ is smooth. These observations combined with the procedure of taking the limit (thermodynamic limit) lead us to

THEOREM 4.3. *Assume the strict-convexity (4.5) of U. Then,*

(i) *$\mu_{1,0}$ is the unique reversible measure of the SPDE (1.2).*

(ii) *The convex combination of $\{\mu_{1,\lambda};\ \lambda \in \mathbf{H}_e \cap C^\infty(\mathbf{R}^d),\ \Delta\lambda = 0\}$ makes a family of reversible probability measures of the SPDE (1.1).*

We remark that the assumption (4.5) combined with the technical condition (3.3) implies the uniqueness of $\mathcal{H}$-Gibbs states satisfying the natural integrability condition.

5. THE HYDRODYNAMIC LIMIT

Let us introduce the hydrodynamic space–time scaling $x \mapsto x/\varepsilon$ and $t \mapsto t/\varepsilon^2$ for the TDGL eq. (1.1) of conservative type and investigate the asymptotic behavior of solutions as $\varepsilon \downarrow 0$. The scaled process $S_t^\varepsilon(x) = S_{t/\varepsilon^2}(x/\varepsilon)$ satisfies

$$dS_t^\varepsilon(x) = \frac{1}{2}\Delta\{\gamma + (-\varepsilon^2\Delta)^m\}S_t^\varepsilon(x)dt + \frac{1}{2}\Delta\{V'(S_t^\varepsilon(x))\}dt + \varepsilon^{d/2}d\{\operatorname{div} \mathbf{w}_t(x)\}. \tag{5.1}$$

The result can be formulated, assuming the condition (4.5) on U, as the law of large numbers.

THEOREM 5.1. *Suppose the initial value $S_0^\varepsilon = S_0 \in C_b^3(\mathbf{R}^d)$ is independent of ε. Then $\langle S_t^\varepsilon, \varphi\rangle$ converges to $\langle \rho_t, \varphi\rangle$ in probability as $\varepsilon \downarrow 0$ for every $t > 0$ and $\varphi \in C_0^\infty(\mathbf{R}^d)$, where we denote*

$$\langle S_t^\varepsilon, \varphi\rangle = \int_{\mathbf{R}^d} S_t^\varepsilon(x)\varphi(x)\, dx$$

and so on. The non-random limit $\rho_t(x)$ is the unique solution of the following nonlinear PDE:

$$\frac{\partial}{\partial t}\rho_t(x) = \frac{1}{2}\text{div}\{D(\rho_t(x))\nabla\rho_t(x)\}\,,\; x \in \mathbf{R}^d\,;\; \rho_0 = S_0. \tag{5.2}$$

In order to explain the diffusion coefficient $D(\rho)$, $\rho \in \mathbf{R}$, appearing in (5.2), we define the pressure (specific free energy):

$$P(\lambda) = \lim_{n\to\infty}\frac{1}{|\Lambda_n|}\log Z_{\Lambda_n}(\lambda),\;\Lambda_n = (-n,n)^d,\;\lambda\in\mathbf{R}, \tag{5.3}$$

where

$$Z_\Lambda(\lambda) = \mathbf{E}^\nu \exp\left\{\int_\Lambda (-V(S(x)) + \lambda S(x))\,dx\right\}.$$

Recall (4.3) for the definition of ν. The variational formula for $P(\lambda)$ is known:

$$P(\lambda) = \sup_{\mu\in P_\theta(C(\mathbf{R}^d))}\{\mathbf{E}^\mu(-V(S(0)) + \lambda S(0)) - H(\mu;\nu)\}\,.$$

Here $P_\theta(C(\mathbf{R}^d))$ denotes the class of all translation-invariant probability measures on $C(\mathbf{R}^d)$ and $H(\mu;\nu)$ is the entropy:

$$H(\mu;\nu) = \lim_{n\to\infty}\frac{1}{|\Lambda_n|}\mathbf{E}^\mu\left[\log\frac{d\mu|_{B(\Lambda_n)}}{d\nu|_{B(\Lambda_n)}}\right]$$

if the limit in the right hand side exists. We simply write μ_λ, $\lambda \in \mathbf{R}$, for the measure $\mu_{1,\lambda(\cdot)}$ with $\lambda(\cdot) \equiv \lambda$ (const) determined in Sect. 4. Note that these μ_λ's are translation-invariant reversible measures of (1.1). After showing some uniform decay property of correlations of functionals with respect to $\mu_{g,\lambda}$, it can be verified that P is smooth in λ, $P'(\lambda) = \rho(\lambda) = \mathbf{E}^{\mu_\lambda}[S(0)]$ the mean spin function and $0 < c_1 \leqslant P''(\lambda) \leqslant c_2 < \infty$, $\lambda \in R$, with some constants c_1 and c_2. The diffusion coefficient is now given by $D(\rho) = 1/P''(\lambda(\rho))$, $\lambda(\rho)$ being the inverse function of $\rho = \rho(\lambda)$. The proof of Theorem 5.1 is carried out by employing the similar method used in (Fritz, 1989) and (Funaki, 1989).

Acknowledgment.
The author is indebted to Professor J. Fritz for stimulating discussions.

REFERENCES

Fritz, J. (1989). On the hydrodynamic limit of a Ginzburg–Landau lattice model. The law of large numbers in arbitrary dimensions. *Probab. Theory Rel. Fields* **81**, 291–318.

Funaki, J. (1989). Derivation of the hydrodynamical equation for one–dimensional Ginzburg—Landau model. *Probab. Theory Rel. Fields* **82**; 39–93.

Prob. Theory and Math. Stat., Vol. 1, pp. 391–400
B.Grigelionis *et al.* (Eds.)
1990 VSP/Mokslas

ALMOST EVERYWHERE CONVERGENCE OF SUBSEQUENCES OF POWERS

V.F. GAPOSHKIN and J.M. ROSENBLATT*
Moscow Institute of Transportation Engineering,
Obrazcova Street, 15, 103055 Moscow, USSR
Mathematics Department, Ohio State University,
Columbus, Ohio 43210, USA

ABSTRACT

In several recent papers by Bellow, Jones and Rosenblatt it was shown that for some classes of contractions T on $L_2(X)$ there exists an increasing sequence (n_m) such that for all $f \in L_2(X)$, $\lim\limits_{m\to\infty} T^{n_m} f(x)$ exists a.e. In this article the rate of growth of (n_m) is considered.
In particular, let μ be a strictly aperiodic probability measure on $\mathbf{Z}$ with finite moment $m_\sigma(\mu)$ for some $\sigma > 0$, and let $\mu f(x) = \sum\limits_{-\infty}^{\infty} \mu(k) f(\tau^k x)$, where τ is an invertible measure preserving transformation on X. Then for all $f \in L_2(X)$ $\lim\limits_{m\to\infty} \mu^{n_m} f(x)$ exists a.e., if (n_m) satisfies $n_{m+1} \geqslant n_m^\alpha$ for some fixed $\alpha > 1$.

Let (X, β, m) be a probability space and let T be a contraction on the Lebesgue space $L_2(X, \beta, m)$. In several recent papers (Bellow et al., 1988, 1989a, 1989b, 1990), one of the problems considered is the almost everywhere convergence of subsequences of $(T^n f)$. The problem is to find an increasing sequence $(n_m, m = 1, 2, 3, \ldots)$ such that for all $f \in L_2(X)$, $\lim\limits_{m\to\infty} T^{n_m} f(x)$ exists a.e. In these papers various positive and negative results about specific subsequence and operators are proved. These types of subsequence convergence results, where the full sequence (T^n) may not have such properties, are new discoveries in operator theory and ergodic theory. In this article, we will illustrate such theorems and consider the behavior of (n_m), especially its rate of growth, necessary to obtain such conclusions.
First, we need the basic property of the spectrum of T, denoted $\mathrm{sp}(T)$, which allows such results.

1. *Definition.* A contraction T on a Hilbert space H is *strictly aperiodic* if $\mathrm{sp}(T) \cap \{\lambda \in \mathbf{C} : |\lambda| = 1\} \subset \{1\}$.

2. THEOREM. *If T is a strictly aperiodic normal contraction on a Hilbert*

* Partially supported by NSF Grant 8802126.

space H, then for all $f \in H, f^* = \lim_{n\to\infty} T^n f$ exists in norm and f^* is a T-invariant function.

Proof. First, H is the orthogonal direct sum of the T-invariant functions and the closed linear span of functions f_1 of the form $f_1 = f - Tf$, $f \in H$. This is clear once one knows that $f \in H$ is T-invariant if and only if f is T^*-invariant. See (Riesz and Sz.-Nagy, 1941–1943).
Therefore, to prove the theorem, it suffices to show $\lim_{n\to\infty} \|T^n(I-T)f\| = 0$ for all $f \in H$, because (T^n) is a uniformly bounded sequence of operators. Indeed, $\lim_{n\to\infty} \|T^n(I-T)\| = 0$. To see this, we apply the spectral theorem to the normal operator T. This shows

$$\|T^n(I-T)\| = \sup_{\lambda \in \mathrm{sp}(T)} |\lambda^n(1-\lambda)|.$$

But for any $\delta > 0$, there is an $\epsilon, 1 > \epsilon > 0$, such that $|\lambda| \leqslant 1-\epsilon$ if $|\lambda - 1| \geqslant \delta$ and $\lambda \in \mathrm{sp}(T)$, because T is strictly aperiodic. Thus,

$$\sup\{|\lambda^n(1-\lambda)| : \lambda \in \mathrm{sp}(T),\ |\lambda - 1| \geqslant \delta\} \leqslant 2(1-\epsilon)^n.$$

So if $\lambda \in \mathrm{sp}(T)$, $|\lambda - 1| \geqslant \delta$, and $n \geqslant \log(\delta/2)/\log(1-\epsilon)$, then $|\lambda^n(1-\lambda)| \leqslant \delta$. But also if $|\lambda - 1| \leqslant \delta$, then $|\lambda^n(1-\lambda)| \leqslant \delta$ if $|\lambda| \leqslant 1$, what is more if $\lambda \in \mathrm{sp}(T)$. Therefore, for any $\delta > 0$, there is $N \geqslant 1$ such that for $n \geqslant N$, if $\lambda \in \mathrm{sp}(T)$, then $|\lambda^n(1-\lambda)| \leqslant \delta$.

Remark. The normality of T is important in the calculation above, as it will be in the ones to follow. This will leave unresolved whether these theorems remain true without the normality hypothesis. For example, if T is a strictly aperiodic contraction, then is $\lim_{n\to\infty} \|T^n(I-T)\| = 0$? This is indeed the case in general. See (Katznelson and Tzafriri, 1986) for this and other facts about $0-1$ laws. As a result, Theorem 2 is also true without the normality hypothesis.
It follows from the above that if T is a strictly aperiodic contraction on $L_2(X)$, then there is a subsequence (n_m) such that $\sum_{m=1}^{\infty} \|T^{n_m}(I-T)\|^2 < \infty$. Then for all $f \in L_2(X)$, $\sum_{m=1}^{\infty} \|T^{n_m}(f - Tf)\|^2 < \infty$ and so $\lim_{m\to\infty} T^{n_m}(f-Tf)(x) = 0$ a.e. It follows that $\lim_{n\to\infty} T^{n_m} f(x)$ exists a.e. for all $f = f_1 + f_2 - Tf_2$ where $f_1, f_2 \in L_2(X)$ and f_1 is T-invariant. Therefore, to find a subsequence (T^{n_m}) such that the same is true for all $f \in L_2(X)$ would only require proving $\sup_{m\geqslant 1} |T^{n_m} f(x)| < \infty$ a.e. for all $f \in L_2(X)$. However, it turns out that the existence of such a subsequence is linked to the a.e. convergence of the Cesaro averages $\frac{1}{N}\sum_{k=1}^{N} T^k f(x)$. This is the reason that the following result is actually a better formulation.

3. *Definition.* Let T be a contraction on a Hilbert space H. For $\delta > 0$, let $\omega_\delta(T) = \sup\{|\lambda| : \lambda \in \mathrm{sp}(T),\ |\lambda - 1| \geqslant \delta\}$. When T is fixed, $\omega_\delta(T)$ is denoted by ω_δ.

In general, we should assume in the sequel that $\lim_{\delta\to 0+} \omega_\delta = 1$ i.e. $1 \in \mathrm{sp}(T)$. Otherwise, the spectral radius of T is less than 1 and so there is $\rho < 1$ and $N \geqslant 1$ such that for $n \geqslant N, \|T^n\| \leqslant \rho^n$. Hence, for all $f \in L_2(X)$, $\sum_{n=1}^{\infty} \|T^n f\|^2 < \infty$ and $\lim_{n\to\infty} T^n f(x) = 0$ a.e. On the other hand, the assumption that T is strictly aperiodic means exactly that $\omega_\delta < 1$ for any $\delta > 0$. This shows that a subsequence (n_m) as in the next theorem always exists.

4. THEOREM. *Let T be a strictly aperiodic normal contraction on $L_2(X)$. Let (n_m) be an increasing sequence such that for some constant C, if we let $\delta_{m+1} = C\left(\sum_{k=1}^{m} n_k^2\right)^{-1/2}$ for $m \geqslant 1$, then*

$$\Sigma_1 = \sum_{m=1}^{\infty} \frac{1}{n_{m+1}^2 \delta_{m+1}^2} < \infty \tag{1}$$

and

$$\Sigma_2 = \sum_{m=1}^{\infty} \omega_{\delta_{m+1}}^{2n_{m+1}} < \infty. \tag{2}$$

Then for all $f \in L_2(X)$, the deviation operator $D_n f = T^n f - \frac{1}{n}\sum_{k=1}^{n} T^k f$, $f \in$ $\in L_2(X)$, satisfies

$$\sum_{m=1}^{\infty} \|D_{n_m} f\|^2 < \infty$$

and $\lim_{m\to\infty} D_{n_m} f(x) = 0$ a.e. for all $f \in L_2(X)$.

Proof. We need the estimate

$$d_n(\lambda) \equiv \left|\lambda^n - \frac{1}{n}\sum_{k=1}^{n} \lambda^k\right| \leqslant \frac{n-1}{2}|\lambda - 1|.$$

Indeed, if $n \geqslant 2$,

$$\left|\lambda^n - \frac{1}{n}\sum_{k=1}^{n}\lambda^k\right| = \frac{1}{n}\left|\sum_{k=1}^{n-1}(\lambda^n - \lambda^k)\right| \leqslant \frac{1}{n}\sum_{k=1}^{n-1}|\lambda^k||\lambda - 1|\left|\sum_{j=0}^{n-k-1}\lambda^j\right| \leqslant$$

$$\leqslant \frac{|\lambda - 1|}{n}\sum_{k=1}^{n-1}(n-k) = \frac{|\lambda - 1|}{n}\cdot\frac{n(n-1)}{2} = \frac{n-1}{2}|\lambda - 1|.$$

Now for $f \in L_2(X)$, the normality of T allows one to use the spectral theorem to choose a positive Borel measure v_f on $\mathrm{sp}(T)$ such that $\|v_f\|_1 = \|f\|_2^2$ and for all polynomials $P(T)$,

$$\|P(T)\|_2^2 = \int_{\mathrm{sp}(T)} |P(\lambda)|^2 dv_f(\lambda).$$

So for any $M \geqslant 1$,

$$\sum_{m=1}^{M} \|D_{n_m} f\|_2^2 = \int_{\mathrm{sp}(T)} \sum_{m=1}^{M} |d_{n_m}(\lambda)|^2 dv_f(\lambda).$$

Thus, to prove the theorem requires only showing that there exists a constant K such that $\sum_{m=1}^{\infty} |d_{n_m}(\lambda)|^2 \leqslant K^2$ for all $\lambda \in \mathrm{sp}(T)$.

Fix $M \geqslant 1$. Then for any λ,

$$\sum_{m=1}^{M+1} |d_{n_m}(\lambda)|^2 \leqslant \sum_{m=1}^{M} |d_{n_m}(\lambda)|^2 + |d_{n_{M+1}}(\lambda)|^2. \tag{3}$$

We can always make the trivial estimate $|d_n(\lambda)|^2 \leqslant 4$ if $|\lambda| \leqslant 1$. So if $|\lambda| \leqslant 1$ and $|\lambda - 1| \leqslant \delta_{M+1}$, we use (3) to get

$$\sum_{m=1}^{M+1} |d_{n_m}(\lambda)|^2 \leqslant 4 + \sum_{m=1}^{M} |d_{n_m}(\lambda)|^2 \leqslant 4 + \sum_{m=1}^{M} \frac{(n_m - 1)^2}{4} |\lambda - 1|^2 \leqslant$$
$$\leqslant 4 + \frac{1}{4}\delta_{M+1}^2 \sum_{m=1}^{M} n_m^2 \leqslant 4 + \frac{1}{4}C^2.$$

On the other hand, (3) says for $\lambda \in \mathrm{sp}(T)$, $|\lambda - 1| \geqslant \delta_{M+1}$,

$$\sum_{m=1}^{M+1} |d_{n_m}(\lambda)|^2 \leqslant \sum_{m=1}^{M} |d_{n_m}(\lambda)|^2 + 2|\lambda^{n_{M+1}}|^2 + \frac{2}{n_{M+1}} \left| \sum_{k=1}^{n_{M+1}} \lambda^k \right|^2 \leqslant$$
$$\leqslant \sum_{m=1}^{M} |d_{n_m}(\lambda)|^2 + 2\omega_{\delta_{M+1}}^{2n_{M+1}} + \frac{8}{n_{M+1}^2 |\lambda - 1|^2} \leqslant$$
$$\leqslant \sum_{m=1}^{M} |d_{n_m}(\lambda)|^2 + 2\omega_{\delta_{M+1}}^{2n_{M+1}} + \frac{8}{n_{M+1}^2 \delta_{M+1}^2}.$$

Since $\delta_1 \geqslant \delta_2 \geqslant \cdots \geqslant \delta_{M+1}$, by induction it follows that for $\lambda \in \mathrm{sp}(T)$,

$$\sum_{m=1}^{M+1} |d_{n_M}(\lambda)|^2 \leqslant |d_{n_1}(\lambda)|^2 + \frac{1}{4}C^2 + 4 + \sum_{m=2}^{M+1} (2\omega_{\delta_m}^{2n_m} + \frac{8}{n_m^2 \delta_m^2}) \leqslant$$
$$\leqslant 8 + \frac{1}{4}C^2 + \sum_{m=2}^{M+1} (2\omega_{\delta_m}^{2n_m} + \frac{8}{n_m^2 \delta_m^2}) \leqslant 8 + \frac{1}{4}C^2 + 2\Sigma_1 + 8\Sigma_2.$$

Since $M \geqslant 1$ was arbitrary, this gives the results with $K^2 = 8 + \frac{1}{4}C^2 + 2\Sigma_1 + 8\Sigma_2$.

Question. Can we show the existence of a sequence (n_m) with $\sum_{m=1}^{\infty} \|D_{n_m} f\|_2^2 <$ $< \infty$ for all $f \in L_2(X)$ as above without the normality hypothesis?

This result shows that if T also satisfies the ergodic theorem on $L_2(X)$, then $\lim_{m\to\infty} T^{n_m}f(x)$ exists a.e. for all $f \in L_2(X)$. Exact criteria for T to satisfy the ergodic theorem in terms of the behavior of the spectral measure of T near 1 are given in (Gaposhkin, 1981). Perhaps the best known examples are when T is also an $L_1 - L_\infty$ contraction. This gives the following general result.

5. COROLLARY. *If T is an $L_1 - L_\infty$ contraction which is a strictly aperiodic normal operator on $L_2(X)$, then for an increasing sequence (n_m) satisfying (1) and (2), $\lim_{m\to\infty} T^{n_m}f(x)$ exists a.e. for all $f \in L_2(X)$.*

Remark. In (Bellow et al., 1988) the question is raised whether the result in Corollary 5 holds for (some) (n_m) simultaneously on all $L_p, 1 \leqslant p \leqslant \infty$ or $1 < p \leqslant \infty$. In some special cases that are discussed below, this is possible.
Various examples in (Bellow et al., 1988, 1989a, 1989b, 1990; Gaposhkin, 1984) have shown that the rate of which $\omega_\delta \to 1$ as $\delta \to 0^+$ determines the rate of growth of (n_m) necessary for $(T^{n_m}f)$ to satisfy an a.e. convergence theorem. This is also reflected in Theorem 4 since the value of $\omega_{\delta_{M+1}}$ determines how large n_{m+1} must be for Σ_2 to converge. For instance, we have this typical result.

6. COROLLARY. *Let T be a strictly aperiodic normal contraction on $L_2(X)$ such that for some $\rho > 0$ and $\delta_0 > 0$, $\omega_\delta \leqslant 1 - \delta^\rho$ for all δ, $0 \leqslant \delta \leqslant \delta_0$. Then for any sequence (n_m) such that for some $\alpha > 1$, $n_{m+1} \geqslant n_m^\alpha$ for all $m \geqslant M_0$, we have $\lim_{m\to\infty} D_{n_m}f(x) = 0$ a.e. for all $f \in L_2(X)$.*

Proof. First, the norm convergence of $(T^n f)$ for $f \in L_2(X)$ identifies the limit of $(T^{n_m}f)$, so if we partition (n_m) into finitely many subsequences and prove a.e. convergence along each of these subsequences, then the full sequence $(T^{n_m}f)$ converges a.e. too. But if $J \geqslant 1$ is fixed, consider the subsequences $S_j = (n_{m(s)} : m(s) = j + sJ,\ s = 1, 2, 3, \ldots)$ for each $j = 0, \ldots, J-1$. We have $n_{m(s+1)} \geqslant n_{m(s)}^{\alpha^J}$ for all sufficiently large s. These observations show that we may assume that $\rho \geqslant 1, n_1 \geqslant 2$ and $n_{m+1} \geqslant n_m^{\rho+1}$ for all $m \geqslant 1$.
Now $\omega_\delta \leqslant 1 - \delta^\rho$ shows that $\omega_{\delta_m}^{2n_m} \leqslant (1 - \delta_m^\rho)^{2n_m} \leqslant e^{-2n_m\delta_m^\rho}$. Thus, (2) holds if $n_m\delta_m^\rho$ eventually increases linearly to ∞ as $m \to \infty$. This would also be good enough for (1) to hold. However, here with $C = 1$,

$$\delta_m = \left(\sum_{k=1}^{m-1} n_k^2\right)^{-1/2} \geqslant (m-1)^{-1/2} n_{m-1}^{-1}.$$

So $n_m\delta_m^\rho \geqslant n_m/((m-1)^{\rho/2}n_{m-1}^\rho) \geqslant n_{m-1}/(m-1)^{\rho/2}$ because $n_m \geqslant n_{m-1}^{\rho+1}$. However, $n_m \geqslant n_{m-1}^{\rho+1} \geqslant n_{m-1}^2, n_1 \geqslant 2$, shows that $n_m \geqslant 2^{2^{m-1}}$. Thus, $n_m\delta_m^\rho \geqslant$ $\geqslant 2^{2^m}/(m-1)^{\rho/2}$ which certainly increases linearly to ∞ as $m \to \infty$.
Before giving more applications of Theorem 4, with some specific estimates of the growth needed in (n_m), it would be good to comment generally on the full sequence (T^n) in case T is an L_2 contraction or an $L_1 - L_\infty$ contraction. Gaposhkin (Gaposhkin, 1984) showed using the general principles from (Gaposhkin, 1977),

that if T is a strictly aperiodic normal operator and the spectrum of T satisfies the Stolz condition, $\sup\{|\lambda-1|/(1-|\lambda|) : \lambda \in \mathrm{sp}(T), \lambda \neq 1\} < \infty$, then for all $f \in L_2(X)$, $\lim_{n\to\infty} D_n f(x) = 0$ a.e. So if T also satisfies the ergodic theorem, then $\lim_{n\to\infty} T^n f(x)$ exists a.e. for all $f \in L_2(X)$. This generalizes Stein's Theorem for symmetric operators. Also, Bellow, Jones and Rosenblatt (1988) generalized Stein's method and showed that if T is an $L_1 - L_\infty$ contraction which is a strictly aperiodic normal operator on L_2, whose spectrum satisfies a Stolz condition, then for all $f \in L_p(X), 1 < p \leqslant \infty$, $\lim_{n\to\infty} T^n f(x)$ exists a.e. The question of extending this result to $L_1(X)$ is a difficult one; see (Oseledets, 1965) for an interesting special case.

One application of these theorems on (T^n) is the following. Let μ be a probability measure on $\mathbb{Z}$ and let T be a contraction on $L_2(X)$. Let $\mu(T)$ denote $\Sigma\mu(k)T^k$ where we assume T is invertible if $\mathrm{supp}(\mu)$ contains negative integers. We say μ is ***strictly aperiodic*** if $\mathrm{supp}(\mu)$ is not contained in a proper coset in $\mathbb{Z}$ (cf. (Bellow, 1989a; Katznelson and Tzafriri, 1986)). If μ is strictly aperiodic, then $|\hat{\mu}(\lambda)| < 1$ for $|\lambda| = 1$, $\lambda \neq 1$. Hence, if T is a normal contraction on $L_2(X)$ and μ is strictly aperiodic, the operator $\mu(T)$ is a strictly aperiodic normal contraction on $L_2(T)$. Also, $\omega_\delta(\mu(T))$ is related to $\sup\{|\hat{\mu}(\lambda)| : |\lambda| = 1, |\lambda - 1| \geqslant \delta\}$ and the contact of $\mathrm{sp}(\mu(T))$ with $\{\lambda : |\lambda| = 1\}$ is determined by the graph of $\hat{\mu}$ on $\{\lambda : |\lambda| = 1\}$. For these reasons, Theorem 4 gives this result.

7. COROLLARY. *Let μ be a strictly aperiodic probability measure on $\mathbb{Z}$ and let T be a normal contraction on $L_2(X)$ such that $\mu(T)$ is well-defined. Then there is a sequence (n_m) determined by μ alone such that for all $f \in L_2(X)$,*

$$\lim_{m\to\infty} \mu(T)^{n_m} f(x) - \frac{1}{n_m}\sum_{k=1}^{n_m} \mu(T)^k f(x) = 0 \quad a.e.$$

If in addition $\{\hat{\mu}(\lambda) : |\lambda| = 1\}$ satisfies a Stolz condition, we could improve the above by taking (n_m) to the full sequence $(1, 2, 3, \ldots)$. For example if μ has finite second moment $m_2(\mu) = \sum_{k=-\infty}^{\infty} k^2\mu(k)$ and the expectation of $\mu, \mathbf{E}(\mu) = = \sum_{k=-\infty}^{\infty} k\mu(k)$, is zero, then this Stolz condition holds. See (Bellow et al., 1988, 1989b). In any case, if $\mathrm{supp}(\mu)$ is not a single point, it is easy to show for some $C > 0, |\hat{\mu}(\lambda)| \leqslant 1 - C|\lambda - 1|^2$ as $\lambda \to 1, |\lambda| = 1$. See (Petrov, 1975). Therefore, Corollary 6 gives this explicit result.

8. COROLLARY. *Let μ be a strictly aperiodic probability measure on $\mathbb{Z}$ with finite moment $m_\sigma(\mu) = \sum_{k=-\infty}^{\infty} |k|^\sigma\mu(k)$ for some $\sigma > 0$. Assume that T is a normal contraction on $L_2(X)$ such that $\mu(T)$ is well-defined. Fix any $\alpha > 1$ and a sequence (n_m) with $n_{m+1} \geqslant n_m^\alpha$ for large m. Then for all $f \in L_2(X)$,*

$$\lim_{m\to\infty} \mu(T)^{n_m} f(x) - \frac{1}{n_m}\sum_{k=1}^{n_m} \mu(T)^k f(x) = 0 \quad a.e.$$

Proof. Without loss of generality, $m_\sigma(\mu) < \infty$ for some σ, $0 < \sigma < 1$. Then $\mu(e^{i\theta})$ is Lipschitz of order σ, and so for some constant $K < \infty$, $|\hat{\mu}(e^{i\theta})-1| \leqslant K\theta^\sigma$ for all θ. The result mentioned above in (Petrov, 1975) says that $|\hat{\mu}(e^{i\theta})| \leqslant 1 - C\theta^2$ for all θ if C is a suitable constant, because μ is also strictly aperiodic. Thus, if $|\hat{\mu}(e^{i\theta}) - 1| \geqslant \delta$, then $|\hat{\mu}(e^{i\theta})| \leqslant 1 - C\theta^2 \leqslant 1 - C(\delta/K)^{2/\sigma}$. So for any $\rho > 2/\sigma$, $|\hat{\mu}(e^{i\theta})| \leqslant 1 - \delta^\rho$ when $|\hat{\mu}(e^{-i\theta}) - 1| \geqslant \delta$ and δ is sufficiently small. Therefore, $\omega_\delta(\mu(T))$ satisfies the condition in Corollary 6.
If we choose T to be induced by an invertible measure–preserving transformation, then $\mu(T)$ satisfies the ergodic theorem. Therefore we have

9. Corollary. *Let μ be a strictly aperiodic probability measure on $\mathbb{Z}$ with finite moment $m_\sigma(\mu)$ for some $\sigma > 0$. Assume that τ is an invertible measure preserving transformation on X and let $\mu f(x) = \sum_{k=-\infty}^{\infty} \mu(k)f(\tau^k x)$ for $x \in X$. Then if (n_m) satisfies $n_{m+1} \geqslant n_m^\alpha$ for some fixed $\alpha > 1$ and m large, we have for all $f \in L_2(X)$, $\lim_{m\to\infty} \mu^{n_m} f(x)$ exists a.e.*

Remark. a) Using the arguments in (Bellow et al., 1989a), one can show here that for all $f \in L_p(X)$, $1 < p \leqslant \infty$, $\lim_{m\to\infty} \mu^{n_m} f(x)$ exists a.e. too.

b) Under what conditions on μ, e.g. a moment condition, do we have $|\hat{\mu}(\lambda)| \leqslant 1 - |\lambda - 1|^\rho$ as $\lambda \to 1$, for some fixed $\rho > 0$? For this class of probability measures, Corollary 9 and Remark a) hold without change of argument.
Of course, if supp(μ) is finite, the second moment is finite and the above holds. Also a typical example of sequences (n_m) satisfying the growth condition above is given by $n_m = \{\beta^{\alpha^m}\}$ for some fixed $\beta, \alpha > 1$. This gives, for instance, the special version of this theorem which was stated in (Bellow et al., 1989a, 1989b, 1990). Let $bf(x) = (f(x) + f(\tau(x)))/2$. Then if $n_m = 2^{2^m}, m \geqslant 1$, for all $f \in L_2(X)$, $\lim_{m\to\infty} b^{n_m} f(x)$ exists a.e. Actually in (Bellow et al., 1990) it is shown that the same is true for any $f \in L_p(X)$, $1 \leqslant p \leqslant \infty$.
This binomial example can be used to clarify the limitations of the theorems previously discussed. For example, for any ergodic transformation τ, there exists $E \in \beta$ such that $\lim_{n\to\infty} b^n \mathbf{1}_E(x)$ fails to exists at almost every point; indeed, we can have $\limsup_{n\to\infty} b^n \mathbf{1}_E(x) = 1$ a.e. and $\liminf_{n\to\infty} b^n \mathbf{1}_E(x) = 0$ a.e. Therefore, some condition on sp(T) like the Stolz condition, is needed for an a.e. convergence theorem along all powers of T^n.
But the binomial example also can be used to clarify what type of growth in (n_m) one can really expect. To understand this best, let us summarize the work in (Bellow et al., 1990) where the general case of moving averages was investigated. Let Ω be a sequence of points (k, n) where $k \in \mathbb{Z}$ and $n \in \mathbb{Z}^+ = \{n \in \mathbb{Z} : n > 0\}$. Let $M_\Omega f(x)$ denote the maximal function given by

$$M_\Omega f(x) = \sup_{(k,n)\in\Omega} \frac{1}{n}\left|\sum_{j=0}^{n} f(\tau^{k+j}x)\right|.$$

Given $\alpha > 0$, let

$$\Omega_\alpha = \{(z, s) \in \mathbb{Z} \times \mathbb{Z}^+ : |z - k| \leqslant \alpha(s - n) \text{ for some } (k, n) \in \Omega\}.$$

For $s > 0$, let $\Omega_\alpha(s)$ denote the cross-section $\{k : (k,s) \in \Omega_\alpha\}$.

10. THEOREM. a) *If there exists constants $A < \infty$ and $\alpha > 0$ such that $|\Omega_\alpha(x)| \leqslant As$ for all $s > 0$, then M_Ω is weak type (1,1) and strong type (p,p) for $1 < p < \infty$.*

b) *If M_Ω is weak type (p,p) for some $p, 0 < p < \infty$, then for all $\alpha > 0$, there exists $A < \infty$ such that $|\Omega_\alpha(s)| \leqslant As$ for all $s > 0$.*

Once this maximal estimate is known, a variety of theorems about block averages can easily be proved, including the following general fact.

11. THEOREM. *If $\Omega = \{(k_m, n_m),\ m = 1,2,3,\ldots\}$ where $\lim_{m\to\infty} n_m = \infty$, then the condition in Theorem 1 is necessary and sufficient for the almost everywhere convergence of the averages $(1/n_m)\sum_{j=0}^{n_m} f(\tau^{k_m+j}x)$ as m tends to ∞ whenever $f \in L_1(X)$.*

The *cone condition* in Theorem 10 was used in (Bellow et al., 1990) to prove the following.

12. THEOREM. *If τ is any ergodic invertible measure preserving transformation of X, then the following are equivalent for an increasing sequence (n_m):*

1) *for all $f \in L_1(X)$, $\lim_{m\to\infty} b^{n_m} f(x)$ exists a.e.,*
2) *for some p, $1 \leqslant p \leqslant \infty$, for all $f \in L_p(X)$, $\lim_{m\to\infty} b^{n_m} f(x)$ exists a.e.,*
3) *for all $E \in \beta$, $\lim_{m\to\infty} b^{n_m} \mathbf{1}_E(x)$ exists a.e.,*
4) *if $\Omega = \{(n_m, \{\sqrt{n_m}\}),\ m = 1,2,3,\ldots\}$, then for some $A < \infty$, $\alpha > 0$, $|\Omega_\alpha(x)| \leqslant As$ for all $s > 0$.*

Remark. Similar results are true for other probability measures μ, with $\mathbf{E}(\mu) \neq$ $\neq 0$, in place of b. Also, in general if $\mathbf{E}(\mu) \neq 0$ and $m_2(\mu) < \infty$, the sequence $(\mu^n \mathbf{1}_E)$ will fail to converge a.e. for some $E \in \beta$.

This result and the discussions in (Bellow et al., 1989a, 1990) show that for any increasing sequence (n_m) such that $n_m \leqslant R^m$ for some $R < \infty$, then there is some $E \in \beta$ such that $\lim_{m\to\infty} b^{n_m} \mathbf{1}_E$ fails to exist a.e. Indeed, even a sequence like $n_m = 2^{m^2}$ fails to have properties 1)–4) in Theorem 12. Thus, a hypergeometric growth condition like $n_{m+1} \geqslant n_m^\alpha$ for some $\alpha > 1$ is in some sense necessary to obtain the type of a.e. convergence theorems for (T^{n_m}) discussed previously. This rapidity of growth is necessary and not an accident of the method of argument used in Theorem 4.

Finally, let us state what can be said along the lines of Theorem 4 for a sequence (T_n) of contractions in place of (T^n). We assume that these contractions are commuting and have a common spectral representation $T_n f = \int_\Delta h_n(\lambda) dE_f(\lambda)$ over some compact set $\Delta \subset \{\lambda; |\lambda| \leqslant 1\}$. Then we assume that (T_n) has ***asymptotically trivial representations*** i.e. for all $\delta > 0$, $(h_n(\lambda) : n = 1,2,3,\ldots)$ converges uniformly to zero on $\{\lambda \in \Delta : |\lambda - 1| \geqslant \delta\}$. We let $C_n f = \int_{\Delta_n} dE_f(\lambda)$ where

$C_n = \{\lambda \in \Delta : |\lambda - 1| \leqslant 1/n\}$. By an argument analogous to the one in Theorem 4, we can prove

13. THEOREM. *If (T_n) is a sequence of contractions on $L_2(X)$ with asymptotically trivial representations, then there is an increasing sequence (n_m) such that for all $f \in L_2(X)$,*

$$\lim_{m\to\infty} T_{n_m} f(x) - C_{n_m} f(x) = 0 \quad a.e.$$

This is the principle underlying the following theorem on probability measures which appears in (Bellow et al., 1989a). Let (μ_n) be a sequence of probability measures on $\mathbb{Z}$. We say (μ_n) has *asymptotically trivial transforms* if $\hat{\mu}_n(\lambda)$ converges to 0 uniformly on $\{\lambda : |\lambda| = 1, |\lambda - 1| \geqslant \delta\}$ for all $\delta > 0$. For some invertible measure-preserving transformation τ of X, let $\mu_n f(x) = \sum_{k=-\infty}^{\infty} \mu_n(k) f(\tau^k x)$ for $x \in X$. Bellow, Jones and Rosenblatt (1989a) show that

14. THEOREM. *If (μ_n) is a sequence of probability measures on $\mathbb{Z}$ with asymptotically trivial transforms, then there exists an increasing sequence (n_m) such that for all ergodic invertible measure-preserving transformations τ of X, and all $f \in L_p(X)$, $1 < p \leqslant \infty$,*

$$\lim_{m\to\infty} \mu_{n_m} f(x) = \int f dm \quad a.e.$$

Acknowledgments.
The authors would like to thank B. Mityagin for helpful discussions. Also, they appreciate the help of A. Bellow and R. Jones whose work is so often referenced here. The authors also wish to thank the Academy of Sciences of the Lithuanian SSR for the invitations to attend the Fifth Conference on Probability and Statistics held in the summer of 1989 in Vilnius.

REFERENCES

Bellow, A., Jones, R. and Rosenblatt, J. (1988). Almost everywhere convergence of powers. *Preprint*, Ohio State University.

Bellow, A., Jones, R. and Rosenblatt, J. (1989a). Almost everywhere convergence of weighted averages. *Preprint*, pp. 38.

Bellow, A., Jones, R. and Rosenblatt, J. (1989b). Almost everywhere convergence of convolution powers. *Preprint*, pp. 18.

Bellow, A., Jones, R. and Rosenblatt, J. (1990). Convergence of moving averages. *Preprint*, pp. 33.

Gaposhkin, V.F. (1977). A theorem on the convergence almost everywhere of a sequence of measurable functions, and its applications to sequences of stochastic integrals. *Math. USSR Sbornik* **33**, 1–17.

Gaposhkin, V.F. (1981). Individual ergodic theorem for normal operators in L_2. *Funct. Analysis and Appl.* **15**, 14–18.

Gaposhkin, V.F. (1984). Tauberian ergodic theorem for normal contraction operators. *Math. Notes* **35**, 208–212.

Katznelson, Y. and Tzafriri, L. (1986). On power bounded operators. *J. Funct. Analysis* **68**, 313–328.

Petrov, V.V. (1975). *Sums of independent random variables. Ergebnisse der Math.* **82**, Springer, Berlin–Heidelberg etc.

Oseledets, V.I. (1965). Markov chains, skew products and ergodic theorem for "general" dynamical systems. *Theory Probab. Appl.* **10**, 499–504.

Riesz, F. and Sz.-Nagy, B. (1941–1943). Über Kontraktionen des Hilbertschen Raumes. *Acta Sci. Math. Szeged* **10**, 202–205.

Prob. Theory and Math. Stat., Vol. 1, pp. 401–408
B.Grigelionis *et al.* (Eds.)
1990 VSP/Mokslas

ON SHOT NOISE PROCESSES WITH LONG RANGE DEPENDENCE

L. GIRAITIS and D. SURGAILIS
Institute of Mathematics and Cybernetics, 232600 Vilnius, Lithuania

ABSTRACT

We discuss the convergence of finite dimensional distributions of integrated shot noise process X_t to that of a H–self similar with stationary increments process $Z_t, t \geqslant 0$ $(0 < H < 1)$, where Z_t is either a fractional Brownian motion, or a stochastic integral with respect to a Poisson random measure on $\mathbf{R} \times \mathbf{R} \times \mathbf{R}_+$.

1. INTRODUCTION AND THE MAIN RESULTS

Consider the shot noise process

$$X_t = \sum_j G((t - \tau_j)V_j)U_j, \quad t \in \mathbf{R}, \tag{1.1}$$

where $G(x), x \in \mathbf{R}$, is a real non–random function, $(\tau_j)_{j \in \mathbf{Z}}$ is the Poisson point process on $\mathbf{R}$ with intensity $\lambda > 0$, and $(U_j, V_j)_{j \in \mathbf{Z}}$ is an i.i.d sequence of $\mathbf{R} \times \mathbf{R}_+$–valued random vectors, independent of $(\tau_j)_{j \in \mathbf{Z}}$ and distributed according to

$$F = F(du, dv) = \mathbf{P}(U_0 \in du, V_0 \in dv). \tag{1.2}$$

The process X_t (1.1) comes as a natural generalization of the simplest shot noise process $X_t = \sum_j G(t - \tau_j)$, with U_j playing the role of random "amplitude" of the "G–shaped response" generated at time $t = \tau_j$, and V_j playing the role of random time scale. However, (1.1) is a rather special case of general shot noise processes (Rice, 1977; Westcott, 1976). On the other hand, shot noise processes of the type (1.1) (and their multiparameter counterparts) were found useful in describing disordered media (Bulinskii and Molchanov, 1989).

In this paper we study limit theorems for the integrated shot noise process

$$S_t = \int_0^t X_s ds, \quad t \geqslant 0. \tag{1.3}$$

In particular, we are interested in conditions on G and F under which

$$A_T^{-1} S_{Tt} \overset{\Delta}{\Rightarrow} Z_t \, (T \to \infty), \tag{1.4}$$

where $Z_t, t \geqslant 0$, is a *H-self similar process with stationary increments* (H-sssi), $0 < H < 1$, and $\overset{\Delta}{\Rightarrow}$ (resp., $\triangleq$) denotes the weak convergence (resp., the equality) of finite dimensional distributions. The self-similarity of Z_t means that for each $a > 0$,

$$Z_{at} \triangleq a^H Z_t. \tag{1.5}$$

An example of H-sssi process is the *fractional Brownian motion* (FBM), which is Gaussian with zero mean and covariance

$$\mathbf{E} Z_t Z_s = \frac{1}{2}(|t|^{2H} + |s|^{2H} - |t-s|^{2H}).$$

We assume below that

$$G \in L^2(\mathbf{R}), \tag{1.6}$$

$$\int_{\mathbf{R}\times\mathbf{R}_+} u^2 v^{-1} dF < \infty, \tag{1.7}$$

and, moreover,

$$\int_{\mathbf{R}} uF(du|v) = 0 \tag{1.8}$$

for each $v \in \mathbf{R}_+ = (0, \infty)$, where $F(du|v) = \mathbf{P}(U_0 \in du | V_0 = v)$. *Under conditions* (1.6)—(1.8), *the series* (1.1) *converge a.s. and in* $L^2(\Omega)$, *for each* $t \in \mathbf{R}$, *and define a strictly stationary process with mean zero and spectral density:*

$$f(x) = \int_{\mathbf{R}\times\mathbf{R}_+} |\widehat{G}(x/v)|^2 u^2 v^{-2} dF, \quad x \in \mathbf{R}, \tag{1.9}$$

where

$$\widehat{G}(x) = \frac{1}{2\pi}\int_{\mathbf{R}} G(t)\exp(-itx)dt$$

is the Fourier transform.

We omit the proof of the statement above, which is elementary (see also (Rice, 1977; Westcott, 1976)). It is well-known, too, that X_t (1.1) has the representation:

$$X_t \triangleq \int_{\mathbf{R}\times\mathbf{R}\times\mathbf{R}_+} uG((t-s)v)dq, \tag{1.10}$$

where $q = q(ds, du, dv)$ is a centered Poisson random measure on $\mathbf{R}\times\mathbf{R}\times\mathbf{R}_+$ with variance

$$\mu \doteq \mu(ds, du, dv) = \lambda ds F(du, dv). \tag{1.11}$$

To formulate our main result, set

$$D_T(x) = \int_0^T e^{itx} dt = (e^{iTx} - 1)/ix. \tag{1.12}$$

Write $A_N \sim B_N$ if $\lim A_N/B_N = 1$.

THEOREM 1. *Assume that*

$$\int_{\mathbf{R}} |\widehat{G}(x)D_T(x)|^2 dx \sim CT^{2H} \tag{1.13}$$

and

$$C_H \equiv \int_{\mathbf{R}\times\mathbf{R}_+} u^2 v^{2H-3} dF < \infty \tag{1.14}$$

for some $C > 0$ *and* $H \in (0,1)$. *Then the convergence* (1.4) *holds, where* Z_t *is a* H*–sssi FBM, and*

$$A_T^2 = \operatorname{Var} S_T \sim CC_H T^{2H}. \tag{1.15}$$

Remark 1. For $H \neq 1/2, H \in (0,1)$, (1.13) follows from

$$G(x) \sim x^{H-3/2}(x \to \infty), \quad G(x) = 0 \quad (x \leqslant 0),$$

the boundedness of $G(x)$ on each finite interval and, for $H \in (0,1/2)$, the relation $\int_{\mathbf{R}} G(x)dx = 0$ (Taqqu, 1979). In the case $H = 1/2$, (1.13) follows from $G \in L^1(\mathbf{R})$ and $\int_{\mathbf{R}} G(x)dx \neq 0$ (Ibragimov and Linnik, 1971).

Remark 2. In a parallel paper (Giraitis and Surgailis, 1989) we prove a similar theorem about the convergence of an integrated shot noise process to a H–sssi α–stable fractional Lévy motion, $\alpha \in (1,2), H \in (0,1), H \neq 1/\alpha$.

The proof of Theorem 1 is based on the estimation of semiinvariants of S_T which can be evaluated explicitly on the ground of the stochastic integral representation (1.10), or the Lévy—Khintchine representation (2.5) of the characteristic function of S_T.
Roughly speaking, condition (1.14) restricts the concentration of F near $v = 0$. One can surmise that if this concentration is sufficiently large, then the limiting behaviour of S_T may depend on the values of $G(x)$ on the whole line rather than on its asymptotics as $|x| \to \infty$. The possibility of such a behaviour is established in Theorem 2 below. Moreover, one can show that (1.14) is exact in the following sense. For any $0 < H' < H < H'' < 1$ there exist G and F satisfying (1.13) and (1.14) with H replaced by H' and H'' respectively, and such that $A_T^{-1} S_{Tt}$ converges in distribution to a non–Gaussian limit; see Remark 4.

In Theorem 2 we assume that F is absolutely continuous having the density

$$f(u,v) = f_1(|u|) f_2(v|u|^{-1/(1-H)}), \tag{1.16}$$

where $f_i \geqslant 0, i = 1,2$, satisfy the conditions:

$$f_1(u) \sim u^{-1} \quad (u \downarrow 0), \tag{1.17}$$

$$\sup\{u f_1(u) : u \in \mathbf{R}_+\} < \infty \tag{1.18}$$

and

$$\int_{\mathbf{R}_+} f_2(v)(1 + v^{2H-3})dv < \infty. \tag{1.19}$$

THEOREM 2. *Let F and $H \in (0,1)$ satisfy conditions* (1.16)—(1.19) *above and, moreover,*

$$\int_{\mathbf{R}} |\widehat{G}(x)|^2(1 + |x|^{2(H-1)})dx < \infty. \tag{1.20}$$

Then

$$A_T^2 = \operatorname{Var} S_T \sim K_1^2 T^{2H} \tag{1.21}$$

and

$$A_T^{-1} S_{Tt} \stackrel{\Delta}{\Rightarrow} K_1^{-1} Z_t, \tag{1.22}$$

where $K_1^2 = \mathbf{E} Z_1^2$,

$$Z_t = \int_{\mathbf{R}\times\mathbf{R}\times\mathbf{R}_+} \left[u \int_0^t G((\tau - s)v)d\tau\right] dq_0, \tag{1.23}$$

and $q_0 = q_0(ds, du, dv)$ *is a centered Poisson random measure on* $\mathbf{R}\times\mathbf{R}\times\mathbf{R}_+$ *with variance*

$$\mu_0 = \lambda|u|^{-1} f_2(v|u|^{-1/(1-H)})dsdudv. \tag{1.24}$$

Remark 3. The class of H–sssi processes Z_t (1.23) depending on the functional "parameter" G is very large. A related class of H–sssi processes subordinated to a Poisson random measure was studied in (Surgailis, 1981).
Let us remark, finally, that there exists a considerable literature on shot noise processes and their applications; see e.g. (Hsing and Teugels, 1989) and the references therein. General limit theorems for shot noise processes were obtained in (Lane, 1984). It seems, however, that shot noise processes with long range dependence were not discussed previously. The case $V_j \equiv$ const corresponds to continuous time moving average processes. The central limit theorem for discrete time moving averages with long range dependence was discussed in (Davydov, 1970; Gorodetskii, 1977).

2. PROOF OF THEOREM 1

Below we identify X_t with the stochastic integral on the right hand side of (1.10). To simplify the notation, we put $\lambda = 1$ and do not indicate the domain of integration (either $\mathbf{R}\times\mathbf{R}\times\mathbf{R}_+$, or $\mathbf{R}\times\mathbf{R}_+$ or $\mathbf{R}$) when it is clear from the context.
Let us prove first (1.15). According to (1.9) and the well–known formula for the variance of an integrated stationary process (Ibragimov and Linnik, 1971, Eq. (18.3.2)),

$$A_T^2 T^{-2H} = \int u^2 v^{2H-3} dF \left[(Tv)^{-2H} \int |\widehat{G}(x) D_{Tv}(x)|^2 dx\right]. \tag{2.1}$$

By (1.13) the expression in the square brackets tends to C uniformly in $v \in (\varepsilon/T, \infty)$ for each $\varepsilon > 0$, and is bounded in $\{T > 0, v \in (0, \varepsilon/T)\}$ as $|D_T(x)| \leqslant T$ and $H < 1$. Hence one can pass to the limit as $T \to \infty$ under the sign of the first integral. This proves (1.15).

Note that it suffices to prove Theorem 1 in the case when

$$\operatorname{supp} F \subset \{(u,v) \in \mathbf{R} \times \mathbf{R}_+ : \varepsilon < |u| < \varepsilon^{-1}, \quad \varepsilon < v < \varepsilon^{-1}\} \tag{2.2}$$

for some $\varepsilon > 0$. Indeed, write D' for the set on the right hand side of (2.2), $D'' = \mathbf{R} \times \mathbf{R}_+ \backslash D', F' = F\mathbf{1}_{D'}, F'' = F\mathbf{1}_{D''}$, where $\mathbf{1}$ is the indicator function. Write also X_t', X_t'' for the corresponding integrals on the right hand side of (1.10), with q replaced by $q' = q\mathbf{1}_{D'}$ and $q'' = q\mathbf{1}_{D''}$, respectively. Then $X_t = X_t' + X_t''$, where X_t' and X_t'' are independent. Setting

$$S_T'' = \int_0^T X_s'' ds$$

we have by (1.15), (1.14) that

$$\lim_{T\to\infty} T^{-2H} \operatorname{Var} S_T'' = C \int u^2 v^{2H-3} dF'' \to 0$$

as $\varepsilon \to 0$. Hence the contribution of S_T'' to (1.4) can be made arbitrary small.

According to (1.10),

$$S_T = \int \phi_T \, dq, \tag{2.3}$$

where

$$\phi_T = \phi_T(s,u,v) = \int_0^T uG((t-s)v)dt. \tag{2.4}$$

By the well-known properties of Poisson stochastic integrals

$$\mathbf{E} \exp\left\{ i \sum_{j=1}^n \theta_j S_{Tt_j} \right\} \doteq \exp\left\{ \int (e^{i\psi} - 1 - i\psi) d\mu \right\}, \tag{2.5}$$

where

$$\psi = \sum_1^n \theta_j \phi_{Tt_j}, \quad t_1, \ldots, t_n \in \mathbf{R}_+, \quad \theta_1, \ldots, \theta_n \in \mathbf{R}$$

and $\mu = \mu(ds, du, dv)$ is given by (1.11), with $\lambda = 1$. To prove Theorem 1, it suffices to show that

$$Y_T = A_T^{-1} \sum_{j=1}^n \theta_j S_{Tt_j} \stackrel{\Delta}{\Rightarrow} \mathcal{N}(0, \sigma^2), \tag{2.6}$$

where

$$\sigma^2 = \frac{1}{2}\sum_{i,j=1}^{n}(|t_i|^{2H} + |t_j|^{2H} - |t_i - t_j|^{2H}). \tag{2.7}$$

Now, (2.6) follows from

$$\operatorname{Var} Y_T \to \sigma^2 \tag{2.8}$$

and

$$\chi_k(Y_T) \to 0 \quad \forall k \geqslant 3, \tag{2.9}$$

where χ_k denotes the k–th semiinvariant.
Here, (2.8) follows from (1.15) and the equality $\mathbf{E}S_{t_1}S_{t_2} = \frac{1}{2}(\mathbf{E}S_{t_1}^2 + \mathbf{E}S_{t_2}^2 - -\mathbf{E}S_{t_2-t_1}), 0 \leqslant t_1 \leqslant t_2$. Let us prove (2.9) for $n = \theta_1 = t_1 = 1$ as the general case can be considered analogously. According to (2.5),

$$\chi_k(Y_T) = A_T^{-k}\int \phi_T^k d\mu. \tag{2.10}$$

By the Hausdorff—Young ineguality,

$$\left|\int\left[\int_0^T G((t-s)v)dt\right]^k ds\right| = \left|v^{-1-k}\int\left[\int_0^{Tv} G(t-s)dt\right]^k ds\right| \leqslant$$

$$\leqslant \operatorname{const} v^{-1-k}\left[\int\left|\widehat{G}(x)D_{Tv}(x)\right|^p dx\right]^{k/p} \tag{2.11}$$

where $p = k/(k-1) \in (1,2]$. Write

$$\int\left|D_{Tv}(x)\widehat{G}(x)\right|^p dx = \int_{|x|<\delta} + \ldots + \int_{|x|>\delta} \ldots \equiv I_1 + I_2.$$

Here,

$$I_1 \leqslant (2\delta)^{(2-p)/2}\left(\int\left|\widehat{G}(x)D_{Tv}(x)\right|^2 dx\right)^{p/2} \tag{2.12}$$

and

$$I_2 \leqslant \left(\int\left|\widehat{G}(x)\right|^2 dx\right)^{p/2}\left(\int_{|x|>\delta}|D_{Tv}(x)|^{2p/(2-p)}\right)^{(2-p)/2} \tag{2.13}$$

The last integral on the right hand side of (2.13) is finite for each $\delta > 0$ and $p > 2/3$. Combining (2.11)—(2.13), we obtain

$$\left|\int \phi_T^k d\mu\right| \leqslant \int |u|^k v^{-1-k}\left[c_1(\delta)(vT)^{kH} + c_2(\delta)\right] dF,$$

where $c_1(\delta) \to 0 \quad (\delta \to 0)$ and $c_2(\delta) < \infty$ for each $\delta > 0$ ($c_2(\delta)$ does not depend on T, v). With (1.15), (2.2) and (2.10) in mind, this concludes the proof of Theorem 1.

3. PROOF OF THEOREM 2

Let us check first that Z_t (1.23) is well-defined. By elementary computations

$$\mathrm{Var}\, Z_T = \int \phi_T^2 d\mu_0 = K_1^2 T^{2H},$$

where ϕ_T is given by (2.4) and

$$K_1^2 = \int \phi_1^2 d\mu_0 = (1-H) \int_{\mathbf{R}_+} f_2(z) z^{2H-3} dz \int_{\mathbf{R}} |\widehat{G}(x)|^2 |x|^{2H-2} dx \times$$

$$\times \int_{\mathbf{R}} |D_1(x)|^2 |x|^{1-2H} dx < \infty$$

according to (1.19) and (1.20).
The self-similarity relation (1.5) for Z_t (1.23) follows from the following invariance property of the Poisson random measure q_0:

$$q_0(a\, ds, a^{H-1} du, a^{-1} dv) \triangleq q_0(ds, du, dv)$$

(which is a consequence of the corresponding invariance property of the measure μ_0 (1.24); see (Surgailis, 1981), and the scaling property of the kernel ϕ_T (2.4):

$$\phi_T(as, a^{H-1}u, a^{-1}v) = a^H \phi_T(s,u,v) \quad \forall a \in \mathbf{R}_+.$$

As in the proof of Theorem 1, we shall discuss the convergence of one-dimensional distributions only, i.e.

$$A_T^{-1} S_T \overset{\triangle}{\Rightarrow} K_1^{-1} Z_1. \tag{3.1}$$

Let us prove first (1.21). We have

$$A_T^2 = \int \phi_T^2 d\mu = T^{2H} \int \phi_1^2 \rho_T d\mu_0, \tag{3.2}$$

where

$$\rho_T = \rho_T(u) = |u| T^{H-1} f_1(|u| T^{H-1}) \to 1 \quad (T \to \infty),$$

for each $u \in \mathbf{R}\backslash\{0\}$ (see (1.17)), and ρ_T is bounded in T, u by (1.18). Hence

$$\int \phi_1^2 \rho_T d\mu_0 \to \int \phi_1^2 d\mu_0 = K_1^2,$$

which proves (1.21). Set $\Xi(x) = e^{ix} - 1 - ix \quad (x \in \mathbf{R})$. According to (2.5),

$$\mathbf{E} \exp\{i\theta A_T^{-1} S_T\} = \exp\left\{\int \Xi(\theta A_T^{-1} \phi_T) d\mu\right\} = \exp\left\{\int \Xi(\theta A_T^{-1} T^H \phi_1) \rho_T d\mu_0\right\} \to$$

$$\to \exp\left\{\int \Xi(\theta K_1^{-1} \phi_1) d\mu_0\right\} = \mathbf{E} \exp\left\{i\theta K_1^{-1} Z_1\right\}$$

for each $\theta \in \mathbf{R}$. Theorem 2 is proved.

Remark 4. Fix any $0 < H' < H < H'' < 1$ and consider the distribution $F(du, dv)$ with density $f(u,v)$ (1.16), where $f_1(u) = |u|^{-1}\mathbf{1}_{(|u|<1)}, f_2(v) = (3v^2/(1-H))\mathbf{1}_{(0<v<1)}$. Then

$$C_{H''} = \int u^2 v^{2H''-3} dF = \frac{6}{1-H} \int_0^1 v^{2H''-1} dv \int_0^1 u^{1+2(H''-1)/(1-H)} du < \infty$$

as $H < H''$. Let $G \in L^2, G(x) = 0 \quad (x < 0)$ be defined by $|\widehat{G}(x)|^2 = |x|^{1-2H'}$ if $|x| < 1, |\widehat{G}(x)|^2 = |x|^{-2}$ if $|x| \geqslant 1$ (such G exists as $\int_{-\infty}^{\infty} \left|\log|\widehat{G}(x)|^2\right| (1 + x^2)^{-1} dx < \infty$). It is easy to show that

$$\int_{-\infty}^{\infty} |\widehat{G}(x)|^2 |D_T(x)|^2 dx \sim \text{const}\, T^{2H'},$$

i.e., F and G satisfy conditions (1.14) and (1.13) of Theorem 1, with H replaced by H'' and H', respectively. On the other hand, F and G satisfy the conditions of Theorem 2, and consequently $T^{-H}S_T$ converges in distribution to a non-Gaussian limit.

REFERENCES

Bulinskii, A. V. and Molchanov, S.A. (1990). Asymptotic Gaussian behaviour of the solution of Burgers equation with random initial data. *Theory Probab. Appl.* (to appear).

Davydov, Yu. A. (1970). The invariance principle for stationary processes. *Theory Probab. Appl.* **15**, 487–498.

Giraitis, L. and Surgailis, D. (1990). On shot-noise processes attracted by fractional Lévy motion. (to appear).

Gorodeckii, V. V. (1977). Convergence to semistable Gaussian processes. *Theory Probab. Appl.* **22**, 498–508.

Hsing, T. and Teugels, J. L. (1989). Extremal properties of shot-noise processes. *Adv. Appl. Prob.* **21**, 513–525.

Ibragimov, I.A. and Linnik, Yu. V. (1971). *Independent and stationary sequences of random variables.* Groningen, Walters-Noordhoff.

Lane, A. J. (1984). The central limit theorem for the Poisson shot noise processes. *J. Appl. Prob.* **21**, 287–301.

Rice, J. (1977). On generalized shot-noise. *Adv. Appl. Prob.* **9**, 553–565.

Surgailis, D. (1981). On infinitely divisible self-similar random fields. *Z. Wahr. verw. Geb.* **58**, 453–477.

Taqqu, M. S. (1979). Convergence of integrated processes of arbitrary Hermite rank. *Z. Wahr. verw. Geb.* **50**, 53–83.

Westcott, M. (1976). On the existence of a generalized shot-noise processes. In: *Studies in Probability and Statistics: Papers in Honour of E.J.G.,* Pitman, ed. E.J. Williams. North-Holland, Amsterdam.

Prob. Theory and Math. Stat., Vol. 1, pp. 409–423
B.Grigelionis *et al.* (Eds.)
1990 VSP/Mokslas

G–ESTIMATES OF EIGEN–VALUES OF COVARIANCE MATRICES

V.L. GIRKO
Kiev State University, Vladimirskaya str. 64, 252017 Kiev, USSR

ABSTRACT

From (Anderson, 1958) it follows that the maximum–likelihood estimates of the eigen–values $\mu_1(R) > \ldots > \mu_m(R)$ and the eigen–vectors $h_1(R), \ldots, h_m(R)$ of a covariance matrix R of order m are equal to $\mu_i(\hat{R}), h_i(\hat{R})$ correspondingly, where

$$\hat{R} = (n-1)^{-1} \sum_{k=1}^{n} (x_k - \hat{x})(x_k - \hat{x})', \quad \hat{x} = n^{-1} \sum_{k=1}^{n} x_k$$

and $x_1, \ldots, x_n$ are independent observations.
Let be independent vector–observations over a random normally $N(a, R)$ distributed vector ξ. It seemes that if one knows the asymptotic properties then the problem of asymptotic estimation of the values μ_i and the veetors h_i is solved. It is so indeed if m is small and does not depend on n. For large m the bias of these estimates can be substantial. Also, in many practical problems it is not necessary to know all the eigen–values and eigen–vectors of a matrix R. For example, to prove the singularity of a matrix R it is sufficient to calculate the eigen–value μ_m. In the main components method it is necessary to know some eigen–values $\mu_1 > \ldots > \mu_s$. In the planning of experiments, the least squares method in some cases we express the errors of an estimate by extreme eigen–values μ_1 and μ_m.
In this paper the methods of G–analysis (Girko, 1987) are used to find consistent estimates of a finite number of eigen–values μ_i, which do not depend on n. We discuss this now briefly. First of all, we impose the G–condition

$$\limsup_{n\to\infty} m_n = \infty, \quad \limsup_{n\to\infty} m_n n^{-1} < \infty. \tag{1}$$

Using this condition and the main limit theorems of the theory of probability and random matrices we prove that

$$p\text{-}\lim_{n\to\infty} \left[\mu_1(\hat{R}) - g(\mu_1(R), R)\right] = 0,$$

where g is some nonrandom function. Then we take the solution of the equation $g(x, R) = \mu_1(\hat{R})$ as an estimate of μ_1. The matrix R is still unknown in this equation. With the help of G_2–estimate we can approximately replace this equation by $\varphi(x, \hat{R}) = \mu_1(\hat{R})$, where φ is some nonrandom function. The estimates obtained

in such a way are better than standard ones, since the latter estimates do not converge to the true values if the G–condition holds.
Note that in this paper we use the idea of G–consistency instead of the idea of consistency of estimates: a sequence of estimates $G(n,m)$ of μ_1 is G–consistent, if G–condition is valid

$$\lim_{n\to\infty} [G(n,m) - \mu_1] = 0.$$

We relate the G–estimates of the values μ_i to the class of G_{26}– estimates. The book (Girko, 1988a) tells about other G_i–estimates. To simplify formulas we omit sometimes indices n, m. The constants are denoted by c_i, c.
A G_{26}–estimate of μ_1 we define as a measurable real solution θ of the equation

$$\mu_1(\hat{R}) = \theta(1-\gamma) + \gamma\theta^2 \text{Re}\left[f(\theta+i\varepsilon)(\theta+i\varepsilon)^{-1} - 1 + \gamma\right](\theta+i\varepsilon)^{-1}, \qquad (2)$$

where $f(\theta + i\varepsilon)$ is an analytic function, which is a solution of the equation

$$1 - \gamma + \gamma f m^{-1}\text{Sp}\left[I(f + i\varepsilon \,\text{sign}\,\text{Re} f) - \hat{R}\right]^{-1} = (\theta + i\varepsilon)^{-1} f, \qquad (3)$$

where $\gamma = m_n n^{-1}$, $\varepsilon \neq 0$.
Note that G_{26} as $\gamma \to 0$ tends to the maximum–likelihood estimate.

THEOREM 1. *Let condition* (1) *hold and*

$$\mu_1 > \mu_2 + \tau, \quad \tau > 0, \quad \mu_{m_n} > 0, \qquad (4)$$

$$\lim_{\varepsilon\to 0} p\text{–}\lim_{n\to\infty} \left| n^{-1}\sum_{p=2}^{m} \mu_p^2 |\mu_1 - \mu_p|^{-1} |\theta + i\varepsilon - \mu_p|^{-1} - 1 \right| > 0,$$

$$\liminf_{n\to\infty} \left[1 - \gamma + 2\gamma\mu_1 m^{-1} \sum_{p=2}^{m} (\mu_1 - \mu_p)^{-1} - \right.$$

$$\left. -\gamma\mu_1^2 m^{-1} \sum_{p=2}^{m} (\mu_1 - \mu_p)^{-2} \right] > 0. \qquad (5)$$

Then

$$\lim_{\varepsilon\to 0} p\text{–}\lim_{n\to\infty} [G_{26} - \mu_1] = 0. \qquad (6)$$

In the proof of Theorem 1 we use asymptotic expressions for the traces of resolvents of matrices $\hat{R}$.

LEMMA 1. *Let conditions* (1), (4) *hold,* ε_n *be a sequence of numbers such that*

$$\lim_{n\to\infty} n^{-\beta} |\varepsilon_n|^{-t} = 0, \qquad (7)$$

where $\beta > 0$ *and* t *is a positive integer.*
Then

$$\frac{\partial}{\partial x} \sum_{k=1}^{m} \mathbf{P}\{\mu_k(\hat{R}) + \varepsilon_n \eta < x\} =$$

$$= -\pi^{-1}\mathrm{Im}\left\{mb(z)+\varphi(z)\right\} + \varkappa_n(x)n^{-\beta}|\varepsilon_n|^{-t},$$

where $z = x + i\varepsilon_n$, $b(z)$ is analytic if $\varepsilon_n \neq 0$ and satisfy the equation

$$b(z) = m^{-1}\sum_{p=1}^{m} b_p, \quad b_p = \left[z - \mu_p(1-\gamma+\gamma z b(z))\right]^{-1},$$

$$\varphi(z) = \frac{1}{2}\frac{\partial}{\partial z}\ln\beta(z) + \frac{3}{2}\frac{\partial}{\partial z}\ln\left[1-\gamma+\gamma z b(z)\right],$$

$$\beta(z) = 1 - m^{-1}\sum_{p=1}^{m} b_p^2\mu_p\gamma z, \tag{8}$$

$\varkappa_n(x)$ is a function such that

$$\sup_{n\geqslant n_0}\int\left|\varkappa_n(x)\right|dx < \infty,$$

η is a random variable not depending on the matrix $\hat{R}$ and distributed according to the Cauchy law.

Proof. From (Anderson, 1958) it follows that $\hat{R}$ is distributed as $R^{\frac{1}{2}}BB'n^{-1}R^{\frac{1}{2}}$, where B is a rectangular $(m_n \times n)$—matrix whose entries are independent and normally $N(0,1)$ distributed. Furthermore, using the invariance of the distribution of B relative to the orthogonal transformations we transform the trace of the resolvent of $\hat{R}$:

$$a(z) = m^{-1}\mathbf{M}\,\mathrm{Sp}(Iz-\hat{R})^{-1} = m^{-1}\mathbf{M}\,\mathrm{Sp}(Iz-HH')^{-1},$$

where $H = \Lambda B_n^{-1/2}$, $\Lambda = (\mu_i^{1/2}\delta_{ij})_{i,j=1}^{m}$, and $\delta_{i,j}$ is the Kronecker symbol. For $a(z)$ we have (Girko, 1980, p.54)

$$a(z) = m^{-1}\sum_{p=1}^{m}\mathbf{M}[z - h_ph_p' - h_pH'\Gamma_pH_ph_p']^{-1} =$$

$$= m^{-1}\sum_{p=1}^{m}[z - \mu_p - \mu_pn^{-1}\mathrm{Sp}\Gamma_pH_pH_p' + \varepsilon_{1p}]^{-1}, \tag{9}$$

where h_p is the p-th vector-row of the matrix H, H_p is the matrix obtained by the deletion of the p-th row of H,

$$\Gamma_p = (Iz - H_pH_p')^{-1},$$

$$\varepsilon_{1p} = -h_ph_p' + \mu_p + \mu_pn^{-1}\mathrm{Sp}\Gamma_pH_pH_p' - h_pH_p'\Gamma_pH_ph_p'.$$

From (9) it is easy to deduce that

$$a(z) = m^{-1}\sum_{p=1}^{m} a_p,$$

$$a_p = \mathbf{M}\left[z - \mu_p(1-\gamma+\gamma z a(z)) + \varepsilon_{2p}\right]^{-1},$$

$$\varepsilon_{2p} = \varepsilon_{1p} - z\mu_p n^{-1}\mathrm{Sp}\Gamma_p + z\mu_p\gamma a(z). \tag{10}$$

Now (10) is equivalent to the equation

$$y = \mathbf{M}m^{-1}\sum_{p=1}^{m} b_p^2(y\mu_p\gamma z - \varepsilon_{2p}) + \mathbf{M}m^{-1}\sum_{p=1}^{m} b_p^3(y\mu_p\gamma z - \varepsilon_{2p})^2 + \tau,$$

$$y = (a(z) - b(z)), \quad \tau = m^{-1}\sum_{p=1}^{m}\mathbf{M}(y\mu_p\gamma z - \varepsilon_{2p})^2 b_p^3(a_p - b_p).$$

This implies the following equation for y

$$y^2 r_1 - y r_2 + \delta_2 = 0,$$

where

$$r_1 = m^{-1}\sum_{p=1}^{m} b_p^3\mu_p^2\gamma^2 z^2, \quad r_2 = 1 - m\sum_{p=1}^{m} b_p^2\mu_p\gamma z,$$

$$\delta_n = \tau - \mathbf{M}m^{-1}\sum_{p=1}^{m} b_p^2\varepsilon_{2p} + \mathbf{M}m^{-1}\sum_{p=1}^{m} b_p^3\varepsilon_{2p}^2 -$$

$$-2\mathbf{M}m^{-1}\sum_{p=1}^{m} b_p^3 y\mu_p\gamma z\varepsilon_{2p}.$$

The solution of this equation is

$$y = (2r_1)^{-1}\left[r_2 \pm (r_2^2 - 4\delta_n r_1)^{1/2}\right].$$

Since $a(z) \to 0$ as $z \to \infty$, choosing sign minus, we obtain

$$ma(z) = mb(z) + 2m\delta_n\left[r_2 + (r_2^2 - 4\delta_n r_1)^{1/2}\right]^{-1}. \tag{11}$$

For further analysis of this expression we need the following auxiliary statements.

LEMMA 2. *If $\gamma < 1$ then $b(z)$ may be represented as*

$$b(z) = \int_{\alpha_1}^{\alpha_2}(z-u)^{-1}p(u)\,du,$$

where $p(u)$ is a bounded density corresponding to a distribution concentrated on the interval (α_1, α_2), $\alpha_1 > 0$, $\alpha_2 < c < \infty$.

Proof. For

$$\beta_1 = \mathrm{Re}\, b(z), \quad \beta_2 = \mathrm{Im}\, b(z)$$

from (8) we get

$$x\beta_1 - \varepsilon\beta_2 + i(\varepsilon\beta_1 + x\beta_2) = m^{-1}\sum_{k=1}^{m} |d_k|^{-2}(\nu_k + i\lambda_k), \tag{12}$$

where

$$d_k = \nu_k + i\lambda_k, \quad \nu_k = 1 - \mu_k(1-\gamma)x|z|^{-2} - \gamma\mu_k\beta_1,$$
$$\lambda_k = \mu_k(1-\gamma)\varepsilon|z|^{-2} - \gamma\mu_k\beta_2.$$

Suppose that $x \neq 0$ and calculate the limit

$$p(x) = -\pi^{-1}\lim_{\varepsilon\to 0}\beta_2.$$

Note that if $p(x) = \infty$ for some values x, then (12) implies that $x\beta_1 - \varepsilon\beta_2 \to 0$, $\beta_1 + x\beta_2 \to 0$ as $\varepsilon \to 0$. But then $\beta_2 \to 0$ if $\varepsilon \to 0$, $x \neq 0$, and we come to a contradiction. Consequently, the function $p(x)$ is bounded for all values $x \neq 0$. Consider the case $x = 0$. Denoting $\theta_1 = \beta_1\varepsilon$, $\theta_2 = \beta_2\varepsilon$ we get

$$|\theta_1| = \left|\int u\varepsilon(u^2+\varepsilon^2)^{-1}\,dF(u)\right| \leqslant \frac{1}{2},$$

$$|\theta_2| = \left|\int \varepsilon^2(u^2+\varepsilon^2)^{-1}\,dF(u)\right| \leqslant 1, \tag{13}$$

where $F(u)$ is some distribution function. From (12) we obtain if $x = 0$

$$-\beta_2 + i\beta_1 = m^{-1}\sum_{k=1}^{m}\left[(\varepsilon - \gamma\mu_k\theta_1)^2 + \mu_k^2(1-\gamma-\gamma\theta_2)^2\right]^{-1}\times$$

$$\times\left\{\varepsilon - \gamma\mu_k\theta_1 - i\mu_k(1-\gamma-\gamma\theta_2)\right\}.$$

Multiplying this equality by ε we get by (13)

$$\lim_{\varepsilon\to 0}\left[|\theta_1| + |\theta_2|\right] = 0,$$

if $\gamma < 1$ and $\mu_k > 0$, since $\theta_2 < 0$.
Consequently

$$\lim_{\varepsilon\to 0}\beta_1 = -m^{-1}\sum_{k=1}^{m}\mu_k^{-1}(1-\gamma)^{-1}, \quad \lim_{\varepsilon\to 0}\beta_2 = 0.$$

Thus $p(0) = 0$. Since $b(z)$ is the Stieltijes transformation of some spectral function and the solution of the equation (8) exists and is unique in the class of analytic functions if $\varepsilon_n \neq 0$, $\operatorname{Im} b(z)\varepsilon < 0$ (Girko, 1988, p. 215), then the function $p(x)$ is unique, so that $p(u) = 0$ if $u < 0$.

Let $p(x) > 0$ for some $x \neq 0$. From (8) it follows that $p(x)$ satisfies the equation

$$p(x) = m^{-1} \sum_{p=1}^{m} \frac{\mu_p \gamma x p(x)}{(x - \mu_p(1-\gamma) - \mu_p \gamma x q(x))^2 + (\mu_p \gamma x p(x) \pi)^2}, \tag{14}$$

where $q(x) \in Q$ and Q is a the set of all limit values of functions β_1 as $\varepsilon \to 0$, $x \neq 0$, so that $q(x) \neq \infty$ and satisfies the equation

$$q(x) = m^{-1} \sum_{p=1}^{m} \frac{x - \mu_p(1-\gamma) - \mu_p \gamma x q(x)}{(x - \mu_p(1-\gamma) - \mu_p \gamma x q(x))^2 + (\mu_p \gamma x p(x) \pi)^2}. \tag{15}$$

Denote $\alpha_2 = \sup\{x\colon\ p(x) > 0\}$.
Passing to the limit as $x \to \alpha_2 + 0$ in (14) and (15) we get the system of equations

$$1 = m^{-1} \sum_{p=1}^{m} \mu_p \gamma \alpha_2 \big[\alpha_2 - \mu_p(1-\gamma) - \mu_p \gamma \alpha_2 y\big]^{-2},$$

$$y = m^{-1} \sum_{p=1}^{m} \big(\alpha_2 - \mu_p(1-\gamma) - \mu_p \gamma \alpha_2 y\big)^{-1}, \tag{16}$$

where y belongs to the set Q_1 of all limit values of the function $q(x)$ as $x \to \alpha_2 + 0$. Denote $\alpha_2(1 - \gamma + \gamma \alpha_2 y)^{-1} = \nu$. Then from (16) we obtain

$$\alpha_2 = \nu - \gamma\nu + \gamma\nu^2 m^{-1} \sum_{p=1}^{m} (\nu - \mu_p)^{-1},$$

$$\alpha_2 = \gamma \nu^2 m^{-1} \sum_{p=1}^{m} \mu_p (\mu_p - \nu)^{-2}. \tag{17}$$

The second equation has bounded solutions ν. Therefore $|\alpha_2| < c < \infty$. Lemma 2 is proved.

LEMMA 3. *If conditions* (1), (4) *hold, then*

$$\max\Big\{|a_p - b_p|,\ p = 1, \dots, m,\ |a(z) - b(z)|\Big\} \leqslant$$

$$\leqslant (c_1|x| + c_2)|\varepsilon_n|^{-5} n^{-1/2}. \tag{18}$$

Proof. As in the proof of (11) we get the equation

$$u^2 \tau_1 - u \tau_2 + \tau_3 = 0,$$

where

$$\tau_1 = m^{-1} \sum_{p=1}^{m} b_p^2 a_p (\mu_p \gamma z)^2, \quad u = a(z) - b(z),$$

$$\tau_2 = 1 - m^{-1}\sum_{p=1}^{m}\mu_p\gamma z b_p^2, \quad \tau_3 = -\mathbf{M}m^{-1}\sum_{p=1}^{m} a_p b_p \varepsilon_{2p} +$$

$$+u\mathbf{M}m^{-1}\sum_{p=1}^{m} a_p b_p^2 \mu_p \gamma z \varepsilon_{2p}.$$

Thus

$$a(z) - b(z) = 2\tau_3\left[\tau_2 + (\tau_2^2 - 4\tau_3\tau_1)\right]^{-1}. \tag{19}$$

Obviously

$$\frac{\partial b(z)}{\partial z} = -m^{-1}\sum_{p=1}^{m} b_p^2 + \theta\left[\frac{\partial b(z)}{\partial z} + \frac{b(z)}{z}\right],$$

where

$$\theta = m^{-1}\sum_{p=1}^{m} b_p^2 \mu_p \gamma z.$$

Also

$$zm^{-1}\sum_{p=1}^{m} b_p^2 = b(z) + \theta(\gamma z)^{-1}(1 - \gamma + \gamma z b(z)).$$

From these equalities we get

$$r_2^{-1} = \tau_2^{-1} = (1-\theta)^{-1} = -z\left(\frac{\partial}{\partial z}\right)\ln\left[1 - \gamma + \gamma z b(z)\right] + 1.$$

Therefore

$$|r_2|^{-1} \leqslant \left[1 - \gamma + \left|\gamma z^2 \frac{\partial b}{\partial z}\right|\right]\left|1 - \gamma + \gamma z b(z)\right|^{-1} \leqslant$$

$$\leqslant |\varepsilon_n|^{-2}\left[\left(1 + \gamma\int_{\alpha_1}^{\alpha_2} u(x-u)[(x-u)^2 + \varepsilon_n^2]^{-1} p(u)\,du\right)^2 +\right.$$

$$\left. +\gamma^2\varepsilon_n^2\left(\int_{\alpha_1}^{\alpha_2} up(u)[(x-u)^2 + \varepsilon_n^2]^{-1}\,du\right)^2\right]^{-1/2} \tag{20}$$

This implies

$$|r_2|^{-1} \leqslant c|\varepsilon_n|^{-3}. \tag{21}$$

We need the inequality

$$\mathbf{M}|\varepsilon_{2p}|^2 \leqslant 4(\mu_p - h_p h_p')^2 + 4\mu_p^2\mathbf{M}|n^{-1}\mathrm{Sp}\Gamma_p H_p H_p' -$$

$$-h_p H_p' \Gamma_p H_p h_p'|^2 + 4|z|^2\mu_p^2|n^{-1}\mathbf{M}\,\mathrm{Sp}\Gamma_p - \gamma a(z)|^2 +$$

$$+4|z|^2\mu_p^2\mathbf{M}|n^{-1}\mathrm{Sp}\Gamma_p - n^{-1}\mathbf{M}\,\mathrm{Sp}\Gamma_p|^2. \tag{22}$$

Note that

$$\mathbf{M}n^{-1}\mathrm{Sp}\Gamma_p - \gamma a(z) = \mathbf{M}\left(\frac{\partial}{\partial z}\right)\ln\left[\int (z-u)^{-1}\,d\mu(u)\right],$$

where

$$\mu(u) = \sum_{k=1}^{m} t_{pk}^2 \chi(\mu_k < u),$$

t_{pk} are the components of the eigen-vector t_p of $\hat{R}$, corresponding to the eigen-value μ_k.
It is easy to check that (see (Girko, 1980, 1987, 1988a, 1988b), where such inequalities are used frequently)

$$\left|\left(\frac{\partial}{\partial z}\right) \ln\left[\int (z-u)^{-1}\, d\mu(u)\right]\right| \leqslant |\varepsilon_n|^{-1},$$

$$\mathbf{M}\,|\mathrm{Sp}\Gamma_p - \mathbf{M}\,\mathrm{Sp}\Gamma_p|^2 \leqslant c|\varepsilon_n|^{-2} n.$$

Using these inequalities we get from (22) that

$$\mathbf{M}|\varepsilon_{2p}|^2 \leqslant c|\varepsilon_n|^{-2} n^{-1}. \tag{23}$$

Furthermore (21), (23) and the inequality $|\tau_1| \leqslant c|\varepsilon_n|^{-2}$ we get that $|a(z) - b(z)|$ satisfies (18). Analogously we prove this inequality for the differences $|a_p - b_p|$. Lemma 3 is proved.

LEMMA 4. *The equality*

$$m\delta_n = \varphi(z) + \varkappa(z) n^{-\beta} |\varepsilon_n|^{-t}$$

holds.

Proof. It is obvious that $|b_p| \leqslant |\varepsilon_n|^{-1}$. Using Lemma 3, we get (see (11))

$$m|\tau| \leqslant |\varkappa(x)| n^{-\beta} |\varepsilon_n|^{-q}. \tag{24}$$

By $|\mathbf{M}\varepsilon_{2p}| \leqslant cn^{-1}$ we have

$$\left|2\mathbf{M}\sum_{p=1}^{m} b_p^3 (a(z) - b(z)) \mu_p \gamma z \varepsilon_{2p}\right| \leqslant mn^{-3/2} \max_p |b_p|^3 \leqslant |\varkappa(x)| n^{-\beta} |\varepsilon_n|^{-t}. \tag{25}$$

Let us calculate the values $\mathbf{M}\varepsilon_{2p}$ and $\mathbf{M}\varepsilon_{2p}^2$. Using (2.5.7) from (Girko, 1980) we get the equalities

$$n\mathbf{M}\varepsilon_{2p} = -z\mu_p \left(\frac{\partial}{\partial z}\right) \ln[z - h_p h_p' - h_p H_p' \Gamma_p H_p h_p']^{-1}.$$

Taking into account (23) in the some way as we obtained (10) we have

$$m\mathbf{M}\varepsilon_{2p} = -z\mu_p \gamma \partial/\partial z \ln b_p + o(1) \tag{26}$$

Omitting elementary transformations we represent $n\mathbf{M}\varepsilon_{2p}^2$ as

$$n\mathbf{M}\varepsilon_{2p}^2 = n\mathbf{M}\varepsilon_{1p}^2 + o(1) = 2\mu_p^2[n^{-1}\sum_{i=1}^{n}(1+2\mathbf{M}q_{ii}) + n^{-1}\mathrm{Sp}\mathbf{M}Q^2] + o(1), \quad (27)$$

where

$$Q = (q_{ij}) = H_p'\Gamma_p H_p.$$

Let us find the limits of $\mathbf{M}q_{ii}$, $\mathbf{M}q_{ii}^2$, $n^{-1}\mathrm{Sp}\mathbf{M}Q^2$ as $n \to \infty$. Without loss of generality we consider the matrix H instead of H_p. For q_{ii} we have

$$q_{ii} = \mathrm{Sp}\, p_i(Iz - \sum_{l=1}^{n} p_l)^{-1} = L_1 + L_2,$$

where

$$L_1 = \mathrm{Sp}\, p_i(Iz - \sum_{l\neq i}^{n} p_l)^{-1},$$

$$L_2 = \mathrm{Sp}\{(Iz - \sum_{l=1}^{n} p_l)^{-1} - (Iz - \sum_{l\neq i}^{n} p_l)^{-1}\}, \quad p_i = \eta_i\eta_i',$$

and η_i are vector–columns of H. For L_1 as in the proof of (9) we obtain

$$p\text{-}\lim_{n\to\infty}[L_1 - \varrho] = 0, \quad (28)$$

where

$$\varrho = n^{-1}\sum_{k=1}^{m} b_k\mu_k.$$

Furthermore

$$L_2 = \eta_i' B p_i B\eta_i(1 - \eta_i' B\eta_i)^{-1},$$

where

$$B = (Iz - \sum_{l\neq i} p_l)^{-1}.$$

We also have

$$p\text{-}\lim_{n\to\infty}[L_2 - \varrho^2(1-\varrho)^{-1}] = 0. \quad (29)$$

Combining (28) and (29) we get

$$q_{ii} = \varrho(1-\varrho)^{-1} + \nu_n, \qquad p\text{-}\lim_{n\to\infty}\mathbf{M}\nu_n^2 = 0.$$

It is easy to see that

$$m^{-1}\mathbf{M}\mathrm{Sp}\, Q^2 = m^{-1}\mathbf{M}\mathrm{Sp}(I - z(Iz - H_pH_p')^{-1})^2 = 1 - 2zb(z) - z^2\partial b(z)/\partial z + o(1).$$

Thus we obtain

$$m\mathbf{M}\varepsilon_{2p}^2 = 2\mu_p^2\gamma[1 - \gamma - \gamma z^2\partial b(z)/\partial z] + o(1). \quad (30)$$

Substituting (26) and (30) in the expression for $m\delta_n$ and taking into account the inequalities (24) and (25) we get

$$m\delta_n = \beta^{-1}(z)m^{-1}\Big\{\sum_{p=1}^{m} b_p^2(z)\mu_p z\gamma(\partial/\partial z)\ln b_p(z)+$$

$$+\sum_{p=1}^{m} b_p^3(z)2\mu_p^2\gamma(1-\gamma-\gamma z^2\partial b(z)/\partial z)\Big\} + \varkappa(x)n^{-\beta}|\varepsilon_n|^{-t} =$$

$$= [(\partial/\partial z)\ln\beta(z) + 3z^{-1}(1-\beta^{-1}(z))]/2 + \varkappa(x)n^{-\beta}|\varepsilon_n|^{-t}.$$

Using now (20) we come to the assertion of Lemma 4.
Return to (11). Applying Lemmas 2—4 and the equality

$$(\partial/\partial x)\sum_{k=1}^{m} p\{\mu_k(\hat{R}) + \varepsilon\eta < x\} = -\pi \mathrm{Im}\, ma(z)$$

we complete the proof of Lemma 1.

LEMMA 5. *If conditions* (1), (4) *and* (7) *hold, then*

$$p\text{-}\lim_{n\to\infty}(\mu_m(\hat{R}) - \alpha_2) = 0, \qquad p\text{-}\lim_{n\to\infty}(\mu_1(\hat{R}) - \alpha_1) = 0,$$

where

$$\alpha_i = \nu_i(1-\gamma) + \gamma\nu_i^2 m^{-1}\sum_{p=1}^{m}(\nu_i - \mu_p)^{-1}, \qquad \gamma = m_n n^{-1}$$

and ν_1, ν_2 *are the maximal and minimal solution of the equation*

$$1 - \gamma + 2\gamma\nu m^{-1}\sum_{p=1}^{m}(\nu-\mu_p)^{-1} = \gamma\nu^2\sum_{p=1}^{m}(\nu-\mu_p)^{-2}.$$

Proof. Multiplying the equation, obtained in Lemma 1 by $\exp[-isx + |\varepsilon_n s|]\psi(s)$ and then integrating by x ($x \in (-\infty,\infty)$) and further by s ($|s| \leqslant |\varepsilon_n|^{-1}$) we get

$$\int_{-\varepsilon_n^{-1}}^{\varepsilon_n^{-1}} \sum_{k=1}^{m} \mathbf{M}\exp[-is\mu_k(\hat{R})]\psi(s)ds =$$

$$= \int_{-\varepsilon_n^{-1}}^{\varepsilon_n^{-1}} \int \exp(-isx)p(x)dx\,\psi(s)ds + I_n + O(1), \tag{31}$$

where

$$I_n = \int\Big[\int_{-\varepsilon_n^{-1}}^{\varepsilon_n^{-1}} [-\pi^{-1}\mathrm{Im}\,\varphi(x+i\varepsilon_n)]\exp(-isx + |s|\varepsilon_n)\varphi(s)ds\Big]dx,$$

$$\psi(s) = \int \exp(isx)g(x)dx,$$

and $g(x)$ is a bounded density such that $|g'(x)| < c < \infty$, $g(x)$ is equal to zero on

$$G(\varepsilon_n, \tau, \delta) = \{x : \alpha_1 - \tau < x < \alpha_2 + \tau, |\mathrm{Im} \ln \beta(z)| > \delta,$$

$$|\mathrm{Im} \ln(1 - \gamma + \gamma z b(z))| > \delta\}, \qquad 0 < \varepsilon_n,\ \tau,\ \delta.$$

Also

$$\int |s|^q |\psi(s)| ds \leqslant c < \infty, \tag{32}$$

for some $q > 0$.
Using the inverse formula for characteristic function, from (31) we deduce

$$\sum_{k=1}^{m} \int g(x) dp\{\mu_k(\hat{R}) < x\} = \sum_{i=1}^{5} A_i, \tag{33}$$

where

$$A_1 = (2\pi)^{-1} \int_{|s|>|\varepsilon_n|^{-1}} \sum_{k=1}^{m} e^{-is\mu_k(\hat{R})} \psi(s) ds,$$

$$A_2 = -(2\pi)^{-1} \int_{|s|>|\varepsilon_n|^{-1}} \int p(x) e^{-isx} dx \psi(s) ds,$$

$$A_3 = -(2\pi)^{-1} \int [-\pi^{-1} \mathrm{Im} \varphi(x + i\varepsilon_n)] \int e^{-isx} \psi(s) ds dx,$$

$$A_4 = -(2\pi)^{-1} \int \left[\int_{|s|<\varepsilon_n^{-1}} [-\pi^{-1} \mathrm{Im} \varphi(x + i\varepsilon_n)] e^{-isx} [e^{|s|\varepsilon_n} - 1] \psi(s) ds \right] dx,$$

$$A_5 = -(2\pi)^{-1} \int [-\pi^{-1} \mathrm{Im} \varphi(x + i\varepsilon_n)] \left[\int_{|s|>\varepsilon_n^{-1}} e^{-isx} \psi(s) ds \right] dx.$$

It is easy to verify (see (21)) that for sufficiently large x

$$|\varphi(x + i\varepsilon_n)| \leqslant c\varepsilon_n^{-r} [(x - \alpha_1)^2 + \varepsilon_n^2]^{-1}, \tag{34}$$

where r is a positive number.
Using this inequality, (32), and choosing a proper q we get

$$\lim_{n \to \infty} [|A_1| + |A_2| + |A_5|] = 0. \tag{35}$$

Applying the formula of integration by parts to A_4 we obtain

$$|A_4| = |(4\pi^2)^{-1} \int \int_{|s|<\varepsilon_n^{-1}} [\mathrm{Im} \ln \beta(z) + 3 \ln(1 - \gamma + \gamma z b(z))] \times$$

$$\times(-is)e^{-isx}[e^{|s|\varepsilon_n}-1]\psi(s)dsdx| \leqslant c|\varepsilon_n|[I_1+I_2], \quad (36)$$

where

$$I_1=\int_{|x|>c}[|\operatorname{Im}\ln\beta(z)|+3|\operatorname{Im}\ln(1-\gamma+\gamma z b(z))|]dx,$$

$$I_2=\int_{|x|\leqslant c}[|\operatorname{Im}\ln\beta(z)|+3|\operatorname{Im}\ln(1-\gamma+\gamma z b(z))|]dx.$$

Using (21) and the inequality $\ln|y| \leqslant \delta^{-1}|y|^{\delta}$ for $|y| \geqslant 1$, $\delta > 0$ we find

$$I_2 \leqslant \varepsilon_n^{-\delta_1}c, \qquad \delta_1 < 1. \quad (37)$$

For sufficiently large x

$$|\varrho_1| \leqslant c_1|z|^{-1} < 1, \qquad |x| > c, \qquad |\varrho_2| \leqslant c_2|z|^{-1} < 1,$$

$$|\operatorname{Im}\varrho_1| \leqslant c_3|x|^{-2}, \qquad |\operatorname{Im}\varrho_2| \leqslant c_4|x|^{-2},$$

where

$$\varrho_1 = m^{-1}\sum_{p=1}^{m} b_p^2\mu_p\gamma z, \qquad \varrho_2 = -\gamma+\gamma z b(z).$$

By these inequalities

$$I_1 \leqslant c\sum_{i=1}^{2}\left\{\int_{|x|>c}|\operatorname{Im}(\ln(1-\varrho_i)+\varrho_i)|dx + \int_{|x|>c}|\operatorname{Im}\varrho_i|dx\right\} \leqslant c_1.$$

This inequality together with (37) implies

$$\lim_{n\to\infty}|A_4| = 0. \quad (38)$$

For A_3 we have

$$A_3 = \int \operatorname{Im}[-\pi^{-1}\{\ln\beta(z)+3\ln(1-\gamma+\gamma z b(z))\}]g'(x)dx,$$

where integration is over $x \notin G_n(\varepsilon_n,\tau,\delta)$. Since for every $x \notin G_n$

$$\lim_{n\to\infty}\operatorname{Im}\varphi(x+i\varepsilon_n)=0,$$

we see that

$$\lim_{n\to\infty}A_3 = 0. \quad (39)$$

By (35), (38), (39), from (33) it follows that

$$\lim_{n\to\infty}\sum_{k=1}^{m}\int g(x)dp\{\mu_k(\hat{R}) < x\} = 0.$$

Therefore for every $d > 0$

$$\mathbf{P}\{d > \mu_1(\hat{R}) > \alpha_2 + \tau\} = \int_L g(x) dp\{\mu_1(\hat{R}) < x\} + o(1),$$

$$L = \{x : |\mathrm{Im} \ln \beta(z)| > \delta, |\mathrm{Im} \ln(1 - \gamma + \gamma z b(z))| > \delta\}. \tag{40}$$

Since (see (20))

$$\ln \beta(z) = -\ln[1 - z(\frac{\partial}{\partial z}) \ln(1 - \gamma + \gamma z b(z))]$$

and for $x \notin (\alpha_1 - \tau, \alpha_2 + \tau)$

$$\lim_{n \to \infty} [|\mathrm{Im} b(z)| + |\mathrm{Im} b'(z)|] = 0, \qquad |\mathrm{Re} b(z)| + |\mathrm{Re} b'(z)| \leqslant c < \infty$$

we have for $x \notin (\alpha_1 - \tau, \quad \alpha_2 + \tau)$

$$\lim_{n \to \infty} |\mathrm{Im} \ln b(z)| = 0, \qquad \lim_{n \to \infty} |\mathrm{Im} \ln(1 - \gamma + \gamma z b(z))| = 0.$$

Also

$$\mathrm{Re}(1 - \gamma + \gamma z b(z)) = 1 + \int_{\alpha_1}^{\alpha_2} u(x - u)[(x - u)^2 + \varepsilon^2]^{-1} p(u) du,$$

$$\liminf_{n \to \infty} \mathrm{Re}[1 - \gamma - z^2 \gamma \partial b(z)/\partial z] > 0.$$

Consequently, from (40) we get

$$\lim_{n \to \infty} \mathbf{P}\{\alpha_2 + \tau < \mu_1(\hat{R}) < d\} = 0, \tag{41}$$

where $d > 0$ is some constant.

To finish the proof of Lemma 5 we use the results of (Girko, 1988b).

Consider equations for α_2, we obtained in Lemma 5. Putting $\nu = \mu_1 + t$ we transform these

$$\alpha_2 = (\mu_1 + t)(1 - \gamma) + \gamma(\mu_1 + t)^2 m^{-1} \sum_{p=2}^{m} (\mu_1 + t - \mu_p)^{-1} + \gamma(\mu_1 + t)^2 m^{-1} t^{-1},$$

$$\alpha_2 = \gamma(\mu_1 + t)^2 m^{-1} \sum_{p=1}^{m} \mu_p(\mu_1 + t - \mu_p)^{-2} + \gamma(\mu_1 + t)^2 m^{-1} t^{-2} \mu_1. \tag{42}$$

Note that $\nu > \mu_1$, therefore $t > 0$. If $t_n \to c > 0$ as $n \to \infty$ then from (42) it is possible to find $\mu_1 + c$ (but not μ_1). If $m t_n^2 \to \infty$ so that $t_n \to 0$, then under the condition $\mu_1 > \mu_2 + \tau$, $\tau > 0$ we get from (42)

$$\alpha_2 = \mu_1(1 - \gamma) + \gamma \mu_1^2 m^{-1} \sum_{p=2}^{m} (\mu_1 - \mu_p)^{-1} + o(1),$$

$$\alpha_2 = \gamma\mu_1^2 m^{-1}\sum_{p=2}^{m}(\mu_1-\mu_p)^{-2}\mu_p + o(1).$$

In general, this system of equations for μ_1 is unsolvable. But it is possible to solve it when $mt_n^2 \to c$, $c > 0$, as $n \to \infty$. Then under the condition $\mu_1 > \mu_2 + \tau$, $\tau > 0$ the system (42) takes on the form

$$\alpha_2 = \mu_1(1-\gamma) + \gamma\mu_1^2 m^{-1}\sum_{p=2}^{m}(\mu_1-\mu_p)^{-1} + o(1),$$

$$\alpha_2 = \gamma\mu_1^2 m^{-1}\sum_{p=2}^{m}(\mu_1-\mu_p)^{-2}\mu_p + \gamma\mu_1^3 c^{-1} + o(1). \qquad (43)$$

We can find c and μ_1 from this system of equations if n is sufficiently large and

$$1-\gamma-\gamma\mu_1^2 m^{-1}\sum_{p=2}^{m}(\mu_1-\mu_p)^{-2} + 2\gamma\mu_1 m^{-1}\sum_{p=2}^{m}(\mu_1-\mu_p)^{-1} > \tau_1 > 0. \qquad (44)$$

Using the G_2 – estimate of the Stieltijes transform

$$\varphi(z, R_{m_n}) = m^{-1}\mathrm{Sp}(Iz - R_{m_n})^{-1}$$

of the spectral function (Girko, 1988b) replace the first equation in (43) by

$$\alpha_2 = \mu_1(1-\gamma) + \gamma\mu_1^2\mathrm{Re}[f(\mu_1+i\varepsilon)(\mu_2+i\varepsilon)^{-1} - 1 + \gamma](\mu_1+i\varepsilon)^{-1}, \qquad (45)$$

where $f(\mu_1+i\varepsilon)^{-1}$ is an analytic solution of the equation

$$1-\gamma+\gamma f m^{-1}\mathrm{Sp}(If-\hat{R})^{-1} = (\mu_1+i\varepsilon)^{-1}f,$$

so that $f(\mu_1) < 0$. It follows (Girko,1988b) that for every z, $\mathrm{Im}z > 0$,

$$p\text{-}\lim_{n\to\infty}[G(z)-\varphi(z,R)] = 0.$$

But since the functions $G(z)$ and $\varphi(z,R)$ are uniformly continous on the domain $\mathrm{Im}z > 0$, we have for every $c < \infty$

$$p\text{-}\lim_{n\to\infty}\sup_{|z|<c,\mathrm{Im}z>0}|G(z)-\varphi(z,R)| = 0. \qquad (46)$$

From (45) we obtain the main equation (2) for G–estimate of the eigen–value μ_1. Let us prove the main assertion of the theorem. Let Θ be arbitrary measurable solution of equation (2). Using Theorem 3.II.2 from (Girko, 1988b), (5), (43), (46), Lemma 5, transform equation (2):

$$\mu_1(1-\gamma) + \gamma\mu_1^2 m^{-1}\sum_{p=2}^{m}(\mu_1-\mu_p)^{-1} =$$

$$= \Theta(1-\gamma) + \gamma\Theta^2 \mathrm{Re}\, m^{-1} \sum_{p=2}^{m} (\Theta - \mu_p + i\varepsilon)^{-1} + \delta_n,$$

where $p - \lim_{n\to\infty} \delta_n = 0$.
It follows that

$$\mu_1 - \Theta = \gamma m^{-1} \sum_{p=2}^{m} \mathrm{Re}\{(\Theta - \mu_p + i\varepsilon)^{-1}(\mu_1 - \mu_p)^{-1}(\mu_p^2(\mu_1 - \Theta) + \varkappa_{np})\} + \delta_n,$$

where

$$\varkappa_{n_p} = -2i\varepsilon\mu_p\mu_1 + i\varepsilon\mu_p^2 - \varepsilon^2\mu_1 + \varepsilon^2\mu_p.$$

This together with (4) and (5) imply Theorem 1.

REFERENCES

Anderson, T.W. (1958). *An Introduction to Multivariate Statistical Analysis*. Wiley, New York.

Girko, V.L. (1980). *Theory of Random Determinants*. Vishcha shkola, Kiev.

Girko, V.L. (1987). Introduction in general statistical analysis. *Theory Probab. Appl.* **32**, 229—242..

Girko, V.L. (1988). *Multivariate statistical analysis*. Vishcha shkola, Kiev.

Girko, V.L. (1988). *Spectral Theory of Random Matrices*. Nauka, Moscow.

Prob. Theory and Math. Stat., Vol. 1, pp. 424–432
B.Grigelionis *et al.* (Eds.)
1990 VSP/Mokslas

ON LARGE DEVIATIONS FOR TRIMMED SUMS OF INDEPENDENT RANDOM VARIABLES

V.V. GODOVAN'CHUK and V.V. VINOGRADOV
Department of Mathematics and Mechanics,
Moscow State University, Lenin Hills, Moscow, 119899 GSP, USSR

ABSTRACT

Consider i.i.d.r.v. $\{X_n,\ n \geqslant 1\}$ with power tails, and let $S_{n,y}(t) = (X_1+\ldots+X_{[nt]})/y$. We investigate the character of the paths $S_{n,y}(\cdot)$ making the main contribution to the probabilities of large deviations (in the full range). Asymptotics of those probabilities is studied and applied to large deviations for trimmed sums when order statistics are based on absolute values. The interplay between the theory of large deviation probabilities for the families of stochastic processes and that for the summation of independent random variables is demonstrated.

INTRODUCTION

Consider a sequence of independent and identically distributed random variables (i.i.d.r.v.) $\{X_n, n \geqslant 1\}$ with distribution function $F(x)$. Assume that their tails have the asymptotics of power type:

$$\left.\begin{array}{r} 1 - F(x) \sim c_+ x^{-\alpha} \\ F(-x) \sim c_- x^{-\alpha} \end{array}\right\} \qquad (0)$$

as $x \to \infty$, where $c_+ > 0, c_- > 0, \alpha > 0$. We also assume, without loss of generality, that each X_i has zero mean when $\alpha > 1$ and unit variance when $\alpha > 2$. For the sake of simplicity we exclude the cases when α equals 1 or 2.

An important problem is the investigation of the limit behaviour of the sums of independent random variables without several maximum summands; the study of the asymptotics of their large deviation probabilities (LDP) is the subject of our work (Godovan'chuk and Vinogradov, 1989). At the same time in (Maller, 1982) it was announced more important, in view of possible statistical applications, to consider the similar problem when a finite number of extremes determined by ordering the moduli of the terms in the sum is removed. The weak convergence and the optimal convergence rate estimates in the weak law of large numbers for such sums have been obtained in (Arov and Bobrov, 1960; Hall, 1978; Teugels, 1981) and (Hatori et al., 1980) respectively, whereas their LDP have not been studied.

Let us introduce some notation. Denote

$$S_n = X_1 + \ldots + X_n, \quad S_0 = 0.$$

Define

$$u_1 = \max_{1\leqslant i\leqslant n} |X_i|, \quad i_1 = \min(i : |X_i| = u_1, \quad 1 \leqslant i \leqslant n), \quad X_1(n) = X_{i_1},$$

$$u_2 = \max_{1\leqslant i\leqslant n, i\neq i_1} |X_i|, i_2 = \min(i : |X_i| = u_2, 1 \leqslant i \leqslant n, i \neq i_1), X_2(n) = X_{i_2},$$

$\ldots$, e.t.c.

The following result (Heyde, 1968) describes the behaviour of LDP for $|S_n|$.

PROPOSITION. *Let conditions* (0) *be satisfied with* $\alpha \in (0,1) \cup (1,2)$ *and let* $\mathbf{M}X_i = 0$ *if* $\alpha \in (1,2)$. *Then*

$$\mathbf{P}\{|S_n| > y\} \sim n\mathbf{P}\{u_1 > y\} \tag{1}$$

as $n \to \infty$, $y/n^{1/\alpha} \to \infty$.

It follows from (1) that a large deviation of the sum occurs due to one large summand. For this reason, when studying some problems connected with the investigation of the absolute values of the sums of a sample of growing size it would be natural to replace them by $X_1(n)$. Therefore we need estimates of errors resulting from such replacements.

Note that in the present paper we call $S_n - X_1(n) - \ldots - X_k(n)$, where $k \geqslant 1$ is any fixed integer, trimmed sums, whereas the term "lightly trimmed sums" is used more frequently.

It turns out that results close to the one mentioned above could be readily obtained by employing theorems from (Godovan'chuk, 1978, 1982) and (Pinelis, 1982). To do this we define the random step-function (r.s.f.) $S_{n,y}$ by $S_{n,y}(t) = S_{[nt]}/y$, where $0 \leqslant t \leqslant 1$. Let $\mathbf{D}[0,1]$ be the CADLAG space equipped with the uniform metrics ρ. Then probabilities related to the distributions of S_n, $X_1(n), \ldots$, e.t.c., may be represented as the probabilities that $S_{n,y}$ belongs to certain subsets of $\mathbf{D}[0,1]$ (see Godovan'chuk, 1978) for details). In this paper the results by Godovan'chuk and by Pinelis related to the distributions of trimmed sums for special subsets of $\mathbf{D}[0,1]$ are refined. This allows us to cover the full range of large deviations for these sums not only in the case $\alpha < 2$, but for $\alpha > 2$ as well.

On the other hand, out results may be interesting not only in view of possible applications, but they also provide a better understanding of the nature of large deviations, i.e. what are the typical paths of $S_{n,y}$ if an event of small probability (a large deviation) has occurred. It is well known that if in the construction of a r.s.f. $S_{n,y}$ random variables with finite exponential moments were used, then its large deviation is mainly contributed by close–to–continuous (or even close–to–smooth) paths (see (Wentzell, 1986), Introduction). Below we shall refer to these paths as paths of the first type. Similar results for sums of order statistics corresponding to

a sample of variables with finite exponential moments have been recently obtained (Bentkus and Zitikis, 1989) (see Ref. therein).
On the other hand, if the distribution function $F(x)$ belongs to the domain of attraction of a non-normal stable law then large deviations occur mostly via almost piecewise constant paths, which have one or several big jumps (below we shall refer to these paths as paths of the second type). This fact has been established in ((Godovan'chuk, 1978), Theorems 1, 2) for the full range of large deviations. For the case of attraction to the normal law (i.e. when $\alpha > 2$) the paths of both types can give comparable contributions to a large deviation. Note that it is possible in the case $\alpha > 2$ and in a very narrow range of large deviations only (see (Pinelis, 1982)) when $S_{n,y}$ has not more than one big jump. Otside this area, if $y \geqslant n^{1/2+\varkappa}$, events of small probability occur mostly due to the second type paths (here $\varkappa > 0$ is any real number). This result follows from (Godovan'chuk, 1982, Theorems $1', 2'$ and Remark 4).
It seems reasonable to assume that large deviations sometimes may occur via the paths that have one or several big jumps (up or down) and are close–to–smooth functions between them. This fact is established below for one subset of $\mathbf{D}[0, 1]$ connected with a certain trimmed sum (see Theorem 2′). Namely, it is shown that in the range of deviations $(cn\ln n)^{1/2} \leqslant y \leqslant (Cn\ln n)^{1/2}$ the event

$$\{S_n - X_1(n) > ay, \quad S_n > by\}$$

occurs mostly due to the paths which have one big jump and are close–to–smooth functions before and after the jump (here c, C, a and b are some positive constants). We supplement this theorem with an analogous (but simpler) result for the case $\alpha < 2$ (see Theorem 2) to emphasize the difference between the cases $\alpha > 2$ and $\alpha < 2$.
Asymptotic behaviour of the LDP for $S_n - X_1(n)$ is also found (see Theorem 1). In this connection see also Corollary 2 of (Pinelis, 1982). Finally we would like to remark that for right–hand large deviations for sums without upper k and lower l order statistics the results are the same as for sums without upper k ones only (of course we mean the ordering by the moduli of r.v.). Note that it is not so for the case of normal deviations (see (Arov and Bobrov, 1960; Hall, 1978) and (Teugels, 1981)).
This paper was motivated by and is close to ((Wentzell, 1986), Chapter 6), where the technique of reducing problems from the real axis to the CADLAG space was applied to the investigation of the asymptotic behaviour of LDP for sums of i.i.d.r.v. In the present paper we demonstrate that this technique is very powerful in the investigation of the asymptotic behaviour of the LDP for certain sums of order statistics as well.

RESULTS.

Let us introduce some notation.
Denote $\mathbf{P}\{S_n - X_1(n) - \ldots - X_k(n) > y\}$ by $P_{n,y}(k)$, $\mathbf{P}\{S_n - X_1(n) > ay| |S_n > by\}$ by $P_{n,y}(a, b)$, the Laplace function by $\Phi(\cdot)$. Let $m_\alpha(a, b)$ be equal to $m_\alpha(a, b) = a^{-2\alpha}$ if $b < 2a$ and to

$$a^{-\alpha}(b-a)^{-\alpha}+\int_{a}^{b-a}(b-x)^{-\alpha}d(-x^{-\alpha})$$

if $b \geqslant 2a$, where a and b are any fixed positive numbers. Fix $t_1,\dots,t_k$: $0<t_1<\dots<t_k\leqslant 1$ and real numbers $x_1,\dots,x_k$. We denote, following (Wentzell, 1985) the function which is equal to zero if $s\in[0,t_1)$, to x_1 if $s\in[t_1,t_2),\dots,$ to x_k if $s\in[t_k,1]$ by $O^{t_1t_2\dots t_k}_{x_1\dots x_{k-1}x_k}(s)$. The set of all such functions $\{O^{t_1t_2\dots t_k}_{x_1\dots x_{k-1}x_k}(s)\}$ will be denoted by B^k, where $s\in[0,1]$, and $k\geqslant 0$ is any fixed integer. In particular, B^0 consists of only one function which is equal to zero identically. Denote the standard univariate Wiener process by $w(s)$, $w(0)=0$; $\max(a,b)$ by $a\vee b$, $\min(a,b)$ by $a\wedge b$, ∂A and $(A)_\varepsilon$ will be the boundary of the set A and its ε–neighbourhood respectively.

THEOREM 1. *Let $k\geqslant 1$ be any fixed integer. Then as $n\to\infty$, $y/n^{1/\min(2,\alpha)}\to\infty$*

a) $P_{n,y}(k)\sim C_n^{k+1}c_+(c_++c_-)^k y^{-(k+1)\alpha}$ *if* $\alpha<2$,

b) $P_{n,y}(k)\sim 1-\Phi(y/n^{1/2})+C_n^{k+1}c_+(c_++c_-)^k y^{-(k+1)\alpha}$ *if* $\alpha>2$.

THEOREM 2. *Let $\alpha<2$, a and b be any fixed positive numbers. Then*

$$P_{n,y}(a,b)\sim(n/2)c_+y^{-\alpha}b^{\alpha}m_\alpha(a,b)$$

as $n\to\infty$, $y/n^{1/\alpha}\to\infty$.

THEOREM 2$'$. *Let $\alpha>2$; a and b be any fixed positive numbers. Then as $n\to\infty$, $y/n^{1/2}\to\infty$*

a) $P_{n,y}(a,b)\sim(1-\Phi(ay/n^{1/2})+C_n^2c_+^2y^{-2\alpha}a^{-2\alpha})/(1-\Phi(by/n^{1/2})+nc_+(by)^{-\alpha})$, *if* $a>b$,

b) $P_{n,y}(a,b)\sim(1-\Phi(by/n^{1/2})+c_+ny^{-\alpha}(b-a)^{-\alpha}(1-\Phi(ay/n^{1/2}))+$
$+C_n^2c_+^2y^{-2\alpha}m_\alpha(a,b))/(1-\Phi(by/n^{1/2})+nc_+(by)^{-\alpha})$ *if* $a\leqslant b$.

To proof a) of Theorems 1 and 2 one may use Theorems 1–2 from (Godovan'chuk, 1978). The proof of Theorem 2$'$ is based on a slight modification of that of Theorem 1 of (Godovan'chuk and Vinogradov, 1989) where the probability of event $\{S_n-X_1(n)>ay, S_n>by\}$ was considered along with the subsequent application of Corollary 2 from (Pinelis, 1982) to get the asymptotic behaviour of $\mathbf{P}\{S_n>y\}$. The details are omitted.
In proving b) of Theorem 1 we consider the simplest case $k=1$.
Rewrite the event $\{S_n-X_1(n)>y\}$ as $\{S_{n,y}\in A\}$, where

$$A = \{f \in \mathbf{D}[0,1] : f(1) = \sup_{t\in[0,1]}(f(t) - f(t-)) > 1\}.$$

Let $\beta = \beta(n)$ be a bounded sequence of positive numbers such that $\beta(n)n^\delta \to \infty$ as $n \to \infty$ for every $\delta > 0$. Let $r = \rho(A, B^1)$. Obviously, $r > 0$. Take an arbitrary

$$\gamma \in (0, (a/8) \wedge (r/40) \wedge (b-a)2).$$

Let us introduce the following quantities by analogy with (Pinelis, 1982):
$\Pi_1 = \mathbf{P}\{S_{n,y} \in A; |X_i| \leqslant \beta n/y, i = 1,\dots,n\}$;
$\Pi_2 = \mathbf{P}(\bigcup_{i=1}^{n} \{|X_i| > \beta n/y\})$;
$\Pi_3 = \sum_{i=1}^{n} \mathbf{P}\{S_{n,y} \in A, |X_i| > \beta n/y, |X_j| \leqslant \beta n/y \quad \forall j \neq i\}$;
$\Pi_4 = \sum_{1\leqslant i\neq j\leqslant n} \mathbf{P}\{S_{n,y} \in A; |X_i| > \beta n/y, |X_j| > \beta n/y, |X_k| < \gamma y \quad \forall k \neq i\}$;
$\Pi_5 = \sum_{1\leqslant i<j\leqslant n} \mathbf{P}\{S_{n,y} \in A; |X_i| > \gamma y, |X_j| > \gamma y\}$;
$\Pi_6 = \sum_{i=1}^{n} \mathbf{P}\{S_{n,y} \in A; |X_i| > \gamma y, |X_j| \leqslant \gamma y \quad \forall j \neq i\}$;
$\Pi_7 = \mathbf{P}\left\{\bigcup_{1\leqslant i<j<k\leqslant n} \{|X_i| \wedge |X_j| \wedge |X_k| > \gamma y\right\}\}$;
$\Pi_8 = \mathbf{P}\{S_{n,y} \in A; |X_i| \leq \gamma y, i = 1,\dots,n\}$.

LEMMA 1. (see (Pinelis, 1982), Lemma 9). *For any $M > 0$ there exists $c > 0$ such that $\Pi_2 = O(e^{My^2/n})$ as $n \to \infty$ for $n^{1/2} \leqslant y \leqslant (cn\ln n)^{1/2}$.*

LEMMA 2. (see (Pinelis, 1982), Lemma 4). *There exist $c_0 > 0$ and $\delta > 0$ such that for any subset U (of the CADLAG space) for which the probability of $\{S_{n,y} \in U\}$ is defined for all $y \leqslant n^{1/2+\delta}$*

$$\mathbf{P}\{S_{n,y} \in U, |X_i| \leqslant \beta n/y, i = 1,\dots,n\} =$$

$$= \mathbf{P}\{n^{1/2}w(\cdot)/y \in U\}(1 + O(n^{-\delta}))+$$

$$+O(P\{n^{1/2}w(\cdot)/y \in (\partial U)\beta\} + O(n^{-c_0 y^2/n})$$

as $n \to \infty$, $y/n^{1/2} \to \infty$.

LEMMA 3. (cf. with (Pinelis, 1982)). *Choose arbitrary $0 < a_1 < a_2 < \infty$. Then if $\eta = o(u^{-2})$*

$$\mathbf{P}\{u(a-\eta) \leqslant w(1) \leqslant u(a+\eta)\} = o(P\{w(1) \leqslant ua\}) \text{ as } u \to \infty$$

uniformly with respect to a $a \in [a_1, a_2]$.

The following five technical lemmas make it possible to neglect the quantities $\Pi_3, \Pi_4, \Pi_6, \Pi_7, \Pi_8$ (as compared to Π_1 and Π_5) in the proof of the theorem.

Lemma 4. *Choose arbitrary $0 < c < C < \infty$ and $\beta(n) = o(1/\ln n)$ as $n \to \infty$. If $(cn\ln n)^{1/2} \leqslant y \leqslant (Cn\ln n)^{1/2}$ then*

$$\Pi_3 = o(P\{n^{1/2}w(1) > y\}).$$

Proof. It follows from the definition of Π_3 that

$$\Pi_3 = nP\{|X_1| > \beta n/y\}P\{S_{n-1} > y, |X_j| \leqslant \beta n/y, \qquad j = 1, \dots, n-1\}.$$

By Lemmas 2 and 3 we obtain that the latter sum is

$$o(P\{n^{1/2}w(1) > y\}).$$

Lemma 5. *There exists $C > 0$ such that if $y \geqslant (Cn\ln n)^{1/2}$, then $\Pi_8 = o(n^2y^{-2\alpha})$ as $n \to \infty$.*

Proof. Obviously

$$\Pi_8 = \mathbf{P}\{S_n - X_1(n) > y, |X_i| \leqslant \gamma y, i = 1, \dots, n\} \leqslant$$
$$\leqslant \mathbf{P} \max_{1 \leqslant i \leqslant n} |S_i| > y;\ |X_i| \leqslant \gamma y, i = 1, \dots, n\}.$$

Note that if $\gamma \in (0, a/(3b-1))$ the right-hand side of this inequality does not exceed $2^l(\max_{1 \leqslant i \leqslant n} \mathbf{P}\{|S_i| > y/(3l-1)\})^l$ for any integer $n \geqslant 1$ and any real $y > 0$. The proof of this estimate is similar to that of ((Godovan'chuk, 1978), Lemma 2) and, therefore, is omitted. Let $l = 3$. Then

$$\Pi_8 \leqslant 8(\max_{1 \leqslant i \leqslant n} \mathbf{P}\{|S_i| > y/8\})^3. \tag{2}$$

Employing Corollary 2 from (Pinelis, 1982), we deduce that

$$\mathbf{P}\{|S_n| > y\} \sim 2(1 - \Phi(y/n^{1/2})) + n(c_+ + c_-)y^{-\alpha}$$

as $n \to \infty$, $y/n^{1/2} \to \infty$. Hence for any $\varkappa > 0$

$$\mathbf{P}\{|S_n| > y\} \leqslant c(\varkappa)ny^{-\alpha} \tag{3}$$

as $n \to \infty$ for $y \geqslant ((\alpha - 2 + \varkappa)n\ln n)^{1/2}$. Note that (3) can be deduced directly from the Nagaev–Fuk inequalities (see (S.Nagaev, 1979), Theorem 1, 3). The inequalities (upper bounds) are often used to prove the asymptotic relations like (1).

The assertion of Lemma 5 follows immediately now from (2) and (3).

Lemma 6. *Let $n \geqslant 1$ be any integer and $y > 0$ be a positive number. Then*

$$\Pi_7 = O(n^3y^{-3\alpha}).$$

The proof is obvious.

LEMMA 7. *There exists $C > 0$ such that*

$$\Pi_6 = o(n^2 y^{-2\alpha})$$

as $n \to \infty$ for $y \geqslant (Cn\ln n)^{1/2}$.

Proof. Recall that $\rho(A, B^1) = r > 0$. Thus for every $i = 1, \dots, n$ the $r/2$–neighbourhood of the function $O^{i/n}X_i/y$ does not have common points with A. Hence

$$\{S_{n,y} \in A; |X_i| > \gamma y, |X_j| \leqslant \gamma y \qquad \forall j \neq i\}$$

implies either
$\max_{1 \leqslant k < i} \left| \sum_{j=1}^{k} X_j/y \right| > r/4$ or $\max_{i < k \leqslant n} \left| \sum_{j=i+1}^{k} X_j/y \right| > r/4$.
Therefore

$$\Pi_6 \leqslant 2n\mathbf{P}\{|X_1| > \gamma y\}\mathbf{P}\{\max_{1 \leqslant i \leqslant n-1} |S_i| > ry/4;$$
$$|X_j| \leqslant \gamma y, j = 1, \dots, n-1\}.$$

The last factor has been already investigated in Lemma 5 and we obtain the desired estimate.

LEMMA 8. *Choose arbitrary $0 < c < C < \infty$. Then $\Pi_4 = o(n^2 y^{-2\alpha})$ as $n \to \infty$ for $(cn\ln n)^{1/2} \leqslant y \leqslant (Cn\ln n)^{1/2}$.*

The proof is analogous to that of Lemma 7.
The following (and the last) lemma is especially important.

LEMMA 9. $\Pi_5 \sim C_n^2 c_+(c_+ + c_-)y^{-2\alpha}$ *as $n \to \infty$, $y/n^{1/2} \to \infty$.*

Proof. Obviously

$$\Pi_5 = \sum_{1 \leqslant i \neq j \leqslant n} \mathbf{P}\{S_n - X_1(n) > y, \gamma y < X_i \leqslant |X_j|\} +$$
$$+ \sum_{1 \leqslant i \neq j \leqslant n} \mathbf{P}\{S_n - X_1(n) > y, |X_i| \leqslant |X_j|, X_i < -\gamma y\}. \tag{4}$$

It is easily seen that the second sum on the right–hand side of (4) does not exceed

$$2 \sum_{1 \leqslant i \neq j \leqslant n} \mathbf{P}\{S_n - X_i - X_j > y, |X_i| \wedge |X_j| > \gamma y\} + O(n^3 y^{-3\alpha}) =$$
$$= 2C_n^2 \mathbf{P}\{S_{n-2} > y\}\mathbf{P}\{|X_1| > \gamma y\}^2 + O(n^3 y^{-3\alpha}) = o(n^2 y^{-2\alpha}). \tag{5}$$

The first sum on the right–hand side of (4) is equal to

$$\sum_{1 \leqslant i \neq j \leqslant n} \mathbf{P}\{S_n - X_1(n) > y, X_i > \gamma y, X_i \leqslant |X_j|, i_1 \in \{i, j\}\} +$$

$$+\sum_{1\leqslant i\neq j\leqslant n}\mathbf{P}\{S_n-X_i(n)>y, X_i>\gamma y, X_i\leqslant |X_j|, i_1\neq i\neq j\}. \qquad (6)$$

Obviously, the latter sum does not exceed

$$O(n^3y^{-3\alpha}).$$

On the other hand, the first summand in (6) is equal to

$$C_n^2\mathbf{P}\{S_{n-2}>y(1-2\gamma)\}P\{|X_1|>\gamma y\}P\{X_1>\gamma y\}+$$
$$+C_n^2\int_{-\infty}^{(1-2\gamma)y}\mathbf{P}\{X_{n-1}\wedge|X_n|>y-z\}\,dF_{S_{n-2}}(z).$$

Since the first term of the latter sum is $o(n^2y^{-2\alpha})$ as $n\to\infty, y/n^{1/2}\to\infty$, the first sum on the right-hand side of (4) is equal to

$$C_n^2\mathbf{P}\{X_{n-1}\wedge|X_n|>y\}F_{S_{n-2}}((1-2\gamma)y)+$$
$$+C_n^2\int_{-\infty}^{(1-2\gamma)y}(\mathbf{P}\{X_{n-1}\wedge|X_n|>y-z\}-$$
$$-\mathbf{P}\{X_{n-1}\wedge|X_n|>y\})dF_{S_{n-2}}(z)+o(n^2y^{-2\alpha}). \qquad (7)$$

It is not difficult to show that the absolute value of this integral is $o(y^{-2\alpha})$ as $y\to\infty$. Similar estimates can be found in ((Wentzell, 1986), Section 6,3). Now (0) implies that the first summand is equivalent to

$$C_n^2c_+(c_++c_-)y^{-2\alpha} \qquad (8)$$

as $n\to\infty, y/n^{1/2}\to\infty$.
Combining (4)–(8) we get the desired estimate.
Let us finish the proof of Theorem 1. Note that as $n\to\infty, y/n^{1/2}\to\infty$ we have

a) $\mathbf{P}\{S_{n,y}\in A\}=\Pi_1+O(\Pi_2)$. Thus it follows from Lemmas 1, 2 and 3 that there exists $c>0$ such that the assertion of the theorem holds for $y\leqslant(c\ln n)^{1/2}$.

b) $\mathbf{P}\{S_{n,y}\in A\}=\Pi_1+\Pi_3+O(\Pi_4)+\Pi_5+O(\Pi_7)$. Thus it follows from Lemmas 2, 3, 4, 6, 8 and 9 that for arbitrary $0<c<C<\infty$ the assertion of the theorem holds for $(cn\ln n)^{1/2}\leqslant y\leqslant(Cn\ln n)^{1/2}$.

c) $\mathbf{P}\{S_{n,y}\in A\}=\Pi_5+O(\Pi_6)+O(\Pi_7)+O(\Pi_8)$.
Thus it follows from Lemmas 5, 6, 7 and 9 that there exists $C>0$ such that the assertion of the theorem holds for $y\geqslant(Cn\ln n)^{1/2}$.
It remains only to combine a), b), c) together.
In conclusion it should be pointed out that the comparison of Theorems 2 and 2′ with the corresponding results from (Godovan'chuk, 1978) and from (Godovan'chuk and Vinogradov, 1989) concerning $S_n-\max(X_1,\ldots,X_n)$ reveals that for the conditional probabilities of the right-hand large deviations of sums without summands with maximum absolute value, the contribution of event $|\min(X_1,\ldots,X_n)|>|\max(X_1,\ldots,X_n)|$ is negligible.

Acknowledgement

We are grateful to A.D.Wentzell for the helpful discussions.

REFERENCES

Arov, D.Z. and Bobrov, A.A. (1960). The extreme terms of a sample and their role in the sum of independent variables. *Theory Probab. Appl.* **5**, 415–435.

Bentkus, V. and Zitikis, R. (1989). *Probabilities of large deviations for L–statistics.* Prepr. No. 8. Inst. Math. and Cybern., Vilnius.

Godovan'chuk, V.V. (1978). Probabilities of large deviations for sums of independent random variables belonging to the domain of attraction of a stable law. *Theory Probab. Appl.* **23**, 624–630.

Godovan'chuk, V.V. (1982). Asymptotic probabilities of large deviations due to large jumps of a Markov process. *Theory Probab. Appl.* **26**, 314–327.

Godovan'chuk, V.V., Vinogradov, V.V. (1989). On large deviations of sums of independent random variables without several maximal summands. *Theory Probab. Appl.* **34**, 569–571.

Hall, P. (1978). On the extreme of a sample from the domain of attraction of a stable law. *J. London Math. Soc.* **18**, 181–191.

Hatori, H. *et al.* (1980). Supplement to the law of large numbers when extreme terms are excluded. *Z. Wahr. verw. Geb.* **52**, 229–234.

Heyde, C.C. (1968). On large deviation probabilities in the case of attraction to a non–normal stable law. *Sankhya* **A30**, 253–258.

Maller, R.A. (1982). Asymptotic normality of lightly trimmed sums – a converse. *Math. Proc. Cambridge Phil. Soc.* **92**, 535–545.

Nagaev, S.V. (1979). Large deviations of sums of independent random variables. *Ann. Probab.* **7**, 745–789.

Pinelis, I.F. (1982). A problem of large deviations in a space of trajectories. *Theory Probab. Appl.* **26**, 69–84.

Teugels, J.L. (1981). Limit theorems on the order statistics. *Ann. Probab.* **9**, 868–880.

Wentzell, A.D. (1985). Infinitesimal characteristics of Markov processes in a function space, which describe the past. *Theory Probab. Appl.* **30**, 625–639.

Wentzell, A.D. (1986). *Limit Theorems on large Deviations for Markov Stochastic Processes.* Nauka, Moscow.

Prob. Theory and Math. Stat., Vol. 1, pp. 433–445
B.Grigelionis *et al.* (Eds.)
1990 VSP/Mokslas

RANDOM OPERATOR FUNCTIONS DEFINED BY STOCHASTIC DIFFERENTIAL EQUATIONS

N.YU. GONCHARUK and YU.L. DALETSKY
Department of Mathematics Methods of System Analysis,
Kiev Polytechnic Institute, Pobedy 37, 252056 Kiev, USSR

ABSTRACT

A Cauchy problem for an abstract parabolic equation in a Banach space is considered. It is ill–posed because of the fact that the unbounded semigroup generator of the problem is multiplied by the negative coefficient. The problem is regularized by the adding the white noise to the coefficient. The obtained stochastic differential equation is considered.

1. Consider a linear differential equation

$$\frac{\partial \xi(t)}{\partial t} = \alpha A\xi(t), \quad t \in \mathbf{R}^1, \tag{1.1}$$

$$\xi(t_0) = u_0,$$

in a Banach space $\mathcal{B}$, where A is an unbounded generator of the analytic semigroup e^{tA} in $\mathcal{B}$, $\alpha \in \mathbf{R}$, $u_0 \in \mathcal{B}$. As it is known if $\alpha(t-t_0) > 0$ then there exists the solution of the Cauchy problem for the equation (1.1) which is given by the formula

$$\xi(t) = e^{\alpha(t-t_0)A}u_0,$$

and if $\alpha(t-t_0) < 0$ then the corresponding Cauchy problem is ill–posed.
Let $w(t)$ be a standard Wiener process. Adding the white noise $\dot{w}(t)$ with the coefficient $\beta \in \mathbf{R}^1$ to the coefficient α we obtain the equation

$$\frac{\partial \xi(t)}{\partial t} = (\alpha + \beta\dot{w}(t))A\xi(t), \quad \xi(t_0) = u_0,$$

which has a meaning as a stochastic differential equation in $\mathcal{B}$:

$$d\xi(t) = \alpha A\xi(t)dt + \beta A\xi(t)dw(t), \quad t \in \mathbf{R}^1, \tag{1.2}$$

$$\xi(t_0) = u_0. \tag{1.3}$$

It will be shown below that the Cauchy problem (1.2), (1.3) is well–posed for $\beta \not\equiv 0$ and for arbitrary sign of $\alpha(t-t_0)$. Moreover its solution is given by the formula

$$\xi(t) = V_\beta(t,t_0)u_0 = \exp\{\alpha\Delta tA - \beta^2|\Delta t|A^2/2 + \beta\Delta wA\}u_0, \quad t \in \mathbf{R}^1, \tag{1.4}$$

where $\Delta t = t - t_0$, $\Delta w = w(t) - w(t_0)$. This formula may be received from Ito's formula if the operator exponential has a sense. In this case it is possible to speak about the ***stochastic regularization*** of the ill-posed Cauchy problem.
As an example let us consider the Cauchy problem for the partial differential equation with A being a second order elliptic operator

$$A = \sum_{k,j=1}^{n} \frac{\partial}{\partial x_k} a_{kj}(\mathbf{x}) \frac{\partial}{\partial x_j} + \sum_{k=1}^{n} b_k(\mathbf{x}) \frac{\partial}{\partial x_k}. \tag{1.5}$$

Here $\mathbf{x} = (x_1, ..., x_n) \in \mathbf{R}^n$; ξ, u_0 are scalar functions. In the case when the Cauchy problem for the nonperturbed equation is well-posed the Cauchy problem (1.2), (1.3) with the operator of the type (1.5) with a_{kj} and b_k being functions of t only is considered in (Gikhman and Mestechkina, 1987).
We consider two examples of the operator exponential from (1.4) in Sections 2 and 3, respectively.

2. Let $-A$ be a sectorial operator in $\mathcal{B}$ satisfying the following properties
1) there exist $a \in \mathbf{R}^1$ and $\varphi \in (0, \pi/4)$ such that

$$S_{a,\varphi} = \{\mu \in \mathbf{C} | \varphi \leqslant |\arg(\mu - a)| \leqslant \pi,\ \mu \neq a\} \subset \rho_{-A},$$

where ρ_{-A} is a resolvent set of the operator $-A$, $\rho_{-A} \neq \varnothing$;
2) there exists $M \geqslant 1$ such that for any $\mu \in S_{a,\varphi}$

$$\|(\mu I + A)^{-1}\| \leqslant \frac{M}{\|\mu - a\|}, \tag{2.1}$$

with I being the identity operator in $\mathcal{B}$.
To simplify the further consideration we set $a = 0$.
It is known (see, for example, (Henry, 1981)) that the operator A generates an analytic semigroup

$$e^{tA} = \frac{1}{2\pi i} \int_{\gamma} e^{t\lambda} (\lambda I - A)^{-1} d\lambda, \quad t \geqslant 0,$$

in $\mathcal{B}$ with γ being a sectionally smooth curve in ρ_A consisting or rays

$$\{re^{-i\theta} : 1 \leqslant r < \infty\}, \qquad \{re^{i\theta} : 1 \leqslant r < \infty\}$$

and the unit circle

$$\operatorname{arc}\{e^{i\delta} : -\theta \leqslant \delta \leqslant \theta\}, \qquad 3\pi/4 < \theta < \pi.$$

Let D_A be a domain of the operator A, $U_c(A)$ be a set of entire vectors for the operator A (see (Goodman, 1969)), $\langle \cdot, \cdot \rangle$ be a pairing in $(\mathcal{B}, \mathcal{B}^*)$.

THEOREM. *For $\beta \neq 0$ the random function (1.4) with $V_\beta(t, t_0)$ being defined by the formula*

$$V_\beta(t,t_0) = \frac{1}{2\pi i}\int\limits_\gamma \exp\{\alpha\Delta t\lambda - \beta^2\lambda^2|\Delta t|/2 + \beta\lambda\Delta w\}(\lambda I - A)^{-1}d\lambda, \quad t \neq t_0, \tag{2.2}$$

possesses a stochastic differential and satisfies the equation (1.2). In addition the family of operators $V_\beta(t, t_0)$ possesses the following properties a. s.:

a) $V_\beta(t,\tau) = V_\beta(t,s)V_\beta(s,\tau)$, $t > s > \tau$, $t < s < \tau$; $V_\beta(t,\tau)V_\beta(\tau,t) = e^{-\beta^2|t-\tau|A^2}$ *is nonrandom;*

b) $\|V_\beta(t,t_0)\| < \infty$;

c) $V_\beta(t,t_0)$ *is a unique evolutionary family for the Cauchy problem* (1.2), (1.3).

d) *for any* $x \in D_A$

$$\lim_{t\to t_0} \|V_\beta(t,t_0)x - x\| = 0;$$

e) *for any* $x \in U_c(A)$

$$\lim_{\beta\to 0} \|V_\beta(t,t_0)x - e^{\alpha\Delta t A}x\| = 0; \tag{2.3}$$

moreover, for $\alpha\Delta t > 0$ equality (2.3) is valid for any $x \in \mathcal{B}$;

f) *for $\alpha\Delta t > 0$ and for any $x \in \mathcal{B}$, $f \in \mathcal{B}^*$*

$$\mathbf{E}\langle V_\beta(t,t_0)x, f\rangle = \langle e^{\alpha\Delta t A}x, f\rangle.$$

We'll reduce the proof of the theorem to the proof of the sequence of propositions.

PROPOSITION 1. *For $\beta \neq 0$ the integral (2.2) converges absolutely a.s., moreover, for almost every trajectory of $w(t)$ the following estimates hold: for any $\varepsilon > 0$ and for any Δt such that $0 < |\Delta t| < T$ there exist positive constants $a_1, \ldots, a_{10}$, $b_1, \ldots, b_4$ depending on M, θ, α, T and ε only, such that*

a) *for* $\alpha\Delta t < 0$

$$\|V_\beta(t,t_0)\| \leqslant \frac{a_1}{|\beta|\sqrt{|\Delta t|}} \exp\left\{\left(a_2\frac{|\Delta t|^{1/2}}{\beta} + a_3|\Delta t|^{-\varepsilon}\right)^2\right\} + b(\beta, |\Delta t|), \tag{2.4}$$

b) *for* $\alpha\Delta t > 0$

$$\|V_\beta(t,t_0)\| \leqslant \frac{a_4}{|\Delta t|} \exp\{-a_5|\Delta t| + a_6|\Delta t|^{-2\varepsilon}\} + b(\beta, |\Delta t|), \tag{2.5}$$

c) *for $\alpha\Delta t > 0$ there exists $\beta_0 > 0$ such that for any β, $0 < |\beta| < \beta_0$,*

$$\|V_\beta(t,t_0)\| \leqslant \frac{a_7}{|\Delta t|} \exp\{a_8(\beta^2 - a_9)|\Delta t| + a_{10}|\beta||\Delta t|^{1/2-\varepsilon}\} + b(\beta, |\Delta t|), \tag{2.6}$$

where

$$b(\beta, |\Delta t|) = b_1 \exp\{b_2(b_9 + \beta^2)|\Delta t| + b_4|\beta||\Delta t|^{1/2-\varepsilon}\}.$$

Proof. Calculating the line integral in (2.2) and estimating its norm using the boundness condition for the resolvent of the operator A along the curve γ which follows from the condition (2.1) we find

$$\|V_\beta(t,t_0)\| \leqslant \frac{M}{\pi}\Bigg[\int_1^\infty \exp\{(\alpha\Delta t + \beta\Delta w)r\cos\theta - (\beta^2 r^2|\Delta t|\cos 2\theta)/2\}\frac{dr}{r} +$$

$$+ \int_0^\theta \exp\{(\alpha\Delta t + \beta\Delta w)\cos\delta - (\beta^2|\Delta t|\cos 2\delta)/2\}d\delta\Bigg]. \tag{2.7}$$

Denote the first and the second integrals in (2.7) by I_1 and I_2, respectively. It is evident that

$$I_2 \leqslant \pi\exp\{|\alpha\Delta t| + |\beta||\Delta w| + \beta^2|\Delta t|/2\}. \tag{2.8}$$

Estimating the integral I_1 we find

$$I_1 < \left(\frac{4\pi}{\beta^2|\Delta t|\cos 2\theta}\right)^{1/2} \exp\left\{\frac{cos^2\theta}{2\cos 2\theta}\left(\left|\frac{\alpha}{\beta}\right|\sqrt{|\Delta t|} + \frac{|\Delta w|}{\sqrt{|\Delta t|}}\right)\right\}. \tag{2.9}$$

For $\alpha\Delta t > 0$ we can improve the estimate (2.9). Indeed, analyzing the function

$$\varphi_1(r) = \exp\{\beta\Delta w r\cos\theta - (\beta^2 r^2|\Delta t|\cos 2\theta)/2\}$$

on the maximum in $[1,\infty)$ we have

$$\max_{r\geqslant 1}\varphi_1(r) \leqslant \max_{r\in\mathbf{R}^1}\varphi_1(r) = \exp\left\{\frac{\cos^2\theta}{\cos 2\theta}\frac{\Delta^2 w}{|\Delta t|}\right\}.$$

Then

$$I_1 = \int_1^\infty \varphi_1(r)e^{\alpha\Delta t r\cos\theta}\frac{dr}{r} \leqslant \max_{r\geqslant 1}\varphi_1(r)\int_1^\infty e^{\alpha\Delta t r\cos\theta}\frac{dr}{r} \leqslant$$

$$\leqslant -\frac{1}{\alpha\Delta t\cos\theta}\exp\left\{\alpha\Delta t\cos\theta + \frac{\cos^2\theta}{2\cos 2\theta}\frac{\Delta^2 w}{|\Delta t|}\right\}. \tag{2.10}$$

Analyzing the function $\varphi_2(r) = \exp\{(\alpha\Delta t r\cos\theta)/2\}\varphi_1(r)$ we find that it attains its global maximum in $(-\infty,\infty)$ at the point

$$r_0 = \frac{\cos\theta}{\beta^2|\Delta t|\cos 2\theta}\left(\frac{1}{2}\alpha\Delta t + \beta\Delta w\right).$$

We note that $r_0 < 1$ for $\alpha\Delta t > 0$ and sufficiently small β and $\varphi_2(r)$ has no other extremums in $(-\infty,\infty)$. Thus, for $\alpha\Delta t > 0$ there exists $\beta_0 > 0$ such that for any $\beta : 0 < |\beta| < \beta_0$,

$$\max_{r\geqslant 1}\varphi_2(r) = \varphi_2(1) = \exp\left\{(\alpha\Delta t/2 + \beta\Delta w)\cos\theta - (\beta^2|\Delta t|\cos 2\theta)/2\right\}$$

and

$$I_1 = \int_1^\infty \varphi_2(r)\exp\left\{(\alpha\Delta tr\cos\theta)/2\right\}\frac{dr}{r} \leqslant \varphi_2(1)\int_1^\infty \exp\left\{(\alpha\Delta tr\cos\theta)/2\right\}\frac{dr}{r} =$$

$$= \frac{2}{\alpha\Delta t\cos\theta}\exp\left\{\alpha\Delta t\cos\theta - (\beta^2|\Delta t|\cos 2\theta)/2 - |\beta||\Delta w|\cos\theta\right\}. \qquad (2.11)$$

Estimates (2.4)—(2.6) follow from inequalities (2.7)—(2.11) and Hölder's condition for the increment of the Wiener process with any exponent less then 1/2 (a.s.), that is the fact that for any $\varepsilon > 0$ for almost every trajectory of $w(t)$ there exists $c_{\varepsilon,t} > 0$ such that

$$|\Delta w| < c_{\varepsilon,t}|\Delta t|^{1/2-\varepsilon}.$$

PROPOSITION 2. *For $\beta \neq 0$ the family (2.2) possesses the evolutionary property for $t > s > \tau$ or $t < s < \tau$ a. s., that is*

$$V_\beta(t,\tau) = V_\beta(t,s)V_\beta(s,\tau), \quad t > s > \tau \quad or \quad t < s < \tau,$$

and

$$V_\beta(t,\tau)V_\beta(\tau,t) = \exp\{-\beta^2|t-\tau|A^2\}$$

is nonrandom.

Proof. Since

$$V_\beta(t,t_0) = f(A,t,t_0) = \frac{1}{2\pi i}\int_\gamma f(\lambda,t,t_0)(\lambda I - A)^{-1}d\lambda$$

with

$$f(\lambda,t,t_0) = \exp\{\alpha\lambda(t-t_0) - (\beta^2\lambda^2|t-t_0|)/2 + \beta\lambda(w(t) - w(t_0))\}, \quad \lambda \in \gamma,$$

then the proposition follows immediately from the condition

$$f(\lambda,t,\tau) = f(\lambda,t,s)f(\lambda,s,\tau) \text{ for } \quad t > s > \tau \text{ or } \quad t < s < \tau$$

and the equality

$$f(\lambda,t,\tau)f(\lambda,\tau,t) = \exp\{-\beta^2\lambda^2|t-\tau|\}.$$

PROPOSITION 3. *For $\beta \neq 0$ the family of operators $V_\beta(t,t_0)$ defined by (2.2) keeps the set D_A invariant a.s., and the solution of the Cauchy problem (1.2), (1.3) exists and is given by the formula (1.4).*

Proof. Consider the scalar random process

$$\eta_\lambda(t) = \alpha\Delta t - \beta^2|\Delta t|\lambda/2 + \beta\Delta w, \quad \lambda \in \gamma,$$

and the scalar function

$$\varphi(\eta_\lambda(t)) = \exp\{\lambda\eta_\lambda(t)\} = \exp\{\alpha\Delta t\lambda - \beta^2\lambda^2|\Delta t|/2 + \beta\lambda\Delta w\}, \quad \lambda \in \gamma. \quad (2.12)$$

The fact that the family of operators $V_\beta(t, t_0)$ keeps the domain of the operator A invariant a.s. follows from the absolute convergence of the integral

$$\frac{1}{2\pi i}\int_\gamma \lambda\varphi(\eta_\lambda(t))(\lambda I - A)^{-1}d\lambda$$

defining for every $t \neq t_0$ and $\beta \neq 0$ the operator–valued function $AV_\beta(t, t_0)$, for the norm of which for almost every trajectory of $w(t)$ estimates of the form (2.4)—(2.6) hold.

According to Ito's formula for the differential of a scalar random process we have

$$d\varphi(\eta_\lambda(t)) = \left(\alpha - \frac{1}{2}\lambda\beta^2\text{sign}\Delta t\right)\varphi'_x(\eta_\lambda(t))dt+$$

$$+\frac{1}{2}\beta^2\text{sign}\Delta t\varphi''_{xx}(\eta_\lambda(t))dt + \beta\varphi'_x(\eta_\lambda(t))dw(t) =$$

$$= (\alpha - \lambda\beta^2\text{sign}\Delta t)\lambda\varphi(\eta_\lambda(t))dt + \frac{1}{2}\beta^2\text{sgn}\Delta t\lambda^2\varphi(\eta_\lambda(t))dt + \beta\lambda\varphi(\eta_\lambda(t))dw(t) =$$

$$= \lambda\varphi(\eta_\lambda(t))(\alpha dt + \beta dw(t))$$

or according to the definition of the stochastic differential

$$\varphi(\eta_\lambda(t)) - \varphi(\eta_\lambda(\tau)) = \alpha\lambda\int_\tau^t \varphi(\eta_\lambda(s))ds + \beta\lambda\int_\tau^t \varphi(\eta_\lambda(s))dw(s). \quad (2.13)$$

Calculating the increment $\xi(t) - \xi(\tau)$ using the formula (2.13) and the representation

$$\xi(t) = \frac{1}{2\pi i}\int_\gamma \varphi(\eta_\lambda(t))(\lambda I - A)^{-1}u_0 d\lambda$$

which follows from (1.4) and (2.2) we obtain

$$\xi(t) - \xi(\tau) = \frac{1}{2\pi i}\int_\gamma [\varphi(\eta_\lambda(t)) - \varphi(\eta_\lambda(\tau))](\lambda I - A)^{-1}u_0 d\lambda =$$

$$= \frac{1}{2\pi i}\int_\gamma \alpha\lambda\left[\int_\tau^t \varphi(\eta_\lambda(s))ds\right](\lambda I - A)^{-1}u_0 d\lambda+$$

$$+\frac{1}{2\pi i}\int_\gamma \beta\lambda\left[\int_\tau^t \varphi(\eta_\lambda(s))dw(s)\right](\lambda I - A)^{-1}u_0 d\lambda. \quad (2.14)$$

Changing the order of integration we have

$$\xi(t)-\xi(\tau)=\alpha\int_{\tau}^{t}\frac{1}{2\pi i}\int_{\gamma}\lambda\varphi(\eta_{\lambda}(s))(\lambda I-A)^{-1}u_0d\lambda ds+$$

$$+\beta\int_{\tau}^{t}\frac{1}{2\pi i}\int_{\gamma}\lambda\varphi(\eta_{\lambda}(s))(\lambda I-A)^{-1}u_0d\lambda dw(s). \tag{2.15}$$

But according to the definition of $\xi(s)$

$$\frac{1}{2\pi i}\int_{\gamma}\lambda\varphi(\eta_{\lambda}(s))(\lambda I-A)^{-1}u_0d\lambda=AV_{\beta}(s,t_0)u_0=A\xi(s). \tag{2.16}$$

Substituting (2.16) to (2.15) we obtain

$$\xi(t)-\xi(\tau)=\alpha\int_{\tau}^{t}A\xi(s)ds+\beta\int_{\tau}^{t}A\xi(s)dw(s).$$

Hence, $\xi(t)$ possesses a stochastic differential and satisfies the equation (1.2).

PROPOSITION 4. *The evolutionary family* (2.2) *defined for* $\beta\neq 0$ *is the unique a.s. evolutionary family for the Cauchy problem* (1.2), (1.3).

Proof. Suppose there exists another different from $V_{\beta}(t,t_0)$ evolutionary family of the Cauchy problem (1.2), (1.3). Denote it by $U(t,t_0)$. Then for any $y\in\mathcal{B}$ the function $U(t,t_0)y$ satisfies the equation

$$dU(t,t_0)y=\alpha AU(t,t_0)ydt+\beta AU(t,t_0)ydw(t). \tag{2.17}$$

Let us consider the random function $\xi(s)=V_{\beta}(t,s)U(s,t_0)x$, $x\in\mathcal{B}$, and calculate its stochastic differential according to the formula of differentiation taking into account (2.17):

$$d\xi(s)=(dV_{\beta}(t,s))U(s,t_0)x+\alpha V_{\beta}(t,s)AU(s,t_0)xds+\beta V_{\beta}(t,s)AU(s,t_0)xdw(s). \tag{2.18}$$

According to Ito's formula

$$(dV_{\beta}(t,s))y=-\alpha AV_{\beta}(t,s)yds-\beta AV_{\beta}(t,s)ydw(s).$$

Having calculated differential $(dV_{\beta}(t,s))y$ we substitute it to (2.18) and taking into account the fact that $V_{\beta}(t,s)$ commutates with A on D_A we obtain

$$d\xi(s)=0\quad\text{(a.s.)}.$$

Thus, $\xi(s)$ doesn't depend on s (a.s.). But

$$\xi(t) = V_\beta(t,t)U(t,t_0)x = U(t,t_0)x$$

and

$$\xi(t_0) = V_\beta(t,t_0)U(t_0,t_0)x = V_\beta(t,t_0)x \quad \text{for any } x \in \mathcal{B} \quad \text{(a.s.).}$$

Hence, $U(t,t_0) = V_\beta(t,t_0)$ (a.s.).

Remark. Under some additional conditions on operator A spectrum the solution (1.4) of the Cauchy problem (1.2), (1.3) is unique a.s.

PROPOSITION 5. *For any $x \in D_A$ and $\beta \neq 0$*

$$\lim_{t\to t_0} \|V_\beta(t,t_0)x - x\| = 0 \quad (a.s.).$$

Proof. Let $x \in D_A$ and $\varphi(\eta_\lambda(t))$ be the random function defined by (2.12). Then for $\beta \neq 0$

$$V_\beta(t,t_0)x - x = \frac{1}{2\pi i}\int\limits_\gamma \varphi(\eta_\lambda(t))[(\lambda I - A)^{-1} - (\lambda I)^{-1}]x d\lambda =$$

$$= \frac{1}{2\pi i}\int\limits_\gamma \varphi(\eta_\lambda(t))(\lambda I - A)^{-1}Ax\frac{d\lambda}{\lambda}$$

and

$$\|V_\beta(t,t_0)x - x\| \leqslant \left\|\frac{1}{2\pi i}\int\limits_\gamma \varphi(\eta_\lambda(t))(\lambda I - A)^{-1}\frac{d\lambda}{\lambda}\right\| \ \|Ax\|. \tag{2.19}$$

Making the change of variables $\mu = \sqrt{|\Delta t|}\lambda$ in the integral in (2.19) we have

$$J_1 = \frac{1}{2\pi i}\int\limits_\gamma \varphi(\eta_\lambda(t))(\lambda I - A)^{-1}\frac{d\lambda}{\lambda} =$$

$$= \frac{1}{2\pi i}\int\limits_{\gamma'} \exp\left\{(\alpha\Delta t + \beta\Delta w)\frac{\mu}{|\Delta t|^{1/2}} - \frac{\beta^2\mu^2}{2}\right\}\left(\frac{\mu}{|\Delta t|^{1/2}}I - A\right)^{-1}\frac{d\mu}{\mu}, \tag{2.20}$$

$$\gamma' = \left\{\mu \in \mathbb{C}\left|\frac{\mu}{|\Delta t|^{1/2}} \in \gamma\right.\right\}.$$

If we observe that for $|\Delta t| < 1$ according to the Cauchy theorem the integration along γ' in (2.20) can be replaced by the integration along γ and using the boundedness condition for the operator $((\mu I/|\Delta t|) - A)^{-1}$ when $\mu \in \gamma$ we obtain

$$\|I_1\| \leqslant \frac{M}{\pi}\sqrt{|\Delta t|}\left[\int\limits_1^\infty \exp\left\{(\alpha\Delta t + \beta\Delta w)\frac{r\cos\theta}{|\Delta t|^{1/2}} - \frac{1}{2}\beta^2 r^2\cos 2\theta\right\}\frac{dr}{r^2} + \right.$$

$$+\int_0^{\theta} \exp\left\{(\alpha\Delta t+\beta\Delta w)\frac{\cos\delta}{|\Delta t|^{1/2}}-\frac{1}{2}\beta^2\cos 2\delta\right\}d\delta\Bigg] <$$

$$< \frac{M}{\pi}\sqrt{|\Delta t|}\Bigg[\Big(\frac{4\pi}{\beta^2\cos 2\theta}\Big)^2\exp\Big\{\frac{\cos^2\theta}{2\cos 2\theta}\Big(\frac{|\alpha|\sqrt{|\Delta t|}}{|\beta|}+\frac{|\Delta w|}{|\Delta t|^{1/2}}\Big)^2\Big\}+$$

$$+\pi\exp\left\{|\alpha|\sqrt{|\Delta t|}+\frac{1}{2}\beta^2+\frac{|\Delta w|}{|\Delta t|^{1/2}}\right\}\Bigg]. \tag{2.21}$$

From the law of the iterated logarithm for the increment of the Wiener process (see, for example (Skorokhod, 1986)) it follows that

$$\lim_{\Delta t\to 0}\sqrt{|\Delta t|}\exp\left\{\delta_1\frac{\Delta_w^2}{|\Delta t|}+\delta^2|\Delta w|\right\}=0 \quad \text{(a.s.)}$$

and

$$\lim_{\Delta t\to 0}\sqrt{|\Delta t|}\exp\left\{\delta_3\frac{\Delta w}{|\Delta t|^{1/2}}\right\}=0 \quad \text{(a.s.)},$$

where δ_1, δ_2 and δ_3 are positive constants.
Thus $\lim_{\Delta t\to 0}\|J_1\|=0$ (a.s.) and

$$\lim_{\Delta t\to 0}\|V_\beta(t,t)x-x\|=\lim_{\Delta t\to 0}\|J_1\|\|Ax\|=0 \quad \text{(a.s.)}.$$

Proposition 6. *For any $x\in U_c(A)$ and for any $x\in\mathcal{B}$ when $\alpha\Delta t>0$*

$$\lim_{\beta\to 0}\|V_\beta(t,t_0)x-e^{\alpha\Delta tA}x\|=0 \quad (a.s.).$$

Proof. For $\alpha\Delta t>0$ the family of operators $e^{\alpha\Delta tA}$ is defined in $\mathcal{B}$ and is given by the formula

$$e^{\alpha\Delta tA}=\frac{1}{2\pi i}\int_\gamma e^{\alpha\Delta t\lambda}(\lambda I-A)^{-1}d\lambda. \tag{2.22}$$

For $\alpha\Delta t<0$ the family $e^{\alpha\Delta tA}$ is defined on a set $U_c(A)$ which is dense in $\mathcal{B}$ as an inverse to the operator $e^{-\alpha\Delta tA}$ (see (Gorbachuk and Knyazyuk, 1989)).
Let $x\in U_c(A)\subset D_{Ae^{\alpha\Delta tA}}$, $\beta\neq 0$. Then

$$\|V_\beta(t,t_0)x-e^{\alpha\Delta tA}x\|=\Big\|\frac{1}{2\pi i}\int_\gamma\exp\Big\{-\frac{1}{2}\beta^2\lambda^2|\Delta t|+$$

$$+\beta\lambda\Delta w\Big\}[(\lambda I-A)^{-1}-(\lambda I)^{-1}]e^{\alpha\Delta tA}x\,d\lambda\Big\|\leqslant$$

$$\leqslant\Big\|\frac{1}{2\pi i}\int_\gamma\exp\{-\beta^2\lambda^2|\Delta t|/2+\beta\lambda\Delta w\}(\lambda I-A)^{-1}\frac{d\lambda}{\lambda}\Big\|\,\|Ae^{\alpha\Delta tA}x\|. \tag{2.23}$$

Making the change of variables $\mu = |\beta|\lambda$ in (2.23) and using the boundedness condition for the operator $((\mu I/|\beta|) - A)^{-1}$ for $\mu \in \gamma$, $|\beta| < 1$ and Hölder's condition for the increment of the Wiener process for almost all trajectories we obtain that of $w(t)$ for any $\varepsilon > 0$ and for any Δt such that $0 < |\Delta t| < T$ there exists the constant $C > 0$ depending on M, θ, α, T and ε only such that

$$\|V_\beta(t,t_0)x - e^{\alpha\Delta t A}x\| \leqslant C|\beta|\,\|Ae^{\alpha\Delta t A}x\|.$$

Thus, for any $x \in U_c(A)$

$$\lim_{\beta\to 0}\|V_\beta(t,t_0)x - e^{\alpha\Delta t A}x\| = 0 \quad \text{(a.s.)}.$$

The second part of the proposition follows from the facts that $U_c(A)$ is dense in $\mathcal{B}$ and for $\alpha\Delta t > 0$ there exists $\beta_0 > 0$ such that for any β, $0 < |\beta| < \beta_0$ the estimate (2.6) for the norm of $V_\beta(t,t_0)$ holds.

Remark. For $\alpha\Delta t < 0$ and $\beta = 0$ the operator (2.2) is unbounded and is defined on the set $U_c(A)$ which is dense in $\mathcal{B}$ as a limit of $V_\beta(t,t_0)$ when β tends to 0:

$$e^{\alpha\Delta t A}u_0 = \lim_{\beta\to 0} V_\beta(t,t_0)u_0, \quad u_0 \in U_c(A).$$

PROPOSITION 7. *For $\alpha\Delta t > 0$ and for any $x \in \mathcal{B}$, $f \in \mathcal{B}^*$*

$$\mathbf{E}\langle V_\beta(t,t_0)x, f\rangle = \langle e^{\alpha\Delta t A}x, f\rangle.$$

Proof. Let $\alpha\Delta t > 0$. Then the family $e^{\alpha\Delta t A}$ can be defined by the formula (2.22). Calculating the expectation of the random variable $\langle V_\beta(t,t_0)x, f\rangle$ we find

$$\mathbf{E}\langle V_\beta(t,t_0)x, f\rangle = \mathbf{E}\Big\langle \frac{1}{2\pi i}\int_\gamma \varphi(\eta_\lambda(t))(\lambda I - A)^{-1}x\,d\lambda, f\Big\rangle =$$

$$= \mathbf{E}\frac{1}{2\pi i}\int_\gamma \varphi(\eta_\lambda(t))\langle(\lambda I - A)^{-1}x, f\rangle d\lambda =$$

$$= \int_{-\infty}^{\infty}\Big[\frac{1}{2\pi i}\int_\gamma \exp\{\alpha\Delta t\lambda - \beta^2\lambda^2|\Delta t|/2 + \beta\lambda\tau\}\langle(\lambda I - A)^{-1}x, f\rangle d\lambda\Big]\times$$

$$\times\frac{1}{\sqrt{2\pi|\Delta t|}}e^{-\tau^2/(2|\Delta t|)}d\tau.$$

Changing the order of integration we have

$$\mathbf{E}\langle V_\beta(t,t_0)x, f\rangle = \frac{1}{2\pi i}\int_\gamma \exp\{\alpha\Delta t\lambda - \beta^2\lambda^2|\Delta t|/2\}\times$$

$$\times\left\{\int_{-\infty}^{\infty}\frac{1}{\sqrt{2\pi|\Delta t|}}e^{\beta\lambda\tau-\tau^2/2|\Delta t|}d\tau\right\}\langle(\lambda I-A)^{-1}x,f\rangle d\lambda=$$

$$=\frac{1}{2\pi i}\int_{\gamma}e^{\alpha\Delta t\lambda}\langle(\lambda I-A)^{-1}x d\lambda,f\rangle=\langle e^{\alpha\Delta tA}x,f\rangle.$$

3. Let $\mathcal{B}=\mathcal{H}$ be a Hilbert space

$$A=\int_{-\infty}^{\sigma}\lambda E(d\lambda) \tag{3.1}$$

be a selfadjoint upperbounded operator in $\mathcal{H}$. Here $E(\Delta)$ is a spectral measure in $\mathcal{H}$, defined on Borel sets Δ in $\mathbf{R}^1$.
This example is the special case of the previous one since $-A$ is a sectorial operator in $\mathcal{B}=\mathcal{H}$ (see, for example, (Henry, 1981)). Thus, for the evolutionary family $V_\beta(t,t_0)$ defined in Section 2 as a function of the operator $-A$ the representation in the form of the spectral integral

$$V_\beta(t,t_0)=\int_{-\infty}^{\sigma}\exp\left\{\alpha\Delta t\lambda-\frac{1}{2}\beta^2\lambda^2|\Delta t|+\beta\lambda\Delta w\right\}E(d\lambda),\quad \alpha,\beta\in\mathbf{R}^1,\quad \beta\neq 0, \tag{3.2}$$

is valid, and all the results of Section 2 concerning the solution of the Cauchy problem (1.2), (1.3) and properties of the family $V_\beta(t,t_0)$ are carried completely to the case of selfadjoint operators in $\mathcal{H}$. But for the operator A considered here one can obtain some additional results which we formulate without proof.

Remark 1. For $\beta\neq 0$ and $t\neq t_0$ the integral (3.2) converges absolutely a.s. Moreover, estimates more simple than given in Proposition 2.1 hold. Namely, for almost every trajectory of $w(t)$ for any $\varepsilon>0$ there exists $C_{\varepsilon,t}>0$ such that

a) for $\alpha\Delta t<0$

$$\|V_\beta(t,t_0)\|\leqslant\exp\left\{\frac{1}{2}\left(\left|\frac{\alpha}{\beta}\right|\sqrt{|\Delta t|}+C_{\varepsilon,t}|\Delta t|^{-\varepsilon}\right)^2\right\}.$$

b) for $\alpha\Delta t>0$

$$\|V_\beta(t,t_0)\|\leqslant\exp\left\{\alpha\Delta t\sigma-\frac{1}{2}C^2_{\varepsilon,t}|\Delta t|^{-2\varepsilon}\right\}.$$

c) for $\alpha\Delta t>0$ there exists $\beta_0>0$ such that for any β, $0<|\beta|<\beta_0$,

$$\|V_\beta(t,t_0)\|\leqslant\exp\left\{\alpha\Delta t\sigma-\beta^2\sigma^2|\Delta t|/2+|\beta\sigma|C_{\varepsilon,t}|\Delta t|^{1/2-\varepsilon}\right\}.$$

Remark 2. The solution (1.4) of the Cauchy problem (1.2), (1.3) in $\mathcal{B}=\mathcal{H}$ with $V_\beta(t,t_0)$ defined by the formula (3.2) is unique a.s.

Remark 3. Let $\mathcal{D} = \{E(\Delta)h$ for any $h \in \mathcal{H}$ with Δ being the arbitrary bounded interval from $(-\infty, \infty)\}$. Then $\mathcal{D} \subset U_c(\mathcal{A})$ and $\mathcal{D}$ is dense in $\mathcal{B}$.

4. The more general case appears when one consider a stochastic differential equation

$$d\xi(t,\lambda) = a(t,\lambda,\xi(t,\lambda))dt + b(t,\lambda,\xi(t,\lambda))dw(t), \tag{4.1}$$

$$\xi(t_0) = x,$$

in a Hilbert space $\mathcal{H}$ with $\lambda \in \mathbf{R}^1$, $x \in \mathcal{H}$. Under usual conditions which guarantee the solvability of the Cauchy problem (4.1) (see, for example (Daletsky and Fomin, 1983)) we can define the random function of a selfadjoint operator A by

$$\xi_f(t,A) = f(\xi(t,A)) = \int\limits_{\mathbf{R}^1} f(\xi(t,\lambda))E(d\lambda), \tag{4.2}$$

with $f : \mathcal{H} \to \mathbf{R}^1$ and using Ito's formula for $f(\xi(t,\lambda))$ calculate its stochastic differential.

From the other hand the derivative of the scalar function of the selfadjoint operator

$$A(t) = \int\limits_{\mathbf{R}^1} \lambda E_t(d\lambda)$$

is calculated according to Yu. Daletsky and S. Krein formula for differentiation of functions of selfadjoint operators on parameter (see (Daletsky and Krein, 1956)):

$$\frac{d}{dt} f(A(t)) = \iint\limits_{\mathbf{R}} \frac{f(\lambda) - f(\mu)}{\lambda - \mu} E_t(d\lambda)A'(t)E_t(d\mu).$$

Under certain conditions combining these two formulas (Ito's formula and formula of Yu. Daletsky and S. Krein) we can calculate the stochastic differential of the random function $\xi_f(t,A) = f(\xi(t,A))$ namely:

$$df(\xi(t,A(t))) = \Big(f'_x(\xi(t,A(t))), \Big\{a(t,A(t),\xi(t,A(t)))+$$

$$+\iint\limits_{\mathbf{R}} \frac{\xi(t,\lambda) - \xi(t,\mu)}{\lambda - \mu} E_t(d\lambda)A'(t)E_t(d\mu)\Big\}\Big)dt+$$

$$+\frac{1}{2}\Big(f''_{xx}(\xi(t,A(t)))b(t,A(t),\xi(t,A(t))), b(t,A(t),\xi(t,A(t)))\Big)dt+$$

$$+\Big(f'_x(\xi(t,A(t))), \quad \Big\{a(t,A(t),\xi(t,A(t)))+$$

$$+\iint\limits_{\mathbf{R}} \frac{\xi(t,\lambda) - \xi(t,\mu)}{\lambda - \mu} E_t(d\lambda)A'(t)E_t(d\mu)\Big\}\Big)dw(t).$$

REFERENCES

Daletsky, Yu.L. and Fomin, S.V. (1983). *Measures and Differential Equations in Infinite Dimensional Spaces.* Nauka, Moscow.

Daletsky, Yu.L. and Krein, S.G. (1956). Integration and differentiation of Hermitian operators functions and applications to the perturbation theory. In: *Proc. Seminar on Func. Anal.*, Voronezh, **1**, 81–105.

Gikhman, Il.I. and Mestechkina, T.M. (1987). The Cauchy problem for the parabolic equation with "white noise" type coefficients. *Theory Stochast. Proc.* **15**, 19–28.

Goodman, R. (1969). Analytic and entire vectors for representations on Lie groups. *Trans. Amer. Math. Soc.* **143**, 55–76.

Gorbachuk, V.I. and Knyazyuk, A.V. (1989). Boundary values of the differential-operator equation solutions. *Uspekhi Mat. Nauk* **44**, N 3(267), 55–91.

Henry, D. (1981). *Geometric theory of semilinear parabolic equations.* Lect. Notes in Math. **840**, Springer, Berlin—Heidelberg etc.

Skorokhod, A.V. (1986). *Random Processes with Independent Increments.* Nauka, Moscow.

Prob. Theory and Math. Stat., Vol. 1, pp. 446–454
B.Grigelionis *et al.* (Eds.)
1990 VSP/Mokslas

HELLINGER INTEGRALS AND HELLINGER PROCESSES FOR SOLUTIONS OF MARTINGALE PROBLEMS

B. GRIGELIONIS
Institute of Mathematics and Cybernetics,
Akademijos 4, 232600 Vilnius, Lithuania

ABSTRACT

The evaluation of the Hellinger integrals for unique solutions of the defined martingale problems is considered. The notions of the Hellinger processes as well as the martingale representation property are exploited.

1. Let $(\Omega, \mathcal{F})$ be a measurable space endowed with two probability measures $\mathbf{P}$ and $\mathbf{P}'$. The Hellinger integral $H(\mathbf{P}, \mathbf{P}')$ of $\mathbf{P}$ and $\mathbf{P}'$ is the nonnegative number

$$H(\mathbf{P}, \mathbf{P}') = \mathbf{E}_{\mathbf{Q}}\left(\sqrt{\frac{d\mathbf{P}}{d\mathbf{Q}}\frac{d\mathbf{P}'}{d\mathbf{Q}}}\right),$$

where $\mathbf{Q}$ is any probability measure such that $\mathbf{P} \ll \mathbf{Q}$ and $\mathbf{P}' \ll \mathbf{Q}$, and $\mathbf{E}_{\mathbf{Q}}$ denotes the expectation with respect to $\mathbf{Q}$. Starting with the famous S. Kakutani's paper (Kakutani, 1948) the Hellinger integrals and the Kakutani–Hellinger distance $\rho(\mathbf{P}, \mathbf{P}')$ defined by means of the formula

$$\rho^2(\mathbf{P}, \mathbf{P}') = 1 - H(\mathbf{P}, \mathbf{P}'),$$

play an increasing role in the theory of statistical inference. We recall that the Kakutani–Hellinger distance defines the same topology as the variation metric $\| \mathbf{P} - \mathbf{P}' \|$ on the space of all probability measures on $(\Omega, \mathcal{F})$ which follows from the inequalities (see, e.g. (Jacod and Shiryaev, 1987; Liese and Vajda, 1987))

$$2\rho^2(\mathbf{P}, \mathbf{P}') \leqslant \| \mathbf{P} - \mathbf{P}' \| \leqslant 2\sqrt{2}\rho(\mathbf{P}, \mathbf{P}').$$

This distance is naturally used to define regularity properties of statistical experiments and asymptotical behaviour of the maximal likelihood estimators of the unknown parameters or error probabilities of the optimal statistical tests (see, e.g. (Liese and Vajda, 1987; Jacod, 1989a; Vajda, 1988; Ibragimov and Hasminsky, 1981)).
Introducing the Hellinger integral of complex valued order z by

$$H(z; \mathbf{P}, \mathbf{P}') = \mathbf{E}_{\mathbf{Q}}\left[\left(\frac{d\mathbf{P}}{d\mathbf{Q}}\right)^z \left(\frac{d\mathbf{P}'}{d\mathbf{Q}}\right)^{1-z}\right], \quad 0 \leqslant \mathrm{Re}\, z \leqslant 1,$$

under assumption that $\mathbf{P} \sim \mathbf{P}'$ we easily find a formula for the characteristic function of the log-likelihood ratio in terms of the related Hellinger integrals (Liese and Vajda, 1987)

$$\mathbf{E}_{\mathbf{P}'} \exp\left\{iv \ln \frac{d\mathbf{P}}{d\mathbf{P}'}\right\} = \lim_{u \downarrow 0} H(u + iv; \mathbf{P}, \mathbf{P}').$$

Let us remark finally, that the important properties of continuity and entire separation of two sequences $\{\mathbf{P}'^n, n \geqslant 1\}$, $\{\mathbf{P}^n, n \geqslant 1\}$ of probability measures on measurable spaces $(\Omega^n, \mathcal{F}^n)$, respectively, can be expressed in terms of Hellinger integrals of order α, $0 < \alpha < 1$. Namely (see, e.g. (Jacod and Shiryaev, 1987; Oosterhoff and Van Zwet, 1979)), $\{\mathbf{P}'^n,\ n \geqslant 1\} \lhd \{\mathbf{P}^n, n \geqslant 1\}$ iff

$$\lim_{\alpha \downarrow 0} \liminf_{n \to \infty} H(\alpha; \mathbf{P}^n, \mathbf{P}'^n) = 1,$$

and $\{\mathbf{P}'^n, n \geqslant 1\} \lhd \{\mathbf{P}^n, n \geqslant 1\}$ iff for some $\alpha \in (0,1)$

$$\liminf_{n \to \infty} H(\alpha; \mathbf{P}^n, \mathbf{P}'^n) = 0.$$

In the case when $(\Omega^n, \mathcal{F}^n, \mathbf{P}^n, \mathbf{P}'^n) = (\Omega, \mathcal{F}, \mathbf{P}, \mathbf{P}')$ for all $n \geqslant 1$, we have the criteria for absolute continuity and singularity of measures $\mathbf{P}$ and $\mathbf{P}'$ in terms of the Hellinger integrals.

In this paper we shall consider the problem of evaluation of the Hellinger integrals for solutions of defined martingale problems on filtered probability spaces. The notions of the Hellinger measures and the Hellinger processes initiated by C.M. Newman (Newman, 1972), will be exploited.

2. Let $(\Omega, \mathcal{F}, \mathbf{F})$ be a filtered measurable space with a family $\{\mathbf{P}^\theta, \theta \in \Theta\}$ of probability measures on $\mathcal{F}$.

For the sake of simplicity assume that there exists a probability measure $\mathbf{Q}$ on $\mathcal{F}$ such that for each $t > 0$, $\theta \in \Theta$

$$\mathbf{P}_t^\theta = \mathbf{P}^\theta \mid \mathcal{F}_t \sim \mathbf{Q}_t = \mathbf{Q} \mid \mathcal{F}_t \tag{1}$$

and $\mathbf{P}_0^\theta = \mathbf{Q}_0$.

Denote

$$Z_t^\theta = \frac{d\mathbf{P}_t^\theta}{d\mathbf{Q}_t}, \quad t \geqslant 0, \quad \theta \in \Theta,$$

$$\alpha = (\alpha_1, \ldots, \alpha_k), \quad 0 < \alpha_j < 1, \quad \sum_{j=1}^k \alpha_j = 1, \quad k \geqslant 2,$$

$$I = (\theta_1, \ldots, \theta_k), \quad \theta_j \in \Theta, \quad j = 1, \ldots, k,$$

$$Y_t(\alpha; I) = \prod_{j=1}^k (Z_t^{\theta_j})^{\alpha_j}, \quad t \geqslant 0.$$

Because Z^θ is a strictly positive $(\mathbf{Q},\ \mathbf{F})$-martingale for each $\theta \in \Theta$ according to the Hölder inequality we have for any $0 \leqslant s < t$ (cf. (Jacod, 1989b))

$$\mathbf{E}_{\mathbf{Q}}\left[\frac{Y_t(\alpha; I)}{Y_s(\alpha; I)} \mid \mathcal{F}_s\right] = \mathbf{E}_{\mathbf{Q}}\left[\prod_{j=1}^k \left(\frac{Z_t^{\theta_j}}{Z_s^{\theta_j}}\right)^{\alpha_j} \mid \mathcal{F}_s\right] \leqslant \prod_{j=1}^k \left[\mathbf{E}\left(\frac{Z_t^{\theta_j}}{Z_s^{\theta_j}} \mid \mathcal{F}_s\right)\right]^{\alpha_j} = 1.$$

Thus $Y(\alpha; I)$ is a strictly positive $(\mathbf{Q}, \mathbf{F})$–supermartingale. Using the Ito–Watanabe's multiplicative decomposition theorem (see, e.g. (Liptser and Shiryaev, 1986)) we find that there exist a unique up to $\mathbf{Q}$–indistinguishability decreasing $\mathbf{F}$–predictable process $D(\alpha; I), D_0(\alpha; I) = 1$, and a $(\mathbf{Q}, \mathbf{F})$–local martingale $L(\alpha; I)$, $L_0(\alpha; I) = 1$, such that

$$Y_t(\alpha; I) = D_t(\alpha; I) L_t(\alpha; I), \quad t \geqslant 0. \tag{2}$$

Denote

$$h_t(\alpha; I) = -\int_0^t \frac{dD_s(\alpha; I)}{D_{s-}(\alpha; I)}, \quad t \geqslant 0. \tag{3}$$

Definition 1 (cf. (Jacod, 1989b)). We call $h(\alpha; I)$ the Hellinger process of order α for the family $\{\mathbf{P}^\theta, \ \theta \in I\}$.

Definition 2. The measures

$$H_t(\alpha; I)(A) = \mathbf{E}_\mathbf{Q}[Y_t(\alpha; I) \mathbf{1}_A], \quad A \in \mathcal{F}_t, \quad t \geqslant 0, \tag{4}$$

will be called the Hellinger measures of order α for the family $\{\mathbf{P}^\theta, \ \theta \in I\}$.

Under the assumption that $L(\alpha; I)$ is a $(\mathbf{Q}, \ \mathbf{F})$–martingale, i.e.

$$\mathbf{E}_\mathbf{Q} L_t(\alpha; I) \equiv 1, \tag{5}$$

we define the probability measures

$$\mathbf{P}_t(\alpha; I)(A) = \mathbf{E}_\mathbf{Q}[L_t(\alpha; I) \mathbf{1}_A], \quad A \in \mathcal{F}_t, \quad t \geqslant 0. \tag{6}$$

From (2) - (6) we obtain that for each $t \geqslant 0$

$$H_t(\alpha; I) \sim \mathbf{P}_t(\alpha; I)$$

and

$$\frac{dH_t(\alpha; I)}{d\mathbf{P}_t(\alpha; I)} = \mathcal{E}_t(-h(\alpha; I)), \tag{7}$$

where $\mathcal{E}_t(\cdot)$ denotes the Doléans–Dade exponential. Bellow we shall identify the measures $\mathbf{P}_t(\alpha; I)$ and the Hellinger processes $h_t(\alpha; I)$ in the case when the probability measures $\mathbf{P}^\theta, \theta \in \Theta$, and $\mathbf{Q}$ are unique solutions to the defined martingale problems. Martingale representation theorems and properties of the Doléans–Dade exponentials will be applied.

3. Assume that $\mathbf{F} = \mathbf{F}^X$, where $\mathbf{F}^X$ denotes the filtration generated by some m–dimensional càdlàg process X, the measures $\mathbf{P}^\theta, \ \theta \in \Theta$, and $\mathbf{Q}$ are uniquely characterized by their restrictions $\mathbf{P}^\theta_0, \ \theta \in \Theta, \ \mathbf{Q}_0$ and the properties that X is a $(\mathbf{P}^\theta, \ \mathbf{F})$–semimartingale and a $(\mathbf{Q}, \mathbf{F})$–semimartingale with the triplets $(\delta^\theta, B, \Pi^\theta)$ and (δ, B, Π) of predictable characteristics (for terminology and definitions see, e.g. (Jacod and Shiryaev, 1987; Liptser and Shiryaev, 1986)), respectively, such that

$$\delta^\theta_t = \delta_t + \int_0^t g^\theta(s) B'(s) d\beta_s + \int_0^t \int_{|x| \leqslant 1} x(\psi^\theta(s, x) - 1) \Pi(ds, dx),$$

$$\Pi^\theta([0,t]\times\Gamma) = \int_0^t \int_\Gamma \psi^\theta(s,x)\Pi(ds,dx), \theta \in \Theta,$$

$$B_t = \| \beta_t^{jk} \|_{j,k=1}^m, \ \beta_t = \sum_{j=1}^m \beta_t^{jj},$$

$$B'(t) = \| \frac{d\beta_t^{jk}}{d\beta_t} \|_{j,k=1}^m, t \geqslant 0,$$

Θ is an arbitrary set.
Let $p([0,t]\times\Gamma) = \sum_{s\leqslant t} \mathbf{1}_\Gamma(\Delta X_s)$ be the jump measure of X, $q([0,t]\times\Gamma) = = p([0,t]\times\Gamma) - \Pi([0,t]\times\Gamma), \Gamma \in \mathcal{B}(\mathbf{E}_m), \mathbf{E}_m = \mathbf{R}^m\{0\}$, and X^c be the continuous local martingale part of the $(\mathbf{Q},\mathbf{F})$–semimartingale X.
Denote $a(t) = \Pi(\{t\}\times\mathbf{E}_m)$, $a^\theta(t) = \Pi^\theta(\{t\}\times\mathbf{E}_m)$,

$$a_I(t) = (a^{\theta_1}(t), \ldots, a^{\theta_k}(t)), \quad \psi_I(t,x) = \left(\psi^{\theta_1}(t,x), \ldots, \psi^{\theta_k}(t,x)\right),$$

$$\hat{u}(t) = \int_{\mathbf{E}_m} u(t,x)\Pi(\{t\}\times dx),$$

$$\psi_{\alpha,I}(t,x) = \prod_{j=1}^k \left(\psi^{\theta_j}(t,x)\right)^{\alpha_j}, \quad a_{\alpha,I}(t) = \widehat{\psi_{\alpha,I}}(t)$$

$$W^\theta_{\alpha_j}(t,x) = (\psi^\theta(t,x))^{\alpha_j} - 1 + \frac{\widehat{(\psi^\theta)^{\alpha_j}}(t) - a(t)}{1-a(t)} \mathbf{1}_{\{a(t)<1\}},$$

$$W_{\alpha,I}(t,x) = \psi_{\alpha,I}(t,x) - 1 + \frac{a_{\alpha,I}(t) - a(t)}{1-a(t)} \mathbf{1}_{\{a(t)<1\}},$$

$$g_{\alpha,I}(t) = \sum_{j=1}^k \alpha_j g^{\theta_j}(t),$$

$$\varphi_\alpha(y) = \sum_{j=1}^k \alpha_j y_j - \prod_{j=1}^k (y_j)^{\alpha_j}, \quad y = (y_1, \ldots, y_k).$$

Assume that for each $t > 0$ and $\theta \in \Theta$ we have $P_t^\theta \sim \mathbf{Q}_t$,

$$W^{\theta_j}_{\alpha_j} \in \mathcal{G}_{\mathrm{loc}}(\Pi,\mathbf{F}), j = 1, \ldots, k, \quad W_{\alpha,I} \in \mathcal{G}_{\mathrm{loc}}(\Pi,\mathbf{F}),$$

$$\{a=1\} \subseteq \{a_{\alpha,I} = 1, \quad a^{\theta_j} = 1, \quad \widehat{(\psi^{\theta_j})^{\alpha_j}} = 1, \quad j = 1, \ldots, k\}.$$

THEOREM 1. (cf. (Jacod and Shiryaev,1987; Jacod,1989b)). *Under above assumptions*

$$L_t(\alpha; I) = \mathcal{E}_t(N(\alpha, I)), \quad t \geqslant 0,$$

where

$$N_t(\alpha; I) = \int_0^t g_{\alpha,I}(s)\,dX_s^c + \int_0^t \int_{\mathbf{E}_m} W_{\alpha,I}(s,x) q(ds,dx),$$

and

$$h_t(\alpha; I) = \frac{1}{2}\Big(\sum_{j=1}^k \alpha_j \int_0^t (g^{\theta_j}(s),\ B'(s) g^{\theta_j}(s))\,d\beta_s -$$

$$- \int_0^t (g_{\alpha,I}(s),\ B'(s) g_{\alpha,I}(s))\,d\beta_s\Big) +$$

$$+ \int_0^t \int_{\mathbf{E}_m} \varphi_\alpha\Big(\psi_I(s,x)\Big)\Pi(ds,\ dx) + \sum_{s\leqslant t} \varphi_\alpha\Big(\mathbf{1}_k - a_I(s)\Big);\quad \mathbf{1}_k = (\underbrace{1,\ldots,1}_{k}).$$

Proof of Theorem 1 is based on the explicit formulas for the local densities Z_t^θ (see (Jacod and Shiryaev,1987)), the properties of the stochastic integrals and the Lemma below.

Let $(\mathfrak{X}, \mathcal{B})$ be a Blackwell space, $(\Omega, \mathcal{F}, \mathbf{F}, \mathbf{P})$ be a stochastic basis, μ be an integer-valued random measure on $(\mathbf{R}_+ \times \mathfrak{X},\ \mathcal{B}(\mathbf{R}_+) \otimes \mathcal{B})$ with the $(\mathbf{P}, \mathbf{F})$–compensator ν, $\varkappa(t) = \nu(\{t\} \times \mathfrak{X})$,

$$\hat{V}(t) = \int_{\mathfrak{X}} V(t,x) \nu(\{t\} \times\ dx),$$

$$W(t,x) = V(t,x) - 1 + \frac{\hat{V}(t) - \varkappa(t)}{1 - \varkappa(t)} \mathbf{1}_{\{\varkappa(t)<1\}},\quad V(t,x) \geqslant 0,$$

$$W_\gamma(t,x) = V^\gamma(t,x) - 1 + \frac{\widehat{V^\gamma}(t) - \varkappa(t)}{1 - \varkappa(t)} \mathbf{1}_{\{\varkappa(t)<1\}}.$$

Under the assumptions that

$$W \in \mathcal{G}_{\mathrm{loc}}(\nu, \mathbf{F}),\quad W_\gamma \in \mathcal{G}_{\mathrm{loc}}(\nu, \mathbf{F}), 0 < \gamma < 1,\quad Y^c \in \mathcal{M}_{\mathrm{loc}}^c(\mathbf{P}, \mathbf{F}),$$

let us denote

$$Y_t = Y_t^c + \int_0^t \int_X W(s,x)(\mu - \nu)(ds, dx),$$

$$Y_t^{(\gamma)} = \gamma Y_t^c + \int_0^t \int_X W_\gamma(s,x)(\mu - \nu)(ds, dx),$$

$$h_t(\gamma) = \frac{1}{2}\gamma(1-\gamma)\langle Y^c\rangle_t + \int_0^t \int_X \varphi^{(\gamma)}(V(s,x), 1)\nu(ds, dx) +$$

$$+\sum_{s\leqslant t}\varphi^{(\gamma)}(1-\hat{V}(s),1-\varkappa(s)),$$

where

$$\varphi^{(\gamma)}(u,v)=\gamma u+(1-\gamma)v-u^{\gamma}v^{1-\gamma}.$$

LEMMA. *If* $\{\varkappa=1\}\subseteq\{\hat{V}=1,\quad \hat{V^{\gamma}}=1\}$, *then the following decomposition holds:*

$$\mathcal{E}_t^{\gamma}(Y)=\mathcal{E}_t(Y^{(\gamma)})\mathcal{E}_t(-h(\gamma)),\quad t\geqslant 0.$$

Proof. Observe that under the assumptions of our Lemma we have that

$$(1+\Delta Y_t)^{\gamma}=1+\Delta Y_t^{(\gamma)},\quad t>0,$$

and complete the proof by standard calculations.

THEOREM 2. *If additionally to the assumptions of Theorem 1*

$$\mathbf{E}_{\mathbf{Q}}\mathcal{E}_t(N(\alpha;I))\equiv 1,$$

then $\mathbf{P}_t(\alpha;I)=\mathbf{P}(\alpha;I)\mid\mathcal{F}_t,\quad t\geqslant 0$, *and* $\mathbf{P}(\alpha;I)$ *is uniquely characterized by the properties that* $\mathbf{P}_0(\alpha;I)=\mathbf{Q}_0$ *and* X *is a* $(\mathbf{P}(\alpha;I),\ \mathbf{F})$*-semimartingale with the triplet* $(\delta(\alpha;I),B,\Pi(\alpha;I))$ *of predictable characteristics, where*

$$\delta_t(\alpha;I)=\delta_t+\int_0^t g_{\alpha,I}(s)B'(s)\,d\beta_s+$$

$$+\int_0^t\int_{|x|\leqslant 1}x(\psi_{\alpha,I}(s,x)-1)\Pi(ds,\ dx),$$

$$\Pi(\alpha;I)([0,t]\times\Gamma)=\int_0^t\int_\Gamma\psi_{\alpha,I}(s,x)\Pi(ds,\ dx).$$

Proof of this statement follows directly from the Theorem 1 and the Girsanov theorem for semimartingales.

4. We shall consider now some examples as illustrations.

Example 1 (exponential class of processes with the independent increments). Assume that

$$\frac{d\mathbf{P}_t^{\theta}}{d\mathbf{Q}_t}=\exp\{(\theta,X_t)-t\mathcal{X}(\theta)\},\quad t\geqslant 0,\quad \theta\in\Theta,$$

where $\mathcal{X}(\theta)=\ln\mathbf{E}_{\mathbf{Q}}\exp\{(\theta,\ X_1)\}$, $\Theta=\{\theta\in\mathbf{R}^m:\mathcal{X}(\theta)<\infty\}$.
After some simple calculations, we find that

$$\mathbf{P}(\alpha;I)=\mathbf{P}^{\theta(\alpha;I)},\quad \Theta(\alpha;I)=\sum_{j=1}^{k}\alpha_j\theta_j,$$

$$h_t(\alpha; I) = t\left(\sum_{j=1}^{k} \alpha_j \mathcal{X}(\theta_j) - \mathcal{X}(\theta(\alpha; I))\right),$$

and

$$L_t(\alpha; I) = \exp\{(\theta(\alpha; I), X_t) - t\mathcal{X}(\theta(\alpha; I))\},$$

where $(\cdot, \cdot)$ denotes the inner product in $\mathbf{R}^m$. We remark that general theory of exponential processes is reviewed in (Küchler and Sorensen, 1989).

Example 2 (m–dimensional counting processes). Assume that

$$\delta_t^\theta = \int_0^t \lambda^\theta(s)\, ds, \quad \lambda^\theta(t) = (\lambda_1^\theta(t), \ldots, \lambda_m^\theta(t)),$$

$B_t \equiv 0$, Π^θ and Π are concentrated at the points

$$e_j = (\underbrace{0, \ldots, 0,}_{j-1}\ 1,\ 0, \ldots, 0), \quad j = 1, \ldots, m, \quad \Pi^\theta([0, t] \times \{e_j\}) =$$

$$= \int_0^t \lambda_j^\theta(s)\, ds, \quad \delta_t = t\,\mathbf{1}_m, \quad \Pi([0, t] \times \{e_j\}) = t, \quad j = 1, \ldots, m, \quad t \geqslant 0,$$

and there exist constants $0 < \underline{\lambda}_j^\theta < \bar{\lambda}_j^\theta$ such that

$$\underline{\lambda}_j^\theta \leqslant \lambda_j^\theta(t) \leqslant \bar{\lambda}_j^\theta, \quad t \geqslant 0, \quad j = 1, \ldots, m.$$

From the known properties of the counting processes and Theorems 1 and 2 we have that

$$N_t(\alpha; I) = \sum_{l=1}^{m} \int_0^t \left(\prod_{j=1}^{k} (\lambda_l^{\theta_j}(s))^{\alpha_j} - 1\right) d(X_s^l - s), \quad X_t = (X_t^1, \ldots, X_t^m),$$

$$h_t(\alpha; I) = \sum_{l=1}^{m} \int_0^t \varphi_\alpha(\lambda_{l,I}(s))\, ds, \quad \lambda_{l,I}(t) = (\lambda_l^{\theta_1}(t), \ldots, \lambda_l^{\theta_k}(t)),$$

and $\mathbf{P}(\alpha; I)$ is the probability measure, corresponding to the m–dimensional counting process with the conditional intensity vector

$$\lambda(\alpha; I)(t) = \left(\prod_{j=1}^{k} (\lambda_1^{\theta_j}(t))^{\alpha_j}, \ldots, \prod_{j=1}^{k} (\lambda_m^{\theta_j}(t))^{\alpha_j}\right).$$

Example 3 (m–dimensional diffusion type processes).
Assume that

$$\delta_t^\theta = \int_0^t b^\theta(s)\, ds, \quad b^\theta(t) = (b_1^\theta(t), \ldots, b_m^\theta(t)),$$

$$\beta_t^{jl} = t\delta_{jl}, \quad j, l = 1, \ldots, m, \quad \Pi^\theta \equiv 0, \quad \delta_t \equiv 0,$$

$\Pi \equiv 0$, and the functions $b^{\theta_j}(t)$, $j = 1, \ldots, m$, are uniformly bounded; δ_{jl} denotes the Kronecker symbol.
Denoting

$$b(\alpha; I)(t) = \sum_{j=1}^{k} \alpha_j b^{\theta_j}(t), \quad t \geqslant 0,$$

and applying the Theorems 1 and 2 we find that

$$N_t(\alpha; I) = \int_0^t b(\alpha; I)(s)\, dX_s,$$

$$h_t(\alpha; I) = \frac{1}{2} \int_0^t \left(\sum_{j=1}^{k} \alpha_j \mid b^{\theta_j}(s) \mid^2 - \mid b(\alpha; I)(s) \mid^2 \right) ds,$$

$\mathbf{P}(\alpha; I)$ is the probability measure, corresponding to the diffusion type process with the drift vector $b(\alpha; I)$ and the unit matrix of diffusion coefficients.
Assuming that $b^\theta(t) = b^\theta(X(t))$, where the functions $b^\theta(x)$, $x \in \mathbf{R}^m$, $\theta \in \Theta$, are non–random and sufficiently smooth in x, and that the initial distributions are Dirac measures $\varepsilon_x(\cdot)$, $x \in \mathbf{R}^m$, we have that for each α and I the corresponding Hellinger measures $H_{t,x}(\alpha; I)(A)$, $A \in \mathcal{F}_t$, generate the Schrödinger semigroup of linear operators

$$T_t(\alpha; I)(f)(x) = \mathbf{E}_{\mathbf{P}_{t,x}(\alpha; I)} e^{-h_t(\alpha; I)} f(X_t),$$

$$f \in C(\mathbf{R}^m), \quad t \geqslant 0, \quad x \in \mathbf{R}^m.$$

Denoting

$$b(\alpha; I)(x) = \sum_{j=1}^{k} \alpha_j b^{\theta_j}(x)$$

and

$$u(t, x) = T_t(\alpha; I)(f)(x),$$

we obtain that

$$\frac{\partial u(t,x)}{\partial t} = \frac{1}{2} \Delta u(t,x) + (b(\alpha; I)(x), \ \nabla u(t,x)) -$$

$$-\frac{1}{2} \left(\sum_{j=1}^{k} \alpha_j \mid b^{\theta_j}(x) \mid^2 - \mid b(\alpha; I)(x) \mid^2 \right) u(t,x),$$

$$u(0, x) = f(x).$$

If we denote by $T_t(\theta)$, $t \geqslant 0$, $\theta \in \Theta$, the semigroup of linear operators generated by the diffusion process with the drift vector $b^\theta(x)$ and the unit diffusion matrix, then we can write symbolically that (cf.(Newman,1972))

$$T_t(\alpha; I) = \prod_{j=1}^{k} T_t(\theta_j)^{\alpha_j}.$$

Let us remark finally, that the domination assumption (1) in Theorems 1 and 2 can be relaxed, but not removed. Necessary and sufficient conditions are found in (Newman,1972) for the probability measures, corresponding to quasi–left continuous processes with independent increments ($m = 1$, $k = 2$, $\alpha_1 = \alpha_2 = 1/2$).

REFERENCES

Ibragimov, I.A. and Hasminsky, R.Z. (1981). *Statistical Estimation: Asymptotic Theory.* Springer, Berlin–Heidelberg etc.

Jacod, J. (1989a). Une application de la topologie d'Emery: le processus information d'un modèle statistique filtré. *Lect. Notes in Math.* **1372**, 448–474.

Jacod, J. (1989b). Filtered statistical models and Hellinger processes. *Stochast. Proc. Appl.* **32**, 3–45.

Jacod, J. and Shiryaev, A.N. (1987). *Limit Theorems for Stochastic Processes.* Springer, Berlin–Heidelberg etc.

Kakutani, S. (1948). On equivalence of infinite product measures. *Ann. Math.* **49**, 214–224.

Liese, F. and Vajda, I. (1987). *Convex Statistical Distances.* Teubner–Texte zur Mathematik, **95**, Leipzig.

Küchler, U. and Sorensen, M. (1989). Exponential families of stochastic processes: a unifying semimartingale approach. *Intern. Statist. Rev.* **57**, 123–144.

Liptser, R.S. and Shiryaev, A.N. (1986). *Theory of Martingales.* Nauka, Moscow.

Newman, C.M. (1972). The inner product of path space measures corresponding to random processes with independent increments. *Bull. Amer. Math. Soc.* **78**, 268–272.

Oosterhoff, J. and Van Zwet, W.R. (1979). A note on contiguity and Hellinger distance. In: *Contributions to Statistics.* Reidel, Dordrecht.

Vajda, I. (1988). *Theory of Statistical Inference and Information.* Reidel, Dordrecht.

Prob. Theory and Math. Stat., Vol. 1, pp. 455–462
B.Grigelionis *et al.* (Eds.)
1990 VSP/Mokslas

ON CONNECTION BETWEEN THE THEORY OF BRANCHING PROCESSES AND QUEUEING THEORY

S.A. GRISHECHKIN

Department of Mathematics, Gubkin Institute of Oil and Gas,
Leninskiy prospekt 65, GSP-1, 117917 Moscow, USSR

ABSTRACT

$M \mid G \mid 1$ queue with various disciplines (repeated calls, random selection, processor sharing, shortest residual processing time) is studied by means of branching processes. A number of theorems are formulated including investigation of heavy traffic and supercritical queues.

The theory of branching processes allows to derive many new results in queueing theory. In this paper we will present some of them. The following four queueing systems will be considered mainly. Let us suppose that customers arrive according to a Poisson process with intensity 1. Each customer has random processing time l, $B(\cdot)$ is the distribution function of l, b is the mean of l, $b(u)$ is the Laplace transform of l.

(a) The first system is the $M \mid G \mid 1 \mid$ RC system with repeated calls. If the arriving customer finds the server free its service begins. Otherwise it stands in the queue. Each customer from the queue independently from each other repeats its calls at intervals which are exponentially distributed with mean μ^{-1}. The moment the customer finds the server free its service begins. This system was studied by Falin, Kulkarni and other authors (Falin, 1986; Yang and Templeton, 1987).

(b) If $\mu = \infty$, we deal with the classical $M \mid G \mid 1 \mid$ SIRO system with random discipline. In this case the moment the server becomes free one customer from the queue is randomly selected and this customer starts to be served.

(c) The third system is the $M \mid G \mid 1 \mid$ PS system with processor sharing. In this system all customers from the queue (say N in number) are served simultaneously with rate $1/N$. This system was studied by many authors, for example (Yashkov, 1983; Schassberger, 1984; Ott, 1984).

(d) If the distribution function B is exponential, then we have the $M \mid M \mid 1$ system with processor sharing.

The first part of this paper is devoted to the sojourn time in these queueing systems. For a $M \mid G \mid 1 \mid$ PS queue let us denote by $W(l_1, \ldots, l_N)$ the sojourn

time of the tagged customer provided that there were N other customers with residual processing times $l_1, \dots, l_N$ in the system at the moment when the tagged customer arrived. Analogously, for a $M \mid G \mid 1 \mid$ RC queue let us denote by $W(N)$ the sojourn time of the tagged customer provided that there were N other customers in the queue at the moment when the tagged customer arrived in the system and the server was free at this moment.
The problems we want to discuss are the properties of the random variable W. The question that we must ask ourselves in the beginning is the following. Is the random variable W finite almost surely or not? For a $M \mid G \mid 1 \mid$ RC queue the answer is the following.

THEOREM 1 (Grishechkin, 1988a, 1990). *The random variable $W(N)$ is finite almost surely if and only if*

$$\int_0^1 (1 - b(u))^{-1} du = \infty.$$

The investigation of such questions is most frequently connected with the explosive condition for the relative branching process.
The theory of branching processes allows to find the expectations $\mathbf{E} \exp(-uW(N))$, $\mathbf{E} \exp(-uW(l_1, \dots, l_N))$, $u > 0$. We can therefore find the moments of integer orders of the random variable W. But more interesting is the problem of finding conditions of finiteness of $\mathbf{E}W^\gamma$, where γ is not nessesary an integer. For example, for $M \mid G \mid 1 \mid$ RC system the following theorem is true.

THEOREM 2 (Grishechkin, 1990). *Let $\gamma \geqslant 1$. Then for $M \mid G \mid 1 \mid$ RC queue $\mathbf{E}W^\gamma(N) < \infty$ if and only if $\gamma(b-1) < 1$, $\mathbf{E}l^\gamma < \infty$.*

Let us note that if the traffic intensity in the system $b > 1$, then the random variable $W(N)$ can have moments only of order less then $1/(b-1)$. Therefore it is natural to ask the question about the asymptotic behavior of the tail of $W(N)$. The answer gives the following theorem.

THEOREM 3 (Grishechkin, 1990). *Let $b > 1$ and $\mathbf{E}l^{\max(1,\ 1/(b-1))+\varepsilon} < \infty$, where $\varepsilon > 0$. Then*

$$\mathbf{P}(W(N) \geqslant x) \sim x^{-1/(b-1)} \mathbf{E}Y_N^{1/(b-1)},$$

where N is fixed, $x \to \infty$ and Y_N is some random variable.

Analogous theorems are valid also for $M \mid G \mid 1 \mid$ PS queues. For example, we have the following theorem.

THEOREM 4. *Let $b > 1$, α be the root of the equation*

$$\int_0^\infty e^{-\alpha u}(1 - B(u))du = 1$$

and the function $R(x) = 1 - B(-\log x)$, $x > 0$, *be regularly varying as* $x \to 0$ *with index* $k \in [0, \infty)$. *Then for the* $M \mid G \mid 1 \mid$ PS *queue*

$$\mathbf{P}(W(l_1, \ldots, l_N) \geqslant x) \sim cR(x^{-1/\alpha}),$$

where l_i, $i = 1, \ldots, N$ *are fixed,* $x \to \infty$ and $c \in (0, \infty)$ *is some constant.*

To conclude this part of the paper we shall remark that the proofs of theorems 2 — 4 are closely connected with the convergence of $Z^{\varphi}(t)e^{-mt}$ almost surely and in

L^p–space, where $Z^{\varphi}(t)$ is some branching process connted by the random characteristic (Nerman, 1981; Jagers and Nerman, 1984).
The next part of this paper is connected with limit theorems for sojourn times. A typical result is the following.

THEOREM 5 (Grishechkin, 1990). *For a* $M \mid G \mid 1 \mid$ RC *queue*

$$W(N)/N \to (b/(b-1))(e^{(b-1)\tau} - 1) \quad a.s., \quad N \to \infty,$$

where τ *is an exponentially distributed random variable with mean* 1.

To formulate an analogous result for the system with processor sharing we shall introduce some new notation. Let $W_L(N)$ be the sojourn time of the tagged customer provided that at the moment of its arrival there were N other customers in a $M \mid M \mid 1 \mid$ PS queue and the tagged customer has processing time L. The following theorem describes the limit behavior of $W_l(N)$ if L or N is large.

THEOREM 6 (Grishechkin, 1988b). *Let us consider a* $M \mid M \mid 1 \mid$ PS *queue.*
(a) *If* $b > 1$, N *is fixed,* $L \to \infty$ *then*

$$W_L(N)e^{L(1-b)} \to Y + X_1 + \cdots + X_N \quad a.s.,$$

where Y, X_i, $i = 1, \ldots, N$ *are independent random variables and* X_i, $i = 1, \ldots, N$ *are identically distributed.*
(b) *If* $b > 1$, $N \to \infty$, $\liminf\ L > 0$, *then* $W_L(N)$ *is asymptoticaly normal.*
(c) *If* $b < 1$, $L + N \to \infty$, $\liminf\ L > 0$, *then* $W_L(N)$ *is asymptoticaly normal.*
(d) *If* $b = 1$, $L \to \infty$, $L/N \to p \in (0, \infty]$, *then*

$$\mathbf{E}\exp(-uL^{-2}W_L(N)) \to \exp\left[\frac{-u^{1/2}(1 - \exp(-2u^{1/2}))}{p(1 + \exp(-2u^{1/2}))}\right] \mathrm{sech}(u^{1/2}).$$

(e) *If* $b = 1$, $\liminf\ L > 0$, $L/N \to 0$ *then* $W_L(N)$ *is asymptoticaly normal.*

In (Grishechkin, 1988b) many other theorems of the same type are formulated. In particular, the case when traffic intensity $b \to 1$ is investigated in detail.
The theory of branching processes allows to derive also more complicated limit theorems. For instance, let us formulate the following result.

THEOREM 7. *Let $V(T)$ be the time from T till the moment when all customers present in a $M \mid G \mid 1 \mid$ PS queue at the moment T are served. Let $b > 1$, $K = \sup(x : B(x) < 1)$, α is the root of the equation*

$$\int_0^\infty e^{-\alpha u}(1 - B(u))du = 1.$$

Then as $T \to \infty$ the following assertions are true.
(a) *If $K < \infty$, then*

$$V(T)/T \to e^{\alpha K} - 1 \quad a.s.$$

(b) *If the function $G(x) = 1/(1 - B(\log x))$ is rapidly varying as $x \to \infty$, then*

$$V(T)/(Ta(T)) \xrightarrow{d} 1,$$

where $a(T) = (G^*(T))^\alpha$ and $G^*(T)$ is the inverse function of $G(T)$.
(c) *If the function $G(x)$ is regularly varying as $x \to \infty$ with index $\gamma \in (0, \infty)$, then*

$$\mathbf{P}(V(T)/Ta(T) \leqslant x) \to \exp(-x^{-\gamma/\alpha}\alpha(\alpha + \gamma)^{-1}).$$

(d) *If the function $G(x)$ is slowly varying as $x \to \infty$, then*

$$\mathbf{P}(T^{-1}G(T^{-1}V(T)) \leqslant x) \to \exp(-1/x), \quad x > 0.$$

Analogous results, concerning a $M \mid G \mid 1 \mid$ RC queue, can be found in (Grishechkin, 1990).
The next part of this paper concerns supercritical ($b > 1$) queueing systems. Supercritical queues were hardly overconsidered in queueing theory. It is believed that there are not interesting and nontrivial results for such queues. But it is not true. Let us consider, for example, the problem of asymptotic behavior of the queue length $S(T)$, as $T \to \infty$. If we deal with the $M \mid G \mid 1$ system with FIFO or SIRO discipline, then $S(T)/T \to 1 - b^{-1}$ a.s. This is evident. But in the case of $M \mid G \mid 1$ system with more complicated discipline that conclusion false.
For example, the following theorem is true.

THEOREM 8. *For a supercritical $M \mid G \mid 1 \mid$ PS queue*

$$S(T)/T \to \alpha \quad a.s., \quad T \to \infty,$$

where α is the root of the equation

$$\int_0^\infty e^{-\alpha u}(1 - B(u))du = 1.$$

The proof of this result is based on deep theorems from the theory of supercritical branching processes, which were derived in (Nerman, 1981; Jagers and Nerman, 1984).

Many other problems concerning supercritical queues can be formulated. We want to mention only one direction of investigation. It concerns the so-called age distribution. The following theorem is true.

THEOREM 9 (Grishechkin, 1988a, 1990). *Let* $S(t, u)$ *be the number of customers arrived in a* $M \mid G \mid 1 \mid$ SIRO *queue after* u *and were not served by the moment* t. *Then*

$$S(t, ct)/S(t) \to 1 - c^{1+1/(b-1)} \quad a.s., \quad t \to \infty$$

for every constant $c \in (0, 1)$.

Thus, this theorem describes the age structure of the queue if t is large. Analogous but more complicated theorems are true also for the $M \mid G \mid 1 \mid$ PS system.
The last part of this paper is devoted to some quite traditional problems in queueing theory. They can be studied with the aid of branching processes and they have not been solved by another methods till now.
The first problem is concerned with W_L in heavy traffic case $b \uparrow 1$ in the $M \mid G \mid 1 \mid$ PS system. Here W_L is the sojourn time of a customer with the processing time L in the steady state.
Although the Laplace transform of W_L is known (Yashkov, 1983), this transform has a very complicated form and the following theorem has not been proved till now.

THEOREM 10. *Let us suppose that customers arrive in a* $M \mid G \mid 1 \mid$ PS *queue according to a Poisson process with intensity* λ. *Then*

$$\mathbf{P}(W_L(1 - b\lambda)/L \leqslant x) \to 1 - e^{-x}, \quad x \geqslant 0,$$

as $\lambda \uparrow b^{-1}$ *and* L *is fixed.*

The proof of this theorem, based on the branching processes theory, is probabilistic and comparatively simple.
The second range of questions concerns $M \mid G \mid 1$ queues with repeated calls and several types of customers. It is described in the following way. The customers of N various types arrive in the system. The customer of type i arrives according to a Poisson process with intensity λ_i. Each customer of type i has a random processing time distribution function $B_i(x)$,

$$b_i = \int_0^\infty x dB_i(x).$$

If at the moment of the customer arrival the server is free, then the service of this customer begins immeidately, otherwise it joins the queue. Every customer from the queue independently from each other repeats it's call after some exponentially distributed delay. The intensity of the recall for the i-th type is μ_i. This queueing system is studied in (Falin, 1988; Kulkarni, 1986). Let us denote by $S = (S_1, \ldots, S_N)$ the queue length in the steady state and let W_i be the virtual waiting time of the i-th type customer in the steady state. Only $\mathbf{E}S_i$ and $\mathbf{E}W_i$ were known till now. The theory of branching processes allows to find the Laplace transform of the virtual waiting time and the generating function of the queue length. Moreover, we can prove limit theoreme for the heavy traffic case.

THEOREM 11. *Let* $\sum_{i=1}^{N} \lambda_i b_i \uparrow 1$ *and* $\int_0^\infty x^2 dB_i(x) < \infty, \quad i = 1, \ldots, N.$ *Then*

$$S_i/S_j \xrightarrow{d} \lambda_i \mu_j / \lambda_j \mu_i$$

and

$$\mathbf{E}\exp(-u(1 - \sum_{i=1}^{N} \lambda_i b_i)\mu_k \lambda_k^{-1} S_k) \to (1 + uc_1)^{-c_2}, \quad u \geqslant 0,$$

where c_1, c_2 *are some constants.*

THEOREM 12. *Let* $\sum_{i=1}^{N} \lambda_i b_i \uparrow 1$ *and* $\int_0^\infty x^2 dB_i(x) < \infty, \quad i = 1, \ldots, N.$ *Then*

$$\mathbf{E}\exp(-u(1 - \sum_{i=1}^{N} \lambda_i b_i) W_k) \to \mu_k \int_0^\infty e^{-\mu_k t}(1 + uc_3 t)^{-c_2} dt, \quad u \geqslant 0.$$

It is nessesary to note that the proofs of these theorems are based on Chistyakov's research concerning transition phenomena in Markov branching processes with several types of particles (Chistyakov, 1961, 1972).

The last result concerns $M \mid G \mid 1$ queues with discipline SRPT. In this system at every moment only one customer is served which have the smallest residual processing time. This discipline was defined in (Schrage and Miller, 1966). It distinguishes from all other possible disciplines by the fact that it have the minimal queue length in the steady state.

Let us consider a $M \mid G \mid 1$ system with batch arrival, i.e. the number of customers arrived in the system in interval $(0, t)$ have generating function $e^{\lambda(h(s)-1)t}$. For each customer A we consider a pair $(l_A, r_A(\cdot))$, where l_A, $\mathbf{P}(l_A \leqslant x) = B(x)$, is the processing time of A, and $r_A(\cdot)$ is some random process. Let us suppose that the pairs $(l_A, r_A(\cdot))$ are independent and identically distributed for of all customers A, $(l_A, r_A(\cdot)) \stackrel{d}{=} (l, r(\cdot))$. We require the following conditions.

(a) $r(t) \geqslant 0$ and $r(t) = 0$ if $t < 0$ or $t \geqslant l$.

(b) The process $r(t)$ is right-continuous, has left limits and the number of points of discontinuity is finite.

Assume that there are n customers $A_1, \ldots, A_n$ in the queue at a moment t and their received amounts of service are equal to $V_{A_1}, \ldots, V_{A_n}$. We want to find the stationary distribution of

$$S^r(t) = \sum_{i=1}^{n} r_{A_i}(V_{A_i})$$

as $t \to \infty$ (this distribution exists iff $\rho = h'(1)b\lambda < 1$). For instance, if $r(t) = \mathbf{1}(t \in [0,1))$, then $S^r(t)$ coincides with the queue length at the moment t. A more complicated example is the following. Let M_t be the random measure

$$M_t(D) = \sum_{i=1}^{n} \mathbf{1}(V_{A_i} \in D).$$

Then $M_t \xrightarrow{d} M, \quad t \to \infty$, and

$$\mathbf{E}\exp(-\int_0^\infty f(x)M(dx)) = \lim_{t\to\infty} \mathbf{E}\exp(-S^r(t)),$$

where $r(t) = f(t) \cdot \mathbf{1}(t \in [0,1))$ and f is a non-random function.

Theorem 13. *If $\rho < 1$ and the function $B(\cdot)$ is continuous then*

$$S^r(t) \xrightarrow{d} S^r, \quad t \to \infty,$$

and for any fixed $v \geqslant 0$

$$\mathbf{E}\exp(-vS^r) = \lambda(1-\rho)(\lambda^{-1} + \int_0^\infty h'(F(v,x))dY(v,x))$$

where the function $Y(v,x)$ is the unique solution (in the class of non-decreasing, continuous functions) of the equation

$$Y(v,t) = b(t) + \int_0^t h'(F(v,u))a(t,u)dY(v,u),$$

where

$$b(t) = \mathbf{E}\Big[\int_0^l e^{-vr(l-y)} \exp\Big\{\int_y^l \lambda(h(F(v,x))-1)dx\Big\}dy, \quad l \leqslant t\Big],$$

$$a(t,u) = \lambda\int_0^u x\exp\Big\{\lambda x(h(F(v,u))-1)\Big\}dB(x)+$$

$$+\lambda\int_u^t u\exp\Big[\lambda\Big\{uh(F(v,u)) + \int_u^x h(F(v,y))dy - x\Big\}\Big]dB(x)+$$

$$+\lambda\mathbf{E}\Big[\int_u^l e^{-vr(1-y)} \exp\Big\{\lambda\int_y^l (h(F(v,w))-1)dw\Big\}dy, \quad u < l < t\Big].$$

The function $0 \leqslant F(v,t) \leqslant 1$ *is the unique solution of the equation*

$$F(v,t) = \mathbf{E}(e^{-vr(0)},\ l > t) + \int_0^t \exp\left\{\lambda u(h(F(v,t)) - 1)\right\} dB(u).$$

Let us note that till now only the mean of the queue length for a $M \mid G \mid 1 \mid$ SRPT system has been known in a special case $h(s) = s$.

REFERENCES

Chistyakov, V.P. (1961). Transition phenomena in branching processes with n types of particles. *Theory Probab. Appl.* **6**, 31–46.

Chistyakov, V.P. (1972). On transition phenomena in branching processes with several types of particles. *Theory Probab. Appl.* **17**, 669–673.

Falin, G.I. (1986). On the waiting–time process in a single–line queue with repeated calls. *J.Appl. Probab.* **23**, 185–192.

Falin, G.I. (1988). On a multiclass batch arrival retrial queue. *Adv. Appl. Probab.* **20**, 483–487.

Grishechkin, S.A. (1988a). On the investigation of queueing systems with random selectionby branching processes methods. *Sov. Math. Dokl.* **300.**

Grishechkin, S.A. (1988b). One–channel systems with round robion or processor sharing and branching processes. *Math. Notes.* **44.**

Grishechkin, S.A. (1990). Branching processes and systems with repeated calls or random discipline. *Theory Probab. Appl.* **35 (to appear).**

Jagers, P. and Nerman, O. (1984). The growth and composition of branching populations. *Adv. Appl. Probab.* **16**, 221–259.

Kulkarni, V.G. (1986). Expected waiting times in a multiclass batch arrival retrial queue. *J.Appl. Probab.* **23**, 144–154.

Nerman, O. (1981). On the convergence of supercritical general (C–M–J) branching processes. *Z. Wahr. verw. Geb.* **57**, 365–395.

Ott, T. (1984). Theb sojourn–time distribution in the $M \mid G \mid 1$ queue with processor sharing. *J.Appl. Probab.* **21**, 360–378.

Schrage, L. and Miller, L. (1966). The queue $M \mid G \mid 1$ with the shortest remaining processing time discipline. *Operat. Research.* **14**, 670–684.

Schassberger, R.A. (1984). A new approach to the $M \mid G \mid 1$ processor–sharing queue. *Adv. Appl. Probab.* **16**, 202–213.

Yang Tao and Templeton, J.G.C. (1987). A survey on retrial queues. *Queueing Syst.* **2**, 201–233.

Yashkov, S.F. (1983). A derivation of response time distribution for a $M \mid G \mid 1$ processor sharing queue. *Probl. Contr. and Inform. Theory.* **12**, 133–148.

Prob. Theory and Math. Stat., Vol. 1, pp. 463–469
B.Grigelionis *et al.* (Eds.)
1990 VSP/Mokslas

ESTIMATION OF MANOVA EIGENVALUES

A.K. GUPTA
Department of Mathematics and Statistics, Bowling Green State University, Bowling Green, Ohio 43403–0221, USA

ABSTRACT

In this paper we consider the problem of estimating the eigenvalues ($\delta_1 \geqslant \delta_2 \geqslant \ldots \geqslant \delta_p > 0$) of $\Sigma_1\Sigma_2^{-1}$ from the corresponding sample eigenvalues of $S_1S_2^{-1}$, where S_1 and S_2 are independent Wishart matrices with $S_i \sim W_p(n_i, \Sigma_i)$, $i = 1, 2$.

1. INTRODUCTION

Suppose that S_1 and S_2 are independent $p \times p$ Wishart random matrices with $S_i \sim W_p(n_i, \Sigma_i)$, where Σ_i is a positive definite matrix and $n_i \geqslant p$ is degrees of freedom, $i = 1, 2$. Let $l_1, l_2, \ldots, l_p$ $(l_1 > l_2 > \ldots > l_p > 0)$ denote the eigenvalues $S_1S_2^{-1}$. We consider here the problem of estimating eigenvalues $\delta_1, \delta_2, \ldots, \delta_p$ $(\delta_1 \geqslant \delta_2 \geqslant \ldots \geqslant \delta_p > 0)$ of $\Sigma_1\Sigma_2^{-1}$. In multivariate analysis under the random effects model, the between and within–group sums of squares and cross–products matrices S_1 and S_2 arise naturally. We need to estimate these eigenvalues in certain distributional problems, for example, in the problem of testing $\Sigma_1 = \Sigma_2$ against $\Sigma_1 \neq \Sigma_2$. The power function of any test statistic which is a function of $l_1, \ldots, l_p$, depends on Σ_1 and Σ_2 only through $\delta_1, \delta_2, \ldots, \delta_p$.

The main purpose of this paper is to review some of the work done for estimating the MANOVA eigenvalues. In Section 2 we first consider the decision theoretic approach under the loss functions

(i) $L_1(\widehat{\Delta}, \Delta) = \mathrm{tr}(\widehat{\Delta}\Delta^{-1}) - \log \mid \widehat{\Delta}\Delta^{-1} \mid -p$,

(ii) $L_2(\widehat{\Delta}\Delta) = \mathrm{tr}(\widehat{\Delta}\Delta^{-1} - I_p)^2$,

(iii) $L_3(\widehat{\Delta}, \Delta) = \mathrm{tr}(\widehat{\Delta} - \Delta)^2$

(iv) $L_4(\widehat{\delta}, \delta) = \sum_{i=1}^{p}(\widehat{\delta}_i - \delta_i)^2$,

where $\widehat{\Delta}$ and $\widehat{\delta}_i$ denote estimators of Δ and δ_i respectively. The maximum likelihood approach is discussed in Section 3.

2. DECISION THEORETIC APPROACH

To estimate $\delta_1, \delta_2, \ldots, \delta_p$ through a decision theoretic approach, one needs to specify a loss function in terms of δ_i's and $\widehat{\delta}_i$'s, where $\widehat{\delta}_i$ is an estimator of

δ_i $(i = 1, 2, \ldots, p)$. However, for this loss, computing the risk of an estimator under the joint distribution of $l_1, l_2, \ldots, l_p$ does not seem feasible due to the complexity of the distribution of the ordered eigenvalues derived in (Khatri, 1967). Instead in (Muirhead and Verathaworn, 1985) an alternate method is described. Define a random matrix $F = V_1^{1/2} V_2^{-1} V_1^{1/2}$, where

$$V_1^{1/2} = \Sigma_2^{-1/2} S_1 \Sigma_2^{-1/2} \sim W_p(n_1, \Delta), \qquad V_2 = \Sigma_2^{-1/2} S_2 \Sigma_2^{-1/2} \sim W_p(n_2, I)$$

and $\Delta = \Sigma_2^{-1/2} \Sigma_1 \Sigma_2^{-1/2}$. The density function of F is given by

$$\left\{\beta_p\left(\frac{n_1}{2}, \frac{n_2}{2}\right)\right\}^{-1} |\Delta|^{-n_1/2} |F|^{(n_1-p-1)/2} \mid I + \Delta^{-1} F \mid^{-n/2}, \qquad F > 0, \quad (2.1)$$

where $n_1 > p+1$, $n_2 > p+1$, $n = n_1 + n_2$, $\beta_p(\cdot)$ is a p–variate beta function and Δ is a positive definite parameter matrix. The eigenvalues of F are the same as those of $S_1 S_2^{-1}$ and their distribution depends only on the eigenvalues $\delta_1, \ldots, \delta_p$ of $\Sigma_1 \Sigma_2^{-1}$ or equivalently of Δ. We then estimate Δ by $\widehat{\Delta}(F)$ and take the eigenvalues of $\widehat{\Delta}(F)$ as estimators of $\delta_1, \ldots, \delta_p$. Note that F is not observable. In (Gupta and Krishnamoorthy, 1987) Δ is estimated by only orthogonally invariant estimators of the form $\widehat{\Delta}(F) = R\varphi(L)R'$ (where $F = RLR'$ is a spectral decomposition of F, $L = \mathrm{diag}(l_1, \ldots, l_p)$ with $l_1 > l_2 > \ldots > l_p > 0$, and $\varphi(L) = \mathrm{diag}(\varphi_1(L), \ldots, \varphi_p(L))$ then the eigenvalues $\varphi_1(L), \varphi_2(L), \ldots, \varphi_p(L)$ of $\widehat{\Delta}(F)$ are observable and are consided as estimators of $\delta_1, \ldots, \delta_p$).

2.1. *Entropy Loss Function*

Consider the problem of estimating Δ by orthogonally invariant estimators under the loss (Haff, 1982; Leung and Muirhead, 1987; Muirhead, 1987)

$$L_1(\Delta, \widehat{\Delta}) = \mathrm{tr}(\widehat{\Delta}\Delta^{-1}) - \log \mid \widehat{\Delta}\Delta^{-1} \mid -p. \qquad (2.2)$$

The usual unbiased (under the density (2.1)) and also orthogonally invariant estimator of Δ given by

$$\widehat{\Delta}_u(F) = \frac{n_2 - p - 1}{n_1} F, \qquad (2.3)$$

is also the "best" as stated below.

THEOREM 2.1 (Muirhead and Verathaworn, 1985). *Under the loss function L_1, the best (smallest risk) estimator of Δ having the form αF is the unbiased estimator $\widehat{\Delta}_u(F)$ given by (2.3).*

In (Muirhead and Verathaworn, 1985) an estimator

$$\widehat{\Delta}_R = R\varphi(L)R, \qquad (2.4)$$

is derived with

$$\varphi_i(L) = l_i \left[(n_1 - p - 1)/2 + \frac{2(n-p-1)}{n_2(n_2-p-1)} l_i \sum_{j \neq i} (l_i - l_j)^{-1}\right]^{-1}, \qquad (2.5)$$

$i = 1, 2, \ldots, p$. As an estimator of Δ, their Monte Carlo comparison study shows that $\widehat{\Delta}_R(F)$ performs better than $\widehat{\Delta}_u(F)$ when $\Delta \approx cI$ or when Δ has groups of equal eigenvalues. However, Monte Carlo study under the loss

$$L(\Delta_e, \widehat{\Delta}_e) = \operatorname{tr}(\widehat{\Delta}_e \Delta_e^{-1}) - \log | \widehat{\Delta}_e \Delta_e^{-1} | - p \tag{2.6}$$

(where, for a matrix A, $A_e = \operatorname{diag}(\lambda_1, \ldots, \lambda_p)$, λ_i's are the eigenvalues of A) indicates that the eigenvalues of $\widehat{\Delta}_u(F)$ dominate the eigenvalues of $\widehat{\Delta}_R(F)$ when the δ_i's are much dispersed. In (Gupta and Krishnamoorthy, 1987) three orthogonally invariant estimators of Δ are introduced, namely

$$\widehat{\Delta}_i(F) = R\varphi_i^*(L)R' \tag{2.7}$$

where $\varphi_i^*(L) = \operatorname{diag}(d_{i1}, d_{i2}, \ldots, d_{ip})$ and

$$\begin{aligned} d_{1j} &= (n_2 - p - 1)/(n_1 + p + 1 - 2j), \\ d_{2j} &= (n_2 - p - 1 + j)(n_2 - p - 2 + j)/(n_1(n_2 - 1)), \\ d_{3j} &= n_1 d_{1j} d_{2j}/(n_2 - p - 1), \end{aligned}$$

$i = 1, 2, 3$ and $j = 1, 2, \ldots, p$. Note that $0 < d_{i1} < d_{i2} < \ldots < d_{ip}$ for $i = 1, 2, 3$. The constants d_{1j}'s and d_{2j}'s are chosen, respectively, from the best invariant minimax estimators of the normal covariance matrix Σ (James and Stein, 1961) and Σ^{-1} (Krishnamoorthy and Gupta, 1989). It has been shown that $\widehat{\Delta}_1(F)$ dominates $\widehat{\Delta}_u(F)$. No other dominance result has been proved analytically.

THEOREM 2.2 (Gupta and Krishnamoorthy, 1987). *For the loss function L_1, $\widehat{\Delta}_1(F)$ dominates the unbiased estimator $\widehat{\Delta}_u(F)$ for all $n_1, n_2 > p + 1$.*

Another orthogonally invariant estimator is developed following an approach given in (Krishnamoorthy and Gupta, 1989). Muirhead and Verathaworn (1985) have derived the best lower triangular invariant estimator $\widehat{\Delta}_m(F)$ of Δ, for the loss (2.1), which is given by $\widehat{\Delta}_m(F) = TDT'$, where T is a lower triangular matrix with positive diagonal elements such that $TT' = F$, $D = \operatorname{diag}(d_{41}, \ldots, d_{4p})$ and

$$d_{4j} = \frac{(n_2 - p - 1 + j)(n_2 - p - 2 + j)}{(n_1 + 1 - j)(n_2 - p - 1 + j) + (p - j)(n - p - 1)}, \quad j = 1, 2, \ldots, p.$$

An orthogonally invariant estimator derived from $\widehat{\Delta}_m(F)$ is

$$\widehat{\Delta}_4(F) = R\varphi_4^*(L)R',$$

where $\varphi_4^*(L) = \operatorname{diag}(d_{41}l_1, \ldots, d_{4p}l_p)$. It is observed that $\widehat{\Delta}_4(F)$ shrinks the larger eigenvalues and expands the smaller eigenvalues of F towards some central value. Gupta and Krishnamoorthy (1987) carry out a Monte Carlo study to compare the eigenvalues of the new estimators with those of $\widehat{\Delta}_u(F)$ and $\widehat{\Delta}_R(F)$ under the loss (2.6). The eigenvalues of $\widehat{\Delta}_1(F)$ and $\widehat{\Delta}_2(F)$ are uniformly dominated by those of $\widehat{\Delta}_3(F)$. For $p = 2$, the eigenvalues of $\widehat{\Delta}_u(F)$ perform better than those of $\widehat{\Delta}_R(F)$

when one of the eigenvalues of Δ is smaller than half times of the other value. Otherwise the eigenvalues of $\widehat{\Delta}_R(F)$ dominate those of $\widehat{\Delta}_u(F)$. The eigenvalues of $\widehat{\Delta}_3(F)$ and $\widehat{\Delta}_4(F)$ dominate those of $\widehat{\Delta}_u(F)$ uniformly. We also note that when $\Delta \approx cI$, the eigenvalues of $\widehat{\Delta}_3(F), \widehat{\Delta}_4(F)$ and $\widehat{\Delta}_R(F)$ are performing equally, while if the eigenvalues of Δ are very spreadout the eigenvalues of $\widehat{\Delta}_3(F)$ and $\widehat{\Delta}_4(F)$ dominate those of $\widehat{\Delta}_R(F)$. For $p = 3$ and 4, when $\Sigma \approx cI$, the eigenvalues of $\widehat{\Delta}_3(F), \widehat{\Delta}_4(F)$ and $\widehat{\Delta}_R(F)$ are not comparable, i.e., their performance depends on the values of (n_1, n_2). However, the eigenvalues of $\widehat{\Delta}_4(F)$ out perform those of $\widehat{\Delta}_R(F)$ when the eigenvalues of Δ are very spreadout. For all p's, n_1's, and n_2's considered by them, the performance of the eigenvalues of $\widehat{\Delta}_3(F)$ is as good as those of $\widehat{\Delta}_4(F)$.
Konno considered the estimators of the form

$$\widehat{\Delta} = a[F + ut(u)I_p],$$

where a is a constant and $t(u)$ is an absolutely continuous, nonincreasing, and non–negative valued function of $u = 1/\mathrm{tr}(F^{-1})$. He proved the following result.

THEOREM 2.3 (Konno, 1989). *Let $p \geqslant 2$ and $t(u)$ be an absolutely continuous and nonincreasing function bounded by*

$$0 \leqslant t(u) \leqslant \frac{2(p-1)(n_1+n_2-p-1)}{n_1(n_2-2)}.$$

Then the estimator

$$\widehat{\Delta}_5(F) = \frac{n_2-p-1}{n_1}\left(F + \frac{t(u)}{\mathrm{tr}F^{-1}}I_p\right)$$

dominates the unbiased estimator $\widehat{\Delta}_u(F)$ under the loss function L_1.

Again, no other dominance result is known here.

2.2. *Loss Function* $L_2(\widehat{\Delta}, \Delta) = \mathrm{tr}(\widehat{\Delta}\Delta^{-1} - I_p)^2$

Here we consider the problem of estimating Δ by $\widehat{\Delta}(F)$ under the loss function

$$L_2(\widehat{\Delta}, \Delta) = \mathrm{tr}(\widehat{\Delta}\Delta^{-1} - I_p)^2. \tag{2.8}$$

This problem was studied in (Leung and Muirhead, 1988) and (Konno, 1989).

THEOREM 2.4 (Leung and Muirhead, 1988). *Assume that $n_2 > p+3$. Then under the loss function L_2, the best estimator of the form αF is given by $\widehat{\Delta}_6(F) =$ $= aF$ where*

$$a = \frac{(n_2-p)(n_2-p-3)}{(n_1+p+1)(n_2-p-1)+pn_1+2}. \tag{2.9}$$

Let us define

$$\varepsilon = \frac{(n_2-4)\{p(n_1-p-1)+2\}\{p(n_1-p-1)+4\}}{p^3(pn_2-2)(pn_2-4)(n_1-p+3)} \tag{2.10}$$

and

$$\beta = \frac{2(n_2-2)(n_2-4)}{(n_1-p+1)(n_1-p+3)}\left[\frac{(n_1-p+1+2\varepsilon)}{a_2 n_2} - \frac{n_1(n_1-p-1)}{n_2(n_2-p-1)} - \right.$$
$$-\frac{2}{(n_2-p-1)(n_2-2)}\left\{\frac{(n_1+n_2)(n_1-p+1)+n_1(n_2-2)}{n_2}+\right.$$
$$\left.\left. + \frac{(n_1-p+1)(n_1-p+3)}{n_2-4}\right\}\right]. \tag{2.11}$$

THEOREM 2.5 (Konno, 1989). *Assume that* $n_2 > p+3$. *If* $\beta > 0$ *and* $p \geqslant 2$, *then the estimator*

$$\widehat{\Delta}_7(F) = a\left(F + \frac{t}{\mathrm{tr}F^{-1}}I_p\right)$$

where a *and* ε *are defined by* (2.9) *and* (2.10) *and* $0 \leqslant t \leqslant \beta$, *dominate* $\widehat{\Delta}_6(F)$ *under the loss function* L_2.

2.3. *Squared-Error Loss Function* $L_3(\widehat{\Delta},\Delta) = \mathrm{tr}(\widehat{\Delta}-\Delta)^2$

The estimation of the scale matrix Δ under the squared-error loss given by

$$L_3(\widehat{\Delta},\Delta) = \mathrm{tr}(\widehat{\Delta}-\Delta)^2 \tag{2.12}$$

is studied in (Dey, 1989). He obtained bounds of the best multiplier, for the estimate in the class αF, under the loss L_3.

THEOREM 2.6 (Dey, 1989). *For* $n_2 > p+3$, *the best estimate in the class* αF *under the loss function* L_3 *belongs to the interval* $[\alpha_L, \alpha_U]$ *where*

$$\alpha_L = \frac{(n_2-p)(n_2-p-3)}{n_1(n_2-p-2)+2+(p-1)(n-p-1)}, \qquad \alpha_U = \frac{(n_2-p-3)}{n_1+2}.$$

The optimum value of α which minimizes the risk $\mathbf{E}\mathrm{tr}(\alpha F - \Delta)^2$ is

$$\alpha_0 = \mathbf{E}[\mathrm{tr}(\Delta F)]/\mathbf{E}[\mathrm{tr}(F^2)]$$

which is a function of the unknown parameter matrix Δ. Dey also gives a better estimator than $\alpha_0 F$.

THEOREM 2.7 (Dey, 1989). *Under the loss function* L_3 *the estimator* $\widehat{\Delta}(F) = \alpha_0 F - b|F|^{1/p} I_p$ *where* b *is a constant, dominates* $\alpha_0 F$, *if* $d < 0$, *and* $2dpc_{p,n_1+2/p}c_{p,n_2-2/p}c^{-1}_{p,n_1+4/p}c^{-1}_{p,n_2-4/p} < b < 0$ *with*

$$c_{p,n} = 2^{pn/2}\Gamma_p(n/2), \quad \textit{and} \quad d = \alpha_0 \frac{n_1+2/p}{n_2-2/p-p-1} - 1.$$

2.4. *Loss Function* $L_4(\widehat{\delta},\delta) = \sum_{i=1}^{p}(\widehat{\delta}_i - \delta_i)^2$

Dey also considered the loss function

$$L_4(\widehat{\delta},\delta) = \sum_{i=1}^{p}(\widehat{\delta}_i - \delta_i)^2. \tag{2.13}$$

He estimates the δ_i's directly and considers estimators of the form

$$\widehat{\delta}_i^c(F) = cl_i, \qquad i = 1,\ldots,p, \tag{2.14}$$

where c is an appropriate constant depending on n_1, n_2 and p. Then he defines the estimators

$$\widehat{\delta}_i(F) = \widehat{\delta}_i^c - b\prod_{1}^{p} l_j^{1/p}, \qquad i = 1,\ldots,p, \tag{2.15}$$

where b is a constant which can be chosen. He proves the following result.

THEOREM 2.8 (Dey, 1988). *Under the loss function* L_4, *the estimators* (2.15) *dominate* (2.14) *if one of the following conditions hold:*

(i) $d > 0$ *and* $0 < b < 2dc_{p,n_1+2/p}c^{-1}_{p,n_1+4/p}c^{-1}_{p,n_2-4/p}$,

(ii) $d < 0$ *and* $2dc_{p,n_1+2/p}c^{-1}_{p,n_1+4/p}c^{-1}_{p,n_2-4/p} < b < 0$,

where $c_{p,n} = 2^{pn/2}\Gamma_p(n/2)$, *and* $d = c\dfrac{n_1 + 2/p}{n_2 - 2/p - p - 1} - 1, \quad c > 0.$

He also conducts a Monte Carlo study and shows that for $p = 4$, for all choices of Δ, the percentage improvements of $\widehat{\delta}_i(F)$ over $\widehat{\delta}_i^c(F)$, $i = 1,\ldots,p$, are very significant and are largest when $\Delta = I_4$.

3. MAXIMUM LIKELIHOOD ESTIMATION

The maximum likelihood estimators of δ_i's have been considered in (Fatti, 1982). When the eigenvalues δ_i's are distinct, the eigenvalues l_i's and corresponding eigenvectors of $S_1S_2^{-1}$ are jointly maximum likelihood estimators of the eigenvalues δ_i's and corresponding eigenvectors of $\Sigma_1\Sigma_2^{-1}$. However, for a single Wishart matrix, in (James, 1966) it is noted that whereas the sample eigenvalues and eigenvectors are jointly maximum likelihood estimators of their population parameters, the sample eigenvalues by themselves do not maximize the likelihood function of their marginal distribution. Since the joint distribution of l_i's is in terms of a generalized hypergeometric function of matrix arguments (Khatri, 1967), Fatti uses the asymptotic expansion of the joint distribution for large $n_1 + n_2$, as given in (Khatri and Srivastava, 1978). To obtain the maximum likelihood estimators $\widehat{\delta}_i$'s, he maximizes the log likelihood function by an optimization routine. He uses the Newton–Raphson iterative procedure.

REFERENCES

Dey, D.K. (1988). Simultaneous Estimation of Eigenvalues. *Ann. Inst. Statist. Math.* **1**, 137–147.

Dey, D.K. (1989). On estimation of the scale matrix of the multivariate F distribution. *Commun. Statist.-Theory Meth.* **18**, 1373–1383.

Fatti, L.P. (1982). Estimation of the MANOVA eigenvalues under the random effects model. National Research Inst. Math. Sci. CSIR, Pretoria, Technical Report TWISK, 271.

Gupta, A.K. and Krishnamoorthy, K. (1987). Improved estimators of eigenvalues of $\Sigma_1\Sigma_2^{-1}$. Depart. Math. Statist. Bowling Green State Univ., T.R., 87–11.

Haff, L.R. (1982). Solutions of Euler–Lagrange equations for certain multivariate normal estimation problems. Unpublished manuscript.

James, A.T. (1966). Inference on latent roots by calculation of hypergeometric functions of matrix argument. *Multivariate Analysis.* Academic Press, New York.

James, W. and Stein, C. (1961). Estimation with quadratic loss. *Fourth Berkeley Symp. on Math. Statist. and Prob.,* Univ. California Press, Berkeley.

Khatri, C.G. (1967). Some distribution problems associated with the characteristic roots of $S_1S_2^{-1}$. *Ann. Math. Statist.* **38**, 944–948.

Khatri, C.G. and Srivastava, M.S. (1978). Asymptotic expansions for distributions of characteristic roots of covariance matrices. *S.Afr. Statist.J.* **12**, 161–186.

Konno, Y. (1989). Improved estimators for scale matrix of the multivariate distribution F. (Unpublished).

Krishnamoorthy, K. and Gupta, A. K. (1989). Improved minimax estimators of normal precision matrix. *Canad. J. Statist.* **17**, 91–102.

Leung, P.L. and Muirhead, R.J. (1987). Estimation of parameter matrices and eigenvalues in MANOVA and canonical correlation analysis. *Ann. Statist.* **15**, 1651–1666.

Leung, P.L. and Muirhead, R.J. (1988). On estimating characteristic roots in a two sample problem. *Statistical Decision Theory and Related Topics IV,* 397–402.

Muirhead, R.J. (1987). Development in eigenvalue estimation. In: *Advanced in Multivariate Statistical Analysis,* Reidel, Dordrecht, 277–288.

Muirhead, R.J. and Verathaworn, T. (1985). On estimating the latent roots of $\Sigma_1\Sigma_2^{-1}$. *Multivariate Analysis VI.* 431–477. North Holland, Amsterdam.

Prob. Theory and Math. Stat., Vol. 1, pp. 470–479
B.Grigelionis *et al.* (Eds.)
1990 VSP/Mokslas

ON JOINT DISTRIBUTIONS OF RANDOM ELEMENTS WITH GIVEN MUTUAL CONDITIONAL DISTRIBUTIONS

B.M. GUREVICH
Dept. Math. Mech. Moscow State University,
Lenin Hills, 119899 Moscow, USSR

ABSTRACT

The joint distribution P of random variables $\eta_1, \dots, \eta_n$ uniquely determines the conditional distribution P_i of each η_i with respect to $\{\eta_j,\ 1 \leqslant j \leqslant \eta,\ j \neq i\}$. When does the family $P_1, \dots, P_n$, in its turn, determine P? Or, more generally, what is the set of all joint distributions P with given $P_1, \dots, P_n$? In the paper we answer these questions under some general assumptions.

1. INTRODUCTION

The joint distribution of random variables $\eta_i, 1 \leqslant i \leqslant n$, determines uniquely the conditional distribution of every η_i given the others. We consider the inverse problem: under what conditions the above n conditional distributions determine uniquely the joint distribution of $\eta_1, \dots, \eta_n$, and also a more general problem: what is the set of joint distributions with given conditional ones? This can be precisely stated as follows.

Let $(X, \mathbf{A}) = \prod_{k=1}^{n} (X_k, \mathbf{A}_k)$ be the product of measurable spaces $(X_k, \mathbf{A}_k), k = 1, \dots, n$, and $(\hat{X}_i, \hat{\mathbf{A}}_i)$ be the same product with the exception of the i–th factor, $1 \leqslant i \leqslant n$. Assume that for every i we are given a function $P_i : \hat{X}_i \times \mathbf{A}_i \to [0,1]$ measurable with respect to the first variable and being a probability measure with respect to the second one. Call it a transition function.

Denote the natural projection mapping from X to X_i and from X to $\hat{X}_i$ by π_i and $\hat{\pi}_i$ respectively and call a probability measure μ defined on $\mathbf{A}$ a hull of the transition function P_i if for all $A \in \mathbf{A}_i$ and $B \in \hat{\mathbf{A}}_i$

$$\mu((\pi_i^{-1} A) \cap (\hat{\pi}_i^{-1} B)) = \int_B P_i(y, A) \hat{\mu}_i(dy), \tag{1.1}$$

where $\hat{\mu}_i$ is the projection of μ to $\hat{\mathbf{A}}_i$ induced by $\hat{\pi}_i$. Denote by $\mathbf{M}(P_1, \dots, P_n)$ the set of measures which are hulls of all the transition functions $P_i, i = 1, \dots, n$, i.e the set of their common hulls. If the number $|\mathbf{M}(P_1, \dots, P_n)|$ of elements in $\mathbf{M}(P_1, \dots, P_n)$ is not greater than one we say that a common hull is unique.

The uniqueness property of a common hull of some transition functions plays an important role in the theory of Gibbs random fields. A Gibbs field on a countable set S is defined by prescribing a form of its conditional distributions inside every finite set $V \subset S$ given its values outside V. The existence problem for such a field corresponding to a given interaction is generally far from being trivial. A method due to R.L.Dobrushin (Dobrushin, 1968, 1970a, 1970b) allows to construct in some cases a random field whose conditional distributions have the Gibbsian form for every one-point set V. In order to deduce from this the desired Gibbsian form of the conditional distributions for any finite one has to prove the uniqueness of a common hull for some transition functions $P_1, \ldots, P_n, n = |V|$ naturally appearing here.

In this paper we are interested in the structure of the set $\mathbf{M}(P_1, \ldots, P_n)$, especially for the case where the following absolute continuity condition (a.c.c.) holds: there are transition functions $P_i^0 : \hat{X}_i \times \mathbf{A}_i \to [0,1], i = 1, \ldots, n$, such that:

(a) for any $i \in [1,n]$, $y', y'' \in \hat{X}_i$

$$P_i^0(y', \cdot) \sim P_i^0(y'', \cdot); \tag{1.2}$$

(b) for any $i \in [1,n], y \in \hat{X}_i$

$$P_i(y, \cdot) \ll P_i^0(y, \cdot); \tag{1.3}$$

here the signs $\sim$ and $\ll$ respectively denote the mutual absolute continuity of two measures and the absolute continuity of one measure with respect to another; abzac (c)

$$\mathbf{M}(P_1^0, \ldots, P_n^0) \neq \emptyset.$$

In the above mentioned application to Gibbs random fields this condition is always satisfied. This is also the case when all the spaces X_i are at most countable (this situation is considered in (Gurevich, 1990)); for another particular case see (Abraham and Thomas, 1984).

If the a.c.c. holds then for every $i \in [1,n], y \in \hat{X}_i$ there exists the Radon–Nikodym derivative $dP_i(y, \cdot)/dP_i^0(y, \cdot)$ which is an $\mathbf{A}_i$–measurable function on X_i. Moreover assume this function is measurable with respect to the both variables. In other words, identifying each point $x \in X$ with the pair $(\pi_i x, \hat{\pi}_i x)$, we thus assume that for $i = 1, \ldots, n$ there exists an $\mathbf{A}$–measurable function $p_i : X \to \mathbf{R}^+$ such that

$$p_i(\pi_i x, \hat{\pi}_i x) = \frac{dP_i(\hat{\pi}_i x, \cdot)}{dP_i^0(\hat{\pi}_i x, \cdot)}(\pi_i x), x \in X. \tag{1.4}$$

From now on we include the existence of such functions p_i into the a.c.c.

Throughout the paper the following separability condition (s.c.) will be assumed to hold: each σ–algebra $\mathbf{A}_i$ is generated by a countable family of its elements.

The next section contains some auxiliary results, Section 3 is devoted to the study of the set $\mathbf{M}(P_1, \ldots, P_n)$ when the a.c.c. holds, in Section 4 we state without proof some results for the general case. The main results of the paper are Theorems 3.7, 3.11, 3.15, and 4.1.

2. AUXILIARY NOTATIONS AND FACTS

2.1. First of all we note that for a measure $\mu \in \mathbf{M}(P_1, \ldots, P_n)$ the transition function P_i is a regular conditional probability with respect to $\hat{A}_i$ defined on the σ-algebra whose elements are the sets of the form $A \times \hat{X}_i$, $A \in \mathbf{A}_i, i = 1, \ldots, n$, (see (Hennequin and Tortrat,1965) 21.3). One can extend this probability to $\mathbf{A}$ by letting for every $A \in \mathbf{A}$, $y \in \hat{X}_i$

$$A_y = \{x \in X_i : (x, y) \in A\}, \ \tilde{P}_i(y, A) = P_i(y, A_y) \tag{2.1}$$

(the right-hand side of the last equality is well-defined since $A_y \in \mathbf{A}_i$ for all $A \in \mathbf{A}, y \in \hat{X}_i$). The function $\tilde{P}_i : \hat{X}_i \times \mathbf{A} \rightarrow [0, 1]$ just defined has all the properties of P_i with the difference equality (1.1) must be replaced by

$$\mu(D \cap (X_i \times B)) = \int_B \tilde{P}_i(y, D)\mu_i(dy), D \in \mathbf{A}, B \in \hat{\mathbf{A}}_i, \mu \in \mathbf{M}(P_1, \ldots, P_n). \tag{2.2}$$

In what follows we shall also use the following extension of (2.2):

$$\int_X f(x)\mu(dx) = \int_{\hat{X}_i} \left[\int_{X_i} f(u, y) P_i(du) \right] \hat{\mu}_i(dy). \tag{2.3}$$

2.2. LEMMA. *Let Q, Q' be two transition functions defined on $\hat{X}_i \times \mathbf{A}_i$, $1 \leqslant i \leqslant$ $\leqslant n$, and μ a measure on $\mathbf{A}$ which is a hull of both Q and Q'. Then there is a set $Y \in \hat{\mathbf{A}}_i$ such that $\hat{\mu}_i(Y) = 1$ and $Q(y, A) = Q'(y, A)$ for all $y \in Y, A \in \mathbf{A}$.*

This lemma can be easily proved by using the separability of the σ-algebra $\mathbf{A}_i$.

2.3. *Definitions.* Let $(\Omega, \mathbf{F}, P)$ be a probability space and $\mathbf{F}_1, \mathbf{F}_2, \ldots, \mathbf{F}_k$ be σ-subalgebras of $\mathbf{F}$. We call them P-consistent if there exist sets $F \in \mathbf{F}, F_i \in \mathbf{F}_i$, $1 \leqslant i \leqslant k$, such that $0 < P(F) < 1$, $P(F \Delta F_i) = 0$, $1 \leqslant i \leqslant k$. Otherwise, call $\mathbf{F}_1, \ldots, \mathbf{F}_k$ P-inconsistent.
For $i = 1, \ldots, n$ denote by $\mathbf{A}^i$ the σ-algebra consisting of the sets of the form $X_i \times A, A \in \hat{\mathbf{A}}_i$.

2.4. LEMMA. *If $\mu, \tau \in \mathbf{M}(P_1, \ldots, P_n)$, $\mu \sim \tau$, and the σ-algebras $\mathbf{A}^1, \ldots, \mathbf{A}^n$ are μ-inconsistent, then $\mu = \tau$.*

Proof. Let $f = d\tau/d\mu$ (since $\mu \sim \tau$ this function can be thought of as everywhere positive). Then one can define functions f_i, $i = 1, \ldots, n$, by

$$f_i(x) = f(x) / \int_{X_i} f(u, \hat{\pi}_i x) P_i(\hat{\pi}_i x, du), \ x \in X, \ i = 1, \ldots, n, \tag{2.4}$$

and easily check that

$$\hat{\tau}_i(A) = \int_A \left[\int_{X_i} f(u, y) P_i(y, du) \right] \hat{\mu}_i(dy), \ A \in \hat{\mathbf{A}}_i, i = 1, \ldots, n, \tag{2.5}$$

where $\hat{\tau}_i = \hat{\pi}_i \tau$, $\hat{\mu}_i = \hat{\pi}_i \mu$. Fixing for $i = 1, \ldots, n$ an arbitrary probability measure δ_i on $\mathbf{A}_i$ we set

$$P_i'(y, A) = \int_A f_i(u, y) P_i(y, du), A \in \mathbf{A}_i, \tag{2.6}$$

if the denominator in the right-hand side of (2.4) is finite, and set $P_i'(y, \cdot) = \delta_i$, otherwise (from (2.5) it is seen that the former is the case for $\hat{\mu}_i$-almost all $y \in \hat{X}_i$). We thus obtain a transition function P_i' defined on $\hat{X}_i \times \mathbf{A}_i, i = 1, \ldots, n$. Using (2.5) it is easy to show that $\tau \in \mathbf{M}(P_1', \ldots, P_n')$. Since we also have $\tau \in \mathbf{M}(P_1, \ldots, P_n)$. Lemma 2.2 says that $P_i'(y, \cdot) = P_i(y, \cdot)$ for $\hat{\mu}_i$-almost all $y \in \hat{X}_i$, $i = 1, \ldots, n$. Then for μ-almost all $x \in X$ we have $f_i(x) = 1$ (see (2.6) and 2.1). Therefore, for $i = 1, \ldots, n$ the function f coinsides μ-almost everywhere with an $\mathbf{A}^i$-measurable function (see (2.4)). From the μ-inconsistency of $\mathbf{A}^1, \ldots, \mathbf{A}^n$ it follows now that the distribution function of the random variable f takes the values 0 and 1 only, that is $f = \text{const}$ μ-a.s. Since f is a Radon–Nikodym derivative we have $f = 1(\mu\text{-a.s.})$, that is $\tau = \mu$.

3. THE SET OF COMMON HULLS UNDER THE ABSOLUTE CONTINUITY CONDITION

In this section we assume that both the a.c.c. and the s.c. (see Section 1) are satisfied. The former of these conditions implies that there exists a measure $\mu^0 \in \mathbf{M}(P_1^0, \ldots, P_n^0)$ (moreover, such a measure is unique – see Theorem 4.3 and Remark 4.4 below).

3.1. Lemma. *If $\mu \in \mathbf{M}(P_1, \ldots, P_n)$, then $\mu \ll \mu^0$.*

Proof. According to Section 2 denote by $\tilde{P}_i^0$ and by $\tilde{P}_i$ the extentions of P_i^0 and P_i respectively to $\hat{X}_i \times \mathbf{A}$. By (1.3)

$$\tilde{P}_i(y, \cdot) \ll \tilde{P}_i^0(y, \cdot),\ y \in \hat{X}_i,\ i = 1, \ldots, n. \tag{3.1}$$

For any $A \in \mathbf{A}$ define by induction four sequences of sets, B_i^0, B_i, C_i^0 and C_i, where

$$C_0^0 = C_0 = A,\ B_i^0 = \{y \in \hat{X}_i : \tilde{P}_i^0(y, C_{i-1}^0) > 0\},$$

$$B_i = \{y \in \hat{X}_i : \tilde{P}_i(y, C_{i-1}) > 0\},\ C_i^0 = X_i \times B_i^0,$$

$$C_i = X_i \times B_i,\ i = 1, \ldots, n-1.$$

By (3.1) $B_1 \subseteq B_1^0$ and hence $C_1 \subseteq C_1^0$. It is easy to show by induction that

$$C_i \subseteq C_i^0,\ i = 1, \ldots, n-1. \tag{3.2}$$

We now wish to prove that

$$C_j^0 = (X_1 \times \ldots \times X_j) \times D_j^0,\ j = 1, \ldots, n-1, \tag{3.3}$$

where D_j^0 is a set from $\mathbf{A}_{j+1} \times \ldots \times \mathbf{A}_n$.

When $j = 1$, (3.3) is obviously true if we set $D_1^0 = B_1^0$. Suppose (3.3) is true for $j = i - 1$, where $2 \leqslant 1 \leqslant n - 1$, i.e. $C_{i-1}^0 = (X_1 \times \ldots \times X_{i-1}) \times \times D_{i-1}^0, D_{i-1}^0 \in \mathbf{A}_i \times \ldots \times \mathbf{A}_n$.
For every $y = (x_1, \ldots, x_{i-1}, x_{i+1}, \ldots, x_n) \in \hat{X}_i$ we have (see (2.1))

$$\tilde{P}_i^0(y, C_{i-1}^0) = P_i^0(y, \{u \in X_i : (x_1, \ldots, x_{i-1}, u, x_{i+1}, \ldots, x_n) \in$$

$$\in (X_1 \times \ldots \times X_{i-1}) \times D_{i-1}^0\}) =$$

$$= P_i^0(y, \{u \in X_i : (u, x_{i+1}, \ldots, x_n) \in D_{i-1}^0\}).$$

The set $\{u \in X_i : (u, x_{i+1}, \ldots, x_n) \in D_{i-1}^0\}$ belongs to $\mathbf{A}_i$, and since it does not depend on $x_1, \ldots, x_{i-1}$, from property (a) of P_i^0 (see (1.2)) it follows that if $y \in B_i^0$ and if $y' = (x_1', \ldots, x_{i-1}', x_{i+1}, \ldots, x_n)$, then $y' \in B_i^0$ for any $x_k' \in X_k, 1 \leqslant k \leqslant i-1$. Therefore, B_i^0 has the form $(X_1 \times \ldots \times X_{i-1}) \times D_i^0$, where D_i^0 is a set belonging to $\mathbf{A}_{i+1} \times \cdots \times \mathbf{A}_n$, and hence (3.3) holds for $j = i$. We now suppose that $\mu^0(A) = 0$. Taking into account that $\mu \in \mathbf{M}(P_1^0, \ldots, P_n^0)$, it is easy to prove by induction that $\mu^0(C_i^0) = 0$ for all i. In particular, $\mu^0(C_{n-1}^0) = 0$ and hence $P_n^0(y, C_{n-1}^0) = 0$ for μ_n^0–almost all $y \in \hat{X}_n$. Since by (3.3) $C_{n-1}^0 = (X_1 \times \ldots \times X_{n-1}) \times D_{n-1}^0$, where $D_{n-1}^0 \in \mathbf{A}_n$, it follows from (2.1) that $\tilde{P}_n^0(y, C_{n-1}^0) = P_n^0(y, D_{n-1}^0), y \in \hat{X}_n$. The last expression vanishes for $\hat{\mu}^0$–almost all $y \in \hat{X}_n$. Then by (1.3) $P_n(y, D_{n-1}^0) = 0$ for all $y \in \hat{X}_n$; therefore, by (1.2) it does so for all $y \in \hat{X}_n$, which implies that $\mu(C_{n-1}^0) = 0$ and hence $\mu(C_{n-1}) = 0$ (see (3.2)) for every $\mu \in \mathbf{M}(P_1, \ldots, P_n)$. Using (2.2) one easily deduces from this by induction that $\mu(C_{i-1}) = 0$ for every $i \geqslant 1$. In particular, when $i = 1$ we get $\mu(C_0) = \mu(A) = 0$.
3.2. Let

$$W = \{x \in X : p_i(x) > 0 \text{ for } i = 1, \ldots, n\}.$$

Clearly, $W \in \mathbf{A}$.

LEMMA. *If $\mu \in \mathbf{M}(P, \ldots, P)$, then $\mu(W) = 1$.*

Proof. We set $W^i = \{x \in X : p_i(x) = 0\}$, $i = 1, \ldots, n$. By (2.2), (2.1) and (1.4) for any $\mu \in \mathbf{M}(P_1, \ldots, P_n)$

$$\mu(W^i) = \int_{\hat{X}_i} \tilde{P}_i(y, W^i)\hat{\mu}_i(dy) = \int_{\hat{X}_i} P_i(y, W_y^i)\hat{\mu}_i(dy) =$$

$$= \int_{\hat{X}_i} \left[\int_{W_y^i} p_i(u, y) P_i^0(y, du) \right] \hat{\mu}_i(dy) = 0, \; i = 1, \ldots, n.$$

It is now sufficient to observe that $W = (X \backslash W^1) \cap \ldots \cap (X \backslash W^n)$.

3.3. COROLLARY. *If $\mathbf{M}(P_1, \ldots, P_n) \neq \emptyset$ then $\mu^0(W) > 0$.*

This is an immediate consequence of Lemmas 3.1 and 3.2.

3.4. Assuming that $\mu^0(W) > 0$ define on $\mathbf{A}$ a probability measure τ^0 by

$$\tau^0(A) = \mu^0(A/W), \ A \in \mathbf{A}. \tag{3.4}$$

It is easy to check that τ^0 is a common hull of the transition function $Q_i^0, i = 1, \dots, n$, where

$$Q_i^0(y, A) = P_i^0(y, A \cap W_y)/P_i^0(y, W_y), \ A \in \mathbf{A}_i, \tag{3.5}$$

if $y \in X_i$ is such that $P_i^0(y, W_y) > 0$, and where $Q_i^0(y, \cdot)$ is an arbitrary probability measure on $\mathbf{A}_i$ (independent of y), otherwise.
3.5. Denote by $\mathbf{A}(\tau^0)$ the family of all sets $A \in \mathbf{A}$ such that for every $i \in [1, n]$ there is a set $A^i \in \mathbf{A}^i$ with $\tau^0(A \Delta \mathbf{A}^i) = 0$. Obviously $\mathbf{A}(\tau^0)$ is a σ-algebra which is trivial (contains only sets of measure 0 and 1) when $\mathbf{A}^1, \dots, \mathbf{A}^n$ are τ^0-inconsistent.

3.6. LEMMA. *If $\mu \in \mathbf{M}(P_1, \dots, P_n)$, then $\mu \ll \tau^0$ and*

$$\left\{ x \in X : \frac{d\mu}{d\tau^0}(x) > 0 \right\} \in \mathbf{A}(\tau^0). \tag{3.6}$$

Proof. The first assertion immediately follows from Lemmas 3.1 and 3.2. To prove the second one we assume that $\mu^0(W) > 0$ (see Corollary 3.3) and for any $\mu \in \mathbf{M}(P_1, \dots, P_n)$ set:

$$f = d\mu/d\tau^0, \ F = \{x \in X : f(x) > 0\}, \ \bar{F} = W \backslash F, \tag{3.7}$$

$$E_i = \{y \in \hat{X}_i : \tilde{P}_i^0(y, F) < \tilde{P}_i^0(y, W)\}, \quad i = 1, \dots, n,$$
$$G_i = \{y \in \hat{X}_i : 0 < \tilde{P}_i^0(y, F) < \tilde{P}^0(y, W)\}, \quad i = 1, \dots, n.$$

Since $\mu \ll \tau^0$ one may assume that $F \subseteq W$. Obviously $F \in \mathbf{A}$, $E_i \in \hat{\mathbf{A}}_i$, $G_i \in \hat{\mathbf{A}}_i$, and $G_i \subseteq E_i$, $1 \leqslant i \leqslant n$. By (2.2), (2.1)

$$0 = \mu(\bar{F}) = \int_{\hat{X}_i} \tilde{P}_i(y, \bar{F}) \hat{\mu}_i(dy) = \int_{\hat{X}_i} P_i(y, \bar{F}_y) \hat{\mu}_i(dy) =$$

$$= \int_{\hat{X}_i} \left[\int_{\bar{F}_y} p_i(u, y) P_i^0(y, du) \right] \hat{\mu}_i(dy)$$

and it follows that the inner integral in the last expression vanishes for $\hat{\mu}_i$-almost all $y \in \hat{X}_i$. Since by the inclusion $F \subseteq W$ we have $p_i(u, y) > 0$ for all $u \in \bar{F}_y$, the above means that for $\hat{\mu}_i$-almost all $y \in \hat{X}_i$

$$0 = P_i^0(y, \bar{F}_y) = P_i^0(y, W_y) - P_i^0(y, F_y) = \tilde{P}_i^0(y, W) - \tilde{P}_i^0(y, F).$$

Hence $\hat{\mu}_i(E_i) = 0$ and then $\hat{\mu}_i(C_i) = 0$. On the other hand, from the definition of f and from equality (2.3) (with τ replacing μ) combined with (3.4), (3.5) it follows that

$$\hat{\mu}_i(C_i) = \mu(X_i \times G_i) = \int\limits_{X_i \times G_i} f d\tau^0 =$$

$$= \int\limits_{G_i} \left[\int\limits_{W_y} \frac{f(u,y)}{P_i^0(y, W_y)} P_i^0(y, du) \right] \hat{\tau}_i^0(dy).$$

Moreover, by the definition of G_i the inner integral in the last expression is positive for all $y \in G_i$. Therefore, the equality $\hat{\mu}_i(G_i) = 0$ implies that $\hat{\tau}_i^0(G_i) = 0$ and so for all $y \in G_i'$, where $G_i' = \{y \in \hat{X}_i : \tilde{P}_i^0(y, F) > 0\} \in \hat{\mathbf{A}}_i$ we have $\tilde{P}_i^0(y, F) = \tilde{P}_i^0(y, W)$. Then by (2.2) and (3.5)

$$\tau^0(F \cap (X_i \times G_i')) = \int\limits_{G_i'} \frac{\tilde{P}_i^0(y, F)}{\tilde{P}_i^0(y, W)} \hat{\tau}_i^0(dy) =$$

$$= \int\limits_{G_i'} \hat{\tau}_i^0(dy) = \tau^0(X_i \times G_i'),$$

i.e. $\tau^0((X_i \times G_i') \backslash F) = 0$. It is also obvious that $\tau^0(F \backslash (X_i \times G_i')) = 0$. So $\tau^0(F \Delta (X_i \times G_i')) = 0$, $i = 1, \dots, n$, which together with (3.7) gives (3.6).

3.7. THEOREM. *If $\mu^0(W) > 0$ and the σ–algebras $\mathbf{A}^1, \dots, \mathbf{A}^n$ are τ^0–inconsistent (see Definition 2.3), then $|\mathbf{M}(P_1, \dots, P_n)| \leqslant 1$.*

Proof. By virtue of the τ^0–inconsistency of $\mathbf{A}^1, \dots, \mathbf{A}^n$ it follows from Lemma 3.6 that each measure $\mu \in \mathbf{M}(P_1, \dots, P_n)$ is equivalent to τ^0 (τ^0 exists since $\mu^0(W) > 0$). Now the theorem follows from Lemma 2.4.
Let us pass to the case where $\mathbf{A}^1, \dots, \mathbf{A}^n$ are τ^0–consistent.

3.8. *Definition.* Let $(\Omega, \mathbf{F}, P)$ be a probability space. Call a σ–algebra $F' \subset \subset F$ discrete with respect to P (or P–discrete) if there exists an at most countable partition $\beta = (B_1, B_2, \dots)$ of Ω (generating partition) such that (i) $B_j \in \in \mathbf{F}'$, $P(B_j) > 0$, $j = 1, 2, \dots$, and (ii) for any $F \in \mathbf{F}'$ there are $j_1, j_2, \dots$ with $P(F \Delta (B_{j_1} \cup B_{j_2} \cup \dots)) = 0$. Clearly, the atoms of β (i.e. $B_1, B_2, \dots$) are defined uniquely up to a set of zero measure.

3.9. LEMMA. *Assume the σ–algebra $\mathbf{A}(\tau^0)$ (see 3.5) is τ^0–discrete. Let B be an atom of its generating partition and let $\mu(B) > 0$ for some $\mu \in \mathbf{M}(P_1, \dots, P_n)$. Then the restrictions of the measures μ and τ^0 to B are equivalent.*

Proof. Let $\mu|B$ and $\tau^0|B$ be the restrictions under consideration. Let $f = = d\mu/d\tau^0$ (see Lemma 3.6) and $F = \{x \in B : f(x) > 0\}$. From the inequality $\mu(B) > 0$ it follows that $\tau^0(F) > 0$ and by (3.5) we have $F \in \mathbf{A}(\tau^0)$. By the definition of B this means that $\tau^0(B \backslash F) = 0$, i. e. $\mu|B \sim \tau^0|B$.

3.10. *Definition.* Assume that the σ–algebra $\mathbf{A}(\tau^0)$ is τ^0–discrete and let B be an atom of its generating partition. We call B complete if

$$\tau^0(\{x \in B : \tilde{P}_i(\hat{\pi}_i x, B) < 1\}) = 0$$

(recall that $\hat{\pi}_i$ is the projection of X to $\hat{X}_i$).

3.11. THEOREM. *Assume that the σ-algebra $\mathbf{A}(\tau^0)$ is τ^0-discrete and that for some atom B of its generating partition there exists a measure $\mu \in \mathbf{M}(P_1, \dots, P_n)$ with $\mu(B) > 0$. Then* (i) *B is complete:* (ii) *there is exactly one measure $\mu \in$ $\in \mathbf{M}(P_1, \dots, P_n)$ with $\mu(B) = 1$.*

Proof. (i) Let

$$\bar{B} = X \backslash B, C_i = \{y \in \hat{X}_i : \tilde{P}_i(y, \bar{B}) > 0, \tilde{Q}_i^0(y, \bar{B}) = 0\}, \ i = 1, \dots, n.$$

By virtue of the relation of $\tilde{Q}_i^0$ to τ^0 we have for $i = 1, \dots, n$

$$\tau^0(\bar{B} \cap (X_i \times C_i)) = \int_{C_i} \tilde{Q}_i^0(y, \bar{B}) \hat{\tau}_i^0(dy) = 0, \tag{3.8}$$

$$\mu(\bar{B} \cap (X_i \times C_i)) = \int_{C_i} \tilde{P}_i^0(y, \bar{B}) \hat{\mu}_i(dy). \tag{3.9}$$

Since by Lemma 3.6 $\mu \ll \tau^0$ (3.8) implies that $\mu(\bar{B} \cap (X_i \times C_i)) = 0$ and so $\hat{\mu}_i(C_i) = 0$ (the integrand in (3.9) is positive on C_i). Let us show that $\hat{\tau}_i^0(C_i) = 0$ as well.
If $\hat{\tau}_i^0(C_i) > 0$, then we get as above

$$\tau^0(B \cap (X_i \times C_i)) = \hat{\tau}_i^0(C_i) > 0. \tag{3.10}$$

Since $\mu(B) > 0$, from (3.10) and Lemma 3.9 it follows that $\mu(B \cap (X_i \times C_i)) > 0$, and so by (3.9) $\hat{\mu}_i(C_i) > 0$. This leads to a contradiction and we conclude that $\hat{\tau}_i^0(C_i) = 0$.
To prove the completeness of B let us observe that

$$\left\{x \in B : \tilde{P}_i(\hat{\pi}_i x, B) < 1\right\} \backslash (X_i \times C_i) \subseteq B \cap (X_i \times \{y \in \hat{X}_i : \tilde{Q}_i^0(y, B) < 1\}). \tag{3.11}$$

Since $B \in \mathbf{A}(\tau^0)$ there is a set $B_i' \in \hat{\mathbf{A}}_i$ such that $\tau^0(B \,\Delta\, (X_i \times B_i')) = 0$. From this fact it is easy to deduce that the measure τ^0 of the right-hand side of (3.11) is zero. This together with the equality $\hat{\tau}_i^0(C_i) = 0$ implies the completeness of B.

(ii) Since $\mu(B) > 0$ and $\tau^0(B) > 0$ (see Lemma 3.6) one can define two measures μ_B and τ_B^0, on A by

$$\mu_B(A) = \mu(A/B), \ \tau_B^0(A) = \tau^0(A/B), \ A \in \mathbf{A}.$$

It is easily checked that $\mu_B \in \mathbf{M}(P_1, \dots, P_n)$. If $\mu', \mu'' \in \mathbf{M}(P_1, \dots, P_n)$ and $\mu'(B) = \mu''(B) = 1$, then by Lemma 3.9 $\mu' \sim \tau_B^0$, $\mu'' \sim \tau_B^0$, and hence $\mu' \sim \mu''$. By the definition of B and τ_B^0 the σ-algebras $\mathbf{A}^1, \dots, \mathbf{A}^n$ are τ_B^0-inconsistent. So by Lemma 2.4 $\mu' = \mu''$, that is $\{\tilde{\mu} \in \mathbf{M}(P_1, \dots, P_n) : \tilde{\mu}(B) = 1\} = \{\mu_B\}$.

3.12. *An example.* Let $n = 2$, $X_i = [0,1]$, $i = 1, 2$, and define P_1, P_2 as follows: $P_1(y, \cdot)$ for $y \in [0, 1/2]$, $P_1(y, \cdot)$ for $y \in (1/2, 1]$, $P_2(y, \cdot)$ for $y \in$

$\in [0,1/2]$, and $P_2(y,\cdot)$ for $y \in (1/2,1]$ are uniform distributions concentrated on $[0,\ 1]$, $[1/2,\ 1]$, $[0,\ 1/2]$, and $[1/2,\ 1]$ respectively. Note that for $n = 2$, $\hat{X}_1 = X_2$, $\hat{X}_2 = X_1$ and for $i = 1,2$ and for any $y \in \hat{X}_i$ take as $P_i^0(y,\cdot)$ the uniform distribution on $[0,\ 1]$. It is easy to see that τ^0 is the uniform distribution on $B_1 \cup B_2$, where $B_1 = [0,1/2] \times [0,1/2]$, $B_2 = (1/2,1] \times (1/2,1]$. One can take as β the partition of the square $X = X_1 \times X_2$ into B_1 and B_2. Clearly, the atom B_2 of β turns out to be complete, and the uniform distribution μ_{B_2} concentrated on B_2 belongs to $\mathbf{M}(P_1, P_n)$, while the atom B_1 is incomplete. So by Theorem 3.11 $\mathbf{M}\ (P_1, P_2) = \{\mu_{B_2}\}$.

3.13. *Remark.* One can prove that the discreteness condition imposed on the σ-algebra $\mathbf{A}(\tau^0)$ in Theorem 3.11 is always satisfied in the important particular case where X_i is a Euclidian space and $P_i(y,\cdot)$ and absolutely continuous distribution on X_i for $i = 1,\dots,n$ and for every $y \in \hat{X}_i$.

3.14. *Remark.* The size of the paper does not allow to discuss in detail the conditions under which $\mathbf{M}(P_1,\dots,P_n)$ is non–empty. One can state such conditions in the framework of this section generalizing those in (Gurevich, 1990). We here confine ourself to a particular case where a necessary condition (close to a sufficient one) can be stated in a simple form.
Let $X_i = R$, $i = 1,\dots,n$, and let $P_i(y,\cdot)$ be for any i and $y \in \hat{X}_i$ an absolutely continuous distribution with density $p_i(x) = p_i(u,y)$, $u = \pi_i x$, $y = \hat{\pi}_i x$, continuous in $x \in R^n$ and continuously differentiable in $u \in X_i$. On the set W where each p_i is positive define a vector field $\Phi(x) = (\Phi_1(x),\dots,\Phi_n(x))$ by

$$\Phi_i(x) = \frac{\partial}{\partial u}\ln p_i(u,y),\ i = 1,\dots,n,\ x \in W,\ u = \pi_i x,\ y = \hat{\pi}_i x.$$

3.15. THEOREM. *Under the assumptions of* 3.14 *the necessary condition for the set* $\mathbf{M}(P_1,\dots,P_n)$ *to be non–empty is that the circulation of the field* Φ *along any closed contour inside* W *is equal to zero.*

4. UNIQUENESS CONDITIONS IN THE GENERAL CASE

Assuming only the s.c. (see Section 1) to be satisfied we formulate without proof the following result.

4.1. THEOREM. (i) *If for any* $\mu \in \mathbf{M}(P_1,\dots,P_n)$ *the* σ*–algebras* $\mathbf{A}^1,\dots,\mathbf{A}^n$ *are* μ*–inconsistent, then* $|\mathbf{M}(P_1,\dots,P_n)| \leqslant 1$;

(ii) *if there is a measure* $\mu \in \mathbf{M}(P_1,\dots,P_n)$ *such that* $\mathbf{A}^1,\dots,\mathbf{A}^n$ *are* μ*–consistent then there exist two measures* μ', $\mu'' \in \mathbf{M}(P_1,\dots,P_n)$ *which are mutually singular.*

4.2. *Remark.* Certainly, only the necessary uniqueness condition contained in this theorem (assertion (ii) can be regarded as rather efficient. However, using assertion (i) one can come to an efficient sufficient condition being in some sence an extention of that in Theorem 3.7.

REFERENCES

Abraham, J. and Thomas, J.B.(1984) A note on the characterization of bivariate density by conditional densities. *Commun.Statist.Theory and Meth.* **3**, 395–400.

Dobrushin, R.L. (1968). The description of the random field by its conditional distributions and its regularity conditions. *Theory Probab.Appl.* **13**, 201–229.

Dobrushin, R.L. (1970a). Prescribing a system of random variables by conditional distributions. *Theory Probab.Appl.* **15**, 458–486.

Dobrushin, R.L. (1970b). Gibbs random fields for the particles without hard core. *Teor.Mat.Fis.* **4**, 101–118.

Gurevich, B.M. (1990). On joint distribution of random variables with given conditional distributions (discrete case). *Theory Probab.Appl.* (to appear).

Hennequin, P. and Tortrat, A. (1965). *Theorie des Probabilites et Quelques Applications.* Masson, Paris.

Prob. Theory and Math. Stat., Vol. 1, pp. 480–489
B.Grigelionis *et al.* (Eds.)
1990 VSP/Mokslas

IMPROVEMENT AND GENERALIZATION OF THE CRAMÉR—RAO INEQUALITY FOR A FILTERED SPACE

A.A. GUSHCHIN
Steklov Mathematical Institute,
Vavilova 42, 117966 Moscow GSP–1, USSR

ABSTRACT

For a parametric family of probability measures on a measurable space together with a family of sub–σ–fields it is possible to improve the Cramér—Rao inequality for the variance of an estimator. This improvement is considered in sections 1 and 2. In the next two sections we consider another problem connected with the first one. The question is the following generalization of the Cramér—Rao inequality: to find conditions under which the quadratic characteristic of the martingale part of an estimator depending on time has a lower bound similar to that for the variance in the Cramér—Rao inequality. Finally, we formulate a generalization of Girsanov's theorem for a non–locally absolutely continuous change of a measure.

1. Let $x = (x_t,\ 0 \leqslant t \leqslant T)$ be a diffusion–type process:

$$dx_t = \theta a_t\, dt + dw_t, \quad x_0 = 0,$$

where $w = (w_t)$ is a Wiener process, $a = (a_t)$ is a non–anticipating functional on the space $\mathbf{C}\ [0,T]$ of continuous functions, θ is an unknown parameter taking values in an open subset $\Theta \subseteq \mathbf{R}^1$. We suppose that

$$\int_0^T a_t^2 dt < \infty, \quad P^\theta\text{–a.s.},$$

where P^θ is the distribution of the process x. Let $\hat\theta_T$ be any real statistic based on the observations of x, $\mathbf{E}^\theta(\hat\theta_T^2) < \infty$. The following simple method allows us to improve the Cramér—Rao inequality.
Put $m_T(\theta) = \mathbf{E}^\theta(\hat\theta_T)$ and fix a point $\theta \in \Theta$. There exists a predictable function $c = (c_t,\ 0 \leqslant t \leqslant T)$ such that

$$\hat\theta_T = m_T(\theta) + \int_0^T c_t dw_t$$

for the given θ (Liptser and Shiryaev, 1977, Theorem 5.20). By the Cauchy—Schwartz inequality

$$\int_0^T c_t^2 dt \geqslant \left(\int_0^T c_t a_t dt\right)^2 \Big/ \int_0^T a_t^2 dt. \tag{1}$$

Taking the expectations of both sides of (1), we obtain

$$\mathbf{E}^\theta \left(\int_0^T c_t^2 dt\right) \geqslant \mathbf{E}^\theta \left[\left(\int_0^T c_t a_t dt\right)^2 \Big/ \int_0^T a_t^2 dt\right]. \tag{2}$$

However,

$$\mathbf{E}^\theta \left(\int_0^T c_t^2 dt\right) = \mathbf{E}^\theta \left(\left(\int_0^T c_t dw_t\right)^2\right) = \mathbf{D}^\theta(\hat{\theta}_T).$$

On the other hand, the derivative of the logarithm of the likelihood ratio is equal to $\int_0^T a_t dw_t$ at the point θ for this model, so the Fisher information at the point θ is equal to

$$I_T(\theta) = \mathbf{E}^\theta \left(\left(\int_0^T a_t dw_t\right)^2\right) = \mathbf{E}^\theta \left(\int_0^T a_t^2 dt\right),$$

and under regularity conditions needed for the Cramér—Rao inequality to be true we get

$$m_t'(\theta) = \mathbf{E}^\theta \left(\hat{\theta}_T \int_0^T a_t dw_t\right) = \mathbf{E}^\theta \left(\int_0^T c_t dw_t \int_0^T a_t dw_t\right) = \mathbf{E}^\theta \left(\int_0^T c_t a_t dt\right).$$

Hence, applying the Cauchy—Schwartz inequality once again, we have

$$\mathbf{E}^\theta \left[\left(\int_0^T c_t a_t dt\right)^2 \Big/ \int_0^T a_t^2 dt\right] \geqslant \left[\mathbf{E}^\theta \left(\int_0^T c_t a_t dt\right)\right]^2 \Big/ \mathbf{E}^\theta \left(\int_0^T a_t^2 dt\right) = \frac{[m_T'(\theta)]^2}{I_T(\theta)}. \tag{3}$$

Thus, under assumptions of the validity of the Cramér—Rao inequality $\mathbf{D}^\theta(\hat{\theta}_T) \geqslant [m_T'(\theta)]^2 / I_T(\theta)$, inequality (2), which is valid without these assumptions, is its improvement. Unfortunately, we failed to write down an explicit form of inequality (2) for, e.g., the maximum likelihood estimator.

Remark 1. When one considers inequalities (1) and (2) for different T it should be kept in mind that the function (c_t) depends on T.

Remark 2. The process $\int_0^t c_s dw_s$ is a special semimartingale with respect to the measure P^ζ, $\zeta \in \Theta$, with the canonical decomposition

$$\int_0^t c_s dw_s = M_t(\zeta) + A_t(\zeta) = \int_0^t c_s(dx_s - \zeta a_s ds) + (\zeta - \theta)\int_0^t c_s a_s ds,$$

so

$$\int_0^T c_s a_s ds = \left.\frac{dA_t(\zeta)}{d\zeta}\right|_{\zeta=\theta}.$$

2. The same improvement of the Cramér—Rao inequality may be established in a more general situation.

Let $(\Omega, \mathcal{F})$ be a measurable space, $F = (\mathcal{F}_t,\ t \geqslant 0)$ be an increasing and right-continuous family of sub-σ-fields of $\mathcal{F}$, $\mathcal{P} = (P^\theta,\ \theta \in \Theta)$ be a family of probability measures depending on a real parameter θ, belonging to an open subset $\Theta \subseteq \mathbf{R}^1$. For simplicity, we assume that $\mathcal{F}_0 = \{\emptyset, \Omega\}$. Let T be a stopping time, and $\hat\theta_T$ be a $\mathcal{F}_T$-measurable statistic, $\mathbf{E}^\theta(\hat\theta_T^2) < \infty$, $m_T(\theta) = \mathbf{E}^\theta(\hat\theta_T)$, $\theta \in \Theta$. Denote by $\mathfrak{z}(\theta) = (\mathfrak{z}_t(\theta),\ t \geqslant 0)$ the density process of the measure P^θ with respect to some dominating probability measure ν.

Let us fix a point $\theta \in \Theta$. Let V_T be the score function, i.e. V_T is the derivative of $\ln \mathfrak{z}_T(\zeta)$ at $\zeta = \theta$ or a variable with an analogous sense, the second moment of which (relative to P^θ) is equal to the Fisher information $I_T(\theta)$ of the family $\{P_T^\zeta,\ \zeta \in \Theta\}$ at θ (for example, if the function $\mathfrak{z}_T^{1/2}(\zeta)$ is differentiable in $L^2(\nu)$ at $\zeta = \theta$, W_t being the derivative, then $V_T = 2W_T/\mathfrak{z}_T^{1/2}(\theta)$ (Ibragimov and Hasminskii, 1981)). The processes $M = (M_t,\ t \geqslant 0)$ and $V = (V_t,\ t \geqslant 0)$ defined by

$$M_t = \mathbf{E}^\theta(\hat\theta_T - m_T(\theta)|\mathcal{F}_t), V_t = \mathbf{E}^\theta(V_t|\mathcal{F}_t)$$

are P^θ-square integrable martingales, and by the Kunita—Watanabe inequality

$$\langle M, M\rangle_T \geqslant \frac{(\langle M, V\rangle_T)^2}{\langle V, V\rangle_T}.$$

Taking the expectations with respect to P^θ we obtain

$$\mathbf{E}^\theta(\langle M, M\rangle_T) \geqslant \mathbf{E}^\theta\left\{\frac{(\langle M, V\rangle_T)^2}{\langle V, V\rangle_T}\right\}. \tag{4}$$

Inequality (4), as well as its particular case (2), is an improvement of the Cramér—Rao inequality. Indeed, by the definition of M, $\mathbf{E}^\theta(\langle M, M\rangle_T) = \mathbf{E}^\theta(M_T^2) = \mathbf{D}^\theta(\hat\theta_T)$. On the other hand, $\mathbf{E}^\theta(\langle V, V\rangle_T) = \mathbf{E}^\theta(V_T^2) = I_T(\theta)$, and under regularity conditions needed for the Cramér—Rao inequality to be true we get

$$m_T'(\theta) = \mathbf{E}^\theta(\hat\theta_T V_T) = \mathbf{E}^\theta(M_T V_T) = \mathbf{E}^\theta(\langle M, V\rangle_T),$$

so by the Cauchy—Schwartz inequality

$$\mathbf{E}^{\theta}\left\{\frac{(\langle M,V\rangle_T)^2}{\langle V,V\rangle_T}\right\} \geqslant \frac{\left[\mathbf{E}^{\theta}(\langle M,V\rangle_T)\right]^2}{\mathbf{E}^{\theta}(\langle V,V\rangle_T)} = \frac{[m'_T(\theta)]^2}{I_T(\theta)}.$$

Remark 3. Regularity conditions are not sufficient for the validity of Remark 2 in the present situation. Even if the process M is a special semimartingale with respect to P^{ζ}, $\zeta \neq \theta$ (which is not necessarily true), with the canonical decomposition $M = M(\zeta) + A(\zeta)$, we need some additional assumptions (see Theorem 2 below) for the validity of the equality

$$\langle M,V\rangle_T = \left.\frac{dA_T(\zeta)}{d\zeta}\right|_{\zeta=\theta},$$

where the derivative is in a suitable sense. Nevertheless, using Theorem 2, it is possible to prove a weakened version of this statement without additional assumptions. Let us suppose for simplicity that $P_T^{\zeta} \ll P_T^{\theta}$, $\zeta \in \Theta$. Then M is a P^{ζ}-semimartingale with the canonical representation

$$M_t = M_t^c(\zeta) + B_t^a(\zeta) + \int_0^t \int_{0<|x|\leqslant a} x\,d(\mu - \nu^{\zeta}) + \int_0^t \int_{|x|>a} x\,d\mu,$$

and $\langle M,V\rangle_T = \lim_{a\to\infty}\lim_{\zeta\to\theta}(\zeta-\theta)^{-1}B_T^a(\zeta)$, where the limits are in P^{θ}-probability.

3. Let $(\Omega,\mathcal{F},F,\mathcal{P})$ be a filtered statistical experiment, i.e. $(\Omega,\mathcal{F})$ is a measurable space, $F = (\mathcal{F}_t,\ t \geqslant 0)$ is an increasing and right-continuous family of sub-σ-fields of $\mathcal{F}$, $\mathcal{P} = (P^{\theta}, \theta \in \Theta)$ is a family of probability measures depending on a real parameter θ, belonging to an open subset $\Theta \subseteq \mathbf{R}^1$. We suppose for simplicity that all measures P^{θ} coincide on $\mathcal{F}_0$.
Let an adapted, real-valued stochastic process $X = (X_t,\ t \geqslant 0)$ be a locally square integrable semimartingale with respect to P^{θ} for each $\theta \in \Theta$, that is $X = X_0 + M(\theta) + A(\theta)$, $M(\theta) \in \mathcal{M}^2_{\text{loc}}(P^{\theta})$, $A(\theta)$ is a predictable process with finite variation (we refer to (Jacod and Shiryaev, 1987) for all basic definitions and notation).
Let us fix a point $\theta_0 \in \Theta$ playing a specific role below. We shall omit the parameter θ when it takes the value θ_0, for example, $P \equiv P^{\theta_0}$, $A \equiv A(\theta_0)$, $\langle M,M\rangle \equiv {}^{\theta_0}\langle M(\theta_0), M(\theta_0)\rangle$ and so on.
We are interested in the following two questions for the considered model. The first one is the problem of local (in time) differentiability (with respect to θ) of the family $\mathcal{P}$ at $\theta = \theta_0$. For the first time this question was considered in (Jacod, 1989a), where the definition of local differentiability Z1 and the criterion H1 in predictable terms are given. We also consider the weaker property Z2 and its sufficient condition H2 in predictable terms.
The second question consists in finding additional assumptions guaranteeing the validity of the equality

$$\langle M,V\rangle = A' = \lim_{\theta\to\theta_0}(\theta-\theta_0)^{-1}(A(\theta) - A),$$

where a local martingale V is the score process, and the limit is in a suitable sense. If this equality is true, then, according to the Kunita—Watanabe inequality, for all $t \geqslant 0$

$$\langle M, M\rangle_t \geqslant \frac{(A'_t)^2}{\langle V, V\rangle_t} \tag{5}$$

(here and below $0/0 = 0$, $a/0 = +\infty$, $a > 0$). Inequality (5) may be considered as a generalization of the Cramér—Rao inequality for a filtered space.
The statement $\langle M, V\rangle = A'$ is also important for the investigation of the so-called quasi-likelihood estimators (Thavaneswaran and Thompson, 1986; Hutton and Nelson, 1986; Godambe and Heyde, 1987), in particular, for their optimality in Godambe's sense. For this purpose the paper (Jacod, 1989b) where the considered range of problems was studied for a partial likelihood process, is very important.
Let $Q^\theta = (P + P^\theta)/2$, $z(\theta)$ and $z'(\theta)$ be the density processes of P and P^θ with respect to Q^θ, $Z^\theta = z'(\theta)/z(\theta)$ be the generalized density process of P^θ with respect to P. Since Z^θ is a P-supermartingale, it has the Doob—Meyer decomposition:

$$Z^\theta = 1 + N^\theta - C^\theta,$$

where $N^\theta = (N^\theta_t,\ t \geqslant 0)$ is a P-local martingale, $C^\theta = (C^\theta_t,\ t \geqslant 0)$ is an increasing predictable process, $N^\theta_0 = C^\theta_0 = 0$. Let $h^{\theta\varsigma} = (h^{\theta\varsigma}_t,\ t \geqslant 0)$ be any version of the Hellinger process of order $1/2$ between P^θ and P^ς, see (Jacod and Shiryaev, 1987); $h^\theta \equiv h^{\theta_0\theta}$;

$$\xrightarrow{L^2},\quad \xrightarrow{L^1},\quad \xrightarrow{\sigma(L^2)},\quad \xrightarrow{\sigma(L^1)}$$

denote the convergence in the spaces $L^2(P)$, $L^1(P)$, the weak convergence in $L^2(P)$, the convergence in the weak topology $\sigma(L^1(P),\ L^\infty(P))$ of the space $L^1(P)$ respectively.
A localizing family is a family $(T_p,\ T(n,p),\ n,\ p \in \mathbb{N})$ of stopping times with

$$T_p \uparrow \infty \quad (\mathbf{P}\text{-a.s.}) \text{ as } p \to \infty; \quad T(n,p) \leqslant T_p \quad (\forall p,\ n \in \mathbb{N});$$

$$\lim_{n\to\infty} \mathbf{P}(T(n,p) < T_p) = 0 \quad (\forall p \in \mathbb{N}).$$

Condition Z1. There exist a process $V = (V_t,\ t \geqslant 0)$ and, for every sequence $\{\theta_n\} \to \theta_0$, a localizing family $(T_p,\ T(n,p))$, such that $\forall t \geqslant 0$, $\forall p \in \mathbb{N}$

$$\frac{\left(Z^{\theta_n}_{t\wedge T(n,p)}\right)^{1/2} - 1}{\theta_n - \theta_0} \xrightarrow{L^2} \frac{1}{2} V_{t\wedge T_p} \text{ as } n \to \infty,$$

$$\mathbf{E}\left(1 - Z^{\theta_n}_{T(n,p)}\right) = o((\theta_n - \theta_0)^2) \text{ as } n \to \infty.$$

Condition Z2. There exist a process $V = (V_t,\ t \geqslant 0)$ and, for some sequence $\{\theta_n\} \to \theta_0$, a localizing family $(T_p,\ T(n,p))$, such that $\forall t \geqslant 0$, $\forall p \in \mathbb{N}$

$$\frac{\left(Z^{\theta_n}_{t\wedge T(n,p)}\right)^{1/2} - 1}{\theta_n - \theta_0} \xrightarrow{\sigma(L^2)} \frac{1}{2} V_{t\wedge T_p} \text{ as } n \to \infty,$$

$$\mathbf{E}\left(1 - Z^{\theta_n}_{T(n,p)}\right) = O((\theta_n - \theta_0)^2) \text{ as } n \to \infty.$$

If Conditions Z1 or Z2 hold, then $V \in \mathcal{M}^2_{\text{loc}}(P)$.

Condition H1. There exists an increasing process $\Lambda = (\Lambda_t, t \geqslant 0)$ such that $\forall t \geqslant 0$

$$\mathrm{Var}\left(\frac{h^\theta + h^\zeta - h^{\theta\zeta}}{(\theta - \theta_0)(\zeta - \theta_0)} - \frac{1}{4}\Lambda\right)_t \xrightarrow{\mathbf{P}} 0 \quad \text{as } \theta,\ \zeta \to \theta_0,$$

$$(\theta - \theta_0)^{-2} C^\theta_T \xrightarrow{\mathbf{P}} 0 \quad \text{as } \theta \to \theta_0.$$

Condition H2. There exists an increasing process $\Lambda = (\Lambda_t, t \geqslant 0)$ such that for each bounded stopping time T

$$(\theta - \theta_0)^{-2} h^\theta_T \xrightarrow{\mathbf{P}} \Lambda_T/8 \quad \text{as } \theta \to \theta_0. \tag{6}$$

Remark 4. If Condition H2 is fulfilled, then Λ is a predictable process (more precisely, Λ is P–indistinguishable from a predictable increasing process). If one demands the predictability of Λ from the beginning, then it is sufficient to check (6) for each bounded predictable stopping time T.

THEOREM 1. 1) Z1 $\Rightarrow$ Z2, H1 $\Rightarrow$ H2.

2) H1 $\Leftrightarrow$ Z1, *and* $\Lambda = \langle V, V\rangle$.

3) H2 $\Rightarrow$ Z2, *and* $\Lambda - \langle V, V\rangle$ *is an increasing process.*

The first statement of Theorem 1 is trivial, the second one was proved in (Jacod, 1989a), and the proof of the third statement is similar to the proof of the second one.

It should be noted that under Condition H2 the sequence $\{\theta_n\}$ in Condition Z2 may be chosen as a subsequence of any given beforehand sequence tending to θ_0. In this case the process V may depend, generally speaking, on the sequence $\{\theta_n\}$. If $\mathcal{F}_t = \{\emptyset, \Omega\}$ for $t < 1$, $\mathcal{F}_t = \mathcal{F}$ for $t \geqslant 1$, Condition Z1 is equivalent to the regularity of the statistical model $(\Omega, \mathcal{F}, \mathcal{P})$ in LeCam's sense (LeCam, 1986) at θ_0, and Condition H2 is equivalent to the property

$$(\theta - \theta_0)^{-2} \varrho^2(P, P^\theta) \to \Lambda_1/4 \quad \text{as } \theta \to \theta_0$$

(cf. (Holevo, 1973; Pitman, 1979)), where

$$\varrho^2(P, P^\theta) = \int \left({z'_\infty}^{1/2}(\theta) - {z'_\infty}^{1/2}(\theta)\right)^2 dQ^\theta$$

is the square of the Hellinger distance between P and P^θ.

More generally, if the processes $h^{\theta\zeta}$ are deterministic for all θ and ζ, then Condition Z1 is equivalent to the regularity of the statistical models $\mathcal{P}_t = (\Omega, \mathcal{F}_t, (P^\theta_t, \theta \in \Theta))$ in LeCam's sense at θ_0 for all $t \geqslant 0$, the process Λ is deterministic and Λ_t coincides with the Fisher information of $\mathcal{P}_t$ at θ_0; Condition Z2 is equivalent to the property $(\theta - \theta_0)^{-2} \varrho^2(P_t, P^\theta_t) \to \Lambda_t/4$ as $\theta \to \theta_0$, for all $t \geqslant 0$, the process

Λ is again deterministic and Λ_t coincides with the appropriately defined Fisher information (see (Holevo, 1973)) of $\mathcal{P}_t$ at θ_0.
In general it is natural to call the process Λ the Fisher information process at the point θ_0. This process was firstly introduced in (Heyde, 1975; Feigin, 1976).
Now we formulate a number of additional assumptions on the family $\mathcal{P}$ and the process X; ${}^\theta\langle X, X\rangle$ denotes the compensator of the process $[X, X]$ with respect to the measure P^θ, and ${}^\theta\langle M(\theta), M(\theta)\rangle$ denotes the quadratic characteristic of the locally square integrable martingale $M(\theta) \in \mathcal{M}^2_{\mathrm{loc}}(P^\theta)$ (with respect to P^θ).

Condition C1. $|\Delta Z^\theta| \leqslant a$ (P–a.s.), $\forall\theta \in \Theta$, for some $a > 0$, and $Z^\theta_{R(\theta)-} = +\infty$ on the set $\{R(\theta) < +\infty\}$ (P^θ–a.s.), $\forall\theta \in \Theta$, where $R(\theta) = \inf\{t: Z^\theta_t = +\infty\}$.

Condition C2. $|\Delta X| \leqslant a$ (P^θ–a.s.), for some $a > 0$, or $|\Delta M| \leqslant a$, P^θ–a.s., $\forall\theta \in \Theta$ for some $a > 0$.

Condition C3. ${}^\theta\langle X, X\rangle_T \xrightarrow{\mathbf{P}} \langle X, X\rangle_T$ as $\theta \to \theta_0$ for each bounded predictable stopping time T.

Condition C4. ${}^\theta\langle M(\theta), M(\theta)\rangle_T \xrightarrow{\mathbf{P}} \langle M, M\rangle_T$ as $\theta \to \theta_0$ $\forall\theta \in \Theta$, for each bounded predictable stopping time T.

Condition C5. ${}^\theta\langle X, X\rangle_t$ is asymptotically P–tight as $\theta \to \theta_0$, $\forall t \geqslant 0$.

Condition C6. ${}^\theta\langle M(\theta), M(\theta)\rangle_t$ is asymptotically P–tight as $\theta \to \theta_0$, $\forall t \geqslant 0$.

Remark 5. The second part of Condition C2 needs some explanations. We follow definitions given in (Jacod and Shiryaev, 1987), in particular, we suppose that for each $\theta \in \Theta$ the trajectories of the processes $X, M(\theta)$ are càdlàg only for P^θ–almost all ω, but the equality $X = X_0 + M(\theta) + A(\theta)$ is true for all ω, the process $A(\theta)$ is measurable with respect to the predictable σ–field not relative to the P^θ–completion of the filtration F but relative to F properly, and $A(\theta)$ has càdlàg trajectories for all ω. Therefore, on the one hand, the process $\Delta M = \Delta M(\theta) + \Delta A(\theta) - \Delta A$ is defined for P^θ–almost all ω. On the other hand, the second part of Condition C2 may be formulated more precisely in the following way: for some $a > 0$ $|\Delta M| \leqslant a$ (P–a.s.) and $|\Delta M| \leqslant a$ (P^θ–a.s.), $\forall\theta \neq \theta_0$, on the set $\Gamma(\theta) = \{t: Z^\theta_{t-} < +\infty\}$. This agrees with the fact that any two modifications of M satisfying the assumptions above, coincide on $\Gamma(\theta)$ up to P^θ–indistinguishability.

Remark 6. A family $\{\xi(\theta), \theta \in \Theta\}$ of random variables is asymptotically P–tight as $\theta \to \theta_0$ if

$$\lim_{N\to\infty} \limsup_{\theta\to\theta_0} \mathbf{P}\{|\xi(\theta)| > N\} = 0.$$

Obviously, C3 $\Rightarrow$ C5 $\Rightarrow$ C6, C4 $\Rightarrow$ C6. It can be proved that, if for any sequence tending to θ_0 there exists its subsequence $\{\theta_n\}$ such that Condition Z2 is fulfilled for $\{\theta_n\}$ (in particular, if Z1, H1 or H2 hold), then any one of Conditions C1 and C2 implies the validity of Conditions C3—C6, C3 is equivalent to C4, C5 is equivalent to C6.
For the last group of conditions, one needs the following definition. We say that a sequence $\{\xi_n, n \in \mathbb{N})$ of random variables converges locally weakly in $L^1(P)$ as $n \to \infty$ if there exists a family $\{B_p, B(n,p), p, n \in \mathbb{N}\}$ of measurable sets such that

$$B_p \uparrow \Omega \quad \text{as} \quad p \to \infty; \quad B(n,p) \subseteq B_p \quad (\forall p, n \in \mathbb{N});$$

$$\lim_{n\to\infty} \mathbf{P}(B_p \backslash B(n,p)) = 0. \quad (\forall p \in \mathbb{N});$$

the sequence $\xi_n \mathbf{1}_{B(n,p)}$ converges in the weak topology $\sigma(L^1(P), L^\infty(P))$ of the space $L^1(P)$ as $n \to \infty$ $(\forall p \in \mathbb{N})$. It is easy to see that in this case the limit of the sequence $\xi_n \mathbf{1}_{B(n,p)}$ is of the form $\xi \mathbf{1}_{B_p}$, where ξ is a random variable not depending on the choice of a family $(B_p, B(n,p))$; ξ is said to be the limit of the sequence $\{\xi_n\}$. This fact will be denoted by $\xi_n \xrightarrow{l} \xi$. Note that the convergence in probability implies the locally weak convergence in $L^1(P)$ to the same limit.

Condition A1. There exist a predictable process $A' = (A'_t, t \geqslant 0)$ with finite variation and, for every sequence $\{\theta_n\} \to \theta_0$, a localizing family $(T_p, T(n,p))$, such that $\forall t \geqslant 0, \forall p \in \mathbb{N}$

$$\frac{A(\theta_n)_{t\wedge T(n,p)} - A_{t\wedge T(n,p)}}{\theta_n - \theta_0} \xrightarrow{L^1} A'_{t\wedge T_p} \quad \text{as } n \to \infty.$$

Condition A1′. $(\theta - \theta_0)^{-1}(A_t(\theta) - A_t) \xrightarrow{\mathbf{P}} A'_t$ as $\theta \to \theta_0 (\forall t \geqslant 0)$, where $A' = (A'_t, t \geqslant 0)$ is a predictable process with finite variation.

Condition A2. There exist a predictable process $A' = (A'_t, t \geqslant 0)$ with finite variation and, for some sequence $\{\theta_n\} \to \theta_0$, a localizing family $(T_p, T(n,p))$, such that $\forall t \geqslant 0, \forall p \in \mathbb{N}$

$$\frac{A(\theta_n)_{t\wedge T(n,p)} - A_{t\wedge T(n,p)}}{\theta_n - \theta_0} \xrightarrow{\sigma(L^1)} A'_{t\wedge T_p} \quad \text{as } n \to \infty.$$

Condition A2′. $(\theta - \theta_0)^{-1}(A_t(\theta) - A_t) \xrightarrow{l} A'_t$ as $\theta \to \theta_0 (\forall t \geqslant 0)$, where $A' = (A'_t, t \geqslant 0)$ is a predictable process with finite variation.
It is easy to see that A1⇒A1′ ⇒A2′ , and Condition A2 implies that for the same sequence $\{\theta_n\}$ $(\theta - \theta_0)^{-1}(A_t(\theta_n) - A_t) \xrightarrow{l} A'_t$ as $n \to \infty$ $(\forall t \geqslant 0)$.
The main result of this paper is

THEOREM 2. 1) *Let Condition* Z1 *and any one of Conditions* C1—C6 *hold. Then Condition* A1 *holds and*

$$\langle M, V\rangle = A'.$$

2) *Let Condition* Z2 *and any one of Conditions* C1—C4 *hold. Then Condition* A2 *holds (for the same sequence* $\{\theta_n\}$ *as in* Z2*) and*

$$\langle M, V\rangle = A'.$$

Theorems 1 and 2 imply

COROLLARY. 1) *Let Condition* H1 *and any one of Conditions* C1—C6 *hold. Then Condition* A1′ *holds and,* $\forall t \geqslant 0$,

$$\langle M, M\rangle_t \geqslant \frac{A'^2_t}{\Lambda_t}\ (P\text{–a.s.}). \tag{7}$$

2) *Let Conditions* H2, A2′ *and any one of Conditions* C1—C4 *hold. Then,* $\forall t \geqslant 0$,

$$\langle M, M\rangle_t \geqslant \frac{A'^2_t}{\Lambda_t}(P\text{–a.s.}).$$

A particular case of Theorem 2 (under assumptions Z1, C2) was also proved in (Jacod, 1989b) by another method.
In the trivial case ($\mathcal{F}_t = \{\varnothing, \Omega\}$ for $t < 1$, $\mathcal{F}_t = \mathcal{F}$ for $t \geqslant 1$) inequality (7) (at $t = 1$) turns out to be the Cramér—Rao inequality, the conditions of the validity of the Cramér—Rao inequality in (Borovkov, 1984; Ibragimov and Hasminskii, 1981; Pitman, 1979, Chapter 5, Theorem 1) imply the fulfilment of H1 and C5; Conditions H2, A2′, C4 coincide with the conditions formulated in (Holevo, 1973; Pitman, 1979, 5.4) under which the Cramér—Rao inequality is true.

4. It is possible to generalize the results of section 3 to the case of a multidimensional parameter $\theta = (\theta^1, \ldots, \theta^d) \in \Theta \subseteq \mathbf{R}^d$ and a multidimensional semimartingale $X = (X^1, \ldots, X^k)$. If we reformulate conditions of Theorem 2 in such a way that for every direction $\theta^i, 1 \leqslant i \leqslant d$, parallel to the coordinate axes, and for every component $X^j, 1 \leqslant j \leqslant k$, conditions of Theorem 2 hold, then the assertion of Theorem 2 will be true for every direction θ^i and every component X^j. (As an example of such reformulation see a multidimensional version of Conditions Z1 and H1 in (Jacod, 1989a).) After this a multidimensional version of inequality (7) may be established by applying the multidimensional Kunita—Watanabe inequality.

5. The central point in the proof of Theorem 2 is the following generalization of Girsanov's theorem for the case of a non–locally absolutely continuous change of a measure.
Let $(\Omega, \mathcal{F})$ be a measurable space, $F = (\mathcal{F}_t, t \geqslant 0)$ be an increasing and right–continuous family of sub–σ–fields of $\mathcal{F}$. Let P, P', Q be probability measures on $(\Omega, \mathcal{F})$ such that $P \overset{\text{loc}}{\ll} Q$, $P' \overset{\text{loc}}{\ll} Q$. Let z and z' be the density processes of P and P' relative to $Q, Z = z'/z$. Define

$$R = \inf\{t\colon\ z_t = 0\}, \Gamma = \{t\colon\ z_{t-} > 0\ \text{ or } z'_{t-} > 0\}, \quad J = z'_R \mathbf{1}_{[R,\infty[}.$$

Since Γ is a predictable set of type $[0, \cdot||$, see (Jacod, 1979, Chapter V), we can consider stochastic processes on the set Γ. For example, let $\mathcal{T}(\Gamma)$ be the set of all stopping times T such that $[0,T] \subseteq \Gamma$; X is a P–semimartingale on Γ ($X \in \mathcal{S}^\Gamma(P)$) if $X^T \in \mathcal{S}(P)$ for any $T \in \mathcal{T}(\Gamma)$; if $X, Y \in \mathcal{S}^\Gamma(P)$ then the process $[X,Y]$ is well–defined on Γ by $[X,Y]^T = [X^T, Y^T]$ for $T \in \mathcal{T}(\Gamma)$; if $[X,Y] \in \mathcal{A}^\Gamma_{\text{loc}}(P)$ (i.e. $[X,Y]^T \in \mathcal{A}_{\text{loc}}(P)$ for any $T \in \mathcal{T}(\Gamma)$) then its compensator $\langle X,Y\rangle$ with respect to the measure P is a process on Γ such that $\langle X,Y\rangle^T$ is a P–compensator of $[X,Y]^T$ for every $T \in \mathcal{T}(\Gamma)$, and so on.

THEOREM 3. *Let $X = (X_t, t \geqslant 0)$ be a special semimartingale on the set Γ relative to P and P', with canonical decompositions $X = X_0 + M + A = X_0 + M' + A'$ respectively on the set Γ. Then, P–a.s. on the set Γ*

$$A' - A = \frac{1}{Z_-} \circ \langle X, Z\rangle + \frac{1}{z'_-} \circ {}^{Q}\langle X, J\rangle, \tag{8}$$

in particular, all expressions in the right–hand side of (8) are well–defined on Γ: $[X,Z] \in \mathcal{A}^{\Gamma}_{\text{loc}}(P)$, $X \in \mathcal{S}^{\Gamma}(Q)$, $[X,J] \in \mathcal{A}^{\Gamma}_{\text{loc}(P)}(Q)$, i.e. for every $T \in \mathcal{T}(\Gamma)$ there exists an increasing sequence $\{T_n\}$ of stopping times such that $T_n \uparrow \infty$ (P–a.s.) and $[X,J]^{T\wedge T_n} \in \mathcal{A}(Q)$, so there exists a unique (up to P–indistinguishability) process ${}^{Q}\langle X, J\rangle$ on the set Γ such that ${}^{Q}\langle X,J\rangle^{T\wedge T_n}$ is a Q–compensator of the process $[X,J]^{T\wedge T_n}$.

REFERENCES

Borovkov, A.A. (1984). *Mathematical Statistics.* Nauka, Moscow.

Feigin, P.D. (1976). Maximum likelihood estimation for continuous–time stochastic processes. *Adv. Appl. Probab.* **8**, 712–736.

Godambe, V.P. and Heyde, C.C. (1987). Quasi-likelihood and optimal estimation. *Inter. Statist. Review* **55**, 231–244.

Heyde, C.C. (1975). Remarks on efficiency in estimation for branching processes. *Biometrika* **62**, 49–55.

Holevo, A.S. (1973). On a generalization of Rao–Cramér inequality. *Theory Probab. Appl.* **18**, 371–375.

Hutton, J.E. and Nelson, P.I. (1986). Quasi–likelihood estimation for semimartingales. *Stoch. Processes Appl.* **22**, 245–257.

Ibragimov, I.A. and Hasminskii, R.Z. (1981). *Statistical Estimation; Asymptotic Theory.* Springer, Berlin—Heidelberg etc.

Jacod, J. (1979). Calcul stochastique et problèmes de martingales. *Lect. Notes in Math.* **714**.

Jacod, J. (1989a). Une application de la topologie d'Emery: le processus information d'un modèle statistique filtré. In: *Séminaire de Probabilités XXIII,* (Lect. Notes in Math. **1372**,). Springer, Berlin—Heidelberg etc, pp. 448–474.

Jacod, J. (1989b). Regularity, partial regularity, partial information process, for a filtered statistical model. *Preprint.*

Jacod, J. and Shiryaev, A.N. (1987). *Limit Theorems for Stochastic Processes.* Springer, Berlin—Heidelberg etc.

LeCam, L. (1986). *Asymptotic Methods in Statistical Decision Theory.* Springer, Berlin—Heidelberg etc.

Liptser, R.S. and Shiryaev, A.N. (1977). *Statistics of Random Processes.* Springer, Berlin—Heidelberg etc.

Pitman, E.J.G. (1979). *Some Basic Theory for Statistic Inference.* Chapman and Hall/Wiley and Sons, London—New York.

Thavaneswaran, A. and Thompson, M.E. (1986). Optimal estimation for semimartingales. *J. Appl. Probab.* **23**, 409–417.

Prob. Theory and Math. Stat., Vol. 1, pp. 490–503
B.Grigelionis *et al.* (Eds.)
1990 VSP/Mokslas

LIMIT THEOREMS FOR RECORD TIMES

A. GUT
Uppsala University, Sweden

ABSTRACT

Let $X_1, X_2, \ldots$ be independent random variables with a common continuous distribution function. Limit theorems for record times and the associated counting process are presented. The proofs are based on elementary methods; limit theorems for sums of independent random variables, inversion, a representation due to Williams and stopped random walks.

1. INTRODUCTION

Throughout this paper we are given a sequence of independent, identically distributed (i.i.d.) random variables whose distribution function is continuous. The modern way to prove limit theorems for the sequences defined above is to prove invariance principles, possibly in conjunction with continuous mapping theorems, to exploit the fact that the record times are the same for all distributions considered (since the distribution function is continuous), and/or to show that the point process of, suitably normalized, record times is asymptotically a Poisson process and, possibly, to combine this fact with some strong approximation theorem. A recent reference is (Resnick, 1987).

It is, however, this authors opinion that for "elementary" limit theorems, such as a central limit theorem, it is of great interest and importance to have proofs on the same "elementary" level. In addition, there are several results, such as theorems on moment convergence and certain theorems on rates of convergence, which are accessible by elementary methods, but not with the more advanced ones.

The aim of this paper is to point to three such elementary methods (one of which has not been used in this context before) and to use them to derive some new results for the sequence of record times and the associated counting process. For further deatils and references we refer to (Gut, 1988b).

2. PRELIMINARIES

Let $X, X_1, X_2, \ldots$ be i.i.d. random variables. We assume throughout that their distribution function F is continuous. Much attention has been devoted to the sequences of sample maxima, record times, interrecord times and record values, which are defined as follows:

The *sample maxima* are

$$Y_n = \max\{X_1, X_2, \ldots, X_n\}, \quad n \geqslant 1.$$

The *record times* are $L(1) = 1$ and, recursively,

$$L(n) = \min\{k : X_k > X_{L(n-1)}\}, \quad n \geqslant 2.$$

The *interrecord times* are $\Delta(1) = L(1) = 1$ and

$$\Delta(n) = L(n) - L(n-1), \quad n \geqslant 2.$$

The *record values* are

$$X_{L(n)}, \quad n \geqslant 1.$$

The present paper is devoted to the study of the sequence of record times and the associated *counting process* $\{\mu(n),\ n \geqslant 1\}$ defined by

$$\mu(n) = \#\text{records in}[1, n] = \max\{k : L(k) \leqslant n\}.$$

The term counting process is inspired from renewal theory.
We first collect some asymptotic properties of $\mu(n)$.

THEOREM 1. (i) $\mu(n)/\log n \xrightarrow{\text{a.s.}} 1$ *as* $n \to \infty$.

(ii) *The sequence* $\{(\mu(n)/\log n)^r,\ n \geqslant 3\}$ *is uniformly integrable and* $\mathbf{E}(\mu(n)/\log n)^r \to 1$ *as* $n \to \infty$ *for all* $r > 0$.

(iii) $(\mu(n) - \log n)/\sqrt{\log n} \xrightarrow{d} N(0,1)$ *as* $n \to \infty$.

(iv) *The sequence* $\{|(\mu(n) - \log n)/\sqrt{\log n}|^r,\ n \geqslant 3\}$ *is uniformly integrable,*

$$\mathbf{E}|(\mu(n) - \log n)/\sqrt{\log n}|^r \to \frac{1}{\sqrt{2\pi}} \int_{-\infty}^{\infty} |x|^r e^{-x^2/2} dx \quad \textit{for all} \quad r > 0$$

and

$$\mathbf{E}((\mu(n) - \log n)/\sqrt{\log n})^k \to 0 \quad \text{for} \quad k = 1, 3, 5, \ldots \ .$$

(v)

$$\limsup_{n\to\infty} (\liminf_{n\to\infty}) \frac{\mu(n) - \log n}{\sqrt{2 \log n \log\log\log n}} = \underset{(-)}{+} 1 \quad a.s.$$

The odd numbered conlusions are proved in the pioneering paper (Rényi, 1962). The starting point is that if we set, for $k \geqslant 1$,

$$I_k = \begin{cases} 1 & \text{if } X_k \text{ is a record,} \\ 0 & \text{otherwise,} \end{cases}$$

then $\{I_k,\ k \geqslant 1\}$ are independent random variables, such that $\mathbf{P}(I_k = 1) = 1 - \mathbf{P}(I_k = 0) = k^{-1}$ and

$$\mu(n) = \sum_{k=1}^{n} I_k, \quad n \geqslant 1, \tag{2.1}$$

(see also (Resnick, 1987), Section 4.1).
The conclusions follow from classical limit theorems for sums of independent, uniformly bounded random variables (note that $|I_k - 1/k| \leqslant 1$). Furthermore, since the moment generating function of $\mu(n)$, normalized as in (i) and (iii), respectively, converges, it also follows that the normalized sequences of the counting process are uniformly integrable and that the moments converge as described in the even numbered conclusions. These, latter, facts do not seem to have been mentioned explicitly before.
The corresponding result for the sequence of record times is as follows.

THEOREM 2. (i) $(\log L(n))/n \xrightarrow{\text{a.s.}} 1$ *as* $n \to \infty$.
(ii) *The sequence* $\{((\log L(n))/n)^r, \quad n \geqslant 1\}$ *is uniformly integrable and*

$$\mathbf{E}((\log L(n))/n)^r \to 1 \quad as \quad n \to \infty \quad for\ all \quad r > 0.$$

(iii) $(\log L(n) - n)/\sqrt{n} \xrightarrow{d} N(0,1)$ *as* $n \to \infty$.
(iv) *The sequence* $\{|(\log L(n) - n)/\sqrt{n}|^r, \quad n \geqslant 1\}$ *is uniformly integrable,*

$$\mathbf{E}|(\log L(n) - n)/\sqrt{n}|^r \to \frac{1}{\sqrt{2\pi}} \int_{-\infty}^{\infty} |x|^r e^{-x^2/2} dx \quad for\ all \quad r > 0$$

and

$$\mathbf{E}((\log L(n) - n)/\sqrt{n})^k \to 0 \quad \text{for} \quad k = 1, 3, 5, \dots \quad .$$

(v)

$$\limsup_{n\to\infty}(\liminf_{n\to\infty}) \frac{\log L(n) - n}{\sqrt{2n \log\log n}} = \underset{(-)}{+} 1 \quad a.s.$$

Remark 2.1. Again, (i), (iii) and (v) are due to Rényi (Rényi, 1962), see Section 3 below. The results (ii) and (iv), which do not seem to have been mentioned explicity before, follow from Williams' representation, which is presented in Section 5.
More detailed asymptotic expressions for $\mathbf{E}\log L(n)$ and $\operatorname{Var}\log L(n)$ are given in (Pfeifer, 1984, 1987), see also (Pfeifer, 1989). We also remark that $\log L(n)$ has moments of all orders, but $\mathbf{E}L(n) = +\infty$.

3. THE INVERSION METHOD

The first proofs of Theorem 2 (i), (iii), (v) are due to Rényi (Rényi, 1962), who used the corresponding results in Theorem 1 together with inversion. In view of the analogy with renewal theory this is a very natural method. Namely, the relation

$$\{L(n) \geqslant k\} = \{\mu(k) \leqslant n\}, \tag{3.1}$$

shows that the (increasing) record time and counting processes are inverses of each other. The proofs of Theorem 2 (i), (iii), (v) follow from Theorem 1 (i), (iii), (v) together with the relation (3.1) by letting k and n tend to infinity in a suitable manner. For the details we refer to (Rényi, 1962).

4. REMAINDER TERM ESTIMATES

In this section we derive Berry—Esseen estimates for $\{\mu(k),\ k \geqslant 1\}$ and $\{L(n),\ n \geqslant 1\}$. For the former sequence the result follows, essentially, from the remainder term estimate for sums of independent random variables, see (Esseen, 1945); for the constant see (van Beek, 1972). For the latter sequence we use the first result and inversion. The proof is modelled after (Englund, 1980).

THEOREM 3. *For $k \geqslant 2$ we have*

$$\sup_n \left| \mathbf{P}(\mu(k) \leqslant n) - \Phi\left(\frac{n - \log k}{\sqrt{\log k}}\right)\right| \leqslant \frac{1.9}{\sqrt{\log k}}.$$

Remark. Because of degeneracy the case $k = 1$ is excluded.

THEOREM 4. *We have*

$$\sup_k \left| \mathbf{P}(L(n) \geqslant k) - \Phi\left(\frac{n - \log k}{\sqrt{n}}\right)\right| \leqslant \frac{4.3}{\sqrt{n}}.$$

For the proofs we need the following estimates for the standard normal distribution function.

LEMMA. a) $\Phi(-x) \leqslant x^{-2}$ *for* $x > 0$.

b) $|\Phi(x) - \Phi(y)| \leqslant (1/\sqrt{2\pi})|x - y|$ *for* $x, y \in R$.

c) $|\Phi(z/x) - \Phi(z/y)| \leqslant 1.25|x/y - 1|$ *for* $x, y > 0, z \in R$.

d) $$\left|\Phi\left(\frac{x-y}{\sqrt{x}}\right) - \Phi\left(\frac{x-y}{\sqrt{y}}\right)\right| \leqslant \frac{1}{e\sqrt{2\pi}} \frac{1 + \theta_0 y_0^{-1/4}}{(1 - \theta_0 y_0^{-1/4})^{3/2}} \frac{1}{\sqrt{y}} \leqslant$$
$$\leqslant \frac{2}{e\sqrt{2\pi}} \frac{1}{(1 - \theta_0 y_0^{-1/4})^{3/2}} \frac{1}{\sqrt{y}} \quad \textit{for all} \quad x, y > 0 \quad \textit{such that} \quad |x - y| \leqslant \theta_0 y^{3/4},$$
$y > y_0 > \theta_0^4 > 0$.

Proof: a) and b) are standard, c) is in (Englund, 1982, Lemma 2.9, d) is similar to (Englund, 1980, Lemma 2.3) and the proof follows the path taken on p. 1112 there.

Now, recall (formula (2.1)) that $\mu(k) = \sum_{j=1}^{k} I_j$, where I_j are independent, $\mathbf{P}(I_j = 1) = 1 - \mathbf{P}(I_j = 0) = 1/j$ and set

$$m_k = \mathbf{E}\,\mu(k), \quad s_k^2 = \operatorname{Var}\mu(k), \quad \beta_k^3 = \sum_{j=1}^{k} \mathbf{E}|I_j - 1/j|^3,$$

let $\gamma_k = \sum_{j=1}^{k} j^{-1} - \log k$ and note that $1 \geqslant \gamma_1 \geqslant \gamma_2 \geqslant \ldots \geqslant \gamma_k \downarrow \gamma = 0.577\ldots$ is Euler's constant. The following relations are easily verified:

$$m_k = \sum_{j=1}^{k} j^{-1} = \log k + \gamma_k = \log k + \gamma + o(1), \quad \text{as} \quad k \to \infty \tag{4.1}$$

$$s_k^2 = \sum_{j=1}^{k} \frac{1}{j}\left(1 - \frac{1}{j}\right) = \log k + \gamma - \frac{\pi^2}{6} + o(1) \quad \text{as} \quad k \to \infty, \tag{4.2}$$

$$\log k \geqslant s_k^2 \geqslant \log k + \gamma - \pi^2/6 \geqslant \log k - 1.07, \tag{4.3}$$

$$\frac{\beta_k^3}{s_k^3} \leqslant \left(\sum_{j=1}^{k} \frac{1}{j}\left(1 - \frac{1}{j}\right)\right)^{-1/2} = s_k^{-1}. \tag{4.4}$$

Proof of Theorem 3. It suffices to consider k-values such that $1.9/\sqrt{\log k} < 1$, i.e. $k \geqslant [e^{3.61}] + 1 = 37$, since otherwise the bound is trivial. We have

$$\left|\mathbf{P}(\mu(k) \leqslant n) - \Phi\left(\frac{n - \log k}{\sqrt{\log k}}\right)\right| \leqslant$$

$$\leqslant \left|\mathbf{P}(\mu(k) \leqslant n) - \Phi\left(\frac{n - m_k}{s_k}\right)\right| + \left|\Phi\left(\frac{n - m_k}{s_k}\right) - \Phi\left(\frac{n - m_k}{\sqrt{\log k}}\right)\right| +$$

$$+ \left|\Phi\left(\frac{n - m_k}{\sqrt{\log k}}\right) - \Phi\left(\frac{n - \log k}{\sqrt{\log k}}\right)\right| = P_1 + P_2 + P_3.$$

The Berry—Esseen theorem for sums of independent random variables, (4.4) and (4.3) together imply that, for $k \geqslant 33$,

$$P_1 \leqslant 0.7975 \frac{\beta_k^3}{s_k^3} \leqslant \frac{0.7975}{\sqrt{\log k - 1.07}} \leqslant \frac{0.92}{\sqrt{\log k}}.$$

By Lemma, c, and (4.3) it follows that, for $k \geqslant 35$, we have

$$P_2 \leqslant \frac{0.73}{\sqrt{\log k}}.$$

Finally, since $k \geqslant 37$ and $\gamma_k \downarrow$, Lemma, b, yields

$$P_3 \leqslant \frac{1}{\sqrt{2\pi}} \frac{\gamma_{37}}{\sqrt{\log k}} \leqslant \frac{0.25}{\sqrt{\log k}}.$$

Proof of Theorem 4. It is sufficient to consider n-values such that $4.3/\sqrt{n} < 1$, that is $n \geqslant 19$.
We connect k and n through the relation $\log k - n = \theta n^{3/4}$ and consider the three cases $\theta \geqslant \theta_0$, $\theta \leqslant -\theta_0$ and $|\theta| < \theta_0$, where, as it turns out later, a suitable choice of θ_0 will be $\theta_0 = 0.74$.

Thus, we first assume that $\log k - n = \theta n^{3/4}$, $\theta \geqslant \theta_0$. Then, by Chebyshev's inequality, (4.3) and the fact that $n \geqslant 19$ one can show that

$$\mathbf{P}(L(n) \geqslant k) = \mathbf{P}(\mu(k) \leqslant n) \leqslant \frac{\operatorname{Var}\mu(k)}{(\log k - n)^2} \leqslant \left(\frac{1}{\theta_0^2} + \frac{1}{\theta_0 \sqrt[4]{19}}\right)\frac{1}{\sqrt{n}}.$$

Moreover, by Lemma, a,

$$\Phi\left(\frac{n - \log k}{\sqrt{n}}\right) = \Phi(-\theta n^{1/4}) \leqslant \frac{1}{\theta_0^2 \sqrt{n}}.$$

The case $\log k - n = \theta n^{3/4}$, $\theta \leqslant -\theta_0$, is treated (almost) the same way.
Finally, suppose that $|\log k - n| \leqslant \theta_0 n^{3/4}$. It follows from (3.1), Theorem 3 and Lemma, d (with $y_0 = \sqrt[4]{19}$), that

$$\left|\mathbf{P}(L(n) \geqslant k) - \Phi\left(\frac{n - \log k}{\sqrt{n}}\right)\right| \leqslant$$

$$\leqslant \left|\mathbf{P}(\mu(k) \leqslant n) - \Phi\left(\frac{n - \log k}{\sqrt{\log k}}\right)\right| + \left|\Phi\left(\frac{n - \log k}{\sqrt{\log k}}\right) - \Phi\left(\frac{n - \log k}{\sqrt{n}}\right)\right| \leqslant$$

$$\leqslant \left(1.9\frac{1}{\sqrt{1-\theta_0}} + \frac{2}{e\sqrt{2\pi}}\frac{1}{(1 - \theta_0/\sqrt[4]{19})^{3/2}}\right)\frac{1}{\sqrt{n}}.$$

The upper bound thus obtained is

$$\max\left\{\frac{2}{\theta_0^2} + \frac{1}{\theta_0 \sqrt[4]{19}},\ \frac{1.9}{\sqrt{1-\theta_0}} + \frac{2}{e\sqrt{2\pi}}\frac{1}{(1 - \theta_0/\sqrt[4]{19})^{3/2}}\right\}\frac{1}{\sqrt{n}}. \qquad (4.5)$$

Numerical calculations for different values of θ_0 show that for $\theta_0 = 0.74$ both estimates involved are slightly smaller than 4.3, which finishes the proof.

Lars Holst has drawn my attention to the fact that a better rate of convergence than that given in Theorem 3 for $\{\mu(k),\ k \geqslant 1\}$ is obtained by approximating $\mu(k)$ with $V_k \in \operatorname{Po}(m_k)$. More precisely, by the Stein—Chen method one has

$$\sup_{A \subset Z_+} |\mathbf{P}(\mu(k) \in A) - \mathbf{P}(V_k \in A)| \leqslant (1 - e^{-m_k})\sum_{j=1}^{k} j^{-2} \Big/ \sum_{j=1}^{k} j^{-1} \leqslant \frac{\pi^2}{6\log k} \qquad (4.6)$$

and this rate is of the correct order of magnitude; see (Barbour and Hall, 1984). This is interesting for the following reason. Namely, by (4.6) the discrepancy between $\mu(k)$ and the $\operatorname{Po}(m_k)$–distribution, in terms of variational distance, is of the order of magnitude $(\log k)^{-1}$. Now, by the ordinary Berry—Esseen theorem, the discrepancy between the $\operatorname{Po}(m_k)$–distribution and the normal distribution is of order of magnitude $(\log k)^{-1/2}$. The Poisson approximation of the counting process thus is better than the normal approximation, that is, the main contribution in the remainder term estimate in Theorem 3 stems from the normal approximation of the Poisson distribution.

However, in view of (2.1) this is not surprising. Namely, $\mu(k)$ can be interpreted (for k large) as a collection of independent, rare, events, which is precisely what characterizes the Poisson process. Let us also recall that the point process of record times is asymptotically Poisson; see (Resnick, 1987, p. 170).

5. WILLIAMS' REPRESENTATION

Recall that $L(1) = 1$. It is easy to see that

$$\mathbf{P}(L(2) = l_2) = \mathbf{P}(X_2, \ldots, X_{l_2-1} < X_1 < X_{l_2}) = \frac{1}{l_2}\frac{1}{l_2 - 1} \tag{5.1}$$

and, similarly, for arbitrary n, that

$$\mathbf{P}(L(k) = l_k,\ k = 2, \ldots, n) = \frac{1}{l_n(l_n - 1)(l_{n-1} - 1)\ldots(l_2 - 1)}, \tag{5.2}$$

from which it follows that $\{L(n),\ n \geqslant 1\}$ is a time-homogeneous Markov chain with transition probabilities

$$p_{ij} = \mathbf{P}(L(n) = j | L(n-1) = i) = \frac{i}{j(j-1)}, \quad j > i \tag{5.3}$$

(and $p_{ij} = 0$ for $j \leqslant i$).
Williams' representation is the following explicit construction of a sequence of random variables with the same properties; see (Williams, 1973) and (Resnick, 1987, p. 194): Let $\{Z_k,\ k \geqslant 1\}$ be i.i.d. Exp(1)–distributed random variables, set $A_1 = 1$ and

$$A_n = [A_{n-1} \exp\{Z_n\}] + 1, \quad n \geqslant 2. \tag{5.4}$$

Then $\{A_n,\ n \geqslant 1\}$ is a time-homogeneous Markov chain and

$$\mathbf{P}(A_n = j | A_{n-1} = i) = \mathbf{P}(j - 1 \leqslant i \exp\{Z_n\} < j) = \frac{i}{j(j-1)}, \quad j > i.$$

This shows that $\{A_n, n \geqslant 1\}$ is a representation of the sequence of record times. We therefore can (and will) write $L(1) = 1$ and

$$L(n) = [L(n-1) \exp\{Z_n\}] + 1, \quad n \geqslant 2. \tag{5.5}$$

It now follows from (5.5) that $L(k-1)\exp\{Z_k\} \leqslant L(k) < L(k-1)\exp\{Z_k\} + 1$ and thus, since $L(n) \geqslant n$, that

$$\exp\{Z_k\} \leqslant \frac{L(k)}{L(k-1)} < \exp\{Z_k\} + \frac{1}{L(k-1)} \leqslant \exp\{Z_k\} + \frac{1}{k-1}. \tag{5.6}$$

By taking logarithms and noting that $\log L(n) = \sum_{k=2}^{n} \log \frac{L(k)}{L(k-1)}$ we obtain (after some computations)

$$\sum_{k=2}^{n} Z_k \leqslant \log L(n) < \sum_{k=2}^{n} Z_k + \log(n-1) + \gamma_{n-1}. \tag{5.7}$$

However, $\Gamma_n = \sum_{k=2}^{n} Z_k$ has a $\Gamma(n-1,1)$–distribution, whose asymptotics are well known and this, together with (5.7) can now be used in order to derive asymptotics for $\{L(n),\ n \geqslant 1\}$. As an example, let

$$\Gamma_n^* = \frac{\Gamma_n - n}{\sqrt{n}}, \quad n \geqslant 2.$$

By the ordinary central limit theorem we have $\Gamma_n^* \xrightarrow{d} N(0,1)$ as $n \to \infty$ and, since

$$\Gamma_n^* \leqslant \frac{\log L(n) - n}{\sqrt{n}} < \Gamma_n^* + \frac{\log(n-1) + \gamma + o(1)}{\sqrt{n}}, \tag{5.8}$$

it follows that $n^{-1/2}(\log L(n) - n) \xrightarrow{d} N(0,1)$ as $n \to \infty$, which reproves Theorem 2 (iii). However, it is also known that the sequence $\{|\Gamma_n^*|^r,\ n \geqslant 2\}$ is uniformly integrable for all $r > 0$ (for general sums of i.i.d. random variable see e.g. (Gut, 1988a), Theorem I.4.2), which, in view of (5.8) and Theorem 2 (iii), proves Theorem 2 (iv).

6. THE METHOD OF STOPPED SUMS

The similarity between $\{L(n),\ n \geqslant 1\}$ and $\{\mu(n),\ n \geqslant 1\}$ on one hand and a renewal process and its counting process on the other hand has been mentioned before. This, together with this authors background (Gut, 1988a), suggests that (some of) the above results might also be obtainable via the theory of stopped sums.

To show that this is, indeed, the case we now demonstrate how Theorem 2 (i) and (iii) can be deduced from Theorem 1 (i) and (iii). A further application is given in Section 7.

Proof of Theorem 2 (i). From Theorem 1 (i) and the fact that $L(n) \to +\infty$ a.s. as $n \to \infty$ it follows (see e.g. (Gut, 1988a), Theorem 2.1) that

$$\frac{\mu(L(n))}{\log L(n)} \xrightarrow{\text{a.s.}} 1 \quad as \quad n \to \infty.$$

Since $\mu(L(n)) = n$ we are done.

Proof of Theorem 2 (iii). In view of formula (4.2) it follows from Theorems 1 (iii), 2 (i) and a generalization of Anscombe's theorem (see (Csörgő and Rychlik, 1980)) that

$$\frac{\mu(L(n)) - \sum_{k=1}^{L(n)} k^{-1}}{\sqrt{\sum_{k=1}^{L(n)} k^{-1}(1-k^{-1})}} \xrightarrow{d} N(0,1) \quad as \quad n \to \infty,$$

which, together with the fact that $\mu(L(n)) = n$, (4.1) and (4.2), shows that

$$\frac{n - \log L(n)}{\sqrt{\log L(n)}} \xrightarrow{d} N(0,1) \quad as \quad n \to \infty.$$

The conclusion now follows from Theorem 2 (i), Cramér's theorem and the symmetry of the normal distribution.
A similar proof can be constructed for Theorem 2 (v) with the aid of an Anscombe law of the iterated logarithm; see (Gut, 1989).

7. THE CLUSTER SETS OF $\{L(n)\}$ AND $\{\mu(n)\}$

In Theorems 2 (v) and 1 (v) the extreme limit points of the sequences $\{L(n), n \geqslant 1\}$ and $\{\mu(n),\ n \geqslant 3\}$ are determined. In this section we use the method of stopped random walks (together with Williams' representation) to show that the cluster sets (the sets of limit points) of these sequences, with probability 1, consist of all points between the extreme limit points.
Let $C(\{x_n\})$ denote the cluster set of the sequence $\{x_n\}$.

THEOREM 5.

$$C\left(\left\{\frac{\log L(n)-n}{\sqrt{2n\log\log n}},\ n \geqslant e^e\right\}\right) = [-1,1] \quad a.s.$$

THEOREM 6.

$$C\left(\left\{\frac{\mu(n)-\log n}{\sqrt{2\log n \log\log\log n}},\ n \geqslant e^{e^e}\right\}\right) = [-1,1] \quad a.s.$$

Proof of Theorem 5. Set, for $n \geqslant 2$,

$$\Gamma_n^{**} = \frac{\sum\limits_{k=2}^{n} Z_k - n}{\sqrt{2n\log\log n}},$$

where $\{Z_k,\ k \geqslant 1\}$ is defined as in Section 5. Then

$$C(\{\Gamma_n^{**},\ n \geqslant 3\}) = [-1,1] \quad \text{a.s.}$$

(see (de Acosta, 1983) for an elementary proof). This, together with (5.7) and (Gut, 1988a, Lemma III.11.1) proves the conclusion.

Proof of Theorem 6. By Theorem 2 (i) and an Anscombe law of the iterated logarithm (see (Torrång, 1987), Theorem 5.2) we have

$$C(\{\Gamma_{\mu(n)}^{**},\ n \geqslant 3\}) = [-1,1] \quad \text{a.s.}$$

and hence, in view of (5.7) and (Gut, 1988a, Lemma III.11.1) and Theorem 1 (i), that

$$C\left(\left\{\frac{\log L(\mu(n))-\mu(n)}{\sqrt{2\log n\log\log\log(n)}},\ n \geqslant e^{e^e}\right\}\right) = [-1,1] \quad \text{a.s.} \tag{7.1}$$

Next we define, for $n \geqslant 1$, $\nu(n) = \min\{k : L(k) > n\}$. The sequence $\{\nu(n),\ n \geqslant 1\}$ thus is the first passage time process of $\{L(n),\ n \geqslant 1\}$. Since the record times are increasing it follows (as for renewal processes) that $\nu(n) = \mu(n) + 1$. The conclusions of Theorem 1, and, as a consequence, formula (7.1) therefore also hold with $\mu(n)$ replaced by $\nu(n)$. These facts, together with the inequality $L(\mu(n)) \leqslant$ $\leqslant n < L(\nu(n))$, finish the proof.

8. CONVERGENCE RATES

A limit theorem states that convergence holds. Given a limit theorem, a natural question is "with what rate?". For the central limit theorems for $\mu(n)$ and $L(n)$ an answer has been given in Section 4. In this section we present rates of convergence in the law of large numbers and the law of the iterated logarithm.
One way to study the rate of convergence in, say, Theorem 2 (i) is to study the convergence/divergence of $\sum a_n \mathbf{P}(|\log L(n) - n| > \varepsilon n)$, where $\varepsilon > 0$ and a_n is some numerical sequence, typically n raised to some power. Another way is to investigate last exit times and the number of boundary crossings; in the above case this corresponds to studying $T(\varepsilon) = \sup\{n : |\log L(n) - n| > \varepsilon n\}$ and $N(\varepsilon) = \text{Card}\,\{n : |\log L(n) - n| > \varepsilon n\}$ (with $\sup \varnothing = 0$ and $\text{Card}\, \varnothing = 0$). Note that we always have $N(\varepsilon) \leqslant T(\varepsilon)$ and that the a.s. finiteness of $T(\varepsilon)$ for all $\varepsilon > 0$ is equivalent to Theorem 2 (i).
We shall present results on the convergence/divergence of sums of the above kind and provide conditions for existence/nonexistence of moments for the last exit times and the number of boundary crossings.
We begin with the results related to the sequence of record times.

THEOREM 7. (i) a) *For all $\varepsilon > 0$ there exists t_0, $0 < t_0 < 1$, such that for $t < t_0$ we have*

$$\sum\nolimits_1 = \sum_{n=1}^{\infty} e^{tn} \mathbf{P}(|\log L(n) - n| > n\varepsilon) < \infty.$$

b)

$$\sum\nolimits_2 = \sum_{n=1}^{\infty} e^{tn} \mathbf{P}\left(\sup_{k \geqslant n} \left| \frac{\log L(k) - k}{k} \right| > \varepsilon \right) < \infty \quad \text{for} \quad t < t_0.$$

(ii) *Let $T_L(\varepsilon) = \sup\{n : |\log L(n) - n| > n\varepsilon\}$. For all $\varepsilon > 0$, $t < t_0$ we have*

$$\mathbf{E} \exp\{T_L(\varepsilon)\} < \infty.$$

(iii) *Let $N_L(\varepsilon) = \text{Card}\,\{n : |\log L(n) - n| > n\varepsilon\}$. For all $\varepsilon > 0$, $t < t_0$ we have*

$$\mathbf{E} \exp\{N_L(\varepsilon)\} < \infty.$$

Proof. (i) It follows from Williams' representation (and (5.7)) that

$$|\log L(n) - n| \leqslant |\Gamma_n - (n-1)| + \log n + 1, \tag{8.1}$$

where $\Gamma_n \in \Gamma(n-1, 1)$. Now, for $0 < t < 1$, one can show that

$$\mathbf{E} e^{t|\Gamma_{n+1} - n|} \leqslant e^{tn}(1+t)^{-n} + e^{-tn}(1-t)^{-n}.$$

This, together with Markov's inequality, then yields

$$\mathbf{P}(|\Gamma_{n+1} - n| > n\varepsilon) \leqslant (e^{t(1-\varepsilon)}(1+t)^{-1})^n + (e^{-t(1+\varepsilon)}(1-t)^{-1})^n. \tag{8.2}$$

Relation (ia), finally, follows by noting that the sum in the RHS of (8.2) converges for t sufficiently small.

Relation (ib) follows from (ia) and the fact that $\sum_2 \leqslant (1-e^{-t})^{-1}\sum_1$.
(ii) The conclusion follows from the observation that

$$\{T_L(\varepsilon) \geqslant n\} = \left\{\sup_{k\geqslant n}\left|\frac{\log L(k)-k}{k}\right| > \varepsilon\right\}, \tag{8.3}$$

partial summation and (ib) (cf. (Gut, 1980), Section 8).
(iii) Immediate, since $N_L(\varepsilon) \leqslant T_L(\varepsilon)$.
A related result concerning the law of the iterated logarithm is next. The proofs of (i), (ii) and (iii) follow from the corresponding results for random walks and (5.7).

THEOREM 8. (i) *For all $r > 0$ and $\varepsilon > 0$ we have*

$$\sum\nolimits_3 = \sum_{n=3}^{\infty} n^{r-1}\mathbf{P}(|\log L(n) - n| > \varepsilon\sqrt{n\log\log n}) = +\infty.$$

(ii) *For $\varepsilon > \sqrt{2r}$, $r \geqslant 1$, we have*

$$\sum\nolimits_4 = \sum_{n=3}^{\infty}\frac{1}{n}(\log n)^{r-1}\mathbf{P}(|\log L(n) - n| > \varepsilon\sqrt{n\log\log n}) < \infty.$$

(iii) *For $\varepsilon > 2$ we have*

$$\sum\nolimits_5 = \sum_{n=3}^{\infty}\frac{1}{n}\log n\,\mathbf{P}(|\log L(n) - n| > \varepsilon\sqrt{n\log\log n}) < \infty,$$

$$\sum\nolimits_6 = \sum_{n=3}^{\infty}\frac{1}{n}\mathbf{P}\left(\sup_{k\geqslant n}\left|\frac{\log L(k)-k}{\sqrt{k\log\log k}}\right| > \varepsilon\right) < \infty.$$

(iv) *If $\varepsilon \leqslant \sqrt{2r}$ then $\sum_4 = +\infty$ and if $\varepsilon \leqslant 2$ then $\sum_5$ and $\sum_6$ diverge.*
(v) *Let $T_L(\varepsilon) = \sup\{n : |\log L(n) - n| > \varepsilon\sqrt{n\log\log n}\}$. Then*

$$\begin{aligned}
\mathbf{E}(T_L(\varepsilon))^r &= +\infty \quad \textit{for all} \quad \varepsilon > 0 \quad \textit{and} \quad r > 0,\\
\mathbf{E}(\log T_L(\varepsilon))^r &< \infty \quad \textit{for} \quad \varepsilon > \sqrt{2(r+1)}, \quad r > 0,\\
\mathbf{E}(\log T_L(\varepsilon))^r &= +\infty \quad \textit{for} \quad \varepsilon < \sqrt{2(r+1)}, \quad r > 0,\\
\mathbf{E}\log\log T_L(\varepsilon) &< \infty \quad \textit{for} \quad \sqrt{2} < \varepsilon < 2.
\end{aligned}$$

(vi) *Let $N_L(\varepsilon) = \mathrm{Card}\{n : |\log L(n) - n| > \varepsilon\sqrt{n\log\log n}\}$. Then all the conclusions in (v) remain valid with $T_L(\varepsilon)$ replaced by $N_L(\varepsilon)$.*

Proof. (i) By Theorem 4 we have

$$|\mathbf{P}(\log L(n) - n \geqslant \varepsilon\sqrt{n\log\log n}) - \Phi(-\varepsilon\sqrt{\log\log n})| \leqslant \frac{4.3}{\sqrt{n}}. \tag{8.4}$$

Furthermore (see e.g. (Feller, 1968, p. 175), we have, as $n \to \infty$,

$$\Phi(-\varepsilon\sqrt{\log\log n}) = 1 - \Phi(\varepsilon\sqrt{\log\log n}) \sim \varepsilon^{-1}(2\pi\log\log n)^{-1/2}(\log n)^{-\varepsilon^2/2}. \quad (8.5)$$

Since, by (8.4),

$$\sum n^{r-1}|\mathbf{P}(\log L(n) - n > \varepsilon\sqrt{n\log\log n}) - \Phi(-\varepsilon\sqrt{\log\log n})| < \infty,$$

for $r < 1/2$ and all $\varepsilon > 0$ and

$$\sum_{n=3}^{\infty} n^{r-1}\Phi(-\varepsilon\sqrt{\log\log n}) = +\infty$$

for all $r > 0$ and $\varepsilon > 0$ it follows that $\sum_3$ diverges for $r < 1/2$ and all $\varepsilon > 0$ and, hence, since $\sum_3$ increases as r increases, for all $r > 0$ and $\varepsilon > 0$.
(ii) and (iii) Let $\varepsilon > \sqrt{2r}$. By (8.4) (and a similar expression for the other tail) and (8.5) we have

$$\sum\nolimits_4 \leqslant \sum_{n=3}^{\infty} \frac{(\log n)^{r-1}}{n} |\mathbf{P}(|\log L(n) - n| > \varepsilon\sqrt{n\log\log n}) - 2(1 - \Phi(\varepsilon\sqrt{\log\log n}))| +$$

$$+2\sum_{n=3}^{\infty} \frac{(\log n)^{r-1}}{n}(1 - \Phi(\varepsilon\sqrt{\log\log n})) < \infty.$$

For $r = 2$ this yields $\sum_5 < \infty$.
(iv) Suppose that $\varepsilon \leqslant \sqrt{2r}$. The proof of the divergence of $\sum_4$ is similar to that of (i).
Now suppose that $\varepsilon \leqslant 2$. The divergence of $\sum_5$ follows from that of $\sum_4$ (with $r = 2$). The divergence of $\sum_6$ follows from (v) (proved below) with $r = 1$, an obvious modification of (8.3) and partial summation.
(v) The nonexistence of moments follows from (i). For the remainder of (v) in the random walk case, see (Russo, 1988, Theorems 1 and 2) (for $r = 1$ see also (Gut, 1980, Theorem 8.3)). The conclusions thus hold with $\log L(n) - n$ replaced by $\Gamma_n - (n-1)$ and, hence, by Williams' representation (and (8.1)) as desired.
(vi) The divergence parts follow from the corresponding results for random walks, see (Slivka, 1969) and (Russo, 1988), respectively, and (8.1) in the usual manner. Note, however, that, since Lemma 1 of (Slivka, 1969) carries over to the present context, the nonexistence of moments for $N_L(\varepsilon)$ is also a consequence of (i). The convergence parts of (vi) follow from (v), since $N_L(\varepsilon) \leqslant T_L(\varepsilon)$.
The proof of the theorem thus is complete.

Remark 8.1. Alternative proofs of (i)–(iii) can be obtained from the corresponding results for random walks together with Williams' representation.

We close by stating the corresponding results for the counting process. Theorem 9 below is related to the law of large numbers and Theorem 10 to the law of the iterated logarithm. The proofs are based on exponential bounds, Theorem 3 and modifications of the proof of Theorem 7. We refer to (Gut, 1988b) for details and references.

THEOREM 9. (i) *For* $\varepsilon > 0$ *we have*

$$\sum\nolimits_7 = \sum_{n=3}^{\infty} \frac{1}{n} \log n \, \mathbf{P}(|\mu(n) - \log n| > \varepsilon \log n) < \infty,$$

$$\sum\nolimits_8 = \sum_{n=3}^{\infty} \frac{1}{n} \mathbf{P}(\sup_{k \geqslant n} |(\mu(k) - \log k)/\log k| > \varepsilon) < \infty.$$

(ii) *Let* $T_\mu(\varepsilon) = \sup\{n : |\mu(n) - \log n| > \varepsilon \log n\}$. *Then*

$$\mathbf{E} \log T_\mu(\varepsilon) < \infty \quad \textit{for all} \quad \varepsilon > 0.$$

(iii) *Let* $N_\mu(\varepsilon) = \mathrm{Card}\,\{n : |\mu(n) - \log n| > \varepsilon \log n\}$. *Then*

$$\mathbf{E} \log N_\mu(\varepsilon) < \infty \quad \textit{for all} \quad \varepsilon > 0.$$

THEOREM 10. (i) *For all* $\varepsilon > 0$ *and* $r \geqslant 0$ *we have*

$$\sum\nolimits_9 = \sum_{n=27}^{\infty} n^{r-1} \mathbf{P}(|\mu(n) - \log n| > \varepsilon \sqrt{\log n \log\log\log n}) = +\infty.$$

(ii) *For* $\varepsilon > \sqrt{2r}$, $r \geqslant 1$, *we have*

$$\sum\nolimits_{10} = \sum_{n=27}^{\infty} \frac{(\log\log n)^{r-1}}{n \log n} \mathbf{P}(|\mu(n) - \log n| > \varepsilon \sqrt{\log n \log\log\log n}) < \infty.$$

(iii) *If* $\varepsilon \leqslant \sqrt{2}$ *then* $\sum_9 = +\infty$ *and if* $\varepsilon \leqslant \sqrt{2r}$ *then* $\sum_{10} = +\infty$.
(iv) *Let* $T_\mu(\varepsilon) = \sup\{n : |\mu(n) - \log n| > \varepsilon\sqrt{\log n \log\log\log n}\}$. *Then*

$$\mathbf{E}(T_\mu(\varepsilon))^r = +\infty \quad \textit{and} \quad \mathbf{E}(\log T_\mu(\varepsilon))^r = +\infty \quad \textit{for all } r > 0,\ \varepsilon > 0.$$

Moreover, for $\varepsilon \leqslant \sqrt{2r}$, $r \geqslant 1$, *we have*

$$\mathbf{E}(\log\log T_\mu(\varepsilon))^r = +\infty.$$

Remark 8.2. Unfortunately we have no positive result concerning the integrability of some function of $T_\mu(\varepsilon)$.

REFERENCES

Barbour, A.D. and Hall, P. (1984). On the rate of Poisson convergence. *Math. Proc. Camb. Phil. Soc.* **95**, 473–480.

Csörgő, M. and Rychlik, Z. (1980). Weak convergence of sequences of random elements with random indices. *Math. Proc. Camb. Phil. Soc.* **88**, 171–174.

de Acosta, A. (1983). A new proof of the Hartman—Wintner law of the iterated logarithm. *Ann. Probab.* **11**, 270–276.

Englund, G. (1980). Remainder term estimate for the asymptotic normality of the number of renewals. *J. Appl. Probab.* **17**, 1108–1113.

Englund, G. (1982). On the coupon collector's remainder term. *Z. Wahr. verw. Geb.* **60**, 381–391.

Esseen, C.-G. (1945). Fourier analysis of distribution functions. A mathematical study of the Laplace–Gaussian law. *Acta Math.* **77**, 1–125.

Feller, W. (1968). *An Introduction to Probability Theory and Its Applications.* Vol. I, 3rd ed., Wiley, New York.

Gut, A. (1980). Convergence rates for probabilities of moderate deviations for sums of random variables with multidimensional indices. *Ann. Probab.* **8**, 298–313.

Gut, A. (1988a). *Stopped Random Walks. Limit Theorems and Applications.* Springer, Berlin–Heidelberg etc.

Gut, A. (1988b). Limit theorems for record times and the associated counting process. *U.U.D.M. Report* 1988:17.

Gut, A. (1989). Anscombe laws of the iterated logarithm. *U.U.D.M. Report* 1989:14.

Pfeifer, D. (1984). A note on moments of certain record statistics. *Z. Wahr. verw. Geb.* **66**, 293–296.

Pfeifer, D. (1987). On a joint strong approximation theorem for record and inter–record times. *Probab. Th. Rel. Fields* **75**, 213–221.

Pfeifer, D. (1989). *Einführung in die Extremwertstatistik.* Teubner, Stuttgart.

Rényi, A. (1962). On the extreme elements of observations. MTA III, *Oszt. Közl.* **12**, 105–121. In: *Collected works*, Vol. III. 1976, Akadémiai Kiadó, Budapest, pp. 50–66.

Resnick, S.I. (1987). *Extreme Values, Regular Variation and Point Processes.* Springer, Berlin–Heidelberg etc.

Russo, R.R. (1988). On the last exit time and the number of exits of partial sums over a moving boundary. *Statist. Probab. Lett.* **6**, 371–378.

Slivka, J. (1969). On the law of the iterated logarithm. *Proc. Nat. Acad. Sci. USA.* **63**, 289–291.

Torrång, I. (1987). The law of the iterated logarithm - cluster points of deterministic and random subsequences. *Probab. Math. Statist.* **VIII**, 133–141.

van Beek, P. (1972). An application of Fourier methods to the problem of sharpening the Berry–Esseen inequality. *Z. Wahr. verw. Geb.* **23**, 183–196.

Williams, D. (1973). On Rényi's "record" problem and Engel's series. *Bull. London Math. Soc.* **5**, 235–237.

Prob. Theory and Math. Stat., Vol. 1, pp. 504–511
B.Grigelionis *et al.* (Eds.)
1990 VSP/Mokslas

ON STATISTICS OF CONTINUOUS MARKOV PROCESSES: SEMI–MARKOV APPROACH

B.P. HARLAMOV
Leningrad Branch of Blagonravov Machine Research Institute,
Academy of Sciences, V.O. Bolshoi pr., 61, 199178 Leningrad, USSR

ABSTRACT

Two criteria for a semi–Markov process to be Markov are proved. The first one is formulated in terms of the ratio

$$P_x\left(e^{-\lambda\sigma_r}-1\right)/P_x(\sigma_r)$$

as $r \to 0$, and the second in terms of the integral

$$P_x(\exp\left(-\lambda\sigma_r/P_x(\sigma_r)\right))$$

($\lambda > 0$, σ_r is the first exit time from the interval $(x-r,\ x+r)$, $r \to 0$), P_x is the semi–Markov family of distributions of the process

INTRODUCTION

Most of the papers devoted to the problem of testing the hypothesis of Markovness of a random sequence deal with the case of discrete state space (see e.g. (Billingsley, 1961)). Their methods can be successfully applied to the case of continuous time processes $\xi(t)$, $t \in \mathbf{R}$, with discrete state space, but they fail when the state space is continuous too. This follows from the impossibility of statistical estimation of the conditional probability $P(\cdot|\xi(t)=x)$, if $P(\xi(t)=x)=0$ for all t. However, it is possible to test Markovness of a stationary continuous process with one–dimensional state space. This possibility is based on the existence of an infinite sequence of random hitting times of $x \in X$, e.g. of the first exit time point process defined with respect to some finite set $X' \subset X$, $x \in X'$. If $\xi(t)$ is (strong) Markov, then every such marked first exit time point process is semi–Markov. As the first step towards testing of Markovness it is reasonable to test the above mentioned property of the first exit time point process. The class of processes for which every embedded first exit time point process is semi–Markov is called the class of continuous semi–Markov processes. As the second step, one should test the hypothesis of Markovness assuming that the process is continuous semi–Markov. Analysis of jump–type semi–Markov processes is interesting and important in view of many applications. It belongs to the sphere of traditional methods of statistics.

This paper is devoted to testing of Markovness of a continuous one–dimensional stationary semi–Markov process.

1. SEMI–MARKOV PROCESSES

We consider Markov processes as a part of the class of semi–Markov (SM) processes. A continuous semi–Markov process is defined as a family of probability measures (P_x), $x \in X$, on the Borel σ–field $\mathcal{B}(C)$ of subsets of the space C of continuous functions $\xi : \mathbf{R}_+ \to X$, where X is some metric space (Harlamov, 1983). The semi–Markov family (P_x) has the Markov property with respect to the first exit time τ_Δ from each open set $\Delta \subset X$:

$$\tau_\Delta(\xi) = \inf\{t \geqslant 0 \ : \ \xi(t) \notin \Delta\}.$$

The kernel

$$F_\tau(dt, dx_1 \mid x) = P_x(\tau \in dt, \ \pi_\tau \in dx_1)$$

(where $\pi_t\xi = \xi(t)$ and τ is a stopping time) is called a semi–Markov transition function of the SM process, where $P_x(\pi_0 = x) = 1$. It is convenient in the theory of SM processes to use the Laplace transform

$$f_\tau(\lambda, dx_1 \mid x) = \int_0^\infty e^{-\lambda t} F_\tau(dt, dx_1 \mid x) \equiv P_x(e^{-\lambda\tau}, \ \pi_\tau \in dx_1).$$

Let $\mathcal{A}$ be the class of all open sets $\Delta \subset X$. The semi–Markov property of the process is reflected in the equation:

$$(\forall x \in X)\ (\forall \Delta_1, \Delta_2 \in \mathcal{A}, \ \Delta_1 \subset \Delta_2)\ (\forall t > 0)\ (\forall S \in \mathcal{B}(X)),$$

$$F_{\tau(\Delta_2)}([0,t] \times S \mid x) =$$

$$= \int_{[0,t]\times X} F_{\tau(\Delta_1)}(dt_1, dx_1|x) F_{\tau(\Delta_2)}([0, t-t_1] \times S \mid x_1). \tag{1}$$

Let $X = \mathbf{R}^1$, $\Delta = (a,b)$, $a < x < b$. In this case $f_{\tau(\Delta)}(\lambda, dx_1 \mid x)$ can be divided into two parts: $f_{\tau(\Delta)}(\lambda, \{a\} \mid x) \equiv g_\Delta(\lambda, x)$ and $f_{\tau(\Delta)}(\lambda, \{b\} \mid x) \equiv h_\Delta(\lambda, x)$. From the equation (1) we have the following system of equations:

$$g_{\Delta_2}(\lambda, x) = g_{\Delta_1}(\lambda, x) g_{\Delta_2}(\lambda, c) + h_{\Delta_1}(\lambda, x) g_{\Delta_2}(\lambda, d),$$

$$h_{\Delta_2}(\lambda, x) = g_{\Delta_1}(\lambda, x) h_{\Delta_2}(\lambda, c) + h_{\Delta_1}(\lambda, x) h_{\Delta_2}(\lambda, d),$$

where $\Delta_2 = (a,b)$, $\Delta_1 = (c,d) \subset \Delta_2$. Under some regularity conditions (Harlamov, 1982, 1983) the functions g_Δ and h_Δ satisfy the differential equation

$$f'' + 4A(x)f' + 2B(\lambda, x)f = 0, \tag{2}$$

where $A(x)$, $B(\lambda, x)$ are some functions such that $(\forall x \in X)\ B(\cdot, x) \leqslant 0$, $B(\lambda, x)$ is completely monotone in λ and $B(\lambda, x) \to -\infty (\lambda \to \infty)$. Moreover, $g_\Delta(\lambda, a) =$ $= h_\Delta(\lambda, b) = 1$, $g_\Delta(\lambda, b) = h_\Delta(\lambda, a) = 0$ $(\Delta = (a,b))$.

2. CRITERIA OF THE MARKOV PROPERTY FOR SEMI–MARKOV PROCESSES

It is reasonable to expect that semi–Markov transition functions and their Laplace transforms g_Δ and h_Δ in the case of a Markov process have some additional properties.

THEOREM 1 (the criterion of Markovness). LET $P(x)$ BE THE FAMILY OF MEASURES OF A CONTINUOUS SM PROCESS ON R. *Let the functions g_Δ and h_Δ satisfy the differential equation (2) with the corresponding boundary conditions. Let the coefficients $A(x)$ and $B(\lambda, x)$ be continuous in x and $(\forall x \in \mathbf{R})$*

$$-\frac{\partial}{\partial\lambda} B(\lambda, x) \mid_{\lambda=0} \equiv \gamma(x) < \infty,$$

$$\frac{\partial^2}{\partial\lambda^2} B(\lambda, x) \mid_{\lambda=0} \equiv \delta(x) < \infty,$$

where γ and δ are continuous.
Under these conditions, the process is Markov iff

$$(\forall\lambda \geqslant 0)\ (\forall x \in \mathbf{R}) \qquad B(\lambda, x) = -\lambda\gamma(x)$$

(hence $\delta(x) \equiv 0$).

To prove this Theorem the criterion of Markovness in terms of the λ–characteristical operator A_λ of the process is applied, where

$$A_\lambda\varphi(x) = \lim_{r\to 0}(f_{\sigma_r}(\lambda, \varphi \mid x) - \varphi(x))/m_r(x). \tag{3}$$

$\varphi(x)$ is a function from the domain of the operator A_λ,

$$f_{\sigma_r}(\lambda, \varphi \mid x) = \int_X \varphi(x_1) f_{\sigma_r}(\lambda, dx_1 \mid x),$$

$$m_r(x) = P_x(\sigma_r) \equiv \int_0^\infty t\ F_{\sigma_r}(dt, X \mid x),$$

$$\sigma_r(\xi) = \inf\{t \geqslant 0\ :\ |\xi(t) - \xi(0)| \geqslant r\}$$

(Harlamov, 1980).

THEOREM 2. *Under the assumptions of Theorem* 1,

1) $m_r(x)/r^2 \to \gamma(x) \qquad (r \to 0)$.

2) *If there exist the derivatives $\varphi'(x)$ and $\varphi''(x)$ of φ at x, then the limit* (3) *exists and*

$$A_\lambda\varphi(x) = \frac{1}{2\gamma(x)}(\varphi''(x) + 4A(x)\varphi'(x) + 2B(\lambda, x)\varphi(x)).$$

Proof. 1) The solutions g_Δ and h_Δ of the equation (2) are of the form (Harlamov, 1983):

$$g_\Delta(\lambda, x) = k_b(x)/k_b(a), \;\; h_\Delta(\lambda, x) = k_a(x)/k_a(b),$$

where $\Delta = (a, b), \; a \leqslant x \leqslant b$ and

$$k_a = \sum_{k=0}^{\infty} \int_a \left(\int_a A_1 + \int_a B_1 \int_a \right)^k 1,$$

$$k_b = -\sum_{k=0}^{\infty} \int^b \left(-\int^b A_1 + \int^b B_1 \int^b \right)^k 1,$$

where $\mathbf{1}(x) = 1 \quad (\forall x), \; \left(\int_a F \right)(x) = \int_a^x F(s)ds \quad (x \geqslant a),$

$$\left(\int^b F \right)(x) = \int_x^b F(s)ds \; (x \leqslant b), \; A_1 = -4A, \; B_1 = -2B(\lambda, \cdot).$$

We have

$$\left| \frac{\partial}{\partial\lambda} \int_a \left(\int_a A_1 + \int_a B_1 \int_a \right)^k \mathbf{1}(x) \right| \leqslant$$

$$\leqslant \gamma_\lambda \sum_{l=1}^{k} \alpha^{k-l} \beta_\lambda^{l-1} C_k^l l \frac{(x-a)^{k+l+1}}{(k+l+1)!} \leqslant$$

$$\leqslant \gamma_\lambda (\alpha + \beta_\lambda (x-a))^{k-1} (x-a)^{k+2}/(k+2)!,$$

where $\alpha = ||A_1||_a^x$, $\beta_\lambda = ||B_1||_a^x$, $\gamma_\lambda = ||\frac{\partial}{\partial\lambda} B_1||_a^x$ and $||F||_c^d = \sup\{|F(x)| : c \leqslant x \leqslant d\}$.
Hence one can differentiate this series termwise and obtain

$$\frac{\partial}{\partial\lambda} k_a(x) = \int_a^x \int_a^{x_1} \frac{\partial}{\partial\lambda} B_1(x_2)(x_2 - a) dx_2 dx_1 + o(x-a)^3.$$

Hence, if $\frac{\partial}{\partial\lambda} B_1(x)$ is continuous in x, then

$$\frac{\partial}{\partial\lambda} k_a(x) = \frac{\partial}{\partial\lambda} B_1(x)(x-a)^3/6 + o(x-a)^3.$$

Similarly one can derive

$$\frac{\partial}{\partial\lambda} k_b(x) = -\int_x^b \int_{x_1}^b \frac{\partial}{\partial\lambda} B_1(x_2)(b - x_2) dx_2 dx_1 + o(b-x)^3 =$$

$$= -\frac{\partial}{\partial\lambda}B_1(x)(b-x)^3/6 + o(b-x)^3.$$

For $a = x - r$, $b = x + r$ we have $k_a(x) = r + o(r)$, $k_b(x) = -r + o(r)$ and

$$\frac{\partial}{\partial\lambda}f_{\sigma_r}(\lambda, \mathbf{1} \mid x)|_{\lambda=0} = \frac{\partial}{\partial\lambda}\left(k_b(x)/k_b(a) + k_a(x)/k_a(b)\right)|_{\lambda=0} =$$

$$= -\left(\frac{\partial}{\partial\lambda}B_1(x)r^2 + o(r^2)\right)/(2 + o(1))|_{\lambda=0}$$

and $m_r(x) = -\frac{\partial}{\partial\lambda}f_{\sigma_r}(\lambda, \mathbf{1} \mid x)|_{\lambda=0} = \gamma(x)r^2 + o(r^2)$. Hence $m_r(x)/r^2 \to \gamma(x)$ uniformly in each finite interval.

2) Let φ be twice differentiable at the point x. Then using the properties of solutions g_Δ, h_Δ of the equation (2) (Harlamov, 1983) we obtain

$$(h_{\Delta_r}(\lambda, x)\varphi(x+r) + g_{\Delta_r}(\lambda, x)\varphi(x-r) - \varphi(x))/m_r(x) \to$$

$$\to (\varphi''(x)/2 + 2A(x)\varphi'(x) + B((\lambda, x)\varphi(x))/\gamma(x) = A_\lambda\varphi(x),$$

where $\Delta_r = (x - r,\ x + r)$. Theorem 2 is proved.

To prove Theorem 1, one should use the criterion of Markovness of a continuous SM process given in (Harlamov, 1980). One can easily check the assumptions of Theorems 5 and 6 from that paper. In particular, the boundedness of the ratio $P_x(\sigma_r^2)/P_x(\sigma_r)$ follows from the integral representation of g_Δ and h_Δ. This implies

$$P_x(\sigma_r^2) = \frac{\partial^2}{\partial\lambda^2}(g_{\Delta_r}(\lambda, x) + h_{\Delta_r}(\lambda, x))|_{\lambda=0} = \delta(x)r^2 + o(r^2).$$

The above mentioned criterion of Markovness consists in the condition $A_\lambda\mathbf{1}(x) = -\lambda$. Consequently, $B(\lambda, x) = -\lambda\gamma(x)$.

The integral representation of g_Δ and h_Δ gives another form of the criterion of Markovness. Consider the normalized random value $\sigma_r^* = \sigma_r/m$. The Laplace transform of its distribution function is $f^*_{\Delta_r}(\lambda, x) = g^*_{\Delta_r}(\lambda, x) + h^*_{\Delta_r}(\lambda, x)$, where $g^*_{\Delta_r}(\lambda, x) = g_{\Delta_r}(\lambda/m_r, x)$ and $h^*_{\Delta_r}(\lambda, x) = h_{\Delta_r}(\lambda/m_r, x)$.

THEOREM 3 (another form of the criterion). *Under the assumptions of Theorem 1, the SM process is Markov iff*

$$(\forall\lambda \geqslant 0)(\forall x \in \mathbf{R}) \quad f^*_{\Delta_r}(\lambda, x) \to (\mathrm{ch}\sqrt{2\lambda})^{-1} \quad (r \to 0).$$

Remark. The function $(\mathrm{ch}\sqrt{2\lambda})^{-1}$ is the Laplace transform of the distribution of the first exit time of the standard Wiener process from the interval (–1, 1) (Bharucha-Reid, 1960). The distribution function itself is given by

$$G(t) = 1 - 2/\pi \cdot \sum_{k=0}^{\infty}(-1)^k(k+1/2)^{-1}\exp(-(k+1/2)^2\pi^2 t/2).$$

Proof. For $a = x - r,\ b = x + r$ we have

$$k_a(\lambda/m_r, x) = \sum_{k=0}^{\infty} \int_a (\alpha + \beta)^k \mathbf{1}(x),$$

where $\alpha = \int_a A_1,\ \beta = \int_a B_1^*,\ \int_a B_1^* = B_1(\lambda/m_r, \cdot) = -2B(\lambda/m_r, \cdot)$.

Moreover,

$$(\alpha + \beta)^k \mathbf{1}(x) = P_k(\alpha,\ \beta)\mathbf{1}(x) + \beta^k \mathbf{1}(x),$$

where P_k is a homogeneous operator polynomial of degree $k-1$ in β. Let $B(\lambda, x) = -\lambda\gamma(x)$. We have

$$\left(\int_a \beta\right) \mathbf{1}(x) = \int_a^x \int_a^{x_1} B_1^*(x_2) \int_a^{x_2} dx_3 dx_2 dx_1 =$$

$$= \int_a^x \int_a^{x_1} B_1(\lambda/m_r, x_2)(x_2 - a) dx_2 dx_1 \sim B_1(\lambda/m_r, x) \frac{(x-a)^3}{3!} =$$

$$= B_1(\lambda/m_r, x) r^3/3! = -2B(\lambda/m_r, x) r^3/6 = \frac{2\lambda}{m_r} \gamma(x) r^3/6 \sim \lambda r/3.$$

Similarly,

$$\left(\int_a \beta\right) \mathbf{1}(x) \sim B_1^k(\lambda/m_r, x)(x-a)^{2k+1}/(2k+1)! =$$

$$= (-2B(\lambda/m_r, x))^k r^{2k+1}/(2k+1)! =$$

$$= (2\lambda/m_r \gamma(x))^k r^{2k+1}/(2k+1)! \sim (2\lambda)^k r/(2k+1)!.$$

On the other hand, each term of the operator polynomial is of the order r^s as $r \to 0$, where $s \geqslant 2$. For example,

$$\left\|\left(\int_a \alpha\beta^{k-1}\right) \mathbf{1}(x)\right\| \sim |A_1(x)| (B_1^*(x))^{k-1} (x-a)^{2k}/(2k)! =$$

$$= |A_1(x)| (-2B(\lambda/m_r, x))^{k-1} r^{2k}/(2k)! \leqslant$$

$$\leqslant |A_1(x)| (2\lambda)^{k-1} r^2/(2k)!.$$

Hence

$$k_a(\lambda/m_r, x)/r \to \sum_{k=0}^{\infty} (2\lambda)^k/(2k+1)! = \operatorname{sh}\sqrt{2\lambda}/\sqrt{2\lambda}.$$

Similarly we obtain

$$\left(\int_a \beta^k\right) \mathbf{1}(b) \sim B_1^k(\lambda/m_r, x)(b-a)^{2k+1}/(2k+1)! =$$

$$= (-2B(\lambda/m_r, x))^k (2r)^{2k+1}/(2k+1)! =$$
$$= (2\lambda/m_r)^k (\gamma(x))^k (2r)^{2k+1}/(2k+1)! \sim (2\lambda)^k 2^{2k+1} r/(2k+1)!.$$

Consequently as $r \to 0$

$$k_a(\lambda/m_r, b)/r \to \sum_{k=0}^{\infty} (2\lambda)^k 2^{2k+1}/(2k+1)! = \mathrm{sh} 2\sqrt{2\lambda}/\sqrt{2\lambda}$$

and $h^*_{\Delta_r}(\lambda, x) \to \mathrm{sh}\sqrt{2\lambda}/\mathrm{sh}2\sqrt{2\lambda} = (2\mathrm{ch}\sqrt{2\lambda})^{-1}$.
Similarly one can obtain $g^*_{\Delta_r}(\lambda, x) \to (2\mathrm{ch}\sqrt{2\lambda})^{-1}$.
Let $B(\lambda, x) \not\equiv -\lambda\gamma(x)$. The existence of the limit

$$\gamma_1(\lambda, x) = \lim_{r \to 0} (-r^2 B(\lambda/m_r, x))$$

(where $0 \leqslant \gamma_1(\lambda, x) < \lambda$) follows from convexity of the function $-B(\lambda, x)$ in λ. Hence there exist the limits

$$\lim_{r \to 0} h^*_{\Delta_r}(\lambda, x) = \lim_{r \to 0} g^*_{\Delta_r}(\lambda, x) = (2\mathrm{ch}\sqrt{2\gamma_1(\lambda, x)})^{-1}.$$

Now, to obtain the statement about Markovness, we have to use Theorem 1. Theorem 3 is proved.

3. ON STATISTICS OF CONTINUOUS MARKOV PROCESSES

Testing the semi–Markov property of a continuous stationary process depends of the possibility to verify independence of random variables in a stationary sequence. For the given set of levels $X' = \{x_1, \ldots, x_m\} \subset \mathbf{R}$ we consider the point process of the first exit times, as follows. Given a continuous function $\xi = \xi(t)$, $t \geqslant 0$, denote

$$\tau_1(\xi) = \inf\{t \geqslant 0,\ \xi(t) \in X'\},$$
$$\tau_2(\xi) = \inf\{t \geqslant \tau_1(\xi),\ \xi(t) \in X' \backslash \{\xi(\tau_1)\}\}, \ldots,$$
$$\tau_{k+1}(\xi) = \inf\{t \geqslant \tau_k(\xi),\ \xi(t) \in X' \backslash \{\xi(\tau_k)\}\},\ (k \geqslant 1).$$

The sequence $(\tau_k(\xi),\ \xi(\tau_k))_{k=1}^{\infty}$ is called the imbedded point process of the first exits with respect to the set X'. For a continuous function ξ on $\mathbf{R}$ and a point $T \in \mathbf{R}$, consider the first exit time after T:

$$\tau_1^T(\xi) = \inf\{t \geqslant T,\ \xi(t) \in X'\},$$

$$\tau_{k+1}^T(\xi) = \inf\{t \geqslant \tau_k^T(\xi),\ \xi(t) \in X' \backslash \{\xi(\tau_k^T)\}\},\ (k \geqslant 1).$$

It is easy to prove that $T_1 < T_2$ implies $(\tau_k^{T_1}(\xi))_2^{\infty} \supset (\tau_k^{T_2}(\xi))_2^{\infty}$. Therefore one can define the imbedded first exit time point process on $\mathbf{R}$ as

$$\lim_{T \to -\infty} (\tau_k^T(\xi),\ \xi(\tau_k^T))_{k=2}^{\infty}.$$

The semi–Markov property of this point process is equivalent to the property: $(\forall t > 0)$ $(\forall A \in B(X))$ $(\forall k \in \mathbb{Z})$ $(\forall m \in \mathbb{N})$

$$P(\tau_k - \tau_{k-1} < t,\ \pi_{\tau_k} \in A | (\tau_{k-1}, \pi_{\tau_{k-1}}), \ldots, (\tau_{k-m}, \pi_{\tau_{k-m}}) = \\ = P(\tau_k - \tau_{k-1} < t,\ \pi_{\tau_k} \in A | \pi_{\tau_{k-1}}).$$

It is also equivalent to the Markovness of the imbedded chain $(\pi_{\tau_k})_{-\infty}^{\infty}$ and the conditional independence of the sequence $(\tau_k - \tau_{k-1})_{-\infty}^{\infty}$ with respect to any fixed sequence of states $(\pi_{\tau_k})_{-\infty}^{\infty}$. Markovness of a random chain can be analysed in a standard way. Independence of intervals is investigated in many papers on the renewal theory.

For testing the Markovness of a semi–Markov process, one can consider a histogram of the normalized random variables T_k/m_n, where $T_k = \tau_{2k} - \tau_{2k-1}$, $(\tau_k)_{k=1}^{\infty}$ is a sequence of the first exit times corresponding to the set of levels $X' = \{x - r,\ x,\ x + r\}$, where τ_1 is the first hitting time of the point x after 0, and $m_n = \sum_{k=1}^{n} T_k/n$. Testing the hypothesis that G is the common distribution function of these random variables is a standard statistical procedure.

One can be interested in estimation of the parameters of the semi–Markov process. We have

$$\frac{1}{n}\sum_{k=1}^{n} e^{-\lambda T_k} \to f_{\sigma_r}(\lambda,\ 1|x)\ (n \to \infty)\ (\text{in } P_x\text{–probability})$$

and

$$\hat{B}_{(r,n)}(\lambda, x) = \frac{1}{r^2 n}\sum_{k=1}^{n}(e^{-\lambda T_k} - 1) \to B(\lambda, x)\ (\text{in } P_x\text{–probability}).$$

Moreover, we have

$$\frac{1}{r}(h_{\Delta_r}(\lambda, x) - g_{\Delta_r}(\lambda, x)) \to A(x),$$

where the limit does not depend on λ. Hence

$$\hat{A}_{(n,r)}(\lambda, x) = \frac{1}{r^2 n}\sum_{k=1}^{n}(\xi(\tau_{2k}) - x) \to A(x)\ (n \to \infty,\ r \to 0)\ (\text{in } P_x\text{–probability}).$$

If $a(x)$ and $b(x)$ are usual Kolmogorov's diffusion coefficients ($a(x)$ is the drift and $b(x)$ is the local diffusion) then in the Markov case $A(x) = a(x)/(2b(x))$ and $B(\lambda, x) = -\lambda/b(x)$ (i.e. $b(x) = (\gamma(x))^{-1}$).

REFERENCES

Bharucha–Reid, A.T. (1960). *Elements of the theory of Markov processes and their applications.* McGraw–Hill, New–York.

Billingsley, P. (1961). Statistical methods in markov chains. *Ann. Math. Stat.* **32**, 12–40.

Harlamov, B.P. (1980). A criterion of the Markov property for continuous semi–Markov processes. *Theory Probab. Apll.* **25**, 535–548.

Harlamov, B.P. (1982). Outleading sequences and continuous semi–Markov processes on the line. *Notes Sci. Seminars Leningrad Branch of Math. Inst.* **119**, 230–236.

Harlamov, B.P. (1983). Transition functions of a continuous semi–Markov process on the line. *Notes Sci. Seminars Leningrad Branch of Math. Inst.* **130**, 190–205.

Prob. Theory and Math. Stat., Vol. 1, pp. 512–525
B.Grigelionis *et al.* (Eds.)
1990 VSP/Mokslas

QUANTUM STOCHASTIC FLOWS

R.L. HUDSON
Mathematics Department, University of Nottingham,
Nottingham NG7 2RD, United Kingdom

ABSTRACT

The notion of stochastic flow on a manifold can be generalised by making both the manifold and the noise noncommutative, the former by replacing the algebra of smooth functions by a noncommutative $*$–algebra, and the latter by replacing the Ito calculus of Brownian motion by a quantum stochastic calculus. The resulting quantum stochastic flows can be analysed in terms of either Hochschild cohomology or more cogently in terms of a nonabelian or nonlinear cohomology theory. The special cases of deterministic and classical stochastic flows emerge naturally as degeneracies of the latter cohomology theory.

1. STOCHASTICISATION AND QUANTISATION OF CLASSICAL FLOWS

Let V be a differentiable manifold. A *homogeneous flow* on V is a smooth map $\varphi : V \times \mathbf{R}_+ \to V$ with the properties
(a) for each $t \in \mathbf{R}_+$, the map $\varphi_t : V \to V$ given by

$$\varphi_t(m) = \varphi(m,t)$$

is a diffeomorphism of V;
(b) for arbitrary $s, t \in \mathbf{R}_+, m \in V$

$$\varphi(m,t+s) = \varphi(\varphi(m,t),s); \tag{1.1}$$

(c) for arbitrary $m \in V$

$$\varphi(m,0) = m.$$

Since nonhomogeneous flows will not be considered in this article, we speak simply of a flow φ.
Given a flow φ on V, we may construct a one parameter semigroup $j =$ $= (j_t : t \in \mathbf{R}_+)$ of automorphisms of the $*$–algebra $K^\infty(V)$ of complex–valued smooth functions of compact support on V, by defining

$$j_t(f) = f \circ \varphi_t^{-1}.$$

Function j is smooth, in the sense that for each $f \in K^\infty(V)$ and $p \in V$, $t \mapsto j_t(f)(p)$ is a smooth curve. Every smooth one-parameter semigroup j of automorphisms of $K^\infty(V)$ arises in this way from a flow φ; φ_t can be constructed by considering the action of j_t on local coordinate functions.
The *infinitesimal generator* of the flow φ is the vector field defined by

$$\tau(f) = \frac{d}{dt} j_t(f)|_{t=0} = \frac{d}{dt}(f \circ \varphi_t^{-1})|_{t=0}.$$

That τ is indeed a vector field, that is, a derivation of $K^\infty(V)$ follows from Leibniz's rule for differentiating a product; note that it expresses the multiplicativity of the automorphisms j_t. From (1.1) or from the semigroup property of j we have that, for all $t \in \mathbf{R}_+$ and $f \in K^\infty(V)$,

$$\frac{d}{dt} j_t(f) = j_t(\tau(f)), \quad j_0(f) = f. \tag{1.2}$$

We now consider two ways of generalising the notion of flow, starting from (1.2). The notion of *classical stochastic flow* can be obtained by including 'noise' or random terms in (1.2). Thus we replace (1.2) by a system of (Ito) stochastic differential equations

$$dj_t(f) = \sum_{k=1}^{n} j_t(\beta_k(f))\, dB_k + j_t(\tau(f))\, dt, \quad j_0(f) = f. \tag{1.3}$$

Here $B_1, \ldots, B_n$ are the components of an n-dimensional Brownian motion B. Each j_t is now an injective $*$-algebra homomorphism from $\mathcal{A} = K^\infty(V)$ to the C^*-algebra $\widetilde{\mathcal{A}}$ of bounded measurable functions on the measurable space $V \times \Omega$, where Ω is the sample space of the Brownian motion, in which $\mathcal{A}$ is embedded naturally by identifying $f \in \mathcal{A}$ with the element $(p, \omega) \mapsto f(p)$ of $\widetilde{\mathcal{A}}$. The maps β^k, $k = 1, \ldots, n$, and τ from $\mathcal{A}$ to itself are linear, corresponding to the linearity of the j_t. Corresponding to the multiplicativity of the j_t, we have, for arbitrary $f, g \in \mathcal{A}$, using the Ito product formula

$$\sum_{k=1}^{n} j_t(\beta_k(fg))\, dB_k + j_t(\tau(fg))\, dt = dj_t(fg) =$$

$$= d(j_t(f) j_t(g)) = dj_t(f) j_t(g) + j_t(f) dj_t(g) + dj_t(f) dj_t(g) =$$

$$= \left(\sum_{k=1}^{n} j_t(\beta_k(f))\, dB_k + j_t(\tau(f))\, dt \right) j_t(g) +$$

$$+ j_t(f) \left(\sum_{k=1}^{n} j_t(\beta_k(g))\, dB_k + j_t(\tau(g))\, dt \right) + \sum_{k=1}^{n} j_t(\beta_k(f)) j_t(\beta_k(g))\, dt =$$

$$= \sum_{k=1}^{n} j_t(\beta_k(f) g + f \beta_k(g))\, dB_k + j_t \left(\tau(f) g + f \tau(g) + \sum_{k=1}^{n} \beta_k(f) \beta_k(g) \right) dt.$$

Using the independence of the stochastic differentials $dB_1, \ldots, dB_n$, dt and the injectivity of the j_t we deduce that the β_k must be vector fields, or derivations,

$$\beta_k(fg) = \beta_k(f)g + f\beta_k(g), \quad k = 1, \ldots, n, \tag{1.4}$$

but that τ is no longer a vector field but instead satisfies

$$\tau(fg) = \tau(f)g + f\tau(g) + \sum_{k=1}^{n} \beta_k(f)\beta_k(g). \tag{1.5}$$

Note that, for given vector fields $\beta_1, \ldots, \beta_n$, a map τ satisfying (1.5) can always be found, namely

$$\tau_0(f) = -\frac{1}{2}\sum_{k=1}^{n} \beta_k(\beta_k(f)), \tag{1.6}$$

and that the most general such map τ is of the form $\tau = \tau_0 + \beta_0$, where β_0 is a further vector field.

In the generalisation from deterministic to stochastic flow the semigroup property

$$j_{s+t} = j_s \circ j_t \qquad (s, t \in \mathbf{R}_+)$$

is lost; indeed j_t no longer maps $\mathcal{A}$ to itself and the product $j_s \circ j_t$ is no longer meaningful.

Classical stochastic flows may be considered alternatively as stochastic processes taking values in the group Diff(V) of diffeomorphisms of the manifold V, just as deterministic flows arise from deterministic Diff(V)–valued processes. Since this approach does not lend itself so readily to quantum generalisation we shall not pursue it but refer the reader to the literature (Baxendale, 1986; Bismut, 1981; Elworthy, 1982).

We turn to the second generalisation of (1.2), the notion of *quantum flow.* The quantum generalisation arises by replacing the $*$–algebra $K^\infty(V)$ by a noncommutative $*$–algebra $\mathcal{A}$ which is a dense subalgebra of a C^*–algebra $\mathcal{B}$ (which is the noncommutative or quantum generalisation of the algebra $K(V)$ of continuous functions of compact support on V). Given such $\mathcal{A} \subseteq \mathcal{B}$, a (deterministic) *quantum flow* over $(\mathcal{A}, \mathcal{B})$ is a one–parameter semigroup $j = (j_t : t \geqslant 0)$ of automorphisms of $\mathcal{B}$ each leaving $\mathcal{A}$ invariant and such that, for arbitrary $a \in \mathcal{A}$

$$\frac{d}{dt} j_t(a) = j_t(\tau(a)), \quad j_0(a) = a,$$

where τ maps $\mathcal{A}$ to $\mathcal{A}$ and the derivative is formed using the C^*–topology of $\mathcal{B}$. (More sophisticated notions of derivative, for example using a noncommutative C^∞–structure on $\mathcal{A}$, are possible.) As in the classical case we find that τ must be a derivation of the algebra $\mathcal{A}$. We call $\mathcal{A}$ the *smooth system algebra* and $\mathcal{B}$ the *complete system algebra.*

In the case when $\mathcal{A} = \mathcal{B} = B(\mathcal{H}_0)$, the C^*–algebra of bounded operators on a Hilbert space, the automorphisms j_t are necessarily of the form

$$j_t(x) = u_t^{-1} x u_t, \tag{1.7}$$

where the u_t are unitary elements of $B(\mathcal{H}_0)$. For a given flow j, the u_t are each non-unique to within a complex number of unit modulus, but can be chosen to form a uniformly continuous one parameter semigroup of unitaries. Writing this in the form

$$u_t = e^{th} \tag{1.8}$$

where h is a bounded skew self-adjoint operator we find that

$$\tau(x) = -[h, x] \tag{1.9}$$

and we recover a primitive form of the Heisenberg dynamics of quantum theory, in which the Hamiltonian h is unphysically restricted to be bounded. (More realistic dynamics, in which h is a unbounded operator, require us to distinguish between $\mathcal{A}$ and $\mathcal{B} = B(\mathcal{H}_0)$ and use a more subtle topology on $\mathcal{A}$.) Notice that u_t is itself the unique solution of

$$\frac{du_t}{dt} = hu_t, \qquad u_0 = 1.$$

Notice also that h is non-unique to the extent of adding a pure imaginary multiple $i\lambda 1$ and the one-parameter semigroup $(u_t = t \in \mathbf{R})$ is correspondingly non-unique to the extent of the replacement $u_t \mapsto e^{it\lambda}u_t$.
In more general cases one may distinguish among the quantum flows over $(\mathcal{A}, \mathcal{B})$ those which are *inner*, that is given by a conjugation (1.7) by a unitary semigroup belonging to $\mathcal{B}$, and for which the infinitesimal generator is an *inner derivation*, given by (1.9) where h is a self-adjoint element of $\mathcal{B}$. That not all flows are inner in this sense is evident from the case when $\mathcal{B}$ is commutative. The implementing unitary semigroup of an inner flow is non-unique to the extent of multiplying by a semigroup of unitary elements of the centre of $\mathcal{B}$, and the Hamiltonian h correspondingly non-unique to the extent of adding a self-adjoint central element.

2. QUANTUM STOCHASTIC FLOWS

We now combine the two generalisations of stochasticisation and quantisation. It is natural in doing this to quantise, that is allow to become noncommutative, not only the system algebra but also the noise. Thus we replace the Ito calculus of n-dimensional Brownian motion with a quantum stochastic calculus. While other choices are possible and interesting, we shall use the n-dimensional Fock quantum stochastic calculus (Hudson and Parthasarathy, 1984).
Let $(e_j, j = 1, \ldots, n)$ and $(E^j_k, j, k = 1, \ldots, n)$ denote the canonical bases of $\mathbf{C}^n$ and its space of linear transformations (so that $E^j_k e_l = \delta^j_l e_k$) and denote by $\mathfrak{h}$ the Hilbert space of $\mathbf{C}^n$-valued functions

$$\mathfrak{h} = L^2(\mathbf{R}_+; \mathbf{C}^n) = L^2(\mathbf{R}_+) \otimes \mathbf{C}^n.$$

The *Boson Fock space* $\mathcal{H} = \Gamma(\mathfrak{h})$ over $\mathfrak{h}$ may be conveniently characterised as a Hilbert space $\mathcal{H}$ equipped with a family $(\psi(f) : f \in \mathfrak{h})$ of so-called *exponential vectors*, satisfying

$$\langle\psi(f),\,\psi(g)\rangle_{\mathcal{H}} = \exp\langle f,\,g\rangle_{\mathfrak{h}}$$

whose linear span $\mathcal{E} = \mathcal{E}(\mathfrak{h})$ is dense in $\mathcal{H}$. Corresponding to each direct sum decomposition

$$\mathfrak{h} = \mathfrak{h}_t \oplus \mathfrak{h}^t, \quad \mathfrak{h}_t = L^2[0,t], \quad \mathfrak{h}^t = L^2[t,\infty), \tag{2.1}$$

there is a Hilbert space tensor product decomposition

$$\mathcal{H} = \mathcal{H}_t \otimes \mathcal{H}^t, \quad \mathcal{H}_t = \Gamma(\mathfrak{h}_t), \quad \mathcal{H}^t = \Gamma(\mathfrak{h}^t) \tag{2.2}$$

in which the exponential vectors are product vectors, so that we may also write

$$\mathcal{E} = \mathcal{E}_t \underline{\otimes} \mathcal{E}^t \tag{2.3}$$

where $\underline{\otimes}$ denotes the algebraic tensor product. We suppose given an "*initial*" Hilbert space $\mathcal{H}_0$ and define

$$\widetilde{\mathcal{H}} = \mathcal{H}_0 \otimes \mathcal{H}, \quad \widetilde{\mathcal{H}}_t = \mathcal{H}_0 \otimes \mathcal{H}_t, \quad \widetilde{\mathcal{E}} = \mathcal{H}_0 \underline{\otimes} \mathcal{E}, \quad \widetilde{\mathcal{E}}_t = \mathcal{H}_0 \underline{\otimes} \mathcal{E}_t$$

so that, corresponding to (2.1), (2.2) and (2.3) give

$$\widetilde{\mathcal{H}} = \widetilde{\mathcal{H}}_t \otimes \mathcal{H}^t, \quad \widetilde{\mathcal{E}} = \widetilde{\mathcal{E}}_t \otimes \mathcal{E}^t.$$

The *integrator processes* of our stochastic calculus are the $(n+1)^2$ operator–valued functions $\Lambda^\mu_\nu(t),\ t \geqslant 0,\ \mu, \nu = 0, 1, \ldots, n$, whose actions on $\widetilde{\mathcal{E}}$ are determined by

$$\Lambda^0_0(t) = t1 \quad (1 \text{ is the identity operator}),$$

$$\Lambda^j_0(t) u \otimes \psi(f) = \int_0^t \langle e_j,\, f(s)\rangle\, ds\, u \otimes \psi(f), \ u \in \mathcal{H}_0,\ f \in \mathfrak{h},\ j = 1, \ldots, n,$$

$$\Lambda^0_k(t) u \otimes \psi(f) = \frac{d}{d\varepsilon} u \otimes \psi(f + \varepsilon\chi_{[0,t]} e_j)|_{\varepsilon=0},\ k = 1, \ldots, n,$$

$$\Lambda^j_k(t) u \otimes \psi(f) = \frac{d}{d\varepsilon} u \otimes \psi(\exp \varepsilon M_{\chi_{[0,t]}} \otimes E^j_k f)|_{\varepsilon=0},\ j, k = 1, \ldots, n.$$

Here $\chi_{[0,t]}$ is the function which is 1 on $[0,t]$ and 0 on its complement, and M_φ is the operator on $\mathfrak{h}$ of multiplication by the function φ. The notations

$$t = \Lambda^0_0(t), \quad A^j = \Lambda^j_0, \quad A^\dagger_j = \Lambda^0_j, \quad j = 1, \ldots, n, \tag{2.4}$$

are also useful, and when $n = 1$ we dispense with indices altogether and denote the four integrators by $t, A, A^\dagger$ and Λ.
The integrands are also operator–valued processes $E = (E(t),\ t \geqslant 0)$, *adapted* in the sense that each $E(t)$ is an operator on the domain $\widetilde{\mathcal{E}} = \widetilde{\mathcal{E}}_t \underline{\otimes} \mathcal{E}^t$ in $\widetilde{\mathcal{H}} = \widetilde{\mathcal{H}}_t \otimes \mathcal{H}^t$ of the form $E(t) = E_t \underline{\otimes} 1^t$. Given an $(n+1)^2$–tuple $E^\mu_\nu,\ \mu, \nu = 0, \ldots, n.$ of *elementary* adapted processes, of form

$$E_\nu^\mu(t) = \begin{cases} 0, & t \notin [t_1, t_2), \\ E_\nu^\mu(t_1), & t \in [t_1, t_2), \end{cases}$$

their ***stochastic integral***

$$M(t) = \int_0^t \sum_{\mu,\nu=0}^n E_\nu^\mu \, d\Lambda_\mu^\nu$$

is defined to be the adapted process

$$M(t) = \sum_{\mu,\nu=0}^n E_\nu^\mu(t_1)(\Lambda_\mu^\nu(\min\{t, t_1\}) - \Lambda_\mu^\nu(\min\{t, t_2\})).$$

The products of operators here make sense even though $\widetilde{\mathcal{E}}$ is not invariant under all the $\Lambda_\mu^\nu(t)$, being in fact algebraic tensor products of the form $E_{t_1} \underline{\otimes} \Lambda^{t_1}$, by adaptedness of integrands and the fact that increments of the integrators "live in the future". Stochastic integration is extended by linearity and continuity to adapted processes satisfying certain measurability and local square-integrability conditions as in the classical Ito theory. In particular adapted process E which are ***continuous***, in the sense that for each $u \in \mathcal{H}_0$ and $f \in \mathfrak{h}$, $t \mapsto E(t)u \otimes \psi(f)$ is continuous from $\mathbf{R}_+$ to $\mathcal{H}$, are integrable. Each stochastic integral is a continuous process, so repeated integration is allowed. If $T^\dagger$ denotes the restriction to the domain $\mathcal{E}$ of the adjoint of an operator T we have

$$\left(\int_0^t \sum_{\mu,\nu=0}^n E_\nu^\mu \, d\Lambda_\mu^\nu \right)^\dagger = \int_0^t \sum_{\nu,\mu=0}^n (E_\mu^\nu)^\dagger \, d\Lambda_\mu^\nu \tag{2.5}$$

whenever the operators occurring on the right hand side exists. We write

$$dM = \sum_{\mu,\nu=0}^n E_\nu^\mu \, d\Lambda_\mu^\nu$$

to mean that the adapted process M differs by a fixed operator of form $M_0 \otimes 1$ in $\widetilde{\mathcal{H}} = \mathcal{H}_0 \otimes \mathcal{H}$ from $\int_0^t \sum_{\mu,\nu=0}^n E_\nu^\mu \, d\Lambda_\mu^\nu$. Then the *quantum Ito product formula* is

$$d(M_1 M_2) = M_1 \, dM_2 + dM_1 \, M_2 + dM_1 \, dM_2 \tag{2.7}$$

where the right hand side is evaluated by bilinear extension of the role that the differentials $d\Lambda_\nu^\mu$ commute with adapted processes, together with the multiplication table

$$d\Lambda_\nu^\mu \, d\Lambda_\tau^\sigma = (\delta_\tau^\mu - \delta_0^\mu \delta_\tau^0) \, d\Lambda_\nu^\sigma. \tag{2.8}$$

It is valid in particular when the processes M_1 and M_2 together with the coefficient processes of their stochastic differentials are all continuous bounded-operator-valued processes locally bounded in the operator bound norm.

Now let there be given smooth and complete system algebras $\mathcal{A} \subseteq \mathcal{B}$, the latter being a unital C^*-subalgebra of the algebra $B(\mathcal{H}_0)$ of all bouded operators on the initial Hilbert space $\mathcal{H}_0$, and the former a unital norm-dense $*$-subalgebra of the latter. An n-dimensional *quantum stochastic flow* over $(\mathcal{A}, \mathcal{B})$ is a family $j = (j_t : t \in \mathbf{R}_+)$ of injective $*$-homomorphisms from $\mathcal{B}$ into $B(\mathcal{H})$ such that each $(j_t(x) : t \geqslant 0)$, $x \in \mathcal{B}$, is an adapted process, and there are maps λ^μ_ν, $\mu, \nu =$ $= 0, 1, \ldots, n$, from $\mathcal{A}$ into $\mathcal{B}$ such that, for each $x \in \mathcal{A}$, $t \geqslant 0$

$$j_t(x) = x \otimes 1 + \int_0^t j_t(\lambda^\mu_\nu(x))\, d\Lambda^\nu_\mu. \tag{2.9}$$

The λ^μ_ν are called the *structure maps* of the flow j. The definition given here is a refinement due to D.B. Applebaum of an earlier definition in which the roles of $\mathcal{A}$ and $\mathcal{B}$ were not distinguished (Applebaum, 1988).
From the linearity of the homomorphisms j_t, together with their injectivity and the independence of the stochastic differentials $d\Lambda^\nu_\mu$, we see (by comparing two expressions for $j_t(x + \alpha y)$) that the structure maps are linear:

$$\lambda^\mu_\nu(x + \alpha y) = \lambda^\mu_\nu(x) + \alpha\lambda^\mu_\nu(y) \tag{2.10}$$

for arbitrary $x, y \in \mathcal{A}$, $\alpha \in \mathbf{C}$ and $\mu, \nu = 0, \ldots, n$. Similarly from the unitality and self-adjointness of the j_t we obtain that

$$\lambda^\mu_\nu(1) = 0; \qquad \lambda^\mu_\nu(x^*) = \lambda^\nu_\mu(x)^*. \tag{2.11}$$

Finally we exploit the multiplicativity of the j_t together with the quantum Ito formula (2.7), (2.8) to obtain

$$\lambda^\mu_\nu(xy) = \lambda^\mu_\nu(x)(y) + x\lambda^\mu_\nu(y) + \sum_{k=1}^n \lambda^\mu_k(x)\lambda^k_\nu(y). \tag{2.12}$$

The Ito formula is applicable since each j_t, being a C^*-algebra homomorphisms, is automatically contractive.
We call (2.9), (2.10) and (2.12) the *structure relations*. In (Evans, 1989) it is shown that, in the case $\mathcal{A} = \mathcal{B}$, a quantum flow satisfying (2.9) exists corresponding to every set of bounded maps $\lambda^\mu_\nu : \mathcal{A} \to \mathcal{A}$ satisfying the structure relations (2.10), (2.11) and (2.12).

3. COHOMOLOGICAL ANALYSIS OF THE STRUCTURE RELATIONS

3a. *Hochschild cohomology* (Hudson,1987)
We use the notations (2.4) and denote the corresponding structure maps by $\tau = \lambda^0_0$, $\alpha^i = \lambda^i_0$, $\alpha^\dagger_i = \lambda^0_i$ respectively. Then $\alpha^\dagger_i$ is the adjoint map of α^i:

$$\alpha^\dagger_i(x) = \alpha^i(x^*)^*$$

and τ is self-adjoint. The remaining structure maps satisfy

$$\lambda_k^i(xy) = \lambda_k^i(x)y + x\lambda_k^i(y) + \sum_{m=1}^{n} \lambda_m^i(x)\lambda_k^m(y) \tag{3.1}$$

for arbitrary $i, k = 1, \ldots, n$. Adding $\delta_k^i xy$ to both sides of (3.1) we see that the map σ defined by

$$\sigma(x) = [x\delta_k^i + \lambda_k^i(x)]_{i,k=1}^n \quad (x \in \mathcal{A})$$

is a $*$-homomorphism from $\mathcal{A}$ into the algebra $M_n(\mathcal{B})$ of $(n \times n)$-matrices over $\mathcal{B}$. Substituting $\lambda_k^i(x) = \sigma_k^i(x) - \delta_k^i x$ into the structure relations for the α^i, $\alpha_i^\dagger$ we find that

$$\alpha^i(xy) = \alpha^i(x)y + \sum_{k=1}^{n} \sigma_k^i(x)\alpha^k(y) \tag{3.2}$$

together with its adjoint relation

$$\alpha_i^\dagger(xy) = x\alpha_i^\dagger(y) + \sum_{k=1}^{n} \alpha_k^\dagger(x)\sigma_k^k(y). \tag{3.3}$$

Finally the structure relation for τ is

$$\tau(xy) = \tau(x)y + x\tau(y) + \sum_{k=1}^{n} \alpha_k^\dagger(x)\alpha^k(y). \tag{3.4}$$

We regard the space $\mathcal{B}^n$ of $\mathcal{B}$-valued (column) n-vectors as a two-sided $\mathcal{A}$-module by defining the left and right actions

$$(xm)^i = \sum_{k=1}^{n} \sigma_k^i(x)m^k, \quad (mx)^i = m^i x, \quad i = 1, \ldots, n,$$

for $x \in \mathcal{A}$, $m \in \mathcal{B}^n$. We may then construct the corresponding Hochschild cohomology theory (Hochschild, 1945) in which the space C_σ^m of m-cochains consists of all m-linear maps from $\mathcal{A}^m$ into $\mathcal{B}^n$, the coboundary operator δ_σ maps C_σ^m into C_σ^{m+1} by

$$\delta_\sigma\eta(x_0, x_1, \ldots, x_m) = x_0\eta(x_1, \ldots, x_m) +$$
$$+ \sum_{i=1}^{n}(-1)^i\eta(x_0, \ldots, x_{i-1}x_i, \ldots, x_n) + (-1)^{n+1}\eta(x_0, \ldots, x_{n-1})x_n,$$

the space Z_σ^m of m-cocycles is the kernel of $\delta_\sigma|_{C^m}$ and the space B_σ^m of m-coboundaries is the range of $\delta_\sigma|_{C^{m-1}}$. Since $\delta_\sigma^2 = 0$, $B_\sigma^m \subseteq Z_\sigma^m$ and we can define the mth cohomology group H_σ^m as the quotient Z_σ^m / B_σ^m.

$\mathcal{B}$ is also a module for $\mathcal{A}$ under the actions given by ordinary multiplication; we denote by C^m, Z^m, B^m and H^m the spaces of m-cochains, m-cocycles, m-coboundaries and the cohomology groups of the corresponding theory.

PROPOSITION 3.1 (Evans, 1988). *Let σ, Δ be $*$–homomorphisms from $\mathcal{A}$ into $M_n(\mathcal{B})$ such that there exists a unitary element w of $M_n(\mathcal{B})$ for which, for arbitrary $x \in \mathcal{A}$,*

$$\Delta(x) = w^{-1}\sigma(x)w.$$

Then the linear transformation T_w^m from C_σ^m to C_Δ^m given by $\eta \mapsto \tilde{\eta}$ where

$$\tilde{\eta}(x_1, \ldots, x_m) = w^{-1}\eta(x_1, \ldots, x_m)$$

maps Z_σ^m bijectively to Z_Δ^m and B_σ^m bijectively to B_Δ^m. In particular H_σ^m and H_Δ^m are isomorphic.

Proof. It is straightforward to verify that

$$T_w^m \delta_\sigma|_{C_{m-1}} = \delta_\Delta|_{C_{m-1}} T_w^{m-1}.$$

This shows that T_w^{m-1} maps Z_σ^{m-1} to Z_Δ^{m-1}, and T_w^m maps B_σ^m to B_Δ^m. Since T_w^m is clearly bijective, having inverse $T_{w^{-1}}^m$, the result follows.

PROPOSITION 3.2. *Let $\alpha \in C_\sigma^1$ and define $\eta_\alpha \in C^2$ by $\eta_\alpha(x, y) = \sum_{j=1}^n \alpha_j^\dagger(x)\alpha^j(y)$. Then*

$$\text{a) } \alpha \in Z_\sigma^1 \Rightarrow \eta_\alpha \in Z^2,$$

$$\text{b) } \alpha \in B_\sigma^1 \Rightarrow \eta_\alpha \in B^2.$$

Proof. a) Straightforward verification.
b) If $\alpha \in B_\sigma^1$ then there exists $l \in B_\sigma^m$ such that $\alpha(x) = lx - xl$, that is $\alpha(x)^i = l^i x - \sum_{k=1}^n \sigma_k^i(x)l^k$ for all $x \in \mathcal{B}$. Define $\tau \in C^1$ by

$$\tau(x) = -\frac{1}{2}\sum_{i=1}^n (l^i)^* l^i x + \sum_{i,k=1}^n (l^i)^* \sigma_k^i(x) l^k - \frac{1}{2}\sum_{i=1}^n x(l^i)^* l^i. \tag{3.5}$$

Then it may be verified directly that $\delta(-\tau) = \eta_\alpha$.
The structure relations (3.2) and (3.4) may now be understood as follows. (3.2) is the requirement that the map $\alpha : \mathcal{A} \to \mathcal{B}^n$, $(\alpha(x))^i = \alpha^i(x)$ is an element of Z_σ^1. By Proposition 3.2a, η_α is then an element of Z^2. However (3.4) requires that η_α is in fact an element of $B^2 \subseteq Z^2$, since it may be expressed as $\delta\tau = -\eta_\alpha$. Note that by Proposition 3.2b this is automatic if α is inner, that is there exist $l^1, \ldots, l^n \in \mathcal{B}$ such that $\alpha^i(x) = l^i x - \sum_{k=1}^n \sigma_k^i(x)l^k$, and that τ given by (3.5) satisfies (3.4). The general solution in τ for given α is then the sum of this particular solution and a self-adjoint 1–coboundary or derivation from $\mathcal{A}$ into $\mathcal{B}$.
In (Hudson and Robinson, 1988), a topological obstruction to the existence of a quantum flow corresponding to a non-trivial 1–cocycle α is overcome by enlarging the algebra. This is closely related to the quantum Hall effect (Robinson, 1989).
As an example of the Hochschild analysis, let us consider the case when $\mathcal{A} = \mathcal{B} = B(\mathcal{H}_0)$ using one–dimensional quantum stochastic calculus. In this case a flow satisfies

$$j_t(x) = x \otimes 1 + \int_0^t [j_s(\sigma(x) - x)\, d\Lambda + j_s(\alpha(x))\, dA^+ + j_s(\alpha^+(x))\, dA + j_s(\tau(x))\, dt]$$

where σ is an endomorphism of $B(\mathcal{H}_0)$, α is a σ–derivation of $B(\mathcal{H}_0)$, and $\tau : B(\mathcal{H}_0) \to B(\mathcal{H}_0)$ satisfies

$$\tau(xy) = \tau(x)y + x\tau(y) + \alpha^\dagger(x)\alpha(y).$$

Because the cohomology of $B(\mathcal{H}_0)$ is trivial (Evans and Hudson, 1990), α is necessarily inner, that is, there exists $l \in B(\mathcal{H}_0)$ such that

$$\alpha(x) = lx - \sigma(x)l.$$

Then τ is necessarily of the form

$$\tau(x) = [h, x] - \frac{1}{2}(l^* lx - 2l^*\sigma(x)l + xl^*l)$$

where $h = -h^* \in B(\mathcal{H}_0)$. This is the general form of infinitesimal generator of a uniformly continuous 1–preserving one–parameter semigroup of completely positive maps on $B(\mathcal{H}_0)$. Thus we obtain a "stochastic dilation", in so far as

$$e^{t\tau}(x) = \mathbf{E}_0[j_t(x)]$$

when $\mathbf{E}_0 : B(\widetilde{\mathcal{H}}) \to B(\mathcal{H})$ is the vacuum conditional expectation defined by

$$\langle u, \mathbf{E}_0[T]v\rangle = \langle u \otimes \psi(0), Tv \otimes \psi(0)\rangle, \quad T \in B(\widetilde{\mathcal{H}})$$

of such a "quantum dynamical semigroup" (Hudson and Shepperson, 1989). This may be compared with the stochastic dilation constructed using in general infinite dimensional quantum stochastic calculus of (Hudson and Parthasarathy, 1984b). Provided $\mathcal{H}_0$ is infinite dimensional and separable, the general form of endomorphism σ of $B(\mathcal{H}_0)$ is

$$\sigma(x) = w^*(x \otimes 1)w$$

where w is a unitary transformation from $\mathcal{H}_0$ to the tensor product $\mathcal{H}_0 \otimes \mathcal{K}$ of $\mathcal{H}_0$ with an auxilliary Hilbert space $\mathcal{K}$ (Hudson and Shepperson, 1989). Only when $\mathcal{K}$ is one–dimensional is the diffusion j inner, that is, given by conjugation by a unitary process of the form constructed in (Hudson and Parthasarathy, 1984b).

3b. *Nonlinear (nonabelian) cohomology* (Accardi and Hudson, 1989)
We adapt to a $*$–algebra context the nonabelian cohomology of associative algebras of Lue (Lue, 1968). Let $\mathcal{A}$ be a $*$–algebra and let $\mathcal{C}$ be a second $*$–algebra which is also a 2–sided $\mathcal{A}$–$*$–module, so that there are actions $\mathcal{A} \times \mathcal{C}\ (a, c) \mapsto ac \in \mathcal{C}$, $\mathcal{C} \times \mathcal{A} \quad (c, a) \to ca \in \mathcal{C}$ satisfying

$$(ac)^* = c^* a^*$$

in addition to the usual linearity, associativity and compatibility relations of a two–sided action.
By definition, a 0–*chain* is an element w of $\mathcal{C}$ satisfying

$$w^* + w + w^* w = w + w^* + w w^* = 0. \tag{3.6}$$

The 0–chains form a group under the composition

$$w_1 * w_2 = w_1 + w_2 + w_1 w_2. \tag{3.7}$$

A 1–*cocycle* is a linear $*$–map $\lambda : \mathcal{A} \to \mathcal{C}$ satisfying the identity

$$\lambda(xy) = \lambda(x)y + x\lambda(y) + \lambda(x)\lambda(y). \tag{3.8}$$

It may be verified that, given a 0–chain w, the map $\delta w : \mathcal{A} \to \mathcal{C}$ given by $(\delta w)(x) = = w^* x + xw + w^* \times w$ is a 1–cocycle. Denote the set of 0–chains by $C^\circ(\mathcal{A}, \mathcal{C})$ and of 1–cocycles by $Z^1(\mathcal{A}, \mathcal{C})$ and let $B^1(\mathcal{A}, \mathcal{C})$ be the range of the map δ.
$Z^1(\mathcal{A}, \mathcal{C})$ is not a group, but the group $C^\circ(\mathcal{A}, \mathcal{C})$ acts on $Z^1(\mathcal{A}, \mathcal{C})$ through $(\lambda, w) \to \to \lambda^w$, where

$$\lambda^w(x) = \lambda(x) + w^*(x + \lambda(x)) + (x + \lambda(x))w + w^*(x + \lambda(x))w. \tag{3.9}$$

As a first example, take $\mathcal{A}$ to be a smooth system algebra and denote by $\mathcal{C}$ the corresponding complete system algebra $\mathcal{B}$ in which the product has been trivialised, that is all products in $\mathcal{C}$ vanish. It is convenient to realise $\mathcal{C}$ as the tensor product $*$–algebra $\mathcal{C} = \mathcal{B} \otimes \mathcal{F}$, where $\mathcal{F}$ is the one–dimensional trivial algebra $\mathcal{C}\, dt$ generated by a single self–adjoint element dt satisfying $(dt)^2 = 0$. $\mathcal{A}$ acts (non–trivially) on $\mathcal{C}$ by using the original multiplication in $\mathcal{B}$, thus, for $a \in \mathcal{A}$ and $b \in \mathcal{B}$,

$$a(b \otimes dt) = ab \otimes dt, \qquad (b \otimes dt)a = ba \otimes dt.$$

In this example all nonlinear terms in the cohomology theory vanish and it reduces to Hochschild cohomology for the $\mathcal{A}$–module $\mathcal{B}$. Thus $C^\circ(\mathcal{A}, \mathcal{C})$ consists of elements of $\mathcal{C}$ of form $h \otimes dt$ where $h \in \mathcal{B}$ is skew, and the group $C^\circ(\mathcal{A}, \mathcal{C})$ is isomorphic to the additive group of such elements. Writing each element λ as $\lambda(x) = = \lambda_0(x) \otimes dt$, where $\lambda_0 : \mathcal{A} \to \mathcal{B}$, we see that $Z^1(\mathcal{A}, \mathcal{C})$ is coextensive with the set of $*$–derivations from $\mathcal{A}$ to $\mathcal{B}$. In this correspondence elements of $B^1(\mathcal{A}, \mathcal{C})$ correspond to inner derivations. We may now identify dt with the differential of time. Then this example corresponds to deterministic quantum flows, and the possibility that $Z^1(\mathcal{A}, \mathcal{C})$ is larger than $B^1(\mathcal{A}, \mathcal{C})$ (equivalently non–triviality of the first Hochschild cohomology space $H^1(\mathcal{A}, \mathcal{B})$) measures corresponds to the existence of non–inner flows. We shall see that a properly nonlinear cohomology theory enables us to make a similar dichotomy for both quantum and classical stochastic flows.
For this purpose let $\mathcal{F}$ be the finite–dimensional $*$–algebra generated by differentials of integrator processes satisfying a closed Ito multiplication table, for example that given by (2.8) with the involution $(d\Lambda^\mu_\nu)^* = d\Lambda^\nu_\mu$. (A moment's thought or a computation will show that the multiplication is indeed associative.) Again let $\mathcal{A}$

and $\mathcal{B}$ be respectively smooth and complete system algebras and let $\mathcal{C}$ be the tensor product $*$–algebra $\mathcal{C} = \mathcal{B} \otimes \mathcal{F}$ on which $\mathcal{A}$ again acts through the multiplication in $\mathcal{B}$; for all $a \in \mathcal{A}$, $b \in \mathcal{B}$ and $d\Lambda \in \mathcal{F}$

$$a(b \otimes d\Lambda) = ab \otimes d\Lambda, \qquad (b \otimes d\Lambda)a = ba \otimes d\Lambda.$$

By writing each element λ in the form

$$\lambda(x) = \sum_{\mu,\nu=0}^{n} \lambda_\nu^\mu(x) \otimes d\Lambda_\mu^\nu,$$

where each λ_ν^μ maps $\mathcal{A}$ to $\mathcal{B}$, we see that the set $Z^1(\mathcal{A},\mathcal{C})$ of 1–cocycles is co–extensive with the set of solutions to the structure relations (2.9), (2.10) and (2.12); indeed (3.8) is simply a notationally abbreviated form of (2.12). Thus quantum stochastic flows are generated by elements of $Z^1(\mathcal{A},\mathcal{C})$.
To interpret the set of 0–chains, we write an element $w \in \mathcal{C}$ in the form

$$w = \sum_{\lambda,\nu=0}^{n} w_\mu^\lambda \otimes d\Lambda_\lambda^\mu, \quad w_\mu^\lambda \in \mathcal{B}.$$

Then it may be verified that the defining relation (3.6) is equivalent to the following relations among the w_μ^λ:

$$(w_\mu^\lambda)^* + w_\lambda^\mu + \sum_{k=1}^{n}(w_k^\lambda)^* w_k^\mu = w_\mu^\lambda + (w_\lambda^\mu)^* + \sum_{k=1}^{n} w_k^\lambda (w_k^\mu)^* = 0. \tag{3.10}$$

The structure maps corresponding to δw are given by

$$\lambda_\nu^\mu(x) = (w_\mu^\nu)^* x + x w_\nu^\mu + \sum_{k=1}^{n}(w_\mu^k)^* x w_\nu^k. \tag{3.11}$$

Provided each w_μ^λ is bounded, the conditions (3.10) are (Hudson and Parthasarathy, 1984a) necessary and sufficient for the existence of a unique unitary–operator valued process u satisfying the stochastic integral equation

$$u(t) = 1 + \sum_{\mu,\nu=0}^{n} \int_0^t (w_\mu^\nu \otimes 1)\, u(s)\, d\Lambda_\nu^\mu.$$

The inner quantum stochastic flow corresponding to the bounded structure maps (3.11) is then given by

$$j_t(x) = u(t)^* x \otimes 1\, u(t).$$

Thus, exactly as in the deterministic case, quantum stochastic flows are generated by elements of $Z^1(\mathcal{A},\mathcal{C})$ and inner flows by elements of $B^1(\mathcal{A},\mathcal{C})$.
The group action (3.9) may be understood as follows. The flow generated by λ^w is an *inner perturbation* of that generated by λ in the following sense. Denoting the corresponding flows by j^w and j respectively, we have that

$$j^w(x) = u^{(j)^*} j(x) u^{(j)},$$

where $u^{(j)}$ is the solution of

$$u^{(j)}(x) = 1 + \sum_{\mu,\nu=0}^{n} \int_0^t j_s(w_\mu^\nu)\, u(s)\, d\Lambda_\nu^\mu.$$

In the case of bounded w_μ^ν and λ_μ^ν, the existence, uniqueness and unitarity of a solution of this quantum stochastic integral equation were proved in (Evans and Hudson, 1990).
It is amusing to carry out a similar analysis to the above for the case of a classical stochastic flow. Thus, we consider the algebras $\mathcal{A} = K^\infty(V)$ and $\mathcal{C} = \mathcal{A} \otimes \mathcal{F}$ where the Ito algebra $\mathcal{F}$ is now generated by the self-adjoint elements $dB_1, \ldots, dB_n$, dt with multiplication table

$$dB_j\, dB_k = \delta_{jk}\, dt, \qquad dt\, dB_j = dB_j\, dt = (dt)^2 = 0;$$

as usual $\mathcal{A}$ acts on $\mathcal{C}$ using the multiplication in $\mathcal{A}$. The vector fields β_k and the map τ characterising the flow may be aggregated into a 1–cocycle

$$\lambda(x) = \sum \beta_k(x) \otimes dB_k + \tau(x) \otimes dt.$$

However in this case there are no inner flows; although there are plenty of 0–chains, the corresponding generators all vanish; for w satisfying (3.6) we have

$$(\delta w)(x) = w^\dagger x + x w + w^\dagger x w = (w^\dagger + w + w^\dagger w) x = 0$$

by commutativity of $\mathcal{C}$.

REFERENCES

Accardi, L. and Hudson, R. (1989). The nonabelian cohomology of quantun stochastic evolutions. In: *Proc. Quantum Probab.* **V**, Heidelberg (to appear).

Applebaum, D.B. (1989). Quantum diffusions on involutive algebras. In: *Proc. Quantum Probab.* **V**, Heidelberg (to appear).

Baxendale, P.H. (1986). Asymptotic behaviour of stochastic flows of diffeomorphisms. *Lect. Notes in Math.* **1203**, 1–19.

Bismut, J.M. (1981). *Mécanique aléatore* (Lect. Notes in Math. **866**). Springer, Berlin—Heidelberg etc.

Elworthy, K.D. (1982). *Stochastic differential equations on manifolds.* CUP, Berkeley.

Evans, M.P. (1988). Quantum diffusions. *Nottingham thesis.*

Evans, M.P. (1989). Existence of quantum diffusions. *Probab. Theor. Rel. Fields* **81**, 473–483.

Evans, M.P. and Hudson, R.L. (1990). Perturbations of quantum diffusions. *Proc. London Math. Soc.* (to appear).

Hochschild, G. (1945). On the cohomology groups of an associative algebra. *Ann. Math.* **46**, 58–67.

Hudson, R.L. (1987). Quantum diffusions and cohomology of algebras. In: *Proc. First World Congress of Bernoulli Soc.* **1**, VNU Sci. Press, Utrecht, pp. 479–485.

Hudson, R.L. and Parthasarathy, K.R. (1984a). Quantum Ito's formula and stochastic evolutions. *Commun. Math. Phys.* **93**, 301–323.

Hudson, R.L. and Parthasarathy, K.R. (1984b). Stochastic dilations of uniformly continuous completely positive semigroups. *Acta Math. Appl.* **2**, 353–379.

Hudson, R.L. and Robinson, P. (1988). Quantum diffusions and the noncommutative torus. *Lett. Math. Phys.* **15**, 47–53.

Hudson, R.L. and Shepperson, P. (1989). Stochastic dilations of quantum dynamical semigroups using one-dimensional quantum stochastic calculus. In: *Proc. Quantum Probab.*, **V**, Heidelberg (to appear).

Kadison, R.V. and Ringrose, J.R. (1986). *Fundamentals of the Theory of Operator Algebras.* Academic Press, New York.

Lue, A.S. (1968). Nonabelian cohomology of associative algebras. *Quart. J. Math.* **19**, 159–180.

Robinson, P. (1989). Quantum diffusions on the rotation algebra and the quantum Hall effect. In: *Proc. Quantum Probab.*, **V**, Heidelberg (to appear).

Prob. Theory and Math. Stat., Vol. 1, pp. 526–538
B.Grigelionis *et al.* (Eds.)
1990 VSP/Mokslas

CONVERGENCE OF THE STOCHASTIC QUANTIZATION METHOD

I.A. IGNATIUK*, V.A. MALYSHEV* and V. SIDORAVIČIUS**
* Moscow State University,
** Institute of Mathematics and Cybernetics, Vilnius

ABSTRACT

We consider classical and fermionic Gibbsian fields on the lattice. The main result of the present paper is the proof of convergence for infinite system of Langevin's equations in the case of weak interaction.

INTRODUCTION

Gibbsian random fields are defined and investigated usually either by means of the thermodynamic limit transition from finite volume, or by means of Dobrushin—Lanford—Ruelle equations (Dobrushin, 1968). The stochastic quantization method is based on the observation of the fact that Gibbsian fields can be obtained also as the limiting distribution as time goes to infinity of the evolution of systems of stochastic differential equations of special type, namely the Langevin equations associated with the potential of the corresponding Gibbsian field.
This method was first introduced by physicists (Parisi and Wu, 1981). It has not received yet a generally accepted physical interpretation, and at present it is considered either as a theoretical or a computational mean of investigation and approximation of Gibbsian fields. In Langevin equations, time is sometimes called the mechanical time, and it is supposed that the Monte–Carlo simulation of Langevin's equations is more simple than that of Gibbsian fields (Migdal, 1986).
We consider Gibbsian fields on the lattice. Convergence of evolution (ergodicity) of Langevin's equations in finite volume is a common fact from the theory of finite dimensional Ito's equations (Gikhman and Skorokhod, 1972), which can be proved, for instance, by the help of Liapunov's functions.
The main result of the present paper is the proof of convergence for infinite system of Langevin's equations in the case of weak interaction. The class of Gibbsian fields under consideration corresponds to the lattice $P(\varphi)$–models of Euclidean quantum field theory. We consider Grassman fields (in particular, a perturbation of Dirac's field). For Grassman fields we had to carry over all basic definitions of the theory of Ito's equations to the case of Grassman variables in a form which is convenient for us. We could not find appropriate formulations in the literature in spite of a large number of papers on Ito's equations on Clifford algebras.
We prefer to give a more constructive definition of stochastic differential equations

then the classical one. It is still not known whether a full analog of the L_2-definition is possible. Complete proofs will be published in (Ignatiuk et al., 1990).

1. THE CONVERGENCE OF THE METHOD OF STOCHASTIC QUANTIZATION IN THE CLASSICAL CASE

With any finite subset $\Lambda \subset \mathbf{Z}^d$ we associate a polynomial U_Λ depending on variables $\xi_x \in \mathbf{R}$, $x \in \Lambda$, such that

$$Z_\Lambda = \int\limits_{\mathbf{R}^{|\Lambda|}} \exp\{-U_\Lambda\} \prod_{x\in\Lambda} d\xi_x < \infty,$$

and consider on $\mathbf{R}^{|\Lambda|}$ the Gibbsian measure μ_Λ with the potential U_Λ:

$$d\mu_\Lambda = \frac{1}{Z_\Lambda} \exp\{-U_\Lambda\} \prod_{x\in\Lambda} d\xi_x .$$

Definition 1. By the system of Langevin's equations for the measure μ_Λ we call the system of stochastic differential equations:

$$d\xi_{x,\Lambda} = -\frac{1}{2}\frac{\partial U_\Lambda}{\partial \xi_x}\left(\bar{\xi}(t)\right) dt + dw_x(t), \qquad x \in \Lambda, \tag{1}$$

where $\bar{\xi}_\Lambda(t) = \{\xi_{x,\Lambda}\,,\ x \in \Lambda\}$ and $w_x(t)$ are standart Wiener processes independent for different $x \in \Lambda$.
The measure μ_Λ is invariant for the corresponding system (1) of Langevin's equations. Let the limit $\lim\limits_{\Lambda\uparrow \mathbf{Z}^d} \mu_\Lambda = \mu$ exists in the sense of the weak convergence of finite dimensional distributions. Assume also that for any $x \in \mathbf{Z}^d$, $F_x = \partial U_\Lambda / \partial \xi_x$ does not depend on Λ if Λ is sufficiently large.

Definition 2. By the system of Langevin's equations for the limiting measure μ we call the infinite system of stochastic differential equations:

$$d\xi_x = -\frac{1}{2} F_x(\bar{\xi}_t)\, dt + dw_x(t), \qquad x \in \mathbf{Z}^d\,, \tag{1a}$$

where $\bar{\xi}(t) = \{\xi_x(t),\ x \in \mathbf{Z}^d\}$, and $w_x(t)$, $x \in \mathbf{Z}^d$, are standart Wiener processes.
In the sequel we will consider the case, when U_Λ can be represented as a quadratic part plus small perturbation.
Consider translation-invariant Gaussian random field $\{\xi_x,\ x \in \mathbf{Z}^d\}$ on the lattice $\mathbf{Z}^d$ with zero mean $\langle \xi_x \rangle = 0$, $x \in \mathbf{Z}^d$, and the spectral density

$$f(\lambda) = \left[2 \sum_{y\in\mathbf{Z}^d} a_y\, e^{-i(\lambda, y)} \right]^{-1}, \qquad \lambda \in [-\pi, \pi]^d,$$

with coefficients a_y, $y \in \mathbf{Z}^d$, satisfying the following conditions:

a) $a_y \neq 0$ only for a finite number of $y \in \mathbf{Z}^d$ (we denote by Q the set of points $y \in \mathbf{Z}^d$ for which $a_y \neq 0$),

b) $a_y = a_{-y}$ for all $y \in \mathbf{Z}^d$,

c) $a_0 - \sum_{y \neq 0} | a_y | > 0$.

The measure μ_0 corresponding to this Gaussian field can be obtained as the limit $\Lambda \uparrow \mathbf{Z}^d$ in the sense of weak convergence of finite dimensional distributions of the Gibbsian measures $\mu_{\Lambda,0}$, $\Lambda = [-N_\Lambda, N_\Lambda)^d \cap \mathbf{Z}^d$ with the quadratic potential

$$U_\Lambda^0 = \sum_{x \in \Lambda} a_0 \xi_x^2 + \sum_{x \neq y;\, x,y \in \Lambda} a_{x-y}^\Lambda \, \xi_x \, \xi_y;$$

where $a_y^\Lambda = a_y$ for $y \in \Lambda$ and $a_{y+N_\Lambda e}^\Lambda = a_y^\Lambda$ for $y \in \Lambda$, $e \in \mathbf{Z}^d$, $| e | = 1$.
The system of Langevin's equations for μ_0 has the form:

$$d\xi_x(t) = -\sum_{y \in Q} a_y \xi_{x+y}(t)\, dt + dw_x(t). \tag{2}$$

Let the initial field $\xi_x(0)$, $x \in \mathbf{Z}^d$, be Gaussian and translation–invariant with $\langle \xi_x(0) \rangle \equiv 0$, $x \in \mathbf{Z}^d$, and

$$| \langle \xi_0(0)\, \xi_x(0) \rangle | \leqslant c\, e^{-\gamma |x|}, \qquad c, \gamma > 0, \tag{3}$$

for any $x \in \mathbf{Z}^d$. By $\mu_0(0)$ we denote the corresponding Gaussian measure on $\mathbf{R}^{\mathbf{Z}^d}$, and by $\langle \cdot \rangle_\mu$ the expectation corresponding to a measure μ.
It is easy to see that the solution of the system (2) is a Gaussian process $\{\xi_x(t),\ x \in \mathbf{Z}^d\}$, $t \in \mathbf{R}_+$, whose evolution is described by the first and the second moments $\langle \xi_x(t) \rangle$, $\langle \xi_x(t) \xi_y(s) \rangle$, $x, y \in \mathbf{Z}^d$, $t, s \in \mathbf{R}_+$. The proof of the convergence to the invariant distribution μ_0 in this case is not difficult, because the moments can be written in an explicit form.
Consider now a Gibssian modification $\mu_\Lambda, \Lambda \subset \mathbf{Z}^d$, of the Gaussian measure μ_0. Let

$$\frac{d\mu_\Lambda}{d\mu_0} = \frac{1}{Z_\Lambda} \exp\left\{ -\varepsilon \sum_{x \in \Lambda} P(\xi_x) \right\},$$

where $P(\xi)$ is a polynomial, which is bounded from below. The covariance of the measure μ_0 decreases exponentially:

$$| \langle \xi_x \, \xi_y \rangle_{\mu_0} | \leqslant C\, e^{-\gamma |x-y|}; \qquad C, \gamma > 0; \qquad x, y \in \mathbf{Z}^d. \tag{4}$$

This follows from the conditions on the coefficients a_y, $y \in \mathbf{Z}^d$.
It is known (Malyshev and Minlos, 1985) that in this case for sufficiently small $\varepsilon > 0$, $\varepsilon < \varepsilon_0$ there exists the limit $\lim_{\Lambda \uparrow \mathbf{Z}^d} \mu_\Lambda = \mu_\varepsilon$ in the sense of weak convergence of finite dimensional distributions. Let us write the system of Langevin's equations for the measure μ_ε:

$$d\xi_x(t) = -\sum_{y \in Q} a_y \, \xi_{x+y}(t)\, dt - \varepsilon\, p(\xi_x(t))\, dt + dw_x(t), \tag{5}$$

where $p(\xi) = (1/2)dP(\xi)/d\xi,\ x \in \mathbf{Z}^d$.
The system (5) has a solution in the real Hilbert space $H = l_2(\mathbf{Z}^d, e^{-|x|})$ with the scalar product $(f,g)_H = \sum_{x\in\mathbf{Z}^d} f(x)\,g(x)\,e^{-|x|}$ (Rozovsky, 1983).
Let $\{\xi_x(t),\ x \in \mathbf{Z}^d\}$, $t \in \mathbf{R}_+$, be the solution of the system (5). By $\mu_\varepsilon(t)$ we denote the probability measure on $\mathbf{R}^{\mathbf{Z}^d}$, which corresponds to the solution $\xi_x(t)$, $x \in \mathbf{Z}^d$, at the time moment $t \in \mathbf{R}_+$. The following theorem is our main result for the classical case.

THEOREM 1. *Let $\varepsilon > 0$ be sufficiently small. Then the measure μ_ε is invariant for the system (2). Moreover, if the measure μ_ε corresponding to the initial field $\xi_x(0)$, $x \in \mathbf{Z}^d$, is the Gibbsian modification of the Gaussian measure $\mu_0(0)$ with the potential $\varepsilon\sum P_0(\xi_x(0))$, where $P_0(\xi)$ is a polynomial bounded from below such that*

$$P_0(\xi) - \frac{1}{2}P(\xi) \geqslant c$$

for all $\xi \in \mathbf{R}$, then

$$\mu_\varepsilon(t) \to \mu_\varepsilon \quad (t \to \infty)$$

in the sense of the weak convergence of finite dimensional distributions.

2. ITO'S CALCULUS ON GRASSMAN ALGEBRAS

For any $\Lambda \subset \mathbf{Z}^d$ and $t \in \mathbf{R}_+$ we consider four Grassman algebras, namely the Grassman algebra $@_0^\Lambda$ with generators

$$1,\quad \xi^{(\alpha,x)},\quad \bar\xi^{(\alpha,x)},\quad (\alpha,x) \in \{1,\ldots,N\} \times \Lambda;$$

the Grassman algebra W_t^Λ with generators

$$1,\quad \overline{w}^{(\alpha,x)}(t),\quad w^{(\alpha,x)}(t),\quad (\alpha,x) \in \{1,\ldots,N\} \times \Lambda,$$

the Grassman algebra $W_{t]}^\Lambda$ with generators

$$1,\quad \overline{w}^{(\alpha,x)}(t'),\quad w^{(\alpha,x)}(t'),\quad (\alpha,x) \in \{1,\ldots,N\} \times \Lambda,\quad t' \in [0,t],$$

and the Grassman algebra W_∞^Λ with generators

$$1,\quad \overline{w}^{(\alpha,x)}(t),\quad w^{(\alpha,x)}(t),\quad (\alpha,x) \in \{1,\ldots,N\},\quad t \in \mathbf{R}_+,\quad t > 0.$$

Denote

$$@_t^\Lambda = @_0^\Lambda \otimes W_t^\Lambda;\quad @_{t]}^\Lambda = @_0^\Lambda \otimes W_{t]}^\Lambda;\quad @_\infty^\Lambda = @_0^\Lambda \otimes W_\infty^\Lambda.$$

All algebras over complex will be considered. For convenience, we use multiindices $\mu = (\mu^1,\ldots,\mu^k)$, $\mu^j \in \{1,\ldots,N\} \times \mathbf{Z}^d$, $k \in \mathbf{Z}_+$, and write

$$\bar\xi^\mu = \bar\xi^{\mu_1}\cdot\ldots\cdot\bar\xi^{\mu_k};\qquad \xi^\mu = \xi^{\mu_1}\cdot\ldots\cdot\xi^{\mu_k};\qquad \xi^\varnothing = 1;$$

$$\overline{w}^{\mu}(t)=\overline{w}^{\mu_1}(t)\cdot\ldots\cdot\overline{w}^{\mu_k}(t);\qquad w^{\mu}(t)=w^{\mu_1}(t)\cdot\ldots\cdot w^{\mu_k}(t).$$

We assume that $\{1,\ldots,N\}\times\mathbf{Z}^d$ is completely ordered, and for any multiindex $\mu=(\mu^1,\ldots,\mu^k)$, $\mu^1<\ldots<\mu^k$. Let M_Λ denote the set of all multiindices $\mu=(\mu^1,\ldots,\mu^k)$, $k\in\mathbf{Z}_+$.

Definition 3. By the Wiener quasistate on W_∞^Λ we mean the Gaussian quasistate $\langle\cdot\rangle_w$ on W_∞^Λ satisfying the following conditions:

$$\langle\,\overline{w}^{(\alpha,x)}(t)\,\overline{w}^{(\beta,y)}(s)\,\rangle_w=\langle\,w^{(\alpha,x)}(t)\,w^{(\beta,y)}(s)\,\rangle_w=0$$

and

$$\langle\,\overline{w}^{(\beta,y)}(t)\,w^{(\alpha,x)}(s)\,\rangle_w=\delta_{\alpha\beta}\,\delta_{xy}\min(t,s)$$

for any

$$(\alpha,\beta)\in\{1,\ldots,N\};\quad x,y\in\Lambda;\quad t,s\in\mathbf{R}_+.$$

PROPOSITION 1. *For any $w\in W_\infty^\Lambda$ there exists $w'\in W_\infty^\Lambda$ such that*

$$\langle w\,w'\rangle\neq 0.$$

We shall consider only quasistates on $@_\infty^\Lambda=@_0^\Lambda\otimes W_\infty^\Lambda$ of the following form:

$$\omega(\cdot)=\langle\cdot\rangle_0\otimes\langle\cdot\rangle_w,$$

where $\langle\cdot\rangle_0$ is a quasistate on $@_0^\Lambda$. By σ_Λ we denote the set of all such quasistates on $@_0^\Lambda$.
We consider on $@_0^\Lambda$ the system of seminorms $\rho_{\zeta,\omega}(\cdot)$, $\zeta\in @_\infty^\Lambda$, $\omega\in\sigma_\Lambda$:

$$\rho_{\zeta,\omega}(\xi)=|\,\omega(\zeta\xi)\,|\,.$$

This system of seminorms separates points of $@_\infty^\Lambda$, i.e., the condition $\rho_{\zeta,\omega}(\xi)=0$ for all $\xi\in @_\infty^\Lambda$, $\omega\in\sigma_\Lambda$ implies $\xi=0$. The locally convex topology on $@_\infty^\Lambda$ with respect to which all $\rho_{\zeta,\omega}$, $\zeta\in @_\infty^\Lambda$, $\omega\in\sigma_\Lambda$, are continuous and the operation of addition is continuous, is a Hausdorff topology. By $\overline{@_\infty^\Lambda}$ we denote the closure of $@_\infty^\Lambda$ with respect to this topology.
Consider the following Grassman algebras, namely:
$\mathcal{A}_\infty^\Lambda$, with the generators

$$1,\quad\overline{\xi}^{(\alpha,x)}(t),\quad\xi^{(\alpha,x)}(t),\quad(\alpha,x)\in\{1,\ldots,N\}\times\Lambda;\quad t\in\mathbf{R}_+;$$

$\mathcal{A}_t^\Lambda$ with generators

$$1,\quad\overline{\xi}^{(\alpha,x)}(t),\quad\xi^{(\alpha,x)}(t),\quad(\alpha,x)\in\{1,\ldots,N\}\times\Lambda$$

and $\mathcal{A}_{t]}^\Lambda$, $t\in\mathbf{R}_+$, with the generators

$$1,\quad\overline{\xi}^{(\alpha,x)}(t'),\quad\xi^{(\alpha,x)}(t'),\quad(\alpha,x)\in\{1,\ldots,N\}\times\Lambda,\quad t'\in[0,t].$$

Definition 4. By the Grassman Wiener process we mean the homomorphism

$$w : \mathcal{A}_\infty^\Lambda \to @_\infty^\Lambda$$

such that

$$w : \overline{\xi}^{(\alpha,x)}(t) \to \overline{w}^{(\alpha,x)}(t),$$

$$w : \xi^{(\alpha,x)}(t) \to w^{(\alpha,x)}(t), \quad (\alpha,x) \in \{1,\dots,N\}, \quad t \in \mathbf{R}_+.$$

All other Grassman processes which we consider will be constructed from the Grassman Wiener process.
Now we shall define the stochastic integrals:

$$\mathfrak{J}^{\mu\nu}(T_1,T_2) = \int_{T_1}^{T_2} \overline{w}^{\mu}(t) w^{\nu}(t)\, dt,$$

$$\overline{\mathfrak{J}}^{\mu\nu}_{(\alpha,x)}(T_1,T_2) = \int_{T_1}^{T_2} \overline{w}^{\mu}(t) w^{\nu}(t)\, d\overline{w}^{(\alpha,x)}(t),$$

$$\mathfrak{J}^{\mu\nu}_{(\alpha,x)}(T_1,T_2) = \int_{T_1}^{T_2} \overline{w}^{\mu}(t) w^{\nu}(t)\, dw^{(\alpha,x)}(t),$$

for all $\mu,\nu \in M_\Lambda$, $(\alpha,x) \in \{1,\dots,N\} \times \Lambda$, $T_1, T_2 \in \mathbf{R}_+$, $T_1 < T_2$ as limits of the corresponding partial sums:

$$S^{\mu\nu}(t_0,\dots,t_n) = \sum_{j=0}^{n-1} \overline{w}^{\mu}(t_j) w^{\nu}(t_j)(t_{j+1} - t_j),$$

$$\overline{S}^{\mu\nu}_{(\alpha,x)}(t_0,\dots,t_n) = \sum_{j=0}^{n-1} \overline{w}^{\mu}(t_j) w^{\nu}(t_j)(\overline{w}^{(\alpha,x)}(t_{j+1}) - \overline{w}^{(\alpha,x)}(t_j)),$$

when the diameter of the partition $T_1 = t_0 < t_1 < \dots < t_n = T_2$ tends to zero.
It is easy to see that in the topology introduced above, all the limits of these partial sums exist, as well as the limits of the various multiproducts.
Next we define the stochastic integrals:

$$\int_{T_1}^{T_2} w(t)\, dt; \quad \int_{T_1}^{T_2} w(t)\, d\overline{w}^{(\alpha,x)}(t); \quad \int_{T_1}^{T_2} w(t)\, dw^{(\alpha,x)}(t);$$

$(\alpha,x) \in \{1,\dots,N\} \times \Lambda$, for elements $w(t) \in @_t^\Lambda$ of the form

$$w(t) = \sum_{\mu,\nu \in M_\Lambda} c_{\mu\nu} \overline{w}^{\mu}(t) w^{\nu}(t), \quad c_{\mu\nu} \in \mathbf{C},$$

by linearity. Finally, we define

$$\int_{T_1}^{T_2} \xi w(t)\,dt = \xi \int_{T_1}^{T_2} w(t)\,dt;$$

$$\int_{T_1}^{T_2} \xi w(t)\,d\overline{w}^{(\alpha,x)}(t) = \xi \int_{T_1}^{T_2} w(t) d\overline{w}^{(\alpha,x)}(t);$$

$(\alpha, x) \in \{1, \dots, N\} \times \Lambda$ for all $\xi \in @_0^\Lambda$.
We consider the superalgebra $@_\infty^{\Lambda,1}$ with the generators

$$\mathbf{1}, \quad \overline{\xi}^{(\alpha,x)}, \quad \xi^{(\alpha,x)}, \quad \mathfrak{I}^{\mu\nu}(0,t), \quad \overline{\mathfrak{I}}^{\mu\nu}_{(\alpha,x)}(0,t), \quad \mathfrak{I}^{\mu\nu}_{(\alpha,x)}(0,t),$$

$$\mu, \nu \in M_\Lambda, \quad (\alpha, x) \in \{1, \dots, N\} \times \Lambda, \quad t \in \mathbf{R}_+;$$

the superalgebra $@_t^{\Lambda,1}$ with the generators

$$\mathbf{1}, \quad \overline{\xi}^{(\alpha,x)}, \quad \xi^{(\alpha,x)}, \quad \mathfrak{I}^{\mu\nu}(0,t), \quad \overline{\mathfrak{I}}^{\mu\nu}_{(\alpha,x)}(0,t), \quad \mathfrak{I}^{\mu\nu}_{(\alpha,x)}(0,t);$$

and the superalgebra $@_{t]}^{\Lambda,1}$ with the generators

$$\mathbf{1}, \quad \overline{\xi}^{(\alpha,x)}, \quad \xi^{(\alpha,x)}, \quad \mathfrak{I}^{\mu\nu}(0,t'), \quad \overline{\mathfrak{I}}^{\mu\nu}_{(\alpha,x)}(0,t'), \quad \mathfrak{I}^{\mu\nu}_{(\alpha,x)}(0,t'),$$

$$(\alpha, x) \in \{1, \dots, N\} \times \Lambda, \quad \mu, \nu \in M_\Lambda, \quad t' \in [0,t],$$

The following inclusion relations hold between the algebras introduced above:

$$@_\infty^\Lambda \subset @_\infty^{\Lambda,1}; \quad @_t^\Lambda \subset @_t^{\Lambda,1}; \quad @_{t]}^\Lambda \subset @_{t]}^{\Lambda,1}, \quad t \in \mathbf{R}_+,$$

because

$$\overline{w}^{(\alpha,x)}(t) = \int_0^t \mathbf{1}\, d\overline{w}^{(\alpha,x)}(s) = \overline{\mathfrak{I}}^{\varnothing\varnothing}_{(\alpha,x)}(0,t),$$

$$w^{(\alpha,x)}(t) = \int_0^t \mathbf{1}\, dw^{(\alpha,x)}(s) = \mathfrak{I}^{\varnothing\varnothing}_{(\alpha,x)}(0,t),$$

$(\alpha, x) \in \{1, \dots, N\} \times \Lambda, t \in \mathbf{R}_+$.
We extend any quasistate from σ_Λ by continuity on $@_\infty^{\Lambda,1}$ and identify all elements ζ, ξ from $@_\infty^{\Lambda,1}$ such that $\omega(\eta\zeta) = \omega(\eta, \xi)$ for all $\eta \in @_\infty^{\Lambda,1}$ and all $\omega \in \sigma_\Lambda$.
We shall consider on $@_t^{\Lambda,1}$, $t \in \mathbf{R}_+$, the family of homomorphisms $S_\tau^1 : @_t^{\Lambda,1} \to$ $\to @_{t+\tau}^{\Lambda,1}$ $(\tau \in \mathbf{R}_+)$ such that

$$S_\tau^1 \xi = \xi, \quad S_\tau^1 \mathfrak{I}^{\mu\nu}(0,t) = \mathfrak{I}^{\mu\nu}(0,t+\tau),$$

$$S_\tau^1 \overline{\mathfrak{I}}^{\mu\nu}_{(\alpha,x)}(0,t) = \overline{\mathfrak{I}}^{\mu\nu}_{(\alpha,x)}(0,t+\tau), \quad S_\tau^1 \mathfrak{I}^{\mu\nu}_{(\alpha,x)}(0,t) = \mathfrak{I}^{\mu\nu}_{(\alpha,x)}(0,t+\tau),$$

for all $\xi \in @_0^\Lambda$, $\tau \in \mathbf{R}_+$, $(\alpha, x) \in \{1, \ldots, N\} \times \Lambda$, $\mu, \nu \in M_\Lambda$, $t \in \mathbf{R}_+$, $t > 0$.

Definition 5. By Grassman random process on $@_\infty^{\Lambda,1}$ we mean a homomorphism $\Psi : @_\infty^\Lambda \to @_\infty^{\Lambda,1}$ such that

$$\Psi \mid_{\mathcal{A}_t^\Lambda} = \Psi_t : \mathcal{A}_t^\Lambda \to @_t^{\Lambda,1}, \quad t \in \mathbf{R}_+,$$

and

$$S_\tau \Psi_t = \Psi_{t+\tau}$$

for all $t, \tau \in \mathbf{R}_+$, $t > 0$. Let $\mathfrak{A}_1^\Lambda$ denote the set of all random processes on $@_\infty^{\Lambda,1}$. We shall construct by induction superalgebras $@_\infty^{\Lambda,n}$, $@_t^{\Lambda,n}$, $@_{t]}^{\Lambda,n}$, $t \in \mathbf{R}_+$, for any $n \in \mathbf{Z}_+$:

$$@_\infty^{\Lambda,0} = @_\infty^\Lambda \subset @_\infty^{\Lambda,1} \subset \ldots \subset @_\infty^{\Lambda,n},$$

$$@_t^{\Lambda,0} = @_t^\Lambda \subset @_t^{\Lambda,1} \subset \ldots \subset @_t^{\Lambda,n},$$

$$@_{t]}^{\Lambda,0} = @_{t]}^\Lambda \subset @_{t]}^{\Lambda,1} \subset \ldots \subset @_{t]}^{\Lambda,n},$$

and sets of random processes $\mathfrak{A}_n^\Lambda$ on $@_\infty^{\Lambda,n}$, $n \in \mathbf{Z}_+$, as follows. We define conse cutively for any $n \in \mathbf{Z}_+$ the stochastic integrals:

$$\int_{T_1}^{T_2} \Psi[\overline{\xi}^{\mu}(t)\xi^{\nu}(t)]\, dt;$$

$$\int_{T_1}^{T_2} \Psi[\overline{\xi}^{\mu}(t)\xi^{\nu}(t)]\, d\overline{w}^{(\alpha,x)}(t);$$

$$\int_{T_1}^{T_2} \Psi[\overline{\xi}^{\mu}(t)\xi^{\nu}(t)]\, dw^{(\alpha,x)}(t);$$

$\Psi \in \mathfrak{A}_{n-1}^\Lambda$, $\mu, \nu \in M_\Lambda$; $(\alpha, x) \in \{1, \ldots, N\} \times \Lambda$; $T_1, T_2 \in \mathbf{R}_+$, and various multiproducts of these integrals as for $n = 1$ and extend by continuity on $@_\infty^{\Lambda,n}$ all quasistates from σ_Λ.

Definition 6. By Grassman random process on $@_\infty^{\Lambda,n}$ we mean a homomorphism

$$\Psi : \mathcal{A}_\infty^\Lambda \to @_\infty^{\Lambda,n}$$

such that

$$\Psi \mid_{\mathcal{A}_t^\Lambda} = \Psi_t : \mathcal{A}_t^\Lambda \to @_t^{\Lambda,n}, \quad t \in \mathbf{R}_+,$$

and

$$S_\tau^n \Psi_t = \Psi_{t+\tau},$$

for all $t, \tau \in \mathbf{R}_+$, $t > 0$, where S_τ^n, $\tau \in \mathbf{R}_+$, is the family of homomorphisms on $@_t^{\Lambda,n}$, $t \in \mathbf{R}_+$, such that

$$S_\tau^n \mid_{@_t^{\Lambda,n-1}} = S_\tau^{n-1}$$

for all $\tau, t \in \mathbf{R}_+$ and

$$S_\tau^n\left(\int_0^t \Psi'\left[\overline{\xi}^{\,\mu}(s)\xi^\nu(s)\right] ds\right) = \int_0^{t+\tau} \Psi'\left[\overline{\xi}^{\,\mu}(s)\xi^\nu(s)\right] ds,$$

$$S_\tau^n\left(\int_0^t \Psi'\left[\overline{\xi}^{\,\mu}(s)\xi^\nu(s)\right] d\widetilde{w}^{(\alpha,x)}(s)\right) = \int_0^{t+\tau} \Psi'\left[\overline{\xi}^{\,\mu}(s)\xi^\nu(s)\right] d\widetilde{w}^{(\alpha,x)}(s),$$

for all $(\alpha, x) \in \{1, \dots, N\} \times \Lambda$; $\Psi' \in \mathfrak{A}_{n-1}^\Lambda$; $\mu, \nu \in M_\Lambda$; $t, \tau \in \mathbf{R}_+$; $t > 0$, where $\widetilde{w}$ means either $\overline{w}$ or w.
By $\mathfrak{A}_n^\Lambda$ we shall denote the set of random processes on $@_\infty^{\Lambda,n}$.
It is clear that

$$\mathfrak{A}_0^\Lambda \subset \mathfrak{A}_1^\Lambda \subset \dots \subset \mathfrak{A}_n^\Lambda, \quad n \in \mathbf{Z}_+.$$

We shall consider the inductive limits:

$$@_\infty^{\Lambda,\infty} = \bigcup_n @_\infty^{\Lambda,n} \subset \overline{@_\infty^\Lambda};$$

$$@_t^{\Lambda,\infty} = \bigcup_n @_t^{\Lambda,n};$$

$$@_{t]}^{\Lambda,\infty} = \bigcup_n @_{t]}^{\Lambda,n};$$

$$\mathfrak{A}_\infty^\Lambda = \bigcup_n \mathfrak{A}_n^\Lambda.$$

Now we shall define the notion of generalized random process on $@_\infty^{\Lambda,\infty}$. For this we introduce the following topology on $\mathfrak{A}_\infty^\Lambda$:
Let $\Psi \in \mathfrak{A}_\infty^\Lambda$, $n \in \mathbf{Z}_+$ be given. We say that $\Psi_n \to \Psi \in \mathfrak{A}_\infty^\Lambda$ as $n \to \infty$ if for any $T \in \mathbf{R}_+$, $k \in \mathbf{Z}_+$

$$\sup_{t_0,\dots,t_k \in [0,T]} \mid \langle \zeta \Psi_n \overline{\xi}^{\,\mu_0}(t_0) \dots \overline{\xi}^{\,\mu_k}(t_k) \xi^{\nu_0}(t_0) \dots \xi^{\nu_k}(t_k) \rangle -$$

$$- \langle \zeta \Psi \overline{\xi}^{\,\mu_0}(t_0) \dots \overline{\xi}^{\,\mu_k}(t_k) \xi^{\nu_0}(t_0) \dots \xi^{\nu_k}(t_k) \rangle \mid \to 0,$$

as $n \to \infty$.
We denote by $\overline{\mathfrak{A}_\infty^\Lambda}$ the closure of $\mathfrak{A}_\infty^\Lambda$ with respect to this topology.

Definition 7. Elements of the set $\overline{\mathfrak{A}_\infty^\Lambda}$ will be called generalized random processes on $@_\infty^{\Lambda,\infty}$, or random processes on $\overline{@_\infty^{\Lambda,\infty}}$.

For $\Psi \in \overline{\mathfrak{A}_\infty^\Lambda}$ we define stochastic integrals by continuity:

$$\int_{T_1}^{T_2} \Psi \overline{\xi}^{\mu}(t) \xi^\nu(t)\, dt,$$

$$\int_{T_1}^{T_2} \Psi \bar{\xi}^{\mu}(t) \xi^{\nu}(t)\, d\bar{w}^{(\alpha,x)}(t),$$

$$\int_{T_1}^{T_2} \Psi \bar{\xi}^{\mu}(t) \xi^{\nu}(t)\, dw^{(\alpha,x)}(t),$$

$(\alpha, x) \in \{1, \ldots, N\} \times \Lambda$.
Similarly as in the classical theory of stochastic integrals, we write

$$d\zeta(t) = \xi_1(t)\, dt + \xi_2(t)\, d\bar{w}^{(\alpha,x)}(t) + \xi_3(t)\, dw^{(\beta,y)}(t) +$$
$$+ d\bar{w}^{(\alpha',x')}(t)\, \xi_4(t) + dw^{(\beta',y')}(t)\, \xi_5(t)$$

if the equality

$$\zeta(t) = \zeta(0) + \int_0^t \xi_1(s)\, ds + \int_0^t \xi_2(s)\, d\bar{w}^{(\alpha,x)}(s) +$$
$$+ \int_0^t \xi_3(s)\, dw^{(\beta,y)}(s) + \int_0^t d\bar{w}^{(\alpha',x')}(s)\, \xi_4(s) + \int_0^t dw^{(\beta',y')}(s)\, \xi_5(s)$$

holds for any $t \in \mathbf{R}_+$.
The following property of Grassman stochastic integrals is the analogue of the classical Ito's formula. Let following two stochastic integrals be given:

$$d\zeta(t) = \eta_1(t)\, dt + \eta_2(t)\, d\bar{w}^{(\alpha,x)}(t) + \eta_3(t)\, dw^{(\beta,y)}(t),$$

$$d\zeta'(t) = \eta_1'(t)\, dt + d\bar{w}^{(\alpha',x')}\, \eta_2'(t) + dw^{(\beta',y')}(t)\, \eta_3'(t)$$

then

$$d\big(\zeta(t)\zeta'(t)\big) = \big(d\zeta(t)\big)\zeta'(t) + \zeta(t)\, d\zeta'(t) + d\zeta(t)\, d\zeta'(t),$$

where

$$d\zeta(t)\, d\zeta'(t) = \delta_{\alpha\beta'}\, \delta_{xy'}\, \eta_2(t)\, \eta_3'(t) - \delta_{\beta\alpha'}\, \delta_{yx'}\, \eta_3(t)\, \eta_2'(t).$$

The proof of Ito's formula follows easily from the definition of stochastic integrals. Consider the following system of stochastic differential equations:

$$d\bar{\xi}^{(\alpha,x)}(t) = \bar{P}_{\alpha,x}\left(\bar{\xi}^{(\beta,y)}(t),\ \xi^{(\beta,y)}(t),\ (\beta,y) \in \{1,\ldots,N\} \times \Lambda\right) dt + d\bar{w}^{(\alpha,x)}(t),$$

$$d\xi^{(\alpha,x)}(t) = P_{\alpha,x}\left(\bar{\xi}^{(\beta,y)}(t),\ \xi^{(\beta,y)}(t),\ (\beta,y) \in \{1,\ldots,N\} \times \Lambda\right) dt + dw^{(\alpha,x)}(t),$$

$$\bar{\xi}^{(\alpha,x)}(0) = \bar{\xi}^{(\alpha,x)}, \quad \xi^{(\alpha,x)}(0) = \xi^{(\alpha,x)}, \quad (\alpha, x) \in \{1, \ldots, N\} \times \Lambda, \tag{6}$$

where $\bar{P}_{\alpha,x}(\cdot)$, $P_{\alpha,x}(\cdot)$ are arbitrary polynomials of odd degree in variables $\bar{\xi}^{(\beta,y)}(t)$, $\xi^{(\beta,y)}(t)$, $(\beta,y) \in \{1,\ldots,N\} \times \Lambda$.

Definition 8. By a solution of the system (6) we mean the random process $\Psi \in \overline{\mathfrak{A}_{\infty}^{\Lambda}}$ such that

$$\Psi\overline{\xi}^{(\alpha,x)}(t) = \overline{\xi}^{(\alpha,x)} + \int_0^t \Psi\overline{P}_{\alpha,x}\left(\overline{\xi}^{(\beta,y)}(s),\ \xi^{(\beta,y)}(s)\right) ds + \overline{w}^{(\alpha,x)}(t),$$

$$\Psi\xi^{(\alpha,x)}(t) = \xi^{(\alpha,x)} + \int_0^t \Psi P_{\alpha,x}\left(\overline{\xi}^{(\beta,y)}(s),\ \xi^{(\beta,y)}(s)\right) ds + w^{(\alpha,x)}(t)$$

holds for all $t \in \mathbf{R}_+$, $(\alpha, x) \in \{1, \ldots, N\} \times \Lambda$.

PROPOSITION 2. *The system* (6) *has a solution for any finite* $\Lambda \subset \mathbf{Z}^d$ $(\mid \Lambda \mid < < \infty)$.

3. GIBBSIAN QUASISTATES AND THE CORRESPONDING LANGEVIN'S EQUATIONS

Let $\Lambda \subset \mathbf{Z}^d$, $\mid \Lambda \mid < \infty$, be given.

Definition 9. A quasistate $\langle\cdot\rangle_\Lambda$ on $@_0^\Lambda$ is called Gibbsian if there exists a potential $U_\Lambda \in @_0^\Lambda$ such that for any $\xi \in @_0^\Lambda$

$$\langle\xi\rangle_\Lambda = \frac{\langle\xi \exp\{-U_\Lambda\}\rangle_B}{\langle\exp\{-U_\Lambda\}\rangle_B},$$

where $\langle V\rangle_B = \int V\, d\overline{\xi}^{\mu_\Lambda}\xi^{\mu_\Lambda}$ is the Berezin integral (Berezin, 1986) and μ_Λ is the maximal multiindex.
We always assume that the potential U_Λ is even. The Gibbsian quasistate with the potential U_Λ will be denoted by $\langle\cdot\rangle_{U_\Lambda}$.

Definition 10. By the system of Langevin's equations corresponding to the Gibbsian quasistate with potential U_Λ we mean the system of stochastic differential equations:

$$\begin{aligned} d\overline{\xi}^{(\alpha,x)}(t) &= -\frac{1}{2}\frac{\partial U_\Lambda(t)}{\partial_R \xi^{(\alpha,x)}(t)}\, dt + d\overline{w}^{(\alpha,x)}(t), \\ d\xi^{(\alpha,x)}(t) &= -\frac{1}{2}\frac{\partial U_\Lambda(t)}{\partial_L \overline{\xi}^{(\alpha,x)}(t)}\, dt + dw^{(\alpha,x)}(t), \end{aligned} \tag{7}$$

where differentiation is the Berezin differentiation, ∂/∂_L and ∂/∂_R stand for the left and the right derivatives respectively, and $U_\Lambda(t)$ is obtained from U_Λ by changing the variables $\overline{\xi}^{(\alpha,x)}$, $\xi^{(\alpha,x)}$ by $\overline{\xi}^{(\alpha,x)}(t)$, $\xi^{(\alpha,x)}(t)$, $(\alpha, x) \in \{1, \ldots, N\} \times \Lambda$, respectively.
Let $\Psi \in \overline{\mathfrak{A}_{\infty}^{\Lambda}}$ be a solution of the system (7), and let a quasistate $\langle\cdot\rangle \in \sigma^\Lambda$ on $@_{\infty}^{\Lambda,\infty}$ be given. Then Ψ defines a quasistate on the Grassman algebra $\mathcal{A}_{\infty}^{\Lambda}$ with the generators $\overline{\xi}^{(\alpha,x)}(t)$, $\xi^{(\alpha,x)}(t)$, $t \in \mathbf{R}_+$, $(\alpha, x) \in \{1, \ldots, N\} \times \Lambda$ by the relation

$$\langle\overline{\xi}^{\mu}(t)\,\xi^{\nu}(t)\rangle_\Lambda = \langle\Psi\overline{\xi}^{\mu}(t)\,\xi^{\nu}(t)\rangle, \quad \mu, \nu \in M_\Lambda.$$

The evolution of the quasistate $\langle\cdot\rangle_\Lambda$ on $@_0^\Lambda$ is defined by:

$$\langle\xi\rangle_{\Lambda,t} = \langle\xi(t)\rangle_\Lambda, \quad t \in \mathbf{R}_+, \quad \xi \in @_0^\Lambda.$$

PROPOSITION 3. *A Gibbsian quasistate on $@_0^\Lambda$ is invariant under the evolution determined by a solution of the corresponding system of Langevin's equations, i.e.,*

$$\langle\cdot\rangle_{\Lambda,t} = \langle\cdot\rangle_\Lambda = \langle\cdot\rangle_{U_\Lambda}$$

for any $t \in \mathbf{R}_+$.

We shall consider the case, when U_Λ is a small perturbation of a quadratic potential

$$U_\Lambda = U_\Lambda^\varepsilon = U_\Lambda^0 + \varepsilon V_\Lambda,$$

where

$$U_\Lambda^0 = \sum_{\alpha,\beta} \sum_{x,y\in\Lambda} a_\Lambda^{\alpha,\beta}(x-y)\overline{\xi}^{(\alpha,x)}\xi^{(\beta,y)},$$

$$V_\Lambda = \sum_{x\in\Lambda} P\Big(\overline{\xi}^{(\alpha,x)},\ \xi^{(\alpha,x)},\ \alpha\in\{1,\dots,N\}\Big),$$

$P\Big(\overline{\xi}^{(\alpha,x)},\ \xi^{(\alpha,x)},\ \alpha\in\{1,\dots,N\}\Big)$ is a polynomial of even degree in variables $\overline{\xi}^{(\alpha,x)},\ \xi^{(\alpha,x)},\ \alpha\in\{1,\dots,N\}$. The coefficients $a_\Lambda^{\alpha\beta}(y),\ \alpha,\beta\in\{1,\dots,N\},\ y\in\mathbf{Z}^d$, satisfy the following conditions. Let $\Lambda = [-N_\Lambda, N_\Lambda)^d \cap \mathbf{Z}^d$ then

a) for any $y \in \Lambda$ and any $\alpha,\beta\in\{1,\dots,N\}$, $a_\Lambda^{\alpha\beta}(y) = a^{\alpha\beta}(y)$, where $a^{\alpha\beta}(y)$ do not depend on Λ,

$$a^{\alpha\beta}(y) \neq 0, \text{ if } y \in Q, \quad a^{\alpha\beta}(y) = 0, \text{ if } y \notin Q,$$

where $Q \subset \Lambda$ and $|Q| < \infty$,

b) for any $\lambda \in [-\pi,\pi)^d$ the eigenvalues of the matrix $\widetilde{A}(\lambda)$ with entries

$$\widetilde{a}^{\alpha\beta}(\lambda) = \sum_{y\in\mathbf{Z}^d} a^{\alpha\beta}(y)\, e^{-i(\lambda,y)}, \quad \alpha\beta\in\{1,\dots,N\},$$

belong to the halfplane $\{z \in \mathbf{C} \mid \operatorname{Re} z > 0\}$.
It is known that for sufficiently small $\varepsilon > 0$ and for any $F \in @_0^{\mathbf{Z}^d}$ there exists the limit

$$\langle F\rangle_{U_\Lambda^\varepsilon} \to \langle F\rangle_\varepsilon$$

as $\Lambda \uparrow \mathbf{Z}^d$ (Malyshev, 1986).
For any finite cube $\Lambda \subset \mathbf{Z}^d$ the system of Langevin's equations corresponding to the Gibbsian quasistate with potential U_Λ defines the evolution $\langle\cdot\rangle_{t,\varepsilon,\Lambda}$ of a quasistate on $@_0^\Lambda$.

We define the evolution $\langle\cdot\rangle_{t,\varepsilon}$ of the quasistate $\langle\cdot\rangle_\varepsilon$ on $@_0^{\mathbf{Z}^d}$ as

$$\langle F\rangle_{t,\varepsilon} = \lim_{\Lambda\uparrow\mathbf{Z}^d}\langle F\rangle_{t,\varepsilon,\Lambda} \tag{8}$$

if this limit exists for any $F \in @_0^{\mathbf{Z}^d}$. It is easy to see that the quasistate $\langle\cdot\rangle_\varepsilon$ on $@_0^{\mathbf{Z}^d}$ is invariant under this evolution.

THEOREM 2. *Let the initial quasistate* $\langle\cdot\rangle_{0,\varepsilon,\Lambda}$, $\Lambda = [-N_\Lambda, N_\Lambda)^d \cap \mathbf{Z}^d$, *at* $t = 0$ *be Gibbsian with the potential:*

$$\sum_{\alpha,\beta}\sum_{x,y} b_\Lambda^{\alpha,\beta}(x-y)\bar{\xi}^{(\alpha,x)}\xi^{(\beta,y)} + \varepsilon\sum_{x\in\Lambda} P_0\big(\bar{\xi}^{(\alpha,x)},\ \xi^{(\alpha,x)},\ \alpha\in\{1,\ldots,N\}\big),$$

where $P_0\Big(\bar{\xi}^{(\alpha,x)},\ \xi^{(\alpha,x)},\ \alpha\in\{1,\ldots,N\}\Big)$ *is an even degree polynomial in variables* $\bar{\xi}^{(\alpha,x)},\ \xi^{(\alpha,x)},\ \alpha\in\{1,\ldots,N\}$, *and the coefficients* $b_\Lambda^{\alpha\beta}(y)$ *satisfy the conditions* a), b) *for the coefficients* $a_\Lambda^{\alpha\beta}(y),\ \alpha,\beta\in\{1,\ldots,N\},\ y\in\mathbf{Z}^d$, *given above.*
Then for sufficiently small $0<\varepsilon<\varepsilon_0$ *the limit* (8) *exists for any* $F \in @_0^{\mathbf{Z}^d}$, $t\in\mathbf{R}_+$ *and*

$$\langle F\rangle_{t,\varepsilon} \to \langle F\rangle_\varepsilon \qquad (t\to\infty).$$

REFERENCES

Berezin, F.A. (1986). *The Method of the Second Quantization.* Nauka, Moscow.

Dobrushin, R.L. (1968). Description of a random field by means of conditional probabilities and conditions for its regularity. *Theory Probab. Appl.* **13**, 201–229.

Gikhman, G.I. and Skorokhod, A.I. (1972). *Theory of Stochastic Differential Equation.* Nauka, Moscow.

Ignatiuk, I.A., Malyshev, V.A. and Sidoravičius, V. (1990). The convergence of the method of stochastic quantization. *Theory Probab. Appl.* (to appear).

Malyshev, V.A. (1986). The ultraviolet problems in the field theory. In: *Probability Theory. Mathematical Statistics. Theoretical Cybernetics.* **24**, VINITI, Moscow, pp. 111–186.

Malyshev, V.A. and Minlos, R.A. (1985). *Gibbsian Random Fields.* Nauka, Moscow.

Migdal, A.A. (1986). The stochastic quantization in the field theory. *Uspekhi Fiz. Nauk.* **149**, No 1.

Parisi, G. and Wu, Y. (1981). Perturbation theory without gauge fixing. *Scientia Sin.* **24**, 483–496.

Rozovsky, B.R. (1981). *Evolutional Stochastic Systems.* Nauka, Moscow.

Prob. Theory and Math. Stat., Vol. 1, pp. 539–549
B.Grigelionis *et al.* (Eds.)
1990 VSP/Mokslas

ADAPTIVE TESTS FOR MINIMAX TESTING OF NONPARAMETRIC HYPOTHESES

Yu.I. INGSTER
Leningrad Branch Steklov Institute of Mathematics,
Fontanka 27, 191011 Leningrad, USSR

ABSTRACT

Minimax problems of hypotheses testing abuot densities of distribution when the sets of alternative densities are the nonparametric set with eliminated neighbuorghoods of null-hypothesis are investigated. The adaptive classes of tests are constructed. The examples of normality, independence and spherical invariance hypotheses are considered and for some alternative sets the exact in the order and asymptotically exact results are given.

1. We shall consider the minimax problem of hypotheses testing. Let $(\mathfrak{X}, \mathfrak{A})$ be a measurable space with a measure P, Φ_0 and Φ_N be the sets of probability distribution densities on $(\mathfrak{X}, \mathfrak{A}, P)$, $\{x_1, \ldots, x_N\} = X_N$ be a random sample which is distributed according to an unknown density f. Let the null hypothesis H_0 be $f \in \Phi_0$, the alternative hypothesis H_1 be $f \in \Phi_N$. The minimax properties of a test $\Psi_N : \mathfrak{X}^N \longrightarrow [0,1]$ are described by the values

$$\alpha_N(\Psi_N, \Phi_0) = \sup_{f \in \Phi_0} \mathbf{E}_{P_f^N} \Psi_N,$$

$$\beta_N(\Psi_N, \Phi_N) = \sup_{f \in \Phi_N} \mathbf{E}_{P_f^N} (1 - \Psi_N)$$

or

$$\gamma_N(\Psi_N; \Phi_0, \Phi_N) = \alpha_N(\Psi_N, \Phi_0) + \beta_N(\Psi_N, \Phi_N).$$

The values

$$\beta_N(\alpha) = \beta_N(\alpha; \Phi_0, \Phi_N) = \inf_{\Psi_N : \alpha_N(\Psi_N, \Phi_0) \leqslant \alpha} \beta_N(\Psi_N, \Phi_0),$$

$$\gamma_N = \gamma_N(\Phi_0, \Phi_N) = \inf_{\Psi_N} \gamma_N(\Psi_N; \Phi_0, \Phi_N)$$

describe the discernibility of the hypotheses, $\alpha \in (0,1)$.
The object of our interest is to investigate the asymptotic (as $N \to \infty$) of these values and to obtain conditions of the discernibility ($\gamma_N \to 0$) or nondiscernibility ($\gamma_N \to 1$) where Φ_N are nonparametric sets.

2. The case of a simple null hypothesis $\Phi_0 = \{f_0\}$ was considered in (Ingster, 1987, 1988; Ermakov, 1988). Let Φ be a set of probability densities, $f_0 \in \Phi$, ρ_N be a sequence of positive numbers, $p \in [1, \infty]$,

$$\Phi_{\rho_N,p} = \left\{ f \in \Phi : \left\| \frac{f}{f_0} - 1 \right\|_{p,f_0} > \rho_N \right\}$$

where $\|\cdot\|_{p,f}$ is a norm in $L_p(\mathfrak{X}, \mathfrak{A}, P_f)$.
The conditions of discernibility and nondiscernibility in the case $\Phi_N = \Phi_{\rho_N,p}$ were considered in (Ingster, 1987). There we obtained asymptotically exact conditions in the case when $f_0 \equiv 1$ and $\Phi = H_k(r, L, p)$ is a set of r–smooth functions on $[0,1]^k$ in L_p–metric. The exact asymptotics of the values $\beta_N(\alpha)$ and γ_N in the case when $\Phi_N = \Phi_{\rho_N,2}$ and $\Phi = A_2(\lambda, L)$ is the set of densities $f(x)$, $x \in [0,1]$ which admit of periodical analitical extention to the set $\{z = x + iy : x \in R, \mid y \mid \leqslant \lambda\}$ and $\| f(\cdot \pm i\lambda) \|_{L_2(0,1)} < L$, were obtained in (Ingster, 1988). The exact asymptotic of $\beta_N(\alpha)$ when $\Phi_N = \Phi_{\rho_N,2}$ and Φ is an ellipsoid in $L_2[0,1]$ and some assumptions on its eigen–functions and on the rate of decrease of its eigen–values hold was announced in (Ermakov, 1988).
The construction of lower bounds for the values $\beta_N(\alpha)$ and γ_N is based on the investigation of Bayes hypothesis testing problems for special probability measures on special subsets in Φ_N. It is useful to consider special discrete subsets or spheres in Φ_N (Ingster, 1987, 1988) or Gaussian measures on special linear subspaces (Ermakov, 1988).
We denote

$$L_2 = \{\varphi \in L_2(\mathfrak{X}, \mathfrak{A}, P_{f_0}) : (\varphi, 1) = 0\},$$

$$\Phi_{f_0} = \left\{ \frac{f}{f_0} - 1 : f \in \Phi \right\}, \quad \Phi_{f_0,N} = \left\{ \frac{f}{f_0} - 1 : f \in \Phi_N \right\}.$$

Example. Let $\Phi_N = \Phi_{\rho_N,2}$ and r_n be the n–dimensional interior radius of Φ_{f_0} at the point 0,

$$n(\rho) = \max\{n : r_n > \rho\}, \quad \delta_N = N\rho_N^2 / \sqrt{n(\rho_N)}.$$

Then (Ingster, 1987) if $\delta_N = o(1)$ then $\gamma_N \to 1$ and if $\delta_N = O(1)$ then $\liminf \gamma_N > 0$.
The construction of upper bounds for the values $\beta_N(\alpha)$ and γ_N is based on the investigation of minimax properties for special classes of tests. The conditions of the minimax consistency $(\gamma_N(\Psi_N; \Phi_0, \Phi_N) \to 0)$ for χ^2–tests and for their generalizations χ^p–tests, $p > 2$, were obtained in (Ingster, 1987). In some examples these classes of tests are minimax in the order.
Now we shall obtain some extention of the results of (Ingster, 1987) about classes of tests which are based on quadratic forms of increasing dimensions. We shall consider U–statistic (Hoeffding, 1948) tests with kernels of increasing rank.
Let $K_N(x, y)$ be a sequence of symmetric non–negative definite functions on $\mathfrak{X} \times \mathfrak{X}$. Let K_N be the integral operator in $L_2(\mathfrak{X}, \mathfrak{A}, P_{f_0})$ with the kernel $K_N(x, y)$. We assume that $K_N \mathbf{1} = 0$. Let $\| K_N \|_{\infty,f_0}$ and $\| K_N \|_{2,f_0}$ be the uniform norm and the Hilbert — Schmidt norm,

$$n = \| K_N \|_{2,f_0}^2 \; / \; \| K_N \|_{\infty,f_0}^2,$$

$$K_{N,\infty} = \operatorname{ess\,sup} K_N(x,x)/\| K_N\|_{\infty,f_0}, \quad r(N) = N^{-1/2}n^{1/4}.$$

Example. Let $K_N = \mathrm{Pr}_{L,f_0}$ be the orthonormal projection of $L_2(\mathfrak{X}, \mathfrak{A}, P_{f_0})$ onto a linear subspace $L = L_N \subset L_2$. Let $\varphi_1, \ldots, \varphi_m$ be an orthonormal basis in L. Then $n = m$,

$$K_N(x,y) = \sum_{i=1}^{m} \varphi_i(x)\varphi_i(y), \quad K_{N,\infty} = \operatorname{ess\,sup} \sum_{i=1}^{m} \varphi_i^2(x)$$

and

$$K_{N,\infty} = O(n) \quad \text{if} \quad \max_i \|\varphi_i\|_{\infty,f_0} = O(1).$$

We consider generated U–statistics and U_N–tests:

$$U_N = \frac{2}{N} \sum_{1 \leqslant \alpha < \beta \leqslant N} K_N(x_\alpha, x_\beta),$$

$$\Psi_N = \Psi_{N,T_N} = \mathbf{1}_{\{U_N > \sqrt{2}\, \| K_N\|_{2,f_0} T_N\}}$$

where T_N is a sequence of positive values. Let

$$\delta_N = \delta_N(\Phi_N, K_N) = \inf_{\varphi \in \Phi_{f_0,N}} N \frac{(\varphi, K_N\varphi)}{\sqrt{2}\, \|K_N\|_{2,f_0}}.$$

Under some assumptions the values δ_N describe the minimax properties of U_N–tests and we can find the optimal kernels $K_N(x,y)$ by the maximization of the values $\delta_N(\Phi_N, K_N)$. Let

$$\delta(L,\Phi) = \inf_{\varphi \in \Phi} \|\varphi - \mathrm{Pr}_{L,f_0}\varphi\|_{2,f_0}$$

be the accuracy of the approximation of a set $\Phi \subset L_2$ by subspace $L \subset L_2$. Let $L_N(\lambda)$ be the image of the operator $(K_N - \lambda\| K_N\|_{\infty,f_0} I)_+$ where I is the unit operator in L_2, A_+ is the positive part of self–conjugate operator A.

THEOREM 1.

1. *Let* $\delta_N \to \infty$ *and one of the assumptions* a)–c) *holds:*
a) $\|\varphi\|_{\infty,f_0} < B$ *for every* $\varphi \in \Phi_{f_0,N}$ *and some* $B > 0$;
b) $K_{N,\infty} = O(n)$ *and* $\|\varphi\|_{2,f_0} < B$ *for every* $\varphi \in \Phi_{f_0,N}$ *and some* $B > 0$;
c) $K_{N,\infty} = o(N\sqrt{n})$ *and for every* $\varphi \in \Phi_{f_0,N}$ *and some* $b > 0$, $B > 0$

$$(\varphi, K_N\varphi) \,/\, \| K_N\|_{\infty,f_0} \geqslant b\, \| \varphi \|_{2,f_0}^2 - Br^2(N). \tag{1}$$

Then $\gamma_N(\Psi_{N,T_N}; \{f_0\}, \Phi_N) \to 0$ *for any* $T_N \to \infty$.

2. *Let* $n \to \infty$ *and one of the assumptions* a)–b) *holds:*
a) $K_{N,\infty} = o(\sqrt{N\,n})$ *and the assumption* c) *of* n.1;
b) $K_{N,\infty} = O(n)$ *and* $\delta(L_N(\lambda), \Phi_{f_0,N}) \to 0$ *for any* $\lambda \in (0,1)$.

Let $T_\alpha = G^{-1}(1-\alpha)$ *where* G *is the Laplace function* $\alpha \in (0,1)$. *Then*

$$\alpha(\Psi_{N,T_\alpha}, \{f_0\}) = \alpha + o(1), \quad \beta_N(\Psi_{N,T_\alpha}, \Phi_N) = G(T_\alpha - \delta_N) + o(1).$$

The proof of theorem 1 is based on estimation of moments of U–statistics (Hoeffding, 1948) and on the central limit theorem for U–statistics (Hall, 1984).
If $K_N = \Pr_{L_N, f_0}$, $n = \dim L_N$, and

$$\delta(L_N, \Phi_{f_0,N}) = O(r(N)) \tag{2}$$

then (1) holds for some $b > 0$, $B > 0$. The conditions of type (2) hold for asymptotically optimal projecting subspaces L_N (Ingster, 1987, 1988) and for asymptotically optimal kernels $K_N(x,y)$ which correspond to the case considered in (Ermakov, 1988).

Example. Let $\Phi_N = \Phi_{\rho_N, p}$, $p \leqslant 2$ and $\| \varphi \|_{\infty, f_0} < B$ for every $\varphi \in \Phi_{f_0}$ and some $B > 0$. Let d_n be the n–dimensional Kolmogorov's diameter of Φ_{f_0} and $\max\limits_n N(\rho_N^2 - d_n^2)/\sqrt{n} \to \infty$. Then

$$\gamma_N(\{f_0\}, \Phi_N) \longrightarrow 0.$$

3. The object of our interest is to extend results of (Ingster, 1987, 1988, Ermakov, 1988) for the case of a simple null–hypothesis to the case of composite and non–parametric hypothesis Φ_0. Some results are announced in (Ermakov, 1988) but we shall consider another variant of this problem. We shall assume that $\Phi_0 \subset F$, $\Phi_N \subset F$ and that the pair (F, Φ_0) have a fibre structure: the map $g : F \to \Phi_0$ is given, $g(f_0) = f_0$ if $f_0 \in \Phi_0$. The sets $g^{-1}(f_0)$ we can consider as fibres. We define the sets

$$F_{f_0} = \left\{ \frac{f}{f_0} - 1 : f \in F,\ g(f) = f_0 \right\},$$

$$\Phi_{f_0,N} = \left\{ \frac{f}{f_0} - 1 : f \in \Phi_N,\ g(f) = f_0 \right\}.$$

For $\Phi \subset F$ and a sequence ρ_N we denote

$$\Phi_{\rho_N, p} = \left\{ f \in \Phi : \left\| \frac{f}{g(f)} - 1 \right\|_{p, g(f)} > \rho_N \right\}. \tag{3}$$

The fibre structure arises in finite parametric hypothesis testing problems (Borovkov, 1984; Ingster, 1985). We can consider the family of simple hypothesis testing problems $H_0 : f = f_0$, $f_0 \in \Phi_0$ against alternative $H_1 : f \in \Phi_N \cap g^{-1}(f_0)$ and lower bounds may be deduced from the results for a simple null–hypothesis. These bounds are sufficiently precise in many cases, see (Borovkov, 1984; Ingster, 1985). But we need another methods to construct sufficiently good tests. In this and the next sections we consider adaptive U–statistic tests. We divide the sample X_N into two parts $X_{N_1} = \{x_1, \ldots, x_{N_1}\}$ and $X_{N_2} = \{x_{N_1+1}, \ldots, x_N\}$, $N_2 = N - N_1$. Using the first part X_{N_1} we "adapte" the kernels and using the

second part X_{N_2} we construct U-statistics and tests. We consider one class of such "adaptive" tests.
Let $(\Delta, \mathfrak{B}, \lambda)$ be a probability space where Δ is a convex compact set in a Banach space. Assume we have a family of maps $h_{f_0} : \mathfrak{X} \to \Delta$, $f_0 \in \Phi_0$, such that the maps $H_{f_0} : L_2(\Delta, \mathfrak{B}, \lambda) \to L_2(\mathfrak{X}, \mathfrak{A}, P_{f_0})$ are isomorphismes, $(H_{f_0}\varphi)(x) = \varphi h_{f_0}(x)$. We denote

$$L_2 = \{\varphi \in L_2(\Delta, \mathfrak{B}, \lambda) : (\varphi, 1) = 0\},$$

and let $\| \cdot \|_p$ be the norm in $L_p(\Delta, \mathfrak{B}, \lambda)$.
Let a family of sets $V_{f_0,N} \subset L_2$ be given. We assume that $\Phi_{f_0,N} = F_{f_0} \cap H_{f_0} V_{f_0,N}$ or

$$\Phi_N = \left\{ f \in F : H^*_{g(f)}\left(\frac{f}{g(f)} - 1\right) \in V_{f_0,N} \right\} \tag{4}$$

where H^* is the conjugate map. Let $W_{f_0,N} = V_{f_0,N} \cap H^*_{f_0} F_{f_0}$.
If a set $V \subset L_2$ and a sequence ρ_N are given we can consider the sets $V_{\rho_N,p} = \{\varphi \in V : \| \varphi \|_p > \rho_N\}$ and

$$\Phi = \left\{ f \in F : H^*_{g(f)}\left(\frac{f}{g(f)} - 1\right) \in V \right\}. \tag{5}$$

The set $\Phi_{\rho_N,p}$ in (3) has the form of Φ_N in (4) if $V_{f_0,N} = V_{\rho_N,p}$.
It is useful to define the set $\Phi_{0,\Delta}$ which corresponds to Φ_0 in L_2:

$$\Phi_{0,\Delta} = \bigcup_{f_0 \in \Phi_0,\, f_1 \in \Phi_0} H^*_{f_0}\left(\frac{f_1}{f_0} - 1\right) \subset L_2.$$

We shall say that a linear subspace $L_{0,\Delta} \subset L_2$ is tangent for $\Phi_{0,\Delta}$ if there exist $\delta > 0$, $B > 0$ such that for every $\varphi \in \Phi_{0,\Delta}$ $\| \varphi \|_2 < \delta$ we can find $h \in L_{0,\Delta}$ such that $\| \varphi - h \|_2 \leqslant B \| h \|_2^2$.
We consider three classical null-hypotheses.

Example 1. *The hypothesis of normality.* Let $(\mathfrak{X}, \mathfrak{A}) = (\mathbf{R}^k, \mathfrak{B})$, $k \geqslant 1$, P be the Lebesgue measure (as in the other examples), $\Phi_0 = \{N_{m,\Sigma}(x) : m \in \mathbf{R}^k, \Sigma > 0\}$ be the set of Gaussian densities on $\mathbf{R}^k$. The map $g(f) = N_{m(f),\Sigma(f)}$ is defined on the set of densities f with finite second order moments. Let $\Delta = [0,1]^k$, λ be the uniform measure (as in the other examples) and $h_{m,\Sigma}(x) = h(\Sigma^{-1/2}(x - m))$ where $h(x^1, \ldots, x^k) = (G(x^1), \ldots, G(x^k))$ maps the standard Gaussian measure on $\mathbf{R}^k$ into λ.
The tangent space $L_{0,\Delta}$ is generated by statistics $G^{-1}(u^i)$ and $G^{-1}(u^j) \times G^{-1}(u^l)$, $i, j, l = 1, \ldots, k$, where G is the Laplace function.

Example 2. *The hypothesis of independence* (Ingster, 1986). Let $(\mathfrak{X}, \mathfrak{A}) = (\mathbf{R}^k, \mathfrak{B})$, $k \geqslant 2$, Φ_0 be a set of densities $f(x^1, \ldots, x^k) = f_1(x^1) \cdot \ldots \cdot f_k(x^k)$. Let $g(f)$ be the product of marginal densities. We assume that $\Delta = (0,1]^k$ and if $f \in \Phi_0$ then $h_f(x) = (F_1(x^1), \ldots, F_k(x^k))$ where F_i is the distribution function for f_i. The tangent space is

$$L_{0,\Delta} = \left\{ \varphi(u^1, \ldots, u^k) = \sum_i \varphi_i(u^i) : \int_0^1 \varphi_i(u)\, du = 0,\ i = 1, \ldots, k \right\}.$$

Example 3. The hypothesis of spherical invariance. Let $(\mathfrak{X}, \mathfrak{A}) = (\dot{\mathbf{R}}^k, \mathfrak{B})$, $k \geqslant 1$. We consider the spherical coordinates in $\dot{\mathbf{R}}^k$: $r = \| x \|$, $\theta = x / \| x \| \in S^{k-1}$. Let $\Phi_0 = \{f(r,\theta) = f(r)\}$ and $g(f)(r)$ be the mean of $f(r,\theta)$ when $\theta \in S^{k-1}$. Let Δ be the unit ball D^k in $\mathbf{R}^k$ and $h_{f_0}(r,\theta) = (F_{f_0}^{1/k}(r), \theta)$ where F_{f_0} is the distribution function of $\| x \|$ for the density f_0. In this case

$$\Phi_{0,\Delta} \subset L_{0,\Delta} = \Big\{\varphi(r,\theta) = \varphi(r) : \int_0^1 \varphi(r)\, r^{k-1} dr = 0\Big\}.$$

We can assume that $F = \{f : \| f / g(f) - 1 \|_{2,g(f)} < \infty\}$ and $H_{f_0}^* \dot{F_{f_0}} = L_2 \ominus L_{0,\Delta}$ in examples 1—3. Let we have a sequence of symmetric non-negative definite functions $K_N(u,v)$, $u,v \in \Delta$. Let K_N be the operator in $L_2(\Delta, \mathfrak{B}, \lambda)$ with the kernal $K_N(u,v)$, $K_N \mathbf{1} = 0$ the values $\| K_N \|_2$, $\| K_N \|_\infty$, $K_{N,\infty}$, n and $r(N)$ are defined as n.2. Let a sequence $N_1 = N_1(N) < N(N_2 = N - N_1)$, and a sequence of random maps $h_{N_1}^* = h_{N_1, X_{N_1}}^* : \mathfrak{X} \to \Delta$ be given.
In example 1 we consider $h_{N_1}^* = h_{g_{N_1}^*}$ where $g_{N_1}^* = N_{m_{N_1}^*, \Sigma_{N_1}^*}$ $m_{N_1}^*$ and $\Sigma_{N_1}^*$ are the standard estimators of a mean vector and a covariance matrix respectively.
In example 2 we consider $h_{N_1}^*(x) = (F_{1,N_1}(x^1), \ldots, F_{k,N_1}(x^k))$ where F_{i,N_1} is the empirical distribution function corresponding to x^i. In example 3 we consider $h_{N_1}^*(r,\theta) = (F_{N_1}^{1/k}(r), \theta)$ where F_{N_1} is the empirical distribution function corresponding to $\| x \|$.
We consider the random kernels $K_N^*(x,y) = K_N(h_{N_1}^*(x), h_{N_1}^*(y))$ and corresponding U-statistics and tests

$$U_{N,N_1}^* = \frac{2}{N_2} \sum_{N_1 < \alpha < \beta \leqslant N} K_N^*(x_\alpha, x_\beta),$$

$$\Psi_N = \Psi_{N,N_1,T_\alpha} = \mathbf{1}_{\{U_{N,N_1}^* > \sqrt{2}\, \|K_N\|_2 T_N\}}. \tag{6}$$

Let

$$\delta_N = \inf_{f_0 \in \Phi_0,\, \varphi \in W_{N,f_0}} N \frac{(\varphi,\ K_N \varphi)}{\sqrt{2} \, \| K_N \|_2}.$$

We assume in this and the next sections that the following assumptions on W_{N,f_0} and K_N hold. Let $L > 0$ and

$$\mathrm{Lip}_\Delta(L) = \{\varphi : | \varphi(u) - \varphi(v) | \leqslant L \| u - v \|;\ u,v \in \Delta\}.$$

We assume that $W_{N,f_0} \subset \mathrm{Lip}_\Delta(L)$ and that unequality (1) holds for some $b > 0$, $B > 0$ and every $f_0 \in \Phi_0$ and $\varphi \in W_{N,f_0}$.
At first we consider the case when $h_{N_1}^* = h_{g_{N_1}^*}$, where $g_{N_1}^* = g_{N_1, x_{N_1}}^* \in \Phi_0$ is an estimator of the density $g(f)$ (as in example 1).
Let $\xi_{N_1,f_0}(x) = h_{N_1}^*(x) - h_{f_0}(x)$ and $\rho(N)$ be a sequence of positive numbers. We shall say that $h_{N_1}^*$ is $(p, \rho(N))$-consistent if for every $b > 0$

$$\sup_{f \in \Phi_N \cup \Phi_0} P_f^{N_1}\Big(\| \xi_{N_1,f_0} \|_{p,g(f)} > b\rho(N)\Big) \to 0 \tag{7}$$

and it is weakly $(p, \rho(N))$–consistent if for every $\varepsilon > 0$ there exist such $N_0 > 0$ and $b > 0$ that the left part of (7) is less than ε if $N > N_0$.
Let $\xi_{N_1,f_0} = \frac{f_0}{g^*_{N_1}} - 1$. We shall say that $g^*_{N_1}$ is $(p, \rho(N))$–consistent if (7) holds for every $b > 0$. The weak $(p, \rho(N))$–consistency is defined in a similar way.

THEOREM 2.

1. *Let* $N_1 = O(N_2)$, $\delta_N \to \infty$, $K_{N,\infty} = o(N, \sqrt{n})$, $h^*_{N_1}$ *be weakly* $(2, r(N))$–*consistent and one of the assumptions* a)—c) *hold:*
a) $g^*_{N_1}$ *is weakly* $(2, r(N))$–*consistent;*
b) *there is a tangent space* $L_{0,\Delta} \subset \operatorname{Ker} K_N$ *and* $g^*_{N_1}$ *is weakly* $(2, r^{1/2}(N))$ *and* $(\infty, 1)$–*consistent;*
c) $\Phi_{0,\Delta} \subset \operatorname{Ker} K_N$ *and* $g^*_{N_1}$ *is weakly* $(\infty, 1)$–*consistent.*
Then

$$\gamma_N(\Psi_N; \Phi_0, \Phi_N) \to 0$$

for any sequence $T_N \to \infty$.

2. *Let* $N_1 = o(N_2)$, $h^*_{N_1}$ *be* $(2, r(N))$–*consistent,* $K_{N,\infty} = o(\sqrt{N\,n})$ *and one of the assumptions* a)—c) *of section* 1 *with weak consistency replaced by consistency holds.*
Then for $\Psi_N = \Psi_{N,N_1,T_\alpha}$

$$\alpha_N(\Psi_N, \Phi_0) = \alpha + o(1), \quad \beta_N(\Psi_N, \Phi_N) = G\,(T_\alpha - \delta_N) + o(1).$$

One can see that consistency assumptions on $h^*_{N_1}$ and a) in example 1 hold if $N\,n^{-1/2} = o(N_1)$.
4. Now we consider the case when there are not enough good estimators $g^*_{N_1}$ (as in examples 2 and 3) but there are discretezations or sequencies of partitions of Δ.
Let the sequencies of probability spaces (Ξ_N, A_N, μ_N) and maps $d_N : \Delta \to \Xi_N$; $\lambda(d_N^{-1}(A)) = \mu_N(A)$ for $A \in A_N$ be given. We denote

$$\mathbf{D}_N : L_2(\Delta, \mathfrak{B}, \lambda) \to L_2(\Xi_N, A_N, \mu_N)$$

where $(\mathbf{D}_N\varphi)(\alpha) = \mathbf{E}(\varphi \mid d_N(u) = \alpha)$. Let $\Phi_{0,\Xi} = \mathbf{D}_N\Phi_{0,\Delta}$. The tangent space $L_{0,\Xi}$ to $\Phi_{0,\Xi}$ is defined as in section 2.
In example 2 the discretezations are based on the partitions of the cube $(0,1]^k$ into m^k cubes, $m = m_N \to \infty$; $\mathbf{D}_N = \{1, \ldots, m\}^k$ and $d_N(x) = (i_1, \ldots, i_k)$ iff $x^l \in ((i_l - 1)/m,\ i_l/m]$, $l = 1, \ldots, k$.
The tangent space is

$$L_{0,\Xi} = \Big\{\varphi(i_1, \ldots, i_k) = \sum_{l=1}^{k} \varphi_l(i^l) : \sum_{\alpha} \varphi_l(\alpha) = 0, \quad l = 1, \ldots, k\Big\}.$$

In example 3 the discretezations are based on the λ–uniform partitions of $\mathbf{D}^k$ into $m = m_N \to \infty$ rings: $\Xi_N = \{1, \ldots, m\} \times S^{k-1}$ and $d_N(r, \theta) = (\alpha, \theta)$ if $r \in \big[((\alpha - 1)/m)^{1/k},\ (\alpha/m)^{1/k}\big]$;

$$\Phi_{0,\Xi} \subset L_{0,\Xi} = \Big\{\varphi(\alpha, \theta) = \varphi(\alpha) : \sum_{\alpha} \varphi(\alpha) = 0\Big\}.$$

We consider the kernels

$$K^N(\alpha,\beta) = \mathbf{E}\big(K_N(u,v) \mid d_N(u) = \alpha,\ d_N(v) = \beta\big),$$

$$K_N^*(x,y) = K^N\big(d_N h_{N_1}^*(x),\ d_N h_{N_1}^*(y)\big)$$

and then we construct U-statistics U_{N,N_1} and U_N-tests Ψ_{N,N_1,T_N} as in (6). Let $P_{f_0,N}^*(A) = P_{f_0}(h_{N_1}^{-1} d_N^{-1}(A))$ be a random measure on $(\Xi_N, \mathcal{A}_N)$ and assume that there is a random density $p_{f_0,N}^* = dP_{f_0,N}^*/d\mu_N$.
We denote

$$\delta_\alpha = d_N^{-1}(\alpha), \quad \delta_\alpha^* = (h_{N_1}^*)^{-1}(\delta_\alpha), \quad dm_N(\alpha) = \operatorname{diam} \delta_\alpha,$$

$$\xi_{f_0,N}(\alpha) = \sup_{x\in\delta_\alpha^*} \|h_{N_1}^*(x) - h_{f_0}(x)\|, \quad \xi_{N,f_0,p} = \|\xi_{f_0,N}\|_p.$$

We shall say that $(h_{N_1}^*, d_N)$ is $(p, \rho(N))$-consistent if for every $b > 0$

$$\sup_{f\in\Phi_N\cup\Phi_0} P_f^{N_1}(\xi_{N,g(f),p} > b\rho_N) \to 0; \tag{8}$$

the weak consistency is defined in a similar way.
Let $\xi_{N,f_0,p} = \|(p_{f_0,N}^*)^{-1} - 1\|_{p,p_{f_0,N}^*}$. We shall say that $p_{f_0,N}^*$ is $(b,p,\rho(N))$-consistent if (8) holds; it is $(p,\rho(N))$-consistent if (8) holds for every $b > 0$; the weak consistency is defined in a similar way.
Let $\delta^N = \delta_N \parallel K_N \parallel_2 / \parallel K^N \parallel_2$, where K_N is the operator with the kernel $K^N(\alpha,\beta)$. It is clear that $\delta^N \geqslant \delta_N$.

THEOREM 3.

1. *Let* $\delta^N \to \infty$, $N_1 = O(N_2)$, $K_{N,\infty} = o(N\sqrt{n})$, $\|dm_N\|_{\mu_N,2} = O(r(N))$, $(h_{N_1}^*, d_N)$ *is weakly* $(2, r(N))$*-consistent and one of the assumptions* a)—c) *holds:*
a) $p_{f_0,N}^*$ *is weakly* $(2, r(N))$*-consistent;*
b) *there is a tangent space* $L_{0,\Xi} \subset \operatorname{Ker} K^N$ *and* $p_{f_0,N}^*$ *is weakly* $(2, r^{1/2}(N))$ *and* $(\infty, 1)$*-consistent;*
c) $\Phi_{0,\Xi} \subset \operatorname{Ker} K^N$ *and* $p_{f_0,N}^*$ *is weakly* $(\infty,1)$*-consistent.*
Then $\gamma_N(\Psi_N, \Phi_0, \Phi_N) \to 0$ *for any sequence* T_N.

2. *Let* $K_{N,\infty} = o(\sqrt{Nn})$, $N_1 = o(N_2)$, $\parallel dm_N \parallel_{\mu_N,2} = o(r(N))$, $(h_{N_1}^*, d_N)$ *is* $(2, r(N))$*-consistent and one of the assumptions* a)—c) *of section* 1 *with weak consistency replaced by consistency holds. Then*

$$\alpha_N(\Psi_{N,N_1,T_\alpha}; \Phi_0) = \alpha + o(1), \quad \beta_N(\Psi_{N,N_1,T_\alpha}, \Phi_N) = G(T_\alpha - \delta^N) + o(1).$$

Using the properties of beta distributions one can see (Ingster, 1986) that the assumptions of theorem 3 hold in example 2 if

$$N^{1/2} n^{-1/4} \ll m \ll \min(N_1/\ln N_1,\ N_1^{1/2} n^{1/4})$$

and in example 3 if

$$\max(N^{k/(k+2)}, N^{1/2}) \ll m \ll N_1/\ln N_1.$$

Using theorems 2 and 3 and the methods of (Ingster, 1988; Ermakov, 1988) it is possible to extend the results of these papers to the composite hypotheses of the considered type. For example the following theorem is true.

THEOREM 4. *Let* Φ_0 *and* Δ, $L_{0,\Delta}$ *be the sets from examples 1—3,* $\varphi_1, \ldots, \varphi_n, \ldots$ *be an orthonormal sistem in* $L_2 \ominus L_{0,\Delta}$ *and* $\| \operatorname{grad} \varphi_n \|_\infty < B\, n^b$ *for some* $B > 0$, $b > 0$. *Let* $\lambda > 0$, $L > 0$ *and the sequences* a_n, ρ_N *be given,* $a_n \asymp \exp n\,\lambda$. *Let*

$$V = \Big\{ \varphi = \sum_{i=1}^{\infty} \varphi_i \theta_i : \sum_i \theta_i^2 a_i^2 < L \Big\}$$

and $\Phi_N = \Phi_{\rho_N,2}$ *be the sets of the type* (4)—(6). *Then*

$$\beta_N(\alpha) = G\left(T_\alpha - N\rho_N^2 \sqrt{\frac{\lambda}{2 \ln N}} \right) + o(1),$$

$$\gamma_N = G\left(- N\rho_N^2 \sqrt{\frac{\lambda}{8 \ln N}} \right) + o(1).$$

5. In this section we consider adaptive χ^p–tests, $p \geqslant 2$, using the structures of sections 3 and 4. We assume that discretezations d_N, Ξ_N where $\Xi_N = \{1, \ldots, n\}$ are given $n = n(N) \to \infty$ and $\mu_N(\alpha) = n^{-1}$. Let ν^*_{α,N_2} be the empirical frequencies of $\delta^*_\alpha = \delta^*_{\alpha,N_1}$ which are computed by the second part $\mathbf{X}_{N_2}$ of the sample $\mathbf{X}_N$,

$$t_{N,p} = \begin{cases} \sum_\alpha |\nu_{\alpha,N_2} - 1/n|^p, & 2 \leqslant p < \infty, \\ \max_\alpha |\nu_{\alpha,N_2} - 1/n|, & p = \infty; \end{cases}$$

$$\chi_N^p = \chi_{N,N_1,T_N}^p = \mathbf{1}_{\{t_{N,p} > T_N\}}.$$

In this case $p^*_{N,f} = \{n\, P_f(\delta^*_{\alpha,N_1})\}$. We define

$$r_p(N) = (N\, n^{(1/p)-1})^{-1/2} \quad \text{for} \quad p < \infty$$

and $r_\infty(N) = (n \ln n/N)^{1/2}$.
Let

$$L_N^* = \operatorname{Lin}\{\mathbf{1}_{\delta^*_{1,N_1}}, \ldots, \mathbf{1}_{\delta^*_{n,N_1}}\}.$$

THEOREM 5. *Let* $N_1 < A N_2$ *for some* $A > 0$, $\| f / g(f) - 1 \|_{p,g(f)} < B$ *for every* $f \in \Phi_N$ *and some* $B > 0$.

1. *Let* $2 \leqslant p < \infty$, $n = o(N^2)$ *for* $p = 2$ *and* $n = o(N)$ *for* $p > 2$, p^*_{N,f_0} *be weakly* $(2, r_2(N))$ *and* $(p, r_p(N))$*–consistent and* $(b, \infty, 1)$*–consistent for some* $b \in (0,1)$. *Let*

$$\sup_{f \in \Phi_N} P_f^{N_1} \left(\left\| \mathrm{Pr}_{L_N^*, g(f)} \left(\frac{f}{g(f)} - 1 \right) \right\|_{p,g(f)} < C\, r_p(N) \right) \to 0 \qquad (9)$$

for every $C > 0$. Then $\gamma_N(\chi_N^p; \Phi_0, \Phi_N) \to 0$ for any T_N.

2. *Let $p = \infty$, $n = o(N/\ln N)$ and p^*_{N,f_0} be $(b, \infty, r_\infty(N))$–consistent for some $b > 0$. Then we can find $C = C(A, B, b)$ in (9) and a sequence T_N such that $\gamma_N(\chi_N^\infty; \Phi_0, \Phi_N) \to 0$ if (9) holds.*

The proof of theorem 5 is based on the methods of (Ingster, 1986). It is possible to extend the results of (Ingster, 1987) for smooth alternatives to the composite hypotheses in examples 1—3.

THEOREM 6. *Let Φ_0 and $\Delta \subset \mathbf{R}^k$ be the sets from examples 1—3, $V = H_\Delta(r, L, q)$ be the set of r–smooth functions on Δ, ρ_N be a sequence of positive numbers and $\Phi_N = \Phi_{\rho_N,p}$ be a set of the types (4)—(6), $1 \leqslant p \leqslant \infty$. Let $q \geqslant p$ for $p \leqslant 2$ and $q = p$, $4rp > k$ for $p > 2$. Let*

$$\xi_N = \begin{cases} \rho_N N^{2r/(4r+k)}, & p \leqslant 2\,; \\ \rho_N N^{r/(2r+k(1-p^{-1}))}, & 2 < p < \infty; \\ \rho_N (N/\ln N)^{r/(2r+k)}, & p = \infty. \end{cases}$$

Let $p < \infty$. If $\xi_N \to \infty$ then $\gamma_N = \gamma_N(\Phi_0, \Phi_N) \to 0$; if $\xi_N = O(1)$ then $\liminf \gamma_N > 0$ and if $\xi_N \to 0$ then $\gamma_N \to 1$.
Let $p = \infty$. We can find such $\lambda_1 \geqslant \lambda_2 > 0$ that if $\liminf \xi_N > \lambda_1$ then $\gamma_N \to 0$ and if $\limsup \xi_N < \lambda_2$ then $\gamma_N \to 1$.

The proof of theorem 6 is based on the results of (Ingster, 1986, 1987) and on theorem 5. We consider the discretezations which are based on the λ–uniform partitions of $(0, 1]^k$ into cubes (in example 1, 2) and on the partitions of the unit ball into m rings and their λ–uniform partitions into sets with the equal in the order diameters (in example 3). In example 3 for $k = 1$ (the hypothesis of a density symmetry) we consider a different variant of χ^p–tests which is based on the statistic

$$t^p_{N,N_1} = \sum_{\alpha=1}^{n} \left|\nu^+_{\alpha,N_2} - \nu^-_{\alpha,N_2}\right|^p$$

where ν^+_{α,N_2} and ν^-_{α,N_2} are the empirical frequencies of symmetric intervals δ^+_{α,N_1} and δ^-_{α,N_1}.

REFERENCES

Borovkov, A.A. (1984). *Mathematical Statistics.* Nauka, Moscow.

Ermakov, M.C. (1988). Asymptotically minimax testing of non–parametric hypothesis about density function. In: *Notes Sci. Seminars of Leningr. Branch of Math. Inst.* **166**, 44–53.

Hall, P. (1984). Central limit theorem for integrated square error of multivariate nonparametric density estimators. *J. Multivar. Anal.* **14**, 1–16.

Hoeffding, W. (1948). A class of statistics with asymptotically normal distribution. *Ann. Math. Statist.* **19**, 293–325.

Ingster, Yu.I. (1985). Asymptotically optimal statistics for testing of complex hypothesis with finite number of parameters. *Theory Probab. Appl.* **30**, 289–308.

Ingster, Yu.I. (1986). Asymptotically minimax testing of independence hypothesis. In: *Notes Sci. Seminars of Leningr. Branch of Math. Inst.* **153**, 60–72.

Ingster, Yu.I. (1987). Asymptotically minimax testing of nonparametric hypothesis. In: *Proc. Fourth Vilnius Conf., 1985.*, **1**, VNU Sci. Press, Utrecht, pp. 553–573.

Ingster, Yu.I. (1988). Minimax testing of a simple hypothesis against weakly nonparametric sets of alternatives. *Theory Probab. Appl.* **33**, 592–596.

Prob. Theory and Math. Stat., Vol. 1, pp. 550–572
B.Grigelionis *et al.* (Eds.)
1990 VSP/Mokslas

A SURVEY OF THE METRIC THEORY OF CONTINUED FRACTIONS, FIFTY YEARS AFTER DOEBLIN'S 1940 PAPER

M. IOSIFESCU
Centre of Mathematical Statistics, Bd. Magheru 22,
RO-70158 Bucharest 22, Romania

ABSTRACT

This is a survey of the state of research in the metric theory of continued fractions from the perspective of the celebrated Doeblin's 1940 paper "Remarques sur la théorie métrique des fractions continues", *Compositio Mathematica* **7**, 353–371, which can be viewed as having set up a working program, only nowadays near completion.

INTRODUCTION

Let Ω denote the set of irrational numbers in the unit interval $I = [0,1]$. Given $\omega \in \Omega$, let $a_1(\omega), a_2(\omega), \ldots$ be the sequence of the partial quotients of the continued fraction expansion of ω constructed as follows. Define $T : \Omega \to \Omega$ as $T(\omega) =$ fractionary part of $1/\omega$, $\omega \in \Omega$. Then $a_{n+1}(\omega) = a_1(T^n(\omega))$, $n \in \mathcal{N}^* = \{1, 2, \ldots\}$, with $a_1(\omega) = [1/\omega]$ – integer part of $1/\omega$, $\omega \in \Omega$. Let $\mathcal{B}_I$ denote the σ-algebra of the Borel subsets of I. Clearly, the a_n, $n \in \mathcal{N}^*$, are random variables on $(I, \mathcal{B}_I)$. They are defined almost surely (a. s.) with respect to any probability measure on $\mathcal{B}_I$ assigning probability 0 to the set $I - \Omega$ of rational numbers in I. For our purposes here, the most important probability measures of this kind are Lebesgue measure λ and Gauss' measure γ defined as

$$\gamma(A) = \frac{1}{\log 2} \int_A \frac{dx}{1+x}, \quad A \in \mathcal{B}_I.$$

The mean value operators under λ (respectively, γ) will be denoted $\mathbf{E}_\lambda$ (respectively, $\mathbf{E}_\gamma$).
Let us now make more precise the notation we shall use. Notice first that writing

$$\cfrac{1}{a_1 + \cfrac{}{\ddots + \cfrac{1}{a_n}}} = [a_1, \ldots, a_n], \quad n \in \mathcal{N}^*,$$

we have $\omega = \lim_{n\to\infty} [a_1(\omega), \ldots, a_n(\omega)]$, $\omega \in \Omega$, and this is precisely the meaning of the equations

$$\omega = \cfrac{1}{a_1(\omega) + \cfrac{1}{a_2(\omega) + \dots}} = [a_1(\omega), a_2(\omega), \dots].$$

Next, put

$$r_n = a_n + \cfrac{1}{a_{n+1} + \cfrac{1}{a_{n+2} + \dots}} = a_n + [a_{n+1}, a_{n+2}, \dots], \quad n \in \mathcal{N}^*.$$

Clearly, $r_n(\omega) = 1/T^{n-1}(\omega), \quad n \in \mathcal{N}^*$, with $T^0(\omega) = \omega, \quad \omega \in \Omega$. Also, define recursively the sequences $(p_n)_{n \geqslant -1}$ and $(q_n)_{n \geqslant -1}$ by

$$p_n = a_n p_{n-1} + p_{n-2}, \quad q_n = a_n q_{n-1} + q_{n-2}, \quad n \in \mathcal{N}^*,$$

with $p_{-1} = 1, \quad p_0 = 0, \quad q_{-1} = 0, \quad q_0 = 1$. It is well-known (see, e.g., (Billingsley, 1965, pp. 40–46)) that for all $n \in \mathcal{N}^*$ and $\omega \in \Omega$:

$$p_{n-1}q_{n-2} - p_{n-2}q_{n-1} = (-1)^n, \quad \frac{p_n}{q_n} = [a_1, \dots, a_n],$$

$$\omega = \frac{p_{n-1}(\omega)r_n(\omega) + p_{n-2}(\omega)}{q_{n-1}(\omega)r_n(\omega) + q_{n-2}(\omega)}, \quad \frac{1}{2q_{n+1}^2(\omega)} < \left| \omega - \frac{p_n(\omega)}{q_n(\omega)} \right| < \frac{1}{q_n^2(\omega)}, \tag{0.1}$$

$$\left| \log q_n - \sum_{i=1}^{n} \log r_i \right| \leqslant 4. \tag{0.2}$$

Finally, for $n \in \mathcal{N}^*$, put

$$y_n = \frac{q_n}{q_{n-1}} = \begin{cases} a_1, & \text{if } n = 1, \\ a_n + [a_{n-1}, \dots, a_1], & \text{if } n \geqslant 2, \end{cases}$$

$$s_n = \frac{1}{y_n} = [a_n, \dots, a_1],$$

$$u_n(\omega) = \left| \omega - \frac{p_{n-1}(\omega)}{q_{n-1}(\omega)} \right|^{-1} q_{n-1}^{-2}(\omega) = r_n(\omega) + s_{n-1}(\omega),$$

with $s_0(\omega) = 0$. Clearly, $r_n > 1, \quad y_n \geqslant 1, \quad 0 < s_n \leqslant 1, \quad u_n > 1, \quad n \in \mathcal{N}^*$.
The metric theory of continued fractions consists of the study of the random sequence $(a_n)_{n \in \mathcal{N}^*}$ and related sequences. There are two basic facts in this theory, namely:

I. The sequence $(a_n)_{n \in \mathcal{N}^*}$ is strictly stationary under γ. (Equivalently, T preserves γ.) In particular,

$$\gamma(a_n = i) = \frac{1}{\log 2} \log \frac{(i+1)^2}{i(i+2)}, \quad i, n \in \mathcal{N}^*.$$

II. The sequence $(a_n)_{n\in\mathcal{N}^*}$ is exponentially ψ–mixing under γ (and under many other probability measures including λ), that is,

$$|\gamma(A_1 \cap A_2) - \gamma(A_1)\gamma(A_2)| \leqslant aq^n\gamma(A_1)\gamma(A_2), \quad n \in \mathcal{N}^*,$$

for any $A_1 \in \sigma(a_1, \ldots, a_k)$ (the σ–algebra generated by the random variables $a_1, \ldots, a_k$), $A_2 \in \sigma(a_{n+k}, a_{n+k+1}, \ldots)$, and $k \in \mathcal{N}^*$, with suitable positive constants $q < 1$ and a.

The second property is a deep result implied by the solution of a slight generalization of the famous 1812 Gauss' problem: estimate the error

$$e_n(x) = e_n(x, \lambda) = \lambda(r_n^{-1} < x) - \frac{\log(1+x)}{\log 2}, \quad x \in \dot{I}.$$

This was the first problem of the metric theory of continued fractions. It was solved almost simultaneously (by different methods), by R.O.Kuzmin (in 1928) and Paul Lévy (in 1929). As $n \to \infty$, uniformly in $x \in I$, according to the former, $e_n(x) = O(q^{\sqrt{n}})$ for some $0 < q < 1$, while, according to the latter, $e_n(x) = O(q^n)$, with $q < 0.68$. The optimal value of q in Lévy's estimate was determined by E.Wirsing in 1971 (see (Wirsing, 1974)), who found that it equalled $q_0 = 0.303663002898732658 60\ldots$. Note that Babenko (1978) obtained what might be called the "complete" solution of Gauss' problem, namely he was able to express $\lambda(r_n^{-1} < x)$ in terms of the eigenvalues and eigenfunctions of a certain linear operator. Putting

$$\psi(n) = \sup \left| \frac{\gamma(A_1 \cap A_2)}{\gamma(A_1)\gamma(A_2)} - 1 \right|, \quad n \in \mathcal{N}^*,$$

where the supremum is taken over all

$$A_1 \in \sigma(a_1, \ldots, a_k),\ A_2 \in \sigma(a_{k+n},\ a_{k+n+1}, \ldots),\ \text{and } k \in \mathcal{N}^*,$$

we have

$$\psi(n) \leqslant aq_0^n, \quad n \in \mathcal{N}^*, \tag{0.3}$$

for some suitable $a > 0$. Clearly, this is an equivalent statement of property II above. Actually, (0.3) is a consequence of an error estimate

$$\frac{e_n(y,\mu) - e_n(x,\mu)}{\gamma([x,y])} = O(q_0^n), \quad x, y \in I, \quad x < y,$$

for all measures $\mu = \lambda(\cdot \mid a_k = i_k,\ 1 \leqslant k \leqslant m)$, $i_k \in \mathcal{N}^*$, $1 \leqslant k \leqslant m$, $m \in \mathcal{N}^*$, rather than λ. For details and proofs, the reader is referred to (Iosifescu and Grigorescu, 1989, Section 5.2). Note that it is important to have estimates of a, since for some limit theorems it is necessary to know how large is $\psi(1)$. In this respect, Philipp (1988), showed that $\psi(n) \leqslant (0.8)^n$, $n \in \mathcal{N}^*$. A better result has been recently obtained by the present author (Iosifescu, 1989), namely that $\psi(n) \leqslant (3.5 - 2\sqrt{2})^{n-1}\log 2$, $n \in \mathcal{N}^*$. Actually, we even have $\psi(1) \leqslant 2\log 2 - 1 = 0.38629\ldots$ and $\psi(2) \leqslant \zeta(2)\log 2 - 1 = 0.13997\ldots$.

The solution of Gauss' problem provided a strong impetus to the development of the metric theory of continued fractions. Subsequently, important results were obtained in the 1930s by A.Ya. Khinchin and Paul Lévy. See Chapter III of (Khinchin, 1964) (whose first Russian edition was published in 1935), and Chapter IX of (Lévy, 1954) (whose first edition was published in 1937).
However, the most important work of the fourth decade on the metric theory of continued fractions is that of Wolfgang Doeblin, see (Doeblin, 1940). (Although, very likely, written in 1937, as shown by a footnote by Paul Lévy, who handled the paper, this was received by the editorial board of *Compositio Mathematica* in June 1939, i.e., exactly 50 years before the 5th Vilnius Conference.) The name of Wolfgang Doeblin is well-known for his many contributions which pervade the theory of Markovian dependence, in spite of the fact that he was only 25 when he died (on 21st June 1940) in World War II fighting the Nazi invaders of France. Details on Doeblin's life and work can be found in (Lévy, 1955), where, curiously enough, Doeblin's 1940 paper is just mentioned in two lines. In contrast to, we do not hesitate to consider this paper as one of the most significant and influential mathematical papers ever written. As a matter of fact, it can be viewed as a concise picture of the *present* state of the metric theory of continued fractions: the author took up and brought important contributions to all the main topics which were studied further on within the half-century elapsed. Thus, Doeblin's paper can be also considered as suggesting a working program (even if the paper was largely ignored by subsequent authors), and the present paper is about how that program has been carried out.
We should note that Doeblin's paper is far from being what is called a well-written paper. On the contrary, it is full of obscurities, misprints, technical imperfections, and even plain errors. (As to misprints, it is plausible that Doeblin was unable to see the paper through the press.) The proofs are, as a rule, just sketched. Most of them are adapted from or hinted at those of the corresponding results for Markov chains given in Doeblin's 1938 Ph. D. thesis, whose printed version is (Doeblin, 1937). (The reason of this strategy will be explained latter.) Such being the situation, it can but only show once more that new ideas and techniques, even in an imperfect form, are more important than formal correctness.

1. THE GAUSS FORMULA

The first main result of Doeblin's 1940 paper, to which its first section is devoted, is a derivation of the (generalized) Gauss formula by using elements of the theory of dependence with complete connections, just in its inception at the time. In our notation, the formula in question (see page 357 of Doeblin's paper) is

$$\lambda(s < r_{m+n} < t \mid a_1, \ldots, a_m) = = \frac{1}{\log 2}\Big(\log\frac{2t}{t+1} - \log\frac{2s}{s+1}\Big)(1 + \theta q^{\sqrt{n}}) \tag{1.1}$$

for all $m, n \in \mathcal{N}^*$ and $t > s \geqslant 1$, with $\mid \theta \mid \leqslant K$, where $q < 1$ and K are positive constants independent of m, n, s, and t. Note that (1.1) implies that

$$\lim_{n\to\infty} \lambda(r_n < t) = \frac{1}{\log 2}\log\frac{2t}{t+1} = \gamma\Big(\Big[\frac{1}{t}, 1\Big]\Big), \quad t \geqslant 1. \tag{1.2}$$

Doeblin's reasoning is not clear, but what he intends to say is as follows (cf. (Iosifescu, 1974) and (Iosifescu and Grigorescu, 1989, Subsection 5.2.2)). Let us start from the so-called Brodén – Borel – Lévy formula

$$\lambda(r_{m+1} > s \mid a_1, \ldots, a_m) = \frac{y_m + 1}{s y_m + 1}, \quad m \in \mathcal{N}^*, \quad s \geqslant 1,$$

which implies that

$$\lambda(a_{m+1} = i \mid a_1, \ldots, a_m) = \lambda(i < r_m < i + 1 \mid a_1, \ldots, a_m) =$$

$$= \lambda(r_{m+1} > i \mid a_1, \ldots, a_m) - \lambda(r_{m+1} > i + 1 \mid a_1, \ldots, a_m) =$$

$$= \frac{y_m + 1}{i y_m + 1} - \frac{y_m + 1}{(i+1) y_m + 1} =$$

$$= \frac{y_m (y_m + 1)}{(i y_m + 1)((i+1) y_m + 1)}, \quad m, i \in \mathcal{N}^*.$$

The last equation shows that $(y_m)_{m \in \mathcal{N}^*}$ is a homogeneous Markov chain with the following transition mechanism: from state $y \geqslant 1$ the only possible one-step transitions are those to states $y + i$, $i \in \mathcal{N}^*$, with corresponding probabilities $y(y+1)/(iy+1)((i+1)y+1)$, $i \in \mathcal{N}^*$. It is easy to check that the probability measure γ' on $\mathcal{B}_{[1,\infty)}$ (the σ-algebra of Borel subsets of $[1, \infty)$), whose distribution function appears at the middle of (1.2), i.e.,

$$\gamma'(A) = \int_{\{x : x^{-1} \in A\}} \gamma(dx) = \frac{1}{\log 2} \int_A \frac{dt}{t(t+1)}, \quad A \in \mathcal{B}_{[1,\infty)}, \tag{1.3}$$

is a stationary distribution for $(y_m)_{m \in \mathcal{N}^*}$. Let U be the transition operator associated with $(y_m)_{m \in \mathcal{N}^*}$, defined as

$$Uf(y) = \mathbf{E}_\lambda(f(y_2) \mid y_1 = y) =$$

$$= \sum_{i \in \mathcal{N}^*} \frac{y(y+1) f(y+i)}{(iy+1)((i+1)y+1)}, \quad y \geqslant 1,$$

for any measurable real-valued bounded function f on $[1, \infty)$. We have

$$\lambda(r_{m+n} > t \mid a_1, \ldots, a_m) = U^{n-1} f_t(y_m), \quad m, n \in \mathcal{N}^*, \quad t \geqslant 1,$$

with

$$f_t(y) = \frac{y+1}{ty+1}.$$

Doeblin's computation shows that

$$U^{n-1} f_s(y_m) - U^{n-1} f_t(y_m) =$$

$$= \frac{1}{\log 2}\left(\log \frac{2t}{t+1} - \log \frac{2s}{s+1}\right)(1 + \theta q^{\sqrt{n}}) = \gamma'([s,t])(1 + \theta q^{\sqrt{n}})$$

for all $m, n \in \mathcal{N}^*$ and $t > s \geqslant 1$, i.e. (1.1) holds.
As already mentioned, (1.1) is 'almost' the mixing property II stated in the preceding section with mixing rate $O(q^{\sqrt{n}})$ rather that $O(q^n)$.
It is the appropriate place to show that the sequence $(a_n)_{n\in\mathcal{N}^*}$ enjoys another mixing property known as the almost Markov property (a concept introduced by the Lithuanian school – see especially the references to the papers by V.A. Statulevičius and B.Riauba in (Heinrich, 1987 Misevičius, 1971). Let μ be a probability measure on $\mathcal{B}_I$. For $k, n \in \mathcal{N}^*$ define the random variable

$$\alpha_{k,n}(\mu) = \sup \left| \mu\Big(B \mid \sigma(a_1, \ldots, a_{k+n-1})\Big) - \mu\Big(B \mid \sigma(a_{k+1}, \ldots, a_{k+n-1})\Big) \right|,$$

where the supremum is taken over all $B \in \sigma(a_{k+n}, a_{k+n+1}, \ldots)$. Put

$$\chi_\mu(n) = \sup_{k\in\mathcal{N}^*} \operatorname{ess\,sup} \alpha_{k,n}(\mu).$$

Then, as shown in (Heinrich, 1987) (for a slightly weaker form of this result see (Misevičius, 1971)), assuming μ to be absolutely continuous with respect to λ and $f = d\mu/d\lambda$ to be a Lipschitz function, we have

$$\chi_\mu(n) \leqslant 2^{-n+1}(24 + L/M),$$

where $L = \sup_{0<u\neq v<1} \mid f(u) - f(v) \mid / \mid u - v \mid$ and $M = \inf_{0<u<1} f(u)$.
Let us note that Doeblin also argues that the distribution function $\lambda(y_{n+1} < < t \mid y_1)$, $n \in \mathcal{N}^*$, $t \geqslant 1$, converges uniformly in t and y_1, as $n \to \infty$, to

$$\gamma'([1,t]) = \frac{1}{2} \log \frac{2t}{t+1}$$

at an $O(q^{\sqrt{n}})$ rate. His reasoning, although difficult to follow, is essentially correct if properly interpreted. Actually, the convergence rate can be improved to $O(q^n)$. See (Knuth, 1971, pp. 326–327) and (Iosifescu, 1989).

2. CLASSICAL LIMIT THEOREMS

The second section of Doeblin's 1940 paper is devoted to the central limit theorem (CLT), the law of the iterated logarithm (LIL), and the strong law of large numbers (SLLN) in the metric theory of continued fractions. (Notice that Doeblin's predecessors did not go beyond the law of large numbers topic !) Actually, Doeblin is concerned there with proving limit theorems for certain types of weakly dependent random variables including ψ–mixing ones. (It should be stressed that this happened many years before the topic would have become fashionable.) It is apparent that Doeblin fully understood how the methods he had used for Markov chains in his Ph.D. thesis could be adapted to suit other, more general, contexts; and this explains the style of certain proofs in his 1940 paper (cf. the concluding part of our introductory Section).

Doeblin's approach to the CLT in his Ph.D. thesis is based on the now classical paper (Bernstein, 1926). This was, in fact, the first systematic use of Bernstein's blocking technique.

Regarding the LIL, Doeblin's results are stated in a form which might appear as unusual for the today reader. When particularized to, e.g., i.i.d. random variables ξ_i, $i \in \mathcal{N}^*$, with $\mathbf{E}\xi_1 = a$ and $0 \neq \mathbf{E}(\xi_1 - a)^2 = \sigma^2 < \infty$, Doeblin's LIL reads as follows: the random events

$$\liminf_{n\to\infty} \left\{ \frac{\sum_{i=1}^{n} \xi_i - na}{\sigma\sqrt{2n(\log\ \log n + c_1 \log\ \log\ \log n)}} \leqslant 1 \right\},$$

for any $c_1 > 3/2$, and

$$\limsup_{n\to\infty} \left\{ \frac{\sum_{i=1}^{n} \xi_i - na}{\sigma\sqrt{2n(\log\ \log n + c_2 \log\ \log\ \log n)}} > 1 \right\},$$

for any $c_2 \leqslant 1/2$, have both probability 1. Clearly, this implies the usual form of the LIL:

$$\limsup_{n\to\infty} \frac{\sum_{i=1}^{n} \xi_i - na}{\sigma\sqrt{2n \log\ \log n}} = 1$$

with probability 1. Doeblin's inspiration is the works (Lévy, 1931) and (Cantelli, 1933) on the LIL for independent random variables, which slightly generalizes the classical results of A.Ya. Khinchin and A.N. Kolmogorov. As a matter of fact, in the study of the Markovian case, Doeblin (Doeblin, 1937, p. 99) uses some ideas from (Cantelli, 1933). Although, as we have noticed long ago (see (Iosifescu, 1968, p. 321)), Doeblin's proof of the LIL for Markov chains in his Ph.D. thesis has a gap, thus making the result doubtful, the proof can be corrected. Such a correction is implicit in our proof of the LIL for some classes of mixing random variables (Iosifescu, 1968).

Regarding the SLLN, Doeblin should be credited for the idea of using the Birkhoff – Khinchin ergodic theorem in the metric theory of continued fractions, in spite of the rooted opinion that this idea is due to Ryll–Nardzewski (1951). See Subsection 2.8 below.

We are now proceeding to a detailed analysis of the subsections of the second section of Doeblin's 1940 paper.

2.1. *Partial sums* $S_n = \sum_{i=1}^{n} f(a_i)$, $n \in \mathcal{N}^*$

Let $f : \mathcal{N}^* \to \mathbf{R}$ be a real-valued function on the natural integers such that

$$\mathbf{E}_\gamma f^2(a_1) = \frac{1}{\log 2} \sum_{i\in\mathcal{N}^*} f^2(i) \log \frac{(i+1)^2}{i(i+2)} < \infty$$

(or, equivalently, $\sum_{i\in\mathcal{N}^*} i^{-2}f^2(i) < \infty$). Put

$$\mathbf{E}_\gamma f(a_1) = \frac{1}{\log 2}\sum_{i\in\mathcal{N}^*} f(i)\log\frac{(i+1)^2}{i(i+2)} = a$$

and

$$\sigma^2 = \mathbf{E}_\gamma(f(a_1)-a)^2 + 2\sum_{i\in\mathcal{N}^*}\mathbf{E}_\gamma(f(a_1)-a)(f(a_{i+1})-a). \tag{2.1}$$

The ψ-mixing property under γ of $(a_n)_{n\in\mathcal{N}^*}$ implies that the series defining σ^2 is absolutely convergent, $\sigma^2 \geqslant 0$, and $\mathbf{E}_\gamma(S_n - na)^2 = n\sigma^2 + O(1)$ as $n \to \infty$. Assuming $\sigma > 0$ and f to be bounded, Doeblin sketches a proof of the fact that $(S_n - na)/\sigma\sqrt{n}$ is asymptotically $N(0,1)$– the standard normal distribution – under λ with an $O(n^{-c})$ remainder for some $c > 0$, and asserts the validity of the LIL. He also asserts that $\sigma = 0$ iff $f=$ const, hinting at the proof of the corresponding result for Markov chains in his Ph. D. thesis. We note that the first complete proof of the last assertion was given in (Gordin, 1971, Section 3). Doeblin also gives valid hints for a proof of the same result (the CLT with remainder) when f is assumed to satisfy $\mathbf{E}_\gamma f^4(a_1) < \infty$ (equivalently, $\sum_{i\in\mathcal{N}^*} i^{-2}f^4(i) < \infty$) rather than being bounded. He conjectures that $\mathbf{E}_\gamma f^2(a_1) < \infty$ is sufficient for the CLT to hold, while $\mathbf{E}_\gamma \mid f(a_1) \mid^{2+\delta} < \infty$ for some $\delta > 0$ is sufficient for the LIL to hold (both under λ). Both conjectures have proved to be true (even with $\delta = 0$). Actually, the functional versions of both CLT and LIL hold under any probability measure $\mu \ll \lambda$ ($\ll$ means" absolutely continuous with respect to"). See (Billingsley, 1968, §20; Heyde and Scott, 1973; Ibragimov and Linnik, 1971, Chapter 19, §4; Iosifescu, 1968, 1972, 1974; Oodaira and Yoshihara, 1971a,b).

Regarding the functional version of the CLT we shall give an elegant recent result (Samur, 1987). Let us first introduce some notation (cf. (Araujo and Giné, 1980, Chapter 3) and (Billingsley, 1968, Chapter 3)). For a real separable Hilbert space $H, D(I,H)$ denotes the Skorokhod space of H-valued functions on I, and we shall write $D = D(I,\mathcal{R})$. If ν is an infinitely divisible (i.d.) probability measure on H, then $\mathbf{Q}(\nu)$ denotes the law of a $D(I,H)$-valued random function $\xi = (\xi(t))_{t\in I}$ with stationary independent increments such that $\xi(0) = 0$ and $\xi(1)$ has law ν. Finally, $\mathbf{W}$ denotes the Wiener measure on $D(I,H)$, $\mathbf{L}_\mu(\xi)$ denotes the law of a random function ξ defined on the probability space $(I, \mathcal{B}_I, \mu)$, and $\xrightarrow{w}$ denotes weak convergence.

THEOREM 2.1. (i) *Let $(A_n)_{n\in\mathcal{N}^*}$ be a sequence of real numbers and $(B_n)_{n\in\mathcal{N}^*}$ be a sequence of positive numbers with $\lim_{n\to\infty} B_n = \infty$. The following assertions are equivalent.*

(I) *The random functions $\xi_n = (\xi_n(t))_{t\in I}$, $n \in \mathcal{N}^*$, defined by*

$$\xi_n(t) = \frac{S_{[nt]} - [nt]A_n}{B_n}$$

satisfy

$$\mathbf{L}_\lambda(\xi_n) \xrightarrow{w} \mathbf{W} \quad in \quad D. \tag{2.2}$$

(II) *The random variable* $\xi_n(1)$ *is asymptotically* $N(0,1)$ $(n \to \infty)$ *under* λ *and* $\forall r_n \in \mathcal{N}^*,\ r_n \leqslant n,\ n \in \mathcal{N}^*$,

$$\lim_{n\to\infty} \frac{r_n}{n} = 0 \Rightarrow \sum_{i=1}^{r_n} \frac{f(a_i) - A_n}{B_n} \to 0 \text{ in } \lambda - \text{probability}.$$

(ii) *The assertion "there exist a bounded sequence* $(A_n)_{n\in\mathcal{N}^*}$ *of real numbers and a sequence* $(B_n)_{n\in\mathcal{N}^*}$ *of positive numbers with* $\lim_{n\to\infty} B_n = \infty$ *for which* (I) *above holds" is equivalent to either*

$$\lim_{x\to\infty} x^2 \sum_{\{i:|f(i)|>x\}} i^{-2} \Big/ \sum_{\{i:|f(i)|\leqslant x\}} f^2(i)i^{-2} = 0$$

or

$$\lim_{x\to\infty} \frac{F(tx)}{F(x)} = 1 \ \text{ for any } \ t > 0 \tag{2.3}$$

(i.e., F is slowly varying), where

$$F(x) = \frac{1}{\log 2} \sum_{\{i:|f(i)|\leqslant x\}} f^2(i)i^{-2}, \quad x > 0.$$

If (2.3) *holds and* $\lim_{x\to\infty} F(x) = \infty$, *then we can take* $A_n = \mathbf{E}_\gamma f(a_1),\ n \in \mathcal{N}^*$, *and any* $(B_n)_{n\in\mathcal{N}^*}$ *such that* $\lim_{n\to\infty} nB_n^{-2}F(B_n) = 1$.

(iii) *If* (I) *holds for some* $(A_n)_{n\in\mathcal{N}^*}$ *and* $(B_n)_{n\in\mathcal{N}^*}$, *then* (2.2) *holds with* λ *replaced by any probability measure* $\mu \ll \lambda$.

We note that Samur (Samur, 1987) also deals with more general partial sums S_n defined as follows. Let $(f_n)_{n\in\mathcal{N}^*}$ be a sequence of functions from $\mathcal{N}^*$ into a separable Hilbert space H, and put

$$S_n = \sum_{i=1}^{n} f_n(a_i), \quad n \in \mathcal{N}^*.$$

He obtain sufficient conditions for the laws of certain random functions associated with the S_n, $n \in \mathcal{N}^*$, to converge weakly in $D(I, H)$ to an i.d. probability measure on H (see his Propositions 2.2 and 2.4).

2.2 *A special case of partial sums* $S_n = \sum_{i=1}^{n} f_i(a_i),\ n \in \mathcal{N}^*$

Another generalization of the case discussed in the preceding subsection is to consider a sequence of functions $(f_n)_{n\in\mathcal{N}^*}$ from $\mathcal{N}^*$ into $\mathbf{R}$ and define partial sums S_n, $n \in \mathcal{N}^*$, as in the heading above (rather than as at the end of the preceding subsection). Doeblin takes up just a very special case, where $f_n(i) = 1$ or 0

according as $i \geqslant \varphi(n)$ or $i < \varphi(n)$, $i, n \in \mathcal{N}^*$. Here $(\varphi(n))_{n\in\mathcal{N}^*}$ is a sequence of positive numbers with $\lim_{n\to\infty} \varphi(n) = \infty$. In this case, S_n is the number of occurrences of the random events $\{a_i \geqslant \varphi(i)\}$, $1 \leqslant i \leqslant n$. A classical theorem of Bernstein (1911) asserts that $\lim_{n\to\infty} S_n < \infty$ or $= \infty$, λ-a.s. in Ω, according as $\sum_{i\in\mathcal{N}^*} 1/\varphi(i)$ converges or diverges. Doeblin gives valid hints for a proof that, if $\sum_{i\in\mathcal{N}^*} 1/\varphi(i) = \infty$, then $(S_n)_{n\in\mathcal{N}^*}$ obeys the CLT, and also asserts the validity of the LIL (both under λ). More precisely, $(S_n - A_n)/\sqrt{A_n}$ is asymptotically $N(0,1)$ $(n \to \infty)$ under λ, with

$$A_n = \mathbf{E}_\gamma S_n = (\log 2)^{-1} \sum_{i=1}^{n} \log(1 + 1/\varphi(i)), \quad n \in \mathcal{N}^*.$$

Complete proofs for both CLT (under any $\mu \ll \lambda$) and LIL (in the usual form of the latter) were given in (Philipp, 1970). The functional version of the CLT under γ was proved in (Philipp and Webb, 1973). Note that the functional version of the LIL might follow from a more general result mentioned in (Szüsz and Volkmann, 1982, p. 458).

2.3. *Partial sums* $S_n = \sum_{i=1}^{n} f(a_i, \ldots, a_{i+k-1})$, $n \in \mathcal{N}^*$, *for a fixed* $k \in \mathcal{N}^*$

Let $f : (\mathcal{N}^*)^k \to \mathbf{R}$ be a real-valued function on the k-dimensional positive integer lattice such that $\mathbf{E}_\gamma f^2(a_1, \ldots, a_k) < \infty$. Put $\mathbf{E}_\gamma f(a_1, \ldots, a_k) = b$ and define σ^2 similarly to (2.1), replacing there a by b, $f(a_1)$ by $f(a_1, \ldots, a_k)$ and $f(a_{i+1})$ by $f(a_{i+1}, \ldots, a_{i+k})$. Doeblin notes that the case $k > 1$ can be treated similarly to the case $k = 1$. However, he asserts that, when $k > 1$, $\sigma = 0$ is no longer equivalent to $f =$ const. A complete proof of this assertion, including the general possible form of f, was given in (Gordin, 1971, Section 3). Note that the partial sums considered here are a special case of those considered in the next subsection.

2.4. *Partial sums* $S_n = \sum_{i=1}^{n} f(r_i)$, $n \in \mathcal{N}^*$

Let $f : [1, \infty) \to \mathbf{R}$ be a Lipschitz function on $[1, \infty)$ such that

$$\mathbf{E}_\gamma f^4(r_1) = \frac{1}{\log 2} \int_1^\infty \frac{f^4(x)\, dx}{x(x+1)} < \infty.$$

Doeblin asserts that these conditions are sufficient to ensure the existence of a finite $\sigma^2 \geqslant 0$ (given by an expression similar to (2.1), where a should be replaced by $\mathbf{E}_\gamma f(r_1)$, $f(a_1)$ by $f(r_1)$, and $f(a_{i+1})$ by $f(r_{i+1})$) such that $\mathbf{E}_\gamma(S_n - n\mathbf{E}_\gamma f(r_1))^2 = n\sigma^2 + O(1)$ as $n \to \infty$. Next, if $\sigma > 0$, then $(S_n - n\mathbf{E}_\gamma f(r_1))/\sigma\sqrt{n}$ is asimptotically $N(0,1)$ $(n \to \infty)$ under λ, with an $O(n^{-c})$ remainder for some $c > 0$, and the LIL under λ also holds.

We have been unable to follow Doeblin's very sketchy and laconical reasoning. Anyhow, his assertions are true for functions f for which $f(x) = g(1/x)$, $x \geqslant 1$, where g is a real-valued function of bounded variation on I. More precisely, for such a function f the series defining σ^2 is absolutely convergent and $\sigma^2 \geqslant 0$. We have $\sigma = 0$ iff g can be written as $g = h \circ T - h$ (λ-a.s. in Ω) for some real-valued function h of essential bounded variation. Finally, if $\sigma \neq 0$, then the CLT holds under γ with optimal remainder $O(n^{-1/2})$. (Replacing γ by a probability measure $\mu \ll \gamma$ may result in a slower convergence rate.) These results were proved in (Rousseau-Egèle, 1983), in a more general context. Also, the functional versions under γ of both CLT and LIL hold (Hofbauer and Keller, 1982, p.133).
Another case in which both classical and functional CLTs hold under any probability measure $\mu \ll \lambda$ is that where f satisfies again $f(x) = g(1/x)$, $x \geqslant 1$, for a real-valued function g on I such that $\int_0^1 g^2\, d\lambda < \infty$ and

$$\sum_{k \in \mathcal{N}^*} \left(\int_0^1 \left(g - \mathbf{E}_\gamma(g \mid a_1, \ldots, a_k)\right)^2 d\lambda \right)^{1/2} < \infty.$$

See (Billingsley, 1968, §21) and (Ibragimov and Linnik, 1971, Chapter 19, §4). Compare (Misevičius, 1971). Under stronger moment assumptions on g, the functional version of LIL also holds. See (Philipp and Stout, 1975, pp. 80–81).

2.5. *Partial sums* $S_n = \sum_{i=1}^n f(y_i)$, $n \in \mathcal{N}^*$

Let $f : [1, \infty) \to \mathbf{R}$ be a Lipschitz function on $[1, \infty)$ such that

$$\mathbf{E}_\gamma f^4(y_1) = \frac{1}{\log 2} \int_1^\infty \frac{f^4(y)\, dy}{y(y+1)} < \infty. \tag{2.4}$$

Concerning the partial sums S_n, $n \in \mathcal{N}^*$, considered here, Doeblin makes assertions identical with those concerning the partial sums S_n considered in the preceding subsection (replace there r_i by y_i, $i \in \mathcal{N}^*$).
We confess again our inability to follow Doeblin's reasoning. However, his assertions are true for functions f for which $f(y) = g(1/y)$, $y \geqslant 1$, where g is a real-valued Lipschitz function on I. (Of course, this amounts to considering partial sums $T_n = \sum_{i=1}^n g(s_i)$, $n \in \mathcal{N}^*$.) More precisely, for such a function f, the series defining σ^2 is absolutely convergent and $\sigma^2 \geqslant 0$. (We do not know of a necessary and sufficient condition expressed in terms of f and T for $\sigma = 0$.) If $\sigma \neq 0$, then the CLT holds under both λ and γ with optimal remainder $O(n^{-1/2})$. See (Iosifescu, 1987) and (Iosifescu and Grigorescu, 1989, Subsection 4.2.3), where a more general context is considered. Also, the functional versions of both CLT and LIL hold under both λ and γ. See (Heyde and Scott, 1973; Iosifescu and Grigorescu, 1989, Subsection 4.2.4; Popescu, 1976; Scott, 1971).

An important special case, not covered by the results described above, is that where g is defined as $g(x) = \log x$, $x \in I$, which *is not* a Lipschitz function on I. (Note, however, that the corresponding f is defined as $f(y) = g(1/y) = -\log y$, $y \geqslant 1$, which *is* a Lipschitz function on $[1, \infty)$, and also satisfies the moment assumption (2.4)!) In this case, $T_n = -\log q_n = -S_n$, $n \in \mathcal{N}^*$. (Note that, by (0.2), the case of the partial sums $S_n = \log q_n$, $n \in \mathcal{N}^*$, is related to the case discussed in the preceding subsection.) Doeblin asserts on page 364 of his paper that the corresponding σ is greater than 0, and, consistent with himself, concludes that both CLT and LIL hold under λ. More precisely,

$$(\log q_n - n\pi^2/12\log 2)/(\sigma\sqrt{n})$$

is asymptotically $N(0,1)$ $(n \to \infty)$ under λ, with an $O(n^{-c})$ remainder for some $c > 0$, while (in the usual form of the LIL)

$$\lambda\left(\limsup_{n\to\infty} \frac{\log q_n - n\pi^2/(12\log 2)}{\sigma\sqrt{2n\log\log n}} = 1\right) = 1. \tag{2.5}$$

The first complete proof of the fact that $\sigma > 0$ was given in (Gordin, 1971, Section 3). The fact that the CLT holds under λ, with an $O(n^{-1/2}\log n)$ remainder, was proved by Misevičius (Misevičius, 1981). His proof is based on the almost Markov property of the sequence $(a_n)_{n\in\mathcal{N}^*}$. Finally, (2.5) was proved independently in (Gordin and Reznick, 1970) and (Philipp and Stackelberg, 1969). It can be shown, using the results of (Heyde and Scott, 1973) and (Scott, 1971), that the functional versions of both CLT and LIL hold under both λ and γ. The details will be given elsewhere.

Let us note that Doeblin uses the limit theorems stated for $\log q_n$ to derive the corresponding limit theorems for

$$z_n(\omega) = \left|\omega - \frac{p_n(\omega)}{q_n(\omega)}\right|, \quad n \in \mathcal{N}^*, \quad \omega \in \Omega.$$

As stated by Doeblin, the latter are marred by obvious inaccuracies. Their correct statements, easily deduced from the double inequality appearing in (0.1), are as follows:

$$\lim_{n\to\infty} \lambda\left(\exp(2u\sigma\sqrt{n}) < z_n \exp\frac{n\pi^2}{6\log 2} < \exp(2v\sigma\sqrt{n})\right) =$$

$$= \frac{1}{\sqrt{2\pi}} \int_{-v}^{-u} e^{-t^2/2}\,dt$$

for any real numbers $u < v$ (with convergence rate $O(n^{-1/2}\log n)$ uniform in $u, v \in \mathbf{R}$), and

$$\lambda\left(\limsup_{n\to\infty} \frac{\log z_n + n\pi^2/6\log 2}{2\sigma\sqrt{2n\log\log n}} = 1\right) = 1.$$

To conclude this subsection, we wish to emphasize that $(y_n)_{n\in\mathcal{N}^*}$, as well as $(s_n)_{n\in\mathcal{N}^*}$, are Markov chains of a special type, which occurs in a natural manner in the theory of dependence with complete connections. Relevant references are (Grigorescu and Popescu, 1973; Iosifescu, 1987, 1990; Iosifescu and Grigorescu, 1989, Chapter 3 and Section 4.2; Popescu, 1976).

2.6. *A multidimensional central limit theorem*

Consider the random variables

$$\Gamma^{(n)}(i) = \text{card}\,\{1 \leqslant j \leqslant n : \ a_j = i\}, \quad i, n \in \mathcal{N}^*,$$

and put

$$g(i) = \gamma(a_1 = i) = \frac{1}{\log 2} \log \frac{(i+1)^2}{i(i+2)}, \quad i \in \mathcal{N}^*.$$

Doeblin gives valid hints for a proof that, for any fixed $r \in \mathcal{N}^*$ and any fixed pairwise distinct $i_1, \ldots, i_r \in \mathcal{N}^*$, the random vector $n^{-1/2}(\Gamma^{(n)}(i_\alpha) - ng(i_\alpha))_{1 \leqslant \alpha \leqslant r}$ is asymptotically normal $(n \to \infty)$ under λ, without indicating the covariance matrix of the limit law (whose mean vector is $\mathbf{0}$).

A natural generalization of this assertion, which also clarifies the point left unsettled by Doeblin, was proved by Samur (Samur, 1987) as a corollary of his results alluded to at the end of Subsection 2.1. Let l^2 be the Hilbert space of square summable real sequences and let $(e_i)_{i \in \mathcal{N}^*}$ denote its canonical orthonormal basis. Define

$$\Gamma^{(n)} = \left(\Gamma^{(n)}(i)\right)_{i \in \mathcal{N}^*} : \Omega \to l^2, \quad g = \left(g(i)\right)_{i \in \mathcal{N}^*},$$

and consider the random functions $\xi_n = (\xi_n(t))_{t \in I}$, $n \in \mathcal{N}^*$, defined by

$$\xi_n(t) = n^{-1/2}(\Gamma^{[nt]} - [nt]g).$$

Then, for any $\mu \ll \lambda$, ξ_n converges weakly under μ to a law $\mathbf{Q}(\nu)$ on the Skorokhod space $D(I, l^2)$ of l^2-valued functions on I. Here ν is the centered Gaussian probability measure on l^2 (see (Araujo and Giné, 1980, Chapter 3)) whose covariance operator S satisfies

$$\langle Se_i, e_j \rangle = \delta_{ij} g(i) - g(i)g(j) + 2 \sum_{k \in \mathcal{N}^*} \left(\gamma(a_1 = i, \ a_{k+1} = j) - g(i)g(j)\right)$$

for $i, j \in \mathcal{N}^*$, where δ is Kronecker's symbol, and $\langle \cdot, \cdot \rangle$ denotes the inner product of l^2.

2.7. *The limit law of* u_n

Using the Brodén—Borel—Lévy formula, Doeblin sketchily derives the limit distribution function $H(t) = \lim_{n \to \infty} H_n(t)$, where $H_n(t) = \lambda(u_n < t)$, $t \in \mathbf{R}$, $n \in \mathcal{N}^*$, as

$$H(t) = \begin{cases} 0, & t \leqslant 1, \\ \dfrac{1-t}{t\log 2} + \dfrac{\log t}{\log 2}, & 1 \leqslant t \leqslant 2\,, \\ 1 - \dfrac{1}{t \log 2}, & t \geqslant 2. \end{cases}$$

He also asserts that

$$\sup_{t \in \mathbf{R}} \mid H_n(t) - H(t) \mid = O(r^n),$$

as $n \to \infty$ for some $0 < r < 1$.

A detailed proof of the above result, following Doeblin's suggestions, was given in (Samur, 1987, pp. 39–40). The proof also yields

$$\sup_{t \in \mathbf{R}} \mid h_n(t) - h(t) \mid = O(r^n) \tag{2.6}$$

as $n \to \infty$. Here $h_n = H_n'$ is the density function under λ of u_n, $n \in \mathcal{N}^*$, and

$$h(t) = H'(t) = \frac{1}{\log 2} \min\left(\frac{1}{t} - \frac{1}{t^2}, \frac{1}{t^2}\right), \quad t \geqslant 1,$$

$$h(t) = 0, \quad t < 1.$$

Note that (2.6) is just a reformulation of a result of (Knuth, 1984) where the limit law of $1/u_n$ was obtained. (The method used there allows to specify the value of r as $(\sqrt{5}-1)/2 = 0.61803\ldots$.)

What is really surprising in the problem discussed, it arises when Doeblin asserts that his result implies that

$$\lim_{n\to\infty} \frac{1}{n} \text{card}\, \{1 \leqslant i \leqslant n : \; u_i < t\} = H(t), \quad t \in \mathbf{R}, \tag{2.7}$$

λ–a.s. in Ω, and that $n^{-1} \sum_{i=1}^{n} u_i^{-1}$ converges λ–a.s. in Ω to a constant M. Equation (2.7), in the equivalent form involving the u_i^{-1} instead of the u_i, $i \in \mathcal{N}^*$, is precisely the object of a conjecture stated by H.W. Lenstra, Jr. The conjecture (which, clearly, we should call after both Doeblin and Lenstra, Jr.) was proved in (Bosma et al., 1983), where the value of M was also obtained as $1/(4\log 2) =$ $= 0.360673\ldots$. In that paper, similar results were proved for other random variables connected with the continued fraction expansion such as

$$\left|\omega - \frac{p_n(\omega)}{q_n(\omega)}\right| \Big/ \left|\omega - \frac{p_{n-1}(\omega)}{q_{n-1}(\omega)}\right| = \frac{s_n(\omega)}{r_{n+1}(\omega)}$$

or

$$q_{n-1}(\omega) q_n(\omega) \left|\omega - \frac{p_{n-1}(\omega)}{q_{n-1}(\omega)}\right| = \left(1 + \frac{s_n(\omega)}{r_{n+1}(\omega)}\right)^{-1}, \quad n \in \mathcal{N}^*, \quad \omega \in \Omega,$$

i.e., functions depend of s_n and r_{n+1} only. Let us note that the proof of (2.7) is based on an idea of (Nakada, 1981), that we shall discuss in the next subsection. Should we think that Doeblin did intuit it so many years before ?

2.8. *The strong law of large numbers*

In two papers published in 1935 and 1936 in *Compositio Mathematica*, A.Ya.Khinchin proved several results of SLLN type about continued fractions (see Chapter III of his 1964 book). These results concern time averages of the form $\sum_{i=1}^{n} f(a_i, \ldots, a_{i+k-1})/n$ for some fixed $k \in \mathcal{N}^*$. Khinchin's proofs are quite intricate, as he made no use of the Birkhoff–Khinchin (!) ergodic theorem, which provides short and elegant proofs.
Unlike Khinchin, Doeblin does make use of the ergodic theorem. On page 366 of his paper, he proves that the continued fraction transformation T is ergodic (a different proof of the ergodicity of T was given earlier in (Knopp, 1926)). Since T preserves γ, this enables him to write

$$\lim_{n\to\infty} \frac{1}{n} \sum_{i=1}^{n} f(r_i) = \frac{1}{\log 2} \int_1^\infty \frac{f(x)\,dx}{x(x+1)} \tag{2.8}$$

a.s. in Ω, for any integrable real-valued function f on $[1, \infty)$. (Here "a.s." refers to λ or to γ indifferently.) Of course, (2.8) is equivalent to the usual form of the ergodic theorem, i.e.,

$$\lim_{n\to\infty} \frac{1}{n} \sum_{i=1}^{n} g(T^{i-1}(\omega)) = \mathbf{E}_\gamma g\Big(\frac{1}{r_1}\Big) = \frac{1}{\log 2} \int_0^1 \frac{g(x)\,dx}{1+x} \tag{2.9}$$

for almost all $\omega \in \Omega$, for any integrable real-valued function g on I. (Note that both (2.8) and (2.9) still hold when the integrals diverge to either $+\infty$ or $-\infty$.) As is well = known (see, e.g., (Billingsley, 1965, pp. 45–46)), Khinchin's results follow from either (2.8) or (2.9) by suitably choosing f or g.
It is the appropriate place to note that, in spite of the fact that, e.g., Billingsley (Billingsley, 1965, p.49) fully credits Doeblin for the idea leading to (2.8), many authors assert that this idea is due to Ryll–Nardzewski (Ryll–Nardzewski, 1951). For a recent reference see (Jager, 1986, pp. 68–69). The only real advance made after 1940 in using the ergodic theorem in the metric theory of continued fractions originated with Nakada (Nakada, 1981), who introduced, in a more general context, the so-called natural extension τ of T, which is the transformation of $J = \Omega \times I$ into itself defined as

$$\tau(\omega, y) = \Big(T(\omega), \frac{1}{[1/\omega] + y}\Big), \quad (\omega, y) \in \Omega \times I.$$

Hence

$$\tau^n(\omega, y) = \Big(T^n(\omega),\ [a_n(\omega), \ldots, a_2(\omega), a_1(\omega) + y]\Big), \quad n \geqslant 2, \quad y \in I,$$

if $\omega = [a_1(\omega), a_2(\omega), \ldots]$, $\omega \in \Omega$. In particular,

$$\tau^n(\omega, 0) = (T^n(\omega), s_n(\omega)), \quad n \in \mathcal{N}^*, \quad \omega \in \Omega.$$

Consider on the σ-algebra $\mathcal{B}_J$ of the Borel subsets of J the probability measure μ defined as

$$\mu(A) = \frac{1}{\log 2} \iint\limits_A \frac{dx\,dy}{(1+xy)^2}, \quad A \in \mathcal{B}_J.$$

Nakada proved that τ is ergodic and preserves μ. This opened the way to a whole class of results involving the bidimensional sequences $(1/r_{n+1},\ s_n)_{n\in\mathcal{N}^*}$ and $(u_n, u_{n+1})_{n\in\mathcal{N}^*}$ (see (Bosma et al., 1983) and (Jager, 1986)). (Note that $u_{n+1} = r_{n+1} + s_n$, but also $u_n = 1/r_{n+1} + 1/s_n$, $n \in \mathcal{N}^*$.) Such a result is precisely (2.7). How did Doeblin guess it ?

We close this subsection by discussing the SLLN for the sequences $(y_n)_{n\in\mathcal{N}^*}$ and $(s_n)_{n\in\mathcal{N}^*}$, which are not covered by (2.8). Doeblin gives valid hints for a proof (see p. 362 of his paper) of the following result. Let f be a bounded continuous real-valued function on $[1, \infty)$. Then

$$\lim_{n\to\infty} \frac{1}{n} \sum_{i=1}^{n} f(y_i) = \mathbf{E}_\gamma f(y_1) = \frac{1}{\log 2} \int_1^\infty \frac{f(y)\,dy}{y(y+1)}$$

a.s. in Ω. Some details, important for completing the proof, are missing but, however, the statement is true. See (Elton, 1988; Grigorescu and Popescu, 1989). The corresponding result for the sequence $(s_n)_{n\in\mathcal{N}^*}$ $(s_n = y_n^{-1})$ is that, g being a bounded continuous real-valued function on I, we have

$$\lim_{n\to\infty} \frac{1}{n} \sum_{i=1}^{n} g(s_i) = \mathbf{E}_\gamma g(s_1) = \frac{1}{\log 2} \int_0^1 \frac{g(s)\,ds}{1+s}$$

a.s. in Ω. Note that this also follows from a theorem of (Breiman, 1960).

3.THE POISSON LAW AND RELATED LAWS

The third section of Doeblin's 1940 paper contains an attempt to prove the following Poisson law. For any fixed positive number θ we have

$$\lim_{n\to\infty} \lambda(a_k > n\theta \ \text{ for } \ p \ \text{ values } \ k,\ 1 \leqslant k \leqslant n) =$$

$$= \left(\frac{1}{\theta \log 2}\right)^p \frac{1}{p!} \exp\left(-\frac{1}{\theta \log 2}\right), \quad p = 0, 1, 2, \ldots . \tag{3.1}$$

Doeblin's proof of (3.1) contains a gap, as shown in (Iosifescu, 1977), where a correct proof is provided, with the specification that λ can be replaced by any probability measure $\mu \ll \lambda$. (Incidentally, our paper was presented at the 2nd Vilnius Conference.) A discussion of the relationship between the Poisson law

(3.1) and the limit laws as $n \to \infty$ of the order statistics $M_n^{(s)}$ = sth largest of $a_1, \ldots, a_n$, $1 \leqslant s \leqslant n$, can be found in (Iosifescu, 1978).
Doeblin also considers a more general problem. Let f be a non-negative function on $\mathcal{N}^*$ and put

$$S_n = \sum_{i=1}^{n} f(a_i)\chi(a_i > n\theta), \quad n \in \mathcal{N}^*,$$

where χ denotes the indicator function and θ is a fixed positive number. He asserts that S_n converges weakly under λ to a compound Poisson law. This assertion, as stated by Doeblin, is obviously incorrect since his "limit" law does actually depend on n. A correct version of a limit compound Poisson law for S_n was proved by Dziubdziela (Dziubdziela, 1988). However, checking his assumptions seems a difficult task.
Now, we shall state recent results of Samur (Samur, 1987), which generalize Doeblin's Poisson law (3.1) and give correct versions of some limit compound Poisson laws. These results follow from Samur's general results alluded to at the end of Subsection 2.1. For any fixed $\alpha \in \mathbf{R}$ and $\theta > 0$ define the random functions $\xi_n = (\xi_n(t))_{t \in I}$, $n \in \mathcal{N}^*$, by

$$\xi_n(t) = n^{-\alpha} \sum_{1 \leqslant i \leqslant [nt]} a_i^{\alpha} \chi(a_i > \theta n).$$

Denote by δ_x the probability measure with total mass concentrated at $x \in \mathbf{R}$. For any finite measure ρ on the Borel subsets of $\mathbf{R}$ put

$$\text{Pois } \rho = \exp(\rho - \rho(\mathbf{R})\delta_0) = e^{-\rho(\mathbf{R})} \sum_{i \geqslant 0} \frac{\rho^i}{i!},$$

where $\rho^0 = \delta_0$, and ρ^i, $i \in \mathcal{N}^*$, is the ith convolution of ρ with itself, and

$$\text{C}_\varepsilon\text{Pois } \rho = (\text{Pois } \rho) * \delta_{b(\varepsilon)}, \quad \varepsilon > 0,$$

where $b(\varepsilon) = -\int_{-\varepsilon}^{\varepsilon} x\rho(dx)$, and $*$ denotes convolution of measures.
Then, with the notation introduced before the statement of Theorem 2.1, for any probability measure $\mu \ll \lambda$ we have $\mathbf{L}_\mu(\xi_n) \xrightarrow{w} \mathbf{Q}(\nu)$ in D, where ν is defined as follows

(i) if $\alpha > 0$, then $\nu = \text{C}_{\theta^\alpha}\text{Pois } \rho$, with

$$\frac{d\rho}{d\lambda}(x) = \begin{cases} 0, & x \leqslant \theta^\alpha, \\ (\alpha \log 2)^{-1} x^{-1/\alpha - 1}, & x > \theta^\alpha; \end{cases}$$

(ii) if $\alpha < 0$, then $\nu = \delta_z * \text{C}_{\theta^\alpha}\text{Pois } \rho$, with $z = 1/(1-\alpha)\theta^{1-\alpha}\log 2$ and

$$\frac{d\rho}{d\lambda}(x) = \begin{cases} 0, & x \leqslant 0 \text{ or } x \geqslant \theta^\alpha, \\ (-\alpha \log 2)^{-1} x^{-1/\alpha - 1}, & 0 < x < \theta^\alpha; \end{cases}$$

(iii) if $\alpha = 0$, then $\nu = \text{Pois}\ ((\theta \log 2)^{-1}\delta_1)$.
Clearly, (iii) above is a functional version of (3.1). Note also that, as Samur points out, (i) to (iii) above remain valid when a_i is replaced by either r_i or u_i, $i \in \mathcal{N}^*$.

4. CONVERGENCE TO STABLE LAWS

Put

$$S_n = \sum_{i=1}^{n} a_i, \quad n \in \mathcal{N}^*.$$

In the fourth section of his 1940 paper, using ψ–mixing and a truncation argument, Doeblin first derives a result proved in 1935 by A.Ya.Khinchin, namely

$$\frac{S_n}{n \log n} \to \log 2, \quad \text{as} \quad n \to \infty \quad \text{in } \lambda\text{–probability.} \tag{4.1}$$

Doeblin's proof is short and clear unlike Khinchin's, which is long and intricate. Then, using the same tools, Doeblin proves that the series

$$\sum_{i \in \mathcal{N}^*} \left(\frac{1}{S_i} - \frac{\log 2}{i \log i} \right)$$

converges absolutely λ–a.s. This implies at once that $\sum_{i \in \mathcal{N}^*} 1/S_i$ diverges λ–a.s., a result which is due again to Khinchin, who derived it as a corollary to (4.1). Finally, and this is the main result of the section, putting

$$S'_n = \sum_{i=1}^{n} a_i \chi(a_i \leqslant n),$$

Doeblin argues that $(S_n - \mathbf{E}_\lambda S'_n)/n$ converges weakly under λ to a quasi–stable law, and indicates the form of its characteristic function. This is a correct assertion, although Doeblin's sketched proof is not convincing, it being based on the invalid compound Poisson law mentioned in the preceding section.
Samur (Samur, 1987) proved a general theorem, entirely similar to Theorem 2.1, dealing with convergence to the stable law $\nu(\kappa_1, \kappa_2, \beta) = c_1 \text{Pois}\ \mu(\kappa_1, \kappa_2, \beta)$, $\kappa_1, \kappa_2 \geqslant 0$, $\kappa_1 + \kappa_2 > 0$, $0 < \beta < 2$, with Lévy density

$$\frac{d\mu(\kappa_1, \kappa_2, \beta)}{d\lambda}(x) = \begin{cases} \kappa_2 \mid x \mid^{-1-\beta}, & x < 0, \\ \kappa_1 x^{-1-\beta}, & x > 0 \end{cases}$$

(see his Proposition 2.10). We shall give here just some consequences of this theorem. See (Samur, 1987, pp. 23 – 26). Let $f : [1, \infty) \to (0, \infty)$ be bounded on finite intervals and regularly varying with exponent $\alpha = 1/\beta > 1/2$, i.e., f is measurable and $\lim_{x \to \infty} f(tx)/f(x) = t^\alpha$, $t > 0$. Put

$$\nu_\alpha = \begin{cases} \delta_{1/(1-\alpha)\log 2} * \nu\left(\dfrac{1}{\alpha \log 2}, 0, \dfrac{1}{\alpha}\right), & \alpha \neq 1, \\ \nu\left(\dfrac{1}{\log 2}, 0, 1\right), & \alpha = 1 \end{cases}$$

and define the random functions $\xi_n = (\xi_n(t))_{t\in I}$, $n \in \mathcal{N}^*$, by

$$\xi_n(t) = \begin{cases} \sum_{i=1}^{[nt]} \Big(f(a_i) - \mathbf{E}_\gamma f(a_1)\Big)/f(n), & 1/2 < \alpha < 1, \\ \sum_{i=1}^{[nt]} \Bigg(f(a_i) - \sum_{\{j: f(j)\leqslant f(n)\}} f(j)\gamma(a_1 = j)\Bigg)\Big/ f(n), & \alpha = 1, \\ \sum_{i=1}^{[nt]} f(a_i)/f(n), & \alpha > 1. \end{cases}$$

Then, for any probability measure $\mu \ll \lambda$, we have

$$\mathbf{L}_\mu(\xi_n) \xrightarrow{w} \mathbf{Q}(\nu_\alpha) \text{ in } D.$$

This conclusion remains valid, in the case where $\alpha > 1$, when a_i is replaced by either r_i or u_i, $i \in \mathcal{N}^*$. Note that the case $f(x) = x^{1/2}$, $x \geqslant 1$, when $\alpha = 1/2$, falls under the incidence of Theorem 2.1, with $A_n = \mathbf{E}_\gamma(a_1^{1/2})$, $B_n = (n\log n/\log 2)^{1/2}$, $n \in \mathcal{N}^*$. Clearly, the case $f(x) = x$, $x \geqslant 1$, when $\alpha = 1$, corresponds to the case considered by Doeblin.

The result above implies that

$$\frac{1}{n}(S_n - \mathbf{E}_\lambda S'_n) = \xi_n(1) + \frac{1}{n}\sum_{i,j=1}^{n} j(\gamma(a_1 = j) - \lambda(a_i = j))$$

converges weakly under any probability measure $\mu \ll \lambda$ to ν_1 since, as it is easy to see,

$$\lim_{n\to\infty} \frac{1}{n}\sum_{i,j=1}^{n} j(\gamma(a_1 = j) - \lambda(a_i = j)) = 0.$$

Similarly, if we define the random functions $\eta_n = (\eta_n(t))_{t\in I}$, $n \in \mathcal{N}^*$, by

$$\eta_n(t) = \frac{1}{n}\left(S_{[nt]} - \frac{[nt]\log n}{\log 2}\right),$$

then

$$\mathbf{L}_\mu(\eta_n) \xrightarrow{w} \mathbf{Q}(\nu') \text{ in } D,$$

where $\nu' = \delta_a * \nu_1$, with

$$a = \lim_{n\to\infty} \frac{1}{\log 2}\left(\sum_{i=1}^{n} i \log\frac{(i+1)^2}{i(i+2)} - \log n\right) = -\frac{1}{\log 2}.$$

The value of a is not given explicitely by Samur. It can be obtained using some computations in (Heinrich, 1987, p. 160). Note that the latter author was able to estimate the convergence rate of $\mathbf{L}_\gamma(\eta_n(1))$ to its limit law, proving that there exists $C_0 < \infty$ such that

$$\left|\gamma\left(\frac{S_n}{n} - \frac{\log n - g}{\log 2} < x\right) - \delta_{(g-1)/\log 2} * \nu_1((-\infty, x))\right| \leqslant C_0\frac{(\log n)^2}{n}$$

for all $n \in \mathcal{N}^*$ and $x \in \mathbf{R}$, where g is Euler's constant. Heinrich's method of proof is that used in (Misevičius, 1981) to estimate the convergence rate to normality for $\log q_n$. (See our Subsection 2.5.) Note that if $\widetilde{\eta}_n(\bar{\eta}_n)$ denotes the random function obtained by replacing a_i by $r_i(u_i)$, $i \in \mathcal{N}^*$, in the definition of η_n we have $\mathbf{L}_\mu(\widetilde{\eta}_n) \xrightarrow{w} \mathbf{Q}(\widetilde{\nu}')$ and $\mathbf{L}_\mu(\bar{\eta}_n) \xrightarrow{w} \mathbf{Q}(\bar{\nu}')$ in D, where $\widetilde{\nu}' = \delta_{1/\log 2 - 1} * \nu'$ and $\bar{\nu}' = \delta_{2(1/\log 2-1)} * \nu'$. (The weak convergence of $\sum_{i=1}^{n} u_i$ is indicated by Doeblin on page 365 of his paper, as a consequence of the equation $\sum_{i=1}^{n} u_i = S_n + \sum_{i=2}^{n+1} 1/r_i + \sum_{i=1}^{n-1} 1/y_i$, $n > 1$.) For the proofs of these results and of other results concerning random functions similar to ξ_n above, constructed from the $f(u_i)$, $i \in \mathcal{N}^*$, for a certain class of functions f, see (Samur, 1987, pp. 41–46).

5. CONCLUDING REMARKS

I believe that the above analysis supports my assertions concerning the importance of Doeblin's 1940 paper. It is only to be regretted that the paper was ignored by many, which resulted in crediting other authors for ideas and results put forward by Doeblin many years before. May I express the hope that the present paper will put things straight in this respect.

I also think that, fifty years after Doeblin's 1940 paper, the working programme set up by it has been practically completed. Of course, this does not mean that the metric theory of continued fractions has become a closed avenue. In view of the lasting importance of the general theory of continued fractions, new problems of metric character will permanently arise in the course of the development of the subject. At the same time, the metric theory of continued fractions is the prototype of the metric theory of other representations of real numbers, a topic which has been given much attention in recent years. See, e.g., (Nakada, 1981) and the references therein.

I am grateful to the organizers of the 5th International Vilnius Conference for their invitation to attend the Conference. This was the fifth time I addressed it as an invited speaker. I was thus able to register with admiration, on the spot, the changes which have been occurring since 1973 in Lithuania, an opportunity that few foreigners have had.

Finnaly, I wish to dedicate this paper to the memory of Wolfgang Doeblin on the semicentenary of his death.

REFERENCES

Araujo, A. and Giné, E. (1980). *The Central Limit Theorem for Real and Banach Valued Random Variables.* Wiley, New York.

Babenko, K.I. (1978). On a problem of Gauss. *Soviet Math. Dokl.* **19**, 136–140.

Bernstein, F. (1911). Über eine Anwendung der Mengenlehre auf ein aus der Theorie der Säkularstörungen herrührendes Problem. *Math. Ann.* **71**, 417–439.

Bernstein, S. (1926). Sur l'extension du théorème limite du calcul des probabilités aux sommes de quantités dépendantes. *Math.Ann.* **97**, 1–59.

Billingsley,P. (1965). *Ergodic Theory and Information.* Wiley, New York.

Billingsley,P. (1968). *Convergence of Probability Measures.* Wiley, New York.

Bosma,W., Jager,H. and Wiedijk,F. (1983). Some metrical observations on the approximation by continued fractions. *Indag.Math.* **45**, 281–299.

Breiman,L. (1960). A strong law of large numbers for a class of Markov chains. *Ann.Math. Statist.* **31**, 801–803.

Cantelli,F.P. (1933). Considerazioni sulla legge uniforme dei grandi numeri e sulla generalisazzione di un fondamentale teorema del Sig. Paul Lévy. *Giorn.Ist.Ital.Attuari* **4**, 327–350.

Doeblin,W. (1937). Sur les propriétés asymptotiques de mouvements régis par certains types de chaînes simples. *Bull.Math.Soc.Roumaine Sci.* **39**, No 1, 57–115; No 2, 3–61.

Doeblin,W. (1940). Remarques sur la théorie métrique des fractions continues. *Compositio Math.* **7**, 353–371.

Dziubdziela,W. (1988). A compound Poisson limit theorem for stationary mixing sequences. *Rev.Roumaine Math.Pures Appl.* **33**, 39–45.

Elton,H.J. (1987). An ergodic theorem for iterated maps. *Ergodic Theory and Dynamical Systems* **7**, 481–488.

Gordin,M.I. (1971). The behaviour of the dispersion of sums of random variables that generate a stationary process. *Theory Probab.Appl.* **16**, 474–484.

Gordin,M.I. and Reznik,M.H. (1970). The law of the iterated logarithm for the denominators of continued fractions. *Vestnik Leningrad.Univ.* **25**, 28–33.

Grigorescu,S. and Popescu,G. (1973). A central limit theorem for a class of Markov chains. In: *Proc.Fourth.Conf.Probability Theory (Braşov, 1971).* Ed. Academiei R.S.România, Bucureşti, pp. 153–166.

Grigorescu,S. and Popescu,G. (1989). Random systems with complete connections as a framework for fractals. *Stud.Cerc.Mat.* **41.**

Heinrich,L. (1987). Rates of convergence in stable limit theorems for sums of exponentially ψ–mixing random variables with an application to metric theory of continued fractions. *Math. Nachr.* **131**, 149–165.

Heyde,C.C. and Scott,D.J. (1973). Invariance principles for the law of the iterated logarithm for martingales and processes with stationary increments. *Ann.Probab.* **1**, 428–436.

Hofbauer,F. and Keller,G. (1982). Ergodic properties of invariant measures for piecewise monotonic transformations. *Math.Z.* **180**, 119–140.

Ibragimov,I.A. and Linnik,Yu.V. (1971). *Independent and Stationary Sequences of Random Variables.* Wolters–Noordhoff, Groningen.

Iosifescu,M. (1968). La loi du logarithme itéré pour une classe de variables aléatoires dépendantes. *Theory Probab.Appl.* **13**, 304–313.

Iosifescu,M. (1970). La loi du logarithme itéré pour une classe de variables aléatoires dépendantes. *Theory Probab.Appl.* **15**, 160.

Iosifescu,M. (1972). On Strassen's version of the loglog law for a class of dependent random variables. *Z.Wahrsch.verw.Geb.* **24**, 155–158.

Iosifescu,M. (1974). On the application of random system with complete connections to the theory of f–expansions. *Colloq.Math.Soc.János Bolyai* **9**, 335–365.

Iosifescu,M. (1977). A Poisson law for ψ–mixing sequences establishing the truth of a Doeblin statement. *Rev.Roumaine Math.Pures Appl.* **22**, 1441–1447.

Iosifescu,M. (1978). Recent advances in the metric theory of continued fractions. In: *Trans. Eighth Prague Conf.Information Theory, Statist.Decision Functions, Random Processes*

(Prague, 1978), **A**, Reidel,Dordrecht, pp. 27–40.

Iosifescu, M. (1987). A class of Markov chains with optimal convergence rate in the central limit theorem, In: *Proc.Fourth Colloq.Probab.Theory and Operations Research (Craiova)*, Craiova Univ.Press, Craiova pp. 89–94.

Iosifescu, M. (1989). On mixing coefficients for the continued fraction expansion. *Stud.Cerc. Mat.* **41.**

Iosifescu, M. (1990). On U–statistics and von Mises statistics for a special class of Markov chains. In: *Proc. 15th Conf. Math. Statist. (Olsztyn, Poland, 1988), J.Statist. Plann. Inference.* Special Issue *(to appear).*

Iosifescu, M. and Grigorescu, S. (1989). *Dependence with Complete Connection and Its Applications.* Cambridge Univ.Press, Cambridge.

Jager, H. (1986). The distribution of certain sequences connected with the continued fraction. *Indag.Math.* **48**, 61–69.

Khinchin, A.Ya. (1964). *Continued Fractions.* Univ. of Chicago Press, Chicago.

Knopp, K. (1926). Mengentheoretische Behandlung einiger Probleme der diophantischen Approximationen und der transfiniten Wahrscheinlichkeiten. *Math.Ann.* **95**, 409–426.

Knuth, D.E. (1971). *The Art of Computer Programming.* **2**, *Seminumerical Algorithms* . Addison–Wesley, Reading, Mass.

Knuth, D.E. (1984). The distribution of continued fraction approximations. *J.Number Theory* **19**, 443–448.

Lévy, P. (1931). Sur un théorème de M.Khintchine. *Bull.Sci.Math.* (2) **55**, 145–160.

Lévy, P. (1954). *Théorie de l'addition des variables aléatoires.* 2ème éd. Gauthier–Villars, Paris.

Lévy, P. (1955). W.Doeblin (V.Doblin) (1915–1940). *Rev.Histoire Sci.Appl.* **8**, 107–115.

Misevičius, G. (1971). Asymptotic expansions for the distribution functions of sums of the form $\sum_{j=0}^{n-1} f(T^j t)$. *Ann.Univ.Sci.Budapest.Eötvös Sect.Math.* **14**, 77–92.

Misevičius, G. (1981). Estimate of the remainder term in the limit theorem for the denominators of continued fractions. *Lithuanian Math.J.* **21**, 245–253.

Nakada, H. (1981). Metrical theory for a class of continued fraction transformations and their natural extensions. *Tokyo J.Math.* **4**, 399–426.

Oodaira, H. and Yoshihara, K. (1971a). The law of the iterated logarithm for stationary processes satisfying mixing conditions. *Kodai Math.Sem.Rep.* **23**, 311–334.

Oodaira, H. and Yoshibara, K. (1971b). Note on the law of the iterated logarithm for stationary processes satisfying mixing conditions. *Kodai Math.Sem.Rep.* **23**, 335–342.

Philipp, W. (1970). Some metrical theorems in number theory II. *Duke Math.J.* **37**, 447–458; Errata *ibid.* **37**, 788.

Philipp, W. (1988). Limit theorems for sums of partial quotients of continued fractions. *Monatsh. Math.* **105**, 195–206.

Philipp, W. and Stackelberg, O.P. (1969). Zwei Grenzwertsätze für Kettenbrüche. *Math.Ann.* **181**, 152–156.

Philipp, W. and Stout, W. (1975). Almost sure invariance principles for partial sums of weakly dependent random variables. *Mem.Amer.Math.Soc.* **161.**

Philipp, W. and Webb, G.R. (1973). An invariance principle for mixing sequences of random variables. *Z.Wahrsch.verw.Geb.* **25**, 223–237.

Popescu, G. (1976). A functional central limit theorem for a class of Markov chains. *Rev.Roumaine Math.Pures Appl.* **21**, 737–750.

Rousseau–Egèle, J. (1983). Un théorème de la limite locale pour une classe de transformations dilatantes et monotones par morceaux. *Ann. Probab.* **11**, 772–788.

Ryll–Nardzewski, C. (1951). On the ergodic theorems. II. Ergodic theory of continued fractions. *Studia Math.* **12**, 74–79.

Samur, J.D. (1987). On some limit theorems for continued fractions. *Notas de Matématica,* No 47. Univ. Nacional de la Plata, Dep. Mat., La Plata. (To appear in *Trans. Amer. Math. Soc.)*

Scott, D.J. (1971). Central limit theorems for martingales and for processes with stationary increments, using a Skorokhod representation approach. *Adv. Appl. Probab.* **5**, 119–137.

Szüsz, P. and Volkmann, B. (1982). On Strassen's law of the iterated logarithm. *Z. Wahrsch. verw. Geb.* **61**, 453–458.

Wirsing, E. (1974). On the theorem of Gauss—Kusmin—Lévy and a Frobenius-type theorem for function spaces. *Acta Arith.* **24**, 507–528.

Prob. Theory and Math. Stat., Vol. 1, pp. 573–579
B.Grigelionis *et al.* (Eds.)
1990 VSP/Mokslas

BOUNDARY AND ENTROPY OF RANDOM WALKS IN RANDOM ENVIRONMENT

V.A. KAIMANOVICH
Department of Applied Mathematics and Mathematical Modelling,
Leningrad Shipbuilding Institute,
Lotsmanskaya 3, Leningrad 190000, USSR

ABSTRACT

The aim of the present paper is a generalization of the entropic theory of random walks on discrete groups (Derriennic, 1980; Kaimanovich, 1985; Kaimanovich and Vershik, 1983) to the case of random walks in random environment (RWRE). It turns out that the principal ideas from the entropic theory of random walks on discrete groups can be extended to this case without difficulties and the main results (with duly reformulation) still hold true in this more general situation (in contradistinction to the theory of transience for RWRE where new phenomena appear). This paper is an enlarged version of a part of the results announced in the abstract (Kaimanovich, 1989). The main result here is an entropic criterion of boundary triviality for RWRE being completely analogical to that for ordinary random walks. We don't discuss here the time reversion, the analogy with the Brownian motion on foliations (Kaimanovich, 1988) and applications of the entropic approach to the description of the non-trivial Poisson boundary for RWRE (cf. (Kaimanovich, 1985)). A more detailed exposition will appear elsewhere.

1. RANDOM WALKS IN RANDOM ENVIRONMENT

Let G be a countable group, $P(G)$ the space of all probability measures on G and MG the set of all families

$$w = \{(w^x) : x \in G\}$$

of measures $w^x \in P(G)$, indexed by elements of G, i.e. the Cartesian product of $|G|$ copies of the space $P(G)$, or the space of the all configurations on G with values in $P(G)$. The space MG is a Polish space (i.e. separable and complete) in the topology induced by the l^1-metrics in $P(G)$ and is provided with the following natural continuous left action of the group G:

$$(gw)^x = w^{(g^{-1}x)}, \quad x,\, g \in G.$$

To a given point $w \in MG$ there corresponds a time homogeneous Markov chain on the group G with the transition probabilities

$$p(x, xg; w) = w^x(g)$$

(i.e. w^x is the probability distribution of the "right increment" of the chain w at the point $x \in G$, rather than simply the transition probability at this point). Thus we identify the space MG with the space of all Markov chains on G, or with the set of the all "environments" on G, each measure w^x, $x \in G$, being the "state of the environment" at the point x.
One can easily see that

$$p(x, y; w) = w^x(x^{-1}y) = (x^{-1}w)^e(x^{-1}y) = p(gx, gy; gw),$$

where e is the unit element of the group G. The n-step transition probabilities (of passing from a point $x \in G$ to a point $y \in G$) will be denoted by $p_n(x, y; w)$, or simply by $p_n(y; w)$ if the initial state x is e. Denote by P^w_Θ the probability measure on the space G^∞ of the paths $\{y_n\}_{n=0}^\infty$ of the chain w with the initial distribution $\Theta \in P(G)$, and by P^w the measure corresponding to the initial distribution $\Theta = \sigma_e$ concentrated on the unit element e of G.

Definition. A *random environment* on the group G is a Borel probability measure λ on the space MG. A *random walk in a random environment (RWRE)* is a family of time-homogeneous Markov chains with the state space G and the transition probabilities $p(x, y; w)$ depending on the random parameter (the "state of the environment") $w \in MG$ with distribution λ. The RWRE corresponding to a measure λ on MG will be denoted by (G, λ).
The measure λ will be always assumed quasi-invariant and ergodic with respect to the action of the group G (i.e. for all $g \in G$ the measures λ and $g\lambda$ are equivalent and there does not exist a non-trivial subset $A \in MG$ such that $A = gA$ (λ-mod 0) for any $g \in G$). Recall that in our situation (the group G is countable) every non-ergodic quasi-invariant measure can be uniquely (mod 0) decomposed into ergodic quasi-invariant components. The point measure $\lambda = \sigma_w$ is quasi-invariant iff all the measures $w^x \in P(G)$ coincide. In this case we get the usual space-homogeneous random walk on the group G corresponding to the "determinate" environment. We also impose on λ the following degeneracy condition: for λ-a.e. $w \in MG$ all points of the group G are attainable from the point e for the chain w. This condition (which actually could be dropped out) is weaker than the usual ellipticity condition (Kozlov and Molchanov, 1984).

Remark. Usually one defines a random walk in random environment using a probability G-space X with a quasi-invariant measure λ and an equivariant mapping $f : X \to MG$. For our purposes it is more convenient to consider directly the measure λ on the space MG (it is clear that this measure carries all the information about the random walk in random environment) since G is countable. We emphasize that all the measure spaces considered below are Lebesgue ones, hence we can use the standard technique of factor-spaces and conditional probabilities.

Example. Our definition of random walks in random environment enables one to consider conditional random walks arising from usual "determinate" random walks as random walks in random environment. Let μ be a non-degenerated probability measure on the group G, and (Γ, v) be the Poisson boundary of the random walk

(G,μ) provided with a natural ergodic quasi-invariant G-action. Then one can assign a conditional random walk with the transition probabilities

$$p_\gamma(x,y) = (dyv/dxv)(\gamma)\mu(x^{-1}y)$$

to each point $\gamma \in \Gamma$ (Kaimanovich, 1985). Thus we get a probability measure on MG (the image of the measure μ) and can identify the conditional random walk with random walks in random environment.

2. A MARKOV CHAIN CORRESPONDING TO RWRE

The state space G of a random walk in random environment is space-homogeneous, hence one can consiger not only the "fixed coordinate system" with the origin at the point e in which the Markov particle moves along a path $\{y_n\}$ with an "environment state" $w \in MG$ being fixed, but also a "moving coordinate system" in which the Markov particle is situated at the origin whereas the environment state constantly changes. Using the formula

$$p(y_n, y_n x; w) = p(e, x; y_n^{-1} w)$$

we get the random sequence $w_n = y_n^{-1} w$ of the elements of MG. Further more,

$$w_{n+1} = y_{n+1}^{-1} w = x_{n+1}^{-1} y_n^{-1} w = x_{n+1}^{-1} w_n,$$

where $x_{n+1} = y_n^{-1} y_{n+1}$ is the $(n+1)$-th increment of the trajectory $\{y_n\}$ having the distribution w_n^e (w_n is fixed). Thus we get a Markov chain with the state space MG and transition probabilities (cf. (Kozlov and Molchanov, 1984)).

$$\pi(w, g^{-1}w) = p(e, g; w).$$

Below we shall always suppose the following condition to be satisfied.
(S) The measure λ is stationary for the chain $\pi(\cdot,\cdot)$.

PROPOSITION 1. *The condition* (S) *is satisfied iff for* λ-a.e. $w \in MG$ *the function*

$$\varphi(g; w) = dg\lambda/d\lambda(w)$$

is the density (with respect to the counting measure) of an invariant measure of the chain w.

Proof. The measure λ is stationary for the chain $\pi(\cdot,\cdot)$ iff

$$d\lambda(w) = \sum_{w'} \pi(w', w) d\lambda(w') =$$

$$= \sum_g \pi(gw, w) d\lambda(gw) = \sum_g p(e, g; gw)\, d\lambda\,(gw) =$$

$$= \sum_g p(g^{-1}, e; w) dg^{-1}\lambda(w).$$

Thus the condition (S) is equivalent to the equality

$$\sum_g \varphi(g;w)p(g,e;w) = 1$$

being true for a.e. w. Since $\varphi(e;w) = 1$, we see that if φ is the density of a stationary measure for the chain w then (S) is satisfied. Conversely, if (S) is satisfied then using the cocycle identity for the Radon–Nikodym derivatives $\varphi(g;w)$ we get for any point $h \in G$

$$\sum_g \varphi(g;w)p(g,h;w) =$$

$$= \varphi(h;w)\sum_g \varphi(h^{-1}g;h^{-1}w)\ p(h^{-1}g,e;h^{-1}w) =$$

$$= \varphi(h;w),$$

i.e. $\varphi(\cdot;w)$ is the density of an invariant measure for a.e. chain w.
One can easily verify that for the random walks in random environment arising from conditional walks of a determinate random walk (G,μ) the condition (S) is satisfied. In a sense the condition (S) seems to be even more natural than the condition of G–invariance of λ (cf. also (Kozlov and Molchanov, 1984)).

PROPOSITION 2. *The measure λ is ergodic with respect to the G–action iff it is ergodic as a stationary measure of the chain $\pi(\cdot,\cdot)$ on MG.*

Proof. Every trajectory of the chain $\pi(\cdot,\cdot)$ consists of the points lying on a G–orbit in MG. Let $A \in MG$ be a non-trivial (with respect to λ) G–invariant subset. Then, evidently, it is also invariant for the chain $\pi(\cdot,\cdot)$. Conversely, if A is π–invariant, then from the non-degeneracy of λ it follows that A is also G–invariant.

3. TAIL AND INVARIANT σ–ALGEBRAS

Let for any positive k, n and $w \in MG$

$$d_n^k(w) = \|p_n(\cdot;w) - p_{n+k}(\cdot;w)\|,$$

where $\|\cdot\|$ is the variance of a measure.

PROPOSITION 3. *For almost all $w \in MG$ there exists the limit*

$$d^k(w) = \lim d_n^k(w),$$

and for any fixed k either $d^k(w) = 0$ for a.e. w, or $d^k(w) = 2$ for a.e. w.

Proof. Since d_n^k decreases as n increases, it follows that the limits d^k exist. On the other hand,

$$d_n^k(w) = \|p_n(\cdot; w) - p_{n+k}(\cdot; w)\| =$$

$$= \sum_x |p_n(x; w) - p_{n+k}(x; w)| =$$

$$= \sum_x | \sum_z p(z; w)[p_{n-1}(z, x; w) - p_{n+k-1}(z, x; w)]| \leqslant$$

$$\leqslant \sum_x \sum_z p(z; w)|p_{n-1}(z, x; w) - p_{n+k-1}(z, x; w)|.$$

Passing to the limit we get

$$d^k(w) \leqslant \sum_z p(z; w) d^k(z^{-1}w) =$$

$$= \sum_{w'} \pi(w; w') d^k(w'),$$

i.e. the function d^k is a positive subharmonic function for the chain π. Now from the condition (S) and Proposition 2 we get that d^k should be a constant, but from the Derriennic's "zero–two law" (Derriennic, 1976) it follows that in this case it should be identically 0 or 2.

Definition. The smallest positive k such that d^k equals 0 is called the *period* of the random walk in random environment (G, λ).

Recall that for any Markov chain $\{y_n\}$ its *tail (asymptotic, residual)* σ–algebra A is the sub–σ–algebra of its path space which is the limit of the decreasing sequence of the coordinate σ–algebras A_n generated by y_k, $k \geqslant n$. The *invariant (stationary)* σ–algebra S consists of all sets from A which are invariant (mod 0) with respect to the shift of the path space. Factor–spaces of the path space with respect to these σ–algebras (defined as measure spaces provided the measure on the path space corresponding to an initial distribution of the chain is given) are called the *exit boundary* and the *Poisson (or, stationary) boundary* of the chain $\{y_n\}$. The Poisson boundary gives an integral representation of all bounded harmonic functions of the chain $\{y_n\}$ (the Poisson formula), its triviality being equivalent to non–existense of non–constant bounded harmonic functions (the Liouville property) (Revuz, 1984).

THEOREM 1. *For λ–a.e. $w \in MG$ the tail and the stationary σ–algebras of the chain w coincide* (mod 0) *with respect to the measure P^w in the path space which corresponds to the initial distribution σ_e.*

Proof. If the period of the random walk in random environment equals 1, then the tail and the stationary σ–algebras coincide for any initial distribution on G according to Derriennic's zero–two law. If the period d is greater than 1 and finite, then the group G for a.e. $w \in MG$ can be divided into d cyclic subsets, hence applying the previous argument to the aperiodic RWRE $\{y_{nd}\}$ we get the desired coincidence. If the period d is infinite, then the supports of the one–dimensional

distributions $p_n(\cdot; w)$ are mutually disjoint for different n, hence every harmonic sequence can be considered as a harmonic function, and the result follows.
Let Γ^w be the Poisson boundary (coinciding with the exit boundary) of the chain $w \in MG$ with the harmonic measure v^w corresponding to the initial distribution σ_e.

THEOREM 2. *The boundaries (Γ^w, v^w) are trivial or non-trivial for λ-a.e. $w \in MG$ simultaneously.*

Proof. Consider the following functions f^k on MG:

$$f^k(w) = \lim_n \|p_n(\cdot; w) - \sum_x p_{n-1}(x, \cdot; w) w^e(x)\|.$$

Then the boundary (Γ^w, v^w) is trivial iff $f^k(w) = 0$ for all positive k. On the other hand, reasoning as in the proof of Proposition 3 we see that all functions f^k are constant λ-a.e.

Definition. We shall say that the random walk in random environment (G, λ) is *Liouvillean* if λ-a.e. boundaries (Γ^w, v^w) are trivial.
Note that if the measure λ is not invariant, then the RWRE (G, λ) can be Liouvillean even for non-amenable groups (e.g. consider the conditional walks treated above). This is not so in the case of G-invariant λ (see (Kaimanovich, 1983; Sunyach, 1987)).

4. ENTROPY OF RANDOM WALKS IN RANDOM ENVIRONMENT

Introduce the following *condition of entropy finiteness* for the RWRE (G, λ):

$$\int \sum_g p(g; w) \log p(g; w) \, d\lambda(w) \tag{H}$$

is finite.

THEOREM 3. *If the condition* (H) *is satisfied then for λ-a.e. $w \in MG$ there exists the limit*

$$\lim_n \left[- \sum_g p_n(g; w) \log p_n(g; w) \right] / n = h(\lambda)$$

(the entropy of the RWRE) and the RWRE (G, λ) is Liouvillean iff $h(\lambda) = 0$.

Proof. Almost verbally reproduces the proof for the deterministic case (Derriennic, 1980; Kaimanovich, 1983) with duly account of the dynamic system corresponding to the Markov chain $\pi(\cdot, \cdot)$ on MG. One should note that the RWRE is Liouvillean iff the quantity

$$\lim_n \int \sum_{x>y} \log \left(\frac{p_{n-1}(x, y; w)}{p_n(e, y; w)} \right) w^e(x) \, d\lambda(w)$$

is zero and then apply the multiplicative ergodic theorem to prove that this quantity is actually $h(\lambda)$.
As a corollary of this theorem we immediately get the following analogue of the classic Choquet–Deny theorem for random walks in random environment.

THEOREM 4. *Let G be the k-dimensional integer lattice $\mathbb{Z}^k$ and the following moment condition be satisfied:*

$$\int \sum_g |g| p(g; w)\, d\lambda(w)$$

is finite. Then the RWRE (G, λ) is Liouvillean.

Remark. In the case $k = 1$ with some additional assumptions this result was proved in (Andjel, 1988). It is interesting, whether one can drop the finiteness conditions and/or the stationarity condition (S) and get this result in maximal generality (as in the usual "deterministic" Choquet–Deny theorem).

REFERENCES

Andjel, E.D. (1988). A zero or one law for one dimensional random walks in random environments. *Ann. Prob.* **16**, 722–729.

Derriennic, Y. (1976). Lois "zero ou deux" pour les processus de Markov, applications aux marches aleatoires. *Ann. Inst. H. Poincaré, Sect. B.* **12**, 111–129.

Derriennic, Y. (1980). Quelques applications du théoréme ergodique sous additif. *Asterisque* **74**, 183–201.

Kaimanovich, V.A. (1985). An entropy criterion for maximality of the boundary of random walks on discrete groups. *Sov. Math. Dokl.* **31**, 193–197.

Kaimanovich, V.A. (1988). Brownian motion on foliations: entropy, invariant measures, mixing. *Funk. Anal.* **22**, No 4.

Kaimanovich, V.A. (1989). Boundary and entropy of random walks in random environment. In: *Abstr. 5th International Vilnius Conference on Probab. Theory and Math. Statist.* **1**, Inst. Math. and Cybern., Vilnius, pp. 234–235.

Kaimanovich, V.A. and Vershik, A.M. (1983). Random walks on discrete groups: boundary and entropy. *Ann. Prob.* **11**, 457–490.

Kozlov, S.M. and Molchanov, S.A. (1984). On conditions for applicability of the central limit theorem to random walks on a lattice. *Sov. Math. Dokl.* **30**, 410–413.

Revuz, D. (1984). *Markov Chains.* (North-Holland Math. Library, **11**) North-Holland, Amsterdam.

Sunyach, C. (1987). Sur la transience et la recurrence des marches aleatoires en milieu aleatoire. *Ann. Inst. H. Poincaré, Sect. Probab. Statist.* **23**, 613–626.

Prob. Theory and Math. Stat., Vol. 1, pp. 580-589
B.Grigelionis *et al.* (Eds.)
1990 VSP/Mokslas

NON-LINEAR LAW OF LARGE NUMBERS AND GAUSSIAN FLUCTUATION FOR STAR-LIKE NETWORKS

M.Y. KELBERT and Y.M. SUHOV
Institute for Problems of Information Transmission,
Yermolova Street 19, 101447, Moscow, GSP-4, USSR

ABSTRACT

The problem of asymptotic description of stationary and transient regimes for the star-like message switching network is discussed. The transient regime is described by an infinite system of non-linear differential equations. Its stationary solution corresponds to the stationary regime of the network. We describe also Gaussian fluctuations near this deterministic evolution.

1. INTRODUCTION

Queueing networks (or, more generally, service networks) become now an object of an intensive investigation on the level of rigorous mathematics (see, e.g., the review (Kelbert and Suhov, 1988) and the references therein). One of the main directions in studying queueing networks is the asymptotic analysis based on various approximation schemes. This paper is devoted to the so-called Poisson approximation for some concrete type of networks. In analogy with statistical mechanics (which turned out to be very fruitful here) one can use also the term "mean-field approximation". Such an approximation was proposed in (Kleinrock, 1964) and remains until now one of the most powerful tools in evaluating and estimating characteristics of complex networks. The basis for such an approximation is provided by (numerous) "phenomenological" observations that, in a network with a "highly branching" configuration graph, the situations in various nodes (even if they are close to each other) are nearly independent. Moreover, the situation in every single network site is nearly the same as it would be when this site is "pluged" in a Poissonian environment which is developing according to some "selfconsistent" law (in statistical mechanics one uses, in the corresponding context, the term 'Vlasov equation').
In probabilistic terms, the Poisson, or mean-field, approximation is related to the (non-linear) law of large numbers. It is valid under the standard normalisation N^{-1} when N, the branching number, tends to infinity. The Vlasov-type equation which appears here connects the parameters (mean-values) which figure in these approach for different times. In a rigorous mathematical form this assertions were proposed independently (for different network classes) by Whitt (Sigal

and Whitt, 1988), Dobrushin and Rybko (Dobrushin, 1986) and Ziednis (private communication by F.P. Kelly).

Besides the law of large number, we study in this paper the $N^{-1/2}$–fluctuations around the deterministic limiting values. As expected, it turns out that these fluctuations are described by a Gaussian process which is determined by its covariance. The paper is organized as follows. In Section 2 we introduce the network class under consideration and formulate our results. Section 3 is devoted to sketches of the proofs.

The authors would like to thank R.L.Dobrushin, A.N.Rybko and F.P.Kelly for fruitful discussions of the topics of the paper. One of us (Y.M.Suhov) thanks the Statistical Laboratory of the Department of Pure Mathematics and Mathematical Statistics, the University of Cambridge, and particularly, F.P. Kelly for the warm hospitality.

2. PRELIMINARIES AND RESULTS

We consider in this paper the so–called star–like message switching networks. The network has a center, v_0, connected with peripherical nodes $v_1, \ldots, v_N$ by the lines $\alpha_j \sim (v_j \to v_0)$ and $\beta_j \sim (v_0 \to v_j)$. On the input of every line, α_j or β_j, a server is placed which transmits the messages arrived, serving them according to the queueing discipline FCFS, with respect to the moments of arrival. The line capacity allows us to transmit simultaneously at most one message, the interuption is not admitted. At the nodes $v_1, \ldots, v_N$ the external message flows $\xi_j, j = 1, \ldots, N$, arise (i.e., arrive from outside), which are supposed to be stationary and independent from each other. Every ξ_j is a stationary random marked point process (r.m.p.p.) with the Poisson "base" (this means that the arrival times of messages form a stationary (unmarked) Poisson point process) and with i.i.d. marks (l_1, l_2, d) taking values in the space $\mathbf{R}_+^2 \times \{1, \ldots, \hat{j}, \ldots, N\}$ (the symbol $\hat{j}$ means that the element j is omitted). Thereby, the distribution of every r.m.p.p. ξ_j is determined by the intensity a_j and the individual mark distribution (i.m.d.) μ_j. We suppose that $a_j \equiv a > 0$ and μ_j is a product, $\mu_j = F \times H_j$, where F is a probability measure on $\mathbf{R}_+^2$ with exponential marginals $F^{(1)}$ and $F^{(2)}, H_j$ is the uniform distribution on the set $\{1, \ldots, \widehat{j}, \ldots, N\}$. We assume henceforth that the following non-overload bound holds

$$\max[\rho^{(1)}, \rho^{(2)}] < 1, \tag{2.1}$$

where $\rho^{(1)} = ab^{(1)}, \rho^{(2)} = ab^{(2)}$,

$$b^{(1)} = \int_{\mathbf{R}_+^1} F^{(1)}(dl)l, \quad b^{(2)} = \int_{\mathbf{R}_+^1} F^{(2)}(dl)l. \tag{2.2}$$

The network operates as follows. A message which arrives from outside at a time t in an external flow ξ_j and has a mark (l_1, l_2, d) waits first in the buffer of the server at the input of the line α_j. Then it takes this line for the time l_1 and reaches the center v_0 thereafter. At the last moment (when the message is received at the center) it is "switched" on (and thereby arrived at) the input of the line β_d

(according to its address) where it waits again in the corresponding buffer. Then it takes the line β_d for the time l_2 and leaves the network after completing its transmission along this line.

One of the problems which arise in dealing with a queueing network is to prove the existence (and uniqueness, if any) of the stationary working regime. In the situation under consideration (where the processes in the network are non-Markovian) a convenient way to make this notion meaningful is to construct a family of "equipped" r.m.p.p.'s $\eta_j, j = 1, \dots, N$, where any message which arrives in a flow ξ_j is endowed with additional components of the mark which characterize the process of its passing through the net. An appropriate example of these additional components is the pair (w_1, w_2) of waiting times which our messages spend in the two buffers along the paths. Of course, unlike the "original" components (l_1, l_2, d), the waiting times of different messages are not independent: they must be related by a system of network equations which correspond to the switching rule and queueing discipline under consideration. We shall not write these equations in explicit form referring the reader to (Kelbert and Suhov, 1988), where such an approach is discussed in details.

The following statement holds (see, e. g., (Dobrushin and Suhov, 1976)):

PROPOSITION A. *There exists a unique family of stationary r.m.p.p.'s* $\eta_j, j = 1, \dots, N$, *with marks* $(l_1, l_2, d, w, w_2) \in \mathbf{R}^2_+ \times \{1, \dots, \widehat{j}, \dots, N\} \times \mathbf{R}^2_+$ *such that*

(i) *the projection* $(l_1, l_2, d, w_1, w_2) \to (l_1, l_2, d)$ *maps the family* $\{\eta_j\}$ *into the external family* $\{\xi_j\}$ *and is 1–1 with probability one,*

(ii) *the values* w_1 *and* w_2 *give, with probability one, a solution to the system of network equations.*

The family $\{\eta_j, j = 1, \dots, N\}$ determines the probability space on which all the processes under consideration are defined.

Let $q^{(1)}(t,s)$ (respectively, $q^{(2)}(t,s)$) denote the proportion of the numbers j among $1, \dots, N$ such that the queue size on the input of the line α_j (respectively, β_j) equals s at the time $t, s \in \mathbf{Z}^1_+, t \in \mathbf{R}^1$. This means that $Nq^{(1)}(t,s)$ and $Nq^{(2)}(t,s)$ are the numbers of the lines α_j and β_j where the input queue size is equal to s. We obtain the two families, $\{q^{(1)}(t,s), t \in \mathbf{R}^1, s \in \mathbf{Z}^1_+\}$ and $\{q^{(2)}(t,s), t \in \mathbf{R}^1 s \in \mathbf{Z}^1_+\}$, of stationary stochastic processes which are k/N-valued, $k = 0, 1, \dots, N$, with $\sum_s q^{(1)}(t,s) = \sum_s q^{(2)}(t,s) = 1$. Sometimes it will be convenient to speak about stochastic vector processes (s.v.p.'s) $\bar{q}^{(1)}(t)$ and $\bar{q}^{(2)}(t), t \in \mathbf{R}^1$.

THEOREM 1. *The s.v.p.'s* $\bar{q}^{(1)}(t)$ *and* $\bar{q}^{(2)}(t)$ *converge, as* $N \to \infty$, *to non-random vectors* $\bar{r}^{(1)}$ *and* $\bar{r}^{(2)}$ *which are constant in time, with the components*

$$r^{(1)}(s) = \rho_1^s(1-\rho_1), \; r^{(2)}(s) = \rho_2^s(1-\rho_2), \; s \in \mathbf{Z}^1_+. \tag{2.3}$$

Remarks. 1. The RHS's of (2.3) are nothing but the steady-state p.d.'s for the queue size in the stationary $M|M|1|\infty$ queueing model with the loads ρ_1 and ρ_2, respectively.

2. We have in mind in Theorem 1 the component–wise vague convergence wrt Skorokhod's topology. The same meaning is in Theorems 2 – 4 below.
Fix a moment $t_0 \in \mathbf{R}^1$ and consider the conditional probability distribution of the s.v.p.'s $\bar{q}^{(1)}(t)$ and $\bar{q}^{(2)}(t)$ with respect to the σ–algebra generated by $\bar{q}^{(1)}(t_0)$ and $\bar{q}^{(2)}(t_0)$. It is convenient to speak about "conditioned" processes $q^{(1)}(t,s|\bar{q}_0^{(1)},\bar{q}_0^{(2)})$ (which depends in fact on the vector $\bar{q}_0^{(1)}$ only) and $q^{(2)}(t,s|\bar{q}_0^{(1)},\bar{q}_0^{(2)})$ with the deterministic initial data $\bar{q}_0^{(1)} = \{q_0^{(1)}(s), s \in \mathbf{Z}^1_+\}$ and $\bar{q}_0^{(2)} = \{q_0^{(2)}(s), s \in \mathbf{Z}^1_+\}$.

Theorem 2. *Suppose the initial vectors $\bar{q}_0^{(1)}$ and $\bar{q}_0^{(2)}$ are bounded from above by a vector $\bar{u} = \{u(s), s \in \mathbf{Z}^1_+\}$ with $\sum_s u(s) < \infty$, and converge, as $N \to \infty$ to limits $\bar{r}_0^{(1)}$ and $\bar{r}_0^{(2)}$:*

$$\lim_{N\to\infty} q_0^{(1)}(s) = r_0^{(1)}(s), \quad \lim_{N\to\infty} q_0^{(2)}(s) = r_0^{(2)}(s), \quad s \in \mathbf{Z}^1_+.$$

Then the conditioned s.v.p.'s $\bar{q}^{(1)}(t|\bar{q}_0^{(1)})$ and $\bar{q}^{(2)}(t|\bar{q}_0^{(1)},\bar{q}_0^{(2)})$ converge, as $N \to \infty$, to non–random vector functions $\bar{r}^{(1)}(t|\bar{r}_0^{(1)})$ and $r^{(2)}(t|\bar{r}_0^{(1)},\bar{r}_0^{(2)})$, $t \geqslant t_0$. The components $r^{(1)}(t,s|\bar{r}_0^{(1)})$ and $r^{(2)}(t,s|\bar{r}_0^{(1)},\bar{r}_0^{(2)})$, $s \in \mathbf{Z}^1_+$, give the (unique) solution of the following Cauchy problem

$$\frac{d}{dt}r^{(1)}(t,s|\bar{r}_0^{(1)}) = ar^{(1)}(t,s-1|\bar{r}_0^{(1)}) + b^{(1)}r^{(1)}(t,s+1|\bar{r}_0^{(1)}) -$$

$$-(a+b^{(1)})r^{(1)}(t,s|\bar{r}_0^{(1)}), \ t \geqslant t_0, \tag{2.4}$$

$$\frac{d}{dt}r^{(2)}(t,s|\bar{r}_0^{(1)},\bar{r}_0^{(2)}) = b^{(1)}\sum_{k\geqslant 1} r^{(1)}(t,k|\bar{r}_0^{(1)})r^{(2)}(t,s-1|\bar{r}_0^{(1)},\bar{r}_0^{(2)}) +$$

$$+b^{(2)}r^{(2)}(t,s+1|\bar{r}_0^{(1)},\bar{r}_0^{(2)}) - (b^{(1)}\sum_{k\geqslant 1} r^{(1)}(t,k|\bar{r}_0^{(1)}) +$$

$$+b^{(2)})r^{(2)}(t,s|\bar{r}_0^{(1)},\bar{r}_0^{(2)}), \quad t \geqslant t_0, \tag{2.5}$$

$$r^{(1)}(t_0,s|\bar{r}_0^{(1)}) = r_0^{(1)}(s), \ r^{(2)}(t_0,s|\bar{r}_0^{(1)}\bar{r}_0^{(2)}) = r_0^{(2)}(s), \ s \in \mathbf{Z}^1_+. \tag{2.6}$$

Remark. Equation (2.4) describes the time evolution of the probability distribution for the queue size in the model $M|M|1|\infty$ with the input intensity a and the service rate $b^{(1)}$ provided that the initial probability distribution is $\bar{r}_0^{(1)}$. As to equation (2.5), it describes the time evolution of the queue size probability distribution in a model with the variable input intensity rate

$$b^{(1)}\sum_{k\geqslant 1} r^{(1)}(t,k|\bar{r}_0^{(1)})$$

and the service rate $b^{(2)}$ provided that the initial probability distribution is $\bar{r}_0^{(2)}$. Theorems 1 and 2 describe the behavior of some average values related to the queues in our network.

One can also speak about "individual" or "distinguished" message characteristics which may be defined, e. g., via the Palm distribution. On this way we obtain the following assertion which generalizes the result of (Dobrushin and Suhov, 1976).

THEOREM 3. *Under the conditions of this section, the distribution of the quadruple (l_1, l_2, w_1, w_2) associated with the "distinguished" message arisen at a time t_0 in a node v_j and with an address $d = k$ tends, as $N \to \infty$, to*

$$F \times G^{(1)} \times G^{(2)},$$

where $G^{(i)}, i = 1, 2$, is the steady state of the waiting time in the $M|M|1|\infty$ queue with the input intensity a and the service time probability distribution $F^{(i)}$ (the marginal of F).

The next step in studying the situation is to establish a Gaussian correction to the limiting formulas and relations (2.3)—(2.6).
Let us introduce some notation and terminology. We shall refer below to p.d.'s related to the stationary queueing models $M|M|1|\infty$ with the FCFS service discipline, which were mentioned before in Theorems 1 and 3. By $r^{(i)}(\tau, s_1, s_2), i = 1, 2$, we denote the joint probability that, in the $M|M|1|\infty$ queue with the input intensity rate a and the service rate $b^{(i)}$, in the steady-state regime, the queue size at a time t is equal to s_1 and at $(t + \tau)$ is equal to s_2 (this probability does not depend on t). By $r^{(i)}_{[h,l]}(t, s), i = 1, 2$, we denote the Palm probability that, in the corresponding $M|M|1|\infty$ model, the queue size at the time t equals s, under the condition that, at the time h, a customer arrived which required the service time l. Likewise, $r^{(i)}_{[h_1,l_1],[h_2,l_2]}(t_1, s_1, t_2, s_2)$ denotes the joint Palm probability that in the same $M|M|1|\infty$ queue (actually, this probability is used for $i = 2$ only) the queue size at the times t_1 and t_2 equals s_1 and s_2, respectively, under the condition that, at the times h_1 and h_2, two customers arrived and required the service times $l^{(1)}$ and $l^{(2)}$. Finally, we use (in fact, for $i = 1$ only) the Palm p.d.'s $P^{(i)}_{[h,l]}(dw)$ and $P^{(i)}_{[h_1,l^{(1)}],[h_2,l^{(2)}]}(dw_1 \times dw_2)$ for customer's waiting times w (respectively, w_1 and w_2) under the condition that this (these) customer(s) arrived at the time h (respectively, at times h_1, h_2) and required the service time(s) l ($l^{(1)}$ and $l^{(2)}$, respectively).

THEOREM 4. *Under the conditions and in the notation of Theorem 1, let us set*

$$x^{(i)}(t, s) = N^{1/2}\Big(q^{(i)}(t, s) - \mathbf{E}q^{(i)}(t, s)\Big), i = 1, 2, \ t \in \mathbf{R}^1, \ s \in \mathbf{Z}^1_+. \tag{2.7}$$

Then the pair of the s.v.p.'s $\bar{x}^{(i)}(t), i = 1, 2$, converge in distribution to a stationary Gaussian pair of s.v.p.'s $\bar{X}^{(i)}(t), i = 1, 2$, with mean 0 (and $\sum_s X^{(i)}(t, s) = 0$) and with the covariances

$$\mathrm{cov}\Big(X^{(1)}(t_1, s_1), X^{(1)}(t_2, s_2)\Big) = r^{(1)}(t_2 - t_1; s_1, s_2) - r^{(1)}(s_1) r^{(1)}(s_2), \tag{2.8}$$

$$\operatorname{cov}\Big(X^{(2)}(t_1,s_1),X^{(2)}(t_2,s_2)\Big) =$$

$$= r^{(2)}(t_2-t_1;s_1,s_2) - r^{(2)}(s_1)r^{(2)}(s_2) + a^2 \int_{-\infty}^{t_1} dt_1' \int_{\mathbf{R}_+^2} F(dl_1^{(1)}\times dl_2^{(1)})\times$$

$$\times \int_{-\infty}^{t_2} dt_2' \times \int_{\mathbf{R}_+^2} F(dl_1^{(2)}\times dl_2^{(2)})\Big(\int_{\mathbf{R}_+^2} P^{(1)}_{[t_1',l_1^{(1)}],[t_2',l_1^{(2)}]}(dw_1^{(1)}\times dw_1^{(2)}\Big)\times$$

$$\times r^{(2)}_{[t_1'+l_1^{(1)}+w_1^{(1)};l_2^{(1)}],[t_2'+l_1^{(2)}+w_1^{(2)};l_2^{(2)}]}(t_1,s_1,t_2,s_2)-$$

$$-\int_{\mathbf{R}_+^1} P^{(1)}_{[t_1',l_1^{(1)}]}(dw_1^{(1)}) \int_{\mathbf{R}_+^1} P^{(1)}_{[t_2',l_1^{(2)}]}(dw_1^{(2)})\times$$

$$\times r^{(2)}_{[t_1'+l_1^{(1)}+w_1^{(1)};l_2^{(1)}]}(t_1,s_1) r^{(2)}_{[t_2'+l_1^{(2)}+w_1^{(2)};l_2^{(2)}]}(t_2,s_2)), \tag{2.9}$$

$$\operatorname{cov}(X^{(1)}(t_1,s_1),X^{(2)}(t_2,s_2)) =$$

$$= a\int_{-\infty}^{t_1} dt' \int_{\mathbf{R}_+^2} F(dl_1\times dl_2)\times\Big(\int P^{(1)}_{[t',l_1]}(dw_1)(r^{(1)}_{[t',l_1]}(t_1,s_1)\times$$

$$\times r^{(2)}_{[t'+w_1+l_1;l_2']}(t_2,s_2) - r^{(1)}(t_1,s_1)r^{(2)}(t_2,s_2)\Big). \tag{2.10}$$

Remark. It is worthwhile to explain the "physical" meaning of expressions in the RHS's of (2.8)-(2.10). The RHS of (2.8) reflects the fact that the r.m.p.p.'s $\xi_j, 1 \leqslant j \leqslant N$, which form the queues on the inputs of the lines $\alpha_j, 1 \leqslant j \leqslant N$, are independent. Hence, correlations between $X^{(1)}(t_1,s_1)$ and $X^{(1)}(t_2,s_2)$ may be created only in the course of development of the "individual" queue size process related to a single line α_j, which corresponds to $M|M|1|\infty$. On the other hand, the RHS's of (2.9) and (2.10) reflect the facts that the correlations between $X^{(1)}(t_1,s_1)$ or $X^{(2)}(t_1,s_1)$ and $X^{(2)}(t_2,s_2)$ may arise from other "sources". Namely, the third term in the RHS of (2.9) and the whole RHS of (2.10) correspond to correlations between different lines β_{j_1},β_{j_2} and between lines α_{j_1} and β_{j_2} (in the sequel we shall call these terms splitting).

To formulate our non-stationary version of Theorem 4 we have to introduce the conditioned processes

$$x^{(1)}(t,s|\bar{q}_0^{(1)}) = N^{1/2}(q^{(1)}(t,s|\bar{q}_0^{(1)}) - \mathbf{E}q^{(1)}(t,s|\bar{q}_0^{(1)})), \tag{2.11}$$

$$x^{(2)}(t,s|\bar{q}_0^{(1)},\bar{q}_0^{(2)}) = N^{1/2}(q^{(2)}(t,s|\bar{q}_0^{(1)},\bar{q}_0^{(2)}) - \mathbf{E}q^{(2)}(t,s|\bar{q}_0^{(1)},\bar{q}_0^{(2)}). \tag{2.12}$$

THEOREM 5. *Suppose that the sequences of initial stochastic vectors $\bar{q}_0^{(i)}$, $i=1,2$, are majorized by a vector $\bar{v}=(v(s), s\in \mathbf{Z}_+^1)$ such that $\sum_s v(s)<\infty$,* *and converge, as $N \to \infty$, to stochastic vectors $\bar{r}_0^{(i)}, i = 1,2$, respectively. Then the pair of conditioned r.p.'s. $\bar{x}^{(1)}(t|\bar{q}_0^{(1)}, \bar{x}^{(2)}(t|\bar{q}_0^{(1)}, \bar{q}_0^{(2)}), t \geqslant 0$, converges in distribution to a Gaussian pair $X^{(1)}(t|\bar{r}_0^{(1)}), X^{(2)}(t|\bar{r}_0^{(1)}, \bar{r}_0^{(2)})$ with zero initial condition and zero mean.*

The covariance $\mathrm{cov}(X^{(1)}(t_1, s_1|\bar{r}_0^{(1)}, \bar{r}_0^{(2)}), X^{(j)}(t_2, s_2|\bar{r}_0^{(1)}, \bar{r}_0^{(2)}))$ may be computed via formulas which are similar to (2.8)—(2.10). The difference is that the $M|M|1|\infty$ queueing processes on the whole time axis $\mathbf{R}^1$ are replaced now by similar processes on the half-axis $\mathbf{R}_+^1$ with the initial condition obtained by adding, at the moment $t = 0$, ordered groups of $s_0^{(1)}$ and $s_0^{(2)}$ customers, respectively (if $s_0^{(1)}$ (resp., $s_0^{(2)}$) is $\geqslant 1$, then one of the customers, namely, the first one, is supposed to be under service at $t = 0$). However, the structure of splitting terms remains unchanged. For the sake of simplicity, we do not write down here all the corresponding formulas. Just to give the reader an idea, let us write

$$\mathrm{cov}(X^{(1)}(t_1, s_1|\bar{r}_0^{(1)}), X^{(1)}(t_2, s_2|\bar{r}_0^{(1)})) = \sum_{s_0^{(1)} \in \mathbf{Z}_+^1} \bar{r}_0^{(1)}(s_0^{(1)}) B^{(1)}(t_1, s_1; t_2, s_2|s_0^{(1)})$$

where

$$B^{(1)}(t_1, s_1; t_2, s_2|s_0^{(1)}) = r^{(1)}(t_1, s_1; t_2, s_2|\delta_{s_0^{(1)}}) - r^{(1)}(t_1, s_1|\delta_{s_0^{(1)}}) r^{(1)}(t_2, s_2|\delta_{s_0^{(1)}})$$

and (in the case $b^{(1)} = b^{(2)}$)

$$\mathrm{cov}\Big(X^{(2)}(t_1, s_1|\bar{r}_0^{(1)}, \bar{r}_0^{(2)}), X^{(2)}(t_2, s_2|\bar{r}_0^{(1)}, \bar{r}_0^{(2)})\Big) =$$

$$= \mathrm{cov}\Big(X^{(1)}(t_1, s_1|\bar{r}_0^{(1)}), X^{(1)}(t_2, s_2|\bar{r}_0^{(1)})\Big) + \text{splitting term},$$

(the covariance $\mathrm{cov}(X^{(1)}(t_1, s_1|\bar{r}_0^{(1)}), X^{(2)}(t_2, s_2|\bar{r}_0^{(1)}, \bar{r}_0^{(2)}))$ contains a splitting term only).
The proofs of Theorems 1–5 are sketched in Section 3.

3. PROOFS

Proof of Theorem 1. The convergence $\bar{q}^{(1)}(t) \to \bar{r}^{(1)}$ is easy to establish because the queues in the peripherical nodes are independent and each of them is $M|M|1|\infty$. Namely, we apply the law of large number to

$$q^{(1)}(\cdot, s) = N^{-1} \sum_{j=1}^{N} \chi_j^{(1)}(\cdot, s), \tag{3.1}$$

where $\chi_j^{(1)}(t, s) = 1$ if, at the moment $t \in \mathbf{R}^1$, on the input of the line α_j the queue size equals s, and $\chi_j^{(1)}(t, s) = 0$ otherwise, $1 \leqslant j \leqslant N, s \geqslant 0$. This gives $r^{(1)}(s) = \mathbf{E}q^{(1)}(\cdot, s)$, which is just the geometrical distribution $\rho^s(1-\rho), s \geqslant 0$.

To prove the convergence $\bar{q}^{(2)}(t) \to \bar{r}^{(2)}$ one has to prove essentially that

$$\lim_{N\to\infty} \mathbf{P}\Big(\big|q^{(2)}(t,s) - \mathbf{E}q^{(2)}(t,s)\big| > \varepsilon\Big) = \lim_{N\to\infty}\Big(\mathbf{E}q^{(2)}(t,s) - r^{(2)}(s)\Big) = 0. \quad (3.2)$$

The first relation in (3.2) is again a law of large numbers. Its proof proceeds as follows. As before, we introduce the one–zero–valued random variables $\chi_j^{(2)}(\cdot\,,s)$, so that

$$q^{(2)}(\cdot\,,s) = N^{-1}\sum_{j=1}^{N}\chi_j^{(2)}(\cdot\,,s). \quad (3.3)$$

The key point is to check the bound

$$\Big|\mathrm{cov}\big(\chi_{j_1}^{(2)}(t,s),\chi_{j_2}^{(2)}(t,s)\big)\Big| < cN^{-1}.$$

This is based on the facts that (i) the variables $\chi_j^{(2)}(t,s)$ depend "essentially" on the situation in the network on a bounded time interval $(t-T,t]$ only and (ii) the "interference" of different peripherical nodes which may cause dependence between $\chi_{j_1}^{(2)}(t,s)$ and $\chi_{j_2}^{(2)}(t,s)$ occurs with probability $O(N^{-1})$. We omit the details ; similar arguments are given below, in the proof of the second relation in (3.2).

The second relation in (3.2) is equivalent to

$$\lim_{N\to\infty} E\chi_j^{(2)}(t,s) = r^{(2)}(s). \quad (3.4)$$

To prove (3.4) we introduce the approximate random variables $\chi_j^{(2)}(t,s,T), T>0$, setting $\chi_j^{(2)}(t,s,T) = 1$ if, at the time t, on the input of the line β_j, the queue size, computed under the "convention" that messages which have the address j and have arrived in the network before the time $t-T$ are not taken into account, equals s, and $\chi_j^{(2)}(t,s,T) = 0$ if this "cut-off-memory-queue size" differs from s. The equalities $\chi_j^{(2)}(t,s) = \chi_j^{(2)}(t,s,T), s \geqslant 0$, mean that the busy period covering the time t starts not earlier than $t-T$.

We are able to check that the probability

$$\mathbf{P}\{\chi_j^{(2)}(t,s) \neq \chi_j^{(2)}(t,s,T) \quad \text{for at least one } s \geqslant 0\}$$

approaches zero as $T\to\infty$ uniformly in N. Moreover, under our assumptions it goes to zero faster than $O(T^{-\varkappa})$ for any $\varkappa > 0$. This finishes the proof of (3.2) and thereby of Theorem 1.

Proof of Theorem 2 proceeds along the same line of ideas as that of Theorem 1. For the processes $q^{(1)}(\cdot\,,s|\bar{q}_0^{(1)},\bar{q}_0^{(2)})$, due to the law of large numbers, and the assumptions regarding the initial conditions, we have

$$\lim_{N\to\infty} q^{(1)}(t,s|\bar{q}_0^{(1)},\bar{q}_0^{(2)}) = \mathbf{E}\chi^{(1)}(t,s|\bar{r}_0^{(1)}), \quad (3.5)$$

where the indicator $\chi^{(1)}(t,s|\bar{r}_0^{(1)})$ takes the value one, if, at the moment t, the size of the $M|M|1|\infty$ queue equals s, provided that the initial condition is given by the stochastic vector $\bar{r}_0^{(1)}$. In other words, the RHS of (3.5) is merely the probability that the birth-death process of the $M|M|1|\infty$–queue size takes the value s at the moment t under the initial condition $\bar{r}_0^{(1)}$. Equation (2.4) with the corresponding initial data in (2.6) follows then immediately.
Similar (but slightly more complicated) arguments are used to deduce (2.5). Here the key remark is

$$\lim_{N\to\infty} q^{(2)}(t,s|\bar{q}_0^{(1)},\bar{q}_0^{(2)}) = \mathbf{E}\chi^{(2)}(t,s|a^{(1)}(\cdot\,|\bar{r}_0^{(1)}),\bar{r}_0^{(2)}), \tag{3.6}$$

where the indicator $\chi^{(2)}(t,s|a^{(1)},\bar{r}_0^{(2)})$ takes the value one if, at the moment t, the queue size in the model $M|M|1|\infty$ with the time–dependent input intensity $a^{(1)}(\cdot\,|\bar{r}_0^{(1)})$ equals s provided that the initial distribution of the size is given by the stochastic vector $\bar{r}_0^{(2)}$. Here

$$a^{(1)}(\tau|\bar{r}_0^{(1)}) = b^{(1)}\sum_{k\geqslant 1} r^{(1)}(\tau,k|\bar{r}_0^{(1)}), 0\leqslant\tau\leqslant t.$$

Equation (2.5) with the initial data (2.6) again follows immediately.

Proof of Theorem 3 is a slight modification of that of Theorem 2 from (Dobrushin and Suhov, 1976). We omit it as well for the sake of brevity.

Proof of Theorem 4. The convergence of the s.v.p.'s $\bar{x}^{(1)}(t)$ to a Gaussian s.v.p. $\bar{X}^{(1)}(t)$ is a direct corollary of the central limit theorem for i.i.d. random variables $\chi_j^{(1)}(t,s), 1\leqslant j\leqslant N, t\in\mathbf{R}^1$ (see (3.1)). The situation with the convergence $\bar{x}^{(2)}(t)\to\bar{X}^{(2)}(t)_j$ is more complicated. One proceeds along the same line of arguments as in proving the convergence $\bar{q}^{(2)}(t)\to\bar{r}^{(2)}(t)$. That is, we introduce the random variables $x^{(2)}(t,s,\tau)$ by replacing in (2.7) and (3.3) the quantities $\chi_j^{(2)}(t,s)$ by $\chi_j^{(2)}(t,s,T), T>0$. We have for any $t_1,\dots,t_s,\lambda_1,\dots,\lambda_s\in\mathbf{R}^1$ uniformly in $N=1,2,\dots$

$$\lim_{T\to\infty}\Bigg(\mathbf{E}\exp\Big[i\sum_{r=1}^{s}\lambda_r\Big/\sqrt{N}\sum_{j=1}^{N}\Big(\chi_j^{(2)}(t,s)-\mathbf{E}\chi_j^{(2)}(t,s)\Big)\Big]-$$

$$-\mathbf{E}\exp\Big[i\sum_{r=1}^{s}\lambda_r\Big/\sqrt{N}\sum_{j=1}^{N}\Big(\chi_j^{(2)}(t,s,T)-\mathbf{E}\chi_j^{(2)}(t,s,T)\Big)\Big]\Bigg)=0. \tag{3.7}$$

Hence, it is natural to reduce the problem to studying the s.v.p. $\bar{x}^{(2)}(t,T)$. Note that both the families of s.v.p.'s are compact: this may be verified, e.g., via Chentsov's criterion (Chentsov, 1956). The final stage of the proof is to check the convergence of the process $\bar{x}^{(2)}(\cdot,T)$ to a Gaussian s.v.p. $\bar{X}^{(2)}(\cdot,T)$ and then the convergence of $\bar{X}^{(2)}(\cdot,T)$ to the limiting s.v.p. $\bar{X}^{(2)}(\cdot)$ which was identified in Theorem 4. This part is technically cumbersome, and the details will be published elsewhere.
In an analogous way one proceeds to prove Theorem 5.

REFERENCES

Chentsov, N.N. (1956). Weak convergence of random processes whose trajectories have no discontinuities of the second kind *Theory Prob. Appl.* **1**, 154–161.

Dobrushin, R.L. (1986). Switching networks, Markov interaction processes, Gibbsian fields–interconnection. In: *Abstr. First World Congress Bernoulli Soc., Tashkent,* **2**, 666-668.

Dobrushin, R.L. and Suhov, Yu.M. (1976). Asymptotic investigation of star–like message switching networks with a large number of radial rays. *Probl. Inform. Transm.* **12**, 70–94.

Kelbert, M.Ya and Suhov, Yu.M. (1988). Mathematical problems of the queueing networks theory. In: *Itogi Nauki i Techniki. Theory Probab., Math. Statist., Theor. Kybern,* **26**, VINITI, Moscow, pp. 3–96.

Kleinrock, L. (1964). *Communication Nets: Stochastic Message Flows and Delay.* McGraw-Hill, New York.

Sigal, M. and Whitt, W. (1988). A queueing network analyzer for manufacturing. In: *Proc. 12th Intern. Teletraffic Congress,* Torino, Italy.

Prob. Theory and Math. Stat., Vol. 1, pp. 590–601
B.Grigelionis *et al.* (Eds.)
1990 VSP/Mokslas

ASYMPTOTIC SOLUTION OF INVERSE KOLMOGOROV EQUATION FOR DIFFUSION PROCESSES WITH SMALL DIFFUSION

V.M. KHAMETOV
MIEM, Bolshoy Vuzovskij proezd, 3/12, 109028 Moscow, USSR

ABSTRACT

This paper deals with the problem of constructing asymptotics of the Cauchy problem's solution for the inverse Kolmogorov equation, associated with a diffusion process with small diffusion. It yields sufficient conditions, under which two first terms of these asymptotics could be constructed.

INTRODUCTION

Let

$$u^{\varepsilon}(t,x):\ (0,1]\times \mathbf{R}^{+}\times \mathbf{R}^{n}\to \mathbf{R}^{1},\ u^{\varepsilon}(t,x)\in C^{1,2}(\mathbf{R}^{+}\times \mathbf{R}^{n})$$

satisfy a parabolic equation of the second order with a small parameter at the second order derivatives ε^2 and the boundary condition at the right end:

$$u^{\varepsilon}(t,x)|_{t=T}=\exp\{-\Phi(x)/\varepsilon^2\},$$

where $\Phi(x):\mathbf{R}^n\to\mathbf{R}^1$ is a smooth function, $\varepsilon\in(0,1]$ is a small parameter.
It is well-known (Ventzel, 1986; Liptser and Shiryaev, 1974; Gikhman and Skorokhod, 1968; Krylov, 1977) that the function $u^{\varepsilon}(t,x)$ has a simple probability interpretation

$$u^{\varepsilon}(t,x)=\mathbf{E}\exp\{-\Phi(\xi^{\varepsilon}_{t,x}(T))/\varepsilon^2\},$$

where $\xi^{\varepsilon}_{t,x}(T)$ is the value of the diffusion process at T assuming that it started from the point $x\in\mathbf{R}^n$ at t; $\mathbf{E}$ means integration with respect to the probability measure associated with the process $(\xi^{\varepsilon}_{t,x})$.
This problem has the numerous applications. It appears in the theory of limit theorems for Markovian random processes (Ventzel, 1986; Ventzel and Freidlin, 1979), in the theory of controled random processes (Bensoussan, 1986), in the modern quantum field theory and is closely connected with the theory of instantons (Maslov, 1988).
It is well-known that the leading term of the asymptotic expansion is the function $S(t,x)$ which is the solution of the Hamilton—Jacobi equation and is called an "action". The main difficulty arising when we try to construct the successive terms of the asymptotics for $t\in\mathbf{R}^+$ is connected with degeneracy of the Hess matrix

of the function $S(t,x)$ for some $t \in \mathbf{R}^+$. The latter, in its turn, is connected with the existence of focal (adjoint) points or Jacobi points for the corresponding Hamilton system (Maslov, 1988). This paper gives sufficient conditions (in terms of the equation coefficients and initial data) under which the first two terms of the asymptotical expansion of $u^\varepsilon(t,x)$ as $\varepsilon \downarrow 0$ exist for any $t \in \mathbf{R}^+$.

A lot of papers deal with the problem of constructing the asymptotic solution of this problem (and more general problems). In some of them the asymptotic expansion of $u^\varepsilon(t,x)$ as $\varepsilon \downarrow 0$ was constructed by using the probability methods, and in this case it was assumed explicitly or implicity that there were no focal points on a given time interval, or the problem of constructing asymptotics in small was considered. The detailed review of these papers is given in (Ventzel, 1986). Papers (Buslaev, 1967; Molchanov, 1975; Varadhan, 1967; Babich, 1986) deal with asymptotics "in small".

In (Maslov, 1988) the global asymptotics for tunnel type equations is constructed a particular case of which is the problem stated above. Our paper suggests the method of "tunnel canonical operator" based on passing by the focal points which is the Weierstrass transformation. This method allows to construct the global asymptotic solution. The method is different from (Maslov, 1988) and is constructive. It yields sufficient conditions under which we can construct only two first terms of the asymptotics, and, finally, it does not give the ways of passing by the focal points at all. We also note that our technique of constructing the asymptotics is not based on the previous results.

1. STATEMENT OF THE PROBLEM. DEFINITIONS. NOTATION

1.1. A diffusion random process $(\xi^\varepsilon(s), \mathcal{F}_s)_{s\geqslant 0}$ with values in $\mathbf{R}^n$ and the stochastic basis $(\Omega, F, (\mathcal{F}_t)_{t\in\mathbf{R}^+}, P)$ is given satisfying the Ito stochastic equation

$$\xi^\varepsilon(s) = \xi^\varepsilon(t) + \int_t^s a(\tau, \xi^\varepsilon(\tau))\,d\tau + \varepsilon \int_t^s \sigma(\tau, \xi^\varepsilon(\tau))\,dW_\tau, \tag{1.1}$$

where $a(t,x) : \mathbf{R}^+ \times \mathbf{R}^n \to \mathbf{R}^n$ is a vector-function; $\sigma(t,x)$ is a $(n \times n)$-matrix function with elements $\sigma_{ij}(t,x) : \mathbf{R}^+ \times \mathbf{R}^n \to \mathbf{R}^1$, $i,j = 1,\dots,n$; $(W_s, \mathcal{F}_s)_{s\geqslant 0}$ is a n-dimensional Wiener process.

Without loss of generality we assume that

$$\mathcal{F}_t = \sigma\{\xi^\varepsilon_\tau,\ \tau \leqslant t\}.$$

We also assume that $a(t,x)$, $\sigma(t,x)$ satisfy the conditions under which a strong solution of (1.1) exists and is unique.

Let $\Phi(x) : \mathbf{R}^n \to \mathbf{R}^1$ be a smooth function.

We denote

$$u^\varepsilon(t,x) = \mathbf{E}\exp\{-\Phi(\xi^\varepsilon_{t,x}(T))/\varepsilon^2\}, \tag{1.2}$$

where $\xi^\varepsilon_{t,x}(T)$ satisfies the stochastic equation

$$\xi^\varepsilon_{t,x}(T) = x + \int_t^T a(\tau, \xi^\varepsilon_{t,x}(\tau))\,d\tau + \varepsilon \int_t^T \sigma(\tau, \xi^\varepsilon_{t,x}(\tau))\,dW_\tau. \tag{1.1a}$$

It is our aim to construct the asymptotics of $u^\varepsilon(t,x)$ as $\varepsilon \downarrow 0$.
Let the following conditions hold

1) $a_i(t,x)$, $\sigma_{ij}(t,x)$ $(i,j=1,\dots,n)$ are continuous in (t,x), twice differentiable with respect to x and their first and second derivatives in x are bounded and continuous with respect to (t,x) in $\mathbf{R}^+ \times \mathbf{R}^n$.

2) $\Phi(x)$ is twice differentiable and its first and second derivatives are bounded and continuous in $\mathbf{R}^n$.

Then (see (Liptser and Shiryaev, 1974; Gikhman and Skorokhod, 1968; Krylov, 1977)):

1) $u^\varepsilon(t,x)$ is once differentiable with respect to t, is twice differentiable with respect to x, the derivatives are continuous in the strip $[0,T] \times \mathbf{R}^n$,

2) $u^\varepsilon(t,x)$ satisfies the Kolmogorov equation

$$\frac{\partial}{\partial t}u^\varepsilon(t,x) + \sum_{i=1}^n a_i(t,x)\frac{\partial}{\partial x_i}u^\varepsilon(t,x) +$$

$$+\frac{\varepsilon^2}{2}\sum_{i,j=1}^n \sigma_{ij}(t,x)\frac{\partial^2 u^\varepsilon(t,x)}{\partial x_i\,\partial x_j} = 0, \quad (t,x) \in [0,T) \times \mathbf{R}^n;$$

$$u^\varepsilon(T,x) = \exp\{-\Phi(x)/\varepsilon^2\}. \tag{1.3}$$

Thus the problem of constructing the asymptotics of the function $u^\varepsilon(t,x)$ as $\varepsilon \downarrow 0$ reduces to constructing of the asymptotics for parabolic equation (1.3) with a small parameter at the highest derivatives.

1.2. For convenience we introduce the following notation:

$$u_t(t,x) = \frac{\partial}{\partial t}u(t,x); \quad u_{x_i}(t,x) = \frac{\partial}{\partial x_i}u(t,x); \quad \nabla u(t,x) = u_x = (u_{x_1},\dots,u_{x_n}).$$

Denote also for all $x,y \in \mathbf{R}^n$, $(x,y) = \sum_{i=1}^n x_i y_i$,

$$L^\varepsilon u(t,x) = u_t + (a,\nabla u) + \frac{\varepsilon^2}{2}(\sigma\sigma^T\nabla,\nabla u).$$

Let $b=(b_{ij})$ be a $(n \times n)$–matrix; the norm of this matrix we define by

$$\| b \| = (\operatorname{tr} bb^T)^{1/2} = \Big(\sum_{i,j=1}^n (b_{ij})^2\Big)^{1/2}.$$

Let $b_{ij}(x) : \mathbf{R}^d \to \mathbf{R}^1$, $i,j=1,\dots,n$, $b_{ij}(x) \in C^1(\mathbf{R}^d)$ and $b(x)$ be the $(n \times n)$–matrix–function with elements $b_{ij}(x)$. Let $b_x(x) = (b_{ijx_k}(x))$ be $(n \times n \times n)$-"matrix"–function; its norm we define as

$$\Big|\, \| b_x(x) \| \,\Big| = \Big(\sum_{i,j,k=1}^n \big(b_{ijx_k}(x)\big)^2\Big)^{1/2};$$

δ_{ij} is Kronecker delta, $i, j = 1, \dots, n$.

1.3. By Q we denote an open set in $\mathbf{R}^d$, and by $\bar{Q}$ we denote its closure, $D_r = \{x \in \mathbf{R}^n : |x| < r\}$, $C_{TR} = (0,T) \times D_R$ is called a *cylinder*.

1.4. Let us introduce the following notations:
a) $C^l(Q)$ is the Banach space of real functions on Q with finite norm

$$\| u \|_{C^l(Q)} = \sum_{|\gamma|=0}^{l} \sup_{x \in Q} \left| \frac{\partial^{|\gamma|} u(x)}{\partial x_1^{\gamma_1} \dots \partial x_d^{\gamma_d}} \right|.$$

b) $C^\infty(Q)$ is the set of real functions on $\bar{Q}$ infinitely differentiable in Q, continuous and bounded in $\bar{Q}$.
c) $\mathcal{L}_p(Q)$, $p \in [1, \infty]$, is the space of real functions on Q with finite norm

$$\| u \|_{\mathcal{L}_p(Q)} = \begin{cases} (\int_Q | u |^p (x) dx)^{1/p}, & p \in [1, \infty), \\ \operatorname*{ess\,sup}_{x \in Q} | u | (x), & p = \infty. \end{cases}$$

d) $\mathcal{L}_{p\mathrm{loc}}(Q)$ is the space of p–locally summable functions u, i.e., for any bounded set $A \subset \bar{A} \subset Q$ we have $u \in \mathcal{L}_p(A)$.

1.5. By $W_p^k(Q)$, $p \in (1, \infty)$, we denote the set of functions $u : \mathbf{R}^d \to \mathbf{R}^1$, which is the completion of $C^\infty(Q)$ with respect to the norm

$$\| u \|_{W_p^k(Q)} = \sum_{|\gamma|=0}^{k} \left\| \frac{\partial^{|\gamma|} u}{\partial x_1^{\gamma_1} \dots \partial x_d^{\gamma_d}} \right\|_{\mathcal{L}_p(Q)}.$$

Its elements are p–summable functions u with p–summable l–th derivatives, $l \leqslant k$. $W_p^{1,k}(C_{T,R})$ is determined in a similar way.
The expression "almost everywhere" will always mean "almost everywhere with respect to the Lebesgue measure".

2. THE MAIN RESULT

Condition (A)

$$\| \Phi \|_{C^5(\mathbf{R}^n)} \leqslant L_1,$$

where L_1 is a constant.

Condition (B)

$$\| a(t,x) \|_{C^{1,5}(\mathbf{R}^+ \times \mathbf{R}^n)} + \left(\sum_{i,j=1}^{n} \| \sigma_{ij} \|^2_{C^{1,5}(\mathbf{R}^+ \times \mathbf{R}^n)} \right)^{1/2} \leqslant L_2,$$

where L_2 is a constant.

Condition (C). There exist constants c_i, $i = 1,2;$ $0 < c_1 < c_2$ such that for any (t, x) and $\theta \in \mathbf{R}^n$

$$c_1 \mid \theta \mid^2 \leqslant \theta^T (\sigma\sigma^T)(t,x)\theta \leqslant c_2 \mid \theta \mid^2 .$$

THEOREM. *Let conditions* (A), (B), (C) *hold,* $v^\varepsilon(t,x)$ *satisfy* (1.2). *Then there exist functions* $v^0 \in W_p^{1,1}(C_{TR})$, $v^1 \in \mathcal{L}_{2\mathrm{loc}}(C_{TR})$ *and a subsequence* $\{\varepsilon'\}$ *such that for any* $\eta \in C_0^\infty(C_{TR})$

$$\lim_{\varepsilon' \downarrow 0} \int_0^T \int_{\mathbf{R}^n} \eta(s,x)\Big(\ln u^{\varepsilon'}(s,x) \exp\{v^0(s,x)/\varepsilon^2\} \Big) dx\, ds =$$

$$= - \int_0^T \int_{\mathbf{R}^n} \eta(s,x) v^1(s,x) dx\, ds \tag{2.1}$$

and

1) $v^0(t,x)$ *is almost everywhere differentiable with respect to* t, x *and is the unique solution of the equation*

$$v_t^0 + \Big(a - \frac{1}{2}\sigma\sigma^T \nabla v^0, \nabla v^0\Big) = 0, \quad v^0 \mid_{t=T} = \Phi(x). \tag{2.2}$$

Moreover, for any (t,x) *the following estimate holds*

$$\mid v^0(t,x) \mid + \mid v_t^0(t,x) \mid + \mid v_x^0(t,x) \mid \leqslant K_1,$$

where K_1 *is a constant;*

2) $v^1(t,x) \leqslant K_2$ *for any* (t,x), *where* K_2 *is a positive constant.*

3. AUXILLIARY STATEMENTS

3.1. By direct verification we can obtain the following

PROPOSITION. *Let conditions* (A), (B), (C) *hold* u^ε *satisfy* (1.2) *and* $\varepsilon > 0$. *Then there exists a unique function* $v^\varepsilon \in C^{1,4}(\mathbf{R}^+ \times \mathbf{R}^n)$ *satisfying the equation*

$$v_t^\varepsilon + \Big(a - \frac{1}{2}\sigma\sigma^T \nabla v^\varepsilon, \nabla v^\varepsilon\Big) + \frac{\varepsilon^2}{2}\Big(\sigma\sigma^T \nabla, \nabla v^\varepsilon\Big) = 0, \; v^\varepsilon \mid_{t=T} = \Phi(x) \tag{3.1}$$

such that for all $(t,x) \in \mathbf{R}^+ \times \mathbf{R}^n$ *we have*

$$u^\varepsilon(t,x) = \exp\{-v^\varepsilon(t,x)/\varepsilon^2\}. \tag{3.2}$$

3.2. We introduce the following probability measures

$$P_{t,x}^{\varepsilon,1}(A) = \mathbf{E}\chi_A(\omega)\exp\left\{-\frac{1}{\varepsilon}\int_t^T\left(\sigma^T\nabla v^\varepsilon(s,\xi_{t,x}^\varepsilon(s)), dW_S\right)-\right.$$

$$\left.-\frac{1}{2\varepsilon^2}\int_t^T|\sigma^T\nabla v^\varepsilon|^2\,ds\right\},$$

$$P_{t,x}^{\varepsilon,2}(A) = \mathbf{E}\chi_A(\omega)\exp\left\{-\frac{1}{2\varepsilon}\int_t^T\left(\sigma^T\nabla v^\varepsilon(s,\xi_{t,x}^\varepsilon(s)), dW_s\right)-\right.$$

$$\left.-\frac{1}{8\varepsilon^2}\int_t^T|\sigma^T\nabla v^\varepsilon|^2\,ds\right\},$$

where $\varepsilon > 0$, and for any $A \in \mathcal{F}_{t,T} = \sigma\{\xi_\tau^\varepsilon,\ \tau \in (t,T]\}$ and $\chi_A(\omega)$ is the indicator of a set A. The integration operations with respect to the measures $P_{t,x}^{\varepsilon,1}$ and $P_{t,x}^{\varepsilon,2}$ we shall denote by $\mathbf{E}_1$ and $\mathbf{E}_2$, respectively. Let $\left(\xi_{t,x}^{\varepsilon,l}(s),\ \mathcal{F}_s\right)_{s\geqslant t}$ be the l-th component of the process $\left(\xi_{t,x}^\varepsilon(s),\ \mathcal{F}_s\right)_{s\geqslant t}$.

By $(\partial/\partial x_j)\xi_{t,x}^{\varepsilon,l}(s), (\partial^2/\partial x_i\partial x_j)\xi_{t,x}^{\varepsilon,l}(s)$ we denote the first and the second mean square derivatives of the initial data, respectively. Evidently, if condition (B) holds, then such derivatives exist.

THEOREM. *Let conditions* (A), (B), (C) *hold and* $\varepsilon > 0$. *Then there exist constants* c_i, $i = 3,\dots,9$, *such that for any* (ε,t,x) *the following estimates are valid*

1) $|v^\varepsilon(t,x)| \leqslant c_3$,

2) $|\nabla v^\varepsilon(t,x)| \leqslant c_4$,

3) *for any* $i,j = 1,\dots,n$ *and* $s \in \mathbf{R}^+$

$$\mathbf{E}_1\left\|(\partial/\partial x_j)\xi_{t,x}^{\varepsilon,i}(s)\right\|^4 \leqslant c_5,$$

4) *for any* $l,i,j = 1,\dots,n$ *and* $s \in \mathbf{R}^+$

$$\mathbf{E}_1\left|\left\|(\partial^2/\partial x_i\,\partial x_j)\xi_{t,x}^{\varepsilon,l}(s)\right\|\right|^2 \leqslant c_6,$$

5) $\xi^T v^\varepsilon_{xx}(t,x)\xi \leqslant c_7 \mid \xi \mid^2$ for any $\xi \in \mathbf{R}$,

6) $\varepsilon^2 |v^\varepsilon_{x_i x_j}(t,x)| \leqslant c_8$ for any $i,j = 1,\dots,n$,

7) $|v^\varepsilon_t(t,x)| \leqslant c_9$.

Proof. 1) The Ito formula and equation (3.1) yield that

$$v^\varepsilon(t,x) = \mathbf{E}_2 \Phi(\xi^\varepsilon_{t,x}(T)).$$

Thus $|v^\varepsilon(t,x)| \leqslant \sup_{x \in \mathbf{R}^n} |\Phi(x)| \leqslant c_3$.

2) We denote

$$g^\varepsilon(t,x) = \sigma^T(t,x) \bigtriangledown v^\varepsilon(t,x).$$

It is easy to show that $g^\varepsilon(t,x)$ satisfies the equation

$$g^\varepsilon(t,x) = \mathbf{E}_1 \sigma^T(T, \xi^\varepsilon_{t,x}(T)) \bigtriangledown \Phi(\xi^\varepsilon_{t,x}(T)) -$$

$$-\mathbf{E}_1 \int_t^T \left[L^\varepsilon \sigma^T - (\bigtriangledown a)\sigma^T \right] \sigma^{T-1} g^\varepsilon(s, \xi^\varepsilon_{t,x}(s)) ds. \tag{3.3}$$

The solution of (3.3) is costructed by the method of successive approximations

$$g^\varepsilon_{n+1}(t,x) = \mathbf{E}_1 \sigma^T(T, \xi^\varepsilon_{t,x}(T)) \bigtriangledown \Phi(\xi^\varepsilon_{t,x}(T)) -$$

$$-\mathbf{E}_1 \int_t^T \left[L^\varepsilon \sigma^T - (\bigtriangledown a)\sigma^T \right] \sigma^{T-1} g^\varepsilon_n(s, \xi^\varepsilon_{t,x}(s)) ds, \tag{3.4}$$

where $n = 0, 1, 2, \dots,\ g^\varepsilon_0(t,x) \equiv 0$.
According to conditions (A), (B) and the Markovian property of the process $(\xi^\varepsilon_{t,x}(s), \mathcal{F}_s)_{s \geqslant t}$ with respect to the measure $P^{\varepsilon,1}_{t,x}$ approximation (3.4) yields the estimate

$$|g^\varepsilon_{n+1}(t,x) - g^\varepsilon_n(t,x)| \leqslant \sqrt{c_2} \parallel \bigtriangledown \Phi \parallel_{C(\mathbf{R}^n)} \frac{c_{10}(T-t)^n}{n!},$$

where $\|L^\varepsilon \sigma^T\| \, \|\sigma^{T-1}\| + \| \bigtriangledown a\| \leqslant c_{10}$, and c_{10} is a positive constant. This gives the second statement of the theorem.
Statements 3) and 4) of the theorem follows from the Ito formula and Girsanov's theorem and Statement 2) of our theorem.

5) Easily we can show that the solution of (3.1) admits the representation

$$v^\varepsilon(t,x) = -\varepsilon^2 \ln \mathbf{E} \exp\{-\Phi(\xi^\varepsilon_{t,x}(T))/\varepsilon^2\}; \tag{3.5}$$

(3.5) also yields

$$\mathbf{E} \exp\{-[\Phi(\xi^\varepsilon_{t,x}(T)) - v^\varepsilon(t,x)]/\varepsilon^2\} = 1.$$

So due to the Ito formula and equation (3.1) we have

$$\mathbf{E}\exp\{-[\Phi(\xi^{\varepsilon}_{t,x}(T)) - v^{\varepsilon}(t,x)]/\varepsilon^2\} =$$

$$= \exp\Big\{-\frac{1}{\varepsilon}\int_t^T (\sigma^T \nabla v^{\varepsilon}, dW_s) - \frac{1}{2\varepsilon^2}\int_t^T |\sigma^T \nabla v^{\varepsilon}|^2 ds\Big\}. \tag{3.6}$$

Hence according to the conditions of the theorem we obtain

$$v^{\varepsilon}_{x_i}(t,x) = \mathbf{E}_1 \sum_{l=1}^{n} \Phi_{x_l}(\xi^{\varepsilon}_{t,x}(T)) \frac{\partial \xi^{\varepsilon,l}_{t,x}(T)}{\partial x_i}, \tag{3.7}$$

$$v^{\varepsilon}_{x_i x_j}(t,x) = \mathbf{E}_1 \Big\{ \sum_{l,k=1}^{n} \Phi_{x_l x_k}(\xi^{\varepsilon}_{t,x}(T)) \frac{\partial \xi^{\varepsilon,l}_{t,x}(T)}{\partial x_i} \frac{\partial \xi^{\varepsilon,k}_{t,x}(T)}{\partial x_j} +$$

$$+ \sum_{l=1}^{n} \Phi_{x_l}(\xi^{\varepsilon}_{t,x}(T)) \frac{\partial^2 \xi^{\varepsilon,l}_{t,x}(T)}{\partial x_i \partial x_j} - \frac{1}{\varepsilon^2}\Big[\sum_{l=1}^{n} \Phi_{x_l}(\xi^{\varepsilon}_{t,x}(T)) \frac{\partial \xi^{\varepsilon,l}_{t,x}(T)}{\partial x_i} -$$

$$-v^{\varepsilon}_{x_i}(t,x)\Big]\Big[\sum_{k=1}^{n} \Phi_{x_k}(\xi^{\varepsilon}_{t,x}(T)) \frac{\partial \xi^{\varepsilon,k}_{t,x}(T)}{\partial x_j} - v^{\varepsilon}_{x_j}(t,x)\Big]\Big\}. \tag{3.8}$$

Therefore Statement 5) follows from (3.8) and from Statements 2) – 4) of the theorem.
Statement 6) follows from (3.8) and 2) – 4).
The last statement follows from 2), 6) and equation (3.1).

3.3. COROLLARY. *Let conditions of Theorem 3.2 hold. Then there exists a constant $c_{11} > 0$ such that for any $\varepsilon > 0$ the function $c_{11} \mid x \mid^2 - v^{\varepsilon}(t,x)$ is convex with respect to x. Moreover, the second order derivatives $v^{\varepsilon}_{x_i x_j}$ (in sense of the theory of generalized functions) are signed measures of bounded variation on each D_R, $R > 0$, uniformly with respect to ε.*

3.4. PROPOSITION. *Let the conditions of Theorem 3.2 hold and $\varepsilon > 0$. Then $v^{\varepsilon}(t,x)$ is differentiable with respect to ε^2 and for any (t,x) the following estimate holds*

$$\frac{\partial}{\partial \varepsilon^2} v^{\varepsilon}(t,x) \leqslant c_{13},$$

where c_{13} is a positive constant.

Proof. According to condition (B) the process $(\xi^{\varepsilon}_{t,x}(s), \ \mathcal{F}_s)_{s \geqslant t}$ is mean square differentiable with respect to ε, and according to (A) v^{ε} is differentiable with respect to ε^2, $\varepsilon > 0$. We denote

$$\bar{v}^{\varepsilon}(t,x) = \frac{\partial}{\partial \varepsilon^2} v^{\varepsilon}(t,x).$$

It is easy to show that

$$\bar{v}^\varepsilon(t,x) = \frac{1}{2}\mathbf{E}_1 \int_t^T (\sigma\sigma^T \nabla, \nabla v^\varepsilon)(s, \xi^\varepsilon_{t,x}(s))ds. \tag{3.9}$$

This yields our statement.

4. THE LEADING TERM OF THE ASYMPTOTIC EXPANSION

The main result of this section is the following.

THEOREM. *Let conditions 3.2 hold and v^ε satisfies equation (3.1). Then there exists a function $v^0(t,x) \in W^{1,1}_p(C_{TR})$, $p \in (1,\infty)$, which is the unique solution of equation (2.2)*

$$v^0_t + \left(a - \frac{1}{2}\sigma\sigma^T \nabla v^0,\ \nabla v^0\right) = 0, \quad v^0|_{t=T} = \Phi(x),$$

and

1) $|v^0| + |v^0_t| + |\nabla v^0| \leqslant c_{14}$ *for any (t,x), where $c_{14} > 0$ is a constant;*

2) $v^\varepsilon(t,x) \to v^0(t,x)$ as $\varepsilon \downarrow 0$ *pointwise;*

3) *the second order derivatives of the function $v^0(t,x)$ (in the sense of the theory of generalized functions) are signed measures with a bounded variation on each D_R, $R > 0$.*

Proof. Proposition 3.4 yields that $(\partial/\partial\varepsilon^2)v^\varepsilon(t,x) \leqslant c_{13}$. So the uniform boundedness of $v^\varepsilon(t,x)$ yields that for any (t,x) there exist limit $v^0(t,x) = \lim_{\varepsilon\downarrow 0} v^\varepsilon(t,x)$ such that $|v^0(t,x)| \leqslant c_3$.
We have from Theorem 3.2 and the Arzell theorem that $v^0(t,x)$ possesses the Sobolev derivatives with respect to t and x_i, $i = 1,\ldots,n$, and $|v^0_t| + |\nabla v^0| \leqslant c_{15}$, where $c_{15} > 0$ is a constant.
As it is well-known from the theory of convex functions there exists a subsequence $\{v^{\varepsilon'}_{x_i}\}$ such that $v^{\varepsilon'}_{x_i}(t,x)$ tends to $v^0_{x_i}(t,x)$ almost everywhere as $\varepsilon' \downarrow 0$.
Now we consider equation (3.1). Multiplying it by $\eta \in C^\infty_0(C_{TR})$ and integrating in the strip $[0,T] \times \mathbf{R}^n$, we have

$$\int\limits_{[0,T]\times\mathbf{R}^n} \eta\Big[v^\varepsilon_t + \Big(a - \frac{1}{2}\sigma\sigma^T \nabla v^\varepsilon,\ \nabla v^\varepsilon\Big)\Big] dx\, dt =$$

$$= \frac{\varepsilon^2}{2} \sum_{i,j=1} \int\limits_{[0,T]\times\mathbf{R}^n} v^\varepsilon_{x_i}((\sigma\sigma^T)_{ij}\eta)_{x_j}\, dx\, dt. \tag{4.1}$$

Letting $\varepsilon \downarrow 0$ in (4.1) we obtain (2.2) according to the statements given above. Evidently, $v^0|_{t=T} = \Phi(x)$. The uniqueness of the solution of (2.2) follows from the estimate obtained in 1) of Theorem 3.2. The last statement of the theorem holds, because $c_{11}|x|^2 - v^0(t,x)$ is a convex function with respect to x.

5. PROOF OF THE THEOREM

5.1. We denote

$$v^{1\varepsilon}(t,x) = \frac{v^\varepsilon(t,x) - v^0(t,x)}{\varepsilon^2}. \tag{5.1}$$

The results of section 4 yield that $v^{1\varepsilon} \in W_p^{1,1}(C_{TR}),\ \ p \in (1,\infty)$ for any $\varepsilon > 0$. Equations (3.1) and (2.1) yield that $v^{1\varepsilon}$ satisfies the equation

$$\begin{cases} v_t^{1\varepsilon} + (a - \sigma\sigma^T \nabla (v^\varepsilon + v^0)/2,\ \nabla v^{1\varepsilon}) + (\sigma\sigma^T \nabla,\ \nabla v^\varepsilon)/2 = 0, \\ v^{1\varepsilon}|_{t=T} = 0. \end{cases} \tag{5.2}$$

We shall investigate the behaviour of the function $v^{1\varepsilon}$ as $\varepsilon \downarrow 0$.

5.2. THEOREM. *Let conditions* (A), (B), (C) *hold,* $v^{1\varepsilon} \in W_p^{1,1}(C_{TR})$, $p \in (1,\infty)$, *satisfy equation* (5.2). *Then there exists a subsequence* ε' *such that for any* $\eta \in C_0^\infty(C_{TR})$

$$\lim_{\varepsilon' \downarrow 0} \int\limits_{[0,T]\times \mathbf{R}^n} \eta(t,x) v^{1\varepsilon'}(t,x)dx\,dt = \int\limits_{[0,T]\times \mathbf{R}^n} \eta(t,x) v^1(t,x)dx\,dt$$

where $v^1 \in \mathcal{L}_{2\mathrm{loc}}(C_{TR})$.

Proof. Due to Proposition 3.4 and the Newton—Leibnitz formula we have from (5.1)

$$v^{1\varepsilon}(t,x) = \frac{1}{\varepsilon^2} \int\limits_0^{\varepsilon^2} \bar{v}^{\varepsilon'}(t,x)d(\varepsilon')^2. \tag{5.3}$$

Moreover, (3.9) yields that $\bar{v}^\varepsilon(t,x)$ satisfies the equation

$$\bar{v}_t^\varepsilon + (a - \sigma\sigma^T \nabla v^\varepsilon,\ \nabla \bar{v}^\varepsilon) + \frac{\varepsilon^2}{2}(\sigma\sigma^T \nabla,\ \nabla \bar{v}^\varepsilon) + \frac{1}{2}(\sigma\sigma^T \nabla,\ \nabla v^\varepsilon) = 0. \tag{5.4}$$

We denote $v_2^\varepsilon = (\bar{v}^\varepsilon)^2$. By multiplying the left–hand and right–hand sides of (5.4) by $2\omega^\lambda \bar{v}^\varepsilon$, where

$$\omega^\lambda(x) = \exp\big\{ -\lambda\sqrt{|x|^2+1}\big\},\ \ 0 < \lambda < \infty,$$

and integrating in the strip $[t,T] \times \mathbf{R}^n$, we obtain

$$\int\limits_{\mathbf{R}^n} v_2^\varepsilon \omega^\lambda(x)dx = \int\limits_y^T \int\limits_{\mathbf{R}^n} \omega^\lambda(x)\Big\{ \mathrm{div} v_2^\varepsilon (a - \sigma\sigma^T \nabla v^\varepsilon) +$$

$$+v_2^\varepsilon(-\text{div}a + (\text{div}\sigma\sigma^T, \nabla v^\varepsilon) + (\sigma\sigma^T\nabla, \nabla v^\varepsilon))+$$

$$+\frac{\varepsilon^2}{2}\text{div}(\sigma\sigma^T \nabla v_2^\varepsilon) - \varepsilon^2|\sigma^T \nabla \bar{v}^\varepsilon|^2 - \frac{\varepsilon^2}{2}(\text{div}\sigma\sigma^T, \nabla v_2^\varepsilon) + \bar{v}^\varepsilon(\sigma\sigma^T\nabla, \nabla v^\varepsilon)\Big\}dxdt.$$

Thus, according to the conditions of the theorem we have

$$\int\limits_{\mathbf{R}^n} \omega^\lambda v_2^\varepsilon(t,x)dx \leqslant c_{16}\int\limits_t^T\int\limits_{\mathbf{R}^n} \omega^\lambda v_2^\varepsilon dxds+$$

$$+c_{17}\int\limits_t^T\int\limits_{\mathbf{R}^n} \omega^\lambda(x)|(\sigma\sigma^T\nabla, \nabla v^\varepsilon)|(s,x)dxds, \tag{5.5}$$

where $c_{16} > 0,\ c_{17} > 0$ are constants independent of (ε, t).
From (5.5) we obtain the estimate

$$\int\limits_{\mathbf{R}^n} \omega^\lambda(x)|v^{1\varepsilon}|^2(t,x)dx \leqslant c_{18},$$

where $c_{18} > 0$ is a constant independent of (ε, t). Thus, $v^{1\varepsilon} \in \mathcal{L}_{2\text{loc}}(C_{TR})$.

5.3. *Proof of the main theorem.* According to Proposition 3.1 we have $u^\varepsilon(t,x) = \exp\{-v^\varepsilon(t,x)/\varepsilon^2\}$. Thus

$$\ln u^\varepsilon(t,x) + v^0(t,x)/\varepsilon^2 = -v^{1\varepsilon}(t,x). \tag{5.6}$$

Multiplying the left-hand and right-hand sides of (5.6) by an arbitrary $\eta \in C_0^\infty(C_{TR})$ and integrating in the strip $[0,T]\times\mathbf{R}^n$, we obtain

$$\int\limits_{[0,T]\times\mathbf{R}^n} \eta(s,x)\Big[\ln u^\varepsilon(s,x) + v^0(s,x)/\varepsilon^2\Big]dx\,ds =$$

$$= -\int\limits_{[0,T]\times\mathbf{R}^n} \eta(s,x)v^{1\varepsilon}(s,x)dx\,ds.$$

According to Theorem 5.2 we can pass to the limit by some subsequence $\{\varepsilon'\}$ in the last relation. We obtain

$$\lim_{\varepsilon'\downarrow 0}\int\limits_{[0,T]\times\mathbf{R}^n} \eta(s,x)\Big[\ln u^{\varepsilon'}(s,x) + v^0(s,x)/\varepsilon'^2\Big]dx\,ds =$$

$$= -\int\limits_{[0,T]\times\mathbf{R}^n} \eta(s,x)v^1(s,x)dx\,ds.$$

REFERENCES

Azencott, R. (1982). Formulae de Taylor stochastique et development asymptotique d'integrale de Feynmann. *Lect. Notes Math.* **921**, 237–285.

Babich, V.M. (1986). Asymptotic solution of a problem about point wise sourse of diffusion in moving media with small coefficiente of diffusion. *Preprint.* Leningrad Branch of Math. Inst., pp. 3–86.

Bensoussan, A. (1986). Perturbation methods in optimal stochastic control. *Second IFAC Symp. on Stochastic Control* **1**, 349–352. IMC Ac. Sci. Lit.SSR, Vilnius.

Buslaev, A. (1986). Continual integrals and asymptotics of solutions of parabolic equations as $t \to 0$. Applications to difraction. *Problems of Modern Physics* **2**, 85–107. Leningrad. State Univ., Leningrad.

Dubrovsky, V.N. (1976). Exact asymptotic formulas of Laplace's type for Markov processes. *Doklady Acad. Sci. USSR* **226**, 1001–1004.

Fridman, A. (1974). Small random perturbation of dynamical systems and applications to parabolic equations. *Indiana Univ. Math. J.* **24**, 533–553.

Gikhman, I.I. and Skorokhod, A.V. (1972). *Stochastic Differencial Equations.* Springer, Berlin – Heidelberg etc.

Khametov, V.M. (1988). Asymptotics of solution of the inverse Kolmogorov equation for diffusion processes with small diffusion. In: *Probab. Problems of Discrete Math.* MIEM, Moscow, pp. 140–151.

Klepikov, V.N. (1986). Asymptotic expansion of the Feynmann integral. *Theory of Random Processes* **14**, 29–37.

Krylov, N.B. (1977). *Controlled Diffusion Type Processes.* Nauka, Moscow.

Liptser, R.S. and Shiryaev, A.N. (1984). *Statistics of Random Processes.* 2nd print. Springer, Berlin—Heidelberg etc.

Maslov, V.P. (1988). *Asymptotic Methods and Theary of Perturbations.* Nauka, Moscow.

Molchanov, S.A. (1975). Diffusion processes and Riemannian geometry. *Uspekhi Math. Nauk* **30**, 3–59.

Scilder, M. (1966). Some asymptotic formulas for Wiener integrals. *Trans. Amer. Math. Soc.* **125**, 63–85.

Shiryaev, A.N. (1989). *Probability.* Nauka, Moscow.

Varadhan, S.R.S. (1967). Diffusion process in a small time interval. *Commun. Pure and Appl. Math.* **20**, 659–685.

Ventzel, A.D. (1986). *Limit Theorem on Large Deviations for Markov Random Processes.* Nauka, Moscow.

Ventzel, A.D. and Freidlin, M.I. (1979). *Fluctuations in Dinamic Systems under the Action of Small Random Perturbations.* Nauka, Moscow.

Prob. Theory and Math. Stat., Vol. 1, pp. 602–610
B.Grigelionis *et al.* (Eds.)
1990 VSP/Mokslas

ASYMPTOTIC ROBUSTNESS OF DISCRIMINANT PROCEDURES FOR DEPENDENT AND NON–HOMOGENEOUS OBSERVATIONS

Yu. KHARIN and A. MEDVEDEV
Department of Mathematical Modelling and Data Analysis,
Byelorussian State University, Leninskij pr.4, 220080 Minsk, USSR

ABSTRACT

The discriminant analysis problems are considered for the situations when independence and homogeneity assumptions are broken. The asymptotic expansions are fined for the risk (mean losses) of the plug — in decision rules using the minimum contrast estimates. The robust decision rules for nonhomogeneous observations are built.

1. MATHEMATICAL MODEL

The Bayes decision rule (BDR) (Anderson, 1957) is widely used in the discriminant analysis, statistical classification theory and their applications. The BDR provides a minimum risk (mean losses) in making decisions only if the loss matrix, the a priori probabilities of classes and the probability distributions of observations, entering this decision rule (DR), are exactly known and fixed. The real data used to determine the above characteristics cannot be considered absolutely reliable in practice. Besides, these characteristics may change with time. In practice, therefore, a special a priori uncertainty appears: the above characteristics in BDR have some distortions. Neglect of them leads to BDR, whose risk r_0 is minimal for an undistorted model only. However, in treating real observations (in the presence of distortions), the risk of this BDR may essentially exceed r_0, these deviations being non–controllable (Kadane and Chuang, 1978; Randless, 1978; Lawoko and McLachlan, 1986). Thus it is important to study the problems of robust evaluation in discriminant analysis and of synthesizing robust DR, in the presence of distortions due to violation independence and homogeneity of observations. The method suggested in (Kharin, 1983, 1984) has been developed to solve these problems.
Let a parametric family of probability densities

$$Q = \left\{q(x;\theta), x \in \mathbf{R}^N : \theta \in \Theta \subseteq \mathbf{R}^m\right\}$$

be given, $L \geqslant 2$ different density functions $q(\cdot;\theta_1^0), \dots, q(\cdot;\theta_L^0) \in Q$ are fixed and observations $x \in \mathbf{R}^N$ from L classes $\{\Omega_1, \dots, \Omega_L\}$ appear with given probabilities $\pi_1, \dots, \pi_L$ ($\pi_1 + \cdots + \pi_L = 1$). An observation from $\Omega_i (i \in S = \{1, \dots, L\})$ is a random vector $X_i \in \mathbf{R}^N$ with the probability density $q(\cdot;\theta_i^0)$; the parameter

$\theta_i^0 = (\theta_{ij}^0) \in \Theta$ is unknown. A random sample $Z = \bigcup_{i=1}^{L} Z_i$ consisting of L sub-samples is observed; $Z_i = \{z_{it} : z_{it} \in \mathbf{R}^N, \quad t = 1, \ldots, n_i\}$ is a sample of n_i random observations from Ω_i. The random observation $X \in \mathbf{R}^N$ entering one of the $\{\Omega_i\}$ classes and independent of Z must be classified. The $(L \times L)$–loss matrix $W = (w_{ij})$ is given; $w_{ij} \geqslant 0$ is the loss which we have if we classify an observation from Ω_i as belonging to $\Omega_j (i, j \in S)$.

Let us define the DR's used in the discriminant analysis. The $\theta^0 = (\theta_i^0)$ being known, the BRD is of the form

$$d = d_0(x) = \arg\min_{i \in S} f_i(x; \theta^0), \quad x \in \mathbf{R}^N, \quad d \in S, \tag{1}$$

$$f_i(x; \theta^0) = \sum_{j=1}^{L} \pi_j w_{ji} q(x; \theta_j^0), \quad i \in S,$$

where d is the number of the class, which the observation x is assigned to; $\theta^0 \in \mathbf{R}^{Lm}$ is the composite vector of parameters $\theta_1^0, \ldots, \theta_L^0$. BDR (1) ensures a minimum risk in the classification

$$r_0 = \sum_{i=1}^{L} \pi_i \mathbf{E}\left\{w_{i, d_0(X_i)}\right\} > 0.$$

As θ^0 is unknown in practice, the set of so–called plug–in rules (PR)

$$d = d(x; Z) = \arg\min_{i \in S} f_i(x; \hat{\theta}), \quad x \in \mathbf{R}^N, \quad d \in S \tag{2}$$

is used, where $\hat{\theta} \in \mathbf{R}^{Lm}$ is the composite vector of estimates $\hat{\theta}_1, \ldots, \hat{\theta}_L$; $\hat{\theta}_i$ is the minimum contrast (MC) estimate (Pfanzagl, 1969; Chibisov, 1973)

$$\hat{\theta}_i = \arg\min_{\theta \in \Theta} l_i(\theta), \quad l_i(\theta) = n_i^{-1} \sum_{t=1}^{n_i} g(z_{it}; \theta), \tag{3}$$

where $g(z; \theta)$ is the contrast function. In particular, if $g(z; \theta) = -\ln q(z; \theta)$, then $\hat{\theta}_i = (\hat{\theta}_{ik}) \in \mathbf{R}^m$ is the maximum likelihood (ML) estimate. The accuracy of PR (2), (3) will be characterized by the risk:

$$r = r(d) = \sum_{i=1}^{n_i} \pi_i \mathbf{E}\left\{w_{i, d(X_i; Z)}\right\}. \tag{4}$$

As it is known, if Q is regular (Pfanzagl, 1969) and independence condition for $z_{i1}, \ldots, z_{in_i} (i \in S)$ is satisfied, then PR (2), (3) is asymptotically optimal: $r(d) \to r_0$ for increasing sample size $n_0 = \min_i n_i \to \infty$. When the traditional independence and homogeneity conditions are violated, the functional of relative risk increment

$$\varkappa = \varkappa(d) = (r(d) - r_0)/r_0 \geqslant 0$$

will be used as a robustness measure for PR $d(\cdot)$. The smaller the robustness coefficient $\varkappa$, the more stable is PR.

2. GENERAL ASYMPTOTIC RISK EXPANSION FOR DEPENDENT SAMPLES

Let the traditional independence assumption (used to build PR (2), (3) is broken: Z_i is a sample of n_i dependent observations from Ω_i $(i \in S)$ with joint probability density function $p(z_{i1}, \dots, z_{in_i}; \theta_i, \varepsilon_i)$, where $\varepsilon_i \in (-1,1)$ is the parameter characterizing the "dependence level". If $\varepsilon_i = 0$, then $z_{i1}, \dots, z_{in_i}$ are independent; the larger $| \varepsilon_i |$, the "stronger" the dependence is. In this case, the relations of co-ordination with Q are valid:

$$\int_{R^{N(n-1)}} p(x_1, \dots, x_n; \theta_i^0, \varepsilon_i) dx_1 \dots dx_{t-1} dx_{t+1} \dots dx_n = q(x_t; \theta_i^0),$$

$$p(z_{i1}, \dots, z_{in_i}; \theta_i^0, 0) = \prod_{t=1}^{n_i} q(z_{it}; \theta_i^0).$$

Let ∇_θ is the gradient symbol with respect to θ; $g^{(k)}(z;\theta') = (g^{(k)}_{i_1,\dots,i_k}(z;\theta'))$ is the set of m^k partial k^{th}–order derivatives of the contrast function $g(z;\theta)$ with respect to θ at a point $\theta = \theta'$; $\mathbf{1}_m$ is a $(m \times m)$–matrix whose elements are equal to unity; superscript T is the transposition symbol;

$$a_i(\theta) = \mathbf{E}_{\theta_i^0}\{g^{(1)}(z_{it};\theta)\} \quad (i \in S), \tag{5}$$

$$b_{i\tau}(\theta) = \mathrm{Cov}_{\theta_i^0}\{g^{(1)}(z_{it};\theta), g^{(1)}(z_{it+\tau};\theta)\} \quad (\tau = 0, \pm 1, \dots),$$

$v_i = \mathbf{E}_{\theta_i^0}\{(\hat\theta_i - \theta_i^0)(\hat\theta_i - \theta_i^0)^T\}$ is the variance matrix of $\hat\theta_i$, where $\mathbf{E}_{\theta_i^0}\{\cdot\}, \mathrm{Cov}_{\theta_i^0}\{\cdot\}$ are the expectation and the covariance with respect to the distribution $p(\cdot;\theta_i^0,\varepsilon_i)$.

THEOREM 1. *Let the regularity conditions* C1 — C5 $(i \in S)$ *are valid;*
C1. *Θ is compact and θ_i^0 is an interior point of Θ;*
C2. *The density function $q(x;\theta)$ is thrice differentiable with respect to θ;*
C3. *$a_i(\theta_i^0) = 0$, the matrix $J(\theta,\theta_i^0) = \mathbf{E}_{\theta_i^0}\{g_i^{(2)}(X_i;\theta)\}$ is not singular and is continuous in $\theta \in \Theta$; $b_0(\theta_i^0) = J(\theta_i^0,\theta_i^0)$ is the positively definite matrix;*
C4. *$z_{it} (t = 1, 2, \dots)$ is a strongly stationary random sequence such that for $\theta \in \Theta$, $j = 1, 2$, $k = (k_1, \dots, k_j)$, $k' = (k'_1, \dots, k'_j)$, $k_l, k'_l \in \{1, \dots, m\}$ as $n_i \to \infty$,*

$$n_i^{-1} \sum_{\tau=0}^{n_i-1} (1 - \tau/n_i) \mathrm{Cov}_{\theta_i^0}\left\{g_k^{(j)}(z_{i1};\theta), g_{k'}^{(j)}(z_{i\tau+1};\theta)\right\} \to 0,$$

$$n_i^{-2} \sum_{t,u,v=1}^{n_i} \mathbf{E}_{\theta_i^0}\left\{g_j^{(1)}(z_{it};\theta_i^0) g_k^{(1)}(z_{iu};\theta_i^0) g_l^{(1)}(z_{iv};\theta_i^0)\right\} \to 0,$$

$$\sum_{\tau=1}^{n_i-1} (1-\tau/n_i)\mathrm{tr}(b_{i\tau}(\theta_i^0)) < \infty \quad (j,k,l=1,\ldots,m).$$

Then the asymptotic expansions

$$v_i = n_i^{-1}\left(b_0^{-1}(\theta_i^0) + \sum_{\tau=1}^{n_i-1}(1-\tau/n_i)b_0^{-1}(\theta_i^0)\Big(b_{i\tau}(\theta_i^0)+\right.$$

$$\left.+b_{i\tau}^T(\theta_i^0)\Big)b_0^{-1}(\theta_i^0)\right) + o(n_i^{-1})\mathbf{1}_m, \tag{6}$$

$$\mathbf{E}_{\theta_i^0}\left\{(\hat\theta_{ij}-\theta_{ij}^0)(\hat\theta_{ik}-\theta_{ik}^0)(\hat\theta_{il}-\theta_{il}^0)\right\} = o(n_i^{-1})$$

are valid.

Proof relies on a stochastic expansion of the statistic $\hat\theta_i - \theta_i^0$ defined by the equation

$$\nabla_{\theta_i} l_i(\theta_i) = 0 \quad (i \in S).$$

Let δ_{ij} be the Kronecker symbol;

$$c_{ikj} = \pi_i(w_{ik}-w_{ij}), f_{kj}(x;\theta) = f_k(x;\theta) - f_j(x;\theta), \Gamma_{kj}(\theta;v) =$$

$$= \left\{x : f_{kj}(x;\theta) = v\right\} \subset \mathbf{R}^N.$$

$\Gamma_{kj}(\theta^0;0)$ is the Bayes discriminant surface for the discrimination Ω_k,

$$\Omega_j (k \neq j), F_{kj}(x) = \mid \nabla_x f_{kj}(x;\theta^0) \mid^{-1}.$$

The function has the superscript 0 when it is calculated at a true value θ^0, for example, $q_i^{0(1)} = \nabla_{\theta_i^0} q(x;\theta_i^0)$. Also, denote linear combinations of surface integrals $(i \in S, \tau = 1,2,\ldots)$ by

$$\rho_{i\tau} = 0.5(1-\delta_{0\tau}/2)\sum_{j=2}^{L}\sum_{k=1}^{j-1} c_{ikj}^2 \int\limits_{\Gamma_{kj}(\theta^0;0)} q_i^{0(1)T} \times$$

$$\times b_0^{-1}(\theta_i^0)(b_{i\tau}(\theta_i^0) + b_{i\tau}^T(\theta_i^0))b_0^{-1}(\theta_i^0)q_i^{0(1)}F_{kj}(x)dS_{N-1}. \tag{7}$$

THEOREM 2. *Let for each of class Ω_i the conditions C1 — C4 are valid, the integrals in (7) are finite and there exit neighbourhoods $U_1,\ldots,U_L$ of $\theta_1,\ldots,\theta_L$, such that for any $\theta' = (\theta_1',\ldots,\theta_L') \in U_1 \times \cdots \times U_L$*

$$\int\limits_{\Gamma_{kj}(\theta';0)} \left|q_{pst}^{(3)}(x;\theta')\right| q_j^0 F_{kj}(x)dS_{N-1} < \infty,$$

$$\left| \frac{d}{dv} \int\limits_{\Gamma_{kj}(\theta';v)} q_{ps}^{(2)}(x;\theta'_i) q_t^{(1)}(x;\theta'_i) q_j^0 F_{kj}(x) dS_{N-1} \right|_{v=0} < \infty,$$

$$\left| \frac{d^2}{dv^2} \int\limits_{\Gamma_{kj}(\theta';v)} q_p^{(1)}(x;\theta'_i) q_s^{(1)}(x;\theta'_l) q_t^{(1)}(x;\theta'_\nu) q_j^0 F_{kj}(x) dS_{N-1} \right|_{v=0} < \infty,$$

$$(p, s, t = 1, \dots, m; \quad i, j, l, \nu \in S).$$

Then the risk of PR (2), (3) *admits the expansion*

$$r(d) = r_0 + \sum_{i=1}^{L} n_i^{-1} \left(\rho_{i0} + \sum_{\tau=1}^{n_i - 1} (1 - \tau/n_i) \rho_{i\tau} \right) + o(n_0^{-1}). \tag{8}$$

Proof employs the method of (Kharin, 1983, 1984) based on asymptotic expansions of moments of $\hat{\theta}_i - \theta_i^0$ from Theorem 1 ($i \in S$).

COROLLARY 1. *If there is no dependence in the samples* $Z_1, \dots, Z_L$ $(\varepsilon_1 = \varepsilon_2 = = \dots = \varepsilon_L = 0)$, *then*

$$r(d) = r_0 + \sum_{i=1}^{L} \rho_{i0} n_i^{-1} + o(n_0^{-1}), \tag{9}$$

where $\rho_{10}, \dots, \rho_{L0} \geqslant 0$.

COROLLARY 2. *If the observations in* Z_i, *have a Markovian dependence it* ($i \in S$):

$$0.5(b_{i\tau}(\theta_i^0) + b_{i\tau}^T(\theta_i^0)) = \varepsilon_i^{|\tau|} b_0(\theta_i^0), \ \ \tau = 0, \pm 1, \dots,$$

then

$$r(d) = r_0 + \sum_{i=1}^{L} (1 + \varepsilon_i)(1 - \varepsilon_i)^{-1} \rho_{i0} n_i^{-1} + o(n_0^{-1}).$$

COROLLARY 3. *If in Corollary* 2 $\varepsilon_1 = \dots = \varepsilon_L = \varepsilon \in (-1, 1)$, *then the robustness coefficient for* PR (2), (3) *can be expanded as*

$$\varkappa = \varkappa(d; \varepsilon, \{n_i\}) = K(\varepsilon) \varkappa_0(d; \{n_i\}) + o(n_0^{-1}), \tag{10}$$

$$K(\varepsilon) = (1 + \varepsilon)/(1 - \varepsilon) > 0,$$

$$\varkappa_0(d; \{n_i\}) = \sum_{i=1}^{L} \rho_{i0} (r_0 n_i)^{-1} \geqslant 0.$$

Proof is carried out by the substitution of $\{b_{i\tau}(\theta_i^0)\}$ in (7), (8). Note that the expansion (9) coincides with the risk expansion in (Kharin, 1983) under the classical independence condition.

Now, we shall analyse the dependence of the robustness coefficient $\varkappa$ in (10) on ε and $\{n_i\}$. This dependence is factorizable. The factor $\varkappa_0(d;\{n_i\})$ characterizes the PR robustness depending on the sample volumes $\{n_i\}$: the larger n_0 the higher the robustness of PR (2). In (10), the factor $K(\varepsilon)$ characterizes the dependence of the robustness coefficient $\varkappa$ on the "dependence level" ε. As it happens, the dependence of sampling values may decrease as well as increase the coefficient $\varkappa$. If $\varepsilon > 0$, the robustness decreases: the decrease increases with ε; if $\varepsilon < 0$, the robustness increases: the increase becomes larger when $|\varepsilon|$ increases.

Let us define two practically important quantities. The first one is the critical "dependence level" ε_+. If $\varepsilon > \varepsilon_+$, the robustness coefficient $\varkappa$ will exceed the given value δ $(0 < \delta < 1) : \varkappa > \delta$. The second one is the critical sample size n_0^- (at $n_i = n_0, \quad i = 1,\ldots,L$), which ensures the given level δ for the coefficient $\varkappa$. Using the main term in expansion (10) we obtain

$$\varepsilon_+ = \varepsilon_+(\delta,\{n_i\}) = \max\Big\{0, 1/K\Big(\sum_{i=1}^{L}\rho_{i0}/(n_i r_0\delta)\Big)\Big\},$$

$$n_0^- = n_0^-(\delta,\varepsilon) = \Big[K(\varepsilon)\sum_{i=1}^{L}\rho_{i0}/(r_0\delta)\Big] + 1, \tag{11}$$

where $[Z]$ is the integer part of a number Z.

3. ROBUSTNESS FOR GAUSSIAN DEPENDENT SAMPLES

Let us illustrate the results of theorem 1 and 2 applying them to the Gaussian family Q, generally used in the discriminant analysis. Assume there are two $(L = 2)$ equiprobable classes $(\pi_1 = \pi_2 = 0.5)$; X_i has N-dimensional normal distribution $\mathcal{N}(\theta_i^0,\Sigma)$ with expectation $\theta_i^0 = (\theta_{ik}^0) \in \mathbf{R}^N$ and covariance matrix $\Sigma (i = 1,2; k = 1,\ldots,L)$. The loss values are $w_{kl} = 1 - \delta_{kl}$ $(k,l = 1,2)$; then the risk r is the error probability. The sample $Z_i (i = 1,2)$ is an autoregressive sequence in $\mathbf{R}^N$:

$$z_{it+1} = (1-\varepsilon_i)\theta_i^0 + \varepsilon_i z_{it} + \xi_{it+1} \quad (t = 1,2,\ldots), \quad 0 \leqslant |\varepsilon_i| < 1,$$

where z_{i1} has the normal distribution $\mathcal{N}(\theta_i^0,\Sigma)$ and is independent of the sequence $\xi_{i2},\xi_{i3},\ldots$ of independent Gaussian random vectors with the same distribution $\mathcal{N}(0,(1-\varepsilon_i^2)\Sigma)$. In this case, BDR (2), (3) with the ML-estimates $(\hat{\theta}_i)$ is linear:

$$d = d_1(x;Z) = U(b^T x - \gamma) + 1, \quad x \in \mathbf{R}^N, \quad b = \Sigma^{-1}(\hat{\theta}_2 - \hat{\theta}_1), \tag{12}$$

$$\gamma = (\hat{\theta}_2 + \hat{\theta}_1)^T \Sigma^{-1}(\hat{\theta}_2 - \hat{\theta}_1)/2, \quad \hat{\theta}_i = n_i^{-1}\sum_{t=1}^{n_i} z_{it} \quad (i = 1,2),$$

where $U(a) = 1$ if $a \geqslant 0$ and $U(a) = 0$ if $a < 0$. Theorem 2 yields a convenient formula for approximate calculations of the robustness coefficient (at $\varepsilon_1 = \varepsilon_2 = \varepsilon$):

$$\varkappa(d_1;\varepsilon,\{n_i\}) \approx (1+\varepsilon)(N - 1 + \Delta^2/4)(n_1^{-1} + n_2^{-1})(4\Delta\Psi(\Delta/2)(1-\varepsilon))^{-1}, \tag{13}$$

where Ψ (a) is the Mills ratio for the standard normal distribution, $\Delta = ((\theta_2^0 - \theta_1^0)^T \Sigma^{-1} (\theta_2^0 - \theta_1^0))^{1/2}$ is the interclass Mahalanobis distance. Formulas (11), (13) yield convenient formulas for critical values ε_+ and n_0^-:

$$\varepsilon_+ \approx \max\left\{0, 1/K((N-1+\Delta^2/4)/(2\Delta\delta n_0\Psi(\Delta/2)))\right\},$$

$$n_0^- \approx \left[K(\varepsilon)(N-1+\Delta^2/4)(2\Delta\delta\Psi(\Delta/2))^{-1}\right]+1.$$

The table below shows the estimates for the robustness coefficient obtained by (13) (the $\varkappa$ value) and by statistical computer modelling (the $\hat{\varkappa}$ value).

ε	estimate	n_0					
		10	15	20	30	40	50
0	$\varkappa$	0.17	0.11	0.08	0.05	0.04	0.03
	$\hat{\varkappa}$	0.20	0.08	0.06	0.04	0.02	0.02
0.8	$\varkappa$	1.10	0.75	0.55	0.37	0.30	0.22
	$\hat{\varkappa}$	0.99	0.66	0.42	0.30	0.36	0.19

It follows from the table that the parameter ε appreciably effects the robustness of PR (2) and the accuracy of approximations derived from our asymptotic expansions is sufficiently high.

Let us study now the robustness of PR $d = d_2(x; Z)$, obtained by (12) with recurrent estimates $\{\bar{\theta}_i\}$ instead of $\{\hat{\theta}_i\}$

$$\bar{\theta}_i = \bar{\theta}_{i,n_i}, \tag{14}$$

$$\bar{\theta}_{i,t} = (I_N - H_{it})\bar{\theta}_{i,t-1} + H_{it}z_{it} \quad (t = 2, \dots, n_i, i = 1, 2),$$

where $\bar{\theta}_{i,1} = z_{i,1}, I_N$ is the unit $(N \times N)$–matrix, H_{it} is the coefficient matrix. Following (7) the optimum coefficient matrices $\{H_{it}^*\}$ will be chosen from the minimum risk condition:

$$\operatorname{tr}(V\{\bar{\theta}_{it}\}B_i) \to \min_{H_{it}}, \tag{15}$$

where $V\{\bar{\theta}_{it}\}$ is the variance matrix for the estimate $\bar{\theta}_{it}$,

$$B_i = \frac{1}{8} \int_{\Gamma_{21}(\theta^0;0)} q_i^{0(1)} q_i^{0(1)T} F_{21}(x) dS_{N-1} \; (i = 1, 2).$$

THEOREM 3. *If $\{z_{it}\}$ is the above defined autoregressive sequence in $\mathbf{R}^N$, then the optimal coefficient matrices in (14) are*

$$\begin{aligned} H_{it}^* &= h_{it} I_N, \\ h_{it} &= (1-\varepsilon_i)A_i(t-1)/(1+(1-2\varepsilon_i)A_i(t-1)), \\ A_i(t) &= A_i(t-1)(1-\varepsilon_i^2 A_i(t-1))/(1+(1-2\varepsilon_i)A_i(t-1)), \\ A_i(1) &= 1, \end{aligned}$$

and the robustness coefficient in this case is given by

$$\varkappa(d_2; \{\varepsilon_i\}, \{n_i\}) = (N-1+\Delta^2/4)(A_1(n_1)+ \\ +A_2(n_2))/(4\Delta\Psi(\Delta/2)) + o(n_0^{-1}).$$

Proof of theorem 3: one has to solve the extremum problem (15).

4. ROBUSTNESS IN THE CASE OF VIOLATED HOMOGENEITY CONDITION

Let the traditional for BDR (1) assumption that the observations Ω (to be classified) are identically distributed is violated: the observation X_i from Ω_i has the density $p_i(x)$, $x \in \mathbf{R}^N$ belonging to some given set of densities $P_i(\varepsilon_i)$, where $\varepsilon_i \in [0,1]$ is the parameter characterizing the "non-homogeneity level" ($i \in S$). In the case of Tukey non-homogeneities

$$P_i'(\varepsilon_i) = \Big\{p_i(x) : p_i(x) = (1-\varepsilon_i)q(x;\theta_i^*) + \varepsilon_i h_i(x), \\ h_i(x) \geqslant 0, \int h_i(x)dx = 1\Big\};$$

yet another kind of non-homogeneities is generated by the ε_i-neighbourhood of the densities with given "center" $q(x;\theta_i^*)$:

$$P_i''(\varepsilon_i) = \Big\{p_i(x) : p_i(x) \geqslant 0, \int p_i(x)dx = 1, \\ \int (p_i(x) - q(x;\theta_i^*))^2 (q(x;\theta_i^*))^{-1} dx \leqslant \varepsilon_i^2\Big\}.$$

We construct the robust decision rule (RDR) $d = d_*(x)$, $x \in \mathbf{R}^N$, $d \in S$, whose guaranteed risk value is minimal:

$$r(d_*) = \min_{d(\cdot)} \max_{\{p_i \in P_i\}} r(d). \tag{16}$$

THEOREM 4. *For the non-homogeneities determined by $\{P_i'(\varepsilon_i)\}$ the RDR is*

$$d = d_*(x) = \arg\min_{i \in S} \sum_{j=1}^{L} \pi_j (1-\varepsilon_j) w_{ji} q(x;\theta_j^*), \ x \in \mathbf{R}^N.$$

THEOREM 5. *In the case of two classes ($L = 2$) and non–homogeneities determined by $\{P_i''(\varepsilon_i)\}$, the RDR is*

$$d = d_*(x) = U(f_*(x)) + 1, \quad x \in \mathbf{R}^N,$$

where the robust discriminant function $f_*(x)$ is the solution of the extremum problem

$$\int\limits_{f(x)\geqslant 0} \sum_{i=0}^{2}(-1)^i c_{i,3-i,i} q(x;\theta_i^*)\times$$

$$\times\left(1+(-1)^{i+1}\varepsilon_i\left(\Big[\int\limits_{f(x)\geqslant 0} q(x;\theta_i^*)dx\Big]^{-1}-1\right)^{1/2}\right)dx \to \max_{f(\cdot)}.$$

Proof of theorems 4, 5: one has to solve the extremum problem (16).

COROLLARY. *If the density $q(x;\theta_i^*) = n(x \mid \theta_i^*, \Sigma)$ is Gaussian $\pi_1 = \pi_2 = 0.5$, then the robust discriminant function*

$$f_*(x) = (\theta_2^* - \theta_1^*)^T\Sigma^{-1}x + (\theta_1^{*T}\Sigma^{-1}\theta_1^* - \theta_2^{*T}\Sigma^{-1}\theta_2^*)/2 + a_*,$$

where a_* is the limit of recurrent sequence:

$$a_{k+1} = \ln\left(1-\varepsilon_2(\Phi(-\Delta/2 - a_k/\Delta)/\Phi(\Delta/2 + a_k/\Delta))^{1/2}\right)-$$

$$-\ln\left(1+\varepsilon_1(\Phi(\Delta/2 - a_k/\Delta)/\Phi(-\Delta/2 + a_k/\Delta)^{1/2}\right) \quad (k = 1, 2, \ldots),$$

$a_1 = 0, \Phi(\cdot)$ is the standard normal distribution function. Computer modelling of $\varkappa$ estimates and robust decision rules confirms their practical value.

REFERENCES

Anderson, T.W. (1957). *An Introduction to Multivariate Statistical Analysis.* Wiley, New York.

Chibisov, D.M. (1973). Asymptotic expansion for a class of estimators. *Theory Probab. Appl.* **18**, 303–311.

Kadane, J.B. and Chuang, D.T. (1978). Stable decision problems. *Ann. Statist.* **6**, 1095–1110.

Kharin, Yu.S. (1983). Investigation of risk of statistical classifiers using minimum contrast estimators. *Theory Probab. Appl.* **28**, 592–598.

Kharin, Yu.S. (1984). Robustness investigation for the decision rules by risk asymptotic expansion method. In: *Proc. Third Prague Symp. Asympt. Statist.*, Elsevier, New York. pp. 309–316.

Lawoko, C.R.O. and McLachlan, G.J. (1986). Asymptotic error rates of the W and Z statistics when the training observations are depended. *Pattern Recogn.* **18**, 467–471.

Pfanzagl, J. (1969). On the measurability and consistancy of MC estimates. *Metrica* **14**, 149–272.

Randless, R.M. (1978). Generalized linear and quadratic discriminant functions using robust estimates. *J.Amer.Statist. Assoc.* **73**, 564–568.

Prob. Theory and Math. Stat., Vol. 1, pp. 611–617
B.Grigelionis *et al.* (Eds.)
1990 VSP/Mokslas

CHARACTERIZATION OF DISTRIBUTIONS: PROBLEMS, METHODS, APPLICATIONS

L.B. KLEBANOV and A.A. ZINGER
Leningrad Institute of Civil Engineering,
Leningrad Institute for Aviation Instrument Making

ABSTRACT

In the communication is stated a new approach to wide class of characterization problems and once closely connected with them. It is based on systematic application of some new metrics.

The problem of characterization of probability distributions consists in describing all probability laws possessing a certain property $\mathfrak{P}$. In some cases such property can be reduced to some functional equation, then the characterization problem reduces to the describtion of all probability solutions of this equation. We find such approach in the well known book (Kagan et al., 1973), and many other publications devoted to the characterization problem.
There are situations when a class of distribution functions for which the property $\mathfrak{P}$ holds is known, so the problem can be reduced to the uniqueness problem, which allows to use some general results about positive solutions of functional equations. This was stated in (Kakosyan et al., 1984).
To the characterization problems belongs an important class of reconstruction problems in which it is necessary to find the parent distribution by distributions of suitable statistics (Zinger, 1956; Linnik, 1956; Kovalenko, 1960; Prokhorov, 1965; Kotlavski, 1971; Zinger and Kagan, 1973; Bondesson, 1974; and others), or by some functionals of such statistics (see (Braverman, 1985, 1987; Neupokoeva, 1986)). For example, Braverman studied the problem of reconstruction of the distribution of a random variable X_1 by the function

$$U(a_1,\dots,a_n) = \mathbf{E}\Big|\sum_{j=1}^{n} a_j X_j\Big|^r, \quad r>0,\ r\neq 2k,$$

under the condition that $X_1, X_2, \dots, X_n$ are independent identically distributed random variables. One can notice that if conditions of independence and equality of distributions of X_j were omitted, it would be necessary to consider the function

$$\Phi(a_1,\dots,a_n,z) = \mathbf{E}\Big|\sum_{j=1}^{n} a_j X_j - z\Big|^r.$$

This problem consists of the following two problems.

1) reconstruction of the distribution of X by the known function

$$\Psi(z) = \mathbf{E}|X - z|^r,$$

2) reconstruction of the common distribution of vectors $(X_1, \ldots, X_n)$ by the distribution of linear statistics $\sum_{j=1}^{n} a_j X_j$.

Let us consider the first problem which may be regarded as a particular case of the reconstruction of a measure by the potential Ψ. It is closely related to the problem of isometry continuation in L_p–spaces (see (Plotkin, 1970, 1971; Rudin, 1976; Linde, 1982; Koldobsky, 1982; Gorin and Koldobsky, 1987)). The most general results refer to the case when X is an element of a Banach space $\mathbf{B}$ and

$$\Psi(z) = \mathbf{E}\|X - z\|^r, z \in \mathbf{B}. \tag{1}$$

For some special types of $\mathbf{B}$ it was shown in (Gorin and Koldobsky, 1987) that (1) allows to reconstruct the unique measure only if r does not belong to a certain sequence $\{r_j\}_{j=1}^{\infty}$ of indexes. Kakosyan, Klebanov and Zinger in (Kakosyan et al., 1989) proposed a new approach to the reconstruction problem by a potential which is closely connected with probability metrics and characterization problems treated as problems of the description of minima points of some functionals given on a metric space.

This method was stated in Kakosyan, Klebanov, Rachev book (Kakosyan et al., 1988). According to them, stability of the characterization problem means correctness of the minimization problem. Sometimes transformation of the functional under consideration leads to a new unexpected form of the characterization problem. For instance, there is the well- known (Polya, 1923) theorem stating that the following two assumptions are equivalent:

(I) Distributions of X_1 and $(X_1 + X_2)/\sqrt{2}$ coinside and X_1, X_2 are independent identically distributed random variables.

(II) X_1 is normally distributed with the mean 0.

Now let us consider the sequence X_1, X_1', X_2, X_2' of independent equally distributed random variables under conditions such that

$$\mathbf{E}|X_1|^r < \infty$$

for some $r \in (0,2)$. Then two following statements are equivalent.

(III) $2\mathbf{E}\left|X_1 - \frac{1}{\sqrt{2}}(X_1 + X_2)\right|^r - \mathbf{E}\left|X_1 - X_1'\right|^r - \mathbf{E}\left|\frac{X_1+X_2}{\sqrt{2}} - \frac{X_1'+X_2'}{\sqrt{2}}\right|^r = 0.$

(IV) X_1 is normally distributed with the mean 0.

Here the point is that

$$m\ (X,Y) = \left(2\mathbf{E}|X - Y|^r - \mathbf{E}|X - X'|^r - \mathbf{E}|Y - Y'|^r\right)^{1/2}, \tag{2}$$

where X, Y, X', Y' are independent and $X \stackrel{d}{=} X'$, $Y \stackrel{d}{=} Y'$, is a simple metrics (in the sense of Zolotarev) in the space of random variables possessing a finite absolute moment of order r $(0 < r < 2)$.
The assumption (1) is the metric form of the statement 1).
The fact that (2) is a simple metrics follows from the inequality

$$2\,\mathbf{E}|X - Y|^r - \mathbf{E}|X - X'|^r - \mathbf{E}|Y - Y'|^r \geqslant 0 \tag{3}$$

and the equality is true iff $X \stackrel{d}{=} Y$. The question arises whether it is possible to substitute $|X - Y|^r$ in (3) by some function $\mathcal{L}(X, Y)$. This has been investigated in (Kakosyan et al., 1989). It was shown that the inequality

$$2\mathbf{E}\mathcal{L}(X, Y) - \mathbf{E}\mathcal{L}(X, X') - \mathbf{E}\mathcal{L}(Y, Y') \geqslant 0 \tag{4}$$

is true for all independent X, X', Y, Y', where

$$X \stackrel{d}{=} X', \; Y \stackrel{d}{=} Y', \; \mathbf{E}\mathcal{L}(X, X') + \mathbf{E}\mathcal{L}(Y, Y') < \infty$$

iff $\mathcal{L}$ is a *negative definite* (n.d.) kernel. Equality in (4) takes place iff $\mathcal{L}$ is n.d. in the strict sense.
Let $(\mathfrak{X}, d)$ be a metric space, $\mathfrak{H}$ be the σ-algebra of Borel subsets of $\mathfrak{X}$ and $\mathfrak{B}$ be the family of all Borel probability measures on $(\mathfrak{X}, \mathfrak{A})$. The well known theorem due to Shoenberg asserts that $\mathfrak{X}$ is isometric to some subset $\tilde{\mathfrak{X}}$ of a Hilbert space $\mathfrak{H}$ iff $d^2(X, Y) = \mathcal{L}(X, Y)$ is a n.d. kernel.
Now let $\mathcal{L}(x, y)$ be a continuous n.d. kernel. We will denote by $\mathfrak{B}_{\mathcal{L}}$ the subset of all measures $\mu \in \mathfrak{B}$ for which

$$\int_{\mathfrak{X}}\int_{\mathfrak{X}} \mathcal{L}(x, y)d\mu(x)d\mu(y) < \infty$$

holds. For every pair $\mu, \nu \in \mathfrak{B}_{\mathcal{L}}$ we will denote

$$N(\mu, \nu) = -\int_{\mathfrak{X}}\int_{\mathfrak{X}} \mathcal{L}(x, y)dR(x)dR(y), \tag{5}$$

where $R(x) = \mu(x) - \nu(x)$.
If $\mathcal{L}(x, y)$ is the n.d. kernel, then $N(\mu, \nu) \geqslant 0$ for all $\mu, \nu \in \mathfrak{B}_{\mathcal{L}}$ (and vice versa). The map $N(\mu, \nu)$ corresponding to a strictly n.d. $\mathcal{L}(x, y)$ is itself a n.d. kernel on the products $\mathfrak{B}_{\mathcal{L}} \times \mathfrak{B}_{\mathcal{L}}$. In this case $N^{1/2}(\mu, \nu)$ is a metrics on $\mathfrak{B}_{\mathcal{L}}$, and $\mathfrak{B}_{\mathcal{L}}$ is isometric to a convex subset $\tilde{\mathfrak{B}}_{\mathcal{L}}$ of some Hilbert space $\mathfrak{H}$, every point of which has a unique representation as a convex linear combination of points of $\tilde{\mathfrak{X}}$. The distance $N^{1/2}(\mu, \nu)$ between measures μ and ν is equal to the distance between the corresponding baricentres in $\mathfrak{H}$. The points in $\mathfrak{H}$ can be treated as δ-measures, i.e., the constitute a natural imbedding of $\mathfrak{X}$ in $\mathfrak{B}_{\mathcal{L}}$. For δ-measures the distance $N^{1/2}(\delta_x, \delta_y)$ is equal to $\mathcal{L}^{1/2}(x, y)$.
It is easy to see that the reverse statement is true. Thus, if there exists an isometry of $\mathfrak{B}_{\mathcal{L}}$ (with the same distance on $\mathfrak{X}$) to some subset $\tilde{\mathfrak{B}}$ of a Hilbert space $\mathfrak{H}$, $\tilde{\mathfrak{B}}$

being a convex set and the distance between measures is the distance between the proper baricentres in $\mathfrak{H}$, then $\mathcal{L}(x,y)$ is a strictly n.d. kernel on $\mathfrak{X}^2$ and $N(\mu,\nu)$ is calculated by formula (5). Let us note that if the function $\mathcal{L}$ is bounded, then $\mathfrak{B}_{\mathcal{L}} = \mathfrak{B}$, and the value of $N(\mu,\nu)$ is determined for all Borel probability measures μ and ν.

Now we will consider the question about imbedding of the set $\mathfrak{B}$ (or its subset) not in Hilbert, but in some Banach space. Let $\mathfrak{X}$ be again a metric space with metrics d, m be an even integer, and $\mathcal{L}(x_1,\ldots,x_m)$ be a real continuous function on $\mathfrak{X}^m$. We will say that $\mathcal{L}$ is m–n.d. iff for every integer $n \geqslant 1$, every collection of points $x_1,\ldots,x_n \in \mathfrak{X}$ and real numbers $h_1,\ldots,h_n$ if $\sum_{j=1}^{n} h_j = 0$ then

$$(-1)^{m/2} \sum_{i_1=1}^{n} \cdots \sum_{i_m=1}^{n} \mathcal{L}(x_{i_1},\ldots,x_{i_m})\, h_{i_1}\ldots h_{i_m} \geqslant 0. \tag{6}$$

If in (6) equality takes place only with $h_1 = \ldots = h_n = 0$, then $\mathcal{L}$ is called strictly m–n.d. It is evident that $\mathcal{L}$ is m–n.d. iff

$$(-1)^{m/2} \int_{\mathfrak{X}} \cdots \int_{\mathfrak{X}} \mathcal{L}(x_1,\ldots,x_m) h(x_1)\ldots h(x_m) dQ(x_1)\ldots dQ(x_m) \geqslant 0 \tag{7}$$

holds for every measure $Q \in \mathfrak{B}$ and integrable function $h(x)$ with $\int h(x)dQ(x) = 0$. With respect to (7) $\mathcal{L}$ is a strictly m–n.d. kernel, if equality in (7) holds only under the condition $h = 0$ (Q–almost everywhere).

Let $\mathfrak{B}(\mathcal{L})$ denote the set of all measures $\mu \in \mathfrak{B}$ satisfying the condition

$$\int_{\mathfrak{X}} \cdots \int_{\mathfrak{X}} \mathcal{L}(x_1,\ldots,x_m) d\mu(x_1)\ldots d\mu(x_m) < \infty.$$

Put

$$N(\mu,\nu) = (-1)^{m/2} \int_{\mathfrak{X}} \cdots \int_{\mathfrak{X}} \mathcal{L}(x_1,\ldots,x_m) dR(x_1)\ldots dR(x_m), \tag{8}$$

where $R = \mu - \nu$, $\mu,\nu \in \mathfrak{B}(\mathcal{L})$.

One can easily prove that if $\mathcal{L}$ is strictly m–n.d., then $N_m^{1/m}(\mu\ \nu)$ is a metrics on the convex set $\mathfrak{B}(\mathcal{L})$. The following proposition is true.

In order for a metric space $(\mathfrak{X},\ d)$ to be isometric to some subset of the space $L^m(T,\sum,\sigma)$ (m is an even integer), it is necessary and sufficient that for some m–n.d. kernel $\mathcal{L}$

$$d(x,y) = N_m^{1/m}(\delta_x,\delta_y). \tag{9}$$

If in (8) $\mathcal{L}$ is a strictly m–n.d. kernel and (9) holds, then $\mathfrak{B}(\mathcal{L})$ is also isometric to some convex subset of $L^m(T,\sum,\sigma)$, with $N_m^{1/m}$ being equal to the distance between the corresponding baricentres.

Now let us consider in some detail the case $\mathfrak{X} = \mathbf{R}^1$. Here $\mathcal{L}(z)$ will be (strictly) n.d., if $\mathcal{L}(x-y)$ is a (strictly) n.d. kernel.

When $\mathcal{L}(z)$ is continuous symmetric and satisfies the condition $\mathcal{L}(0) = 0$, it is n.d. iff

$$\mathcal{L}(z) = \int_0^\infty \left(1 - \cos(zx)\right) \frac{1+x^2}{x^2} d\theta(x) \tag{10}$$

where $\theta(x)$ is a real bounded nondecreasing function that $\theta(+0) = 0$. The function defined by (10) is strictly n.d., iff supp $\theta = [0, \infty)$. We have for (9)

$$N(\mu, \nu) = c \int_0^\infty \left|f(t) - g(t)\right|^2 \frac{1+t^2}{t^2} d\theta(t).$$

Here f and g are the characteristic functions of the measures μ and ν respectively, and c is an appropriate constant.
Particularly,

$$2\mathbf{E}|X - Y|^r - \mathbf{E}|X - X'|^r - \mathbf{E}|Y - Y'|^r = c_r \int_0^\infty \left|f(t) - g(t)\right|^2 \frac{dt}{t^{1+r}}.$$

In the case $r = 1$ we obtain representation

$$2\mathbf{E}|X - Y| - \mathbf{E}|X - X'| - \mathbf{E}|Y - Y'| = 4 \int_{-\infty}^\infty \left(F(x) - G(x)\right)^2 dx,$$

where F and G denote the distribution functions of the random variables X and Y.
Let us now consider m–n.d. functions in this case. Let

$$\mathcal{L}(x_1, \ldots, x_m) = g(x_1 - x_2 + x_3 - x_4 + \ldots + x_{m-1} - x_m), \tag{11}$$

where g is a continuous symmetric function. A necessary and sufficient condition for $\mathcal{L}$ to be m–n.d. is the representation of g in the form

$$g(u) = \int_0^\infty \left(\sum_{k=0}^{(m-2)/2} (-1)^k u^{2k} x^{2k}/(2k)! - \cos(ux) \right) \frac{1+x^m}{x^m} d\theta(x) + P_m(u^2), \tag{12}$$

where $P_m\ (\cdot)$ is some polynomial of degree less or equal to $(m-2)/2$. We have for $\mathcal{L}$ determined by (11) and (12)

$$N_m(\mu,\ \nu) = c \int_0^\infty \left|f(t) - g(t)\right|^m \frac{1+t^m}{t^m} d\ \theta(t).$$

We can also observe that if $m - 2 < r < m$ then the function

$$\mathcal{L}(x_1, \ldots, x_m) = |x_1 - x_2 + x_3 - x_4 + \ldots + x_{m-1} - x_m|^r$$

with is a strictly m–n.d. kernel. For such function $\mathcal{L}$ we have

$$N_m(\mu,\ \nu) = c_r \int_0^\infty \left|f(t) - g(t)\right|^m \frac{dt}{t^{1+r}}.$$

In close relation with metrics of the $N(\mu,\ \nu)$ type is the above mentioned problem of reconstruction of a measure by its known potential. We will illustrate this in the simpliest case. Let X be a random variable satisfying the condition $\mathbf{E}|X|^r < \infty$ $(0 < r < 2)$. Then, knowing the function

$$\varphi(x) = \mathbf{E}|X - x|^r,\ x \in \mathbf{R}^1,$$

one can uniquely determine the distribution function $F(x)$ of X. Indeed, let Y be a random variable satisfying

$$\mathbf{E}|X - x|^r = \mathbf{E}|Y - x|,\ x \in \mathbf{R}^1. \tag{13}$$

First let us integrate (13) with respect to the measure $dF(x)$. We obtain

$$\mathbf{E}|Y - X|^r = \mathbf{E}|X - X'|^r, \tag{14}$$

where as before X' is independent of and equally distributed with X. At the same time integrating (13) with respect to the measure $dG(x)$, where $G(x)$ denotes the distribution function of the random variable Y, we obtain

$$\mathbf{E}|Y - X|^r = \mathbf{E}|Y - Y'|^r. \tag{15}$$

Here Y' is a random variable independently and equally distributed with Y. From (14) and (15) one immediately obtains

$$2\mathbf{E}|X - Y|^r - \mathbf{E}|X - X'|^r - \mathbf{E}|Y - Y'|^r = 0,$$

and therefore $X \stackrel{d}{=} Y$ (we have the equality in (3)).
In conclusion we will formulate some problems which are closely related to those discussed above. They are:

1) the reconstruction problem for a measure by its values on some families of sets;

2) the problem of extending a multidimensional positive definite in some domain function to the whole space;

3) some aspects of the multidimensional moment problem;

4) different forms of the property of positive definiteness in terms of some generalized expansions;

5) construction of estimators of the parameters of distribution families by means of n.d. "score" functions;

6) application of the results about reconstruction of a measures by its potential to the problems of description of the structure of linear optimal estimators and the study of properties of nonquadratic loss functions (see (Kagan, 1986; Kagan and Zinger, 1987)).

REFERENCES

Bondesson, L. (1973). Characterizations of probability laws through constant regression. *Inst. Mittag-Leffler, Report* 9-75, 1-26.

Braverman, M. Sh. (1985). Characterizing properties of the normal and stable distributions. *Theory Probab. Appl.* **30**, 465-474.

Braverman, M. Sh. (1987). On a method of characterization of probability distributions. *Theory Probab. Appl.* **32**, 552-556.

Gorin, E. A. and Koldobsky, A. L. (1987). On measure potentials in Banach spaces. *Siber. Math. J.* **28**, 65-80.

Kagan, A. M., Linnik, Yu. V. and Rao, C. R. (1973). *Characterization Problems of Mathematical Statistics.* Wiley, New York.

Kagan, A. M., and Zinger, A. A. (1973). Some statistical problems for distribution families, depending of the location and scale parameters. In: *Abstr. of Communic. Intern. Conf. Probab. and Statist,* I, Vilnius.

Kagan, A. M., and Zinger, A. A. (1990). Last square estimator nonquadratic loss function and normal distribution. In: *Theory Probab. Appl,* (to appear).

Kakosyan, A. V., Klebanov, L. B. and Melamed, I. A. (1984). *Characterization of Distributions by the Method of Intensively Monotone Operators.* Springer, Berlin-Heidelberg etc.

Kakosyan, A. V., Klebanov, L. B. and Rachev, S. T. (1988). *Quantitative criteria of the convergence of probability measures.* Aiastan, Yerevan.

Kakosyan, A. V., Klebanov, L. B. and Zinger, A. A. (1989). Characterization of distributions by means of statistics and some probability metrics. In: *Stability of Stochastic Models,* VNIISI, Moscow.

Koldobsky, A. L. (1982). Uniqeness theorem for measures and its application in random processes theory. In: *Notes Sci. Seminars Leningrad Branch of Math. Inst,* **119**,

Kotlarski, I. (1971). On a characterization of probability distributions by the joint distribution of their linear functions. *Sankhya* **A33**, 73-80.

Kovalenko, I. N. (1960). On reconstruction of the additive type of distribution by sequential series of independent trials. In: *Proc. Sov. Symp. Probab. and Statist,* Yerevan.

Linde, W. (1982). Moments of measures on Banach spaces. *Math. Ann.* **258**, 277-287.

Linnik, Yu. V. (1956). On the problem of reconstruction of general distribution by distribution of statistics. *Theory Probab. Appl.* **1**, 466-478.

Neupokoeva, M. V. (1986). *Probability problems of mathematical theory of reliability (Cand. Thesis).* Leningrad.

Plotkin, A. I. (1970). On isometric operators on subspaces of L_p. *Soviet Math. Dokl.* **193**, 537-539.

Plotkin, A. I. (1971). Extending of L_p-isometrics. In: *Notes Sci. Seminars Leningrad Branch of Math. Inst,* **29**,

Polya, G. (1923). Deduction of Gauss probability low from a functional equation. *Math. Z.* **19**,

Prokhorov, Yu. V. (1965). Characterization of the class of distributions by distribution of some statistics. *Theory Probab. Appl.* **10**, 479-487.

Rudin, W. (1976). L_p-isometrics and equimesurability. *Indiana Univ. Math. J.* **25**, 215-228.

Shoenberg, I. J. (1938). Metric spaces and positive definite functions. *Trans. Amer. Math. Soc.* **44**, 522-536.

Zinger, A. A. (1956). On a Kolmogorov problem. *Vestnik. Leningrad Univ.* **1**, 53-56.

Prob. Theory and Math. Stat., Vol. 1, pp. 618–626
B.Grigelionis *et al.* (Eds.)
1990 VSP/Mokslas

THE ENTROPY ORDER OF OPERATORS AND LIMIT THEOREMS IN BANACH SPACES

V.I. KOLCHINSKY
Kiev University, Vladimirskaya, 64, 252017 Kiev, USSR

ABSTRACT

A new class of operators acting between Banach spaces is defined (so called operators with power or logarithmic entropy order). Such operators are closely related to the law of large numbers and the central limit theorem in Banach spaces.

1. INTRODUCTION AND NOTATION

Let E, F be the Banach spaces and let $T : E \to F$ be a bounded linear operator. E^*, F^* will denote the dual spaces of E, F and $T^* : F^* \to E^*$ is the dual operator of T. $S_E, S_F, \ldots$ are the unit spheres and U_E, $U_F, \ldots$ are the unit balls in corresponding Banach spaces.
The subset $\mathfrak{N} \subseteq S_{F^*}$ is called essential if

$$\|x\| = \sup\{|f(x)| : f \in \mathfrak{N}\}, \ x \in F^*.$$

We denote Ess (F^*) the set of all countable essential subsets of S_{F^*} and assume that Ess $(F^*) \neq \varnothing$. In particular, the separable space F satisfies this condition, but also some nonseparable spaces satisfy it too (e.g.,$F = l_\infty$). Let $(\Omega, \Sigma, \mathbf{P})$ be a probability space and let $\xi : \Omega \to F$ be a weakly measurable random element. We denote $(\xi_n)_{n\geqslant 1}$ the sequence of independent copies of ξ.
Assume that $\mathbf{E}|f(\xi)| < \infty$ for all $f \in F^*$ and $\mathbf{E}f(\xi) = f(E\xi)$ with $\mathbf{E}\xi \in F$. We set

$$S_n = \sum_{k=1}^{n}(\xi_k - \mathbf{E}\xi), \quad n \geqslant 1.$$

Let $\mathfrak{B}_F$ be a minimal σ–algebra in F, generated by all balls in F (if F is separable space, then $\mathfrak{B}_F$ is Borel σ–algebra). Given Ess $(F^*) \neq \varnothing$ it is easy to see that $(F, \mathfrak{B}_F)$ is a linear measurable space and ξ, ξ_n, S_n are random elements in $(F, \mathfrak{B}_F)$.

Definition 1. Random element ξ *satisfies the central limit theorem in space* F, or $\xi \in$ CLT (F), if there exists a centered Gaussian Radon measure γ in F such that for every $\Phi : F \to \mathbf{R}^1$ bounded, $\mathfrak{B}_F$–measurable and continuous

$$\mathbf{E}\Phi(n^{-1/2}S_n) \to \int_F \Phi(x)\gamma(dx), \quad n \to \infty.$$

Let $\eta : \Omega \to E$ be weakly measurable random element. Assume that $\|\eta\|$ is a random variable and $\xi = T_\eta$. It is well known, that if E, F are separable Banach spaces, $T : E \to F$ is operator of type 2 and $\mathbf{E}\|\eta\|^2 < \infty$, then $\xi \in$ CLT (F) (Hoffmann–Jorgensen and Pisier, 1976; Hoffmann–Jorgensen, 1977).
In this paper some new classes of operators $T : E \to F$ are considered, such that ξ satisfies the central limit theorem under weaker conditions on $\|\eta\|$, although we need some additional conditions on $\|\xi\|$. In some sense the conditions of the central limit theorem, obtained in the paper, are exact.

Definition 2. Random element ξ satisfies *the Marcinkiewicz strong law of large numbers of order p* $(p \in [1,2])$, or $\xi \in$ SSLN (F,p) if

$$\|S_n\| = o(n^{1/p}), \quad n \to \infty \quad \text{a.s.}$$

If (S,d) is a metric space and $A \subseteq S$ then $N_d(A,\varepsilon)$ will denote the minimal number of ε*-balls*, covering A, $H_d(A,\varepsilon)$ will denote the ε*-entropy* of A with respect to d:

$$H_d(A,\varepsilon) = \ln N_d(A,\varepsilon), \quad \varepsilon > 0.$$

If $(X, \mathfrak{A})$ is a measurable space, then $\mathcal{P}\,(X)$ is the set of all discrete probabilities on X, $\mathcal{P}^*(X)$ is the set of all probabilities on $(X, \mathfrak{A})$.
If $\mu \in \mathcal{P}^*(X)$ and $p \in [1,\infty)$, then $\|\cdot\|_{\mu,p}$ will denote the norm and $d_{\mu,p}$ will denote the metric of space $L_p(X, d\mu)$:

$$\|f\|_{\mu,p} = \left(\int_X |f(x)|^p \mu(dx)\right)^{1/p},$$

$$d_{\mu,p}(f,g) = \|f-g\|_{\mu,p}, f, g \in L_p(X, d\mu)$$

(we use the same notation $\|\cdot\|_{\mu,p}, d_{\mu,p}$ for different measurable spaces).
For $\mathfrak{R} \subset E^*$ and $p \in [1,+\infty)$ we set

$$H_p(\mathfrak{R};\varepsilon) = \sup\{H_{d_{\mu,p}}(\mathfrak{R};\varepsilon) : \mu \in \mathcal{P}(S_E)\}, \varepsilon > 0.$$

Let us denote

$$\rho^* = \begin{cases} \rho/(\rho-1), & \rho > 2; \\ 2, & \rho \in (0,2]. \end{cases} \tag{1}$$

Definition 3. Operator T is called operator of power (respectively, logarithmic) entropy order ρ $(\rho > 0)$ if there exists a set $\mathfrak{N} \in$ Ess (F^*), such that

$$H_{\rho^*}(T^*\mathfrak{N};\varepsilon) = O(\varepsilon^{-\rho}), \quad \varepsilon \to 0$$

(respectively,

$$H_2(T^*\mathfrak{N};\varepsilon) = O\big(\ln^\rho(1/\varepsilon)\big), \varepsilon \to 0).$$

PH(ρ) (respectively, LH(ρ)) will denote the class of operators of power (respectively, logarithmic) order ρ.

If T is operator of type p ($p \in (1,2]$), then $T \in$ PH (q) for $q = p/(p-1)$. If $T \in$ PH (q) ($q \geqslant 2$), then T is operator of type p for every p, $p \in [1, q/(q-1))$ (Kolchinsky, 1988).

2. BASIC RESULTS

The following theorems extend the well-known results for operators of type 2 to the classes of operators PH (ρ) and LH (ρ).

THEOREM 1. *If $T \in$ PH (ρ) for some $\rho \in (0,2)$,*

$$\mathbf{E}\|T\eta\|^r < \infty \quad \text{for some} \quad r > 2 \tag{2}$$

and

$$\mathbf{E}\|\eta\|^s < \infty \quad \text{for some} \quad s > \rho r/(\rho + r - 2), \tag{3}$$

then $T\eta \in$ CLT (F).

Remark . Under the conditions of Theorem 1

$$\rho r/(\rho + r - 2) < 2.$$

So we can assume that $s < 2$ in condition (3).

THEOREM 2. *If $T \in$ LH (ρ) for some $\rho > 0$, η satisfies the condition* (2) *for some $r > 2$ and*

$$\mathbf{E}\ \ln^s\|\eta\| < \infty \quad \text{for some } s > \rho r/(r-2), \tag{4}$$

then $T\eta \in$ CLT (F).

Similarly one can use the operators of classes PH (ρ) and LH (ρ) to obtain the conditions of the strong law of large numbers of Marcinkiewicz type.
Let $p \in (1,2)$, $q = p(p-1)^{-1}$.

THEOREM 3. *If $T \in$ PH (ρ) for some $\rho \in (0, q)$, η satisfies the condition* (2) *for some $r > p$ and η satisfies the condition* (3) *for some $s > \rho r/(\rho + (q-1)(r-p))$, then $T\eta \in$ SLIN (F, p).*

THEOREM 4. *If $T \in$ LH (ρ) for some $\rho > 0$, η satisfies the condition* (2) *for some $r > p$ and η satisfies the condition* (4) *for some $s > \rho r/(q-1)(r-p)$ then $T\eta \in$ SLLN (F, p).*

The above Theorems 1—4 are corollaries of the following more general result. Let $\alpha : [0, \infty) \to (0, \infty)$ be continuous decreasing function, $\alpha(u) \to 0$ as $u \to \infty$ and let $\beta : [0, \infty) \to (0, \infty)$ be continuous increasing function.
We set $\Delta = \beta \circ \alpha^{-1}$. Let $p \in (1,2]$, $q = p/(p-1)$.

THEOREM 5. *If*

$$\mathbf{E}\big[\|T\eta\|/\alpha(\|\eta\|)\big]^p < \infty, \tag{5}$$

$$\mathbf{E}\left[\|\eta\|/\beta(\|\eta\|)\right]^p < \infty \tag{6}$$

and for some $\mathfrak{B} \in$ Ess (F^*) *and* $\delta > 0$

$$\int_0^\delta H_p^{1/q}(T^*\mathfrak{N}; \varepsilon/\Delta(\varepsilon))d\varepsilon < \infty, \tag{7}$$

then

(i) $T\eta \in$ CLT (F) *for* $p = 2$

(ii) $T\eta \in$ SLLN (F, p) *for* $p < 2$.

The main step of the proof of Theorem 5 is the following lemma (Kolchinsky, 1990).

LEMMA. *Let* $\mu \in \mathcal{P}(E)$ *and for some constants* $C_1 > 0,\ C_2 > 0$

$$\int_E \|Tx\|^p/\alpha(\|x\|)^p \mu(dx) \leqslant C_1^p,$$

$$\int_E \|x\|^p/\beta(\|x\|)^p \mu(dx) \leqslant C_2^p.$$

If $\mathfrak{N} \subset S_{F^*}$, *then for sufficiently small* $\varepsilon > 0$

$$H_{d_{\mu,p}}(T^*\mathfrak{N}; \varepsilon) \leqslant H_p(T^*\mathfrak{N}; \varepsilon/(2C_2\Delta(\varepsilon/4C_1))). \tag{8}$$

Proof of the theorem 5. (i) Let ν be the distribution of a random element η, let μ be the distribution of $\xi = T\eta$, let ν_n^* be the empirical measure, based on sample $(\eta_1, \eta_2, \ldots, \eta_n)$ and let μ_n^* be the empirical measure, based on sample $(\xi_1, \xi_2, \ldots, \xi_n)$, where $\xi_j = T\eta_j$.
We set

$$C_1 = \left(2\mathbf{E}\|T\eta\|^2/\alpha(\|\eta\|)^2\right)^{1/2},$$

$$C_2 = \left(2\mathbf{E}\|\eta\|^2/\beta(\|\eta\|)^2\right)^{1/2}$$

and choose $\mathfrak{N} \in$ Ess (F^*), satisfying the condition (7) for $p = q = 2$. If

$$\int_E \|Tx\|^2/\alpha(\|x\|)^2 \nu_n^*(dx) \leqslant C_1^2,$$

$$\int_E \|x\|^2/\beta(\|x\|)^2 \nu_n^*(dx) \leqslant C_2^2, \tag{9}$$

then by Lemma for sufficiently small $\varepsilon > 0$

$$H_{d_{\nu^*_{n,2}}}(T^*\mathfrak{N},\varepsilon) \leqslant H_2(T^*\mathfrak{N},\varepsilon/(2C_2\Delta(\varepsilon/(4C_1)))).$$

Hence under the conditions (9), for small $\delta > 0$ we have

$$\int_0^\delta H^{1/2}_{d_{\nu^*_{n,2}}}(T^*\mathfrak{N},\varepsilon)d\varepsilon \leqslant$$

$$\leqslant \int_0^\delta H_2^{1/2}(T^*\mathfrak{N},\gamma\varepsilon/\Delta(\gamma\varepsilon))d\varepsilon =$$

$$= \gamma \int_0^{\delta/\gamma} H_2^{1/2}(T^*\mathfrak{N},\varepsilon/\Delta(\varepsilon))d\varepsilon. \tag{10}$$

where $\gamma = \min\left(1/2C_2,\ 1/4C_1\right)$.
Using the law of large numbers, by the condition (7) and estimate (10), we obtain

$$\lim_{\delta\to 0}\lim_n \mathbf{P}\left\{\int_0^\delta H^{1/2}_{d_{\nu^*_{n,2}}}(T^*\mathfrak{N};\varepsilon)d\varepsilon \geqslant \sigma\right\} \leqslant$$

$$\leqslant \lim_n \left[\mathbf{P}\left\{\int_F \|Tx\|^2/\alpha(\|x\|)^2\nu_n^*(dx) \geqslant C_1^2\right\} + \right.$$

$$\left. + \mathbf{P}\left\{\int_E \|x\|^2/\beta(\|x\|)^2\nu_n^*(dx) \geqslant C_2^2\right\}\right] = 0, \quad \sigma > 0.$$

Obviously, $d_{\nu^*_{n,2}}(T^*f,T^*g) = d_{\mu^*_{n,2}}(f,g), f,g \in \mathfrak{N}$. Hence

$$\lim_{\delta\to 0}\lim_n \mathbf{P}\left\{\int_0^\delta H^{1/2}_{d_{\mu^*_{n,2}}}(\mathfrak{N};\varepsilon)d\varepsilon \geqslant \sigma\right\} = 0, \quad \sigma > 0. \tag{11}$$

For $p = 2$ condition (5) implies that $\mathbf{E}\|T\eta\|^2 < \infty$. Therefore

$$\mathbf{E}\sup_{f\in\mathfrak{N}}|f(\xi)|^2 = \mathbf{E}\|\xi\|^2 = \mathbf{E}\|T\eta\|^2 < \infty. \tag{12}$$

By the Theorem 8.11 of the paper (Gine and Zinn, 1984), under the conditions (11) and (12) $\mathfrak{N}$ is μ–Donsker class. This implies that $\xi \in$ CLT (F).
The proof of the statement (ii) of the Theorem is similar.

Proof of Theorem 1. We use the statement (ii) of Theorem 1. In this case $p = q = 2$. Since $T \in$ PH (ρ), we obtain that $\mathfrak{N}$ satisfies the condition (7) for $p = q = 2$ if

$$\int_0^\delta \varepsilon^{\rho/2}\Delta(\varepsilon)^{\rho/2}d\varepsilon < \infty. \tag{13}$$

Let $\beta(v) = v^{1-s/2}, \alpha(v) = v^{s(2-r)/(2r)}$, $v > 0$. Then

$$\Delta(\varepsilon) = \beta\big(\alpha^{-1}(\varepsilon)\big) = \varepsilon^{(2-s)r/(s(2-r))}, \ \varepsilon > 0.$$

Under the condition $s > \rho r/(\rho + r - 2)$ the subset $\mathfrak{N}$ satisfies the conditions (13) and (7). On the other hand, we have $v^2/\beta(v)^2 = v^s$, so condition (6) follows from (3). Using Hölder inequality, we obtain

$$\mathbf{E}\|T\eta\|^2/\alpha(\|\eta\|)^2 \leqslant$$

$$\leqslant (\mathbf{E}\|T\eta\|^r)^{2/r}(\mathbf{E}[1/\alpha(\|\eta\|)]^{2r/(r-2)})^{(r-2)/r}.$$

Under the condition (2). In order to obtain (5), it is enough to show, that

$$\mathbf{E}[1/\alpha(\|\eta\|)]^{2r/(r-2)} < \infty. \tag{14}$$

Since $[1/\alpha(v)]^{2r/(r-2)} = v^s, \quad v > 0$, (14) follows from (3).
The proof of Theorems 2, 3, 4 are similar.

3. EXAMPLES OF THE OPERATORS OF CLASSES PH(ρ) AND LH(ρ)

Given a set S, $l^\infty(S)$ will denote the space of uniformly bounded functions $x : S \to \mathbf{R}^1$ with a norm $\|\cdot\|_\infty$,

$$\|x\|_\infty = \sup\{|x(t)| : t \in S\}.$$

If (S, d) is a metric space, then $C_d(S) \subset l^\infty(S)$ is the subspace of uniformly continuous with respect to d and bounded functions. $\mathrm{Lip}_d(S)$ is the space of bounded functions $x : S \to \mathbf{R}^1$, satisfying Lipschitz condition, with a norm

$$\|x\|_{\mathrm{Lip}} = \|x\|_\infty + \sup\{|x(t) - x(s)|/d(t,s) : t \neq s\}.$$

$W(X)$ for a measurable space $(X, \mathfrak{A})$ will denote the space of signed measures with finite total variation on (X,$\mathfrak{A}$). For $\nu \in W(X)\|\nu\|_W$ is a total variation of ν.

(i) Let (S, d) be a compact metric space, $T : \mathrm{Lip}_d(S) \to C_d(S)$ be an identical embeding operator. If for some $\rho > 0$

$$H_d(S, \varepsilon) = O(\varepsilon^{-\rho}), \quad \varepsilon \to 0,$$

then $T \in$ PH (ρ).
If for some $\rho > 0$

$$H_d(S; \varepsilon) = O\big(\ln^\rho(1/\varepsilon)\big), \quad \varepsilon \to 0,$$

then $T \in$ LH (ρ).
This result should be compared with a well-known Zinn's theorem (Zinn, 1977).

(ii) Let $(X,\mathfrak{A})$ be a measurable space and $\mathcal{F}$ be a countable uniformly bounded class of $\mathfrak{A}$–measurable functions $f : X \to \mathbf{R}^1$. We define the operator $T : (W(X), \|\cdot\|_W) \to (l^\infty(\mathcal{F}), \|\cdot\|_\infty)$ by

$$(T\nu)(f) = \int_X f d\nu, f \in \mathcal{F}, \nu \in W(X).$$

If for some $\rho > 0$

$$\sup_{\lambda \in \mathcal{P}^*(X)} H_{d_{\lambda,\rho^*}}(\mathcal{F};\varepsilon) = O(\varepsilon^{-\rho}), \quad \varepsilon \to 0,$$

then $T \in$ PH (ρ) $\quad$ (ρ^* is defined by (1)).
If for some $\rho > 0$

$$\sup_{\lambda \in \mathcal{P}^*(X)} H_{d_{\lambda,\rho^*}}(\mathcal{F};\varepsilon) = O\big(\ln^\rho(1/\varepsilon)\big), \quad \varepsilon \to 0,$$

then $T \in$ LH (ρ).

(iii) Let $\mathfrak{C} \subset \mathfrak{A}$ be a countable class of sets.

$$T : (W(X), \|\cdot\|_W) \to (l^\infty(\mathfrak{C}), \|\cdot\|_\infty) : (T\nu)(C) =$$

$$= \nu(C), \quad C \in \mathfrak{C}, \quad \nu \in W(X).$$

If $\mathfrak{C}$ is Vapnik—Chervonenkis class, then $T \in$ LH(1). This proposition refines the result of (Pisier, 1984). Pisier's theorem states that T is operator of type 2 if and only if $\mathfrak{C}$ is Vapnik—Chervonenkis class. If $\mathfrak{C}$ is not Vapnik—Chervonenkis class, then T is not operator of type p for all $p > 1$.
Let ν be a random signed measure with a.s. finite total variation on $(X,\mathfrak{A})$. It follows from theorem 2, that $\nu \in$ CLT $(l^\infty(\mathfrak{C}))$ if $\mathbf{E}\ln^s \|\nu\|_W < \infty$ for some $s > 1$ and $\mathbf{E}\|\nu\|_\infty^r < \infty$ for all $r > 0$. This proposition improves the result of (Gine and Zinn, 1985) (they proved the central limit theorem under the conditions $\|\nu\|_\infty \leqslant C$ a.s., where C is a constant (not random), and

$$u^2\mathbf{P}\{\ln \|\nu\|_W > u\} \to 0, \; u \to \infty.$$

If $s < 1$ then there exists a random measure ν, such that $\|\nu\|_\infty \leqslant C$ a.s. and $\mathbf{E}\ln^s \|\nu\|_W < \infty$ but the central limit theorem does not hold.

(iv) Let l_∞ be the Banach space of uniformly bounded sequences of real numbers $T : l_\infty \to l_\infty$:

$$Tx = (\lambda_1 x_1, \lambda_2 x_2, \ldots), \quad x = (x_1, x_2, \ldots) \in l_\infty,$$

where $\lambda_n > 0$, $\sup\{\lambda_n : n \geqslant 1\} < \infty$.
We denote

$$\lambda_N^* = \sup\{\lambda_n : n \geqslant N\}.$$

If for some $\rho > 0$

$$\lambda_N^* = O(\ln^{-1/\rho} N), \quad N \to \infty,$$

then $T \in$ PH (ρ).
If for some $\rho > 0$

$$\ln 1/\lambda_N^* = O(\ln^{1/\rho} N), \quad N \to \infty,$$

then $T \in$ LH (ρ).
The proofs of all statements of this assertion are given in (Kolchinsky, 1990).

4.THE EXACTNESS OF THE RESULTS

The following statement shows the exactness of the conditions of Theorem 1.

THEOREM 6. *Let $\rho \in (0,2], r > 2$. There exist a compact space (S,d) and a weakly measurable random element ξ in* $\mathrm{Lip}_d(S)$, *such that*

(i) $H_d(S,\varepsilon) = O(\varepsilon^{-\rho}), \varepsilon \to 0$,
(ii) $\mathbf{E}\|\xi\|_\infty^r < \infty$,
(iii) $\mathbf{E}\|\xi\|_{\mathrm{Lip}}^s < \infty$ *for all* $s < \rho r/(\rho + r - 2)$,
(iv) $\xi \notin \mathrm{CLT}(C_d(S))$.

Proof. Let $s = [0,1]$, $D = \exp\{1/\rho + 1\}$, $d(t,s) = \ln^{-1/\rho} D/|t - s|$, $t \neq s$ and $d(t,s) = 0$, $t = s$. Then (S,d) is a compact metric space and condition (i) is fulfilled. We define the sequence $(f_n)_{n\geqslant 1} \subseteq \mathrm{Lip}_d(S)$:

$$f_n(t) = \cos 3^{n-1}\pi t, \quad t \in [0,1], \quad n \geqslant 1.$$

Then with some constant $A > 0$, we have

$$\|f_n\|_{\mathrm{Lip}} \leqslant A n^{1/\rho}, \quad n \geqslant 1.$$

Let

$$\alpha = (r-2)^{-1}, \lambda = 2(r-1)/(r-2),$$
$$c_n = n^\alpha/\ln n, p_n = Bn^{-\lambda}, \quad n \geqslant 1, \tag{15}$$

where constant $B > 0$ is such that $\sum_{n\geqslant 1} p_n = 1$.
Now we define the random element ξ in $\mathrm{Lip}_d(S)$, such that

$$\mathbf{P}\{\xi = +c_n f_n\} = \mathbf{P}\{\xi = -c_n f_n\} = 0.5 p_n, \quad n \geqslant 1.$$

It is easy to check that $\mathbf{E}\xi = 0$ and ξ satisfies the conditions (ii), (iii).
We set

$$G(t) = \sum_{n\geqslant 1} p_n^{1/2} c_n g_n f_n(t), \quad t \in [0,1],$$

where $\{g_n\}_{n\geqslant 1}$ is the sequence of independent standard normal random variables. Then

$$\mathbf{E}G(t)G(s) = \sum_{n\geqslant 1} p_n c_n^2 f_n(t) f_n(s) = \mathbf{E}\xi(t)\xi(s), \ \ t, s \in [0,1].$$

Under the conditions (15)

$$\sum_{n\geqslant 1} p_n^{1/2} c_n = \infty.$$

Therefore, the Gaussian process G is not bounded a.s. (see Dudley, 1973)). Hence $\xi \notin \mathrm{CLT}(C_d(S))$.
Similarly, the exactness of the conditions of the theorems 2, 3, 4 can be proved.

REFERENCES

Dudley, R.M. (1973). Sample functions of the Gaussian process. *Ann. Probab.* **1**, 66–103.

Dudley, R.M. (1984). A course on empirical processes. *Lect. Notes in Math.* **1097**, 2–242.

Gine, E. and Zinn, J. (1984). Some limit theorems for empirical processes. *Ann.Probab.* **12**, 929–989.

Gine, E. and Zinn, J. (1985). Some results on empirical processes. In: *Abstr. Fourth Intern. Vilnius Conf. on Probab. Theory and Math. Statistics.*, IMC, Vilnius, **4**, pp. 89–91.

Hoffmann-Jørgensen, J. and Pisier, G. (1978). The law of large numbers and the central limit theorem in Banach spaces. *Ann. Probab.* **4**, 587–599.

Hoffmann-Jørgensen, J. (1977). Probability in B-spaces. *Aarhus Univ. Lect. Notes Ser.* **48**, 1–185.

Kolchinsky, V.I. (1988). Operators of type p and metric entropy. *Theory Probab. and Math. Statist.* **38**, 69–76.

Kolchinsky, V.I. (1990). Entropy order of operators in Banach spaces and the central limit theorem. *Theory Probab. Appl.* (to appear).

Pisier, G. (1984). Remarques sur les classes de Vapnik–Chervonenkis. *Ann. Inst. H. Poincare.* **20**, 287–298.

Zinn, J. (1977). A note on the central limit theorem in Banach space. *Ann.Probab.* **5**, 282–289.

Prob. Theory and Math. Stat., Vol. 1, pp. 627–637
B.Grigelionis *et al.* (Eds.)
1990 VSP/Mokslas

INVERSE PROBLEM FOR POTENTIALS OF MEASURES IN BANACH SPACES

A.L. KOLDOBSKII

Department of Mathematics, Leningrad Institute of Finance and Economics, Sadovaja ul. 21, 197183 Leningrad, USSR

ABSTRACT

Let μ be a measure on a Banach space E, $p > 0$, such that the mapping $g: a \to \int_E \| x - a \|^p \, d\mu(x)$ is well-defined from E into $\mathbf{R}$. Then the main question is as follows: for which numbers p does g always describe μ completely? We give a survey of results and methods connected with this problem.

Let E be a Banach space, $p > 0$, $M_p(E)$ be a set of Borel measures μ on E, such that $\int_E (1+ \| x \|^p) \, d\mu(x) < \infty$, $g(a) = \int_E \| x - a \|^p \, d\mu(x)$ be a function on E, which is the potential of some unknown measure $\mu \in M_p(E)$. The problem is to identify μ by means of the function g.

The interest to this problem arose only recently, in spite of its close connection with the classical problems of the potential theory (for instance, if $E = \mathbf{R}^n$ and $-n < p < 0$, we have the inverse problem for Riesz potentials). First results were devoted to the uniqueness problem. In order to describe isometries on some subspaces of L_p A.I.Plotkin (Plotkin, 1970, 1974, 1976) and W.Rudin (Rudin, 1976) proved that in the one-dimensional case (real or complex) the inverse problem has a unique solution for all $p > 0$, except for even integers. The situation turned out to be the same for some other Banach spaces E: the uniqueness holds for all $p > 0$, except may be a countable set, which is called exceptional for the space E. In (Stephenson, 1977; Al–Hisaini, 1978; Koldobskii, 1979) (the finite dimensional case) and (Koldobskii, 1982; Linde, 1986a) (the infinite dimensional case) it was proved that only even integers are exceptional for separable Hilbert spaces. Exceptional for L_q–spaces are p–s such that $p/q \in \mathbf{N}$, and, besides that, in the case of n–dimensional space l_q^n one of the following three conditions must be fulfilled; $p/q < n$, q is an even integer, q and $n+p$ are odd integers (Gorin and Koldobskii, 1985, 1987; Linde, 1986b). For the complex space l_∞^n even numbers are exceptional, and for the real l_∞^n $n+p$ must be odd (Gorin and Koldobskii, 1987). The set of exceptional exponents is empty for spaces $C(Q)$, where Q is an infinite metric compact space without isolated points (Gorin and Koldobskii, 1987) and for spaces $C_0(\Omega)$, where Ω is locally compact but not compact (Linde, 1986). If Q contains isolated points there are no exceptional exponents among numbers $p \notin \mathbf{N}$, and the problem is open for $p \in \mathbf{N}$ (Koldobskii, 1984).

However, the methods used in the proofs of these uniqueness theorems could not give the expression of μ by means of the potential g. The difficulties were in the computation of the Fourier transforms of norms. In the recent paper (Koldobskii, 1990) the Levy representations of norms are used to compute the Fourier transforms of norms and to get the full solution of the inverse problem in some finite dimensional spaces. In particular, the Fourier transforms of l_q^n –norms are expressed in terms of q–stable measures. For the two–dimensional space $E = \operatorname{span}(e_1, e_2)$ the exponent $p > 0$, $p \neq 2, 4, 6, \dots$, is exceptional iff the $(p+1)$–th fracture derivative $(\| e_1 + t\, e_2 \|^p)^{(p+1)}$ vanishes on some open set in $\mathbf{R}$. In (Koldobskii, 1990) the Fourier transform technique is also extended to convolution equations in the infinite dimensional Hilbert spaces, and so the inverse problem in these spaces is completely solved. Some results on the stability of solutions may be found in (Koldobskii, 1987).

In this paper we give a survey of methods used in the proofs of the above mentioned results. These methods are essentially different for finite dimensional and infinite dimensional spaces. Some applications of the inverse problem in statistics and for the description of L_p–isometries one can find in (Klebanov and Zinger, 1989; Plotkin, 1972; Lusky, 1978; Hardin, 1981; Koldobskii, 1989).

1. FINITE DIMENSIONAL SPACES

At first we shall follow the reasoning of the paper (Gorin and Koldobskii, 1987), where uniqueness theorems are proved for some finite dimensional spaces similar to l_q^n.

When E is a finite dimensional space the potential g is considered as a convolution of distributions $\| x \|^p$ and μ over the space $S = S(\mathbf{R}^n)$ of rapidly decreasing functions (as usual we denote the space of distributions by $S' = S'(\mathbf{R}^n)$).

Let R be a subset of S. The R-convolution of distributions $f_1, f_2 \in S'$ is defined in (Gorin and Koldobskii, 1987) as the set of distributions $f \in S'$ such that

$$(f * (\varphi_1 * \varphi_2))(x) = \int_{\mathbf{R}^n} (f_1 * \varphi_1)(\xi)(f_2 * \varphi_2)(x - \xi)\, d\xi$$

for all φ_1, $\varphi_2 \in R$ and $x \in \mathbf{R}^n$ (note, that $f * \varphi \in C^\infty(\mathbf{R}^n)$ if $f \in S'$, $\varphi \in S$).

The convolution defined in this way is naturally connected with the Fourier transform. Let $D(\Omega)^\wedge$ be the space of the Fourier transforms of functions $\varphi \in S$ with supports in an open subset Ω of $\mathbf{R}^n$.

LEMMA 1. *Let f_1, f_2 be distributions such that their Fourier transforms $\widehat{f_1}, \widehat{f_2}$ are L_2–functions on Ω. Then $D(\Omega)$–convolution of these distributions is not empty, and the Fourier transforms of the elements of the convolution coincide on Ω with $\widehat{f_1} \cdot \widehat{f_2}$.*

To apply Lemma 1 to the inverse problem one must compute $(\| x \|^p)^\wedge$, verify that $(\| x \|^p)^\wedge$ is regular and don't vanish on open sets, and put $\widehat{\mu} = \widehat{g}/(\| x \|^p)^\wedge$.

In (Gorin and Koldobskii, 1987) a class of symmetric analytic norms is considered, i.e. norms such that for some basic $e_1, \dots, e_n$ and $\varepsilon > 0$ the function

$f(x_1,\ldots,x_n) = \| x_1e_1 + \ldots + x_ne_n \|$ is even with respect to all coordinates and extends to an analytic function in the cone

$$K_\varepsilon = \{z \in \mathbf{C}^n : | \operatorname{Im} z_k | < \varepsilon | \operatorname{Re} z_k |, \quad k = 1,\ldots,n\}$$

satisfying the condition

$$| f(z) | < A(1 + | z |^\rho), \qquad A, \rho > 0, \qquad z \in K_\varepsilon.$$

It is obvious that the l_q^n–norm is symmetric analytic.

LEMMA 2. *If E is a n–dimensional space with analytic norm and $p > 0$, then $(\| x \|^p)^\wedge$ is a real analytic function outside the coordinate planes in $\mathbf{R}^n$.*

It is clear now that the exponent p is exceptional for a space E with a symmetric analytic norm iff $(\| x \|^p)^\wedge = 0$ outside the coordinate plane. Because of the connection between the Fourier transform and differentiation the last condition is equivalent to the vanishing of some partial derivatives of the function $\| x \|^p$. In the final part of the proof in (Gorin and Koldobskii, 1987) the possibility of this condition for l_q^n–norms is studied, and in this way the uniqueness theorem for l_q^n is obtained. However the attempts to compute $(\| x \|^p)^\wedge$ were unsuccessful (except the case $E = \mathbf{R}^n$), so the full solution of the inverse problem was not obtained in (Gorin and Koldobskii, 1987) even for l_q^n–spaces.
Further we shall follow the paper (Koldobskii, 1990), where the Levy representation of norm is used to compute the Fourier transform of $\| x \|^p$.
Let h be an even continuous function on $\mathbf{R}$ with the power growth at infinity, such that $\widehat{h}$ is a continuous function on $\mathbf{R}\setminus\{0\}$ with the power growth at infinity, ν be a symmetric Borel charge of bounded variation on the unit sphere S^{n-1} in $\mathbf{R}^n$. The distribution $\widehat{h}(t)\,d\nu(\xi)$ is defined by the equation

$$\langle \widehat{h}(t)\,d\nu(\xi), \varphi\rangle = \int\limits_{S^{n-1}} d\nu(\xi) \int\limits_{\mathbf{R}} \widehat{h}(t)\,\varphi(t\xi)\,dt$$

for all $\varphi \in S(\mathbf{R}^n)$ such that $0 \notin \operatorname{supp}\varphi$.

THEOREM 1. *The Fourier transform of the function*

$$f(x) = \int\limits_{S^{n-1}} h(\langle x, \xi\rangle)\,d\nu(\xi)$$

coincides on $\mathbf{R}^n\setminus\{0\}$ with the distribution $\widehat{h}(t)\,d\nu(\xi)$ (here $\langle x, \xi\rangle$ is the scalar product).

COROLLARY. *If $\int\limits_{S^{n-1}} h(\langle x, \xi\rangle)\,d\nu(\xi) = 0$ for all $x \in \mathbf{R}^n$, then either $\nu = 0$ or h is a polynomial containing even powers only.*

This is a generalization of the result for $h(t) = | t |^p$, which was obtained independently in (Kanter, 1973; Linde, 1982; Neyman, 1984).

The computation of $(\| x \|^p)^\wedge$ is based on the representation of $\| x \|^p$ in such a form as in Theorem 1 with $h(t) = | t |^p$. It is the well-known Levy representation of norms.

Let us begin with a n-dimensional subspace of $L_p(\Omega, \sigma)$, where (Ω, σ) is a measure space, $\sigma(\Omega) < \infty$. If $E = \mathrm{span}(f_1, \ldots, f_n) \subset L_p(\Omega, \sigma)$, then for every $x = \sum_1^n x_i f_i \in E$ we have

$$\| x \|^p = \int_\Omega \left| \sum_1^n x_i f_i(\omega) \right|^p d\sigma(\omega) = \int_{\mathbf{R}^n} | \langle x, \xi \rangle |^p \, d\mu(\xi) = \int_{S^{n-1}} | \langle x, \xi \rangle |^p \, d\nu(\xi),$$

where μ is the joint distribution of functions $f_1, \ldots, f_n$ and

$$2\nu(B) = \int_{B \times \mathbf{R}} | x |^p \, d\mu(x)$$

for all Borel sets B in S^{n-1} (here $| x |$ is the eucledian norm,

$$B \times \mathbf{R} = \{ y \in \mathbf{R}^n : \; y / | y | \in B \cup (-B) \};$$

we call ν the p-projection of μ onto S^{n-1}). So if $p > 0$, $p \neq 2, 4, 6, \ldots$, then by Theorem 1 $(\| x \|^p)^\wedge = c_p \, | t |^{-1-p} \, d\nu(\xi)$ on $\mathbf{R}^n \backslash \{0\}$, where $c_p = 2^{p+1} \sqrt{\pi}\, \Gamma((p + +1)/2) / \Gamma(-p/2)$. In particular, the exponent $p \neq 2, 4, 6, \ldots$ is exceptional for E iff μ vanishes on some open cone $B \times \mathbf{R}$ (i.e. $\nu(B) = 0$ for some open set $B \subset S^{n-1}$). In order to obtain the Levy representation for concrete spaces (cf. (Bretagnolle et al., 1966; Misiewicz, 1988)) consider a n-dimensional Banach space E, and suppose, that there exists a function $f \in L_1(R)$ such that $\big(f(\| x \|)\big)^\wedge = = u \in L_1(\mathbf{R}^n)$. One can verify easily that the functions $\langle x, \xi \rangle / \| x \|$ are equimeasurable with respect to the charge $u(\xi)\, d\xi$ for all $x \in E$, $x \neq 0$. Therefore if u_1 is the image of the charge $u(\xi)\, d\xi$ under the mapping $\xi \to \langle x, \xi \rangle / \| x \|$ (u_1 is a charge on $\mathbf{R}$), and the p-th moment of the charge u_1 exists and is not zero, then we get the Levy representation with the charge $u(\xi)\, d\xi$:

$$\| x \|^p = \left(\int_{\mathbf{R}} | t |^p \, du_1(t) \right)^{-1} \int_{\mathbf{R}^n} | \langle x, \xi \rangle |^p \, u(\xi)\, d\xi \, .$$

For instance, if $E = l_q^n$, $f(t) = \exp(- | t |^q)$ and $0 < p < q \leqslant 2$, then $u(\xi) = = \gamma_q(\xi) = \big(\exp(- \| x \|_q^q)\big)^\wedge(\xi)$, and we have

$$\| x \|_q^p = \frac{q\, \Gamma(-p/2)}{2^{p+2} \sqrt{\pi}\, \Gamma((p+1)/2)\, \Gamma(-p/q)} \int_{\mathbf{R}^n} | \langle x, \xi \rangle |^p \, d\gamma_q(\xi).$$

Thus by Theorem 1 $(\| x \|_q^p)^\wedge = (2^{-1} q / \Gamma(-p/q)) \, | t |^{-1-p} \, d\nu(\xi)$ on $\mathbf{R}^n \backslash \{0\}$, where ν is the p-projection of the q-stable measure γ_q onto S^{n-1}.

Now we shall describe a generalized Levy representation, which gives in the two–dimensional case a convenient form of the Fourier transform (for details see (Koldobskii, 1990)).
We say that a n–dimensional space E admits the Levy representation with an exponent p and a distribution $\gamma \in S'(\mathbf{R}^{n-1})$ if

$$\langle \| x \|^p, \varphi \rangle = \Big\langle \gamma(\xi_1, \dots, \xi_{n-1}),$$

$$\int\limits_{\mathbf{R}^n} | x_1\xi_1 - x_2\xi_2 - \dots - x_{n-1}\xi_{n-1} - x_n |^p \varphi(x_1, \dots, x_n)\, dx_1 \dots dx_n \Big\rangle$$

for each function $\varphi \in S(\mathbf{R}^n)$ with the Fourier transform $\widehat{\varphi}$ supported outside of the coordinate planes.
Define the distribution $| t |^{-1-p}\, d\gamma(\xi)$ by the equation

$$\langle | t |^{-1-p}\, d\gamma(\xi), \varphi \rangle =$$

$$= \Big\langle \gamma(\xi_1, \dots, \xi_{n-1}), \int\limits_{\mathbf{R}} | t |^{-1-p} \varphi(t\xi_1, -t\xi_2, \dots, -t\xi_{n-1}, -t)\, dt \Big\rangle$$

for all $\varphi \in S(\mathbf{R}^n)$ supported outside of coordinate planes.

Theorem 2. *Every n–dimensional Banach space E admits the Levy representation with every exponent $p > 0$, $p \neq 2, 4, 6, \dots$, and with the distribution $\gamma \in S'(\mathbf{R}^{n-1})$ such that $(\| x \|^p)^\wedge = c_p \,| t |^{-1-p}\, d\gamma(\xi)$ everywhere outside the coordinate planes. Thus $(\| x \|^p)^\wedge = 0$ on an open set in $\mathbf{R}^n$ iff $\gamma = 0$ on some open set in $\mathbf{R}^{n-1}$.*

If $E = \operatorname{span}(e_1, e_2)$ is a two–dimensional space, the distribution γ can be computed independently.

Theorem 3 (see Theorem 6 from (Koldobskii, 1990)). *If $p > 0$, $p \neq 2, 4, 6, \dots$, then $\gamma = (1/c_p)(\| e_1 + te_2 \|^p)^{p+1}$.*

Here $(\| e_1 + te_2 \|^p)^{p+1} = \big(| t |^{p+1} (\| e_1 + te_2 \|^p)^\wedge(t) \big)^\vee$ is the $(p+1)$–th fracture derivative of the function $\| e_1 + te_2 \|^p$. It follows from Theorems 2 and 3 that an exponent $p > 0$, $p \neq 2, 4, 6, \dots$, is exceptional for a two–dimensional space iff $(\| e_1 + te_2 \|^p)^{p+1} = 0$ on some open set in $\mathbf{R}$. In particular the exponent $p = 1$ is exceptional for a two–dimensional space $E = \operatorname{span}(e_1, e_2)$ iff $\| e_1 + te_2 \|'' = 0$ on an open set, i.e. when E is not a strictly convex space.

Corollary. *An exponent $p > 0$, $p \neq 2, 4, 6, \dots$, is exceptional for a real two–dimensional space with a symmetric analytic norm iff there exists a basis e_1, e_2 in E such that*

$$\| e_1 + te_2 \| = \left(\sum_{k=0}^{[p]} a_k \,| t |^{p-k} (\operatorname{sgn} t)^k + \sum_{k=0}^{[p]} b_k t^k \right)^{1/p}, \qquad \forall t \in \mathbf{R},$$

where a_k, b_k are scalars.

2. INFINITE DIMENSIONAL SPACES

In the infinite dimensional case there is no general approach to the inverse problem because of the absence of the Fourier transform in these spaces. The proofs of the uniqueness theorems are based on the special properties of norms. However, in these proofs there are some general ideas which may be interesting in some other topics.

Let μ be a bounded Borel measure on an infinite dimensional separable Banach space E. Consider the potentials

$$S_n(a) = \int_E \| x - a \|^{-n} \, d\mu(x), \qquad n \in \mathbb{N}.$$

In contrast to the finite dimensional case the set of $a \in E$ such that $S_n(a) < \infty$ for all $n \in \mathbb{N}$ is rather large. In the proofs of the uniqueness theorems for $E = l_2$ and $E = C(K)$ in (Koldobskii, 1982, 1984) it was used that the set of elements $a \in E$ for which $S_n(a) \leqslant n^n$ for all $n \in \mathbb{N}$ is dense in E.

Here we shall follow the general way of estimation of potentials developed in (Gorin and Koldobskii, 1987).

Let φ be a continuous strictly increasing function such that

$$\varphi(0) = 0, \qquad \lim_{t\to\infty} \varphi(t) > \mu(E), \qquad \lim_{t\to\infty} t^{-n}\varphi(t) = \lim_{t\to 0} t^{-n}\varphi(t) = 0$$

for every $n \in \mathbb{N}$. Denote by $B(a,t)$ the closed ball with center at $a \in E$ and radius $t \geqslant 0$, and by μ_a the measure on $[0, +\infty)$ such that $\mu_a([0,t]) = \mu B(a,t)$ for all $t \geqslant 0$.

A point $a \in E$ is said to be regular (with respect to μ and φ) if $\mu B(a,t) < \varphi(t)$ for all $t > 0$.

If $a \in E$ is a regular point, then integrating by parts we get

$$S_n(a) = \int_E \| x - a \|^{-n} \, d\mu(x) = \int_0^\infty y^{-n} \, d\mu_a(y) \leqslant n \int_0^\infty y^{-n-1}\varphi(y) \, dy. \qquad (1)$$

THEOREM 4. *Each ball of radius greater than $2\varphi^{-1}(\mu(E))$ contains a regular point.*

Suppose now that μ is a Borel charge of bounded variation on E, and consider the function

$$\varphi(t) = 2 \mid \mu \mid (E) \exp\left(- \left(\frac{1}{t}\right)^M \right),$$

where $\mid \mu \mid$ is the variation of μ and $M > 1$.

If $a \in E$ is a regular point with respect to $|\mu|$ and φ, then by (1)

$$\int_E \| x-a \|^{-n} \, d|\mu|(x) \leqslant 2|\mu|(E) n \int_0^\infty z^{n-1} \exp\left(-z^M\right) dz =$$

$$= 2|\mu|(E)\frac{n}{M}\Gamma\left(\frac{n}{M}\right). \tag{2}$$

Suppose that $\int_E \| x-a \|^{-n} \, d\mu(x) = 0$ for all $n \in N$.

Representing μ_a as a difference of two measures μ_a^+ and μ_a^- we have

$$u_n = \int_0^\infty y^{-n} \, d\mu_a^+(y) = \int_0^\infty y^{-n} \, d\mu_a^-(y)$$

for all $n \in \mathbb{N}$. It follows from (2) that for some $\alpha > 0$

$$\sum_{n=1}^\infty \frac{1}{\sqrt[n]{u_n}} \geqslant \alpha \sum_{n=1}^\infty \left(\frac{M}{n}\right)^{1/M} = \infty .$$

Thus by the Carleman criterion (Ahiezer, 1961) the moments problem has a unique solution, and $\mu_a^+ = \mu_a^-$. Hence the charge μ vanishes on all balls with center at a.

THEOREM 5. *Let E be a separable Banach space and μ be a Borel charge on E with bounded variation. Suppose that*

$$\int_E \| x-a \|^{-n} \, d\mu(x) = 0$$

for all points $a \in E$ which are regular with respect to $|\mu|$ and $m\varphi$ for some $m \in \mathbb{N}$ and for all $n \in \mathbb{N}$. Then $\mu = 0$.

Indeed by Theorem 4 and by the reasoning preceding Theorem 5 μ vanishes on balls with centers at some dense subset of E. Therefore μ vanishes on all balls in E. Recently in (Preiss and Tiser, 1989) a problem from (Hoffman–Jorgensen, 1975) was solved and it was proved that measures in separable Banach spaces are uniquely determined by their values on balls. Thus $\mu = 0$.

Now it is easy to explain how the uniqueness theorems for l_2 and $C(K)$ were proved in (Koldobskii, 1982, 1984). If $p > 0$, $p \notin \mathbb{N}$, and μ, ν are bounded Borel measures on $E = l_2$ or $E = C(K)$ such that

$$\int_E \| x-a \|^p \, d\mu(x) = \int_E \| x-a \|^p \, d\nu(x) < \infty \tag{3}$$

for all $a \in E$, then, differentiating this equality with respect to a in special directions, we obtain

$$\int_E \| x-a \|^{p-2n} \, d\mu(x) = \int_E \| x-a \|^{p-2n} \, d\nu(x) < \infty$$

for all regular points $a \in E$ and for all $n \in \mathbb{N}$. Now the uniqueness theorem follows from Theorem 5.

The reasoning which was used independently in (Gorin and Koldobskii, 1985, 1987; Linde, 1986b) to prove the uniqueness theorem for the infinite dimensional separable spaces L_q is suitable, apparently, only for these spaces. The proof is based on the following fact, which was for the first time observed in (Linde, 1986a) for the case $q = 2$.

LEMMA 3. *Let e_n, $\| e_n \| = 1$, be a sequence of disjoint functions from the infinite dimensional space L_q, $q \geqslant 1$. Then*

$$\| f + c e_n \|^q \to \| f \|^q + | c |^q$$

as $n \to \infty$ for each function $f \in L_q$ and $c \in \mathbf{R}$.

Suppose now, that μ and ν are bounded Borel measures on L_q satisfying (3) for all $a \in L_q$. Substitute $a + c^{1/q} e_n$ instead of a in (3) and let n tend to the infinity. Then by Lemma 3

$$\int_E (\| x - a \|^q + c)^{p/q} \, d\mu(x) = \int_E (\| x - a \|^q + c)^{p/q} \, d\nu(x), \qquad \forall c > 0.$$

If $p/q \notin N$, one can differentiate infinitely many times this equality with respect to c and use the Carleman criterion to verify, that μ and ν coincide on balls with center at a. Thus $\mu = \nu$, since a is arbitrary.

In (Gorin and Koldobskii, 1987) a new proof of the uniqueness theorem in $C(Q)$ was presented. This proof is suitable only for infinite compact metric spaces Q without isolated points, but it is applicable to a wide class of kernels and to some subspaces of $C(Q)$.

Let E be a closed subspace of $C(Q)$, such that for arbitrary points $s_1, \ldots, s_n \in Q$ and arbitrary constants $\zeta_1, \ldots, \zeta_n$, $| \zeta_i | = 1$, $i = 1, \ldots, n$, there exists a function $a \in E$ such that $a(s_k) = \zeta_k$, $k = 1, \ldots, n$, and $| a(s) | < 1$ for all $s \neq s_k$, $k = 1, \ldots, n$. Let K be a non-negative increasing convex function on the positive semi-axis $t \geqslant 0$, differentiable at $t = 1$ with $K'(1) > 0$.

THEOREM 6. *If μ is a Borel charge of bounded variation on E such that*

$$\int_E K(t \| x \|) \, d | \mu | (x) < \infty$$

for all $t \geqslant 0$ and

$$\int_E K(t \| x - a \|) \, d\mu(x) = 0$$

for all $t \geqslant 0$ and all $a \in E$, then $\mu = 0$.

To prove this result we differentiate the equality with respect to a in some special directions. After that the problem turns out to be reduced to the uniqueness theorem for the finite dimensional space l_∞^n.

An approach which seems to be natural for the infinite dimensional inverse problem was suggested recently in (Koldobskii, 1990) for the case of the infinite dimensional Hilbert space.
Let E be the separable Hilbert space l_2, e_i, $i \in \mathbb{N}$, be the standard basis in l_2, $E_n = \mathrm{span}(e_1, \ldots, e_n)$, $x^{(n)} = \sum_{i=1}^{n} x_i e_i$ be the projection of an element $x = \sum_{i=1}^{\infty} x_i e_i \in l_2$ to E_n, and g_n be the restriction of the potential g to E_n.

THEOREM 7. *If $p \in \mathbb{R}$, $p \neq 0, 2, 4, \ldots$, then for each $\xi \in l_2$, $\xi \neq 0$,*

$$\widehat{\mu}(\xi) = \lim_{n\to\infty} \frac{\widehat{g}_n(\xi^{(n)}) \parallel \xi^{(n)} \parallel^{n+p} \Gamma(-p/2)}{2^{n+p}\pi^{n/2}\Gamma((n+p)/2)}.$$

Here on the right-hand side we have $\widehat{g}_n/(\| x^{(n)} \|^p)^\wedge$, so this formula gives a natural generalization of the finite dimensional Fourier transform technique. The proof is based on the straightforward computation of the Fourier transform of the function $\| x - a \|^p$ of $a \in E_n$ for fixed $x \in l_2$.

3. STABILITY OF THE SOLUTION OF THE INVERSE PROBLEM

Let $M_p(\mathbb{R})$ be the set of Borel measures μ on $\mathbb{R}$ such that

$$\int_{\mathbb{R}} (1 + | x |^p)\, d\mu(x) < \infty.$$

The problem is to define some metrics on the set of functions $| x |^p * \mu$, $\mu \in M_p(\mathbb{R})$, and on the set $M_p(\mathbb{R})$ in such a way, that the proximity of functions $| x |^p * \mu$ and $| x |^p * \nu$ in the first metric ensures the proximity of μ and ν in the second one. We assume, that $p > 0$, $p \neq 2, 4, 6, \ldots$. We shall restrict ourself to the one-dimensional case, although analogous results are valid in more general situations.
We define the metric ρ on the set $M_p(\mathbb{R})$

$$\rho(\mu, \nu) = \sup_{t\in\mathbb{R}} \frac{| (| x |^p * \mu)(t) - (| x |^p * \nu)(t) |}{1 + | t |^p}.$$

This metric reflects the proximity of convolutions.
As the second metric we use the Kantorovich—Rubinshtein metric A_p on $M_p(\mathbb{R})$:

$$A_p(\mu, \nu) = \inf\left\{ \int_{\mathbb{R}^2} | x - y |^p \, d\psi(x, y) \right\},$$

where ψ is a measure on $\mathbb{R}^2$ with marginal distributions μ and ν.

THEOREM 8. *Let μ_n, $\mu \in M_p(\mathbb{R})$ be probability measures, $p > 0$, $p \neq 2, 4, 6, \ldots$. If $\lim_{n\to\infty} \rho(\mu_n, \mu) = 0$, then $\lim_{n\to\infty} A_p(\mu_n, \mu) = 0$.*

The metrics are chosen in such a way, that the result can be used to prove that linear ε-isometries on finite dimensional subspaces of L_p are approximable in the operator norm by linear isometries (Koldobskii, 1987).

To prove Theorem 8 we verify at first that the sequence μ_n weakly converges to μ, and then use the Rachev criterion of convergence in the metric A_p (Rachev, 1984) (for details of the proof see (Koldobskii, 1987)).

REFERENCES

Ahiezer, N.I. (1961). *Classical Problem of Moments*. Fizmatgiz, Moscow.

Al-Husaini, A.L. (1978). Potential operators and equimeasurability. *Pacific J.Math.* **76**, 1–7.

Bretagnolle, J., Dacunha–Castelle, D. and Krivine, J.L. (1966). Lois stables et espaces L_p. *Ann.Inst.H.Poincare B.* **2**, 231–259.

Gorin, E.A. and Koldobskii, A.L. (1985). On potentials, identifying measures in Banach spaces. *Sov.Math.Dokl.* **32**, 659–663.

Gorin, E.A. and Koldobskii, A.L. (1987). On potentials of measures in Banach spaces. *Siberian Math.J.* **28**, 65–80.

Hardin, C.D. (1981). Isometries on subspaces of L_p. *Indiana Univ. Math.J.* **30**, 449–465.

Hoffman–Jorgensen, J. (1975). Measures, which agree on balls. *Math. Scand.* **37**, 319–326.

Kanter, M. (1973). The L_p–norm of sums of translates of a function. *Trans.Amer.Math.Soc.* **179**, 35–47.

Klebanov, L.B. and Zinger, A.A. (1989). Characterization: problems, methods, applications. In: *Abstr. Fifth Inter.Vilnius Conf. Probab.Theory and Math. Statist.* **1**, IMC, Vilnius, pp. 259–260.

Koldobskii, A.L. (1979). On isometric operators in $L_p(X; \mathbf{R}^n)$. In: *Functional Analysis* **12**, Ulyanovsk, pp. 90–99.

Koldobskii, A.L. (1982). On isometric operators in vector–valued L_p–spaces. *Notes Sci.Seminars Leningrad Branch of Math.Inst.* **107**, 198–203.

Koldobskii, A.L. (1984). Uniqueness theorem for measures in $C(K)$ and its application in the theory of random processes. *J.Soviet Math.* **27**, 3095–3102.

Koldobskii, A.L. (1987). Convolution metrics on spaces of measures and ε–isometries in L_p. *Notes Sci.Seminars Leningrad Branch of Math.Inst.* **157**, 151–156.

Koldobskii, A.L. (1989). Isometries in spaces $L_p(X; L_q)$ and equimeasurability. *Izvestija Vyzow. Matematika.* **3**, 25–34.

Koldobskii, A.L. (1990). Convolution equations in certain Banach spaces. *Proc.Amer.Math. Soc. (to appear).*

Linde, W. (1982). Moments of measures on Banach spaces. *Math.Ann.* **258**, 277–287.

Linde, W. (1986a). On Rudin's equimeasurability theorem for infinite dimensional Hilbert spaces. *Indiana Univ.Math.J.* **35**, 235–243.

Linde, W. (1986b). Uniqueness theorems for measures in L_r and $C_0(\Omega)$. *Math.Ann.* **274**, 617–626.

Lusky, W. (1978). Some consequences of Rudin's paper "L_p–isometries and equimeasurability". *Indiana Univ.Math. J.* **27**, 859–866.

Misiewicz, J.K. (1988). On norm–dependent positive definite functions. *Bull.Acad.Sci. Georgian SSR.* **130**, 253–256.

Neyman, A. (1984). Representation of L_p–norms and isometric embedding in L_p–spaces. *Israel J.Math.* **48**, 129–138.

Plotkin, A.I. (1970). On isometric operators on subspaces of L_p. *Soviet Math.Dokl.* **11**, 537–539.

Plotkin, A.I. (1972). Isometric operators in L_p-spaces of analytic and harmonic functions. *Notes Sci.Seminars Leningrad Branch of Math.Inst.* **30**, 130–145.

Plotkin, A.I. (1974). Continuation of L_p-isometries. *J.Soviet Math.* **2**, 1976.

Plotkin, A.I. (1976). An algebra, generated by translation operators, and L_p-norms. In: *Functional Analysis* **6**, Ulyanovsk, pp. 112–121.

Preiss, D. and Tiser, J. (1989). Measures in Banach spaces are determined by their values on balls. *Preprint.*

Rachev, S.T. (1984). Problem of Monge–Kantorovich about the transfer of mass and its application in stochastics. *Theory Probab.Appl.* **29**, 647–676.

Rudin, W. (1976). L_p-isometries and equimeasurability. *Indiana Univ.Math.J.* **25**, 215–228.

Stephenson, K. (1977). Certain integral equalities, which imply equimeasurability of functions. *Canad.J.Math.* **29**, 827–844.

Prob. Theory and Math. Stat., Vol. 1, pp. 638–646
B.Grigelionis *et al.* (Eds.)
1990 VSP/Mokslas

INVESTIGATION OF THE CONVERGENCE OF FUNCTIONS OF SAMPLE MEANS AND COVARIANCES TO LIMIT DISTRIBUTIONS

T. KOLLO
Tartu University, Estonia, USSR

ABSTRACT

The classical method of deriving asymptotic distribution is generalized to the case, when statistics depend on more than one multivariate argument. Asymptotic covariance matrices of discussed statistics are presented.

Parameters of the asymptotic normal law have been found for many different statistics in the multivariate analysis. Usually these statistics are functions of the sample mean vector $\bar{X}$ or the sample covariance matrix S. In this paper we show that a method of deriving asymptotic distributions can be easily generalized from the case where our statistic depends on one asymptotically normal argument ($\bar{X}$ or S, for example) to the case where there are more multivariate arguments.
Assume we have samples $\mathbf{X}_i$ $(i = 1, \ldots, k)$ from p–dimensional populations and let for simplicity the sample size be n in every sample. We suppose that for the vector–valued statistic $T_i = T_i(\mathbf{X}_i)$ $(1 \leqslant i \leqslant k)$ the following convergence in distribution takes place as $n \to \infty$:

$$\sqrt{n}(T_i - \mathbf{E}T_i) \xrightarrow{D} N(0, \Psi_i). \tag{1}$$

For some values of indexes samples $\mathbf{X}_i$ and $\mathbf{X}_j$ may coincide. If we denote

$$Z = \begin{pmatrix} T_1 \\ \ldots \\ T_k \end{pmatrix},$$

we have

$$\sqrt{n}(Z - \mathbf{E}Z) \xrightarrow{D} N(0, \Psi), \tag{2}$$

as $n \to \infty$, where matrices Ψ_i are diagonal blocks in the block–matrix Ψ. From a classical result (see, e.g., (Anderson, 1958)) we know that if a vector–valued function $g(Z)$ is sufficiently smooth in a neighbourhood of $g(\mathbf{E}Z)$ (partial derivatives of $g(Z)$ are continuous), the distribution of the function $g(Z)$ converges to the normal law:

$$\sqrt{n}[g(Z) - g(\mathbf{E}Z)] \xrightarrow{D} N(0, \Omega), \tag{3}$$

where

$$\Omega = \left.\frac{dg(Z)}{dZ}\right|_{Z=\mathbf{E}Z} \Psi \left(\left.\frac{dg(Z)}{dZ}\right|_{Z=\mathbf{E}Z}\right)'. \tag{4}$$

Here $dg(Z)/dZ$ is the matrix derivative. We use the Neudecker's definition (Magnus and Neudecker, 1988): if elements of $(m \times n)$–matrix Y are functions of elements of $(p \times q)$–matrix X then

$$\frac{dY}{dX} = \frac{d}{d\,\mathrm{vec}'X} \otimes \mathrm{vec}\,Y.$$

This means that the partial derivatives $\partial y_{kl}/\partial x_{ij}$ are arranged in the $(mn \times pq)$–matrix dY/dX by the Kronecker product of vectorized matrices Y and X. The vec–operator transforms a matrix into a vector by stacking the columns of the matrix one underneath the other.
We can find the asymptotic covariance matrix Ω using (4) after fixing the vector Z. We have considered two basic cases

$$\text{a)}\quad Z = \begin{pmatrix} \bar{X} \\ \cdots \\ \mathrm{vec}S \end{pmatrix}; \quad \text{b)}\quad Z = \begin{pmatrix} \mathrm{vec}S_1 \\ \cdots \\ \mathrm{vec}S_2 \end{pmatrix}.$$

In the case a) $\bar{X}$ is the sample mean and S is the covariance matrix for the sample $\mathbf{X} = (X_1, \ldots, X_n)$ with population characteristics $\mathbf{E}X_i = \mu$; $\mathbf{D}X_i = \Sigma$. In the case b) S_1 and S_2 are the sample covariance matrices for two different samples with theoretical covariance matrices Σ_1 and Σ_2 respectively. The case a) makes it possible to derive the asymptotic normal distribution for statistics like Hotelling's T^2:

$$g(\bar{X}, S) = (\bar{X} - \mu_0)'S^{-1}(\bar{X} - \mu_0)$$

or the statistic

$$g(\bar{X}, S) = \bar{X}'S^{-1}\bar{X}.$$

In the case b) we get the asymptotic normal distributions for F matrix or MANOVA matrix:

$$g(S_1, S_2) = S_1 S_2^{-1} = F,$$

beta matrix

$$g(S_1, S_2) = S_1(S_1 + S_2)^{-1} = B$$

and for their functions used in estimation theory.
In the case a) we assume that the fourth order moments of the population distribution are finite:

$$M_4 = \mathbf{E}[(X_i - \mu) \otimes (X_i - \mu)' \otimes (X_i - \mu) \otimes (X_i - \mu)'] < \infty.$$

Then the convergence (1) reduces to the following convergences:

$$\sqrt{n}\,(\bar{X} - \mu) \xrightarrow{D} N(0, \Sigma); \tag{5}$$

$$\sqrt{n}\,\mathrm{vec}\,(S - \Sigma) \xrightarrow{D} N(0, \Pi), \tag{6}$$

where

$$\Pi = M_4 - \operatorname{vec}\Sigma \operatorname{vec}'\Sigma.$$

In parallel with the general case we also find the asymptotic covariance matrices for the considered statistics for elliptically and normally distributed samples. Matrices Π in all these cases are presented in Table 1.

Table 1
Asymptotic Covariance Matrices Π of Sample Covariance Matrix S

distribution	Π
general	$M_4 - \operatorname{vec}\Sigma\operatorname{vec}'\Sigma$
elliptical	$(1+k)\,(I_{p^2} + I_{p,p})\,(\Sigma \otimes \Sigma) + k\,\operatorname{vec}\Sigma\,\operatorname{vec}'\Sigma$
normal	$(I_{p^2} + I_{p,p})\,(\Sigma \otimes \Sigma)$

In Table 1 $I_{p,p}$ denote the permutation matrix — $(p^2 \times p^2)$–block–matrix consisting of $(p \times p)$–blocks where in the ji–th block the ij–th element equals to one and all other elements are zeros.
p–vector Y is distributed elliptically with parameters μ and V if the density function of Y is of the form

$$f_Y(x) = c \mid V \mid^{-1/2} g\{(x-\mu)'V^{-1}(x-\mu)\},$$

where μ is a real p–vector, V is a positive definite $(p \times p)$–matrix, c is the normalizing constant and g is a non–negative function. For the elliptical distribution the covariance matrix $\Sigma = \alpha V$ for some constant α and all cumulants of the fourth order are functions of the second order moments and the numerical parameter k:

$$k = \frac{\mathbf{E}(Y_i - \mu_i)^4}{3\sigma_{ii}^2} - 1.$$

In the following the triplet (μ, Σ, k) for indicating an elliptical distribution is used. In the case b) we get from (1) the following convergences:

$$\sqrt{n}\operatorname{vec}(S_1 - \Sigma_1) \xrightarrow{D} N(0, \Pi_1); \tag{7}$$

$$\sqrt{n}\operatorname{vec}(S_2 - \Sigma_2) \xrightarrow{D} N(0, \Pi_2), \tag{8}$$

where

$$\Pi_1 = M_4(1) - \operatorname{vec}\Sigma_1 \operatorname{vec}'\Sigma_1; \tag{9}$$

$$\Pi_2 = M_4(2) - \operatorname{vec}\Sigma_2 \operatorname{vec}'\Sigma_2. \tag{10}$$

$M_4(1), M_4(2)$ are the fourth order moments of the two considered populations respectively. The matrices $\Pi_i\ (i = 1, 2)$ for different distributions are presented in Table 2.

Table 2
Asymptotic Covariance Matrices Π_1, Π_2 of Sample Covariance Matrices S_1, S_2

distribution	Π_1	Π_2
general	$M_4(1) - \text{vec}\Sigma_1 \text{ vec}'\Sigma_1$	$M_4(2) - \text{vec}\Sigma_2 \text{ vec}'\Sigma_2$
elliptical	$(1+k_1)(I_{p^2} + I_{p,p})(\Sigma_1 \otimes \Sigma_1)+$ $+k_1 \text{ vec}\Sigma_1 \text{ vec}'\Sigma_1$	$(1+k_2)(I_{p^2} + I_{p,p})(\Sigma_2 \otimes \Sigma_2)+$ $+k_2 \text{ vec}\Sigma_2 \text{ vec}'\Sigma_2$
normal	$(I_{p^2} + I_{p,p})(\Sigma_1 \otimes \Sigma_1)$	$(I_{p^2} + I_{p,p})(\Sigma_2 \otimes \Sigma_2)$

Statistic $\bar{X}'S^{-1}\bar{X}$

Let us denote $Z = \begin{pmatrix} \bar{X} \\ \cdots \\ \text{vec}S \end{pmatrix}$; $Z_0 = \begin{pmatrix} \mu \\ \cdots \\ \text{vec}\Sigma \end{pmatrix}$. As was shown in (Kollo, 1987), the following convergence takes place if $\mu \neq 0$:

$$\sqrt{n}(\bar{X}'S^{-1}\bar{X} - \mu'\Sigma^{-1}\mu) \xrightarrow{D} N(0, \xi), \tag{11}$$

$$\xi = \left.\frac{d(\bar{X}'S^{-1}\bar{X})}{dZ}\right|_{Z=Z_0} \Psi \left(\left.\frac{d(\bar{X}'S^{-1}\bar{X})}{dZ}\right|_{Z=Z_0} \right)',$$

where

$$\frac{d(\bar{X}'S^{-1}\bar{X})}{dZ} = \left(2\bar{X}'S^{-1} \ \vdots \ -(\bar{X}'S^{-1}) \otimes (\bar{X}'S^{-1}) \right).$$

Thus in the general case we have the asymptotic variance ξ in the following form:

$$\xi = \left(\mu'\Sigma^{-1} \ \vdots \ -(\mu'\Sigma^{-1}) \otimes (\mu'\Sigma^{-1}) \right) \begin{pmatrix} \Sigma & \vdots & M_3' \\ \cdots & \vdots & \cdots \\ M_3 & \vdots & \Pi \end{pmatrix} \times$$

$$\times \begin{pmatrix} \Sigma^{-1}\mu \\ \cdots \\ -(\Sigma^{-1}\mu) \otimes (\Sigma^{-1}\mu) \end{pmatrix}; \tag{12}$$

where M_3 is the third central moment of the vector X_i. If the population distribution is symmetrical, $M_3 = 0$ and

$$\xi = 4\mu'\Sigma^{-1}\mu + (\mu'\Sigma^{-1})^4 \text{vec}\Pi.$$

Here the fourth power of the vector $\mu'\Sigma^{-1}$ means the Kroneckerian power — for a vector X

$$X^n = \underbrace{X \otimes \cdots \otimes X}_{n \text{ times}}.$$

For different populations expressions of ξ are presented in Table 3.

Table 3
Asymptotic Variance ξ of Statistic $\bar{X}'S^{-1}\bar{X}$

distribution	ξ
general	$\left(2\mu'\Sigma^{-1} \vdots -(\mu'\Sigma^{-1}) \otimes (\mu'\Sigma^{-1})\right) \begin{pmatrix} \Sigma & \vdots & M_3' \\ \cdots & & \cdots \\ M_3 & \vdots & \Pi \end{pmatrix} \begin{pmatrix} 2\Sigma^{-1}\mu \\ \cdots \\ -(\Sigma^{-1}\mu)\otimes(\Sigma^{-1}\mu) \end{pmatrix}$
symmetrical ($M_3 = 0$)	$4\,\mu'\,\Sigma^{-1}\mu + 2\,(\mu'\Sigma^{-1})^4\ \mathrm{vec}\Pi$
elliptical	$4\,\mu'\,\Sigma^{-1}\mu + (2+3k)\,(\mu'\,\Sigma^{-1}\mu)^2$
normal	$4\,\mu'\,\Sigma^{-1}\mu + 2\,(\mu'\,\Sigma^{-1}\mu)^2$

F-matrix $S_1 S_2^{-1}$
Let us propose that the sample covariance matrices S_1 and S_2 are calculated from two independent samples. If we denote

$$Z = \begin{pmatrix} \mathrm{vec}S_1 \\ \cdots \\ \mathrm{vec}S_2 \end{pmatrix}; \quad Z_0 = \begin{pmatrix} \mathrm{vec}\Sigma_1 \\ \cdots \\ \mathrm{vec}\Sigma_2 \end{pmatrix},$$

we have the convergence

$$\sqrt{n}\mathrm{vec}(S_1S_2^{-1} - \Sigma_1\Sigma_2^{-1}) \xrightarrow{D} N(0, \Phi), \tag{13}$$

as $n \to \infty$, where

$$\Phi = \left.\frac{d(S_1S_2^{-1})}{dZ}\right|_{Z=Z_0} \Omega \left(\left.\frac{d(S_1S_2^{-1})}{dZ}\right|_{Z=Z_0}\right)'.$$

Here the derivative

$$\frac{d(S_1S_2^{-1})}{dZ} = \left(S_2^{-1} \otimes I_p \;\vdots\; S_2^{-1} \otimes S_1S_2^{-1}\right)$$

and the matrix

$$\Omega = \begin{pmatrix} \Pi_1 & \vdots & 0 \\ \cdots & \vdots & \cdots \\ 0 & \vdots & \Pi_2 \end{pmatrix},$$

where the matrices Π_i are defined by equalities (9)—(10). In the general case we have the following result:

$$\Phi = \left(\Sigma_2^{-1} \otimes I_p \;\vdots\; \Sigma_2^{-1} \otimes \Sigma_1\Sigma_2^{-1}\right) \begin{pmatrix} \Pi_1 & \vdots & 0 \\ \dots & \vdots & \dots \\ 0 & \vdots & \Pi_2 \end{pmatrix} \begin{pmatrix} \Sigma_2^{-1} \otimes I_p \\ \dots \\ \Sigma_2^{-1} \otimes \Sigma_2^{-1}\Sigma_1 \end{pmatrix} =$$

$$= (\Sigma_2^{-1} \otimes I_p)\Pi_1(\Sigma_2^{-1} \otimes I_p) + (\Sigma_2^{-1} \otimes \Sigma_1\Sigma_2^{-1})\Pi_2(\Sigma_2^{-1} \otimes \Sigma_2^{-1}\Sigma_1).$$

In the case of two elliptically distributed populations with parameters (μ_1, Σ_1, k_1) and (μ_2, Σ_2, k_2) respectively we get Φ in the form

$$\Phi = (I_{p^2} + I_{p,p})[(1 + k_1)(\Sigma_2^{-1}\Sigma_1\Sigma_2^{-1} \otimes \Sigma_1) + (1 + k_2)(\Sigma_2^{-1} \otimes \Sigma_1\Sigma_2^{-1}\Sigma_1)] +$$
$$+ (k_1 + k_2)\mathrm{vec}(\Sigma_1\Sigma_2^{-1})\mathrm{vec}'(\Sigma_1\Sigma_2^{-1}).$$

In the normal population case the formula above reduces to

$$\Phi = (I_{p^2} + I_{p,p})(\Sigma_2^{-1}\Sigma_1\Sigma_2^{-1} \otimes \Sigma_1 + \Sigma_2^{-1} \otimes \Sigma_1\Sigma_2^{-1}\Sigma_1).$$

The expressions of the asymptotic covariance matrix Φ are much simpler if $\Sigma_1 = \Sigma_2 = \Sigma$. For this case the results are presented in Table 4.

Table 4
Asymptotic Covariance Matrix Φ of F matrix $(\Sigma_1 = \Sigma_2 = \Sigma)$

distribution	Φ
general	$(\Sigma^{-1} \otimes I_p)\,(\Pi_1 + \Pi_2)\,(\Sigma^{-1} \otimes I_p)$
$k_1 \neq k_2$	$(2 + k_1 + k_2)\,(I_{p^2} + I_{p,p})\,(\Sigma^{-1} \otimes \Sigma) + (k_1 + k_2)\,\mathrm{vec}\, I_p\,\mathrm{vec}'\, I_p$
elliptical	
$k_1 = k_2 = k$	$2\,[(1 + k)\,(I_{p^2} + I_{p,p})\,(\Sigma^{-1} \otimes \Sigma) + k\,\mathrm{vec}\, I_p\,\mathrm{vec}'\, I_p]$
normal	$2\,(I_{p^2} + I_{p,p})\,(\Sigma^{-1} \otimes \Sigma)$

In the matrix Φ in (13) is known, it is easy to get asymptotic normal distributions of functions of F matrix $S_1S_2^{-1}$. Let us consider the determinant $|\, S_1S_2^{-1} \,|$. By (3), (4) and (13) the following convergence takes place

$$\sqrt{n}(|\, S_1S_2^{-1} \,| - |\, \Sigma_1\Sigma_2^{-1} \,|) \xrightarrow{D} N(0, \beta),$$

as $n \to \infty$, where

$$\beta = \left.\frac{d\,|\, S_1S_2^{-1} \,|}{d(S_1S_2^{-1})}\right|_{Z=Z_0} \Phi \left(\left.\frac{d\,|\, S_1S_2^{-1} \,|}{d(S_1S_2^{-1})}\right|_{Z=Z_0} \right)'.$$

Using the formula for a regular $(p \times p)$–matrix X (see (Magnus and Neudecker, 1988))

$$\frac{d \mid X \mid}{dX} = \mid X \mid \mathrm{vec}'(X^{-1}),$$

we get

$$\frac{d \mid S_1 S_2^{-1} \mid}{d(S_1 S_2^{-1})} = \mid S_1 \mid \mid S_2^{-1} \mid \mathrm{vec}'(S_1^{-1} S_2).$$

In the general case

$$\beta = \mid \Sigma_1 \Sigma_2^{-1} \mid^2 (\mathrm{vec}'\Sigma_1^{-1}\Pi_1 \mathrm{vec}\Sigma_1^{-1} + \mathrm{vec}'\Sigma_2^{-1}\Pi_2 \mathrm{vec}\Sigma_2^{-1}).$$

For normal samples we have

$$\beta = 2|\Sigma_1\Sigma_2^{-1}|[(p + \mathrm{tr}(\Sigma_2\Sigma_1^{-1}\Sigma_2^{-1}\Sigma_1))].$$

The asymptotic variance β in the case $\Sigma_1 = \Sigma_2 = \Sigma$ is given in Table 5.

Table 5
Asymptotic Variance β of Determinant of F matrix $S_1 S_2^{-1}$ $(\Sigma_1 = \Sigma_2 = \Sigma)$

distribution	β
general	$\mathrm{vec}'\ \Sigma^{-1}\ (\Pi_1 + \Pi_2)\ \mathrm{vec}\ \Sigma^{-1}$
$k_1 \neq k_2$	$2\,(2 + k_1 + k_2)\,p + (k_1 + k_2)\,p^2$
elliptical	
$k_1 = k_2 = k$	$4\,(1 + k)\,p + 2kp^2$
normal	$4p$

Beta matrix $S_1(S_1 + S_2)^{-1}$
As for F matrix, we assume that both S_1 and S_2 are sample covariance matrices from two independent samples of size n.
Then

$$\sqrt{n}\,\mathrm{vec}[S_1(S_1 + S_2)^{-1} - \Sigma_1(\Sigma_1 + \Sigma_2)^{-1}] \xrightarrow{D} N(0, \Theta)$$

as $n \to \infty$, where

$$\Theta = \left.\frac{d(S_1(S_1 + S_2)^{-1})}{dZ}\right|_{Z=Z_0} \Omega \left(\left.\frac{d(S_1(S_1 + S_2)^{-1}}{dZ}\right|_{Z=Z_0} \right)'.$$

Here the form of the asymptotic covariance matrix Θ is more complicated than in the previous cases. For the derivative we have

$$\frac{d(S_1(S_1 + S_2)^{-1})}{dZ} =$$

$$= \left((S_1+S_2)^{-1} \otimes (S_1(S_1+S_2)^{-1} + I_p) \ \vdots \ (S_1+S_2)^{-1} \otimes (S_1(S_1+S_2)^{-1}) \right).$$

In the general case the form of the matrix Θ cannot be simplified. When $\Sigma_1 = = \Sigma_2 = \Sigma$, the derivative takes on the form

$$\left.\frac{d(S_1(S_1+S_2)^{-1})}{dZ}\right|_{Z=Z_0} = \frac{1}{2}\left(\Sigma^{-1} \otimes I_p \ \vdots \ \Sigma^{-1} \otimes I_p \right).$$

In this case the expressions of the matrix Θ are simpler and we present them in Table 6.

Table 6
Asymptotic Covariance Matrix Θ of Beta Matrix $S_1(S_1+S_2)^{-1}$ ($\Sigma_1 = \Sigma_2 = \Sigma$)

distribution	Θ
general	$\frac{1}{4}(\Sigma^{-1} \otimes I_p)(9\Pi_1 + \Pi_2)(\Sigma^{-1} \otimes I_p)$
elliptical $k_1 \neq k_2$	$\frac{1}{4}[(10 + 9k_1 + k_2)(I_{p^2} + I_{p,p})(\Sigma^{-1} \otimes \Sigma) + +(9k_1 + k_2) \operatorname{vec} I_p \operatorname{vec}' I_p]$
elliptical $k_1 = k_2 = k$	$\frac{5}{2}[(1+k)(I_{p^2} + I_{p,p})(\Sigma^{-1} \otimes \Sigma) + k \operatorname{vec} I_p \operatorname{vec}' I_p]$
normal	$\frac{5}{2}[(I_{p^2} + I_{p,p})(\Sigma^{-1} \otimes \Sigma)]$

Although the forms of the asymptotic covariance matrices presented above in the Tables are relatively simple, one has to be extremely careful in using these results in the estimation problems. All the considered statistics are biased estimators for the corresponding theoretical values of the statistics ($\bar{X}'S^{-1}\bar{X}$ for $\mu'\Sigma^{-1}\mu$; $S_1S_2^{-1}$ for $\Sigma_1\Sigma_2^{-1}$ and $S_1(S_1+S_2)^{-1}$ for $\Sigma_1(\Sigma_1+\Sigma_2)^{-1}$) and before using their asymptotic distributions some additional information about the bias and the rate of convergence is needed. For example, if we have a normal sample $X_i \sim N(\mu, \Sigma)$ with parameters

$$\mu = \begin{pmatrix} 1 \\ 1 \\ 1 \end{pmatrix}; \quad \Sigma = \begin{pmatrix} 1 & 0.1 & 0.2 \\ 0.1 & 1.1 & 0.3 \\ 0.2 & 0.3 & 1.2 \end{pmatrix},$$

the main term of the bias b_n of the statistic

$$Y_n = \sqrt{n}(\bar{X}'S^{-1}\bar{X} - \mu'\Sigma^{-1}\mu)$$

is rather big as can be seen from Table 7.

Table 7
Main Term of Bias b_n of Statistic $\sqrt{n}\ (\bar{X}'S^{-1}\bar{X} - \mu'\Sigma^{-1}\mu)$.

n	100	200	500	1000	2000
b_n	1.116	0.789	0.499	0.353	0.278

Practical usage of a statistic with such systematic error seems to be unreasonable (formulae for calculating bias b_n can be found in (Vilismäe and Kollo, 1988)).

REFERENCES

Anderson, T. W. (1985). *An Introduction to Multivariate Statistical Analysis*. Wiley, New York.

Kollo, T. (1987). Asymptotic distributions of functions of sample means and covariances.In: *Proc. Second International Tampere Conference in Statistics*, pp. 499–510.

Magnus, J.R. and Neudecker, H. (1988). *Matrix Differential Calculus with Applications in Statistics and Economics*. Wiley, New York.

Vilismäe, J. and Kollo, T. (1988). Asymptotic distribution of the function $\bar{X}'S^{-1}\bar{X}$. *Contributions of the Computer Centre of Tartu State Univ.* **55**, 94–105.

Bei Fragen zur Produktsicherheit wenden Sie sich bitte an:
If you have any questions regarding product safety,
please contact:

Walter de Gruyter GmbH
Genthiner Straße 13
10785 Berlin
productsafety@degruyterbrill.com